BASIC SET THEORY

AZRIEL LEVY

Institute of Mathematics
The Hebrew University of Jerusalem

DOVER PUBLICATIONS, INC.
MINEOLA, NEW YORK

Copyright

Published in the United Kingdom by David & Charles, Brunel House, Forde Close, Newton Abbot, Devon TQ12 4PU.

Bibliographical Note

This Dover edition, first published in 2002, is an unabridged republication of the work published by Springer-Verlag, Berlin, Heidelberg, and New York, in 1979 as part of its series Perspectives in Mathematical Logic. Corrections and Additions, Additional Bibliography and a Preface to the Dover Edition have been prepared especially for this edition by the author.

AMS Subject Classification (1970): 04-02, 04-01, 04A15, 04A20, 04A25, 04A30, 02K35

Library of Congress Cataloging-in-Publication Data

Levy, Azriel.
Basic set theory / Azriel Levy.
p. cm.
Originally published: Berlin : New York: 1979, in series: Perspectives in Mathematical Logic.
Includes bibliographical references and indexes.
ISBN 0-486-42079-5 (pbk.)
1. Set theory. I. Title.

QA248 .L398 2002
511.3'22—dc21

2002022292

Manufactured in the United States of America
Dover Publications, Inc., 31 East 2nd Street, Mineola, N.Y. 11501

מוקדש להורי ע"ה
שרה ושלמה לוי

To the memory of my parents
Sarah and Shlomo Levy

ה' בחכמה יסד ארץ,
כונן שמים בתבונה.
משלי ג, יט

The Lord by wisdom founded the earth,
by understanding he established the heavens.
Proverbs 3:19

Preface to the Dover Edition

Much has happened in set theory since the original edition of this book. Naturally, this has happened not in the basic areas covered by the book but in more advanced matters. Therefore this book is almost as up-to-date as it was originally, but the road from it to the current research has become much longer. Among the books which cover parts of this road are Jech 1978 and Kunen 1980 in general set theory, Kanamori 1994 in large cardinals, Moschovakis 1980 in descriptive set theory, and Shelah 1994 on cardinal numbers.

The original edition contained its fair share of misprints and mistakes. All those of which I am aware are listed in the Corrections and Additions appendix, which also contains some additional historical references and a small number of exercises based on the material in the book. An Additional Bibliography also follows the original Bibliography.

The author owes much to all his colleagues who pointed out the improvements needed. Since this new edition is a reprint and not a fully revised, newly edited edition, it has been possible to implement only some of the suggestions. The most useful were supplied by K. Carmodi, W. Hodges, A. Kanamori, and G. Moran, to whom I owe a great debt.

I hope this edition will reach more students and will help them to learn this beautiful area of mathematics.

March 25, 2002 A. Levy
Jerusalem

Preface

Almost all the recently-published books on set theory are of one of the following two kinds. Books of the first kind treat set theory on an elementary level which is, roughly, the level needed for studying point set topology and Steinitz's theorem on the existence of the algebraic closure of a general field. Books of the second kind are books which give a more or less detailed exposition of several areas of set theory that are subject to intensive current research, such as constructibility, forcing, large cardinals and determinacy. Books of the first kind may serve well as an introduction to the subject but are too elementary for the student or the mathematician who wants to gain a deeper understanding of set theory. The books of the second kind usually go hurriedly through the basic parts of set theory in their justified haste to get at the more advanced topics. One of the advantages of writing a book in a series such as the Perspectives in Mathematical Logic is that one is able to write a book on a rather advanced level covering the basic material in an unhurried pace. There is no need to reach the frontiers of the subject as one can leave this to other books in the series. This enables the author to pay close attention to interesting and important aspects of the subject which do not lie on the straight road to the very central topics of current research.

I started writing this book in 1970. During the long period since that time I have been helped by so many people that I cannot name them all here. Several of my colleagues advised me on the material in the book, read parts of the manuscript and made very useful remarks, and taught me new theorems and better proofs of theorems I knew. Many typists typed the numerous versions of the manuscript and bore with admirable patience all my inconsistent instructions. I shall mention in name only Klaus Glöde and Uri Avraham to whom I am most grateful for diligently reading the galley proofs, correcting many misprints and mistakes. This book would not have been written without the initiative and encouragement of my colleagues in the Ω group. I enjoyed very much their company and collaboration.

I shall be most grateful to any reader who will point out misprints, mistakes and omissions and who will supply me with additional bibliographical references. This will hopefully be incorporated in later printings of this book.

Most of this book was written while I stayed as a visitor at Yale University in the academic year 1971–72 and at UCLA in 1976–77. I extend my thanks to the

National Science Foundation of the United States for partially supporting me during those years.

May 12, 1978 A. Levy
Jerusalem

Table of Contents

Introduction

As its name indicates, this book does not contain all of what is known about set theory. This poses the question of what to include in the book. A major guideline was to stop short of those areas where model-theoretic methods are used. Thus constructibility is formally introduced, but the consequences of the axiom of constructibility are not proved; some large cardinals are discussed, but their model-theoretic properties are not investigated; Martin's axiom and Jensen's $\Diamond$ are introduced, but their respective consistencies are not proved.

A basic tacit assumption for almost all the metamathematical results in this book is that the Zermelo–Fraenkel set theory ZF is consistent. For example, when we say that the axiom $V=L$ of constructibility is consistent with ZF we mean that assuming ZF is consistent then ZF with $V=L$ is consistent as well.

This book contains few historical remarks, since the history of the subject alone is material enough for a whole book. Nevertheless, to give the reader some feeling of the history of the subject, the original authors and publications are quoted whenever this information was available. There are of course theorems which were known to their discoverers a long time before they were published, but in most cases all that is given here is the publication reference. "Gödel 1939" refers to the bibliographical item listed as such in the bibliography, but "Gödel in 1939" means that the theorem next to which this is written was proved by Gödel in 1939. Theorems are often given here not in their original form but in a form which comprises improvements by several mathematicians. In some of these cases later contributors are mentioned beside the original one, but no uniformity of reference has been achieved.

When we use in Chapter V, for example, the reference 4.29(i) we refer to part (i) of the 29th subsection of Section 4 of Chapter V, which is the subsection marked **4.29**. When we refer to II.2.4 we mean Subsection 2.4 in Chapter II. When we say inside III.2.35 "by II.2.14(i) and (ii)" we mean "by II.2.14(i) and part (ii) of the present subsection (which is III.2.35(ii))" if we want to refer to parts (i) and (ii) of II.2.14 we say "by II.2.14(i) and II.2.14(ii)".

The end of a proof, or of a theorem which is given without a proof, is usually marked with $\square$. In some places, where a theorem or a proof of a theorem is immediately followed by a discussion which dwells on some points in the theorem, the end of the proof, or the end of the theorem if it is not followed by a proof, is marked with a narrow box ▯.

Most of the exercises supply additional information on the subject matter and are therefore an integral part of the discussion. The exercises are usually followed by hints so as to make their solution not too difficult.

Part A

Pure Set Theory

Chapter I
The Basic Notions

All branches of mathematics are developed, consciously or unconsciously, in set theory or in some part of it. This gives the mathematician a very handy apparatus right from the beginning. The most he usually has to do in order to have his basic language ready is to describe the set theoretical notation he uses. In developing set theory itself we have no such advantage and we must go through the labor of setting up our set theoretical apparatus. This is a relatively long task. Even the question as to which objects to consider, only sets or also classes, is by no means trivial, and its implications will be discussed here in detail. In addition, we shall formulate the axioms of set theory; we shall show how the concepts of ordered pair, relation and function, which are so basic in mathematics, can be developed within set theory, and we shall study their most basic properties. By the end of this chapter we shall be just about ready to begin our real mathematical investigation of the universe of sets.

1. The Basic Language of Set Theory

In the present section and in Sections 3 and 4 we shall thoroughly discuss the language we are going to use for set theory. Usually when one studies a branch of mathematics one is not concerned much with the question as to which exactly is the language used in that branch. The reason why here we must look carefully at the language lies in the difference between set theory and most other branches of mathematics. Most mathematical fields use a relatively "small" fragment of set theory as their underlying theory, and rely on that fragment for the language, as well as for the set theoretical facts. The source of the difficulty we have in set theory with the language is the fact that not every collection of objects is a set (something which will be discussed in detail in the next few sections), but we still have to refer often to these collections, and we have to arrange the language so that we shall be able to do it handily. This difficulty does not come up in the fragments of set theory used for most mathematical theories, since those fragments deal only with a very restricted family of collections of objects, and all these collections are indeed sets.

Our present aim is to obtain for set theory a language which is sufficiently rich and flexible for the practical development of set theory, and yet sufficiently simple so as not to stand in the way of metamathematical investigation of set theory. For this purpose we start by choosing for set theory a very simple basic language. The simplicity of this language will be a great advantage when we wish to discuss set theory from a metamathematical point of view. The only objects of our set theory will be sets. One could also consider atoms, i.e., objects which are not sets and which serve as building blocks for sets, but they are not essential to what we shall do and, therefore, will not be considered in the present book. As a consequence of this decision, we view the sets as follows. We start with the null set 0, from it we obtain the set $\{0\}$, from the two sets 0 and $\{0\}$ we obtain the sets $\{0, \{0\}\}$ and $\{\{0\}\}$, and so on. Much of set theory is concerned with what is meant by this "and so on".

The language which we shall use for set theory will be the first-order predicate calculus with equality. Why first-order? Because a second-order or a higher-order theory admits already a part of the set theory in using its higher order variables. To see, for example, that second-order variables are essentially set variables let us consider the following axiom of second-order logic: $\exists A \forall x(x \in A \leftrightarrow \Phi(x))$, where $\Phi(x)$ is, essentially, any formula. This axiom is read: there exists a *set* A such that for every x, x is a member of A if and only if $\Phi(x)$. It would, of course, change nothing if we would choose another term instead of "set", since "set" is what we mean anyway. When we develop a formal system of set theory it does not seem right to handle sets in two or more parts of the language, i.e., by considering some sets as first-order objects, while having around also second-order objects which are sets. As a consequence of our decision we shall have, in principle, just one kind of variable, lower case letters, which will vary over sets.

The reason why we take up first-order predicate calculus *with equality* is a matter of convenience; by this we save the labor of defining equality and proving all its properties; this burden is now assumed by the logic.

Our basic language consists now of all the expressions obtained from $x=y$ and $x \in y$, where x and y are any variables, by the sentential connectives $\neg$(not), $\rightarrow$ (if...then...), $\vee$ (or), $\wedge$ (and), $\leftrightarrow$ (if and only if), and the quantifiers $\exists x$ (there exists an x) and $\forall x$ (for all x). These expressions will be called *formulas*. For metamathematical purposes we can consider the connectives $\neg$ and $\vee$ as the only primitive connectives, and the other connectives will be considered as obtained from the primitive connectives in the well known way (e.g., $\phi \wedge \psi$ is $\neg(\neg\phi \vee \neg\psi)$, $\phi \rightarrow \psi$ is $\neg\phi \vee \psi$, etc.). For the same reason, we can consider $\exists$ as the only primitive quantifier and the quantifier $\forall$ as defined by means of $\exists$ by taking $\forall x\phi$ to be an abbreviation of $\neg\exists x \neg\phi$. We shall also use the abbreviations $x \neq y$ and $x \notin y$ for $\neg x=y$ and $\neg x \in y$. We shall write $\exists! x\phi$, and read: there is exactly one x such that ϕ, for the formula $\exists y \forall x(x=y \leftrightarrow \phi)$, where y is any variable which is not free in ϕ. Finally, we shall write $(\exists x \in y)\phi$ and $(\forall x \in y)\phi$ for $\exists x(x \in y \wedge \phi)$ and $\forall x(x \in y \rightarrow \phi)$, respectively, and read: "there is an x in y such that ϕ", and "for all x in y, ϕ".

By a *free variable* of a formula we mean, informally, a variable occurring in that formula so that it can be given different values and the formula says something concerning the values of the variable. E.g., x is a free variable in each of the following formulas (which are not necessarily taken from set theory): $x<3x$,

$x^2=y$, x is a real number, $\sin x>\frac{1}{2}$, $\forall y(z<y\rightarrow x<y)$. x is not a free variable in the formulas $\forall x(x^2\geqslant 0)$, $\neg\exists x(x\in y)$, $\int_0^y \sin x\, dx<\frac{1}{2}$. In the latter three formulas x is an auxilliary variable which cannot be given a definite value and which can be replaced throughout each formula by another variable, say z, without changing the meaning of the formula. In these examples x is used as a *bound variable*. Note that $\forall x(x^2\geqslant 0)$ says exactly what $\forall z(z^2\geqslant 0)$ says, while $\sin x>\frac{1}{2}$ does not say the same thing as $\sin z>\frac{1}{2}$; in fact, for appropriate values of x and z $\sin x>\frac{1}{2}$ may be true, while $\sin z>\frac{1}{2}$ may be false. A variable may have both free and bound occurrences in the same formula, even though one would usually try to avoid it; e.g., in $7<z\wedge\exists z(z>x)$, the occurrences of z in $\exists z(z>x)$ are bound, while the occurrence of z in $7<z$ is free (since the quantifier $\exists z$ applies only to $z>x$).

A formula with free variables says something about the values of its free variables. A formula without free variables makes a statement not about the value of some particular variable, but about the universe which the language describes. A formula of the latter kind is called a *sentence*. We shall also refer, informally, to formulas and sentences as *statements*.

Whenever we use a formula with free variables as an axiom or as a theorem we mean to say that the formula holds for all possible values given to its free variables. Thus, if we state a theorem $\exists z(z=x\cup y)$ we mean the same thing as $\forall x\forall y\exists z(z=x\cup y)$.

By a *theory* we mean a set of formulas, which are called the *axioms* of the theory. If T is a theory, we shall write $T\vdash\phi$ for "ϕ is provable from T".

When we refer to a formula as $\phi(x)$ this does not mean that x is necessarily a free variable of $\phi(x)$ nor does it mean that $\phi(x)$ has no free variables other than x; it means that the interesting cases of what we shall say are those where x is indeed a free variable of $\phi(x)$. When we shall mention $\phi(z)$ after we have first mentioned $\phi(x)$, then $\phi(z)$ denotes the formula obtained from $\phi(x)$ by substituting the variable z for the free occurrences of x. (z may also be a bound variable of $\phi(x)$, and then before we substitute z for the free occurrences of x we may have to replace the bound occurrences of z by some other variable.)

2. *The Axioms of Extensionality and Comprehension*

By a set we mean a completely structure-free set, and therefore a set is determined solely by its members. This leads us to the first axiom of set theory.

2.1 Axiom of Extensionality (Frege 1893). $\forall x(x\in y\leftrightarrow x\in z)\rightarrow y=z$.
In words: if y and z have the same members they are equal. The converse, that equal objects have the same members, is a logical truth.

2.2 The Existence of Sets. Now we face the question of finding or constructing the sets. We want any collection whatsoever of objects, i.e., sets, to be a set. This is not a precise idea and therefore we cannot translate it into our language. We must therefore be satisfied with a somewhat weaker stipulation. We shall require that every collection of sets which is "specifiable" in our language is a set; i.e., for every statement of our language the collection of all objects which satisfy it is a set. We shall by no means assume that it is necessarily true that *all* sets are specifiable; moreover, by introducing the axiom of choice we shall require the existence of sets which are not necessarily specifiable. The requirement that all specifiable collections are indeed sets is the following one.

2.3 Axiom of Comprehension (Frege 1893). $\exists y \forall x(x \in y \leftrightarrow \phi(x))$, where $\phi(x)$ is any formula (of the language of set theory) in which the variable y is not free (since if y were free in $\phi(x)$ this would cause a confusion of the y free in $\phi(x)$ with the y whose existence is claimed by the axiom). Our only reason in writing $\phi(x)$ instead of just ϕ is to draw attention to the fact that the "interesting" cases of this axiom schema are those for which the formula ϕ does actually contain free occurrences of the variable x.

The axiom of comprehension is an *axiom schema*, i.e., it is not a single sentence but an infinite set of sentences obtained by letting ϕ vary over all formulas. Any single sentence obtained from 2.3 by choosing a particular formula for ϕ in 2.3 is said to be an *instance* of the axiom schema, and is also called "an axiom of comprehension." □

Those readers who were convinced by the axiom schema of comprehension are now in for a shock; the axiom schema of comprehension is not consistent—Theorem 2.4 below is the negation of one of its instances.

2.4 Theorem (Russell's antinomy—Russell 1903). $\neg \exists y \forall x(x \in y \leftrightarrow x \notin x)$.

Proof. Notice that this theorem is not just a theorem of set theory; it is a theorem of logic, since we do not use in its proof any axiom of set theory. We prove it by contradiction. Suppose y is a set such that $\forall x(x \in y \leftrightarrow x \notin x)$, then, since what holds for every x holds in particular for y, we have $y \in y \leftrightarrow y \notin y$, which is a contradiction. □

Russell's antinomy is the simplest possible refutation of an instance of the comprehension schema. We refer to a refutation of such an instance as an *antinomy.* The first antinomy to be discovered is the Burali–Forti paradox discovered by Cantor and by Burali–Forti in the 1890's; it is given in II.3.6 and II.3.15. Some variants of Russell's antinomy are given in 2.5.

2.5 Exercise. Prove the negation of the instance of the axiom of comprehension where $\phi(x)$ is one of the following formulas:

(a) $\neg \exists u(x \in u \wedge u \in x)$,

(b) $\neg \exists u_1 \ldots \exists u_n(x \in u_1 \wedge u_1 \in u_2 \wedge \cdots \wedge u_{n-1} \in u_n \wedge u_n \in x)$. □

2.6 How to Avoid the Antinomies. One can react to Russell's antinomy in two different ways. One way is to think again of what led us to the axiom of comprehension, and to decide that since a set is something like 0, $\{0\}$, $\{0, \{0\}\}$, etc., we should not have come up with anything like the axiom of comprehension anyway. According to this view, the axiom of comprehension is basically false, since it represents a mental act of "collecting" all sets which satisfy $\phi(x)$, and this cannot be done since we can "collect" only those sets which have been "obtained" at an "earlier" stage of the game. This point of view was suggested first by Russell 1903 as one of the ingredients of his theory of types. The other possible reaction to Russell's antinomy is to continue believing in the essential truth of the axiom schema of comprehension, viewing the Russell antinomy as a mere practical joke played on mankind by the goddess of wisdom. According to this point of view the axiom schema of comprehension is only in need of some tinkering to avoid the antinomies; the guide on how to do it will be the *doctrine of limitation of size*. The doctrine says that we should use the axiom schema of comprehension only in order to obtain new sets which are not too "large" compared to the sets whose existence is assumed in the construction. Also this doctrine, which is already implicit in Cantor 1899, was formulated first by Russell 1906. In our framework of set theory both approaches lead to the same result, and therefore there is no mathematical need to go through the arguments in favor of each one of them. Motivations for the choice of the axioms, from both points of view, are presented in the literature (see, e.g., Fraenkel, Bar–Hillel and Levy 1973 and Scott 1974) and will hopefully be presented in a later book in this series devoted to the axiomatics of set theory. Here we shall mostly rely on the acceptance by the reader of the axioms which we shall introduce as intuitively reasonable axioms. □

Let us still notice one feature of the axiom of comprehension. After the failure of the full axiom of comprehension, we cannot be sure that, given a formula ϕ, there is a set y such that $\forall x(x \in y \leftrightarrow \phi(x))$. However, if there is such a y it is unique, as stated in the next theorem.

2.7 Proposition. *If there is a y such that*

$$\forall x(x \in y \leftrightarrow \phi(x))$$

then this y is unique.

Proof. If y' is also such, i.e., $\forall x(x \in y' \leftrightarrow \phi(x))$, then we have, obviously, $\forall x(x \in y' \leftrightarrow x \in y)$, and by the axiom of extensionality, $y' = y$. □

3. *Classes, Why and How*

As we shall come to see, the main act of generation of set theory is that objects are collected to become a set, which is again an object which can be collected into a new set. We saw above that, because of the antinomies, not every collection of objects which can be specified in our language can be collected to become a "new" object. This is by no means disastrous for mathematics, since, by means of appropriate axioms which we shall introduce, we shall be able to show that sufficiently many of the intuitive collections can indeed be taken as sets to satisfy the mathematical needs. This will enable us to obtain sets such as the set of all real numbers, the set of all countable ordinals, the set of all measures on some given set, etc.

There are many things we can say about an intuitive collection of objects without assuming that the collection is an object itself. Let us see an example. Suppose we want to say

(1) Every non-void subset u of the collection of all sets x such that $x \notin x$ has a member y such that y has no common member with the collection.

This can also be said as

(2) $$\forall u(u \neq 0 \wedge (\forall x \in u) x \notin x \rightarrow (\exists y \in u)\ (\forall x \in y) x \in x).$$

Notice that (2) does not mention the collection mentioned in (1). We could decide never to use (1) and always to use (2) instead; but, as we shall point out now, and as will become even clearer to the reader as he goes on reading this book, this would have required a considerable sacrifice of convenience. Sometimes we want to say about many or all specifiable collections what we said in (1) about one particular collection. We can proceed as in (2) but this requires using an infinite family of formulas. This is illustrated by the following example. Suppose we want to say that for every specifiable collection A

(3) $$\forall u(u \neq 0 \wedge u \subseteq A \rightarrow (\exists y \in u) y \notin A) \rightarrow A \textit{ has at most 10 members}.$$

We can say the same thing also by asserting that for all formulas $\phi(x, x_1, \ldots, x_n)$

(4) $$\forall x_1 \ldots \forall x_n[\forall u(u \neq 0 \wedge (\forall x \in u)\phi(x, x_1, \ldots, x_n) \rightarrow (\exists y \in u) \neg \phi(y, x_1, \ldots, x_n)) \rightarrow \textit{there are at most 10 objects } x \textit{ such that } \phi(x, x_1, \ldots, x_n)].$$

To see that (3) and (4) say the same thing notice that the specifiable collections A are exactly those given as the collection of all objects x such that $\phi(x, x_1, \ldots, x_n)$ holds, for some formula $\phi(x, x_1, \ldots, x_n)$ and for some fixed values of $x_1, \ldots, x_n$. Comparing (3) with (4) shows that (3) is not only shorter but also much easier to comprehend than (4). Thus we see that it is a great advantage to be able to talk about collections as if they were sets even though we know, as a result of Russell's antinomy, that not all of them are sets. A uniform way of talking about sets and collections has also the following advantage. We often come across collections which, at a certain point in the discussion, we do not know whether they are sets or not. Speaking about them in the same way as we speak about sets puts what we say about them in a form which retains its convenience even after the collections turn out to be sets.

We shall now introduce into our language additional notation which will enable us to make our discussion of general specifiable collections look similar to the discussion of sets, while *everything which we say by means of the additional notation can also be said without it*, possibly only in a clumsy way. Thus, we shall be able to talk about general specifiable collections, which we want to do for the sake of convenience, without introducing them as objects of one kind or another.

A general specifiable collection, which may or may not be a set, will be called a *class*. A class is given by a formula $\phi(x)$ as the class of all objects x for which $\phi(x)$ holds. Such a class will be denoted by $\{x \mid \phi(x)\}$. The expression $\{x \mid \phi(x)\}$ will be called a *class term*. The formula $\phi(x)$ may also contain free variables other than x. These other variables are called *parameters*. Different values of the parameters may yield different classes. For example, the class $\{x \mid x \textit{ is a natural number} \wedge x<y\}$ is a class with no members if $y=0$, has a single member if $y=1$, and so on. Note also that since we took all specifiable collections to be classes, sets are classes too. In fact, the set y is the class $\{x \mid x \in y\}$.

3.1 Class Terms. We shall extend our language by incorporating class terms also, but in such a way that every statement containing a class term will stand for some other, possibly longer, statement which does not contain class terms. We shall now write out what kind of statements containing class terms we allow, and for which statements containing no class terms these will stand. Since $\{x \mid \phi(x)\}$ is the class of all x's for which $\phi(x)$ holds, we shall take the statement $y \in \{x \mid \phi(x)\}$ to stand for $\phi(y)$ (where $\phi(y)$ is the formula obtained from $\phi(x)$ by proper substitution of y for x), i.e., "y belongs to the class of all x's such that $\phi(x)$" stands for $\phi(y)$. Since we consider two sets with the same members to be equal, we shall also consider two classes with the same members as equal. Therefore we shall admit the statement $\{x \mid \phi(x)\}=\{y \mid \psi(y)\}$ and let it stand for $\forall z(\phi(z) \leftrightarrow \psi(z))$. Since also the sets are classes, we admit also the statements $x=\{y \mid \phi(y)\}$ and $\{y \mid \phi(y)\}=x$ and let them stand for $\forall z(z \in x \leftrightarrow \phi(z))$. To say that one class is a member of the other will mean that the first class is equal to a set which is member of the other. Accordingly, we admit the statement $\{x \mid \phi(x)\} \in \{y \mid \psi(y)\}$ and let it stand for $\exists z(z=\{x \mid \phi(x)\} \wedge z \in \{y \mid \psi(y)\})$, and similarly we let the statement $\{x \mid \phi(x)\} \in y$ stand for $\exists z(z=\{x \mid \phi(x)\} \wedge z \in y)$.

3.2 The Extended Language. We are now on our way to extending the language by adding notation for classes. This is a rather special way of extending the language. We shall also extend our language later, routinely, by new definitions as is done commonly in the development of any mathematical theory. All the extensions will be such that while they increase the ease of the use of the language, they do not add anything to the absolute expressive power of the language, since everything which can be expressed in the extended language is also expressible in the basic language of set theory. We shall examine this point again later. Let us only decide now to refer at every point throughout our book, to the language, as extended by various agreements and definitions up to that point, as *the extended language*, while by *the basic language* we shall always mean the basic language of set theory as outlined in Section 1. Note that while the term "the basic language" has a fixed meaning, the term "the extended language" does not, since as we go along we enrich the latter by new definitions. Unless otherwise mentioned, or implied, we shall use lower-case Greek letters for formulas of the basic language and capital Greek letters for formulas of the extended language.

We extended our language by adding to the basic language the class terms. We shall now further extend this language in two steps. The first step is to allow as formulas $\phi(x)$ in the class terms $\{x \mid \phi(x)\}$ not only formulas of the basic language, but also formulas of the extended language, i.e., formulas which contain also class terms. We have to tell now what do the formulas which contain the more general class terms $\{x \mid \Phi(x)\}$ stand for. It is easy to see that we can still use the same way of interpreting class terms as above, only that now more than one step may be needed in passing from the formula of the extended language to the corresponding formula of the basic language. Let us consider a simple example. We start with the statement:

(1) $$z = \{x \mid x \in \{y \mid \phi(y)\} \vee x \in \{y \mid \psi(y)\}\}.$$

By what we said in 3.1 about what $z = \{x \mid \phi(x)\}$ stands for, (1) stands for.

(2) $$\forall u(u \in z \leftrightarrow u \in \{x \mid x \in \{y \mid \phi(y)\} \vee x \in \{y \mid \psi(y)\}\}.$$

In (2), $u \in \{x \mid x \in \{y \mid \phi(y)\} \vee x \in \{y \mid \psi(y)\}\}$ stands for $u \in \{y \mid \phi(y)\} \vee u \in \{y \mid \psi(y)\}$, hence (1) stands for

(3) $$\forall u(u \in z \leftrightarrow u \in \{y \mid \phi(y)\} \vee u \in \{y \mid \psi(y)\}).$$

We still have to eliminate in (3) the statements $u \in \{y \mid \phi(y)\}$ and $u \in \{y \mid \psi(y)\}$ which we know stand for $\phi(u)$ and $\psi(u)$, respectively. Thus, (1) stands for

(4) $$\forall u(u \in z \leftrightarrow \phi(u) \vee \psi(u)).$$

This example suffices to show how to interpret the statements of the extended language. Later we will discuss the formal proof that such an elimination exists for every statement. In practice, none of the statements which we will use in this book will have an essentially more complicated nesting of class terms than (1) above, and hence, the reader will have no trouble at all in finding the formula of the basic language for which a given formula in our extended language stands.

It should be noted that even though we have added now new class terms by allowing all the statements of the extended language inside the class terms, we did not add any classes; we have only more class terms referring to the same classes. Since every formula $\Phi(x)$ of the extended language is equivalent to some formula $\phi(x)$ of the basic language, we get that every class $\{x \mid \Phi(x)\}$ is equal to some class $\{x \mid \phi(x)\}$ (since $\{x \mid \Phi(x)\} = \{x \mid \phi(x)\}$ stands for $\forall x(\Phi(x) \leftrightarrow \phi(x))$, which holds since $\Phi(x)$ and $\phi(x)$ are equivalent).

Our second step is to add also class *variables* to our language. We shall use capital Roman letters for the class variables. The class variables will stand for arbitrary class terms and will enable us to make statements about *all* classes, while until now we were able to make statements essentially only about one class at a time or about a single family of classes given by a single class term with parameters. For example, if we state the theorem $A=B \to B=A$ we mean by it that for all $\Phi(x)$ and $\Psi(y)$ we have

(5) $$\{x \mid \Phi(x)\}=\{y \mid \Psi(y)\} \to \{y \mid \Psi(y)\}=\{x \mid \Phi(x)\}.$$

As we mentioned in 3.1, (5) stands for $\forall z(\Phi(z) \leftrightarrow \Psi(z)) \to \forall z(\Psi(z) \leftrightarrow \Phi(z))$, which, in turn, stand for the schema $\forall z(\phi(z) \leftrightarrow \psi(z)) \to \forall z(\psi(z) \leftrightarrow \phi(z))$ of formulas in the basic language. Thus, we saw that $A=B \to B=A$ stands not for a single formula of the basic language, but for a schema of such formulas. Since we allow the occurrence of class terms inside the formula $\Phi(x)$ of a class term $\{x \mid \Phi(x)\}$ we shall also allow the occurrence of class variables inside such a $\Phi(x)$. The role of a class variable in the formula $\Phi(x)$ of a class term $\{x \mid \Phi(x)\}$ is similar to the role of a set variable y which is a parameter of $\{x \mid \Phi(x)\}$.

3.3 Quantification of Class Variables. Note that the class variables occur only as free variables, i.e., we shall not use quantifiers like $\exists X$ or $\forall X$ over these variables. If we were to admit such quantifiers, we would get formulas which do not stand for any formula of the basic language, and our extended language would become a new language, and not just a convenient way of handling the statements of the basic language. However, the way we interpret a statement which contains a class variable, say A, is such that the statement is assumed to hold for all classes A (or, for a "general" A). As a consequence, we shall not hesitate to read $A=B \to B=A$ as "for all A and B, if $A=B$ then $B=A$".

3.4 A Historical and General Remark. The informal idea of using classes in set theory occurred already to Cantor 1899, once he became aware of the antinomies. An axiomatic system, using functions instead of classes, was presented first by Fraenkel 1922. This approach was brought to maturity by von Neumann 1925. An axiomatic system of set theory with classes was given first by Bernays 1937. The particular way in which we look here at classes has been stated explicitly first by Quine 1963, but its ideas were known earlier from the writings of Bernays (particularly Bernays 1958). Our way of looking at classes is by no means the only one. Many mathematicians, beginning with von Neumann, view the classes as objects as "real" as the sets, except that they are not members of other classes (unless they are sets). Systems of set theory based on that point of view are discussed in the literature (see, e.g., Levy 1976). However, the differences between such systems and the system with which we deal here turned out to be, from a mathematical point of view, only of a minor nature. Thus, once one accepts the axioms of ZF about the existence of sets, there is really only one underlying set theory.

4. Classes, the Formal Introduction

In §1 we have explained how to introduce class terms and class variables, considering them as abbreviated notation for formulas of the basic language. While we shall retain, in principle, the attitude that the class notations are just abbreviations for formulas of the basic language, we shall, nevertheless, follow Bernays 1958 and introduce a formal system in which class terms and class variables are expressions in their own right and not abbreviations of other expressions. The reason for doing this is that once we are anyway using in practice an extended language which is not the basic language there is nothing to be gained by not introducing the system we actually use as a formal system. This does not conflict with our attitude of regarding the expressions of the extended language as abbreviations of expressions of the basic language; this attitude will be justified by rules which will tell us how to interpret the formulas of the extended language in the basic language. We shall now first introduce the formal system for the extended language, and then dwell on its relationship to the basic language.

4.1 The Formal System of the extended language, which admits classes as well as sets, is formulated in a two-sorted predicate calculus with equality, having capital letters as variables for classes and lower case letters as variables for sets. Notice that the class variables are now formal variables and not metamathematical symbols standing for class terms as in the last section. *Atomic formulas* are formed from the class terms and the class and set variables by means of the equality and membership relation symbols, i.e., atomic formulas are of the form $\mathfrak{a} \in \mathfrak{b}$ or $\mathfrak{a} = \mathfrak{b}$, where each of $\mathfrak{a}$ and $\mathfrak{b}$ is either a class term or a class variable or a set variable. All other *formulas* are obtained from the atomic formulas by the sentential connectives $\neg$, $\vee$, $\wedge$, $\rightarrow$ and $\leftrightarrow$ and by quantification *over set variables only*. *Class terms* are expression $\{x \mid \Phi(x)\}$ where $\Phi(x)$ is an arbitrary formula. The "logic" of this language consists of the axioms and rules of inference of the first order predicate calculus with equality for the set variables, the axioms and rules of inference of the free variable calculus with equality for the class variables, as well as the following rule of inference.

4.2 The Rule of Substitution of Sets for Classes. From $\Phi(A)$ infer $\Phi(x)$. This rule says, essentially, that all sets are classes (since it says that whatever holds for all classes holds also for all sets).

It is not necessary that the reader should memorize the logical axioms and rules of inference of the extended language. All one has to do is to use ordinary mathematical reasoning in the extended language, remembering that all sets are classes (but classes are not necessarily sets).

4.3 The Basic Extralogical Axioms of the Extended Theory, including the axiom of extensionality, are as follows.

$$y \in \{x \mid \Phi(x)\} \leftrightarrow \Phi(y)$$
$$A = B \leftrightarrow \forall x(x \in A \leftrightarrow x \in B)$$
$$A \in B \leftrightarrow \exists x(x = A \wedge x \in B) \quad \square$$

The statements $A = y \leftrightarrow \forall x(x \in A \leftrightarrow x \in y)$, $y = z \leftrightarrow \forall x(x \in y \leftrightarrow x \in z)$ and $A \in y \leftrightarrow \exists x(x = A \wedge x \in y)$ are consequences of the axioms 4.3 by the rule 4.2. The first and third axioms of 4.3 are not proper axioms of set theory since they convey no information concerning sets. The second axiom of 4.3 is partly a proper axiom of set theory since it includes the axiom of extensionality.

4.4 Terminology. We shall use "*A is a set*" as an abbreviation for $\exists x(x = A)$. We shall say "*A is a proper class*" for the negation of "*A* is a set", i.e., for $\neg\exists x(x = A)$. We shall also write $\{x \in A \mid \Phi(x)\}$ for $\{x \mid x \in A \wedge \Phi(x)\}$. $\square$

Let us denote by P the theory formulated in the basic language, with the axiom of extensionality as its only extralogical axiom. Let P^* be the theory formulated here in the extended language with the axioms 4.3 in addition to the logical axioms. We claimed in §1, informally, that in P^* we are able to do only what we can already do in P, with the difference that in the extended language we can handle by means of one formula infinitely many formulas of the basic language, using class variables. When we say that we are able to do in P^* only what we can do already in P, we mean two things by the word "do", namely, to express and to prove. Our claim is fully established by the following two theorems.

4.5 The Eliminability Theorem. (Comparison of the expressive power of the extended language and the basic language.) *Let $\Phi(A_1, \ldots, A_n)$ be a formula of the extended language with no class variables other than indicated. Let $\{x_1 \mid \Psi_1(x_1)\}$, $\ldots$, $\{x_n \mid \Psi_n(x_n)\}$ be class terms which do not contain class variables. Then there is a formula ϕ of the basic language such that*

$$P^* \vdash \Phi(\{x_1 \mid \Psi_1(x_1)\}, \ldots, \{x_n \mid \Psi_n(x_n)\}) \leftrightarrow \phi.$$

Such a formula ϕ is said to be a basic instance *of $\Phi(A_1, \ldots, A_n)$ corresponding to the class-terms $\{x_i \mid \Psi_i(x_i)\}$, $i = 1, \ldots, n$.*

Remark. We had to require in 4.5 that the class terms $\{x_i \mid \Psi_i(x)\}$ contain no class variables in order that the formula $\Phi(\{x_1 \mid \Psi_1(x_1)\}, \ldots, \{x_n \mid \Psi_n(x_n)\})$ contain no class variables, which it must if it is to be equivalent to a single formula of the basic language.

The eliminability theorem will be proved in the appendix (X.1.3). Let us only note that it says that whatever $\Phi(A_1, \ldots, A_n)$ says about the particular classes $A_i = \{x_i \mid \Psi_i(x_i)\}$, $i = 1, \ldots, n$, can be said by a basic instance ϕ of it, which is in the basic language.

4.6 The Conservation Theorem. (Comparison of the deductive power of corresponding axiom systems in the basic language and in the extended language.)

(i) *Let* ϕ *be any formula of the basic language.* $P^* \vdash \phi$ *iff* $P \vdash \phi$.

(ii) *Let* T *be a theory formulated in the basic language and* T^* *a theory formulated in the extended language such that for every formula* ϕ *of the basic language* $T^* \vdash \phi$ *iff* $T \vdash \phi$. *Then, if* Q *is any set of axioms of the basic language we have, again, for every* ϕ *of the basic language* $T^* \cup Q \vdash \phi$ *iff* $T \cup Q \vdash \phi$.

(iii) *If* T *and* T^* *are as in* (ii), $\Psi(A)$ *is a formula of the extended language having no class variables other than* A, *and* R *is a set of basic instances of* $\Psi(A)$ *(or of formulas equivalent in* T *to basic instances of* $\Psi(A)$*) which contains for each class term* $\{x \mid \chi(x)\}$ *a corresponding basic instance of* $\Psi(A)$ *(or a formula equivalent to it in* T*), then we have for every* ϕ *of the basic language* $T^* \cup \{\Psi(A)\} \vdash \phi$ *iff* $T \cup R \vdash \phi$.

Partial proof. Parts (i) and (iii) of this theorem will be proved in the appendix (X.1.4 and X.1.5). Part (ii) is shown as follows. Suppose that $T^* \cup Q \vdash \phi$, then for some finite subset $\{\chi_1, \ldots, \chi_n\}$ of Q we have $T^* \cup \{\chi_1, \ldots, \chi_n\} \vdash \phi$ (since a proof can use only finitely many axioms). Therefore, we have $T^* \vdash \chi_1 \wedge \cdots \wedge \chi_n \rightarrow \phi$. Since $\chi_1 \wedge \cdots \wedge \chi_n \rightarrow \phi$ is a formula in the basic language (Q being a set of formulas in the basic language) we have, by the hypothesis of (ii), also $T \vdash \chi_1 \wedge \cdots \wedge \chi_n \rightarrow \phi$ and hence $T \cup \{\chi_1, \ldots, \chi_n\} \vdash \phi$ and $T \cup Q \vdash \phi$. The other direction, namely, that if $T \cup Q \vdash$ also $T^* \cup Q \vdash \phi$, is shown in the same way. □

4.7 Applying the Conservation Theorem. Altogether, the conservation theorem tells us that if we start with P and P^* and keep extending them as in 4.6(ii) and 4.6(iii) we always get corresponding theories which are the same as far as the basic language is concerned. The correspondence between the extensions in 4.6(iii), of T^* by $\{\Psi(A)\}$ and of T by R, is indeed natural, since under the interpretation given in the last section the formula $\Psi(A)$ stands for nothing more than what is obtained from it after substituting all class-terms $\{x \mid \chi(x)\}$ for A, and this gives exactly the basic instances of $\Psi(A)$.

4.8 Asserting the Existence of Classes. We shall sometimes formulate theorems of the kind

(1) There exists a class B such that $\Psi(\mathfrak{a}_1, \ldots, \mathfrak{a}_n, B)$,
or, stated somewhat differently,
for all $\mathfrak{a}_1, \ldots, \mathfrak{a}_n$ there exists a class B such that $\Psi(\mathfrak{a}_1, \ldots, \mathfrak{a}_n, B)$,

where $\mathfrak{a}_1, \ldots, \mathfrak{a}_n$ are class or set variables, and Ψ has no free variables other than indicated. This is to be understood as a promise, fulfilled explicitly or implicitly in the proof of the theorem, to provide a class term $\{x \mid \Phi(x, \mathfrak{a}_1, \ldots, \mathfrak{a}_n, y_1, \ldots, y_k)\}$, where $y_1, \ldots, y_k$ are possible additional parameters, and to prove the statement

(2) $\exists y_1 \ldots \exists y_k \Psi(\mathfrak{a}_1, \ldots, \mathfrak{a}_n, \{x \mid \Phi(x, \mathfrak{a}_1, \ldots, \mathfrak{a}_n, y_1, \ldots, y_k)\})$.

(1) is not a statement of the extended language, since the extended language does not admit (existential) quantifiers over class variables. (1) is just a way of telling the reader to expect (2). Note that, on the other hand, "For every class A, $\Psi(A)$" stands directly for the statement $\Psi(A)$ of the extended language.

4.9 Exercises. (i) Formulate the axiom of comprehension (2.3) as a single statement about classes.
(ii) Prove that $\{x \mid x \notin x\}$ is a proper class. □

4.10 Adding Relation and Function Symbols. In the development of set theory, we shall have to add to our extended language, as we should have done in the case of any other mathematical theory, symbols for defined relations and functions. For practical reasons, this is absolutely necessary for going beyond trivialities, since the original language contains only the simple relation-symbols of equality and membership, and to express even relatively simple relations between sets and classes we need very long formulas in the original language. (One can avoid extending the language by regarding definitions as rules for abbreviation, but once one introduces function symbols, and uses them to build up complicated terms, it is difficult to keep track of which formula is abbreviated by a given formula.)

The new symbols thus introduced can be eliminated from the language, and they do not add anything to the expressive or deductive power of the language. Since, unlike the case of the introduction of classes, the introduction of new function and relation symbols is familiar to the reader, being used so often throughout all of mathematics, we shall not give here the corresponding eliminability and conservation theorems.

Let us denote with lower case German letters variables which can be either class variables or set variables. A relation $R(\mathfrak{a}_1, \ldots, \mathfrak{a}_n)$ is defined by means of the statement

(1) $$R(\mathfrak{a}_1, \ldots, \mathfrak{a}_n) \leftrightarrow \Phi(\mathfrak{a}_1, \ldots, \mathfrak{a}_n)$$

where Φ is a formula with no free variables other than $\mathfrak{a}_1, \ldots, \mathfrak{a}_n$.

A function whose values are sets is defined as follows. Let $\Phi(\mathfrak{a}_1, \ldots, \mathfrak{a}_n, y)$ be a formula with no free variables other than $\mathfrak{a}_1, \ldots, \mathfrak{a}_n, y$, such that in whatever theory T we are dealing with at that time we have

$$T \vdash \exists! y \Phi(\mathfrak{a}_1, \ldots, \mathfrak{a}_n, y).$$

We define y as a function $F(\mathfrak{a}_1, \ldots, \mathfrak{a}_n)$ of $a_1, \ldots, \mathfrak{a}_n$ by the statement

(2) $$\Phi(\mathfrak{a}_1, \ldots, \mathfrak{a}_n, F(\mathfrak{a}_1, \ldots, \mathfrak{a}_n))$$

4.11 More on Function Symbols. When we define functions in practice, we will sometimes not exhibit the formula $\Phi(\mathfrak{a}_1, \ldots, \mathfrak{a}_n, y)$ explicitly, but it will be clear which is the appropriate formula $\Phi(\mathfrak{a}_1, \ldots, \mathfrak{a}_n, y)$. Also, in several cases, we will not be interested in all n-tuples $\mathfrak{a}_1, \ldots, \mathfrak{a}_n$ but only in a restricted collection of such n-tuples. In such a case, we shall prove the existence and uniqueness of y only for the relevant $\mathfrak{a}_1, \ldots, \mathfrak{a}_n$. For other $\mathfrak{a}_1, \ldots, \mathfrak{a}_n$ we shall assume, tacitly, that y has some fixed value, for example 0 (which is the null set, to be defined in the next section).

Functions whose values are classes are given by the class terms. We can, of course, abbreviate an expression $\{x \mid \Phi(x, \mathfrak{a}_1, \ldots, \mathfrak{a}_n)\}$ by $F(\mathfrak{a}_1, \ldots, \mathfrak{a}_n)$, but this will be regarded as a mere abbreviation.

4.12 Discussion. When we defined relations and functions whose values are sets we allowed the arguments $\mathfrak{a}_1, \ldots, \mathfrak{a}_n$ to be set or class variables. We shall see that, in principle, there is an advantage to using only class variables for $\mathfrak{a}_1, \ldots, \mathfrak{a}_n$. This would not lead to any difficulty since we can always substitute set variables for the class variables. If we intend the variable $\mathfrak{a}_1$ of $F(\mathfrak{a}_1, \ldots, \mathfrak{a}_n)$, e.g., to vary only over sets, we can still use a class variable for $\mathfrak{a}_1$ and set the value of $F(\mathfrak{a}_1, \ldots, \mathfrak{a}_n)$ in the case where $\mathfrak{a}_1$ is a proper class to be some arbitrary object, such as 0. The advantage of using class variables only is that we shall be able to substitute also class terms and class variables in the expression for the function or the relation, which is very useful since sets are often given by means of class terms. Therefore, we shall assume that whenever a set variable is used as an argument in a definition of a relation or a function, this definition will only be considered as an informal way of writing the true definition, in which all the arguments are class variables. This is embodied in the following agreement.

4.13 Agreement on Notation. A relation defined as $R(A_1, \ldots, A_l, y_1, \ldots, y_m)$ will in fact be $R(A_1, \ldots, A_l, B_1, \ldots, B_m)$, where it will be defined to be false when at least one of the B_i's is a proper class. A function defined as $F(A_1, \ldots, A_l, y_1, \ldots, y_m)$ is indeed $F(A_1, \ldots, A_l, B_1, \ldots, B_m)$ and is defined to have the value 0 when at least one of the B_i's is a proper class. □

The reason why we still intend in practice to use also set variables when we define relations and functions is that often we are interested only in the values of the functions in the case where some fixed arguments are sets; defining the function explicitly also for the case where the arguments are classes may divert the reader's attention from the main point.

Later on we shall mostly use the terms "relation" and "function" for certain kinds of sets and classes. Those notions are somewhat related to the relations and functions mentioned here, but they are still different notions and one has to bear that in mind.

Once we have symbols for defined functions $F(\mathfrak{a}_1, \ldots, \mathfrak{a}_n)$, we get by means of these symbols and by means of the class terms, more complicated expressions which we call *terms*. Terms are the expressions obtained from expressions like $F(\mathfrak{a}_1, \ldots, \mathfrak{a}_n)$ and class terms by repeated substitution of such expressions for variables in such expressions. Examples of terms are

$$F(G(H_1(a), H_2(a), b), c, I(a)), \quad \{\{u\}, \{u, v\}\}, \quad \log(\sin(x)),$$
$$H(\{x \mid \Phi(x)\}, y, A), \quad \{y \mid F(y, z) \in \{u \mid \chi(u)\}\}.$$

Terms other than class terms and class variables, i.e., set variables or terms whose leftmost symbol is a function symbol, will be called set terms. The values of such terms must always be sets, by our definition of the functions. Note that a set term, while not a class term, may have a class term as a part, as in $H(\{x \mid \Phi(x)\}, y, A)$. □

It is now convenient to extend our notation for class terms.

4.14 Agreement on Notation. Let τ be a term (which may be a set term or a class term). Let us pick some variables $x_1, \ldots, x_n$ and call them the *active* variables of τ (for the purposes of the following syntactical construction), while the variables which occur free in τ other than $x_1, \ldots, x_n$ will be called parameters. We shall write τ as $\tau(x_1, \ldots, x_n)$ to emphasize that $x_1, \ldots, x_n$ are the active variables. Let $\Phi(x_1, \ldots, x_n)$ be any formula which, again, may have free variables other than $x_1, \ldots, x_n$. The class of all the *sets* which have the form $\tau(x_1, \ldots, x_n)$ for $x_1, \ldots, x_n$ such that $\Phi(x_1, \ldots, x_n)$, i.e., the class

(1) $$\{x \mid \exists x_1 \ldots \exists x_n(\Phi(x_1, \ldots, x_n) \wedge x = \tau(x_1, \ldots, x_n))\}$$

will be denoted by

(2) $$\{\tau(x_1, \ldots, x_n) \mid \Phi(x_1, \ldots, x_n)\}$$

The parameters of the class term (1) are the free variables of τ and Φ other than $x_1, \ldots, x_n$.

Note that the value of $\tau(x_1, \ldots, x_n)$ for $x_1, \ldots, x_n$ such that $\Phi(x_1, \ldots, x_n)$ can also be a proper class. Such a class will, of course, not be a member of $\{\tau(x_1, \ldots, x_n) \mid \Phi(x_1, \ldots, x_n)\}$. As a matter of practice, we will avoid using expressions of the form (2) unless we know that all the values of $\tau(x_1, \ldots, x_n)$ are sets for $x_1, \ldots, x_n$ such that $\Phi(x_1, \ldots, x_n)$.

We do not introduce a consistent notation to distinguish between the active variables and the parameters of (2); but, unless otherwise mentioned, the active variables will be those which occur free in both τ and ϕ, while all the variables occurring in only one of τ and ϕ are parameters. Examples of this notation are:

$\{2n \mid n \in w\}$—where w is a parameter,

$\{\langle xy \rangle \mid x \in s \wedge y \in t\}$—where s and t are parameters,

$\{\{F(x, y) \mid x \in a\} \mid y \in b\}$—here we have the term $\{F(x, y) \mid x \in a\}$ inside the outer term; $\{F(x, y) \mid x \in a\}$ has y and a as its parameters, $\{\{F(x, y) \mid x \in a\} \mid y \in b\}$ has a and b as its parameters.

4.15. Definition. (i) $\bigcup A = \{x \mid (\exists u \in A)\,(x \in u)\}$.

$\bigcup A$ is the class of all members of the members of A, and is called the *union* of A.

(ii) $\bigcap A = \{x \mid (\forall u \in A)\,(x \in u)\}$.

$\bigcap A$ is the class of all common members of all members of A and is called the *intersection* of A. Note that $\bigcap 0 = V$, where 0 is the class with no members (5.9) and V is the universal class (5.16(iv)), since every set belongs to each member of 0.

(iii) $\bigcup_{\Phi(x_1, \ldots, x_n)} \tau(x_1, \ldots, x_n) = \bigcup\{\tau(x_1, \ldots, x_n) \mid \Phi(x_1, \ldots, x_n)\}$.

Read: The union of all sets $\tau(x_1, \ldots, x_n)$ *such that* $\Phi(x_1, \ldots, x_n)$.

(iv) $\bigcap_{\Phi(x_1, \ldots, x_n)} \tau(x_1, \ldots, x_n) = \bigcap\{\tau(x_1, \ldots, x_n) \mid \Phi(x_1, \ldots, x_n)\}$.

Read: The intersection of all sets $\tau(x_1, \ldots, x_n)$ *such that* $\Phi(x_1, \ldots, x_n)$.

5. *The Axioms of Set Theory*

As we have already made clear, we shall consider set theory as a theory developed, in principle, in the basic language, while in practice we shall use the extended language which admits also class variables. We shall take care to formulate the axioms of set theory in the basic language, making use, if convenient, of some defined relations symbols, but not of class variables. The axioms of set theory which we shall introduce are of two kinds; we shall have several single axioms (such as the axiom of extensionality) and several axiom schemas. As follows from the conservation theorem (4.6(ii)), when we add the single axioms to the logical axioms of the basic language and the axiom of extensionality and, at the same time, also to the logical and basic extralogical axioms of the extended language we get two theories in which exactly the same statements of the basic language are provable. As to the axiom schemas of set theory, we shall find for each one of them a single statement of the extended language, called a *class form* of that axiom schema, such that the axiom schema consists exactly of the basic instances of that statement (or of statements equivalent to them). Part (iii) of the conservation theorem tells us that adding the axiom schemas to the theory we have in the basic language and adding their class forms to the theory we have in the extended language yields again two theories in which exactly the same statements of the basic language are provable. An axiom schema itself will be called the *set form* of the axiom schema, in order to distinguish it from the class form of the axiom schema. The main reason for using in the extended language the class forms rather than the original axiom schemas is not the "economy" obtained by having a single axiom instead of a schema, but the fact that if we tried to use in the extended language the axiom schemas the handling of set theory in the extended language would be very clumsy and the advantage of the extended language over the basic language would be lost. On the other hand, by using for the extended language the natural class forms of the axiom schemas we obtain a system which has all the necessary means for a smooth handling of classes.

Having abandoned the axiom schema of comprehension, we are left only with the axiom of extensionality (2.1). In the present section, we shall go through the other axioms of set theory.

The following axioms of union, power-set and replacement can be regarded as direct instances of the axiom schema of comprehension. As such, the sets whose existence they claim are unique by Proposition 2.7. From the point of view of the limitation of size doctrine the justification of these axioms is rather obvious; the sets whose existence is claimed by these axioms are not much bigger than the sets whose existence is assumed by these axioms.

5.1 Axiom of Union (Cantor 1899, Zermelo 1908). $\forall z \exists y \forall x(x \in y \leftrightarrow \exists u(x \in u \wedge u \in z))$. In words: For every set z there is a set y which consists of the members of the members of z. This can also be written as "$\{x \mid \exists u(x \in u \wedge u \in z)\}$ is a set", or as "$\bigcup z$ is a set".

5.2 Definition. (i) $A \subseteq B \leftrightarrow \forall z(z \in A \to z \in B)$ (read: A is a *subclass* of B, or a *subset* of B if A is a set, or B *includes* A).

(ii) $A \subset B \leftrightarrow A \subseteq B \wedge \exists z(z \in B \wedge z \notin A)$ (read: A is a *proper subclass* of B, or A is a *proper subset* of B if A is a set, or B *properly includes* A).

5.3 Proposition. $A \subset B \leftrightarrow A \subseteq B \wedge A \neq B$. □

5.4 Axiom of Power-Set (Zermelo 1908). $\forall z \exists y \forall x(x \in y \leftrightarrow x \subseteq z)$.
In words: For every set z there is a set y which consists of all subsets of z. This can also be written as "$\{x \mid x \subseteq z\}$ is a set".

5.5 Definition. $\mathbf{P}(A) = \{x \mid x \subseteq A\}$. $\mathbf{P}(A)$ is the class of all subsets of A and is called the *power-class* of A. If A is a set, then by the axiom of power-set, $\mathbf{P}(A)$ also is a set, and it is called the *power-set* of A.

5.6 Axiom (-Schema) of Replacement (Skolem 1923, Fraenkel 1922, with the idea going back to Cantor 1899). We shall present now the set form of this axiom schema; the class form will be given in 6.19.

$$\forall u \forall v \forall w(\psi(u, v) \wedge \psi(u, w) \to v = w) \to \forall z \exists y \forall v(v \in y \leftrightarrow (\exists u \in z)\psi(u, v))$$

where $\psi(u, v)$ is a formula which has no free occurrences of the variables y, z and w (this is required in order to avoid confusion of the y or w free in $\psi(u, v)$ with the y and w mentioned explicitly in the axiom of replacement), and $\psi(u, w)$ is the formula obtained from $\psi(u, v)$ by substituting in it w for v. $\psi(u, v)$ may have free occurrences of z and of other variables.

The meaning of the axiom schema of replacement is as follows. The hypothesis of the axiom is $\forall u \forall v \forall w(\psi(u, v) \wedge \psi(u, w) \to v = w)$, i.e., for every u there is at most one v such that $\psi(u, v)$. Thus, v can be taken to be "a function of u" whenever v exists. In other words, there is a function F not necessarily "defined for each u" such that $\psi(u, F(u))$ holds whenever it is defined. What the axiom claims is that there is a *set* y which consists exactly of the images of the members of z under the function F. If $\psi(u, v)$ has free variables $x_1, \ldots, x_n$ other than u and v we can regard them as parameters and write $\psi(u, v)$ as $\psi(u, v; x_1 \ldots, x_n)$. The v such that $\psi(u, v; x_1, \ldots, x_n)$ depends not only on u but also on $x_1, \ldots, x_n$; for given $x_1, \ldots, x_n$ the set y consists of those v's for which $(\exists u \in z)\psi(u, v; x_1, \ldots, x_n)$.

The axiom of replacement is, as we see, an axiom schema. As shown by Montague 1961, the fact that this axiom cannot be given as a single axiom of the basic language is not an accident but an inherent feature of set theory. This, of course, does not stand in our way when we come to present that axiom as a single statement of the extended language by using class variables, as we shall do in §6.

5.7 Proposition. *Let* $\tau(u, x_1, \ldots, x_n)$ *be a term, then the class* $\{\tau(u, x_1, \ldots, x_n) \mid u \in z\}$ *is a set. As a consequence, also* $\bigcup_{u \in z} \tau(u, x_1, \ldots, x_n)$ *is a set (by the axiom of union).*

Remark. Recall that the class $\{\tau(u, x_1, \ldots, x_n \ldots, x_n) \mid u \in z\}$ contains only those $\tau(u, x_1, \ldots, x_n)$'s which are *sets*.

Proof. We shall prove this theorem now only for terms $\tau(u, x_1, \ldots, x_n)$ with no class variables. For terms with class variables we need the class form of the axiom of replacement to prove our theorem; we shall do it in 6.22. Till then we shall draw the attention of the reader whenever we shall use the present theorem for terms τ with class variables.

To prove our proposition for a term $\tau(u, x_1, \ldots, x_n)$ with no class variables, let $\psi(u, v)$ be a formula of the basic language equivalent to $v = \tau(u, x_1, \ldots, x_n)$. That $\{\tau(u, x_1, \ldots, x_n) \mid u \in z\}$ is a set follows from the axiom of replacement with $\psi(u, v)$ as chosen here. □

None of the axioms we gave until now postulated the *unconditional* existence of a set. The three last axioms said that for every set z there is a set y such that ....., but this does not say that there is any set at all. The axiom of infinity, which will be presented later does postulate the existence of a set. However, since $\exists x(x = x)$ is a theorem of first order logic, and since our only objects here are sets, we can say that we assumed the existence of at least one set when we decided to adopt first order logic here.

5.8 Proposition (Null-set). *There exists a unique set which has no members.*

Proof. Let z by any set (there is a set by the above arguments). Let us take for $\psi(u, v)$ in the axiom of replacement a logically invalid formula (e.g., $u \neq u$), then the set y whose existence is claimed by the axiom of replacement has no members. If y' is another set with no members, then $y' = y$ by the axiom of extensionality. □

5.9 Definition (Boole 1847, 1854). 0 is the set with no member, i.e., $\{x \mid x \neq x\}$ (the *null-set*). Any other class or set, i.e., any class or set which has at least one member is said to be a *non-void*, or *non-empty*, class or set.

5.10 Proposition. $x \subseteq 0 \leftrightarrow x = 0$. □

By 5.10, $\mathbf{P}(0)$ has 0 as its only member. Anticipating notation to be introduced in 5.12, we denote $\mathbf{P}(0)$ also by $\{0\}$. Since $0 \in \{0\}$ but $0 \notin 0$ we have, by the logical rules of equality, $\{0\} \neq 0$. $\mathbf{P}(\{0\})$ consists exactly of the members 0 and $\{0\}$ and we denote it by $\{0, \{0\}\}$. The two members of $\{0, \{0\}\}$ are unequal as we have seen.

5.11 Proposition (Pairing). $\forall s \forall t \exists y \forall x(x \in y \leftrightarrow x = s \vee x = t)$, *i.e., for all s and t there is a set y which consists exactly of s and t.*

Proof. Let us use the axiom of replacement where we take for $\psi(u, v)$ the formula $(u = 0 \wedge v = s) \vee (u = \{0\} \wedge v = t)$ and where we take for z the set $\{0, \{0\}\}$. The set y whose existence is claimed by the axiom of replacement consists of exactly s and t. □

Notice that in the proof of 5.11 we use an instance of the axiom schema of replacement where we take for $\psi(u, v)$ a formula containing also defined symbols.

Even though the $\psi(u, v)$ in the schema is intended to vary only over formulas of the basic language, we can, in fact, take for $\psi(u, v)$ any formula of the extended language which does not mention class variables, since every such formula is indeed equivalent to a formula $\psi'(u, v)$ of the basic language, and the instances of the axiom schema of replacement corresponding to $\psi(u, v)$ and $\psi'(u, v)$ are equivalent.

5.12 Definition. (i) $\{s, t\} = \{x \mid x = s \vee x = t\}$. $\{s, t\}$ is the class which consists of exactly s and t; it is a set by Proposition 5.11; it is called the *pair* s, t.

(ii) $\{s\} = \{s, s\}$. $\{s\}$ consists of the single member s and is called the *singleton of* s.

By our agreement of 4.13, Definition 5.12 is, in fact, a definition of $\{A, B\}$, where the value of $\{A, B\}$ is 0 if one of A and B is a proper class.

5.13 Theorem-Schema (Subsets). $\forall z \exists y \forall x (x \in y \leftrightarrow x \in z \wedge \phi(x))$ *where* y *is not a free variable in* $\phi(x)$, *i.e., for all* z *there is a set* y *which consists exactly of the members of* z *for which* $\phi(x)$ *holds. This can also be written as* "$\{x \in z \mid \phi(x)\}$ is a set."

Proof. Take for $\psi(u, v)$ in the axiom of replacement the formula $\phi(u) \wedge u = v$. Then the set y whose existence is claimed by the axiom of replacement is exactly the set y of the theorem. ☐

The schema of Theorem 5.13 was used by Zermelo 1908 as an axiom of set theory, and is still often used as such today (see 5.14); it is called the *axiom of subsets*, or, the *axiom of separation.* An informal version of this schema was already mentioned by Cantor 1899.

5.14 Exercise. Prove that the instance $\exists y \forall x (x \in y \leftrightarrow \phi(x))$ of the axiom schema of comprehension follows from the statement $\exists y \forall x (\phi(x) \rightarrow x \in y)$ by means of the axiom of subsets. ☐

By the *weak versions* of the axioms of union, power-set and replacement we mean the axioms

$$\begin{aligned} &\forall z \exists y \forall x \forall u (x \in u \wedge u \in z \rightarrow x \in y) \\ (1) \qquad &\forall z \exists y \forall x (x \subseteq z \rightarrow x \in y) \\ &\forall u \forall v \forall w (\psi(u, v) \wedge \psi(u, w) \rightarrow v = w) \rightarrow \forall z \exists y \forall u \forall v (u \in z \wedge \phi(u, v) \rightarrow v \in y). \end{aligned}$$

These axioms differ from the axioms of union, power-set and replacement by claiming only the existence of sets which *include*, but not necessarily coincide with, the sets whose existence is claimed by the respective axioms of union, power-set and replacement. By what is claimed in 5.14 the axioms of union, power-set and replacement can be replaced by their weak versions (1) taken together with the axiom of subsets. (1) will be convenient for the purpose of verifying that certain structures are indeed models of set theory, since it is usually much easier to obtain a set which just includes a desired set than to obtain the desired set itself. The only *exact* construction of a set that will be needed is that of the set required by the axiom of subsets.

5.15 Exercise. Show, without using any axiom of set theory, that the axiom schema of subsets (5.13) is equivalent to the schema $\forall z[\forall x(\phi(x)\rightarrow x\in z)\rightarrow \exists y\forall x(x\in y\leftrightarrow\phi(x))]$, where y is not free in $\phi(x)$.
Hint. z is allowed to be free in $\phi(x)$. □

5.16 Definition (Boole 1874, 1854). (i) $A\cup B=\{x\mid x\in A\vee x\in B\}$.
Read: the *union* of A and B. $A\cup B$ is the class which consists of all the members of A and of all the members of B.

(ii) $A\cap B=\{x\mid x\in A\wedge x\in B\}$.
Read: the *intersection* of A and B. $A\cap B$ is the class which consists of all the common members of A and B.

(iii) $A\sim B=\{x\mid x\in A\wedge x\notin B\}$.
Read: the *difference* of A and B. $A\sim B$ is the class which consists of all the members of A which are not members of B. If $B\subseteq A$ then we call $A\sim B$ also the *complement of B with respect to A*, or just the *complement* of B if it is clear which class A is intended.

(iv) $V=\{x\mid x=x\}$.
V is the class of all objects; it is called the *universal class*.

(v) $\sim A=V\sim A$.
$\sim A$ is the class which consists exactly of all objects which are not members of A. □

The basic properties of these operations, as well as their order of precedence when written without parentheses, are assumed to be well-known to the reader. We shall write $\bigcup_{i=1}^{n}A_i$ and $\bigcap_{i=1}^{n}A_i$ for $A_1\cup A_2\cup\ldots\cup A_n$ and $A_1\cap A_2\cap\ldots\cap A_n$, respectively.

5.17 Class Form of the Axiom of Subsets. $\forall z(z\cap A$ is a set).
The name we gave 5.17 is justified by the following proposition.

5.18 Proposition. *The addition of the class form 5.17 of the axiom of subsets to the extended language without the axiom of replacement yields in the basic language the same theorems as the axiom of subsets 5.13.*

Proof. $\forall z(z\cap A$ is a set) is an abbreviation of $\forall z\exists y(y=z\cap A)$ which is equivalent to $\forall z\exists y(y=\{x\mid x\in z\wedge x\in A\})$. To find the basic instances of 5.17 we substitute $\{x\mid\phi(x)\}$ for A and obtain, by means of the axioms 4.3, the equivalent statement $\forall z\exists y\forall x(x\in y\leftrightarrow x\in z\wedge\phi(x))$ which is a typical instance of the axiom of subsets. Our theorem follows now from the conservation theorem (4.6). □

There is no need to add 5.17 as an axiom of the extended system since it will be a consequence of the class form of the axiom of replacement (6.19 below). In 5.19 we shall use 5.17 in the proofs as a temporary axiom.

5.19 Proposition. (i) *If A is a set and $B\subseteq A$, then B too is a set.*

(ii) *If A is a set, so are $A\cap B$ and $A\sim B$.*

(iii) *If A and B are sets, so is $A\cup B$ $(=\bigcup\{A, B\})$.*

(iv) *V is a proper class.*
(v) *$\sim x$ is a proper class.*
(vi) *If $A \neq 0$, then $\bigcap A$ is a set.*

Hint of proof. (i)–(ii). Use 5.17. (iv). By contradiction, using 4.9(ii) and (i). (v). By (iii) and (iv). □

5.20 Exercise. Prove that 5.7 is in fact equivalent to the axiom schema of replacement.
Hint. If $\phi(u, v; x_1, \ldots, x_n)$ and z are such that $(\exists u \in z)\exists! v\phi(u, v; x_1, \ldots, x_n)$ then let v_0 be such a v for some $u \in z$. Define

> $F(u, x_1, \ldots, x_n)$ = the unique v such that $\phi(u, v; x_1, \ldots, x_n)$ if there is such a unique v, and v_0 if there is no such unique v. □

One of the basic tenets of set theory is the existence of infinite sets. We can actually write down now the statement "there exists an infinite set" as our next axiom, provided we define what we mean by an infinite set. It is, however, more convenient to postulate now the existence of a particular set which will turn out to be infinite when we define what "infinite" means.

5.21 Axiom of Infinity (Zermelo 1908). $\exists z(0 \in z \wedge (\forall x \in z)(\forall y \in z)(x \cup \{y\} \in z))$. □

For the sake of the record, we will mention here also the other axioms, even though we shall discuss them only later.

5.22 Axiom of Schema Foundation (Skolem 1923, von Neumann 1925).

$$\exists x\phi(x) \rightarrow \exists x(\phi(x) \wedge (\forall y \in x) \neg \phi(y)), \text{ where } y \text{ is not free in } \phi(x).$$

We can formulate the axiom of foundation as the following single statement.

5.23 Class Form of the Axiom of Foundation. $A \neq 0 \rightarrow (\exists x \in A)(x \cap A = 0)$.

It is clear that the basic instances of 5.23 are exactly the instances of the axiom of foundation, and therefore, by the conservation theorem (4.6) we can take 5.23 to be the axiom of foundation of the extended language.

5.24 The Zermelo–Fraenkel Set Theory. The system consisting of the axioms of extensionality, union, power-set, replacement, infinity, and foundation is called the *Zermelo–Fraenkel set theory* and is denoted by ZF. An additional axiom is the axiom of choice (formulated in Chapter V but discussed already in II. 2.23. The system of axioms obtained from ZF by adding to it the axiom of choice will be denoted by ZFC. The reason for this segregation of the axiom of choice is not because this axiom is a dubious one. It is because the system ZF is sufficient for many set theoretical purposes. Also, investigations concerning statements of set theory which are contradicted by the axiom of choice turned out to be very interesting and to have considerable applications to the set theory ZFC and its extensions.

5.25 Historical Remark. The first system of axioms along this line was introduced by Zermelo 1908. His system contained essentially all the axioms of ZF, except that he had the axiom of subsets instead of the axiom of replacement, and he did not have the axiom of foundation. Informal versions of the axiom of replacement were suggested by Cantor 1899 and Mirimanoff 1917; formal versions were introduced by Fraenkel 1922 and Skolem 1923. The axiom of foundation was added by Skolem 1923 and von Neumann 1925.

5.26 Use of Axioms in This Book. All theorems in this book will be proved in ZF or in ZFC. Whenever we shall use the axiom of choice in the proof of a theorem we shall write the letters Ac next to the number of the theorem as in 2.25Ac. Also the axiom of foundation will be used quite sparingly and it shall be clear from the proof of each theorem whether this axiom was used in it.

6. *Relations and Functions*

The concepts of set and membership in a set are by no means the only concepts of set theory; there are many additional concepts which are absolutely vital for any useful development of set theory such as the concepts of ordered pair, relation and function. The beauty of the matter is that even though these entities are a priori distinct from the sets (and the classes), they can be represented by sets and classes in a completely satisfactory way. In the present section we shall see how this representation is carried out and investigate the properties of those concepts. This leaves membership of sets as the only fundamental relation of set theory from which all other concepts can be defined, and their properties can be established by means of the axioms which are about sets.

A very useful notion of set theory is the notion of a *pairing function.* A pairing function is a function $\langle x, y\rangle$ (with the arguments x and y) such that

(1) $$\langle x, y\rangle = \langle u, v\rangle \leftrightarrow x = u \wedge y = v.$$

We refer to $\langle x, y\rangle$ as the *ordered pair* x, y; the particular way in which the function $\langle x, y\rangle$ is defined is of no concern for ordinary mathematical purposes.

6.1 Definition (Wiener 1914, Kuratowski 1921). $\langle x, y\rangle = \{\{x\}, \{x, y\}\}$. A set z is said to be an *ordered pair* if for some x and y, $z = \langle x, y\rangle$.

6.2 Proposition. $\langle x, y\rangle = \langle u, v\rangle \rightarrow x = u \wedge y = v.$

We prove first a simple lemma.

6.3 Lemma. *If* $\{u, a\}=\{u, b\}$ *then* $a=b$.

Proof. If $a=u$ then since $b \in \{u, b\}=\{u, a\}=\{u\}$ then $b=u=a$. If $a \neq u$ then since $a \in \{u, a\}=\{u, b\}$ we must have $a=b$. □

Proof of Proposition 6.2 $\{x\} \in \{\{x\}, \{\{x, y\}\}=\langle x, y\rangle=\langle u, v\rangle=\{\{u\}, \{u, v\}\}$, i.e., $\{x\} \in \{\{u\}, \{u, v\}\}$. Thus, either $\{x\}=\{u\}$ and then $u \in \{u\}=\{x\}$ and hence $u=x$, or $\{x\}=\{u, v\}$ and then, again, $u \in \{u, v\}=\{x\}$ and hence $u=x$. We have shown that in either case $u=x$. Since we have now $\{x\}=\{u\}$ and $\{\{x\}, \{x, y\}\}=\{\{u\}, \{u, v\}\}$ we get, by applying Lemma 6.3 once, $\{x, y\}=\{u, v\}$ and by applying it again $y=v$. □

6.4 Definition. Let z be an ordered pair. $1^{st}(z)$ and $2^{nd}(z)$ are the unique (by 6.3) x and y, respectively, such that $\langle x, y\rangle=z$. $1^{st}(z)$ and $2^{nd}(z)$ are called the *first* and *second* components, respectively, of z. We have $1^{st}(\langle x, y\rangle)=x$, $2^{nd}(\langle x, y\rangle)=y$, and if z is an ordered pair then $z=\langle 1^{st}(z), 2^{nd}(z)\rangle$.

6.5 Definition. $A \times B=\{\langle u, v\rangle \mid u \in A \wedge v \in B\}$. $A \times B$ is called the *Cartesian product* of A and B.

6.6 Proposition. *Let* $\langle x, y\rangle$ *be defined in any way such that 6.2 holds, then the Cartesian product of two sets is a set.*

Hint of proof. Use the axiom of replacement first to show that for every x $\{\langle x, y\rangle \mid y \in t\}$ is a set, then use the axiom of replacement and the axiom of union to obtain the required result. □

6.7 Exercises. (i) (Wiener 1914). Prove that the function defined by $\langle x, y\rangle'=\{\{\{x\}, 0\}, \{\{y\}\}\}$ is a pairing function, i.e., it satisfies (1).

(ii) Formulate a requirement analogous to (1) for the notion of an *ordered triple* and define a function which satisfies it.

Hint. (ii) Use the pairing function $\langle x, y\rangle$ or else proceed in a way similar to that used in (i). □

One of the main features of set theory is that in it we handle collections as objects. Notions closely related to the notion of a collection are the notions of a relation and a function. It is also very important for mathematics that we be able to handle relations and functions as objects. One example of this necessity is the frequent use in analysis of topological spaces whose members (points) are functions. It turns out that we do not have to introduce relations and functions as new kinds of objects; the sets and classes which we already have are sufficient for handling these notions. We shall see how to do it in the following definition.

6.8 Definition. A class S is said to be a (binary) *relation* if every member x of S is an ordered pair. We shall write $y \, S \, z$ for $\langle y, z\rangle \in S$.

For example, if $<$ is the natural order relation on the natural numbers (i.e., $\langle x, y\rangle \in <$ if and only if x is less than y), then we write $x<y$ for $\langle x, y\rangle \in <$. The class $S\times T$ is always a relation. Also the null-set 0 is a relation.

Historical Remark. This way of representing relations is essentially due to Hausdorff 1914 who represented order relations in a way similar to this one.

6.9 Definition. Let S be any class. (i) The *domain* of S, $\mathrm{Dom}(S)$, is the class $\{x \mid \exists y(\langle x, y\rangle \in S)\} = \{1^{\mathrm{st}}(z) \mid z \in S \wedge z \textit{ is an ordered pair}\}$, i.e., the class of the first components of the ordered pairs in S.

(ii) The *range* of S, $\mathrm{Rng}(S)$, is the class $\{y \mid \exists x(\langle x, y\rangle \in S)\} = \{2^{\mathrm{nd}}(z) \mid z \in S \wedge z \textit{ is an ordered pair}\}$, i.e., the class of the second components of the ordered pairs in S.

(iii) The *field* of S is the class $\mathrm{Dom}(S) \cup \mathrm{Rng}(S)$, i.e., the class of all objects which occur as a first or second component in an ordered pair in S.

(iv) $S^{-1} = \{\langle y, x\rangle \mid \langle x, y\rangle \in S\}$; S^{-1} is the *inverse relation* of S.

6.10 Proposition. (i) *The domain and the range of a set are sets.* (*This can be shown without using any property of the ordered pair except 6.2.*)

(ii) *If S is a proper class and a relation then at least one of* $\mathrm{Dom}(S)$ *and* $\mathrm{Rng}(S)$ *is a proper class.* (*Is the requirement that S be a relation necessary?*)

Hint of proof. (i) To prove that the domain is a set use the axiom of replacement with a mapping which gives the value x to $\langle x, y\rangle$ and the value 0 to all sets which are not ordered pairs. □

6.11 Exercises. (i) $S \subseteq \mathrm{Dom}(S) \times \mathrm{Rng}(S)$ if and only if S is a relation.

(ii) Under what additional assumption can you conclude from $A\times B = C\times D$ that $A=C$ and $B=D$?

(iii) $(S^{-1})^{-1} = S$ iff S is a relation.

(iv) $\mathrm{Dom}(S^{-1}) = \mathrm{Rng}(S)$, $\mathrm{Rng}(S^{-1}) = \mathrm{Dom}(S)$. □

6.12 Definition. (i) A relation F is said to be a *function* if for all x, y, z if $\langle x, y\rangle \in F$ and $\langle x, z\rangle \in F$ then $y=z$; in other words, if for all $x \in \mathrm{Dom}(F)$ there is a *unique* y such that $\langle xy\rangle \in F$.

(ii) For any class F $F(x)$ will denote the unique y such that $\langle x, y\rangle \in F$ if there is a unique such y, and $F(x)$ will denote 0 otherwise. $F(x)$ is called the *value of F for x*, or *at x*, and also the *image* of x under F. If there is no unique y such that $\langle x, y\rangle \in F$, and in particular if there is no such y at all, we shall say that *F is undefined at x* or that $F(x)$ is undefined (even though $F(x)$ is formally defined as 0).

(iii) We say that F maps *A into B*, or that F is a *mapping* of *A into B*, and we write $F: A \to B$, if F is a function, $\mathrm{Dom}(F) = A$ and $\mathrm{Rng}(F) \subseteq B$.

(iv) If F is a function, $\mathrm{Dom}(F) = A$ and $\mathrm{Rng}(F) = B$ we say that *F is a surjection* of A on B.

(v) A function F is said to be *one-one* if for every $y \in \mathrm{Rng}(F)$ there is a *unique* x such that $F(x) = y$.

(vi) If F is a one-one function mapping A into B it is said to be an *injection* of A into B; if F is also *onto* B it is said to be a *bijection* of A on B.

(vii) Functions of two variables are represented as functions whose domains consist of ordered pairs. Accordingly, we define $F(x, z)=F(\langle x, z\rangle)$. Functions of three variables are given as functions whose domains consist of ordered triples. Therefore we define $F(x, z, u)=F(\langle x, z, u\rangle)$.

6.13 Exercise. The function F is one-one iff F^{-1} is a function. □

6.14 Proposition (The principle of extensionality for functions). *If F and G are functions such that* $\mathrm{Dom}(F)=\mathrm{Dom}(G)$ *and for all* $x\in\mathrm{Dom}(F)$ $F(x)=G(x)$, *then* $F=G$.

Proof. We shall prove $F\subseteq G$, and since the assumptions are symmetrical with respect to F and G we have also $G\subseteq F$, hence, $F=G$ (by 4.3). Let $u\in F$, then since F is a function we have $u=\langle v, w\rangle$, where $v=1^{\mathrm{st}}(u)$, $w=2^{\mathrm{nd}}(u)$. Since $\langle v, w\rangle\in F$ we have $v\in\mathrm{Dom}(F)$ and $w=F(v)$. We assumed $\mathrm{Dom}(G)=\mathrm{Dom}(F)$ and $G(x)=F(x)$ for $x\in\mathrm{Dom}(F)$, hence we have $v\in\mathrm{Dom}(G)$ and $w=F(v)=G(v)$. Since G is a function and $v\in\mathrm{Dom}(G)$ and $w=G(v)$ we have $u=\langle v, w\rangle\in G$. Thus we have shown $F\subseteq G$. □

6.15 Definition (Explicit definition of functions). (i) Let $\tau(x)$ be a term and let A be a class. By "the function F defined by $\mathrm{Dom}(F)=A$ and $F(x)=\tau(x)$ for every $x\in A$" we mean the function $\{\langle x, \tau(x)\rangle \mid x\in A\}$. We shall denote this function by $\langle\tau(x)\mid x\in A\rangle$. This way of obtaining a function from a term and, in particular, this notation for a function is often called *functional abstraction.*

(ii) (Explicit definition by cases.) Let A be a class, let $\tau_i(x)$, $i=1,\ldots,n$, be terms and let $\Phi_i(x)$, $i=1,\ldots,n$, be formulas. By "the function F defined by $\mathrm{Dom}(F)=A$ and $F(x)=\tau_i(x)$ if $\Phi_i(x)$, $i=1,\ldots,n$" we mean the relation $\bigcup_{i=1}^{n}\{\langle x, \tau_i(x)\rangle \mid x\in A\wedge\Phi_i(x)\}$. If we prove, or assume, as we take care always to do in practice, that for each $x\in A$ exactly one of the $\Phi_i(x)$ holds, then F will indeed be a function with $\mathrm{Dom}(F)=A$. In practice, we often say "$F(x)=\tau_i(x)$ if $\Phi_i(x)$, $i=1,\ldots,n-1$, and $F(x)=\tau_n(x)$ *otherwise*" where by "otherwise" we mean, of course, $\bigwedge_{i=1}^{n-1}\neg\Phi_i(x)$.

6.16 Definition. (i) $R\mid A=\{\langle x, y\rangle \mid x\in A\wedge y\in A\wedge xRy\}$ (read: the *restriction* of R to A).

(ii) $F\restriction A=\{\langle x, y\rangle \mid x\in A\wedge xFy\}$; read: the *restriction* of F to A. The slanted line of the symbol $\restriction$ points leftwards since in $F\restriction A$ we leave only those pairs of F whose left component is in A.

(iii) $R[A]=\{y\mid(\exists x\in A)(\langle x, y\rangle\in R)\}$. If F is a function, then $F[A]$ is the class $\{F(x)\mid x\in\mathrm{Dom}(F)\cap A\}$ of all values of the members of A under the mapping F. In this case we shall call $F[A]$ the *image* of A under F.

Note that in the first two parts of Definition 6.16 we use the same verbal names for two different operations. This will not cause any difficulty since we shall usually use the two operations under different circumstances: $R\mid A$ will usually be applied to relations which are not functions. $F\restriction A$ will usually be applied to functions.

Notice also that we use the phrase "image of x under F" both for $F(x)$ (6.12(ii)) and for $F[x]$ (6.16(iii)). This will cause no confusion since we shall refrain from using that phrase whenever it will not be absolutely clear whether we consider x as a subset or as a member of $\mathrm{Dom}(F)$. In fact, many mathematicians use also $F(x)$ for both our $F(x)$ and $F[x]$ since in their discussions they do not encounter an x which is both a subset and a member of the same class A. We cannot go that far because we shall deal a lot with transitive classes A, which are such that every member of A is also a subset of A.

6.17 Definition. (i) $R \circ S = \{\langle x, z\rangle \mid \exists y(xRy \wedge ySz)\}$.

(ii) $FG = G \circ F$ (the *composition* of F and G). We define composition this way so as to get for functions F and G $(FG)(x) = F(G(x))$ (see 6.18(v) below).

(iii) The functions F and G are said to be *compatible* if for every $x \in \mathrm{Dom}(F) \cap \mathrm{Dom}(G)$ we have $F(x) = G(x)$ (i.e., if $F \upharpoonright \mathrm{Dom}(G) = G \upharpoonright \mathrm{Dom}(F)$). If $\mathrm{Dom}(F) \cap \mathrm{Dom}(G) = 0$ then F and G are trivially compatible.

6.18 Proposition. (i) $\mathrm{Dom}(R \circ S) = \mathrm{Dom}(R \cap (V \times \mathrm{Dom}(S))) \subseteq \mathrm{Dom}(R)$.

(ii) $\mathrm{Rng}(R \circ S) = \mathrm{Rng}(S \upharpoonright \mathrm{Rng}(R)) \subseteq \mathrm{Rng}(S)$.

(iii) $(R \circ S)^{-1} = S^{-1} \circ R^{-1}$, *hence*, $(FG)^{-1} = G^{-1}F^{-1}$.

(iv) *If F and G are functions so is FG. If F and G are one-one functions so is FG.*

(v) *If F and G are functions then for all x $(FG)(x) = F(G(x))$, provided that one of the sides of this equality is defined (see 6.12(ii), i.e., if either $x \in \mathrm{Dom}(FG)$ or else $x \in \mathrm{Dom}(G)$ and $G(x) \in \mathrm{Dom}(F)$) and then the other side of this equality is defined too (and the two sides are equal).*

(vi) *Let $I = \{\langle x, x\rangle \mid x = x\} = \langle x \mid x \in V\rangle$. For every relation R $R^{-1} \circ R \subseteq I$ iff R is a function.*

(vii) *The operation $\circ$ is associative, i.e., $R \circ (S \circ T) = (R \circ S) \circ T$. Hence, also the composition (of functions) is associative, i.e., $F(GH) = (FG)H$.* □

Let us observe that we have had here three different notions to which we refer by the term *relation*. The first one is sets which are relations. These are real objects of set theory. The second notion is classes which are relations. These are not, in general, objects of set theory; a single relation of this kind can be handled in the basic language by using, instead of a class-term, the formula which defines the class, but general statements about such relations can be made only by means of class variables in the extended language. As in the case of general classes, where we are able to discuss in the extended language only matters concerning the membership of the class, we can, in the case of relations, discuss only matters concerning the objects which stand in the relations of this kind. Thus, using class variables for relations serves as a convenience, not a necessity. The third notion of a relation is the notion of a syntactic relation $R(\mathfrak{a}_1, \ldots, \mathfrak{a}_n)$ defined in the extended language as in 4.10(1). This is, indeed, no relation at all, but just a relation symbol used essentially as an abbreviation for a formula $\Phi(\mathfrak{a}_1, \ldots, \mathfrak{a}_n)$. Whatever was said here about relations applies as well to functions.

6.19 The Class Form of the Axiom of Replacement. Now that we have at our disposal a notion of a class which is a function we have an easy way of formulating the class form of the axiom of replacement, which is a single statement of the extended language. The axiom is

(1) If F is a function, then for every set z, $F[z]$ is a set.

In order to show that this is indeed a formulation of the axiom schema of replacement, we shall show, employing the conservation theorem (4.6), that the theorems of the basic language added to the extended language by (1) are exactly those that follow from the axiom of replacement. It will be somewhat easier to prove this by considering the following statement (2) which we shall prove to be equivalent to (1).

(2) If F is a class such that $\forall u \forall v \forall w(\langle u, v\rangle \in F \wedge \langle u, w\rangle \in F \rightarrow v=w)$, then for every set z, $F[z]$ is a set.

(2) obviously implies (1) since the hypothesis of (2) is weaker than that of (1). To show that (1) implies (2) let F be a class as required in (2), then while F itself is not necessarily a function $F \cap (V \times V)$ is a function and hence, using (1), we get that for every set z $(F \cap (V \times V))[z]$ is a set. However, as is easily seen, $(F \cap (V \times V))[z] = F[z]$ and thus, (2) is proved from (1).

Having shown that (2) and (1) are equivalent, we shall employ the conservation theorem with respect to (2). Substituting $\{s \mid \chi(s)\}$ for F in (2) and replacing $F[x]$ by its definition, we get

$$\forall u \forall v \forall w(\chi(\langle u, v\rangle) \wedge \chi(\langle u, w\rangle) \rightarrow v=w) \rightarrow \forall z(\{v \mid (\exists u \in z)\chi(\langle u, v\rangle)\} \textit{ is a set}),$$

which is obviously equivalent to

$$(3) \qquad \forall u \forall v \forall w(\chi(\langle u, v\rangle) \wedge \chi(\langle u, w\rangle) \rightarrow v=w) \rightarrow \forall z \exists y \forall v(v \in y \leftrightarrow (\exists u \in z)\chi(\langle u, v\rangle))$$

(3) is an instance of the axiom schema of replacement where we take $\chi(\langle u, v\rangle)$ for $\psi(u, v)$. Thus every basic instance of (2) is equivalent to an instance of the axiom schema of replacement. In the other direction, the instance of the axiom of replacement corresponding to $\psi(u, v)$ is indeed equivalent to (3), and hence equivalent to a basic instance of (2), if we take for $\chi(s)$ in (3) the formula $\exists u \exists v(s=\langle u, v\rangle \wedge \psi(u, v))$. We get in this way the instance of (2) obtained by taking for F the class $\{\langle u, v\rangle \mid \psi(u, v)\}$. The proof of this equivalence requires, though, the assumption that for all u, v the ordered pair $\langle u, v\rangle$ exists. This follows right away from the statement that for all u, v the set $\{u, v\}$ exists. The latter is known as the "axiom of pairing". It is not needed as an axiom of ZF since it follows in ZF from the axiom schema of replacement (5.11). However, if we want, in the extended language, to replace the axiom schema of replacement by the single statement (1) we have to add also the axiom of pairing.

6.20 The Axioms of ZF in the Extended Language, in addition to those of 4.3 (which include extensionality) are now as follows:

Pairing: $\{x, y\}$ is a set.
Union: $\{x \mid \exists u(x \in u \wedge u \in z)\}$ is a set.
Power Set: $\{x \mid x \subseteq z\}$ is a set.
Replacement: If F is a function, then $F[z]$ is a set.
Infinity: $\exists z(0 \in z \wedge (\forall x \in z)(\forall y \in z)(x \cup \{y\} \in z))$.
Foundation: $A \neq 0 \rightarrow (\exists x \in A) x \cap A = 0$.

Historical Remark. The first full-fledged axiom system of set theory with something like classes was given by von-Neumann 1925, who used functions instead of classes. The first axiom system of set theory with classes as described here is due to Bernays 1937.

6.21 Proposition. *The schema of replacement.*

$$\forall u \forall v \forall w(\Psi(u, v) \wedge \Psi(u, w) \rightarrow v = w) \rightarrow \forall z \exists y \forall v(v \in y \leftrightarrow (\exists u \in z)\Psi(u, v))$$

(*where* w, z *and* y *are not free in* $\Psi(u, v)$) *holds for all formulas* $\Psi(u, v)$ *of the extended language, even if they have class variables as parameters.*

Hint of Proof. Take $\{\langle u, v\rangle \mid \Psi(u, v)\}$ for F in (1). □

6.22 Remark. We recall that until now Theorem 5.7 has been proved there only for terms with no class variables as parameters. Using 6.21 instead of the schema of replacement of the basic language (5.6) we can complete the proof of Theorem 5.7 and claim that the class $\{\tau(u, x_1, \ldots, x_n, X_1, \ldots, X_m) \mid u \in z\}$ is a set.

6.23 Proposition (The class form of the axiom of subsets). $\forall z(z \cap A)$ *is a set.*

Hint of Proof. Take for F in 6.19(1) $\{\langle x, x\rangle \mid x \in A\} = \langle x \mid x \in A\rangle$. This is like the proof of the axiom–schema of subsets (5.13) from the set-form of the axiom schema of replacement (5.6). □

Having now proved the class form 5.17 of the axiom of subsets we have available also Proposition 5.19 which was proved by means of 5.17.

6.24 Exercise. If F is a function, then $F \upharpoonright z$ is a set, and hence if $\mathrm{Dom}(F)$ is a set, then F itself is a set. □

As a consequence of 6.24, whenever we define a function by an explicit definition, as in 6.15, on a *set* we know that the function itself is a set, too.

6.25 Proposition. $\mathrm{Dom}(r)$ *and* $\mathrm{Rng}(r)$ *are sets.* □

6.26 Definition. ${}^zA=\{f \mid f\colon z\to A\}$.

6.27 Proposition. *If A is a set, so is ${}^zA$.*

Hint of proof. Compare the members of ${}^zA$ with $z\times A$. □

6.28 Exercises. (i) Give necessary and sufficient conditions on A and z for ${}^zA=0$.

(ii) Define a function F on V such that for all x, $F(x)\notin x$. Try not to use the axiom of foundation.

Hint. (ii) Use a construction like the one in Russell's antinomy 2.4. □

6.29 Definition. The classes A and B are said to be *disjoint* if $A\cap B=0$.

6.30 Proposition. (i) *Let F and G be functions. $F\cup G$ is a function iff F and G are compatible. As a consequence, if F and G have disjoint domains, then $F\cup G$ is a function.*

(ii) *Let T be a class of functions. $\bigcup T$ is a function iff every two members of T are compatible.*

(iii) *Let F and G be compatible functions. $F\cup G$ is a one–one function iff both F and G are one–one functions and* $\mathrm{Rng}(F)\cap\mathrm{Rng}(G)\subseteq F[\mathrm{Dom}(F)\cap\mathrm{Dom}(G)]$ $(=G[\mathrm{Dom}(F)\cap\mathrm{Dom}(G)])$. *In particular, if F and G are one–one functions with disjoint domains, then $F\cup G$ is a one–one function iff F and G have disjoint ranges.* □

Chapter II

Order and Well-Foundedness

After having described the language and the axioms of set theory we start the mathematical investigation of the universe of sets. The first concept which we turn to is the concept of order. This is an important concept of mathematics in general since ordered structures are abundant in mathematics. But this is not the reason for the high preference given to this concept here. We study the concept of order, together with the related concepts of well-order and well-foundedness since these are the concepts which reveal the structure of the universe of sets. The present chapter will tell us, in a sense, what the skeleton of this universe looks like, while other chapters and other books will deal with the flesh and organs.

The study of ordinal numbers is an indispensable part of an investigation of well-order. The ordinal numbers will be introduced and studied in this chapter, but only to the extent needed for the purposes of this chapter; a more detailed study of the ordinal numbers will be carried out in Chapter IV.

Applying the concept of well-foundedness to the membership relation among sets we get the class of all well founded sets. The axiom of foundation asserts that all sets belong to this class. This shows the universe of sets to be a neat cumulative hierarchy, where the null set lies at the bottom and each successive layer of the hierarchy is the power set of the union of all earlier layers.

1. Order

1.1 Definition. (i) "R is *irreflexive*" stands for $\forall x(\langle x, x\rangle \notin R)$.

(ii) "R is *transitive*" stands for $\forall x\forall y\forall z(\langle x, y\rangle \in R \wedge \langle y, z\rangle \in R \rightarrow \langle x, z\rangle \in R)$.

(iii) "R is a *partial ordering relation*" stands for "R is an irreflexive and transitive relation".

(iv) "R is an *order relation* on A" and "R *orders* A" stand for "$R \mid A$ is a partial order relation $\wedge \forall x\forall y(x, y \in A \rightarrow x = y \vee xRy \vee yRx)$". When we shall want to stress that we deal with order rather than with partial order we shall also use the phrase "*total order*" for order, e.g., we shall say R *totally orders* A.

Notice that the present concepts of partial order and order are of the $<$-type (i.e., irreflexive). Later we shall also mention reflexive order, i.e., order of the

$\leqslant$-type. Since we are dealing here with irreflexive partial order, every partial ordering relation can be understood to partially order the whole universe, not just its field (x is not in the field of the partial ordering R just in case x is not "connected", or "compared", by R with any other object y). The case of a total irreflexive order is different; if R orders A then A is a subclass of the field of R, except for the case where A is a singleton in which case every relation, even 0, orders A.

(v) For a relation R, we say that x is an *R-minimal* member of A if $x \in A$ and there is no $y \in A$ such that yRx. If R orders A this means that x is the "least" member of A (according to R). Notice also that every $x \in A$ which is not in the field of R is an R-minimal member of A.

(vi) "R is *left-narrow*" stands for $\forall x(\{y \mid yRx\}$ is a set); "R is *left-narrow* on A" stands for $(\forall x \in A)(\{y \in A \mid yRx\}$ is a set); "R is *right-narrow*" stands for $\forall x(\{y \mid xRy\}$ is a set); and "R is *right-narrow* on A" stands for $(\forall x \in A)(\{y \in A \mid xRy\}$ is a set).

(vii) "R well-orders A" stands for "R orders A, and every non-void subset z of A has an R-minimal member, and R is left-narrow on A" (i.e., for every $x \in A$ the class of all members of A which "come before x" is set).

(viii) We say that R is *symmetric* if $\forall x \forall y(xRy \leftrightarrow yRx)$.

1.2 Examples of Narrow and Non-Narrow Relations. Notice that if A is a set then, by the axiom of subsets, every relation R is both left-narrow and right-narrow on A. Every function is right-narrow, and every one-one function is both right-narrow and left-narrow. The $\in$ relation is left-narrow but not right-narrow.

In our definition 1.1(vii) of the statement "R well-orders A" we require, in addition to what is usually required in the literature, that R be left-narrow on A. As remarked above, if A is a set this requirement holds always and is therefore superfluous. We chose to require left-narrowness because it enables us to give a simple unified treatment of well-ordered sets and classes. Almost all the well-ordering relations which one encounters in set theory are anyway left-narrow.

1.3 Definition. *A (one-binary-relation-) structure* is an "ordered pair" $\langle A, R\rangle$ where A is a class and $R \subseteq A \times A$ (R is binary relation on A). A is said to be the *universe* or the *class* (or the *set*, if appropriate) of the structure $\langle A, R\rangle$. $\langle A, R\rangle$ is said to be a structure *on the class A*. If A is a set and $R \subseteq A \times A$ then R is a set too (by the axiom of subsets, since $A \times A$ is a set by I.6.6), and $\langle A, R\rangle$ is indeed the ordered pair it is claimed to be. If, however, A is a proper class, then the $\langle A, R\rangle$ we defined serves no purpose, since by our agreement of I.4.13 it is the void set and we cannot reconstruct A and R from $\langle A, R\rangle$. Therefore, let us agree to use $\langle A, R\rangle$, if A is a class which is not known to be a set, only as a figure of speech. Whatever we say about "the structure $\langle A, R\rangle$" we shall also be easily able to say by mentioning just A and R, as we shall see, for example, in Definition 1.5 below.

1.4 Exercise. Suppose we defined the notion of an ordered pair $\langle A, B\rangle$, where A and B are arbitrary classes, to be equal to our old $\langle A, B\rangle$ if both A and B are sets, and to $(\{0\} \times A) \cup (\{\{0\}\} \times B)$ otherwise. Prove that $\langle A, B\rangle = \langle C, D\rangle \rightarrow A = C \wedge B = D$. □

We could use the definition of $\langle A, B\rangle$ in 1.4 to serve as a definition of a structure which covers also the case where A of $\langle A, R\rangle$ is a proper class, but this does not seem to serve any real purpose.

1.5 Definition. (i) The structure $\langle B, S\rangle$ is a *substructure* of $\langle A, R\rangle$ if $B \subseteq A$ and $S = R \mid B$ (i.e., for all $x, y \in B$ $xSy \leftrightarrow xRy$).

(ii) F is an *isomorphism* of the structure $\langle A, R\rangle$ *into* the structure $\langle B, S\rangle$ if F is an injection of A into B and for all $x, y \in A$ $xRy \leftrightarrow F(x)SF(y)$. F is an isomorphism of $\langle A, R\rangle$ *onto* $\langle B, S\rangle$ if F is an isomorphism of $\langle A, R\rangle$ *into* $\langle B, S\rangle$ and F is a mapping of A *onto* B.

(iii) The structures $\langle A, R\rangle$ and $\langle B, S\rangle$ are said to be *isomorphic* (in symbols $\langle A, R\rangle \cong \langle B, S\rangle$) if there exists an isomorphism F on $\langle A, R\rangle$ *onto* $\langle B, S\rangle$. (Notice that, by I.6.24 $\langle a, r\rangle \cong \langle b, s\rangle$ can be written as "there is a function f which is an isomorphism of $\langle a, r\rangle$ onto $\langle b, s\rangle$" and therefore $\langle a, r\rangle \cong \langle b, s\rangle$ is a formula of set theory. $\langle A, R\rangle \cong \langle B, S\rangle$, on the other hand, involves the assertion "there exists a function F such that . . ." and therefore $\langle A, R\rangle \cong \langle B, S\rangle$ is not even a statement of the extended language, but, as mentioned in I.4.8, it is a promise to provide such an F by means of a class term.)

(iv) If $\langle A, R\rangle$ is a structure and F is a bijection of A on B then the relation $S = \{\langle F(x), F(y)\rangle \mid x, y \in A \wedge \langle x, y\rangle \in R\}$ on B is called the relation *induced* on B by R and F. F is obviously an isomorphism of $\langle A, R\rangle$ on $\langle B, S\rangle$.

In practice, we refer to $\langle A, R\rangle$ as a *structure* even when R is not a subclass of $A \times A$. In this case, when we mention the structure $\langle A, R\rangle$ we shall mean the structure $\langle A, R \mid A\rangle$, and $R \mid A$ is indeed a subclass of $A \times A$. Accordingly, $\langle A, R\rangle$ and $\langle A, S\rangle$ will be taken to be the same structure in case $R \mid A = S \mid A$. This agreement enables us to represent the substructures of $\langle A, R\rangle$ as $\langle B, R\rangle$, where $B \subseteq A$. Thus, the structure $\langle a, R\rangle$, where a is a set and R is a class is the object $\langle a, R \mid a\rangle$ of set theory. Nevertheless, whenever we write a structure as $\langle A, R\rangle$ and there is no reason to assume otherwise, we assume that $R \subseteq A \times A$.

1.6 Exercises. (i) The class of all one-binary-relation-structures on a is a set.

(ii) The class of all isomorphisms of $\langle a, r\rangle$ onto $\langle b, s\rangle$ is a set.

(iii) The structure $\langle B, S\rangle$ is a substructure of $\langle A, R\rangle$ iff the identity map on B (i.e., the function $I \mid B = \langle x \mid x \in B\rangle$) is an isomorphism of $\langle B, S\rangle$ into $\langle A, R\rangle$. □

1.7 Proposition. *If F is an isomorphism of $\langle A, R\rangle$ into $\langle B, S\rangle$ and G is an isomorphism of $\langle B, S\rangle$ into $\langle C, T\rangle$ then GF is an isomorphism of $\langle A, R\rangle$ into $\langle C, T\rangle$. If F and G are isomorphisms of $\langle A, R\rangle$ and $\langle B, S\rangle$, respectively, onto $\langle B, S\rangle$ and $\langle C, T\rangle$, then F^{-1} is an isomorphism of $\langle B, S\rangle$ onto $\langle A, R\rangle$ and GF is an isomorphism of $\langle A, R\rangle$ onto $\langle C, T\rangle$. Therefore, the relation $\cong$ of isomorphism of one-binary-relation structures is reflexive, symmetric and transitive.* □

1.8 Definition. A structure $\langle A, R\rangle$ is said to be a *partially ordered class* if R is a partial ordering; it is said to be an *ordered class* if R orders A; and a *well ordered class* if R well-orders A. If A is a set then we say that $\langle A, R\rangle$ is a *partially ordered set*, or an *ordered set*, or a *well ordered set* according to which one of the conditions above is satisfied. We shall sometimes use the abbreviation *poset* for partially ordered set.

1.9 Terminology. We shall often refer to the partially ordered class $\langle A, R\rangle$ as "the partially ordered class A", if it is clear which R is intended. For $x, y \in A$, if xRy we say that x *precedes* y (in the partial ordering R) or, even, x is *less than* y and y is *greater than* x. If B is a subclass of A we say that x is the *least*, or *first* (in case of a total ordering), member of B if $x \in B$ and $(\forall y \in B)\,(y \neq x \rightarrow xRy)$. More generally, we shall say that x is the least member of A for which $\Phi(x)$ holds if we have $\Phi(x) \wedge (\forall y \in A)\ (\Phi(y) \rightarrow xRy \vee x = y)$. If $B \subseteq A$ we say that x is the *greatest* or *last* (in case of a total ordering) member of B if $x \in B \wedge (\forall y \in B)(y \neq x \rightarrow yRx)$. It should also be clear by now what we mean by saying, for example, that x is the last member of A for which $\Phi(x)$ holds. Whenever we have no other relation R in mind, expressions like "the least set" or "the greatest set" will refer to the proper inclusion relation $\subset$. Thus by "the least set x such that $\Phi(x)$" we mean the set x such that $\Phi(x) \wedge \forall y(\Phi(y) \rightarrow y \supseteq x)$.

1.10 Examples of Ordered Sets. The best known ordered sets in mathematics are the sets of the natural numbers, the integers, the rational numbers and the real numbers with their natural order. The set of the natural numbers is well-ordered since, by the least number principle, every non-void set of natural numbers has a least member. The other sets mentioned are not well-ordered since for each one of those sets even the set itself has no least member.

1.11 Proposition. (i) *Let $\langle A, R\rangle$ and $\langle B, S\rangle$ be isomorphic structures. If $\langle A, R\rangle$ is a partially ordered class, an ordered class, or a well-ordered class, then so is $\langle B, S\rangle$. This is, essentially, a particular case of a general theorem about properties of structures preserved under isomorphism.*

(ii) *Every substructure of a partially ordered class, an ordered class, or a well-ordered class is a partially ordered class, an ordered class, or a well-ordered class, respectively.* □

1.12 Definition. Let $\langle A, R\rangle$ and $\langle B, S\rangle$ be partially ordered classes, and let F be a mapping of A into B. We say that F is an *increasing function*, or a *strictly monotonic function* (with respect to R and S), if for all $x, y \in A$, $xRy \rightarrow F(x)SF(y)$. In particular every isomorphism of $\langle A, R\rangle$ into $\langle B, S\rangle$ is an increasing function. We say that F is a *monotonic function* if for all $x, y \in A$ $xRy \leftarrow F(x)SF(y) \vee F(x) = F(y)$. (Clearly every strictly monotonic function and every constant function is monotonic.) Whenever we shall not mention or imply what the identity of $\langle A, R\rangle$, or $\langle B, S\rangle$, is we shall intend the respective structure to be $\langle V, \subset\rangle$, i.e., the universal class, partially ordered by proper inclusion. Thus when we say that the power set function P given by $P(x) = \mathbf{P}(x)$ is increasing we mean that it is increasing as a function from $\langle V, \subset\rangle$ to $\langle V, \subset\rangle$.

1.13 Proposition. *If $\langle A, R\rangle$ is an ordered class, $\langle B, S\rangle$ a partially ordered class and F an increasing function on A into B then F is an isomorphism of $\langle A, R\rangle$ into $\langle B, S\rangle$.* □

1.14 Proposition. *If $\langle A, R\rangle$, $\langle B, S\rangle$ are ordered classes such that $A\cap B=0$ (and $R\subseteq A\times A$, $S\subseteq B\times B$), then $\langle A\cup B, R\cup(A\times B)\cup S\rangle$ is an ordered class. (This is an ordered class in which the members of A are ordered as in $\langle A, R\rangle$ and those of B are ordered as in $\langle B, S\rangle$, while all the members of A precede all the members of B.) This ordered class is called the* ordered union *of $\langle A, R\rangle$ and $\langle B, S\rangle$ and is denoted by $\langle A, R\rangle\oplus\langle B, S\rangle$. If $\langle A, R\rangle$ is a well ordered* set *and $\langle B, S\rangle$ is a well ordered class, then their ordered union is a well ordered class. (If also B is a set, then the ordered union is a well ordered set.) In particular, if $\langle a, r\rangle$ is a well ordered set and $z\notin a$, then so is $\langle a\cup\{z\}, r\cup a\times\{z\}\rangle$ (this is the ordered union of $\langle a, r\rangle$ and $\langle\{z\}, 0\rangle$).* □

1.15 Exercises. (i) Is the ordered union of any two well ordered classes a well ordered class?

(ii) Let T be a class and R a relation such that

(a) R orders every member of T, and

(b) for all $x, y\in T$ we have $x\subseteq y$ or $y\subseteq x$, then R also orders $\bigcup T$. □

1.16 Definition. Let $\langle A, R\rangle$ be an ordered class, and let B be a subclass of A.

(i) B is said to be *cofinal* in $\langle A, R\rangle$ if for every member $x\in A$ there is a member $y\in B$ which is greater than or equal to x. (In symbols: $(\forall x\in A)(x\in B\vee(\exists y\in B)xRy)$.)

(ii) The member z of A is a *bound* of B if z is greater than or equal to all the members of B; z is a *strict bound* of B if z is greater than all the members of B.

(iii) B is said to be *bounded* in $\langle A, R\rangle$ if it has a bound z in A. B is said to be *strictly bounded* in $\langle A, R\rangle$ if it has a strict bound in A (i.e., if B is not cofinal in A).

(iv) B, and also the structure $\langle B, R\rangle$, is called an (*initial*) *section* of $\langle A, R\rangle$ if for some member y of A $B=\{x\in A\mid xRy\}$. This is the initial section of $\langle A, R\rangle$ *determined* by y. (If B is a section of $\langle A, R\rangle$, then B is strictly bounded in $\langle A, R\rangle$. Notice also that by our particular definition of a well ordered class, every section of a well ordered class is a set.) Whenever the relation R is assumed to be known we denote the section of $\langle A, R\rangle$ determined by y by A_y.

(v) B is said to be an *initial segment* of $\langle A, R\rangle$ if for every $x, y\in A$ if $x\in B$ and yRx, then also $y\in B$. (Every section of A is an initial segment of A.)

1.17 Exercises. (i) If $B\subseteq A$ then B is cofinal in the ordered class $\langle A, R\rangle$ iff $B\cup R^{-1}[B]=A$.

(ii) Let B be cofinal in the ordered class $\langle A, R\rangle$. B has a last member if and only if A has one, and it is the same member.

(iii) If $C\subseteq B\subseteq A$, B is cofinal in the ordered class $\langle A, R\rangle$ and C is cofinal in $\langle B, R\rangle$, then also C is cofinal in $\langle A, R\rangle$. □

1.18. Proposition. *Different members of an ordered class $\langle A, R\rangle$ determine different sections.* □

1.19. Proposition. *If B and C are initial segments of the ordered class $\langle A, R\rangle$, then B is an initial segment of C or C is an initial segment of B.* $\square$

1.20 Exercise (Hessenberg 1906). Let $\langle a, r\rangle$ be an ordered set, and let t be the set of all subsets of a which are initial segments of $\langle a,r\rangle$. t has the following properties:

(a) $u, v \in t \rightarrow u \subseteq v \vee v \subseteq u$,
(b) $\forall x \forall y(x, y \in a \wedge x \neq y \rightarrow (\exists u \in t)(x \in u \wedge y \notin u \vee x \notin u \wedge y \in u))$,
(c) $s \subseteq t \rightarrow \bigcup s \in t$,
(d) $s \subseteq t \rightarrow a \cap (\bigcap s) \in t$.

Prove that for every $t \subseteq \mathbf{P}(a)$ which satisfies (a) and (b) there is a unique relation $r \subseteq a \times a$ which orders a so that every member of t is an initial segment of $\langle a, r\rangle$. If t satisfies also (c) and (d), then t is equal to the set of all initial segments of $\langle a, r\rangle$. Notice that now we have a new way of representing the ordered set $\langle a, r\rangle$ as $\langle a, t\rangle$ where $t \subseteq \mathbf{P}(a)$ and t satisfies (a) and (b). If we want t to be unique for a given order we require also that it satisfy (c) and (d). What is t for the ordered set whose members are just x and y? Compare your answer with $\langle x, y\rangle = \{\{x\}, \{x, y\}\}$. $\square$

1.21 Historical Remark. In § I.6 we mentioned the program of representing the various notions of set theory by means of sets and classes. In 1.20 we saw the way in which the order of a set was represented first by Hessenberg 1906. This is superseded now by the way of Hausdorff 1914 which serves also to represent all relations (I.6.8).

2. *Well-Order*

We shall soon give an intuitive description of the concept of well-order, but first let us dwell on the following point. In the definition 1.1(vii) of when R well-orders A we required that every non-void *subset* z of A have a least member. We could not require in that definition that every non-void *subclass* Z of A have a least member since this would necessitate quantifying over the class variable Z, which we cannot do in a definition since a definition has to be a single statement. However, we get the stronger requirement we mentioned, without assuming it, by means of the least member principle in the following theorem.

2.1 Theorem (Schema). *Let $\langle A, R\rangle$ be a well ordered class, then:*

(i) The least member principle. *If $(\exists x \in A)\Phi(x)$, then there is a least member t of A for which $\Phi(t)$ holds.*

(ii) Induction. *If $(\forall x \in A)$ $[\forall y(yRx \rightarrow \Phi(y)) \rightarrow \Phi(x)]$ then $(\forall x \in A)\Phi(x)$; i.e., if for every x in A it is the case that if for every y which precedes x $\Phi(y)$ holds then $\Phi(x)$ holds for x too, then $\Phi(x)$ holds for every $x \in A$.*

Remark. When we come to prove $(\forall x \in A)\Phi(x)$, for some Φ, 2.1(ii) tells us that instead of directly proving that $\Phi(x)$ holds for $x \in A$ we can use the assumption that $\forall y(yRx \rightarrow \Phi(y))$ and prove $\Phi(x)$ from that assumption, and this establishes $(\forall x \in A)\Phi(x)$ anyway. This assumption $\forall y(yRx \rightarrow \Phi(y))$, which we get free of charge, is called the *induction hypothesis*.

Proof. (i) Let $B = \{x \in A \mid \Phi(x)\}$; we shall first show that B has a least member t. By our hypothesis that $(\exists x \in A)\Phi(x)$, $B \neq 0$. If we knew that B is a set it would follow directly from the definition of well-order that B has a least member; since we cannot assume that, we have to look for the least member of some suitable set. Since $B \neq 0$ B has a member z. If z is the least member of B, then set $t = z$. Otherwise, let $C = \{y \in B \mid yRz\} \neq 0$. Since $C \subseteq \{y \in A \mid yRz\}$, and $\{y \in A \mid yRz\}$ is a set because R is left-narrow on A, C is a set, and indeed a subset of A. R well-orders A and $C \neq 0$, hence, C has a least member t. We claim now that t is the least member of B. Suppose y is a member of B which precedes t. Since $t \in C$ we have tRz; and since we assumed yRt the transitivity of R implies yRz. But $y \in B$ and yRz imply $y \in C$, thus, y is a member of C which precedes t, contradicting our choice of t. Thus, t is the least member of B. Since $t \in B$ we have $\Phi(t)$. If $y \in A$ and yRt we must have $\neg\Phi(y)$, since if we had $\Phi(y)$ we would have $y \in B$, which contradicts what we showed that t is the least member of B.

(ii) Suppose that for some $x \in A$ we have $\neg\Phi(x)$. Then, by (i), with $\neg\Phi(x)$ replacing $\Phi(x)$, A has a least member t such that $\neg\Phi(t)$. Thus, we have $(\forall y \in A)\,(yRt \rightarrow \Phi(y))$, but since we have also $\neg\Phi(t)$, this contradicts our hypothesis that for every $x \in A$ $(\forall y \in A)\,(yRx \rightarrow \Phi(y)) \rightarrow \Phi(x)$. □

2.2 Proposition (Cantor 1897). *Let $\langle A, R\rangle$ be an ordered class. $\langle A, R\rangle$ is a well ordered class iff*

(a) *every strictly bounded subset u of A has a* least strict bound *in A, or, to use a term which may be even more suggestive, u has an* immediate successor *t in A, i.e., t is the least member of A which is greater than all the members of u, and*

(b) *R is left-narrow on A.*

Proof. Let $\langle A, R\rangle$ satisfy (a) and (b) and let z be a non-void subset of A. To prove that A is well-ordered by R it suffices to show that z has a least member. Since $z \neq 0$ there is an $s \in z$. Let U be the class of all members of A which are less than every member of z, then $U \subseteq A_s$, where A_s is the section of A determined by s. By (b) A_s is a set, hence U is a set too. By (a) U has an immediate successor t. Since every member of z is greater than every member of U there is no member of z below t. On the other hand, since $t \notin U$ we have, by the definition of U, that t is not less than all the members of z; thus $t \in z$ and t is the least member of z.

In the other direction, if R well-orders A then (b) holds, by the definition of well-ordering, and let us prove (a). Let u be a strictly bounded subset of A then it has strict bounds. By the least member principle (2.1(i)) u has a least strict bound t. □

2.3 Remark. The concept of well-order was introduced by Cantor 1883, who defined a set A to be well-ordered by R if every strictly bounded subset of A has an immediate successor in A. We saw in 2.2 that this is essentially equivalent to the present definition 1.1(vii).

Every finite ordered set is well-ordered (III.1.19). Let $\langle a, <\rangle$ be an infinite well-ordered set. Being non-void, a has a first member a_0 (which is the immediate successor of the null-set). a_0 has an immediate successor a_1 and a_1 has an immediate successor a_2. Since a is infinite the sets 0, $\{a_0\}$, $\{a_0, a_1\}$ are *proper* initial segments of a and hence they are strictly bounded and have immediate successors in a. Thus we obtain the infinite sequence $a_0, a_1, a_2, \ldots$. It is possible that this sequence exhausts a and thus $a = \{a_0, a_1, a_2, \ldots\}$. It is also possible that this sequence does not exhaust a and then it is a proper initial segment of a and it has an immediate successor a_0^1. a may go on containing a finite number of additional members and then it will be of the form $\{a_0, a_1, a_2, \ldots, a_0^1, a_1^1, \ldots, a_n^1\}$ for some finite n, but it may also have all the members $a_0, a_1, a_2, \ldots, a_0^1, a_1^1, a_2^1, \ldots$. Sets ordered in this way are the set of all integers ordered in the order $0, 1, 2, 3, \ldots, -1, -2, -3, \ldots$, and the set $\{0, \frac{1}{2}, \frac{3}{4}, \frac{7}{8}, \ldots, 1, 1\frac{1}{2}, 1\frac{3}{4}, 1\frac{7}{8}, \ldots\}$ ordered by the natural order of the real numbers. Our well-ordered set may go on as far as containing all of $a_0, a_1, a_2, \ldots, a_0^1, a_1^1, a_2^1, \ldots, a_0^2, a_1^2, a_2^2, \ldots, a_0^n, a_1^n, a_2^n, \ldots, \ldots$, and even beyond. All well ordered proper classes also "begin like the well-ordered sets", but "go on even more", as will become clear later on in this section.

2.4 Proposition. *If $\langle A, R\rangle$ is a well ordered class then every initial segment B of it is either A itself or else a section of $\langle A, R\rangle$.*

Hint of proof. If $B \neq A$ consider the least member z of A such that $z \notin B$. □

2.5 Exercise. Prove the following converse of 2.4. If $\langle A, R\rangle$ is an ordered class such that every initial segment of it is either A itself or else a set and a section of $\langle A, R\rangle$, then $\langle A, R\rangle$ is a well ordered class. □

2.6 Proposition. *If $\langle A, R\rangle$ and $\langle B, S\rangle$ are well ordered classes and F is an isomorphism of an initial segment of $\langle A, R\rangle$ on an initial segment of $\langle B, S\rangle$, then for every $x \in \mathrm{Dom}(F)$, $F \upharpoonright A_x$ is an isomorphism of the section A_x of A determined by x onto the section $B_{F(x)}$ of B determined by $F(x)$.* □

2.7 Proposition. *Let T be a class and R a relation such that*

(a) *R well-orders every member of T, and*

(b) *for all $x, y \in T$ either x is an initial segment of y or y is an initial segment of x,*

then R well-orders $\bigcup T$ (see 1.15(ii)). □

2.8 Exercise. Will 2.7 stay true if we weaken (b) to require only, as in 1.15(ii), that if $x, y \in T$ then $x \subseteq y$ or $y \subseteq x$? □

One of the most powerful tools of set theory is definition by recursion on well-ordered classes. Suppose $\langle A, R\rangle$ is a well ordered class. We want to define a function F on A by making the value of $F(x)$, for $x \in A$, dependent on the behavior of F

on the section A_x of A determined by x. Translating this to a strict mathematical requirement, we require that

(1) $F(x) = \tau(F1A_x)$

where τ is a set term which prescribes the way in which $F(x)$ should depend on the "behavior" of F on A_x. Notice that A_x is a set, and therefore, by the axiom of replacement $F1A_x$ is a set too (I.6.24). If we wanted to make $F(x)$ also explicitly dependent on x we could require $F(x) = \tau(x, F1A_x)$, but this is no more general than (1) since we can obtain x from $F1A_x$ as the least member of A which is not in the domain of $F1A_x$.

2.9 The Definition by Recursion Theorem (von Neumann 1923 and 1928a). *Let $\langle A, R\rangle$ be a well-ordered class. Given a set term τ, there exists a unique function F on A such that for every $x \in A$*

(1) $F(x) = \tau(F1A_x)$

(*where τ may also have parameters which are class- or set-variables*).

Proof. (1) tells us what the values of the function F should be. If the members of A, as ordered by R, are $a_0, a_1, a_2, \ldots$ then

$$F(a_0) = \tau(F1A_{a_0}) = \tau(0) \quad \text{(since } a_0 \text{ is the first member of } A\text{)},$$
$$F(a_1) = \tau(F1A_{a_1}) = \tau(\{\langle a_0, F(a_0)\rangle\}) = \tau(\{\langle a_0, \tau(0)\rangle\}),$$
$$F(a_2) = \tau(\{\langle a_0, \tau(0)\rangle, \langle a_1, \tau(\{\langle a_0, \tau(0)\rangle\})\rangle\}), \quad \text{and so on.}$$

However, the fact that we know what the values of F should be does not constitute a proof that F exists; the existence of F can be shown only by providing a class term for which we can prove (1). The fact that we know the values of F makes the existence of F very plausible and were it necessary we could introduce such F's into set theory by means of new symbols and new axioms. However, as we shall see, we can actually provide a class term for F. This is not outright obvious; one can actually write an implicit definition of a function F, of a form similar to that of (1), for which there is no class term F which satisfies the implicit definition. (This is the situation we encounter when we come to define a truth function F for set theory.) We begin now proving our theorem.

(a) Let F and F' be two functions whose domains are initial segments of A and such that each one of them satisfies (1), for every x in its domain; then we shall prove that F and F' are compatible. Before proving it, let us note that this establishes the uniqueness of the F of the theorem, since if F and F' are functions on A which satisfy (1), then as we claim here, they are compatible, and since they have the same domain, they are equal, by the principle of extensionality for functions (I.6.14).

Let F and F' be functions on initial segments of A, which satisfy (1). We shall prove by induction that

$$\text{for every } x \in A \text{ if } x \in \mathrm{Dom}(F) \cap \mathrm{Dom}(F') \text{ then } F(x) = F'(x).$$

This will, of course, establish the compatibility of F and F'. Our induction hypothesis is that for every y such that yRx we have $y \in \mathrm{Dom}(F) \cap \mathrm{Dom}(F') \to F(y) = F'(y)$. Assume $x \in \mathrm{Dom}\,(F) \cap \mathrm{Dom}(F')$, then since $\mathrm{Dom}(F)$ and $\mathrm{Dom}(F')$ are initial segments of A, we have for every $y \in A$ such that yRx also $y \in \mathrm{Dom}(F) \cap \mathrm{Dom}(F')$, and, by the induction hypothesis $F(y) = F'(y)$. Therefore, the functions F and F' are the same on A_x, i.e.,

(2) $\qquad F \upharpoonright A_x = F' \upharpoonright A_x.$

Since both F and F' satisfy (1) we get

$$\begin{aligned} F(x) &= \tau(F \upharpoonright A_x), \text{ by (1)}, \\ &= \tau(F' \upharpoonright A_x), \text{ by (2)}, \\ &= F'(x), \text{ by (1)}. \end{aligned}$$

(b) Let T be the class of all functions f (i.e., the functions f which are sets) whose domains are initial segments of A and which satisfy (1), i.e.,

(3) $\qquad x \in \mathrm{Dom}(f) \to f(x) = \tau(f \upharpoonright A_x).$

By what we proved in part (a) every two members of T are compatible, hence by I.6.30(ii) $\bigcup T$ is a function. Let us denote it by F. Our proof of the theorem will be completed once we show that $\mathrm{Dom}(F) = A$ and that F satisfies (1) for every $x \in \mathrm{Dom}(F)$.

(c) F satisfies (1). Let $x \in \mathrm{Dom}(F)$, then since $F = \bigcup T$, $x \in \mathrm{Dom}(f)$ for some $f \in T$. Since $f \in T$ $\mathrm{Dom}(f)$ is an initial segment of A, and because $x \in \mathrm{Dom}(f)$ also $A_x \subseteq \mathrm{Dom}(f)$. Since $F = \bigcup T \supseteq f$ we have $F(x) = f(x)$ and $F \upharpoonright A_x = f \upharpoonright A_x$. By (3) we get $F(x) = f(x) = \tau(f \upharpoonright A_x) = \tau(F \upharpoonright A_x)$, thus establishing (1).

(d) $\mathrm{Dom}(F) = A$. The essential point is to prove that for every $x \in A$ there is an $f \in T$ such that $x \in \mathrm{Dom}(f)$. The idea used to prove this is that (1) shows us how to extend a function in T to functions defined for more and more x's. While keeping this idea in mind, we shall use a somewhat indirect approach. Suppose $\mathrm{Dom}(F) \subset A$. $\mathrm{Dom}(F) = \bigcup\{\mathrm{Dom}(f) \mid f \in T\}$, hence, $\mathrm{Dom}(F)$ is an initial segment of A, being the union of initial segments. Since we assumed that $\mathrm{Dom}(F)$ is a *proper* subclass of A, we get that $\mathrm{Dom}(F)$ is a section of A (2.4); let it be the section determined by a member z of A. F is a set, since its domain is a set, being a section of a well ordered class. Let $h = F \cup \{\langle z, \tau(F)\rangle\}$. h is a function since $z \notin \mathrm{Dom}(F)$ by our choice of z. Let us prove now that $h \in T$. $\mathrm{Dom}(h) = \mathrm{Dom}(F) \cup \{z\}$; since $\mathrm{Dom}(F)$ is the section determined by z, $\mathrm{Dom}(F) \cup \{z\}$ is obviously an initial segment of A. Let us show now that for every $x \in \mathrm{Dom}(h)$ we have $h(x) = \tau(h \upharpoonright A_x)$. If $x \in \mathrm{Dom}(h)$, then $x \in \mathrm{Dom}(F)$ or $x = z$. If $x \in \mathrm{Dom}(F)$, then since $h \supseteq F$ and $\mathrm{Dom}(F) \supseteq A_x$ we have, by (c), $h(x) = F(x) = \tau(F \upharpoonright A_x) = \tau(h \upharpoonright A_x)$. If $x = z$ we have, since $\langle z, \tau(F)\rangle \in h$ and

$h \upharpoonright A_z = F$, $h(x) = h(z) = \tau(F) = \tau(h \upharpoonright A_z) = \tau(h \upharpoonright A_x)$. We have shown that $h \in T$, hence, $h \subseteq \bigcup T = F$. Therefore, $z \in \mathrm{Dom}(h) \subseteq \mathrm{Dom}(F)$, which is a contradiction since our choice of z was such that $z \notin \mathrm{Dom}(F)$. □

2.10 Parameters in Definition by Recursion. In the definition by recursion theorem, 2.9, we mentioned that the set term τ of the recursion equation $F(x) = \tau(F \upharpoonright A_x)$ may contain parameters which are class- or set-variables. In such a case the function F obtained depends on the parameters. For example, if τ has a set parameter u and a class parameter W then F is given explicitly by the class term

$$\{t \mid \exists f(\mathrm{Dom}(f) \text{ is an initial segment of } \langle A, R\rangle \\ \wedge (\forall x \in \mathrm{Dom}(f))(f(x) = \tau(f \upharpoonright A_x, u, W)) \wedge t \in f)\}$$

in which u and W occur as parameters. One can incorporate a set parameter u of τ as a variable rather than a parameter of F; let us call the resulting function of two variables (u and x) F'. F' is the unique function on $V \times A$ such that for all $u \in V$ and $x \in A$

$$(4) \qquad F'(u, x) = \tau(\langle F'(u, y) \mid y \in A \wedge yRx\rangle, u).$$

F' is obviously given by the class term

$$\{\langle\langle u, x\rangle, z\rangle \mid \exists f(\mathrm{Dom}(f) \text{ is an initial segment of } \langle A, R\rangle \\ \wedge (\forall x \in \mathrm{Dom}(f))(f(x) = \tau(f \upharpoonright A_x, u)) \wedge \langle x, z\rangle \in f)\}.$$

Notice that in (4) the value of $F'(u, x)$ depends, via τ, on u and on the values of $F'(u, y)$ for yRx, but not on any value $F(u', y)$ where $u' \neq u$. One can generalize this by making $F(u, x)$ dependent also on all the values $F(u', y)$ where yRx and u' is a member of some set, which may depend on x and y. However, the generalization which makes $F(u, x)$ dependent on the values $F(u', y)$ for y such that yRx and *all* u' is not a theorem schema of set theory.

2.11 Corollary (*Definition of a class by recursion*). *Let $\langle A, R\rangle$ be a well ordered class. Given a formula Φ there exists a unique subclass P of A such that for all $x \in A$*

$$x \in P \leftrightarrow \Phi(A_x \cap P).$$

Outline of proof. The idea is to apply definition by recursion to the characteristic function of P. Thus we take $\tau(f)$ to be the set term defined by cases (I.6.15) as follows: $\tau(f) = 1$ if $\Phi(\{x \mid f(x) = 1\})$ holds and $\tau(f) = 0$ if $\Phi(\{x \mid f(x) = 1\})$ does not hold. Take F to be the unique function on A satisfying $F(x) = \tau(F \upharpoonright A_x)$ for all $x \in A$, and $P = \{x \mid F(x) = 1\}$. □

When we are defining a function by recursion on a well ordered class $\langle A, R\rangle$, often the function we are really after is not a function on the whole of A but on some initial segment of A which cannot be conveniently determined before the function is constructed. In these cases the term τ in the recursion equation $F(x)=\tau(F \upharpoonright A_x)$ is usually such that $\tau(F \upharpoonright A_x)$ has no natural value for x's outside that initial segment. Formally, Theorem 2.9 is sufficient to deal with this case, too, since the term $\tau(f)$ is necessarily defined for every f, even if by the use of the clause "and 0 otherwise" (I.4.11). However, applying terms which are defined artificially outside their natural domain is rather inconvenient. Therefore it is worth while to have the following special theorem which will enable us to define the functions we need on their natural domain.

2.12 Theorem (*Definition by recursion as long as some condition is met*). *Let $\langle A, R\rangle$ be a well ordered class. Given a set term $\tau(f)$ and a formula $\Phi(f)$ there exists a unique initial segment C of $\langle A, R\rangle$ and a unique function H on C such that for every $x \in C$*

(5) $\quad \mathrm{H}(x)=\tau(H \upharpoonright A_x)$

and

(6) $\quad \Phi(H \upharpoonright A_x)$

and if C is a proper initial segment of A (i.e., if $C \subset A$) and z is the least member of A not in C then

(7) $\quad \neg\Phi(H \upharpoonright A_z)$, i.e., $\neg\Phi(H)$.

Notice that $\Phi(f)$ is the formula that tells us how far is H to be defined: $H(x)$ is to be defined exactly as long $\Phi(H \upharpoonright A_x)$ holds.

Proof. Let F be the function defined on the whole of A by

(1) $\quad F(x)=\tau(F \upharpoonright A_x)$

as in the definition by recursion theorem (2.9). Let C be the subclass of A consisting of all the members z of A such that $\Phi(F \upharpoonright A_x)$ holds for every xRz as well as for $x=z$. C is obviously an initial segment of A. We have for every $x \in C$

(8) $\quad \Phi(F \upharpoonright A_x)$

and if C is a proper initial segment of A and z is the least member of $A \sim C$ then

(9) $\quad \neg\Phi(F \upharpoonright A_z)$

since if we had also $\Phi(F \upharpoonright A_z)$ z were, trivially, a member of C. Let $H=\mathrm{F} \upharpoonright C$. Since C is an initial segment (1) implies (5) and (8) implies (6), and by our choice of z also (9) implies (7).

The proof of the uniqueness of C and H is left to the reader. The idea of the proof is that as in the proof of part (a) of 2.9 any two such H's must give the same values as long as both are defined, and the place where they stop being defined is the same since it is determined by the values of the H's which are the same. □

We shall often denote general order relations by $<$, without intending this to be ordering by magnitude. In these cases, we shall write $x \leqslant y$ for $x = y \vee x < y$. The properties of the relation $\leqslant$ are well known. We shall even use $<$ for several order relations discussed at the same time if no confusion will arise. We shall denote with $>$ the inverse relation of $<$, i.e., $> = <^{-1}$. $\geqslant$ will have the obvious meaning.

2.13 Theorem. *If $\langle A, <\rangle$ is a well ordered class, and F is an increasing function on $\langle A, <\rangle$ into itself, then for all $x \in A$ we have $x \leqslant F(x)$.*

Proof. We shall prove $x \leqslant F(x)$ by induction. We assume, as our induction hypothesis, that for all $y \in A$ such that $y < x$ we have $y \leqslant F(y)$. Since F is increasing we have for all $y < x$ $y \leqslant F(y) < F(x)$, thus $F(x)$ is greater than all y's which are less than x hence $F(x) \geqslant x$. □

2.14 Corollary (Cantor 1897). *If $\langle A, <\rangle$ is a well ordered class and B is a strictly bounded subclass of A (and, in particular, if B is a section of A), then $\langle A, <\rangle$ is not isomorphic to $\langle B, <\rangle$.*

Proof. Suppose F is an isomorphism of $\langle A, <\rangle$ on $\langle B, <\rangle$, then F is an isomorphism of $\langle A, <\rangle$ into itself. If y is a strict bound of B, then by 2.13 $F(y) \geqslant y$; hence, $F(y) \notin B$, contradicting our assumption that $\mathrm{Rng}(F) = B$. □

2.15 Corollary. (i) *If $\langle A, <\rangle$ is a well ordered class, then the identity map on A $I_A = \langle x \mid x \in A\rangle$ is the only automorphism of $\langle A, <\rangle$, i.e., the only isomorphism of $\langle A, <\rangle$ on itself.*

(ii) *If $\langle A, <\rangle$ and $\langle B, <\rangle$ are isomorphic well ordered classes, then there is a unique isomorphism of $\langle A, <\rangle$ on $\langle B, <\rangle$.*

Proof. (i) Let F be an automorphism of $\langle A, <\rangle$. By 2.13, we have $x \leqslant F(x)$ for every $x \in A$. F^{-1} is also an automorphism of A (1.7), hence, again by 2.13, $y \leqslant F^{-1}(y)$ for every $y \in A$. Substituting $F(x)$ for y we get $F(x) \leqslant F^{-1}(F(x)) = x$. We have shown that $x \leqslant F(x)$ and $F(x) \leqslant x$ for every $x \in A$, hence $F(x) = x$ for every $x \in A$, and $F = I_A$.

(ii) Let F and G be two isomorphisms of $\langle A, <\rangle$ on $\langle B, <\rangle$. Then G^{-1} is an isomorphism of $\langle B, <\rangle$ on $\langle A, <\rangle$ (1.7) and hence, $G^{-1}F$ is an automorphism of $\langle A, <\rangle$. By (i), we have $G^{-1}F = I_A$, from which we get immediately $F = G$. □

2.16 Theorem (Cantor 1897). *If $\langle A, <\rangle$ and $\langle B, <\rangle$ are well ordered classes then they are isomorphic or else one of them is isomorphic to a section of the other.*

Proof. The intuitive idea of the construction of the isomorphism H required in this theorem is that we let H map the first member of A on the first member of B, then the next member of A on the next member of B, and so on, until one of the classes A and B, or both, run out. Accordingly, the version of the theorem on definition by recursion which we shall use here is the theorem on definition by recursion as long as some condition is met (2.12).

We define now a function H on an initial segment C of A such that for every $x \in C$ $H(x)$ is the least member of B which is not an image of members of A smaller than x, i.e.,

(1) $\quad H(x) = \text{the least member of } B \sim H[\{y \in A \mid y < x\}]$ (in the ordering $<$)

and $\quad x \in C$ as long as B is not exhausted, i.e., as long as

(2) $\quad B \sim H[y \in A \mid y < x\}] \neq 0$

This is a valid definition by recursion as given in 2.12, where τ is the term given by

(3) $$\tau(f) = \begin{cases} \text{the least member of } B \sim \operatorname{Rng}(f) & \text{if } \quad B \sim \operatorname{Rng}(f) \neq 0 \\ 0 & \text{if } \quad B \sim \operatorname{Rng}(f) = 0 \end{cases}$$

and $\Phi(f)$ is

(4) $\quad B \sim \operatorname{Rng}(f) \neq 0,$

since by substituting $H \upharpoonright \{y \in A \mid y < x\}$ for f we get from (3) the right-hand side of (1) and from (4) we get (2).

We shall first prove that H is an isomorphism of C on an initial segment D of B. Let $x, z \in C$, $x < z$. $H(z)$ is the least member of $B \sim H[A_z]$, hence $H(z) \notin H[A_z]$, and also, since $H(x) \in H[A_z]$, $H(z) \neq H(x)$. Since $H(z) \notin H[A_z] \supseteq H[A_x]$ we have $H(z) \in B \sim H[A_x]$. By definition of $H(x)$ it is the *least* member of the set $B \sim H[A_x]$ of which $H(z)$ is a member, hence, since $H(x) \neq H(z)$, $H(x) < H(z)$. Thus we have shown that if $x < z$ then $H(x) < H(z)$ which establishes that H is increasing and hence, by 1.13, it is an isomorphism. Let us prove now that $D = \operatorname{Rng}(H)$ is an initial segment of B. Let u, $v \in B$, $u \in D$ and $v < u$, we have to show that also $v \in D$. Since $u \in D$ $u = H(x)$ for some x. Since $v < u$ and u $(= H(x))$ is, by (1), the *least* member of $B \sim H[A_x]$ v is not a member of that set, i.e., $v \in H[A_x] \subseteq \operatorname{Rng}(H) = D$.

If $\operatorname{Dom}(H) = A$ then H is an isomorphism of A on an initial segment of B, which is, by 2.4, B itself or a section of B. If $C = \operatorname{Dom}(H) \subset A$ then C is a proper initial segment of A and, by 2.12 (7), (4) fails for $f = H$, i.e., $B \sim \operatorname{Rng}(H) = 0$, hence $\operatorname{Rng}(H) = B$, and thus C, which is a section of A by 2.4, is isomorphic to B. □

We shall now strengthen Theorem 2.16 by the following proposition which adds strong uniqueness statements to 2.16.

2.17 Proposition. *If $\langle A, <\rangle$ and $\langle B, <\rangle$ are well ordered classes, then exactly one of the following holds.*

(a) *$\langle A, <\rangle$ and $\langle B, <\rangle$ are isomorphic.*
(b) *$\langle A, <\rangle$ is isomorphic to a unique section of $\langle B, <\rangle$.*
(c) *$\langle B, <\rangle$ is isomorphic to a unique section of $\langle A, <\rangle$.*

Proof. Theorem 2.16 asserts that at least one of (a), (b), (c) holds, without the uniqueness conditions of (b) and (c).

In each one of the cases where (a) and (b) hold at the same time, or where (a) and (c) hold at the same time, or when the section mentioned in (b) or in (c) is not unique, we have a situation where a well ordered class $\langle C, <\rangle$ is isomorphic to two different initial segments, say D' and D'' of a well ordered class $\langle D, <\rangle$. Since of any two initial segments one of them includes the other, we can assume that D' is an initial segment of D'', and since $D' \neq D''$, D' is a section of D'' (by 2.4). Thus, we have $\langle D', <\rangle \cong \langle C, <\rangle \cong \langle D'', <\rangle$, which is a contradiction since the well ordered class $\langle D', <\rangle$ cannot be isomorphic to a section of itself (2.14).

Dealing with the only case left, let us suppose that (b) and (c) hold at once. Then there are isomorphisms F of $\langle A, <\rangle$ on a section B' of $\langle B, <\rangle$, and G of $\langle B, <\rangle$ on a section of $\langle A, <\rangle$. By 2.6, $G \upharpoonright B'$ is an isomorphism of B' on a section A' of A. Thus, we have $\langle A, <\rangle \cong \langle B', <\rangle \cong \langle A', <\rangle$, and A is isomorphic to a section of itself, contradicting 2.14. □

Remark. A somewhat shorter, but slightly less intuitive, proof of 2.16 is carried out as follows. Let H be the relation given by

> $y\ H\ z$ iff $y \in A$, $z \in B$ and y, z determine isomorphic sections of A and B, respectively.

It is easily seen that H is a function as required by 2.16. 2.17 can be proved along the same lines by showing that every isomorphism as in 2.17 must be included in H, and hence equal to H.

2.18 Corollary. *If $\langle A, R\rangle$ is a well ordered* proper *class, and $\langle B, <\rangle$ is a well ordered class, then B is a set iff $\langle B, <\rangle$ is isomorphic to a section of A, and B is a proper class iff $\langle B, <\rangle$ is isomorphic to $\langle A, <\rangle$. Thus, all well ordered proper classes are isomorphic.*

2.19 Remark. There are many non-isomorphic structures $\langle A, R\rangle$, where

(a) R orders A, and
(b) every non-void subset of A has an R-minimal member,

but R does not well order A because R is not left-narrow on A. Let us call structures which satisfy (a) and (b) *weakly well ordered classes.* Weakly well ordered classes which are not well-ordered must of course be proper classes. Examples of such classes can be obtained as follows. If $\langle B, <\rangle$ is a well ordered proper class (the existence of such a class is established in 3.12 and 3.15) and $\langle C, <\rangle$ is any non-void well ordered class where $C \cap B = 0$ then the ordered union (1.14) of $\langle B, <\rangle$ and $\langle C, <\rangle$ is a weakly well ordered class which is not well ordered, and two such ordered unions are isomorphic only if the corresponding $\langle C, <\rangle$'s are isomorphic. The last part of 2.18 asserts that as a result of the additional requirement on a well ordered class A that every section of A be a set there is just one well ordered proper class, up to isomorphism, among the many weakly well ordered proper classes. We shall discuss again the weakly well ordered classes in 7.9.

Proof of 2.18. By Theorem 2.16, there is a function F which is an isomorphism of one of $\langle A, <\rangle$ and $\langle B, <\rangle$ on the other or on a section of the other. Since A is a proper class it cannot, by the axiom of replacement, be isomorphic to a section of B, which must be a set (by Definition 1.1(vii)). Hence, F can be assumed to be an isomorphism of B on A or on a section of A. If F is an isomorphism of B on A then, by the axiom of replacement, B must be a proper class too; if F is an isomorphism of B on a section of A, which must be a set by Definition 1.1(vii), B must be a set too. □

2.20 Proposition. *If $\langle A, <\rangle$ is a well ordered class and $B \subseteq A$, then $\langle B, <\rangle$ is isomorphic to $\langle A, <\rangle$ or to a section of $\langle A, <\rangle$.*

Proof. If $\langle B, <\rangle$ is not isomorphic to $\langle A, <\rangle$ or to a section of $\langle A, <\rangle$ then, by 2.16, $\langle A, <\rangle$ is isomorphic to a section $\langle B_u, <\rangle$ of $\langle B, <\rangle$ determined by a member u of B. B_u is a subset of A strictly bounded by u, and hence, by 2.14, $\langle A, <\rangle$ cannot be isomorphic to $\langle B_u, <\rangle$, contradicting what we claimed above. □

2.21 The Role of Well-Ordering. Let us conclude the present section with a preview on the role of well-ordering in set theory and mathematics. The two main tools which we have for dealing with well ordered classes are the least member principle (or proof by induction, which is immediately equivalent to it) and definition by recursion. It turns out that these are also very useful tools for dealing with an "unordered" class, i.e., where no particular well-ordering of the class is related to whatever we are doing with the class. We shall now give a more detailed explanation of what we can do in such cases with the least member principle and with definition by recursion.

2.22 Using the Indefinite Article in Definitions. In defining a function mathematicians often use the indefinite article in the defining clause. For example, it is standard mathematical usage to define a function F on the closure $\bar{A}$ of a set A of real numbers by

(5) $F(x) =$ a sequence $\langle a_n \mid n < \omega\rangle$ of members of A such that $\lim_{n\to\infty} a_n = x$.

The reason why one does not specify in the right-hand side of (5) which of the possibly many sequences in A converging to x one wants $F(x)$ to be is as follows. It is possible that the right-hand side of (5) gives one all the information about $F(x)$ that he needs to know, and there is no reason for him to specify $F(x)$ further, but it is also possible that one is *unable* to specify a particular sequence of members of A which converges to x. (5) is not a valid definition of F according to our rule I.6.15 for explicit definition of functions. The main trouble is not that (5) does not specify F completely but that (5) is just a statement of what we want F to be without a guarantee that there is such an F. This contrasts with the case of a valid explicit definition (which uses, essentially, the definite article) where one can immediately write down the class term F required (see I.6.15).

The well-ordering comes into the problem of using the indefinite article as follows. Suppose that $<$ is a relation which well-orders a class A. If we use an indefinite article in the expression "a y such that $\phi(x, y)$", where the x's are taken from some class W, and it is the case that for all $x \in W$ there is a $y \in A$ such that $\phi(x, y)$, then the expression "*a* y such that $\phi(x, y)$" can be interpreted as "the least member of y of A, in the sense of $<$, such that $\phi(x, y)$". Since all we want is some y such that $\phi(x, y)$ it does not matter which well-ordering $<$ of A we take as long as we have one. Notice that what we use here is the least member principle.

The other important aspect of well ordering an "unordered" class is the following. Some mathematical constructions are such that one goes through a set A in some order and carries out the construction for the members of A one at a time so that whatever is constructed for x depends on what has already been constructed for the preceding members of the class. The definition by recursion theorem asserts that such a construction can be carried out once we have a well-ordering $<$ of A.

Having seen to what extent the existence of a well-ordering of a class can be helpful, we ask whether every class can be well ordered. We shall discuss the possibility of well ordering all classes in § V.4; here we shall deaí only with sets. It will sufficiently serve our present purposes if every set can be well-ordered, since, after all, the objects we are really dealing with are just the sets, and proper classes are rarely encountered in mathematics. In ZF it cannot be proved that all sets can be well-ordered. However, once we admit the axiom of choice (thereby passing to the theory ZFC) we get the following theorem, which is equivalent to the axiom of choice (V.1.6).

2.23Ac The Well-Ordering Theorem (*of* ZFC). *For every set a there is a relation r which well-orders a.*

We shall use this theorem from now on, even though it will be proved only in V.1.6. The reason why we do not take the well-ordering theorem itself to be the axiom, even though it is perhaps the most useful form of the axiom of choice, is that despite its usefulness the well-ordering theorem does not have the same ring of truth about it as the principal version (V.1.1) of the axiom of choice. A mathematician is usually willing to accept the well-ordering theorem only after he has seen some proof of it from the axiom of choice.

We still want to carry out most of our development in ZF, i.e. without using the axiom of choice, yet we want to enjoy the benefits of the existence of a well-ordering whenever possible. Therefore the following definition will be useful.

2.24 Definition. A set a is said to be *well-orderable* if there is a relation r which well-orders a. If we assume the axiom of choice then every set is well-orderable.

2.25 More on Using the Indefinite Article. Notice that by our discussion in 2.22 we can use the indefinite article freely with respect to a well-orderable set a, i.e., whenever there is a fixed-well orderable set a such that the objects denoted by the indefinite article are taken to be in a. When we shall use the indefinite article for the first time we shall go over the matter carefully, but after that we shall use the indefinite article as a matter of course.

3. Ordinals

An idea used frequently in mathematical construction is the idea of abstraction. We take a property common to a set or a class of objects and regard it as a new object. In fact, the very notion of set is obtained in this way, to some extent, since in mathematics one deals almost exclusively with sets with some sort of structure on them, such as an order relation, rather than with plain sets. The common property to all the structures with the same underlying set is abstracted and called a set. Now we wish to get an object to represent the common property of isomorphic structures (which are sets); such an object will be called an isomorphism type. For this purpose, we need a function G as in the following assumption.

3.1 Temporary Assumption (Isomorphism type). We assume the existence of a function G such that $\mathrm{Dom}(G)$ contains every structure, and such that for all structures $\langle a, r\rangle$ and $\langle b, s\rangle$

(1) $$G(\langle a, r\rangle) = G(\langle b, s\rangle) \leftrightarrow \langle a, r\rangle \cong \langle b, s\rangle.$$

We shall show later (in 7.13; see also 3.24 and 3.25) that there is indeed such a function G.

3.2 Temporary Definition. $G(\langle a, r\rangle)$ will be called the *isomorphism type* of $\langle a, r\rangle$. The isomorphism types of ordered sets will be called *order types* (Cantor 1895). The order types of well ordered sets will be called *ordinals*. Ord will denote the function $G \upharpoonright \{x \mid x \textit{ is an ordered set}\}$. This is only a temporary definition of the notion of an order type, an ordinal, and the function Ord. We shall give the permanent definitions of these notions in 3.8, IV.1.1 and 3.24.

3.3 Temporary Definition (Cantor 1897). The order relation $<$ among ordinals is defined by: $x<y$ if x is the ordinal of a section b of a well-ordered set a of the ordinal y, i.e.,

$$< = \{\langle x, y\rangle \mid x \textit{ and } y \textit{ are ordinals} \wedge \exists a \exists r \exists b(\langle a, r\rangle \textit{ is a well-ordered set} \wedge y = \mathrm{Ord}(\langle a, r\rangle) \wedge b \textit{ is a section of } \langle a, r\rangle \wedge x = \mathrm{Ord}(\langle b, r\rangle))\}$$

(the permanent definition of $<$ is given in 3.8(iv)).

3.4 Proposition. *If* $\langle a, r\rangle$ *is a well ordered set,* $\mathrm{Ord}(\langle a, r\rangle) = x$ *and* $y < x$, *then* $\langle a, r\rangle$ *has a section b such that* $\mathrm{Ord}(\langle b, r\rangle) = y$.

Proof. Since $y<x$ there is, by Definition 3.3, a well ordered set $\langle a', r'\rangle$ and a section b' of $\langle a', r'\rangle$ such that $\mathrm{Ord}(\langle a', r'\rangle)=x$ and $\mathrm{Ord}(\langle b', r'\rangle)=y$. b' is determined by some member u of a'. Since $\mathrm{Ord}(\langle a', r'\rangle)=x=\mathrm{Ord}(\langle a, r\rangle)$ there is an isomorphism f of $\langle a', r'\rangle$ onto $\langle a, r\rangle$. Let b be the section of $\langle a, r\rangle$ determined by the member $f(u)$ of a. By 2.6 $f\restriction b'$ is an isomorphism of $\langle b', r'\rangle$ on $\langle b, r\rangle$. Thus, we have $\mathrm{Ord}(\langle b, r\rangle)=\mathrm{Ord}(\langle b', r'\rangle)=y$. □

3.5 Theorem. (i) $<$ *well-orders the class of all ordinals.*

(ii) *If y is an ordinal, then the class $\{x \mid x<y\}$ is a well ordered set by* (i), *and its ordinal, under the relation $<$, is y.*

Proof. $<$ is irreflexive, since by 2.17 a section of a well ordered set cannot have the same ordinal as the set itself.

To show that $<$ is transitive, suppose $x<y$ and $y<z$. Then, by Definition 3.3, there is a well ordered set $\langle c, t\rangle$ and a section b of it such that $\mathrm{Ord}(\langle c, t\rangle)=z$ and $\mathrm{Ord}(\langle b, t\rangle)=y$. By 3.4, $\langle b, t\rangle$ has a section a such that $\mathrm{Ord}(\langle a, t\rangle)=x$. a is a section of $\langle b, t\rangle$ which is a section of $\langle c, t\rangle$, hence a is a section of $\langle c, t\rangle$. By Definition 3.3, $x=\mathrm{Ord}(\langle a, t\rangle)<\mathrm{Ord}(\langle c, t\rangle)=z$.

It is an immediate consequence of Theorem 2.16 that for every two ordinals x, y $x=y$ or $x<y$ or $y<x$. This completes the proof that $<$ orders the class of all ordinals.

Now let us show that every non-void set w of ordinals has a least member. Let $z\in w$. If z is the least member of w then we have already what we wanted to get. Otherwise, let $\langle a, r\rangle$ be a well-ordered set such that $\mathrm{Ord}(\langle a, r\rangle)=z$; since w has members which are less than z there are, by 3.4, members x of a such that the section a_x of $\langle a, r\rangle$ determined by x fulfils $\mathrm{Ord}(\langle a_x, r\rangle)\in w$. Let x be the least such member of a. We claim that $\mathrm{Ord}(\langle a_x, r\rangle)$ is the least member of w; otherwise let $u<\mathrm{Ord}(\langle a_x, r\rangle)$, and $u\in w$, then $\langle a_x, r\rangle$ has a section a_y such that $\mathrm{Ord}(\langle a_y, r\rangle)=u\in w$ but, since $y\in a_x$ we have $y<x$ and this contradicts the minimality of x.

We shall conclude the proof of (i) and prove (ii) by showing that $\{y \mid y<x\}$ is a set and its ordinal is x. Let $\langle a, r\rangle$ be a well-ordered set such that $\mathrm{Ord}(\langle a, r\rangle)=x$. Let f be the function on a given by $f(u)=\mathrm{Ord}(\langle a_u, r\rangle)$. By Definition 3.3 and Proposition 3.4 $\mathrm{Rng}(f)=\{y \mid y<x\}$. Since $\mathrm{Dom}(f)=a$ is a set, $\mathrm{Rng}(f)=\{y \mid y<x\}$ is a set, too. We still have to prove that if $u, v\in a$ and urv, then $f(u)<f(v)$. This will establish that f is an isomorphism of $\langle a, r\rangle$ on $\langle\{y \mid y<x\}, <\rangle$. Since urv, a_u is a section of $\langle a_v, r\rangle$ and hence, by Definition 3.3 $f(u)=\mathrm{Ord}(\langle a_u, r\rangle)<\mathrm{Ord}(\langle a_v, r\rangle)=f(v)$. Since f is an isomorphism of $\langle a, r\rangle$ on $\langle\{y \mid y<x\}, <\rangle$ we have $\mathrm{Ord}(\langle\{y \mid y<x\}, <\rangle)=\mathrm{Ord}(\langle a, r\rangle)=x$. □

3.6 Theorem (Burali–Forti 1897, Cantor 1899). *The class of all ordinals is a proper class.*

Proof. If this class A were a set then, by 3.5, $\langle A, <\rangle$ would be a well-ordered set. Let $x=\mathrm{Ord}(\langle A, <\rangle)$, then since also $x=\mathrm{Ord}(\langle\{y \mid y<x\}, <\rangle)$, by Theorem 3.5(ii) we have, by 3.1, $\langle A, <\rangle\cong\langle\{y \mid y<x\}, <\rangle$; but, this is an isomorphism of a well ordered set on a section of itself, contradicting 2.14. □

3.6 is essentially what is known as the Burali–Forti paradox. If we were to adopt the comprehension schema (I.2.3) as an axiom we should have that every class is a set, and, in particular, the class of all ordinals would be a set, contradicting Theorem 3.6. Thus, the Burali–Forti paradox is a way of obtaining a contradiction from the schema of comprehension which is different from that of proving Russell's antinomy. □

We had to postulate in 3.1 the existence of a function G which assigns to each structure $\langle a, r\rangle$ its isomorphism type, rather than prove the existence of such a function. Suppose that for all structures $\langle a, r\rangle$ we are able to identify among the structures isomorphic to $\langle a, r\rangle$ a particular one, which can be called the *canonical* structure isomorphic to $\langle a, r\rangle$. Then we could *define* the function G by taking $G(\langle a, r\rangle)$ to be the canonical structure isomorphic to $\langle a, r\rangle$. As a consequence of Theorem 3.5(ii) we do have, in the case of a well ordered set $\langle a, r\rangle$, a canonical structure isomorphic to $\langle a, r\rangle$, namely the structure $\langle\{x \mid x<\mathrm{Ord}(\langle a, r\rangle)\}, <\rangle$, and therefore, we can define the function Ord by setting $\mathrm{Ord}(\langle a, r\rangle)=\langle\{x \mid x< \mathrm{Ord}(\langle a, r\rangle)\}, <\rangle$. There is no need for each ordinal to carry with it (as the second component) the order relation $<$ on $\{x \mid x<\mathrm{Ord}(\langle a, r\rangle)\}$ since this relation is anyway determined by the *set* $\{x \mid x<\mathrm{Ord}(\langle a, r\rangle)\}$ and by the order relation $<$ on the class of all ordinals. Therefore we prefer to define

(1) $\quad \mathrm{Ord}(\langle a, r\rangle)=\{x \mid x<\mathrm{Ord}(\langle a, r\rangle)\}$

i.e., an ordinal is the set of all smaller ordinals. (1) is not an explicit definition of the notion of an ordinal since this notion occurs also in the right-hand side of (1). We can, however, "compute" the various ordinals from (1). The first ordinal (which is the ordinal of $\langle 0, <\rangle$) must be the null-set 0. The next ordinal must be the set consisting of 0 only, namely $\{0\}$, the next one is $\{0, \{0\}\}$, and so on. In order to make our treatment of ordinals independent of the notion of the isomorphism type we shall extract from our assumptions on ordinals and from (1) some basic properties, which we shall then take for a definition of the notion of an ordinal.

3.7 Proposition. *Assuming our temporary definition of an ordinal as an isomorphism type of a well ordered set, and assuming that every ordinal is equal to the set of all smaller ordinals we have, for every ordinal x:*

(i) *x is a transitive set, namely if $z \in y \in x$ then also $z \in x$.*

(ii) *The membership relation $\in$ well-orders x, where $\in$ is the relation given by $\in=\{\langle y, z\rangle \mid y \in z\}$.*

Proof. (i) If $y \in x$, then since x is the set of all smaller ordinals, then y is also an ordinal and $y<x$. Similarly, $z \in y$ implies that z is an ordinal and $z<y$. Since $<$ orders the ordinals (by 3.5) $z<y$ and $y<x$ imply $z<x$. Since z is an ordinal smaller than x we have $z \in x$, because x is the set of all smaller ordinals.

(ii) By 3.5 the relation $<$ well-orders the set $\{y \mid y<x\}$. By our present assumption $\{y \mid y<x\}=x$, hence, $<$ well-orders x. Let $y, z \in x$. As we saw in part (i) y and z are ordinals and y is the set of all ordinals smaller than y. Thus, $z \in y$ if and only if $z<y$, i.e., the $\in$ relation coincides with $<$ on x ($\in\cap x\times x=<\cap x\times x$). Since $<$ well-orders x, also $\in$ well-orders x. □

We shall now use the properties of the ordinals given in Proposition 3.7 as our permanent definition of the notion of an ordinal.

3.8 Definition (Zermelo in 1916—see von Neumann 1928, Mirimanoff 1917, von Neumann 1923). (i) A class X is called *transitive* if every member of a member of X is a member of X, i.e., if every member of X is also a subset of X. Thus, X is transitive iff $\forall y\forall z(z\in y\wedge y\in X\rightarrow z\in X)$, iff $\bigcup X\subseteq X$. (This definition collides with our definition of a transitive relation in 1.1(ii). Whenever it may not be clear which is the notion which we intend, we shall speak of a *transitive relation* when we refer to the notion of 1.1, and of a *transitive class* (or *set*) when we refer to the notion of our present definition.)

(ii) A set x is called an *ordinal* if it is transitive and it is well ordered by the $\in$-relation.

(iii) $\mathrm{On}=\{x\mid x\textit{ is an ordinal}\}$; On is the class of all ordinals.

(iv) $<=\{\langle x,y\rangle\mid x,y\in\mathrm{On}\wedge x\in y\}$. This is the order relation on the ordinals, and we have $<=\in\cap\mathrm{On}\times\mathrm{On}$.

(v) $>=<^{-1}=\{\langle x,y\rangle\mid x,y\in\mathrm{On}\wedge y\in x\}$.

3.9 Proposition. *If every member of a class X is transitive then also $\bigcup X$ and $\bigcap X$ are transitive.* □

3.10 Exercise. For every class X, its power-class $\mathbf{P}(X)$ is transitive if and only if X is transitive. □

We shall use the lower-case Greek letters α, β, γ, δ, ξ, η, ζ for ordinals, i.e., whenever we speak about α it is to be understood as if "α is an ordinal" has been inserted in the appropriate place. For example, $\forall\alpha\Phi(\alpha)$ stands for $\forall\alpha(\alpha$ *is an ordinal* $\rightarrow\Phi(\alpha))$ and $\exists\alpha\Phi(\alpha)$ stands for $\exists\alpha(\alpha$ *is an ordinal* $\wedge\ \Phi(\alpha))$.

3.11 Proposition. *Every member of an ordinal is an ordinal.*

Proof. Let x be an ordinal and let $y\in x$. We have to prove that y is transitive and that $\in$ well-orders y. The latter is obvious, since by the transitivity of x we have $y\subseteq x$, and $\in$ well-orders x therefore it also well-orders y (1.11(ii)). We still have to prove that y is transitive. Assume $u\in z\in y$. We shall prove that $u\in y$ thereby establishing the transitivity of y. Since $y\subseteq x$ and $z\in y$ we have $z\in x$. This implies, by the transitivity of x, $u\in x$. Since $\in$ orders x and u,z,y are members of x, $u\in z\in y$ implies $u\in y$. □

3.12 Theorem. $\in$ *well-orders the class* On *of all ordinals.*

Proof. We prove first that $\in$ orders On, and we start with $\alpha\notin\alpha$. Suppose $\alpha\in\alpha$, then α is not an ordinal since then $\in$ does not order α, because the member α of α

stands in the relation $\in$ to itself. The fact that $\in$ is "transitive on On", i.e., $\alpha\in\beta\wedge\beta\in\gamma\rightarrow\alpha\in\gamma$ follows directly from the transitivity of the set γ. We conclude this part of the proof by showing that for every α and β we have $\alpha\in\beta$ or $\alpha=\beta$ or $\beta\in\alpha$. Consider $\alpha\cap\beta$, it is transitive since it is the intersection of two transitive sets (3.9), and therefore, by the definition of an initial segment, it is an initial segment of the ordered set $\langle\alpha,\in\rangle$. Therefore $\alpha\cap\beta$ is, by 2.4, either α or a section of α determined by some $\gamma\in\alpha$, i.e., $\alpha\cap\beta=\{\delta\in\alpha\mid\delta\in\gamma\}=\alpha\cap\gamma$; but since in the latter case $\gamma\subseteq\alpha$ because $\gamma\in\alpha$ we have in this case $\alpha\cap\beta=\alpha\cap\gamma=\gamma\in\alpha$. Thus, we have shown that $\alpha\cap\beta$ is either α itself, or else a member of α. Similarly, $\alpha\cap\beta$ is β itself or else a member of β. If $\alpha\cap\beta$ is equal to both α and β, or if $\alpha\cap\beta$ is equal to one of α and β and is a member of the other, then we have obtained one of the alternatives we set out to prove. We are left, thus, with the case where $\alpha\cap\beta$ is a member of both α and β; we shall see that this case cannot happen. In this case $\alpha\cap\beta$ is also a member of the intersection $\alpha\cap\beta$ of α and β, hence $\alpha\cap\beta\in\alpha\cap\beta$. But $\alpha\cap\beta$, being an ordinal (since it is a member of α), cannot be a member of itself, as we saw above.

To show that $\in$ *well*-orders On we prove first that every non-void subset a of On has a least member. Let $0\neq a\subseteq$ On and let $\alpha\in a$. If α is not the least member of a, then let β be the least member of the subset $\alpha\cap a$ of α (there is such a β because a has members which are less than α, and because $\in$ well-orders α). β is the least member of a, since if $\gamma\in a$ and $\gamma<\beta$ we have $\gamma\in\beta\in\alpha$, hence $\gamma\in\alpha$ and therefore $\gamma\in\alpha\cap a$, contradicting our assumption that β is the least member of $\alpha\cap a$. In any case, either α or β is the least member of a. We still have to show that the second requirement of well-ordering holds, namely that $\{y\mid y\in x\}$ is a set for every $x\in$ On, but this is obvious since this class is x itself. □

3.13 Exercise. Give an alternative proof of 3.12 as follows: After proving that $\in$ is irreflexive and transitive on On prove:

(a) Every subclass A of On has an $\in$-minimal member (1.1(v)).

(b) For all $\alpha,\beta\in$ On $\alpha=\beta$ or $\alpha\in\beta$ or $\beta\in\alpha$.

Hint. If (b) does not hold take α to be an $\in$-minimal member of the class $\{\alpha\mid\neg\forall\beta(\alpha=\beta$ or $\alpha\in\beta$ or $\beta\in\alpha)\}$ and choose β similarly to obtain a contradiction. □

3.14 Proposition. *If A is a non-void class of ordinals, then $\bigcap A$ is the least member of A.* □

3.15 Proposition. *The class* On *of all ordinals is a proper class.*

Remark and hint of proof. This is another form of the "Burali–Forti paradox", but is even easier to prove than 3.6. Show that if On is a set, then On $\in$ On. □

3.16 Proposition. *Every transitive set of ordinals is an ordinal. In particular, the null-set* 0 *is an ordinal.*

Hint of proof. Use 3.8(ii) and 3.12. □

3.17 Proposition. *If x is a set of ordinals, then $\bigcup x$ is an ordinal. $\bigcup x$ is the least upper bound of x. Thus every* set *of ordinals is bounded, and an unbounded class of ordinals is a proper class.*

Hint of proof. Use 3.16. □

3.18 Proposition. *For every ordinal α, also $\alpha\cup\{a\}$ is an ordinal, and this is the* successor *of α, i.e., $\alpha<\alpha\cup\{\alpha\}$ and there is no β such that $\alpha<\beta<\alpha\cup\{\alpha\}$.*

Hint of proof. For the first part use 3.16. □

3.19 Definition. 1 will denote $\{0\}$, 2 will denote $\{0, 1\}=\{0, \{0\}\}$, 3 will denote $\{0, 1, 2\}$, and so on up to 9.

By 3.16 and 3.18 it is obvious that the least ordinal is 0, the next ordinal is 1, then comes 2, and so on. □

Specializing to On Theorems 2.1 and 2.9 and Remark 2.10, which establish induction and recursion for arbitrary well-ordered classes, we get the following Theorems 3.20–3.22.

3.20 Theorem (The least ordinal principle). *If $\exists\alpha\Phi(\alpha)$ then there is a least ordinal α such that $\Phi(\alpha)$.* □

3.21 Theorem (Induction). *If for all α we have*

$$(\forall\beta<\alpha)\Phi(\beta)\rightarrow\Phi(\alpha)$$

then we have $\Phi(\alpha)$ for all ordinals α. □

3.22 Theorem (Definition by recursion—von Neumann 1923, 1928a). (i) *Given a set term $\tau(x)$, there is a unique function F on* On *such that for all α*

(1) $$F(\alpha)=\tau(F\upharpoonright\alpha).$$

(ii) *Given a set term $\tau(u, x)$ there is a unique function F on $V\times$*On *such that for all u and α*

(2) $$F(u, \alpha)=\tau(u, F_u\upharpoonright\alpha)$$

where F_u is the function on On *such that $F_u(\alpha)=F(u, \alpha)$ for all α* (i.e., $F_u=\langle F(u, \alpha)\mid\alpha\in\mathrm{On}\rangle$). □

3.23 Theorem (Mirimanoff 1917). *For every well ordered set $\langle a, r\rangle$ there is a unique ordinal α such that $\langle a, r\rangle\cong\langle\alpha, \in\rangle$. For every well ordered proper class $\langle A, R\rangle$ we have $\langle A, R\rangle\cong\langle\mathrm{On}, \in\rangle$. In both cases the isomorphisms are unique.*

Proof. Let $\langle a, r\rangle$ be a well ordered set. Since On is a proper class (3.15) $\langle a, r\rangle$ is isomorphic, by 2.18, to a section of On determined by some $\alpha \in \mathrm{On}$, but this section is obviously α itself. The rest of the theorem follows directly from 3.15, 2.18 and 2.15(ii). ☐

3.24 Definition. Ord is a function defined on the class of all well ordered sets by

$$\mathrm{Ord}(\langle a, r\rangle) = \text{ the unique } \alpha \text{ such that } \langle a, r\rangle \cong \langle \alpha, \in\rangle$$

(see 3.23). $\mathrm{Ord}(\langle a, r\rangle)$ is called *the ordinal of* $\langle a, r\rangle$. For a set a of ordinals we shall also write $\mathrm{Ord}(a)$ for $\mathrm{Ord}(\langle a, \in\rangle)$.

3.25 Corollary. *If* $\langle a, r\rangle$ *and* $\langle b, s\rangle$ *are well ordered sets, then* $\langle a, r\rangle \cong \langle b, s\rangle$ *iff* $\mathrm{Ord}(\langle a, r\rangle) = \mathrm{Ord}(\langle b, s\rangle)$. *This justifies our calling* $\mathrm{Ord}(\langle a, r\rangle)$ *the ordinal of* $\langle a, r\rangle$ (*compare with* 3.1 *and* 3.3). ☐

3.26 Corollary. *If* $\langle a, r\rangle$ *and* $\langle b, s\rangle$ *are well ordered sets, then* $\mathrm{Ord}(\langle a, r\rangle) < \mathrm{Ord}(\langle b, s\rangle)$ *iff* $\langle a, r\rangle$ *is isomorphic to a section of* $\langle b, s\rangle$.

Hint of proof. Use 2.6 and the properties of the ordinals. ☐

3.27 Proposition. *If* $a \subseteq \alpha$, *then* $\mathrm{Ord}(\langle a, <\rangle) \leqslant \alpha$.

Hint of proof. Use 2.20. ☐

3.28 Definition. Let x be a set of ordinals. $\sup x$ will denote the least upper bound *of* x (by 3.17 $\sup x = \bigcup x$), and $\sup^+ x$ will denote the least strict upper bound of x. If A is a non-void class of ordinals then $\min A$ will denote the least member of A (by 3.14 $\min A = \cap A$). We shall write $\max(\alpha_1, \ldots, \alpha_n)$ for the largest ordinal among $\alpha_1, \ldots, \alpha_n$. If $\tau(x)$ is a term such that $\tau(x)$ is an ordinal for every $x \in u$ we write also $\sup_{x\in u} \tau(x)$ for $\sup\{\tau(x) \mid x \in u\}$ and $\sup^+_{x\in u} \tau(x)$ for $\sup^+\{\tau(x) \mid x \in u\}$.

3.29 Proposition. *If* u *is a set and* $\tau_1(x)$ *and* $\tau_2(x)$ *are terms such that for all* $x \in u$ $\tau_1(x)$ *and* $\tau_2(x)$ *are ordinals and* $\tau_1(x) \leqslant \tau_2(x)$ *then* $\sup_{x\in u}\tau_1(x) \leqslant \sup_{x\in u}\tau_2(x)$ *and* $\sup^+_{x\in u}\tau_1(x) \leqslant \sup^+_{x\in u}\tau_2(x)$. ☐

3.30 Definition. (i) α is a *successor ordinal* if there is an ordinal β such that $\alpha = \beta \cup \{\beta\}$; this β, which is unique by 3.18, is called the *predecessor* of α, and is denoted by $\alpha - 1$.

(ii) α is a *limit ordinal* if $\alpha \neq 0$ and α is not a successor.

3.31 Proposition. *If a set* x *of ordinals has no largest member then* $\sup^+ x = \sup x$ *and it is a limit ordinal or zero; if* x *has a largest member* α *then* $\sup x = \alpha$ *and* $\sup^+ x = \alpha \cup \{\alpha\}$. ☐

As of now, we have not yet shown that there are limit ordinals. Note that now we have divided the ordinals to three mutually exclusive classes, one consists only of the ordinal 0, the second one consists of the successor ordinals and the third one consists of the limit ordinals. This is the major classification of the ordinals; when we shall prove theorems about all ordinals α we shall often give different proofs for three cases where $\alpha=0$, α is a successor, and α is a limit ordinal. A corresponding form of the induction theorem is 3.34 below.

3.32 Proposition. *If α is a limit ordinal and $\beta<\alpha$, then there is an ordinal γ such that $\beta<\gamma<\alpha$, and in particular $\beta<\beta\cup\{\beta\}<\alpha$.* □

3.33 Proposition. *If α is* 0 *or a limit ordinal then* $\sup\alpha=\alpha$. *If α is a successor ordinal then* $\sup\alpha$ *is the predecessor of α. (What is* $\sup^{+}\alpha$*?)* □

3.34 Theorem (Induction). *If we have*

(a) $\Phi(0)$,

(b) $\forall\beta(\Phi(\beta)\rightarrow\Phi(\beta\cup\{\beta\}))$, and

(c) *for every limit number* λ $\forall\gamma(\gamma<\lambda\rightarrow\Phi(\gamma))\rightarrow\Phi(\lambda)$,

then we have $\forall\alpha\Phi(\alpha)$.

Proof. In proving $\Phi(\alpha)$ we assume, as an induction hypothesis,

(1) $$\forall\gamma(\gamma<\alpha\rightarrow\Phi(\gamma)).$$

If $\alpha=0$ we have $\Phi(\alpha)$ by (a). If α is a successor ordinal then let β be the predecessor of α. By the induction hypothesis (1) we have, in particular, $\Phi(\beta)$, hence by (b) we have $\Phi(\beta\cup\{\beta\})$, i.e., $\Phi(\alpha)$. If α is a limit number we have $\Phi(\alpha)$ by (1) and (c). □

3.35 Exercise. Formulate and prove a version of the definition by recursion theorem which is related to 3.22 in the same way in which 3.34 is related to 3.21. □

4. *Natural Numbers and Finite Sequences*

4.1 Definition. α is a *finite ordinal*, or a *natural number*, if $\alpha=0$, or α is a successor and every ordinal $\beta<\alpha$ is 0 or a successor.

In the following, the letters i, j, k, l, m, n will be, unless otherwise indicated, variables for natural numbers. I.e., "for all n..." or $\forall n$... stands for "for all n if n is a natural number, then...", etc.

4.2 Proposition. (i) 0 *is a natural number.*

(ii) *If n is a natural number so is also $n\cup\{n\}$.*

(iii) *If n is a natural number and $\alpha<n$ then so is also α.*

Partial proof. (ii) We have to prove that $n\cup\{n\}$ and each one of its members is 0 or a successor. $n\cup\{n\}$ is a successor. A member of $n\cup\{n\}$ is either a member of n

or n itself, and in either case it is 0 or a successor (by 4.1) since n is a natural number. □

4.3 Theorem (Induction on the natural numbers). *If* $\Phi(0)$ *and*

(1) $\forall n(\Phi(n) \to \Phi(n \cup \{n\}))$ (*this will become later* $\forall n(\Phi(n) \to \Phi(n+1))$)

then $\forall n \Phi(n)$.

Proof. We shall prove that for every ordinal

(2) if α is a natural number then $\Phi(\alpha)$,

by induction on α. If α is a natural number then either $\alpha = 0$, in which case $\Phi(\alpha)$ holds by the hypothesis of 4.3, or else $\alpha = \beta \cup \{\beta\}$ for some β. By 4.2(iii) β is also a natural number. By the induction hypothesis that (2) holds for natural numbers $< \alpha$ we have $\Phi(\beta)$, and by (1) we have $\Phi(\beta \cup \{\beta\})$, i.e., $\Phi(\alpha)$. □

4.4 Theorem (Definition by recursion on the natural numbers). *For every pair* H, J *of functions there is a unique function* F *on* $V \times \{n \mid n$ *is a natural number*$\}$ *such that for all* u *and* n

$$F(u, 0) = H(u)$$

$$F(u, n \cup \{n\}) = J(u, n, F(u, n)) \text{ (this will become later } F(u, n+1) = J(u, n, F(u, n))\text{)}.$$

Hint of proof. Use the theorem on definition by recursion on the ordinals (3.22(ii)) with the term $\tau(u, f)$, which establishes the value of $F(u, \alpha)$ in terms of the parameter and the sequence $f = \langle F(u, \beta) \mid \beta < \alpha \rangle$ of the "earlier" values of F, given by

$$\tau(u, f) = \begin{cases} H(u) & \text{if } f = 0 \\ J(u, \beta, f(\beta)) & \text{if Dom}(f) \text{ is a successor ordinal and } \beta \text{ is its predecessor} \\ 0 & \text{otherwise.} \end{cases}$$ □

4.5 Definition (Addition of natural numbers). We define by recursion on the natural numbers

$$m + 0 = m$$

$$m + (n \cup \{n\}) = (m + n) \cup \{m + n\}.$$

4.6 Proposition. $m + 1 = m \cup \{m\}$. □

4.7 Remark. By 4.6, we shall use from now on $m+1$ for $m \cup \{m\}$ which is the successor of m. Therefore, we can rewrite the second part of the definition of addition as

$$m+(n+1)=(m+n)+1.$$

4.8 Proposition (Associative law of addition). $(k+m)+n=k+(m+n)$.

Proof. By induction on n. For $n=0$, we have, by the definition of addition, $(k+m)+0=k+m=k+(m+0)$. Assuming $(k+m)+n=k+(m+n)$ as the induction hypothesis, let us prove 4.8 for $n+1$.

$$\begin{aligned}(k+m)+(n+1)&=((k+m)+n)+1 &&\text{by 4.7}\\ &=(k+(m+n))+1 &&\text{by the induction hypothesis}\\ &=k+((m+n)+1) &&\text{by 4.7}\\ &=k+(m+(n+1)) &&\text{by 4.7}\end{aligned}$$

which is what we set out to prove. □

4.9 Exercises. (i) (Commutativity of addition). $m+n=n+m$.
(ii) If $l\neq 0$, then $n+l>n$.
(iii) If $n>k$, then there is an $l\neq 0$ such that $n=k+l$.

Hints. (i) Prove $1+n=n+1$ by induction on n, then prove $m+n=n+m$ by induction on n. (ii) Use induction on l. (iii) Use induction on n. □

We shall assume that all other facts about the order of the natural numbers and their addition and subtraction are known to the reader. In particular, for $n\neq 0$, the difference $n-1$ is indeed the predecessor of n, and thus our notation of $\alpha-1$ as the predecessor of the successor ordinal α is compatible with our use of $-$ for subtraction.

4.10 Definition. (i) f is a *sequence* if f is a function and $\mathrm{Dom}(f)$ is an ordinal; $\mathrm{Dom}(f)$ is also called the *length* of f and is also denoted with $\mathrm{Length}(f)$. Sometimes we shall also informally refer to a function whose domain is a set of ordinals which is not necessarily an ordinal as a sequence. If f is a sequence whose domain is a finite ordinal then we say that f is a *finite sequence*. A finite sequence of length n is also called an *ordered n-tuple* or just an *n-tuple*. In the case of sequences we shall also write f_α for $f(\alpha)$.

(ii) 0 is a finite sequence of length 0 and is denoted also by $\langle\ \rangle$. $\langle a\rangle$ will denote the sequence $\{\langle 0, a\rangle\}$ of length 1. $\langle a, b\rangle$ will denote the finite sequence $\{\langle 0, a\rangle, \langle 1, b\rangle\}$ of length 2 (where $\langle 0, a\rangle=\{\{0\}, \{0, a\}\}$ and $\langle 1, b\rangle=\{\{1\}, \{1, b\}\}$) as in Definition I.6.1; this makes the term $\langle a, b\rangle$ ambiguous, having been defined differently here and in I.6.1; we shall discuss this problem below. In general, if k is one of 0, 1, 2, 3, ... and $a, b, \ldots, t$ are k terms then $\langle a, b, \ldots, t\rangle$ denotes the finite sequence $\{\langle 0, a\rangle, \langle 1, b\rangle, \ldots, \langle k-1, t\rangle\}$ of length k. For a function F, $F(a, b, \ldots, t)$ denotes $F(\langle a, b, \ldots, t\rangle)$.

(iii) If f is a finite sequence of length n, then for $i<n$ the sets $f(i)$ are called the *terms* of the sequence f. $f(0)$ is called the *first term* of f and $f(n-1)$ is called the *last term* of f. □

We chose to write $\langle x, y\rangle$ for both $\{\{x\}, \{x, y\}\}$ and $\{\langle 0, x\rangle, \langle 1, y\rangle\}$ (where inside the latter expression $\langle\ ,\ \rangle$ has the same meaning as in the former expression), because both notions are ordinarily known as ordered pairs, and both satisfy the requirement

$$(3) \qquad \langle x, y\rangle=\langle u, v\rangle \rightarrow x=u \wedge y=v$$

of I.6.2. Since it is only on very rare occasions that we need any property of $\langle x, y\rangle$ beyond (3), there is usually no need to distinguish between the two notions. In the few cases where a distinction is needed it will be clear from the context which of the two notions is the intended one.

4.11 On the Use of Subscripts. In 4.10 we introduced the use of *subscripts* as function arguments, writing f_α for $f(\alpha)$. Alongside this usage we shall continue our old habit of using subscripts as integral parts of the symbols. Thus x_1 and x_2 can denote just two different (and unrelated) variables, and they can denote two terms of the sequence x. This ambivalence is customary in mathematics; in our book it will always be clear, hopefully, which is the intended meaning.

4.12 Definition. For finite sequences the *concatenation* $f^\frown g$ of f and g is $f\cup\{\langle \mathrm{Dom}(f)+i, g(i)\rangle \mid i<\mathrm{Dom}(g)\}$, i.e., if f is $\langle a, b, \ldots, c\rangle$ and g is $\langle d, e, \ldots, j\rangle$ then $f^\frown g$ is $\langle a, b, \ldots, c, d, e, \ldots, j\rangle$.

4.13 Proposition. *If f and g are finite sequences, then $f^\frown g$ is a finite sequence whose length is the sum of the lengths of f and g.* □

4.14 Definition. (i) For a relation R let us say that a finite sequence f of length n is an *R-chain from x to y* if $n\geq 2$, the first term of f is x, the last term of f is y and R holds between any two consecutive terms of f (i.e., $(\forall i<n-1) f_i R f_{i+1}$).

(ii) For a relation R we define R^n, for $n\geq 1$, to be the relation, with n as a parameter, given by

$$\{\langle x, y\rangle \mid \textit{there is an R-chain of length } n+1 \textit{ from } x \textit{ to } y\}.$$

R^0 is defined to be the identity relation $\{\langle x, x\rangle \mid x\in V\}$.

(iii) For a relation R let

$$R^*=\{\langle x, y\rangle \mid \text{there is an R-chain from } x \text{ to } y\}.$$

R^* is called the *ancestral* of R. If xRy is the relation "x is a parent of y" then xR^*y is the relation "x is an ancestor of y".

4.15 Proposition. *$R^1=R$; for $n\geq 0$, $R^{n+1}= R^n\circ R$.* □

For a relation r which is a set we can take 4.15 to be a definition by recursion of r^n; but for a class R this would not be covered by our theorem, 4.4, on definition by recursion, since that theorem enables us to obtain only set-valued functions and for a class R R^n is a class-valued function. Therefore, we had to give an explicit definition of R^n in 4.14(ii).

4.16 Proposition. *For $m, n \geqslant 0$, $R^{m+n} = R^m \circ R^n$.* □

4.17 Proposition. *For $n > 0$ $\mathrm{Dom}(R^n) \subseteq \mathrm{Dom}(R)$, $\mathrm{Rng}(R^n) \subseteq \mathrm{Rng}(R)$; if $\mathrm{Rng}(R) \subseteq \mathrm{Dom}(R)$ then $\mathrm{Dom}(R^n) = \mathrm{Dom}(R)$.*

Hint of proof. Use I.6.18. □

4.18 Proposition. *If F is a function then, for $n \geqslant 0$, F^n is a function, too. If F is a one–one function and $n \geqslant 0$ then F^n is also a one–one function.* □

4.19 Remark. $xR^*y \leftrightarrow (\exists n > 0)xR^ny$. Hence we can write, informally, $R^* = \bigcup_{n>0} R^n$.

4.20 Theorem. *For a relation R, R^* is a transitive relation which includes R and is indeed the least transitive relation which includes R.*

Proof. $R \subseteq R^*$ is obvious, since if xRy then $\langle x, y\rangle$ is an R-chain. To see that R^* is transitive, assume xR^*y and yR^*z. Let $\langle f_0, \ldots, f_{n-1}\rangle$ be an R-chain from x to y and let $\langle g_0, \ldots, g_{m-1}\rangle$ be an R-chain from y to z. Then, as easily seen, the finite sequence $\langle f_0, \ldots, f_{n-2}, g_0, \ldots g_{m-1}\rangle$ $(=(f\upharpoonright(n-1))^\frown g)$ is an R-chain from x to z, and we have xR^*z, which establishes the transitivity of R^*.

To see that R^* is the least transitive relation which includes R, we assume that S is a transitive relation which includes R and prove that $R^* \subseteq S$. Let xR^*y, then there is an R-chain $f = \langle f_0, \ldots, f_{n-1}\rangle$ from x to y. We prove that f_0Sf_i for every $0 < i < n$ by induction on i. For $i = 1$ we have f_0Rf_1, since f is an R-chain and hence, since $S \supseteq R$, f_0Sf_1. Assume now that f_0Sf_i for $i < n-1$, then since f is an R-chain and $R \subseteq S$ we have also f_iSf_{i+1}. Since S is transitive, f_0Sf_i and f_iSf_{i+1} imply f_0Sf_{i+1}, which is what we had to show in our induction step. In particular, for $i = n-1$ we have f_0Sf_{n-1}, i.e., xSy; thus, $R^* \subseteq S$. □

4.21 Exercises. (i) For a relation R, $R^* \circ R = R \circ R^* \subseteq R^*$.

(ii) If R and S are relations such that $S \circ R \subseteq S$ then $S \circ R^* \subseteq S$.

(iii) $(R^{-1})^* = (R^*)^{-1}$.

(iv) $R^* = R$ *iff* R is a transitive relation.

(v) $R^{**} = R^*$. □

We have defined the notion of a *finite* ordinal, but until now we have not proved the existence of an infinite ordinal, or, for that matter, of any infinite set. As we have already mentioned, the existence of infinite sets and ordinals is proved by means of the axiom of infinity, an axiom which we have not yet used. The theory of integers and rational numbers can be developed without the axiom of infinity, since the integers can be defined as ordered pairs of natural numbers and the rational numbers can be defined as ordered pairs of integers, and the order and the addition and multiplication of integers and rational numbers can be defined in terms of the order and the addition and multiplication of the natural numbers. However, the real numbers cannot be represented by means of finite sets only, hence we need the axiom of infinity already in order to develop the theory of real numbers in set theory. We shall now prove the existence of an infinite ordinal since we need this fact now in order to get certain results about relations (mainly Theorem 4.32). We shall have a further discussion of the notions of finiteness and infinity in § III.1.

4.22 Proposition. (Existence of ω.) *The class of all finite ordinals is a set; it is the least infinite* (*i.e., non-finite*) *ordinal.*

Proof. Let z be a set such as required by the axiom of infinity (I.5.21). Since $0 \in z$ and for every x also $x+1 = x \cup \{x\} \in z$ we have, by induction, that every finite ordinal is a member of z. Thus, the class N of all finite ordinals is a subclass of the set z, and hence, by the axiom of subsets, N is a set. N is a transitive set of ordinals, by 4.2, and hence, by 3.16, N is an ordinal. N is greater than every finite ordinal since every finite ordinal is a member of N; N is the least infinite ordinal since all the ordinals smaller than N are members of N, and hence finite. □

4.23 Definition. ω will denote the least infinite ordinal. By 4.22 ω is the set of all finite ordinals.

4.24 Proposition. *ω is a limit ordinal; it is the least limit ordinal.* □

Our first application of the existence of the ordinal ω will be in 4.29 and 4.33 below where we prove the existence, for every set a, of a set b which is the closure of a under certain functions and relations. We shall first investigate this concept of closure.

4.25 Definition. For a relation R, a class A is said to be *R-closed*, or *closed under R*, if whenever $x \in A$ and xRy then also $y \in A$ (i.e., $R[\mathrm{A}] \subseteq A$).

4.26 Discussion. There are two ways which one can follow in defining the class B which is the R-closure, or the closure under the class R, of a class A. The first way, which an experienced mathematician chooses instinctively, is to take for B the least R-closed class with includes A, i.e., B is an R-closed class which includes A, and every R-closed class C which includes A includes also B. The second way, is to put first in B all the members of A, to add then to B all the objects which stand in the relation R to the members of A (i.e., the members of $R[A]$), and to continue, indefinitely if necessary, with a process of adding to B at each stage all the objects which stand in the relation R to the objects which are already in B. In our case this means, disregarding formal difficulties, to define $C_0 = A$, for $\alpha > 0$ $C_\alpha = \bigcup_{\beta<\alpha}(C_\beta \cup R[C_\beta])$ and $B = \bigcup_{\alpha \in \mathrm{On}} C_\alpha$. This second way is basically a more straightforward way; both ways lead always to the same class B. The reason why the first way is usually preferred is that it makes it easy to prove, without having to use induction, that all the members of the R-closure B of A have certain properties. In order to prove that every $x \in B$ satisfies $\Phi(x)$ it is sufficient to prove that the class $\{x \mid \Phi(x)\}$ is an R-closed class which includes A, thus all one has to prove is that $x \in A \rightarrow \Phi(x)$ and $\Phi(x) \wedge xRy \rightarrow \Phi(y)$. Since we deal here with classes we have also to make sure that we do not use, in the definition of R-closure, quantifiers over class variables. Therefore we shall adopt, in Definition 4.27 below the second way just mentioned. We can use a definition simpler than that of $B = \bigcup_{\alpha \in \mathrm{On}} C_\alpha$ above since in the present case no new members are obtained after ω steps. Proposition 4.28(iii) below will justify Definition 4.27 by showing that the ω steps of 4.27 were sufficient in order to get an R-closed class. That proposition will also establish the equivalence of both ways of defining the R-closure of A in our case.

4.27 Definition. Let R be a relation. The *R-closure* of A is defined to be the class $A \cup R^*[A]$, (i.e., by 4.19, $A \cup \bigcup_{1 \leqslant n < \omega} R^n[A]$).

4.28 Proposition. (i) 0 *and* V *are R-closed.*

(ii) *A is R^*-closed if A is R-closed.*

(iii) *The R-closure of a class A is an R-closed class which includes A, and is the least such class. (i.e., if B is an R-closed class which includes A then B includes the R-closure of A.)* □

4.29 Proposition. *Let $F_1, \ldots, F_k$ be functions. For every set a the set $b = a \cup \bigcup_{1 \leqslant n < \omega} (F_1 \cup \ldots \cup F_k)^n[a]$ is the closure of a under $F_1, \ldots, F_k$, i.e., b is the least set which includes a and is closed under each one of the functions $F_1, \ldots, F_k$.* □

The assumption in 4.29 that $F_1, \ldots, F_k$ are functions is too strong. We shall see that we can replace this assumption by the weaker assumption that $F_1, \ldots, F_k$ are right-narrow relations (see Definition 1.1(vi)).

4.30 Proposition. *The relation R is right-narrow iff $R[x]$ is a set for every set x.*

Hint of proof and a remark. Use the axioms of replacement and union. Notice that the "only if" direction of this lemma is a generalization of the axiom of replacement (or I.6.19). □

4.31 Proposition. *If $R_1, \ldots, R_n$ are right-narrow relations, so is $\bigcup_{i=1}^n R_i$.* □

4.32 Proposition. *If R is a right-narrow relation so is its ancestral R^*; if R is left-narrow so is R^*.*

Proof. Let R be right-narrow and let x be an object. We define a function F on ω by recursion as follows. $F(0) = \{x\}$, $F(n+1) = R[F(n)]$; $R[F(n)]$ is a set by 4.30. Since $\mathrm{Dom}(F) = \omega$ is a set, also $\mathrm{Rng}(F)$ is a set. Let $u = \bigcup \mathrm{Rng}(F)$. We shall show that if xR^*y then $y \in u$, hence $R^*[\{x\}] \subseteq u$, and R^* is right-narrow. If xR^*y then there is an R-chain $f = \langle f_0, \ldots, f_{n-1} \rangle$ from x to y, $n \geqslant 2$. $f_0 = x \in \{x\} = F(0)$; it follows easily, by induction on i, that $f_i \in F(i)$ for $i<n$, hence in particular, $y=f_{n-1} \in F(n-1) \subseteq u$, which is what we had to show. The case of a left-narrow R is treated similarly. □

4.33 Theorem. (i) *If R is a right-narrow relation then the R-closure of a set is a set.*

(ii) *If $R_1, \ldots, R_k$ are right-narrow relations then for every set a the $\bigcup_{i=1}^k R_1$-closure b of a is also the closure of a with respect to $R_1, \ldots, R_k$, i.e., b is the least class which includes a and is closed under $R_1, \ldots, R_k$. This is the generalization of* 4.29 *to right-narrow relations.* □

Proposition 4.29 and Theorem 4.33(ii) which deal with unary functions and binary relations can easily be generalized to give the same result for functions and relations with an arbitrary finite, or even infinite, number of arguments (see III.4.25, III.4.26, IV.4.29 and IV.4.30).

5. *Well-Founded Relations*

5.1 Definition (Zermelo 1935). A relation R is said to be *well-founded* if

(a) Every non-void set y has an R-minimal member x, i.e., there is a member x of y such that for no $z \in y$ does zRx hold, and

(b) R is left-narrow, i.e., for every y the class $\{x \mid xRy\}$ is a set.

Remark. The notion of a well founded relation is a generalization of that of a well-ordering relation; a relation $R \subseteq A \times A$ well-orders the class A exactly when it is a well-founded order relation on A. We shall see that many of the theorems which we proved about well ordered classes can be generalized to well-founded relations, and this is exactly what makes these relations so important. As in the case of a well-ordering relation, the condition of the left-narrowness of R is often not taken in the literature as a part of the definition of a well-founded relation. As we have remarked in 1.2, if R is a set then this condition is fulfilled automatically.

5.2 Proposition. *If R is a well-founded relation and $S \subseteq R$ then also S is a well-founded relation.* □

5.3 Proposition. (i) *If R is a well-founded relation* (*or even if R just satisfies part* (a) *of* 5.1) *then*

(1) *there is no sequence x of length ω such that $x_{n+1} R x_n$ for all $n<\omega$.*

(ii) *If R is a relation for which* (1) *holds then every non-void* well-orderable *set y has an R-minimal member. As a consequence, if we assume the axiom of choice, in which case every set is well-orderable, then* (1) *can replace* (a) *in the definition* 5.1 *of well-foundedness.*

Proof. (i) If x is as in (1) then the set $\{x_n \mid n<\omega\}$ has obviously no R-minimal member. (ii) Let y be a non-void well-orderable set. Suppose that y has no R-minimal member. Since no $u \in y$ is R-minimal in y there is for every such u a $v \in y$ such that vRu. We define now a sequence x of length ω by

$$x_0 = \text{a member of } y$$
$$x_{n+1} = \text{a member } z \text{ of } y \text{ such that } zRx_n.$$

Since y is well-orderable our uses of the indefinite article in the defining clauses of x_0 and x_{n+1} are legitimate since they can be interpreted as follows. Let $<$ be a well-ordering of y. By our agreement in 2.25, the above definition is interpreted as

$$x_0 = \text{the } <\text{-least member of } y$$
$$x_{n+1} = \text{the } <\text{-least member of the set } \{z \in y \mid zRx_n\}.$$

The sequence x thus obtained obviously contradicts (1). □

5.4 Corollary. *If R is a well-founded relation then there are no R-cycles, i.e., there is no finite sequence $\langle x_1, \ldots, x_n \rangle$, $n \geq 1$, such that $x_n R x_1 R x_2 R \ldots R x_{n-1} R x_n$. In particular, the relation R is irreflexive.* □

5.5 Corollary. *If R is a relation which orders A so that every section of A is a well-orderable set, and if* (1) *holds for R then R well-orders A.* □

5.6 Theorem. *Let R be a well-founded relation.*

(i) The minimal member principle. *If $\exists x \Phi(x)$ then there is an R-minimal object x for which $\Phi(x)$ holds.*

(ii) Induction. *If $\forall x[\forall y(yRx \to \Phi(y)) \to \Phi(x)]$ then $\forall x \Phi(x)$; i.e., if for any given x, if $\Phi(y)$ holds for every y such that yRx then $\Phi(x)$ holds also for x, then $\Phi(x)$ holds for every x. (In applying induction, the assumption $\forall y(yRx \to \Phi(y))$ which we are allowed to assume in proving $\Phi(x)$, is called the* induction hypothesis.)

Proof. (i) Let $B = \{x \mid \Phi(x)\}$; since we assume $\exists x \Phi(x)$ we have $B \neq 0$. We want to prove that the non-void class B has an R-minimal member, while the definition of well-foundedness requires only that every non-void *set* has an R-minimal member. Therefore, it pays to find a non-void set u such that every R-minimal member of u is also an R-minimal member of B; to obtain such a set we use the left-narrowness of R as follows. Let z be such that $z \in B$. Let $u = B \cap \{x \mid xR^*z\} \cup \{z\}$; the right-hand side is indeed a set since R^* is left-narrow, by 4.32. Since R is well founded and $u \neq 0$ (because $z \in u$) u has an R-minimal member x. Since $u \subseteq B$, $x \in B$. Let us prove that x is an R-minimal member of B, and hence, by the definition of B, R-minimal among the objects satisfying $\Phi(x)$. Suppose there is a $y \in B$ such that yRx. Since $x \in u$ we have, by the definition of u, xR^*z or $x = z$. Combining this with yRx we get yR^*z, i.e., $y \in \{x \mid xR^*z\}$. Since also $y \in B$ we have, by the definition of u, $y \in u$. Now we have $y \in u$ and yRx, contradicting our choice of x as an R-minimal member of u.

(ii) Suppose that for some x we have $\neg\Phi(x)$. Then, by part (i) with $\neg\Phi(x)$ replacing $\Phi(x)$, there is an object x which is R-minimal among the objects satisfying $\neg\Phi(x)$. For such an x we have both $\neg\Phi(x)$ and $\forall y(yRx \rightarrow \Phi(y))$, which contradicts the assumption of (ii). □

We are proving here theorems which are natural generalizations of the theorems we proved about well ordered classes; note that the generalization of the notion of an initial segment of a well ordered class is the notion of an R^{-1}-closed class.

5.7 Theorem (Definition by recursion on a well-founded relation—Tarski 1955, Montague 1955). *Let R be a well founded relation. Given a set term τ, there exists a unique function F on V such that for every x*

(2) $$F(x)=\tau(x, F \upharpoonright \{y \mid yRx\}).$$

(Notice that $\{y \mid yRx\}$ is a set since R is left-narrow.)

Proof. Our proof is an almost verbal repetition of the proof of the theorem 2.9 on definition by recursion on a well ordered class, with the notion of an initial segment replaced by the notion of an R^{-1}-closed class.

(a) Let F and F' be two functions on R^{-1}-closed classes such that each one of them satisfies (2) for every x in its domain; then we shall prove that F and F' are compatible. Before proving it, let us note that this establishes the uniqueness of the F of the theorem, since any two compatible functions on V are identical.

Let F and F' be as above. We shall prove by induction on R, that

$$\text{if } x \in \mathrm{Dom}(F) \cap \mathrm{Dom}(F') \text{ then } F(x)=F'(x).$$

Assume $x \in \mathrm{Dom}(F) \cap \mathrm{Dom}(F')$ and the induction hypothesis that for every y such that yRx, if $y \in \mathrm{Dom}(F) \cap \mathrm{Dom}(F')$ then $F(y)=F'(y)$. Let y be any object such that yRx. Since $\mathrm{Dom}(F)$ and $\mathrm{Dom}(F')$ and R^{-1}-closed classes, we have also $y \in \mathrm{Dom}(F) \cap \mathrm{Dom}(F')$ and hence, by the induction hypothesis $F(y)=F'(y)$. Thus we have $F \upharpoonright \{y \mid yRx\} = F' \upharpoonright \{y \mid yRx\}$. By (2), we have $F(x)=\tau(x, F \upharpoonright \{y \mid yRx\})=\tau(x, F' \upharpoonright \{y \mid yRx\})=F'(x)$. By the theorem on proof by induction on R (5.6(ii)) we have that for all x if $x \in \mathrm{Dom}(F) \cap \mathrm{Dom}(F')$ then $F(x)=F'(x)$, i.e., F and F' are compatible.

(b) Let T be the class of all functions f (i.e., all functions f which are sets) whose domains are R^{-1}-closed sets and which satisfy (2), i.e.,

(3) $$x \in \mathrm{Dom}(f) \rightarrow f(x)=\tau(x, f \upharpoonright \{y \mid yRx\}).$$

By what we proved in part (a) every two members of T are compatible, hence (by I.6.30(ii)) $\bigcup T$ is a function, let us denote it by F. Our proof of the theorem will be completed once we show that $\mathrm{Dom}(F)=V$ and that F satisfies (2).

(c) F satisfies (2). Let $x \in \mathrm{Dom}(F)$; since $F=\bigcup T$, $x \in \mathrm{Dom}(f)$ for some $f \in T$. Since $f \in T$, $\mathrm{Dom}(f)$ is R^{-1}-closed, and since $x \in \mathrm{Dom}(f)$ we have $\{y \mid yRx\} \subseteq \mathrm{Dom}(f)$. Since $F=\bigcup T \supseteq f$ we have $F(x)=f(x)$ and $F \upharpoonright \{y \mid yRx\} = f \upharpoonright \{y \mid Rx\}$. By (3) we get $F(x)=f(x)=\tau(x, f \upharpoonright \{y \mid yRx\})=\tau(x, F \upharpoonright \{y \mid yRx\})$, thus establishing (2).

(d) $\text{Dom}(F)=V$, since from $\text{Dom}(F)\subset V$ we shall derive a contradiction. Using the minimal member principle, let z be an R-minimal member of $V\sim\text{Dom}(F)$. The rough idea of the proof is to extend F to z by means of the recursion equation (2) and thereby to contradict the maximality of F. To carry out the proof we notice first that $\text{Dom}(F)$ is R^{-1}-closed, since $F=\bigcup T$ implies $\text{Dom}(F)=\bigcup\{\text{Dom}(f)\mid f\in T\}$ and for each $f\in T$ $\text{Dom}(f)$ is R^{-1}-closed. For the z we chose let

(4) $$f=F\upharpoonright\{y\mid yR^*z\}\cup\{\langle z,\tau(z,F\upharpoonright\{y\mid yRz\})\rangle\}.$$

The right-hand side of (4) is indeed a set since R* is left-narrow by 4.32. We shall obtain a contradiction once we show that $f\in T$, since then $z\in\text{Dom}(f)\subseteq\text{Dom}(F)$, contradicting our choice of z such that $z\notin\text{Dom}(F)$. f is a function since F is one and $z\notin\text{Dom}(F)$. Now let us prove that $\text{Dom}(f)$ is R^{-1}-closed. Let

$$w=\text{Dom}(F\upharpoonright\{y\mid yR^*z\})=\text{Dom}(F)\cap\{y\mid yR^*z\};$$

w is R^{-1}-closed, being the intersection of two R^{-1}-closed classes. In order to prove that also $\text{Dom}(f)=w\cup\{z\}$ is R^{-1}-closed it is enough to show that $\{y\mid yRz\}\subseteq w$. We have, obviously, $\{y\mid yRz\}\subseteq\{y\mid yR^*z\}$; also since z is an R-minimal member of $V\sim\text{Dom}(F)$ we have $\{y\mid yRz\}\subseteq\text{Dom}(F)$; therefore

$$\{y\mid yRz\}\subseteq\text{Dom}(F)\cap\{y\mid yR^*z\}=w,$$

which establishes that $\text{Dom}(f)$ is R^{-1}-closed. In order to prove $f\in T$ we still have to show that (3) holds for f. We saw above that w is R^{-1}-closed and that $\{y\mid yRz\}\subseteq w$, hence if $x\in\text{Dom}(f)=w\cup\{z\}$ then $\{y\mid yRx\}\subseteq w$. Therefore by the definition (4) of f $f\upharpoonright\{y\mid yRx\}=F\upharpoonright\{y\mid yRx\}$. Let us now continue separately with the cases where $x\in w$ and $x=z$. If $x\in w$ then, by the definition of f, $f(x)=F(x)$ and hence, since we know already that F satisfies (2),

$$f(x)=F(x)=\tau(x,F\upharpoonright\{y\mid yRx\})=\tau(x,f\upharpoonright\{y\mid yRx\}),$$

i.e., (3) holds. If $x=z$, then, by definition of f,

$$f(z)=\tau(z,F\upharpoonright\{y\mid yRz\})=\tau(z,f\upharpoonright\{y\mid yRx\}),$$

and (3) holds. □

5.8 Exercise. Prove that if R is a well founded relation then so is also R^*.

Hint. Use 4.32, then prove that if $u\neq 0$ then u has an R^*-minimal member. To do this extend u to u' by adding to u exactly the objects necessary to guarantee that every R-minimal member of u' be an R^*-minimal member of u' and of u. Alternatively, prove by R-induction on x that every set which contains x has an R^*-minimal member. □

5.9 Definition. We say that the relation R is well-founded on the class B if the restriction $R \mid B$ is a well founded relation. $\langle B, R\rangle$ is a *well founded structure* if $\langle B, R\rangle$ is a structure and R is well-founded on B.

5.10 Proposition. *A relation R is well-founded on a class A iff every non-void subset y of A has an R-minimal member and for every $x \in A$ $\{z \in A \mid zRx\}$ is a set.* □

5.11. Proposition. *If the relation R is well-founded on A, and $B \subseteq A$, then R is well-founded also on B.* □

For a well founded relation we shall now define the notion of the R-rank $\rho_R(x)$ of an object x. The R-minimal objects x, i.e., the objects x for which there is no y with yRx, will have R-rank 0. The objects x which are not R-minimal but which are such that yRx holds only for R-minimal y will have R-rank 1, and so on.

5.12 Definition. (The rank with respect to a well founded relation—Zermelo 1935.) Let R be a well founded relation. The function ρ_R on V is defined by recursion on R as follows:

$$\rho_R(x) = \sup{}^+\{\rho_R(y) \mid yRx\}$$

(where $\sup^+ z$ is the least *strict* upper bound of z). Notice that since R is left-narrow $\{y \mid yRx\}$ is a set and hence $\{\rho_R(y) \mid yRx\}$ is a set. One can easily verify by induction on R that $\rho_R(x)$ is an ordinal for every x.

5.13 Proposition. *If R is well-founded, then*

(5) $$xRy \rightarrow \rho_R(x) < \rho_R(y). \quad \square$$

5.14 Exercises. (i) If R is a well founded relation then $xR^*y \rightarrow \rho_R(x) < \rho_R(y)$.

(ii) Let R be a left-narrow relation and let $H: V \rightarrow \mathrm{On}$ be a function which satisfies (5), i.e., for all x, y $xRy \rightarrow H(x) < H(y)$, then R is well founded.

(iii) If R is a well founded relation so is also R^*. (Give a proof different from that of 5.8.)

(iv) If R is a well founded relation and $S \subseteq R$, then for every x $\rho_S(x) \leqslant \rho_R(x)$.

For a relation R, a function $H: V \rightarrow \mathrm{On}$ is called a *rank function* for R if it satisfies (5), i.e., if for all x $\{H(y) \mid yRx\}$ is a set and $H(x) = \sup^+\{H(y) \mid yRx\}$.

(v) For a relation R there is at most one rank function.

(vi) If ρ_R is a rank function for R and A is an R^{-1}-closed class then $\rho_R[A]$ is either On or an ordinal.

Hints. (ii) For a set z, a member u of z with least $H(u)$ is an R-minimal member of z. (iv) Use induction on x with respect to the relation R. (vi) If $\rho_R[A] \neq \mathrm{On}$ let α be the least member of $\mathrm{On} \sim \rho_R[A]$. If $\rho_R[A] \neq \alpha$ obtain a contradiction from the existence of a least member β of $\rho_R[A] \sim \alpha$. □

6. *Well-Founded Sets*

Let us recall that the $\in^{-1}$-closed classes were called transitive classes in 3.8(i). Accordingly, we define:

6.1 Definition. The *transitive closure* of a class A is the $\in^{-1}$-closure of the class, i.e., $A \cup (\in^{-1})^*[A]$; we denote it with $\mathrm{Tc}(A)$.

6.2 Corollary. (i) *For every class A, $\mathrm{Tc}(A)$ is a transitive class, and is the least transitive class which includes A. For every set x, $\mathrm{Tc}(x)$ is a set.*

(ii) $B \subseteq \mathrm{Tc}(A) \to \mathrm{Tc}(B) \subseteq \mathrm{Tc}(A)$.

(iii) $\mathrm{Tc}(A) = A \cup \bigcup_{x \in A} \mathrm{Tc}(x)$.

Hint of proof. (i) Use 4.28(iii) for the first part and 4.32 and 4.30 for the second part. (ii) Use (i). (iii) Use (i) and (ii). □

6.3 Definition (Mirimanoff 1917). $\mathrm{Wf} = \{x \mid \in \text{ is well-founded on } \mathrm{Tc}(x)\}$. The members of Wf are called the *well founded sets*.

6.4 Theorem. Wf *is a transitive class on which* $\in$ *is well-founded, and is the maximal such class.*

Proof. First let us show that Wf is transitive. If $y \in x$ then, by 6.2(ii), $\mathrm{Tc}(y) \subseteq \mathrm{Tc}(x)$. If $x \in \mathrm{Wf}$ then $\in$ is well founded on $\mathrm{Tc}(x)$ and hence, by 5.11, also on $\mathrm{Tc}(y)$ and thus $y \in \mathrm{Wf}$ and Wf is transitive. Now let us prove that $\in$ is well-founded on Wf. Let u be a non-void subset of Wf; we have to show that u has an $\in$-minimal member. Take an $x \in u$. If $x \cap u = 0$ then x is an $\in$-minimal member of u. If $x \cap u \neq 0$ then $\mathrm{Tc}(x) \cap u \supseteq x \cap u$ is a non-void subset of $\mathrm{Tc}(x)$ and since $\in$ is well founded on $\mathrm{Tc}(x)$ $\mathrm{Tc}(x) \cap u$ has an $\in$-minimal member z. z is obviously also an $\in$-minimal member of u.

Finally let us assume that C is a transitive class on which $\in$ is well-founded, and prove $C \subseteq \mathrm{Wf}$. Let $x \in C$, then $x \subseteq C$; since C is transitive also $\mathrm{Tc}(x) \subseteq C$, by 6.2. Since $\in$ is well-founded on C, $\in$ is also well-founded on $\mathrm{Tc}(x)$ (by 5.11), hence $x \in \mathrm{Wf}$. Thus $C \subseteq \mathrm{Wf}$. □

6.5 Proposition. (i) $x \subseteq \mathrm{Wf} \to x \in \mathrm{Wf}$.

(ii) $x \in \mathrm{Wf} \wedge y \subseteq x \to y \in \mathrm{Wf}$.

(iii) $x \in \mathrm{Wf} \to \bigcup x \in \mathrm{Wf} \wedge \mathbf{P}(x) \in \mathrm{Wf}$.

(iv) $\mathrm{On} \subseteq \mathrm{Wf}$.

Hint of proof. (i) By 6.2 and 6.4 $\mathrm{Tc}(x) \subseteq \mathrm{Wf}$; by 6.4 and 5.11 $\in$ is well-founded on $\mathrm{Tc}(x)$. (ii) Use 6.4 and (i). (iii) Use directly the definitions of Wf and On, or else use induction and 6.5(i). □

6.6 Definition (Mirimanoff 1917). The *rank* $\rho(x)$ of $x \in \mathrm{Wf}$ is the $\in \restriction \mathrm{Wf}$-rank of x, i.e., $\rho(x) = \rho_{\in \restriction \mathrm{Wf}}(x)$. $\rho(x)$ is an ordinal for every $x \in \mathrm{Wf}$.

6.7 Corollary. $x \in \mathrm{Wf} \to \rho(x) = \sup^+\{\rho(y) \mid y \in x\}$ (by 5.12); $y \in \mathrm{Wf} \wedge x \in y \to x \in \mathrm{Wf} \wedge \rho(x) < \rho(y)$. □

6.8 Proposition. *For every ordinal* α, $\rho(\alpha) = \alpha$.

Hint of proof. Use induction on α. □

6.9 Exercise. (i) Let x be a transitive well-founded set, then $\{\rho(y) \mid y \in x\} = \rho(x)$, i.e., x has members of every rank $< \rho(x)$.

(ii) $x \in \mathrm{Wf} \leftrightarrow \forall z(x \in z \to (\exists y \in z)(y \cap z = 0))$.

Hint. (ii) If $x \in \mathrm{Wf}$ choose for y an $\in$-minimal member of $z \cap \mathrm{Wf}$. If the right-hand side holds consider $z = \mathrm{Tc}(\{x\}) \sim \mathrm{Wf}$, and use the right-hand side to show that $x \notin z$. □

6.10 Proposition (Mirimanoff 1917). $\{x \in \mathrm{Wf} \mid \rho(x) < \alpha\}$ *is a set.*

Proof. By induction on α. If $\alpha = 0$ then this class is 0. If $\alpha = \beta \cup \{\beta\}$ then by the induction hypothesis $\{x \in \mathrm{Wf} \mid \rho(x) < \beta\}$ is a set. By 6.7,

$$y \in \mathrm{Wf} \wedge \rho(y) \leqslant \beta \to y \subseteq \{x \in \mathrm{Wf} \mid \rho(x) < \beta\},$$

hence

$$\{x \in \mathrm{Wf} \mid \rho(x) < \alpha\} \subseteq \mathbf{P}(\{x \in \mathrm{Wf} \mid \rho(x) < \beta\}),$$

and $\{x \in \mathrm{Wf} \mid \rho(x) < \alpha\}$ is a set by the power-set axiom. If α is a limit number then, by 3.32,

$$\{x \in \mathrm{Wf} \mid \rho(x) < \alpha\} = \bigcup_{\beta < \alpha} \{x \in \mathrm{Wf} \mid \rho(x) < \beta\},$$

and the right-hand side is a set by the axioms of union and replacement (I.5.7). □

6.11 Definition. R is a function on On given by $R(\alpha) = \{x \in \mathrm{Wf} \mid \rho(x) < \alpha\}$ (the right-hand side is a set by 6.10).

6.12 Proposition. (i) $R(\alpha)$ *is a transitive set.*

(ii) $\mathrm{Wf} = \bigcup_{\alpha \in \mathrm{On}} R(\alpha)$.

(iii) $x \subseteq R(\alpha) \leftrightarrow x \in \mathrm{Wf} \wedge \rho(x) \leqslant \alpha$.

(iv) $x \in \mathrm{Wf} \to \rho(x) =$ *the least* α *such that* $x \subseteq R(\alpha)$.

(v) $R(\alpha) = \bigcup_{\beta < \alpha} \mathbf{P}(R(\beta))$.

(vi) $\alpha < \beta \to R(\alpha) \subseteq R(\beta)$.

(vii) $R(\alpha + 1) = \mathbf{P}(R(\alpha))$.

(viii) *If* α *is a limit ordinal, then* $R(\alpha) = \bigcup_{\beta < \alpha} R(\beta)$.

Hint of proof. (i) Use 6.7. (iii) Use 6.11, 6.5(i) and 6.7. (iv) Use (iii). (v) Use 6.11 and (iii). (vii) Use (v) and (vi). □

We came to the $R(\alpha)$'s and to the notion of the rank $\rho(x)$ of a set after a rather long discussion of well founded relations. It is, however, possible to define directly the function R on On by recursion, with 6.12(v) as the defining clause. Having thus defined the function R we define Wf by 6.12(ii) and we define the function ρ on Wf by 6.12(iv).

6.13 The Comulative Hierarchy. Let us try to gain an intuitive understanding of what the $R(\alpha)$'s stand for. $R(0)$ is just the void set. By 6.12(v) $R(\alpha+1)$ can be regarded as the set obtained from $R(\alpha)$ by adding to it all the subsets of $R(\alpha)$, i.e., $R(\alpha+1)=R(\alpha)\cup\mathbf{P}(R(\alpha))$. For a limit number α $R(\alpha)$ is just the union of the earlier $R(\beta)$'s. Thus, the "sequence" of the $R(\alpha)$'s is what we get when we carry on a process of construction of sets which starts with no sets at all, and at every step adds, to the collection of the already obtained sets, all the subsets of this collection. The steps of this process come one after the other in the same fashion as the ordinal numbers. This process yields exactly all the well founded sets. The reader can now show directly the following facts for every well founded set x: (i) $x \notin x$; (ii) there is no finite sequence $\langle y_1, \ldots, y_n\rangle$ such that $x \in y_n \in y_{n-1} \in \ldots \in y_2 \in y_1 \in x$; and (iii) there is no sequence y of length ω such that $\ldots \in y_{n+1} \in y_n \in \ldots \in y_1 \in y_0 \in x$. (i)–(iii) are also direct consequences of 6.7. □

We shall often deal with structures $\langle A, R\rangle$ where R is a well founded relation and $\langle A, R\rangle$ satisfies some or all of the axioms of ZF. Almost always the structures we shall use satisfy the axiom of extensionality, i.e., for all x, $y \in A$ if for all z, zRx iff zRy, then $x=y$. Accordingly, we define:

6.14 Definition. A structure $\langle A, R\rangle$ is said to be *extensional* if it satisfies the axiom of extensionality, i.e., for all $x, y \in A$, if $(\forall z \in A)(zRx \leftrightarrow zRy)$ then $x=y$, or, equivalently, for all x, y in A, if $x \neq y$ then there is a $z \in A$ which stands in the relation R to one of x and y but not to the other.

6.15 Exercise. For every ordinal α the structure $\langle \alpha, \in\rangle$ is extensional. □

Now we formulate and prove the analogue of Theorem 3.23 (that every well ordered set is isomorphic to an ordinal).

6.16 The Representation Theorem (Mostowski 1949). *Let $\langle A, R\rangle$ be an extensional well-founded structure; there is a unique transitive subclass T of* Wf *and a unique isomorphism F of $\langle A, R\rangle$ on $\langle T, \in\rangle$.*

Proof. For a function F from A to T to be an isomorphism as required we must have, for all $x, y \in A$:

(1) $F(x)=F(y) \rightarrow x=y$

(2) $yRx \rightarrow F(y) \in F(x)$

(3) $F(y) \in F(x) \rightarrow yRx.$

We shall now prove that an F as in the theorem must satisfy, for all $x \in A$:

(4) $\qquad F(x) = \{F(y) \mid y \in A \wedge yRx\}.$

By (2) $F(x) \supseteq \{F(y) \mid y \in A \wedge yRx\}$. To prove the other inclusion let $s \in F(x)$. Since $F(x) \in \mathrm{Rng}(F) = T$ and T is transitive we have $s \in T = \mathrm{Rng}(F)$, hence $s = F(z)$ for some $z \in A$. We have $F(z) = s \in F(x)$, hence, by (3), zRx, and therefore $s = F(z) \in \{F(y) \mid y \in A \wedge yRx\}$, which completes our proof of (4).

(4) can now be taken as a definition of F by recursion on the well founded relation $R \mid A$. Since (4) determines F uniquely (by 5.7) this establishes the uniqueness of F and of $T = \mathrm{Rng}(F)$. To prove that F, as defined by (4), is indeed an isomorphism we have to show that it satisfies (1)–(3). Since (2) follows directly from (4) we are left to prove (1) and (3).

We prove (1), for all y, by induction on x with respect to R. For the induction step we assume $F(x) = F(y)$ and prove $x = y$. Since $\langle A, R\rangle$ is extensional it suffices to prove, for every $s \in A$, that if sRx then sRy and if sRy then sRx. If sRx then, by (4), $F(s) \in F(x) = F(y)$, and by $F(s) \in F(y)$ and the definition (4) of $F(y)$ we get $F(s) = F(t)$ for some $t \in A$ such that tRy. Since sRx we can apply the induction hypothesis to s and get $s = t$ from $F(s) = F(t)$. tRy and $s = t$ imply sRy, which is what we had to show. Similarly, if we assume sRy we get $F(s) \in F(y) = F(x)$, hence $F(s) = F(t)$ for some $t \in A$ such that tRx. Applying the induction hypothesis to t we get $s = t$ and hence sRx.

To prove (3) let $F(y) \in F(x)$, then, by the definition (4) of $F(x)$, there is a $t \in A$ such that tRx and $F(t) = F(y)$. By (1) $t = y$, hence yRx, which is what we had to show.

Put $T = F[A]$. T is transitive since by (4) $F(x) = \{F(y) \mid y \in A \wedge yRx\} \subseteq T$. $\in$ is well-founded on T since $\langle T, \in\rangle$ is isomorphic to $\langle A, R\rangle$, which is a well founded structure. By 6.4 Wf is the maximal transitive class on which $\in$ is well founded, hence $T \subseteq \mathrm{Wf}$. □

6.17 The Compression Theorem (Gödel 1939). *If A is a subclass of* Wf *and $\langle A, \in\rangle$ is an extensional structure then there is a unique transitive subclass T of* Wf *and a unique function F on A such that F is an isomorphism of $\langle A, \in\rangle$ on $\langle T, \in\rangle$. If B is a transitive subclass of A then F is the identity map on B, i.e., $F(x) = x$ for every $x \in B$. (The intuitive picture here is that A may have gaps, i.e., there may be members of* $\mathrm{Tc}(A)$ *which are not in A, and as a result F may move members of A downward, i.e., to sets of lower rank. If $B \subseteq A$ is transitive then B is already densely packed and cannot be compressed any more.)*

Proof. Since $A \subseteq \mathrm{Wf}$, $\in$ is well founded on A, by 6.4 and 5.11. By the representation theorem 6.16 applied to $\langle A, \in\rangle$ there is a unique transitive subclass T of Wf and a unique isomorphism F of $\langle A, \in\rangle$ on $\langle T, \in\rangle$. If B is a transitive subclass of A we prove $x \in B \rightarrow F(x) = x$ by induction on $\in$ which is a well founded relation on A. Let $x \in B$; since B is transitive $x \subseteq B \subseteq A$. By (4), with $\in$ instead of R we have $F(x) = \{F(y) \mid y \in A \wedge y \in x\} = \{F(y) \mid y \in x\}$. Since $x \subseteq B$ we have, by the induction hypothesis $F(y) = y$ for every $y \in x$, hence $F(x) = \{y \mid y \in x\} = x$. □

7. The Axiom of Foundation

The versions of the axiom of foundation which we have looked at (in I.5.22 and I.5.23) are as follows.

7.1 Axiom Schema of Foundation. $\exists x\phi(x) \rightarrow \exists x(\phi(x) \wedge (\forall y \in x) \neg \phi(y))$, where y is not free in $\phi(x)$.

7.2 Class Form of the Axiom of Foundation. $A \neq 0 \rightarrow (\exists x \in A)\,(x \cap A = 0)$. □

We shall now consider the following statement, and we shall also justify its title.

7.3 The Local Version of the Axiom of Foundation (Bernays and Gödel—see Bernays 1941). $\forall a(a \neq 0 \rightarrow (\exists x \in a)\,(x \cap a = 0))$.

7.4 The Equivalence of the Global and Local Versions. 7.3 is a particular case of 7.2 where we replace the class A by the set a. It is thus equivalent to an instance of 7.1 (where we take $x \in a$ for $\phi(x)$). Since the $\in$ relation (i.e., the class $\{\langle x, y\rangle \mid x \in y\}$) is obviously left-narrow, 7.3 asserts that the $\in$ relation is well-founded (see Definition 5.1 of well-foundedness). Once we know that the $\in$ relation is well-founded then 7.1 and 7.2 are immediate consequences of the minimal member principle for well founded relations (5.6(i)), whose proof did not use the axiom of foundation. We have thus seen that 7.3 which is a consequence of each one of 7.1 and 7.2 does also imply both and is hence equivalent to each one of them. Thus, unlike the axiom schema of replacement, the axiom schema of foundation is equivalent to the single instance 7.3 of itself.

7.5 Proposition. *The axiom of foundation is equivalent to each one of the following statements.* (a) $V = \mathrm{Wf}$; (b) $\forall x \exists \alpha(x \in R(\alpha))$; (c) $\forall x \exists \alpha(x \subseteq R(\alpha))$.

Hint of proof. As mentioned in the discussion above, 7.3 is trivially equivalent to the statement that the $\in$ relation is well-founded. The present proposition follows now by means of 6.4 and 6.12. □

As a consequence of 7.5, we can say that the axiom of foundation asserts that all the sets are obtained by the process associated with the $R(\alpha)$'s which we have described in 6.13. Many logicians take up the view that this process is the true basis of our intuition about sets. Accordingly, the full axiom of comprehension, I.2.3, was a mistake anyway because it is a result of attempting to construct a new set while assuming that all the sets are already at hand. According to this view, Russell's antinomy is not a bad surprise, but an obstacle which may naturally occur when one follows an unsound way. One can formulate an axiom system of set theory which is based directly on the idea that the whole universe of sets is obtained by the process associated with the $R(\alpha)$'s, and which turns out to be equivalent to ZF (Scott 1974).

During the rest of this section we shall prove theorems using the axiom of foundation, thereby exhibiting its role in set theory.

7.6 Proposition. (i) *There is no sequence z of length* ω *such that for all i,* $z_{i+1} \in z_i$.

(ii) *If we assume the axiom of choice then* (i) *is equivalent to the axiom of foundation.*

(iii) *For no x does* $x \in x$; *for no x, y we have both* $x \in y$ *and* $y \in x$.

Hint of proof. Use 5.3 and 5.4. □

7.7 Cutting Down Classes to Sets (Tarski 1955). It seems that the strongest technical consequence of the axiom of foundation is that it enables us to obtain a set term $\tau(A)$ such that

(1) $$\tau(A) \subseteq A \wedge (A \neq 0 \rightarrow \tau(A) \neq 0).$$

$\tau(A)$ is taken to be the subclass of A consisting of all the members of A of least rank, i.e.,

$$\tau(A) = \{x \in A \mid (\forall y \in A)\,(\rho(x) \leqslant \rho(y))\}.$$

$\tau(A)$ obviously satisfies (1) (by the least ordinal principle, 3.20). We shall now see that $\tau(A)$ is always a set. If $A=0$ then $\tau(A)=0$ and $\tau(A)$ is a set. If $A \neq 0$ let α be the least ordinal which is the rank of some member of A. By the definition of $\tau(A)$, all the members of $\tau(A)$ are of the rank α and hence $\tau(A) \subseteq R(\alpha+1)$ (as a matter of fact $\tau(A) = A \cap (R(\alpha+1) \sim R(\alpha)))$ and $\tau(A)$ is a set by the axiom of subsets.

An important consequence of 7.7 is the following:

7.8 Proposition. *For every relation R there is a right-narrow relation* $S \subseteq R$ *such that* $\mathrm{Dom}(S) = \mathrm{Dom}(R)$, *and a left-narrow relation* $T \subseteq R$ *such that* $\mathrm{Rng}(T) = \mathrm{Rng}(R)$.

Hint of proof. Take $S = \{\langle x, y\rangle \mid y \in \tau(\{y \mid xRy\})\}$, i.e.,

$$S = \{\langle x, y\rangle \mid xRy \wedge \forall z(xRz \rightarrow \rho(z) \geqslant \rho(y))\}. \quad \square$$

As a consequence of 7.8 we are able to generalize in 7.9 the minimal member principle and the theorem on proof by induction for well founded relations (5.6) by dropping the assumption that R is left-narrow.

7.9 Theorem *Let R be a relation which satisfies only the first part of the definition of well-foundedness, i.e., every non-void set y has an R-minimal member, then:*

(i) The minimal member principle. *If* $\exists x \Phi(x)$ *then there is an R-minimal object x for which* $\Phi(x)$ *holds.*

(ii) Induction. *If* $\forall x[\forall y(yRx \rightarrow \Phi(y)) \rightarrow \Phi(x)]$ *then* $\forall x \Phi(x)$.

Proof. Since (ii) was proved in 5.6 from (i) without using the left-narrowness of R we have here only to prove (i). Let $B=\{x \mid \Phi(x)\}$; our aim is to show that B has an R-minimal member. By 7.8 there is a left-narrow relation S such that $S \subseteq R \upharpoonright B$ and $\mathrm{Rng}(S)=\mathrm{Rng}(R \upharpoonright B)$. Since $S \subseteq R \upharpoonright B$ and S is left-narrow S is, by the proof of 5.2, a well founded relation. By the minimal member principle 5.6, the class B has an S-minimal member x; we claim that x is also an R-minimal member of B. Suppose there is a $y \in B$ such that yRx. Then, obviously $x \in \mathrm{Rng}(R \upharpoonright B)=\mathrm{Rng}(S)$, and hence there is a y' such that $y'Sx$. $y' \in B$ since $y' \in \mathrm{Dom}(S) \subseteq \mathrm{Dom}(R \upharpoonright B) \subseteq B$. Thus $y'Sx$ contradicts our choice of x as an S-minimal member of B. □

In 4.29 and 4.33 we proved that if $R_1, \ldots, R_k$ are functions or even only right-narrow relations then for every set a there is a set b which includes a and such that b is R_i-closed for $1 \leqslant i \leqslant k$. Since the axiom of foundation allows us to "convert" relations into narrow ones (7.8) we can use it to obtain the following generalization of 4.33.

7.10 Definition. Let R be a relation. A set a is said to be *nearly R-closed* if for every $x \in a$ if there is a y such that xRy, then there is such a $y \in a$. (If a is R-closed, then a is nearly R-closed, if F is a function and a is nearly F-closed then a is F-closed.)

7.11 Exercise. Let $R_1, \ldots, R_k$ be relations. For every set a there is a set b which includes a and such that b is nearly R_i-closed for every $1 \leqslant i \leqslant k$. □

An idea which occurs often in mathematics is the idea of abstraction. This idea has already been encountered and discussed in § 3, where we introduced temporarily the notion of an isomorphism type. There we had to add a new function symbol G to the language and a new axiom to set theory in order to obtain the isomorphism types. Now, by means of the axiom of foundation, we can carry out formally the idea of abstraction without adding any new symbols to the language.

7.12 Definition. (i) A relation R is said to be *reflexive* on the class A if for every $x \in A$ xRx (i.e., if $I=\{\langle x, x\rangle \mid x \in V\}$ then $I \upharpoonright A \subseteq R$).

(ii) R is said to be an *equivalence relation* on A if $R \subseteq A \times A$, R is reflexive on A, symmetric and transitive (1.1).

7.13 Theorem (The existence of equivalence types—Frege 1884, Russell 1903, Scott 1955). *Let R be an equivalence relation on a class A. There is a function F on A such that*

(2) *for all* $x, y \in A$, $\quad F(x)=F(y) \leftrightarrow xRy$.

Remark. The classes $\{u \mid uRx\}$ are called the *equivalence classes* of the relation R. If R is such that all its equivalence classes are sets, then we can define $F(x)=\{u \mid uRx\}$ and it is easily seen that (2) holds. However, in the general case the equivalence classes of R are proper classes, and yet we often want to consider the equivalence classes as sets. The values $F(x)$ of our theorem are indeed sets and can be regarded as the representatives of the equivalence classes; $F(x)$ is indeed the result of applying the τ of 7.7 to the equivalence class $\{u \mid uRx\}$.

Proof. We define the function F on A by setting $F(x)$ to be the class of all objects u of least rank such that uRx, i.e.,

$$(3) \qquad F(x) = \{u \mid uRx \wedge \forall v(vRx \rightarrow \rho(u) \leqslant \rho(v))\}.$$

For $x \in A$ the class $\{u \mid uRx\}$ is non-void since it contains x, by the reflexivity of R on A. Therefore, by the least ordinal principle, this class contains a member w of least rank, and thus $w \in F(x)$ and $F(x) \neq 0$.

Now let us prove (2). Assume xRy, then, by the symmetry and transitivity of R, we have for all u uRx *iff* uRy, and hence, by the definition (3) of F, $F(x) = F(y)$. Assume now $x, y \in A$ and $F(x) = F(y)$. Since, as we have shown above, $F(x) \neq 0$ there is a $w \in F(x) = F(y)$. By the definition of F we have wRx and wRy and hence xRy. □

Chapter III
Cardinal Numbers

The present chapter is devoted to a quantitative study of sets. The cardinal numbers are, naturally, the focal point of this study. In the last chapter we have developed the theory of well-order and ordinals, and we have used it there to obtain information about the structure of the universe of sets. The study of the sets from a quantitative point of view was carried out, historically, before the study of the structure of the universe, and is also of much more importance to most other branches of mathematics. This is the case because branches of mathematics other than set theory are not concerned with the structure of the universe, but they often need quantitative information about the sets with which they deal.

The theory of well-order and ordinals will also be a very important tool throughout the present chapter. It plays the same role in the study of infinite sets as that played by induction in the study of finite sets, i.e., it enables us to deal with an infinite well ordered set by looking at one member at a time. Accordingly we shall see that we know much more about the cardinals of well-orderable sets than about the cardinals of sets which cannot be well-ordered. As a consequence, once we assume the axiom of choice, which implies that every set can be well-ordered, the theory of cardinals will be considerably simplified.

1. Finite Sets

1.1 Definition (Bolzano 1851, Cantor 1878). The sets a and b are said to be *equinumerous* if there is a bijection of a on b. We write in this case $a \approx b$.

1.2 Proposition. *The relation of equinumerosity is an equivalence relation on* V.

Hint of Proof. Use I.6.13 and I.6.18(iv). □

Our intuitive idea of the notion of a finite set is that it is a set obtained from the null set by adding to it one element at a time finitely many times. This does not directly yield a good definition of this notion since we used the notion "finitely many times" in the defining clause. However, we shall now give a definition of the notion of a finite set which will define a set to be finite if it can be reached from the null set by the process of adding single elements. The theory developed on the basis of this definition will show that we chose the right definition.

1.3 Definition (Whitehead and Russell 1912). Given a set a and a subset u of the power set $\mathbf{P}(a)$ of a, we say that u is an *inductive family of subsets of a* if $0 \in u$ and if for every $x \in u$ and $y \in a$ also $x \cup \{y\} \in u$. a is a *finite set* if a is a member of every inductive family of its subsets. A set which is not finite is said to be *infinite*.

1.4 Corollary. *0 is a finite set.* □

1.5 Proposition. *If a is a finite set, so is* $a \cup \{z\}$.

Proof. Let u be an inductive family of subsets of $a \cup \{z\}$; we have to show that $a \cup \{z\} \in u$. $u \cap \mathbf{P}(a)$ is obviously an inductive family of subsets of a. Since a is finite $a \in u \cap P(a) \subseteq u$. Since u is inductive and $a \in u$ also $a \cup \{z\} \in u$. □

1.6 Theorem. (Proof by induction on finite sets—Zermelo 1909). *If $\Phi(0)$ and for every finite a and for every $z \notin a$ if $\Phi(a)$ then $\Phi(a \cup \{z\})$, then we have $\Phi(a)$ for every finite set a.*

Proof. Let a be a finite set; we shall prove $\Phi(a)$. Let $u = \{b \mid b \subseteq a \wedge b$ is finite $\wedge \Phi(b)\}$. By 1.4, 1.5 and the hypothesis of our theorem, u is an inductive family of subsets of a, hence $a \in u$ and $\Phi(a)$ holds. □

Remark. 1.6 is very useful for proving the properties of the finite sets.

1.7 Proposition. *If a is a finite set and F is a function such that* $a \subseteq \text{Dom}(F)$ *then* $F[a]$ *is a finite set too. In particular, if a set b is equinumerous to a then b is a finite set, too.*

Proof. By induction on a (1.6). If $a=0$ then also $F[a]=0$ and $F[a]$ is finite. Assume now that the theorem is true for a, i.e., that $F[a]$ is finite. $F[a \cup \{z\}] = F[a] \cup \{F(z)\}$, hence $F[a \cup \{z\}]$ too is finite by 1.5. □

1.8 Proposition. *Every natural number is a finite set.*

Proof. By induction. By 1.4, 0 is a finite set. If n is a finite set, then $n+1 = n \cup \{n\}$ is a finite set by 1.5. □

1.9 Proposition. *Every finite set is equinumerous to some natural number.*

Proof. By induction on the finite set a. 0 is itself a natural number. By the induction hypothesis there is a bijection f of a on a natural number n. Let $z \notin a$ then $f \cup \{\langle z, n\rangle\}$ is a bijection of $a \cup \{z\}$ on the natural number $n+1$, hence also $a \cup \{z\}$ is equinumerous to a natural number. □

1.10 Theorem. *A set a is finite iff it is equinumerous to some natural number.*

Hint of proof. Use 1.9, 1.7, and 1.8. □

1.10 shows that we could have *defined* that a set a be finite if it is equinumerous to a natural number.

1.11 Proposition. *Let $<$ be a partial ordering relation, and let b be a non-void finite set, then b has a minimal and a maximal member with respect to $<$. (x is a maximal member of b if $(\forall y \in b) \neg y > x$.)*

Proof. We shall prove that b has a minimal member; a similar proof shows that b has a maximal member. The proof will be by induction on b. For 0 the lemma holds vacuously. Assuming the lemma for b we consider $b \cup \{z\}$ for $z \notin b$. If $b=0$ then z is a minimal member of $b \cup \{z\}$. If $b \neq 0$ then, by the induction hypothesis, b has a minimal member u. If $z<u$ then z is a minimal member of $b \cup \{z\}$, if it is not the case that $z<u$ then u is also a minimal member of $b \cup \{z\}$. □

1.12 Proposition. *ω is an infinite set.*

Proof. $\in | \omega$ is a (partial) ordering relation. In this ordering ω has no maximal member, by II.4.2(ii). Hence, by 1.11, it is an infinite set. □

1.13 Proposition. *Every subset of a finite set a is finite.*

Proof. By induction on the finite set a. The lemma is trivial for the null set 0. We assume that every subset of a is finite. Let $z \notin a$ we shall prove that also every subset of $a \cup \{z\}$ is finite. Let $b \subseteq a \cup \{z\}$. If $z \notin b$ then $b \subseteq a$ and b is finite by our hypothesis. If $z \in b$ then $b = (b \sim \{z\}) \cup \{z\}$; $b \sim \{z\} \subseteq a$ and hence $b \sim \{z\}$ is finite; $b = (b \sim \{z\}) \cup \{z\}$ is now finite by 1.5. (Can you prove 1.13 from 1.7?) □

1.14 Definition (Bolzano 1851, Cantor 1878). A set is said to be *denumerable* if it is equinumerous with ω. A set is said to be *countable* if it is either equinumerous with ω or finite.

1.15 Proposition. *A set which is denumerable, or which includes a denumerable set, is infinite.*

Hint of proof. Use 1.12, 1.7, and 1.13. □

1.16 Remark. By 1.15 all the ordinals $\geqslant \omega$ are infinite sets. This, and 1.10, shows that we were consistent in II.4.1 when we called the natural numbers, and only them, finite ordinals.

1.17 Proposition. *For well-orderable sets also the converse of 1.15 holds, i.e., every infinite well-orderable set includes a denumerable subset.*

Proof. Let a be an infinite set and let $<$ be a relation which well-orders a. Let α be the ordinal of $\langle a, <\rangle$; by II.3.23 there is an isomorphism F of $\langle\alpha, \in\rangle$ on $\langle a, <\rangle$. If $\alpha<\omega$ then, by 1.10, a is finite since it is equinumerous to $\alpha<\omega$; hence $\alpha\geqslant\omega$. $F[\omega]$ is clearly a denumerable subset of a. □

1.18Ac Corollary. *Assuming the axiom of choice, a set is infinite iff it includes a denumerable subset.* □

1.19 Proposition. *Every finite ordered set is well-ordered and its ordinal is a natural number.*

Proof. Let $\langle a, <\rangle$ be a finite ordered set. Every subset b of a is finite by 1.13, thus every non-void subset b of a has a least member, by 1.11, and $\langle a, <\rangle$ is well-ordered. If α is the ordinal of $\langle a, <\rangle$ then $\langle a, <\rangle\cong\langle\alpha, \in\rangle$. By 1.7 and 1.15, α cannot be an infinite ordinal, hence it is a natural number. □

1.20 Theorem (Dedekind 1888). *A set a is equinumerous to some proper subset b of itself iff a includes a denumerable subset.*

Proof. Assume that a includes a denumerable subset c. Let f be a bijection of ω on c. We define a function g on a as follows:

$$g(x)=\begin{cases} f(n+1) & \text{if} \quad x=f(n),\ n<\omega \\ x & \text{if} \quad x\in a\sim c \end{cases}$$

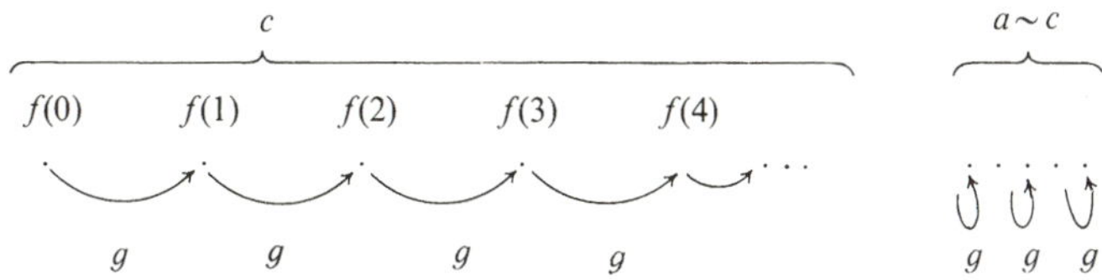

Fig. 1

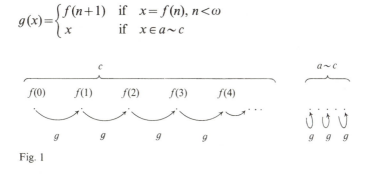

It is easily seen that g is an injection of a into a. Thus, a is equinumerous to Rng(g) which is a proper subset of a since $f(0)\notin$ Rng(g).

Assume now that a is equinumerous to a proper subset b of a. Let g be a bijection of a on b. Let us look at the iterations g^n of g. By II.4.17 Dom(g^n)$=a$ for every $n>0$. Since $b\subset a$ there is a $z\in a\sim b$. We define a function f on ω by $f(n)=g^n(z)$ ($f(0)=g^0(z)=z$). Obviously, Rng(f)$\subseteq a$. We shall now prove that f is one-one, and hence Rng(f) is a denumerable subset of a. Suppose f is not one-one, then there are natural numbers n and m such that $n<m$ and $f(n)=f(m)$. By the least ordinal principle, let n be the least natural number for which there is such an m. n cannot be 0 since $f(0)=z$ was chosen outside b, while for $m>0$ $f(m)$ was defined to be in Rng(g)$=b$. Since now $m>n>0$ we have $f(m)=g^m(z)=g(g^{m-1}(z))$ and $f(n)=g^n(z)=g(g^{n-1}(z))$. Since $f(n)=f(m)$ we have $g(g^{n-1}(z))=g(g^{m-1}(z))$. g is one-one hence $g^{n-1}(z)=g^{m-1}(z)$, i.e., $f(n-1)=f(m-1)$, contradicting our choice of n as the least natural number for which there is an $m>n$ such that $f(n)=f(m)$. □

1.21 Corollary. *By* 1.15 *and* 1.20, *no finite set is equinumerous to a proper subset of itself. In particular, no two different natural numbers are equinumerous, since one of them is less than the other, and is hence a proper subset of the other.* □

1.22 Exercise. Give a more direct proof, by induction on a, that a finite set a is not equinumerous to a proper subset of itself. □

1.23 Remark. Combining 1.18 and 1.20 we get that if the axiom of choice holds then a set a is finite iff it is not equinumerous to any proper subset of itself. This property of finite sets was used by Peirce 1885 and Dedekind 1888 to define the concept of a finite set. Definition 1.3 is preferable to that of Peirce and Dedekind since it allows for the development of the full theory of finite sets without ever using the axiom of choice. At the end of § 2 we shall briefly study the sets which are not equinumerous to any of their proper subsets.

1.24 Proposition. *If a and b are finite sets, so is $a\bigcup b$.*

Hint of proof. By induction of the finite set b. □

1.25 Proposition. *If a is a finite set and all the members of a are finite sets, then $\bigcup a$ is a finite set.*

Hint of proof. Use induction on a. □

1.26 Proposition. *If a is a finite set so is its power set* $\mathbf{P}(a)$.

Hint of proof. By induction on a. If $z\notin a$ then $\mathbf{P}(a\cup\{z\})=\mathbf{P}(a)\cup\{t\cup\{z\}\mid t\in\mathbf{P}(a)\}$. □

1.27 Proposition. *If e is an infinite set, then every finite set a is equinumerous to a subset of e. (If the axiom of choice holds this follows trivially from* 1.9 *and* 1.18.)

Hint of proof. Use induction on a, using 1.7. □

1.28 Proposition (Tarski 1924a). *A set a is infinite iff the power set of its power set,* $\mathbf{P}(\mathbf{P}(a))$ *includes a denumerable set. (Contrast this with* 1.17 *where we proved, using the axiom of choice, that if a is infinite than a itself includes a denumerable set; without the axiom of choice one cannot even prove that* $\mathbf{P}(a)$ *includes a denumerable set.)*

Proof. If a is finite then so is $\mathbf{P}(\mathbf{P}(a))$ by 1.26, and hence, by 1.15, $\mathbf{P}(\mathbf{P}(a))$ does not include a denumerable set. If a is infinite, we define a function f on ω as follows:

$$f(n)=\{b\mid b\subseteq a \text{ and } b \text{ is equinumerous to } n\}.$$

By 1.27, $f(n)\neq 0$ for every $n<\omega$. Since different natural number m and n are not equinumerous, by 1.21, $f(m)$ and $f(n)$ have no member in common and hence $f(m)\neq f(n)$ and f is one-one. Therefore, $\mathrm{Rng}(f)\subseteq\mathbf{P}(\mathbf{P}(a))$ is denumerable. □

When we introduced the axiom of infinity, I.5.21, we said that the formulation given there is a formally simpler way of saying "there exists an infinite set". Now after we have defined the notion of an infinite set, let us prove that the two formulations are equivalent in the presence of the other axioms of set theory.

1.29 Theorem. *In the presence of the other axioms of* ZF *the axiom of infinity*

(1) $\exists z(0\in z\wedge(\forall x\in z)\,(\forall y\in z)\,(x\cup\{y\}\in z))$

is equivalent to the statement "there is an infinite set".

Proof. We used (1) to prove the existence of ω and we proved that ω is an infinite set. As a matter of fact, one can easily show that a set z as in (1) must include $R(\omega)$ (where R is the function of II.6.11), that $R(\omega)$ is infinite (since $R(\omega)$ itself is an inductive family of subsets of $R(\omega)$ which does not contain $\mathrm{R}(\omega)$), and hence such a z must be infinite.

Let us assume now the existence of an infinite set a. Let N be the class of all natural numbers. Let F be the function such that $\mathrm{Dom}(F)=N$ and $F(n)=\{b\mid b\subseteq a$ and b is equinumerous to $n\}$. This is the same as the function f of the proof of 1.28, except that we do not claim now that it is a set. As in the proof of 1.28 F is one-one function. Its range is a set, being a subclass of $\mathbf{P}(\mathbf{P}(a))$. We have $N=\mathrm{Dom}(F)=F^{-1}[\mathrm{Rng}(F)]$ and by the axiom of replacement (I.6.19) N is a set and we can denote it with ω. The set $R(\omega)$ satisfies all the requirements on z in (1), by the properties of the function R (II.6.12). □

1.30 Exercise. A set a is finite iff every non-void set u of subsets of a has a maximal member b (i.e., for no $c\in u$ does $b\subset c$). □

1.31 Exercise. The following three conditions on a are equivalent:

(a) a is finite.
(b) There is a relation r which orders a, and every relation s which orders a does in fact well-order a.
(c) There is a relation r such that both r and r^{-1} well-order a. □

1.32 Proposition. *Every bounded subset of ω is finite, every cofinal subset of ω is denumerable.*

Proof. If $a\subseteq\omega$ and n is a bound of a then $a\subseteq n\cup\{n\}$, hence a is finite by 1.8 and 1.13. If a is a finite subset of ω then a is bounded since a finite set has a maximal member (1.11). By II.3.27, if $a\subseteq\omega$ then $\mathrm{Ord}(\langle a,<\rangle)\leqslant\omega$. If a is a cofinal subset of ω we cannot have $\mathrm{Ord}(\langle a,<\rangle)<\omega$ because then a is finite, by 1.7, and as we saw, a

must be bounded. Therefore, we have $\mathrm{Ord}(\langle a, <\rangle)=\omega$ and thus a is denumerable. □

1.33 Proposition. *Every finite subset of $R(\omega)$ is a member of $R(\omega)$. (R is the function of* II.6.11.)

Proof. Let y be a finite subset of $R(\omega)$. The set $\{\rho(x) \mid x \in y\}$ is finite, being an image of the finite set y (1.7) and hence, by 1.32, $\{\rho(x) \mid x \in y\}$ is a bounded subset of ω; let n be a bound of it. Then $y \subseteq R(n+1)$, $y \in R(n+2) \subseteq R(\omega)$. □

1.34 Definition. For a formula $\Phi(x)$ we say that a set y has *hereditarily* the property expressed by $\Phi(x)$ if y itself and every member of its transitive closure has this property, i.e., $\{y\} \cup Tc(y) \subseteq \{x \mid \Phi(x)\}$. The reader can easily verify that y is such iff y is a member of a transitive subset of $\{x \mid \Phi(x)\}$. For example, we say that y is *hereditarily finite* if every member of $\{y\} \cup Tc(y)$ is finite.

1.35 Proposition. *$R(\omega)$ is the class of all hereditarily finite well-founded sets; hence, by the axiom of foundation, $R(\omega)$ is the class of all hereditarily finite sets.*

Outline of proof. Prove by induction on $n<\omega$, using II.6.12(vii) and 1.26, that $R(n)$ is finite. Hence, by II.6.12(v) every member of $R(\omega)$ is finite and since $R(\omega)$ is transitive also every member of $R(\omega)$ is hereditarily finite. In the other direction, if there is a hereditarily finite well-founded set x not in $R(\omega)$ let x be an $\in$-minimal such set. Every $y \in x$ is also hereditarily finite; therefore, by the minimality of x, every such y is in $R(\omega)$ and $x \subseteq R(\omega)$. x is thus a finite subset of $R(\omega)$ and is hence, by 1.33, a member of $R(\omega)$. This contradicts our choice of x. □

Remarks. An example of a hereditarily finite set which is not well founded is a set x such that $x=\{x\}$ (we can refute the existence of such a set only by means of the axiom of foundation—see II.7.6(iii)). For a slightly different proof of the second direction of 1.35 proceed as follows. Let x be a hereditarily finite well-founded set. If $\rho(x) \geqslant \omega$ then by II.6.9(i) there is a $y \in \{x\} \cup \mathrm{Tc}(x)$ such that $\rho(y)=\omega$. y is a finite subset of $R(\omega)$ hence, by 1.33, $y \in R(\omega)$, contradicting $\rho(y)=\omega$.

1.36 Proposition. *$R(\omega)$ is a denumerable set.*

Proof (Ackermann 1937). This proposition can be easily proved by the methods of cardinal arithmetic (§ 4) but we shall establish here a simple bijection f of $R(\omega)$ on ω. We define f on $R(\omega)$ by recursion on $\in$ as follows: $f(x)=\Sigma_{y \in x} 2^{f(y)}$. Since x is a finite set the right-hand side is a natural number. (We use here familiar facts which are not proved in this book.) We shall now prove that f is an injection. Suppose not, then let n be the least natural number for which there are $x \neq z$ such that $n=f(x)=f(z)$. Then

(2) $$n=\Sigma_{y \in x} 2^{f(y)}=\Sigma_{y \in z} 2^{f(y)}.$$

If $y \in x$ or $y \in z$ we have, by (2), $f(y) < n$ and by the minimality of n for two different such y's the corresponding $f(y)$'s are different. Thus the sums in (2) are sums of different powers of 2 and by the theorem on the uniqueness of the binary expansion of a natural number we get $\{f(y) \mid y \in x\} = \{f(y) \mid y \in z\}$. Since for those y's different y's yield different $f(y)$'s we get $x = z$, contradicting our assumption that $x \neq z$. Finally we shall show that f is a bijection on ω. Suppose $\text{Rng}(f) \subset \omega$ then let n be the least member of $\omega \sim \text{Rng}(f)$. Let $\Sigma_{k \in t} 2^k$ be the binary expansion of n, where t is a subset of n. By the minimality of n, $t \subseteq n \subseteq \text{Rng}(f)$. Let $s = f^{-1}[t]$, then $n = \Sigma_{k \in t} 2^k = \Sigma_{x \in s} 2^{f(x)} = f(s)$, contradicting $n \notin \text{Rng}(f)$. □

Remark. 1.36 can be proved by constructing the bijection $g = f^{-1}$ of ω on $R(\omega)$ as follows. For $k, n \in \omega$ define $k \,\tilde{\in}\, n$ if 2^k occurs in the binary expansion of n. Prove that $\tilde{\in}$ is a well founded extensional relation on ω and hence, by II.6.16, there is a unique bijection g of ω on a transitive well-founded set t such that for all $k, n \in \omega$, $f(k) \in f(n) \leftrightarrow k \,\tilde{\in}\, n$. $t = R(\omega)$ is shown without trouble.

2. The Partial Order of the Cardinals

2.1 Proposition. (i) *If a and b are equinumerous sets then a is well-orderable iff b is.*
(ii) *A set a is well-orderable iff it is equinumerous to some ordinal.* □

2.2 Definition. The *cardinal* of x, or, synonymously, the *cardinality* of x, which we shall denote by $|x|$, is

(a) the least ordinal α equinumerous to x, if x is well-orderable, and
(b) the set of all sets y of least rank which are equinumerous to x, otherwise.

The idea of the cardinal numbers is due to Cantor 1878. The origins of our particular definition of this concept and the reasons for choosing it are discussed in 2.5 below.

Calling $|x|$ the cardinal of x will be justified once we prove, in 2.4 below, that for all x, y, $|x| = |y|$ iff x is equinumerous to y. An object a is a *cardinal* if it is the cardinal of some set x. Cn denotes the class of all cardinals. We shall use lower case German letters for cardinals.

2.3 Lemma. *$|x|$ is an ordinal iff x is well-orderable.*

Proof. If x is well-orderable then $|x|$ is an ordinal by Definition 2.2. If x is not well-orderable then $|x|$ is the set of all y's of least rank equinumerous to x. We have then $|x| \neq 0$ (since there is at least one such y), hence if $|x|$ were an ordinal we would have $|x| > 0$, i.e., $0 \in |x|$. Since all members of $|x|$ are equinumerous to x we have $0 \approx x$, i.e., $x = 0$. This contradicts our assumption that x is not well-orderable, hence $|x|$ is not an ordinal. □

2.4 Proposition. *For all sets* $x, y, |x| = |y|$ *iff* $x \approx y$.

Proof. If $x \approx y$ then $|x| = |y|$ follows directly from 2.1(i) and the transitivity of the relation of equinumerosity. If $|x| = |y|$ then we distinguish two cases. If $|x| = |y|$ is an ordinal then, by 2.3 and 2.2, it is the least ordinal equinumerous to each of x and y and thus $x \approx y$. If $|x| = |y|$ is not an ordinal then, by 2.3, neither x nor y is well-orderable. In this case, by 2.2, if $z \in |x| = |y|$ then $x \approx z \approx y$, hence $x \approx y$. Such a z does indeed exist, since $|x| = |y| \neq 0$ (because $|x| = |y|$ is not an ordinal). □

2.5 Discussion. By 2.3, for every set x, $|x|$ is an ordinal iff x is well-orderable. We shall therefore refer to the cardinals which are also ordinals as the *well ordered cardinals*. It follows from the axiom of choice that every set is well-orderable, hence all the cardinals are ordinals. Thus in the presence of the axiom of choice the cardinal $|x|$ of x is always the least ordinal equinumerous to x and is hence a canonical set equinumerous to x (it is canonical because it does not depend on x, only on the equinumerosity class of x). This is how von Neumann 1928 defined the concept of the cardinal numbers (he assumed the axiom of choice throughout). Having the cardinal itself as the canonical set of that cardinality is a very convenient feature. This is the reason why in the absence of the axiom of choice we define the cardinals by using von Neumann's definition for clause (a) in 2.2 and the Frege–Russell–Scott definition for clause (b) in 2.2, rather than the straight Frege–Russell–Scott definition of equivalence types (II.7.13). Our definition makes x an ordinal and a canonical set equinumerous to x for all well-orderable sets x, which is still quite good. On the basis of ZF alone one cannot do the same thing for all sets x, i.e., we cannot prove the existence of a function F such that for all x and y

(a) $F(x) = F(y)$ iff x and y are equinumerous, and
(b) $F(x)$ is equinumerous to x (Pincus 1974).

2.6 Definition. We define a relation $\preccurlyeq$ on V as follows: $a \preccurlyeq b$ if there is an injection of a into b.

2.7 Proposition. (i) *The relation* $\preccurlyeq$ *is reflexive and transitive.*

(ii) *If* $a \approx b$ *then* $a \preccurlyeq b$.

(iii) *If* $a \subseteq b$ *then* $a \preccurlyeq b$.

(iv) $a \preccurlyeq b$ *iff there is a subset* c *of* b *such that* $a \approx c$.

Proof. (i) follows from I.6.18. (ii) is obvious. For (iii), the identity function on a is the required injection. (iv) Let f be an injection of a in b, then f is a bijection of a on $f[a]$, which is a subset of b. The other direction is equally trivial. □

An important, and non trivial, feature of the relation $\preccurlyeq$ is given by the following theorem.

2.8 The Cantor Bernstein Theorem (Proved by Dedekind in 1887—see Dedekind 1932, p. 447, conjectured by Cantor 1895, proved by Bernstein in 1898—see Borel 1898). *If $a \preccurlyeq b$ and $b \preccurlyeq a$ then a and b are equinumerous.*

Proof. Since $b \preccurlyeq a$, a has, by 2.7(iv), a subset c such that $b \approx c$. Since $a \preccurlyeq b \approx c$ we have, by 2.7(i) and 2.7(ii), $a \preccurlyeq c$ and, again by 2.7(iv), c has a subset d such that $d \approx a$. Since $b \approx c$ our proof is done once we prove that $c \approx a$. The only assumptions which we shall use during the rest of the proof are that $d \subseteq c \subseteq a$, and that $d \approx a$ via a bijection h of a on d. We shall show that a set c which is "sandwiched" between the equinumerous sets $d \subseteq a$ is equinumerous with them. To prove $a \approx c$ we construct a bijection j of a on c. If we read $j(x)=y$ as "j moves x to y" (and thus if $j(x)=x$ then "j does not move x") then we shall construct j so that j moves only those members of a which have to be moved, and those members x of a which j moves are moved by means of h (i.e., $j(x)=h(x)$).

Since the members of $a \sim c$ are not in c, which is the intended range of j, they have to be moved. The set $a \sim c$ goes over to $h[a \sim c]$. Now the members of $h[a \sim c]$ have to be moved too if j is to be one-one, and they get moved to the members of $h[h[a \sim c]] = h^2[a \sim c]$, and so on. This suggests that it is exactly the members of the sets $h^n[a \sim c]$, for $n \geqslant 0$, which have to be moved, where h^0 denotes the identity function on a ($h^0 = \langle x \mid x \in a \rangle$), and thus $h^0[a \sim c] = a \sim c$.

Accordingly we define the function j on a by

$$j(x)=\begin{cases} h(x) & \text{if } \ x \in \bigcup_{n<\omega} h^n[a \sim c] \quad \text{(i.e., } x \in (a \sim c) \cup h^*[a \sim c]) \\ x & \text{if } \ x \in a \sim \bigcup_{n<\omega} h^n[a \sim c] \end{cases}$$

Fig. 2 illustrates the action of j on a.

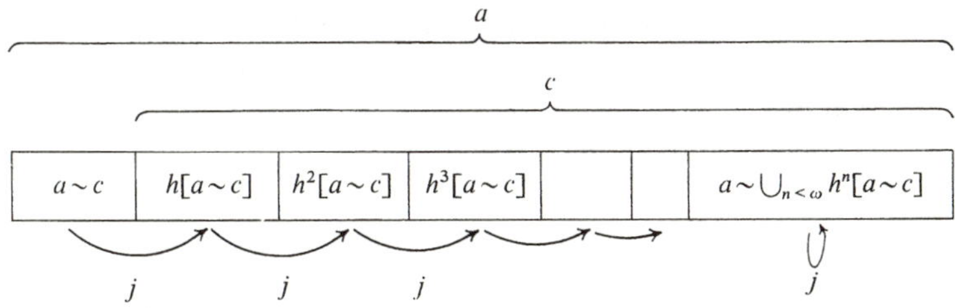

Fig. 2

$$\begin{aligned} \text{Rng}(j) &= j[a] = j[\bigcup_{n<\omega} h^n[a \sim c]] \cup j[a \sim \bigcup_{n<\omega} h^n[a \sim c]] \\ &= \bigcup_{n<\omega} j[h^n[a \sim c]] \cup (a \sim \bigcup_{n<\omega} h^n[a \sim c]) \\ &= \bigcup_{n<\omega} h[h^n[a \sim c]] \cup (a \sim \bigcup_{n<\omega} h^n[a \sim c]), \end{aligned}$$

since j coincides with h on $h^n[a \sim c]$,

$$= \bigcup_{n<\omega} h^{n+1}[a \sim c] \cup (a \sim \bigcup_{n<\omega} h^n[a \sim c])$$

$= a \sim h^0[a \sim c]$, since $h^0[a \sim c] = a \sim c$ is disjoint from each $h^{n+1}[a \sim c]$ and each $h^{n+1}[a \sim c]$ is a subset of a,

$$= a \sim (a \sim c) = c.$$

Once we prove that j is one–one this will show that $a \approx c$. $j \upharpoonright \bigcup_{n<\omega} h^n[a \sim c]$ is one–one since it is equal to $h \upharpoonright \bigcup_{n<\omega} h^n[a \sim c]$ and h is a bijection. $j \upharpoonright (a \sim \bigcup_{n<\omega} h^n[a \sim c])$ is one–one since it is an identity function. We shall complete proving that j is one–one by showing that $j[\bigcup_{n<\omega} h^n[a \sim c]]$ and $j[a \sim \bigcup_{n<\omega} h^n[a \sim c]]$ are disjoint (I.6.30(iii)). As we have seen when we computed Rng(j) these sets are $\bigcup_{n<\omega} h^{n+1}[a \sim c]$ and $a \sim \bigcup_{n<\omega} h^n[a \sim c]$, respectively, and hence obviously disjoint. □

From the relation $\preccurlyeq$ we pass in a natural way to a partial ordering of the cardinals as given in the next definition.

2.9 Definition (The natural order of the cardinals—Cantor 1878). We define relations $\leqslant$ and $<$ on the class Cn of all cardinals as follows: For all cardinals $\mathfrak{a}$, $\mathfrak{b}$, $\mathfrak{a} \leqslant \mathfrak{b}$ if there are sets x, y such that $|x| = \mathfrak{a}$, $|y| = \mathfrak{b}$, and $x \preccurlyeq y$; $\mathfrak{a} < \mathfrak{b}$ if $\mathfrak{a} \leqslant \mathfrak{b}$ and $\mathfrak{a} \neq \mathfrak{b}$.

We have denoted with $<$ the natural order of the ordinals. Now we denote with $<$ a relation on the cardinals. We shall show in 2.20 that these two notations do not clash.

2.10 Proposition. *If $x \subseteq y$ then $|x| \leqslant |y|$. If $\mathfrak{a} \leqslant |y|$ then there is a subset x of y such that $|x| = \mathfrak{a}$.*

Hint of proof. Use 2.7. □

2.11 Proposition. *For all sets u, v, $u \preccurlyeq v$ iff $|u| \leqslant |v|$.*

Proof. If $u \preccurlyeq v$, then $|u| \leqslant |v|$ by the definition of $\leqslant$. If $|u| \leqslant |v|$ then, by Definition 2.9, there are sets x, y such that $|x| = |u|$, $|y| = |v|$ and $x \preccurlyeq y$. We have to show $u \preccurlyeq v$. Let f be a bijection of x on u, g a bijection of y on v and h an injection of x in y. Then ghf^{-1} is an injection of u into v, hence $u \preccurlyeq v$. □

2.12 Theorem. *The relation $<$ on* Cn *is a partial ordering relation.*

Proof. By definition of $<$, $\mathfrak{a} < \mathfrak{a}$ never holds. To prove the transitivity of Cn we assume $\mathfrak{a} < \mathfrak{b} < \mathfrak{c}$ and prove $\mathfrak{a} < \mathfrak{c}$. By 2.10 there are sets x, y, z such that $x \subseteq y \subseteq z$ and $|x| = \mathfrak{a}$, $|y| = \mathfrak{b}$, and $|z| = \mathfrak{c}$, and hence, again by 2.10, $\mathfrak{a} \leqslant \mathfrak{c}$. We shall now see that $\mathfrak{a} \neq \mathfrak{c}$. If $\mathfrak{a} = \mathfrak{c}$, then in addition to $\mathfrak{a} < \mathfrak{b}$ we have also $\mathfrak{b} < \mathfrak{a}$ (since $\mathfrak{b} < \mathfrak{c} = \mathfrak{a}$) and hence, by the Cantor–Bernstein theorem (2.8), $\mathfrak{a} = \mathfrak{b}$, contradicting $\mathfrak{a} < \mathfrak{b}$. □

The Cantor–Bernstein theorem is the only non-trivial fact involved in the proof that $<$ is a partial ordering relation. That theorem can be formulated as $\mathfrak{a} \leqslant \mathfrak{b} \wedge \mathfrak{b} \leqslant \mathfrak{a} \rightarrow \mathfrak{a} = \mathfrak{b}$, which is the requirement that the relation $\leqslant$ be antisymmetric. (It follows trivially from its definition that it is reflexive and transitive.) □

We shall now prove two theorems which will give us some information about how to obtain larger and larger cardinals.

2.13 Cantor's Theorem (Cantor 1892). *For every set z,* $|z|<|\mathbf{P}(z)|$. *Therefore, for every cardinal* $\mathfrak{a}$ *there is a cardinal greater than* $\mathfrak{a}$.

Proof. Let f be the function on z given by $f(x)=\{x\}$. f is an injection of z into $\mathbf{P}(z)$, hence $|z|\leqslant|\mathbf{P}(z)|$. Assume now that $|z|=|\mathbf{P}(z)|$, and let g be a bijection of z on $\mathbf{P}(z)$. Let

$$u=\{x\in z\mid x\notin g(x)\}.$$

We claim that $u\notin \mathrm{Rng}(g)$, contradicting our assumption that g is a bijection on $\mathbf{P}(z)$. Suppose $u\in \mathrm{Rng}(g)$ then $u=g(t)$ for some $t\in z$. If $t\in u$ then, by the definition of u, $t\notin g(t)=u$; if $t\notin u=g(t)$ then, by definition of u, $t\in u$; in either case we have both $t\in u$ and $t\notin u$, which is a contradiction. □

The proof of Cantor's theorem is very similar to the proof of Russell's antinomy (I.2.4). In fact, if we apply Cantor's theorem to the universal class V, which is a set if the axiom of comprehension I.2.3 is assumed, we get the following antinomy. By Cantor's theorem the universal set V, like every other set, has more subsets than members, but this is a contradiction since every subset of V is obviously also a member of V. A variant of this antinomy was discovered by Cantor 1899. By analyzing its proof Russell arrived at his antinomy, which is of a simpler form.

2.14 Proposition. *Every set p of cardinals has a bound, i.e., there is a cardinal* $\mathfrak{c}$ *such that for every* $\mathfrak{a}\in p$, $\mathfrak{a}\leqslant\mathfrak{c}$.

Proof. Let F be the function on p defined by

$$F(\mathfrak{a})=\text{the rank of a set } x \text{ of least rank among the sets } x \text{ with } |x|=\mathfrak{a}.$$

Let x be as in the definition of $F(\mathfrak{a})$, then $|x|=\mathfrak{a}$ and $\rho(x)=F(\mathfrak{a})$ and we have, by II.6.7 and II.6.11, $x\subseteq R(F(\mathfrak{a}))$, and hence

(1) $\quad \mathfrak{a}=|x|\leqslant|R(F(\mathfrak{a}))|.$

Since p is a set so is $\mathrm{Rng}(F)$; let $\alpha=\sup \mathrm{Rng}(F)$; Then we have, for every $\mathfrak{a}\in p$, $\mathfrak{a}\leqslant|R(F(\mathfrak{a}))|\leqslant|R(\alpha)|$, by (1) and II.6.12(vi). Thus $|R(\alpha)|$ is a bound as required. □

2.15 Corollary. *The class* Cn *of all cardinals is a proper class.*

Hint of proof. Use 2.14 and 2.13. □

Now we shall look more closely at the well ordered cardinals (2.5).

2.16 Proposition. *If* $\mathfrak{a}$ *is a well ordered cardinal and* $\mathfrak{b}$ *is a cardinal such that* $\mathfrak{b}<\mathfrak{a}$ *then* $\mathfrak{b}$ *too is a well ordered cardinal.*

Hint of proof. Use 2.10 and II.2.20. □

2.17 Proposition. *For every ordinal* α:

(i) $|\alpha|$ *is an ordinal and* $|\alpha| \leqslant \alpha$.

(ii) α *is a cardinal iff* α *is not equinumerous to any smaller ordinal, iff* $|\alpha| = \alpha$.

Proof. (i) is trivial. (ii) If α is not equinumerous to any smaller ordinal, then α is the least ordinal equinumerous to α and hence, by Definition 2.2, $|\alpha| = \alpha$, hence $\alpha \in \mathrm{Cn}$. If α is a cardinal, and in particular if $|\alpha| = \alpha$, then $\alpha = |x|$ for some x. Since $|x|$ is an ordinal we know, by 2.3 that case (a) of the definition of $|x|$ (2.2) holds for x and hence $x \approx \alpha$. If $\alpha \approx \beta$ for some ordinal β smaller than α, then $x \approx \alpha \approx \beta$ and α is not the least ordinal equinumerous with x, contradicting $|x| = \alpha$. □

2.18 Corollary. *Every natural number is a cardinal.*

Hint of proof. Use 1.21. □

2.19 Corollary. ω *is a cardinal.*

Hint of proof. Use 1.12, 1.7 and 1.8. □

2.20 Proposition. *The order relations* $<$ *on the cardinals and on the ordinals coincide for the cardinals which are also ordinals, i.e., for the well ordered cardinals; in other words, the restrictions of both relations to* $\mathrm{Cn} \cap \mathrm{On}$ *are identical.*

Proof. To distinguish between the two relations $<$ let us write, for the purpose of this proof only $<_0$ for the order relation on the ordinals (i.e., $<_0 = \in | \mathrm{On}$) and $<_c$ for the partial ordering of the cardinals defined in 2.9.

Let $\alpha, \beta \in \mathrm{Cn} \cap \mathrm{On}$. Assume first that $\alpha <_0 \beta$, then $\alpha \subseteq \beta$, and hence $|\alpha| \leqslant_c |\beta|$. Since by 2.17 $|\alpha| = \alpha$ and $|\beta| = \beta$ we have $\alpha \leqslant_c \beta$. We do not have $\alpha = \beta$ since $\alpha <_0 \beta$, therefore $\alpha <_c \beta$.

Assume now $\alpha <_c \beta$ then $\alpha <_0 \beta$, since otherwise $\beta \leqslant_0 \alpha$, and this would imply $\beta \subseteq \alpha$ and, by 2.10 and 2.17, $\beta = |\beta| \leqslant_c |\alpha| = \alpha$, contradicting $\alpha <_c \beta$. □

2.21 Is $<$ a Total Ordering of Cn? We have shown in 2.12 that $<$ is a partial ordering of the class Cn. We cannot prove that this is a total ordering without assuming the axiom of choice (since we shall show in V.1.8 that the statement that $<$ is a total ordering of Cn is equivalent to the axiom of choice, and the axiom of choice is known not to be provable in ZF). 2.20 shows that the subclass of Cn of all well ordered cardinals is totally ordered and even well ordered by $<$, since on this subclass $<$ is the natural order of the ordinals. If we assume the axiom of choice then, as we mentioned in 2.5, all cardinals are well ordered cardinals and hence the class Cn of all cardinals is totally ordered, and even well ordered, by $<$.

2.22 Corollary. (i) *If* $\alpha \leqslant \beta$ *then* $|\alpha| \leqslant |\beta|$, *hence if* $|\alpha| < |\beta|$ *then* $\alpha < \beta$.

(ii) *For every ordinal* α, $|\alpha|$ *is the largest cardinal which is* $\leqslant \alpha$.

Proof. (i) *If* $\alpha \leqslant \beta$ *then* $\alpha \subseteq \beta$ *and hence, by* 2.10, $|\alpha| \leqslant |\beta|$. (ii) $|\alpha|$ *is obviously a cardinal. Let* β *be a cardinal which is* $\leqslant \alpha$. *By* 2.17(ii) *and* (i) $\beta = |\beta| \leqslant |\alpha|$, *which shows that every cardinal* β *which is* $\leqslant \alpha$ *is also* $\leqslant |\alpha|$. □

2.23 Proposition. *If* α *is a cardinal and* $a \subseteq \alpha$ *then* $|a| = \alpha$ *iff* $\mathrm{Ord}(\langle a, \in \rangle) = \alpha$ *(otherwise* $|a| < \alpha$*).* □

2.24 Exercise. $|x|$ is an ordinal iff $|x| \approx x$ (i.e., iff $||x|| = |x|$). If $\mathfrak{a}$ is not a well ordered cardinal then $|\mathfrak{a}| > \mathfrak{a}$.

Hint. Use the fact that for infinite α $R(\alpha) \times 2 \approx R(\alpha)$, thus $R(\alpha)$ includes two disjoint copies of itself. □

By Cantor's theorem we know that for every given cardinal there is a bigger cardinal. We shall now prove a theorem about the existence of large *well ordered* cardinals.

2.25 Theorem (Hartogs 1915). *For every set* u *there is an ordinal* α *such that* $\alpha \preccurlyeq \mathbf{P}(\mathbf{P}(u \times u))$ *and* $\alpha \not\preccurlyeq u$ *(i.e., not* $\alpha \preccurlyeq u$*).*

Proof. Let $t = \{r \mid r$ is a reflexive well-ordering of a subset of $u\}$, where by r being a *reflexive well-ordering* of a set v we mean that $r \subseteq v \times v$ and r is the $\leqslant$ relation corresponding to a well-ordering $<$ of v (i.e., $r = \{<x, y> \mid x, y \in v \wedge (x < y \vee x = y)\}$). We use here reflexive well-orderings rather than the usual (irreflexive) well-orderings since for a reflexive well-ordering r of a set v one can recover v from r by $v = \mathrm{Dom}(r)$ while from an irreflexive well-ordering of v one cannot recover v if v is a singleton. Obviously $t \subseteq \mathbf{P}(u \times u)$.

We define a function f on t by setting, for $r \in t$,

$$f(r) = \mathrm{Ord}(\langle \mathrm{Dom}(r), r \sim \{\langle x, x \rangle \mid x \in \mathrm{Dom}(r)\} \rangle).$$

Since t is a set f and $\mathrm{Rng}(f)$ are sets; we shall see that $\mathrm{Rng}(f)$ is a transitive set of ordinals and hence an ordinal (by II.3.16). If $r \in t$, $\beta = f(r)$ and $\gamma < \beta$ then the ordered set $\langle \mathrm{Dom}(r), r \sim \{\langle x, x \rangle \mid x \in \mathrm{Dom}(r)\} \rangle$, which is of the order type β has a section w of the ordinal γ. $r|w$ is obviously a member of t and $f(r|w) = \gamma$, hence $\gamma \in \mathrm{Rng}(f)$. Thus we have indeed shown that $\mathrm{Rng}(f)$ is an ordinal α. We shall now see that there is no injection of α into u. Suppose that g is such an injection. Then the reflexive well ordering $s = \{\langle g(\beta), g(\gamma) \rangle \mid \beta \leqslant \gamma < \alpha\}$ of $g[\alpha]$ induced by g is clearly a member of t and $f(s) = \alpha$. This yields $\alpha \in \mathrm{Rng}(f) = \alpha$ which is a contradiction, arising from our assumption that α can be injected into u. $\alpha \preccurlyeq \mathbf{P}(\mathbf{P}(u \times u))$ since we get an injection h of α into $\mathbf{P}(t) \subseteq \mathbf{P}(\mathbf{P}(u \times u))$ by setting, for $\beta < \alpha$, $h(\beta) = \{r \in t \mid f(r) = \beta\} = f^{-1}[\{\beta\}]$. □

2.26 Definition. For every set u, u^+ will denote the least ordinal β such that $\beta \not\preccurlyeq u$; the existence of such a β follows from 2.25.

2.27 Proposition. *For every set* u, u^+ *is a well ordered cardinal, and is the least well ordered cardinal* $\mathfrak{a}$ *such that* $\mathfrak{a} \not\leqslant |u|$. $u^+ \leqslant |\mathbf{P}(\mathbf{P}(u \times u))|$.

Proof. If $\beta < u^+$ then, by the definition of u^+, $\beta \preccurlyeq u$. If we had $u^+ \approx \beta$ then $u^+ \approx \beta \preccurlyeq u$ would yield $u^+ \preccurlyeq u$, contradicting the definition of u^+. Thus u^+ is not equinumerous with any smaller ordinal, i.e., u^+ is a cardinal. The rest of 2.27 follows immediately from 2.26 and 2.25. □

2.28 Remark. Formulating 2.25 in terms of cardinals we can say that for every cardinal $\mathfrak{a}$ there is well ordered cardinal $\mathfrak{b}$ such that $\mathfrak{b} \nleqslant \mathfrak{a}$.

2.29 Corollary. *For every well ordered cardinal* $\mathfrak{a}$ *there are well ordered cardinals* $\mathfrak{b} > \mathfrak{a}$. *The least such* $\mathfrak{b}$ *is* $\mathfrak{a}^+$. □

2.30 Corollary. *There is no cardinal* $\mathfrak{a}$ *which is greater than all the well ordered cardinals.* □

2.31 Proposition. *For every set p of well ordered cardinals there is a well ordered cardinal which is greater than all the members of p. Therefore the class of all well ordered cardinals is a proper class.*

Hint of proof. Use either 2.14 and 2.28 or prove that for every set p of ordinals which are cardinals $\sup p$ is also a cardinal, and if $\sup p \in p$ take $(\sup p)^+$. □

2.32 Definition (Cantor 1895 and 1899). The class A of all *infinite* well-ordered cardinals is an unbounded class of ordinals (since ω is a cardinal and by 2.31). Therefore, this class, too, is a proper class and by II.3.23 there is an isomorphism of $\langle \mathrm{On}, \in \rangle$ on $\langle A, \in \rangle$, we denote this isomorphism with $\aleph$ (*aleph*). It is customary to write $\aleph_\alpha$ for $\aleph(\alpha)$.

2.33 Corollary. (i) *Each* $\aleph_\alpha$ *is an infinite well-ordered cardinal, and each infinite well-ordered cardinal is* $\aleph_\alpha$ *for some* α.

(ii) $\aleph_0 = \omega$

(iii) $\alpha < \beta$ iff $\aleph_\alpha < \aleph_\beta$.

(iv) *If* $\mathfrak{a}$ *is a cardinal such that* $\aleph_\alpha < \mathfrak{a} < \aleph_\beta$ *then* $\mathfrak{a} = \aleph_\gamma$ *for some* γ *such that* $\alpha < \gamma < \beta$; *hence* $\aleph_{\alpha+1} = (\aleph_\alpha)^+$.

(v) *The class* $\{\aleph_\alpha \mid \alpha \in \mathrm{On}\}$ *is a proper class.*

Hint of proof. (ii) By 2.19, 1.12 and 1.8. (iv) Use 2.16. □

2.34 Definition. (i) We shall call the natural numbers also *finite cardinals* since they are the cardinals of the finite sets (by 1.10, 2.18 and 2.17(ii)). The other cardinals are called *infinite cardinals*. The infinite well ordered cardinals are called *alephs* (since they are exactly the $\aleph_\alpha$'s for $\alpha \in \mathrm{On}$).

(ii) Cardinals which are of the form $\mathfrak{a}^+$, for a well ordered cardinal $\mathfrak{a}$, are called *successor cardinals*, $\mathfrak{a}^+$ is called the *cardinal successor* of $\mathfrak{a}$, and $\mathfrak{a}$ is called the *cardinal predecessor* of $\mathfrak{a}^+$. Cardinals of the form $\aleph_\lambda$, where λ is a limit ordinal or 0 are called *limit cardinals*. Obviously the successor cardinals are exactly the nonzero finite cardinals and the alephs of the form $\aleph_{\alpha+1}$ (IV.1.9).

2.35 Proposition. *If n is a finite cardinal and $\mathfrak{a}$ an infinite cardinal, then $n<\mathfrak{a}$ (by* 1.27 *and* 1.7). □

2.36 Proposition. *If $\mathfrak{a}<\aleph_0$ then $\mathfrak{a}$ is a finite cardinal (by* 2.16 and 2.34). □

2.37 Proposition. *Let f be a function such that* $\mathrm{Dom}(f)$ *is well-orderable, then there is a subset a of* $\mathrm{Dom}(f)$ *such that $f \upharpoonright a$ is a bijection of a on* $\mathrm{Rng}(f)$. *As a consequence* $|\mathrm{Rng}(f)| \leqslant |\mathrm{Dom}(f)|$ *and* $\mathrm{Rng}(f)$ *is well-orderable too.*

Proof. We define a function g on $\mathrm{Rng}(f)$ by setting

$$g(y) = \text{an } x \in \mathrm{Dom}(f) \text{ such that } f(x) = y.$$

This definition is legitimate because Dom (f) is well-orderable (II.2.25). We have, obviously, for every $y \in \mathrm{Rng}(f)$, $g(y) \in \mathrm{Dom}(f)$ and $f(g(y)) = y$. This implies immediately that $f \upharpoonright \mathrm{Rng}(g)$ is a one–one function (since if $f(g(y_1)) = f(g(y_2))$ then $y_1 = y_2$ and hence $g(y_1) = g(y_2)$), and that $f[\mathrm{Rng}(g)] = \mathrm{Rng}(f)$. Let $a = \mathrm{Rng}(g)$, then $f \upharpoonright a$ is a bijection of $a \subseteq \mathrm{Dom}(f)$ onto $\mathrm{Rng}(f)$. As a consequence $|\mathrm{Rng}(f)| = |a| \leqslant |\mathrm{Dom}(f)|$; by 2.16 $\mathrm{Rng}(f)$ is well-orderable too. □

2.38 Definition (Tarski in 1926—see Lindenbaum–Tarski 1926). $\preccurlyeq *$ is the relation on V given by $u \preccurlyeq * w$ if $u = 0$ or if there is a function f with $\mathrm{Dom}(f) = w$ and $\mathrm{Rng}(f) = u$. $\leqslant *$ is the relation on the cardinal numbers given by: $\mathfrak{a} \leqslant * \mathfrak{b}$ if there are sets a, b such that $|a| = \mathfrak{a}$, $|b| = \mathfrak{b}$ and $a \preccurlyeq * b$. As easily seen, for all sets x, y, $x \preccurlyeq * y$ iff $|x| \leqslant * |y|$. $* \geqslant$ is the inverse of $\leqslant *$, i.e., $\mathfrak{a} * \geqslant \mathfrak{b}$ iff $\mathfrak{b} \leqslant * \mathfrak{a}$. $\leqslant *$ is a transitive relation (by I.6.18).

2.39 Proposition. (i) *If $\mathfrak{a} \leqslant \mathfrak{b}$ then $\mathfrak{a} \leqslant * \mathfrak{b}$.*

(ii) *If $\mathfrak{a} \leqslant * \mathfrak{b}$ and $\mathfrak{b}$ is a well ordered cardinal then $\mathfrak{a} \leqslant \mathfrak{b}$.*

(iii) *If the axiom of choice holds then $\mathfrak{a} \leqslant * \mathfrak{b}$ iff $\mathfrak{a} \leqslant \mathfrak{b}$.*

Hint of proof. (ii) Use 2.37. (iii) Use (i) and (ii). □

By 2.39 the relation $\leqslant *$ gives nothing new if we assume the axiom of choice. However, in set theory without the axiom of choice this is an interesting relation; we shall soon see examples where $\mathfrak{a} \leqslant * \mathfrak{b}$ but $\mathfrak{a} > \mathfrak{b}$! Those examples show, of course, that the requirements in 2.37 that Dom (f) be well-orderable and in 2.39(ii) that $\mathfrak{b}$ be a well ordered cardinal cannot be dropped.

2.40 Definition. A set a which does not include a denumerable subset is said to be *Dedekind-finite*. The cardinal $|a|$ of such a set is said to be a *Dedekind-finite cardinal.* (See 1.23 for the origin of this concept.)

2.41 Proposition. (i) *Every finite set is Dedekind-finite.*

(ii) *No infinite well-orderable set is Dedekind-finite (and hence if we assume the axiom of choice then no infinite set is Dedekind finite and the Dedekind-finite sets are exactly the finite sets).*

(iii) *A set a is Dedekind-finite iff there is no bijection of a on a proper subset of itself.*

(iv) *A subset of a Dedekind-finite set is also Dedekind-finite.*

Hint of proof. (i) Use 1.15. (ii) Use 1.17. (iii) Use 1.20. □

2.42 Exercise. For a set a let $a^{(n)}$ denote the set of all ordered n-tuples without repetitions of members of a (i.e., the set of all injections of n into a). For a Dedekind-finite infinite set a prove the following facts.

(i) $\bigcup_{n\in\omega} a^{(n)}$ is a Dedekind-finite infinite set, and so is also every set $\bigcup_{n\in t} a^{(n)}$ where $t \subseteq \omega$, $t \sim \{0\} \neq 0$.

(ii) For all infinite s, $t \subseteq \omega$, $|\bigcup_{n\in s} a^{(n)}| *\geqslant |\bigcup_{n\in t} a^{(n)}|$.

(iii) If $s \subset t \subseteq \omega$ then $|\bigcup_{n\in s} a^{(n)}| < |\bigcup_{n\in t} a^{(n)}|$ (while $|\bigcup_{n\in s} a^{(n)}| *\geqslant |\bigcup_{n\in t} a^{(n)}|$ for infinite s!).

(iv) There is a set w of Dedekind-finite infinite cardinals such that $\langle w, < \rangle$, where $<$ is the natural partial order of the cardinals, is isomorphic to the set $\mathbb{R}$ of all real numbers in its natural order, while for all $\mathfrak{a}, \mathfrak{b} \in w$ $\mathfrak{a} \leqslant * \mathfrak{b}$.

Hints. (i) Prove that if $\langle b_0, b_1, b_2, \ldots \rangle$ is an infinite sequence of distinct members of $\bigcup_{n\in\omega} a^{(n)}$ we can obtain from it an infinite sequence of distinct members of a. (ii) Let $t = \{t_0, t_1, \ldots\}$, let $\langle r_0, r_1, \ldots \rangle$ be an increasing sequence of members of s such that $r_i \geqslant t_i$ for $i \in \omega$, let u be any member of $\bigcup_{n\in t} a^{(n)}$, and let $f(\langle x_0, \ldots, x_{r_i - 1} \rangle) = \langle x_0, \ldots, x_{t_i - 1} \rangle$ for $i \in \omega$, $x_0, \ldots, x_{r_i - 1} \in a$, and $f(\langle x_0, \ldots, x_{n-1} \rangle) = u$ for $n \in s \sim \{r_0, r_1, \ldots\}$. f is a surjection of $\bigcup_{n\in s} a^{(n)}$ onto $\bigcup_{n\in t} a^{(n)}$. (iii) Use (i) and 1.20. (iv) There is an injection g of the set $\mathbb{R}$ of all real numbers into the set of all infinite subsets of ω such that for all real numbers p, q if $p < q$ then $g(p) \subset g(q)$ (we get it by taking a bijection h of the set $\mathbb{Q}$ of all rational numbers on ω (VII.1.4) and defining $g(p) = \{h(r) \mid r \in \mathbb{Q} \wedge r < p\}$). We take $w = \{|\bigcup_{n\in g(p)} a^{(n)}| \mid p \in \mathbb{R}\}$. □

3. The Finite Arithmetic of the Cardinals

3.1 Lemma. *Let x, y, x', y' be sets such that $x \approx x'$, $y \approx y'$, $x \cap y = 0$, and $x' \cap y' = 0$ then $x \cup y \approx x' \cup y'$.*

Proof. Let f be a bijection of x on x', and let g be a bijection of y on y', then $f \cup g$ is a bijection of $x \cup x'$ on $y \cup y'$ (I.6.30(iii)). □

3.2 Definition (Addition of cardinals—Cantor 1887). $\mathfrak{a} + \mathfrak{b}$ is the unique cardinal $\mathfrak{c}$ such that if x and y are sets with $|x| = \mathfrak{a}$, $|y| = \mathfrak{b}$, and $x \cap y = 0$ then $|x \cup y| = \mathfrak{c}$.

To see that there is always such a cardinal we proceed as follows. Given any cardinals $\mathfrak{a}$ and $\mathfrak{b}$ there are sets u, v such that $|u| = \mathfrak{a}$, $|v| = \mathfrak{b}$. Obviously, $u \times \{0\}$ and $v \times \{1\}$ are equinumerous to u and v, respectively. Denote $x = u \times \{0\}$, $y = v \times \{1\}$ then $|x| = |u| = \mathfrak{a}$, $|y| = |v| = \mathfrak{b}$, and $x \cap y = 0$, hence there are x and y as required in the definition of + above. The cardinal $|x \cup y|$ is unique (i.e., independent of our choice of x and y) by 3.1.

We refer to $\mathfrak{a}+\mathfrak{b}$ as $\mathfrak{a}$ *plus* $\mathfrak{b}$ or as the *sum* of $\mathfrak{a}$ and $\mathfrak{b}$.

3.3 Proposition (Basic properties of addition—Cantor 1887).
(i) Associativity. $(\mathfrak{a}+\mathfrak{b})+\mathfrak{c}=\mathfrak{a}+(\mathfrak{b}+\mathfrak{c})$.
(ii) Commutativity. $\mathfrak{a}+\mathfrak{b}=\mathfrak{b}+\mathfrak{a}$.
(iii) $\mathfrak{a}+0=0+\mathfrak{a}=\mathfrak{a}$.
(iv) The relation of order to addition. $\mathfrak{a}\geqslant\mathfrak{b} \leftrightarrow \exists\mathfrak{c}(\mathfrak{a}=\mathfrak{b}+\mathfrak{c})$.
(v) Monotonicity of addition. $\mathfrak{a}\leqslant\mathfrak{a}' \wedge \mathfrak{b}\leqslant\mathfrak{b}' \rightarrow \mathfrak{a}+\mathfrak{b}\leqslant\mathfrak{a}'+\mathfrak{b}'$.

Hint of proof. (i) Take sets u, v, w such that $|u|=\mathfrak{a}$, $|v|=\mathfrak{b}$ and $|w|=\mathfrak{c}$. Carry out the proof by working with $u\times\{0\}$, $v\times\{1\}$, $w\times\{2\}$. (iv) Use 2.10. □

3.4 Proposition. *For all sets u and v, not necessarily disjoint,* $|u\cup v|\leqslant|u|+|v|$. □

3.5 Proposition. *The operation of addition of natural numbers defined in* II.4.5 *coincides on the natural numbers with the operation of addition of cardinals. In particular, the cardinal sum of any two finite cardinals is a finite cardinal.*

Proof. For the purpose of this proof let us denote with $+_c$ the addition of cardinals defined in 3.2. If m and n are natural numbers so is $m+_c n$, since the union of finite sets is finite (1.24). We shall show first that for every natural number n, $n\cup\{n\}=n+_c 1$. Since $n\notin n$, n and $\{n\}$ are disjoint, hence by the definition of addition of cardinals

(1) $$|n\cup\{n\}| = |n| +_c |\{n\}|.$$

Since for every well ordered cardinal α we have $|\alpha|=\alpha$ and since for every x $|\{x\}|=1$ we get from (1) $n\cup\{n\}=n+_c 1$.

Our next step is to show that the function $+_c\upharpoonright(\omega\times\omega)$ of cardinal addition of natural numbers satisfies the recursion equations $m+0=m$ and $m+(n\cup\{n\})=(m+n)\cup\{m+n\}$ of the addition + of natural numbers defined in II.4.5.

By 3.3(iii) $m+_c 0=m$. By the associativity of cardinal addition (3.3(i)) and what we have just shown $m+_c(n\cup\{n\})=m+_c(n+_c 1)=(m+_c n)+_c 1=(m+_c n)\cup\{m+_c n\}$. Thus, the function, $+_c\upharpoonright(\omega\times\omega)$, satisfies the recursion equations of +, and because of the uniqueness of functions defined by recursion, we get that the addition + of natural numbers coincides with $+_c\upharpoonright(\omega\times\omega)$. □

3.6 Proposition (Cantor 1895). *For every finite cardinal n,* $\aleph_0+n=\aleph_0$.

Proof. Since n is finite we have, by 2.35, $n<\aleph_0$. Therefore, by 3.3(iv) there is a cardinal $\mathfrak{c}$ such that $\aleph_0=n+\mathfrak{c}$. By 3.3(iv) $\mathfrak{c}\leqslant\aleph_0$ and hence, by 2.36 $\mathfrak{c}=\aleph_0$ or $\mathfrak{c}$ is a finite cardinal. If $\mathfrak{c}$ were a finite cardinal, then also $n+\mathfrak{c}$ were finite, by 3.5, contradicting $n+\mathfrak{c}=\aleph_0$. Hence, $\mathfrak{c}=\aleph_0$ and we have $n+\aleph_0=\aleph_0$. □

3.7 Exercise. Give a direct proof of 3.6 by constructing particular disjoint sets of cardinalities $\aleph_0$ and n and showing that the cardinality of their union is also $\aleph_0$. □

3.8 Proposition. *For every cardinal* $\mathfrak{a} \geqslant \aleph_0$ *and for every finite cardinal* n *we have* $\mathfrak{a}+n=\mathfrak{a}$. *In particular*, $\aleph_\alpha+n=\aleph_\alpha$.

Proof. If $\mathfrak{a} \geqslant \aleph_0$ then, by 3.3(iv), there is a cardinal $\mathfrak{c}$ such that $\mathfrak{a}=\aleph_0+\mathfrak{c}$ and hence $\mathfrak{a}+n=(\aleph_0+\mathfrak{c})+n=(\mathfrak{c}+\aleph_0)+n=\mathfrak{c}+(\aleph_0+n)=\mathfrak{c}+\aleph_0=\mathfrak{a}$, by 3.6. □

3.9 Proposition. *For every cardinal* $\mathfrak{a}$, $\mathfrak{a}+1=\mathfrak{a}$ *iff* $\mathfrak{a} \geqslant \aleph_0$.

Proof. If $\mathfrak{a} \geqslant \aleph_0$ then $\mathfrak{a}+1=\mathfrak{a}$ by 3.8. On the other hand, if $\mathfrak{a}+1=\mathfrak{a}$, let y be a set of cardinality $\mathfrak{a}$ and let u be an object not in y. $|y \cup \{u\}| = |y|+|\{u\}|=\mathfrak{a}+1=\mathfrak{a}=|y|$. Thus, $y \cup \{u\}$ is equinumerous to its proper subset y. By 1.20 $y \cup \{u\}$ includes a denumerable subset, hence, $\mathfrak{a}=\mathfrak{a}+1=|y \cup \{u\}| \geqslant \aleph_0$. □

3.10 Proposition. *Every* $\aleph_\alpha$ *is a limit ordinal.*

Proof. $\aleph_0=\omega$ is a limit ordinal by II.4.24. For $\alpha>0$ assume $\aleph_\alpha=\beta \cup \{\beta\}$, for some ordinal β. Then, since by 2.33(iii) $\aleph_\alpha>\aleph_0$, we have $\beta \geqslant \aleph_0$ (by 3.5). Thus, $|\beta| \geqslant \aleph_0$ (since $\beta \supseteq \aleph_0$) and, therefore, by 3.8, $\aleph_\alpha=|\beta \cup \{\beta\}|=|\beta|+1=|\beta|$. Now the cardinal $\aleph_\alpha$ is equinumerous to $\beta<\aleph_\alpha$, which contradicts 2.17. □

3.11 Definition. The relation R on On × On defined by

$$\langle \alpha, \beta \rangle R \langle \alpha', \beta' \rangle \quad \text{iff} \quad \alpha<\alpha', \quad \text{or} \quad \alpha=\alpha' \quad \text{and} \quad \beta<\beta'$$

is called the (*left*) *lexicographic ordering* of On × On. The relation S on On × On defined by

$$\langle \alpha, \beta \rangle S \langle \alpha', \beta' \rangle \quad \text{iff} \quad \beta<\beta', \quad \text{or} \quad \beta=\beta' \quad \text{and} \quad \alpha<\alpha'$$

is called the *right lexicographic ordering* of On × On.

3.12 Proposition. *Let* u *and* v *be sets of ordinals. The lexicographic ordering of* $u \times v$ *is a well-ordering relation.*

Proof. Since $u \times v$ is a set we have only to prove that every non-void subset y of $u \times v$ has a least member. We shall prove it only for the left lexicographic ordering; the proof for the right lexicographic ordering is the same. Let y be a non-void subset of $u \times v$, then $\mathrm{Dom}(y) \subseteq u$ is a non-void set, hence it has a least member α. $\alpha \in \mathrm{Dom}(y)$ hence the set $\{\gamma \mid \langle \alpha, \gamma \rangle \in y\}$ is non-void, let β be least member of this set. It is now easily seen that $\langle \alpha, \beta \rangle$ is the least member of $u \times v$. □

3.13 Theorem (Hessenberg 1906). $\aleph_\alpha+\aleph_\alpha=\aleph_\alpha$.

Proof. By induction on α. The set $\aleph_\alpha \times 2 = \aleph_\alpha \times \{0\} \cup \aleph_\alpha \times \{1\}$ has the cardinality $\aleph_\alpha + \aleph_\alpha$, by the definition of the addition of cardinals. Let r be the left lexicographic ordering of $\aleph_\alpha \times 2$. By 3.12 $\langle \aleph_\alpha \times 2, r \rangle$ is a well-ordered set and by II.3.23 there is an ordinal β such that $\langle \aleph_\alpha \times 2, r \rangle \cong \langle \beta, \in \rangle$. Once we prove that $\beta \leqslant \aleph_\alpha$ then we have, since $\beta \subseteq \aleph_\alpha$, $\aleph_\alpha + \aleph_\alpha = |\aleph_\alpha \times 2| = |\beta| \leqslant \aleph_\alpha$. Combining this with $\aleph_\alpha \leqslant \aleph_\alpha + \aleph_\alpha$ (3.3(iv)) we get by the Cantor–Bernstein theorem $\aleph_\alpha + \aleph_\alpha = \aleph_\alpha$. (We use here the Cantor–Bernstein theorem out of laziness. It is really easy to show directly that $\beta \geqslant \aleph_\alpha$.) Assume that $\beta > \aleph_\alpha$. Let f be the isomorphism of $\langle \beta, \in \rangle$ on $\langle \aleph_\alpha \times 2, r \rangle$; since $\beta > \aleph_\alpha$, $\aleph_\alpha \in \mathrm{Dom}(f)$. Let $f(\aleph_\alpha) = \langle \gamma, i \rangle$, where $\gamma < \aleph_\alpha$, $i < 2$. Thus, $f \upharpoonright \aleph_\alpha$ is an injection of $\aleph_\alpha$ into $(\gamma \cup \{\gamma\}) \times 2$, and $|(\gamma \cup \{\gamma\}) \times 2| \geqslant \aleph_\alpha$. Since, by 3.10, $\aleph_\alpha$ is a limit number also $\gamma \cup \{\gamma\} < \aleph_\alpha$. If $\gamma \cup \{\gamma\}$ is finite then $(\gamma \cup \{\gamma\}) \times 2$ is finite too, being the union of two finite sets, contradicting $\aleph_\alpha \leqslant |(\gamma \cup \{\gamma\}) \times 2|$. If $\gamma \cup \{\gamma\}$ is infinite then, since $\gamma \cup \{\gamma\} < \aleph_\alpha$, $|\gamma \cup \{\gamma\}| = \aleph_\delta$ for some $\aleph_\delta < \aleph_\alpha$, by 2.33(i) and 2.17(i), and hence $\delta < \alpha$, by 2.33(iii). The cardinal of the set $(\gamma \cup \{\gamma\}) \times 2$ is $\aleph_\delta + \aleph_\delta$, being the union of two disjoint sets of cardinality $\aleph_\delta$. By the induction hypothesis we have $\aleph_\delta + \aleph_\delta = \aleph_\delta$, since $\delta < \alpha$, hence $|(\gamma \cup \{\gamma\}) \times 2| = \aleph_\delta$ contradicting $|(\gamma \cup \{\gamma\}) \times 2| \geqslant \aleph_\alpha$. □

3.14 Proposition. $\aleph_\alpha + \aleph_\beta = \aleph_{\max(\alpha, \beta)}$.

Proof. Assume, without loss of generality, $\alpha \geqslant \beta$. Then, by 3.13, $\aleph_\alpha \leqslant \aleph_\alpha + \aleph_\beta \leqslant \aleph_\alpha + \aleph_\alpha = \aleph_\alpha = \aleph_{\max(\alpha, \beta)}$, hence $\aleph_\alpha + \aleph_\beta = \aleph_{\max(\alpha, \beta)}$. □

3.15 Proposition. *If* $\mathfrak{a}, \mathfrak{b}$ *are cardinals such that* $\mathfrak{a} + \mathfrak{b} \geqslant \aleph_\alpha$, *then* $\mathfrak{a} \geqslant \aleph_\alpha$ *or* $\mathfrak{b} \geqslant \aleph_\alpha$. *As a consequence, the sum of two Dedekind-finite cardinals is Dedekind-finite.* □

3.16 Lemma. *If* x, y, x', y' *are sets such that* $x \approx x'$, $y \approx y'$ *then* $x \times y \approx x' \times y'$. □

3.17 Definition (Multiplication of cardinals—Cantor 1887). $\mathfrak{a} \cdot \mathfrak{b}$ (or $\mathfrak{ab}$) is the unique cardinal $\mathfrak{c}$ such that if x and y are sets with $|x| = \mathfrak{a}$, $|y| = \mathfrak{b}$, then $|x \times y| = \mathfrak{c}$. This $\mathfrak{c}$ is always unique by 3.16.

3.18 Proposition (Basic properties of multiplication—Cantor 1887).

(i) Associativity. $\mathfrak{a} \cdot (\mathfrak{b} \cdot \mathfrak{c}) = (\mathfrak{a} \cdot \mathfrak{b}) \cdot \mathfrak{c}$.

(ii) Commutativity. $\mathfrak{a} \cdot \mathfrak{b} = \mathfrak{b} \cdot \mathfrak{a}$.

(iii) $\mathfrak{a} \cdot 0 = 0 \cdot \mathfrak{a} = 0$.

(iv) $\mathfrak{a} \cdot 1 = 1 \cdot \mathfrak{a} = \mathfrak{a}$.

(v) Distributivity of multiplication over addition. $\mathfrak{a} \cdot (\mathfrak{b} + \mathfrak{c}) = \mathfrak{a} \cdot \mathfrak{b} + \mathfrak{a} \cdot \mathfrak{c}$.

(vi) Monotonicity of multiplication. $\mathfrak{a} \leqslant \mathfrak{a}' \wedge \mathfrak{b} \leqslant \mathfrak{b}' \rightarrow \mathfrak{a} \cdot \mathfrak{b} \leqslant \mathfrak{a}' \cdot \mathfrak{b}'$.

Hint of proof. (i) For all sets x, y, z there is a natural bijection of $x \times (y \times z)$ on $(x \times y) \times z$. (ii) For all sets x, y there is a natural bijection of $x \times y$ on $y \times x$. (v) If x, y, z are sets such that y and z are disjoint then $x \times y$ and $x \times z$ are disjoint and $x \times (y \cup z) = x \times y \cup x \times z$. □

3.19 Proposition. *If* m *and* n *are finite cardinals then* $m \cdot n$ *is also a finite cardinal. On the finite cardinals the operation of cardinal multiplication is the ordinary* *multiplication of natural numbers, since the former satisfies the recursive definition of the latter, namely*

$$m \cdot 0 = 0, \quad \textit{and} \quad m \cdot (n+1) = m \cdot n + m.$$

Hint of proof. To get the first part prove by induction on the finite set b that if a and b are finite sets, then also $a \times b$ is finite. □

3.20 Proposition. *Let R be the relation on* On × On *defined by* $\langle\alpha, \beta\rangle R \langle\alpha', \beta'\rangle \leftrightarrow \max(\alpha, \beta) < \max(\alpha', \beta') \vee \max(\alpha, \beta) = \max(\alpha', \beta') \wedge (\alpha < \alpha' \vee \alpha = \alpha' \wedge \beta < \beta')$. *$R$ well-orders the class* On × On, *and will be called the* canonical ordering *of* On × On.

Remarks. Let us represent the pairs $\langle\alpha, \beta\rangle$ as points in a plane, with α as the x-coordinate and β as the y-coordinate. Then the ordering can be pictured as in Figure 3.

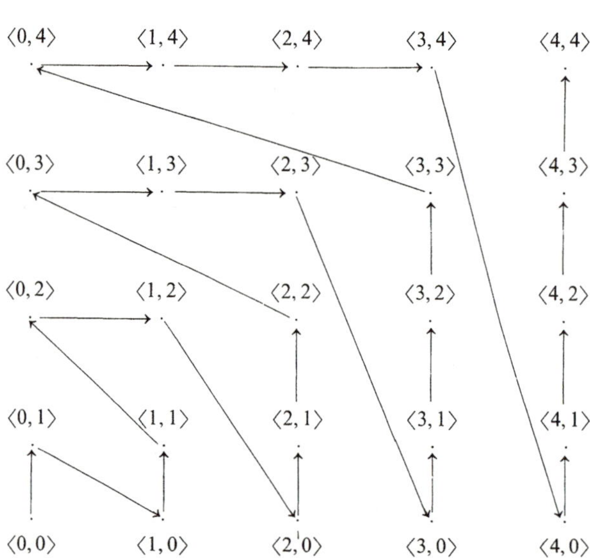

Fig. 3. The ordering R of On × On

The canonical ordering of On × On is useful for many purposes for which the lexicographic ordering of On × On is of no avail. For one thing, the lexicographic ordering of On × On is not a well-ordering in our sense as it is not left-narrow; e.g., the section determined by $\langle 1, 0\rangle$ in the left lexicographic ordering is $\{0\} \times$ On, which is clearly a proper class. The canonical ordering, on the other hand, handles first all the pairs $\langle\beta, \gamma\rangle$ with $\beta, \gamma < \alpha$ before it takes on pairs with one or two components $\geqslant \alpha$. As a result, not only do we get a well-ordering of On × On but also each set $\alpha \times \alpha$ is a section of the canonical well-ordering (determined by

$\langle 0, \alpha \rangle$). This property of the canonical ordering will enable us to use it to prove in 3.22 that $\aleph_\beta \cdot \aleph_\beta = \aleph_\beta$ for all β.

Proof of 3.20. Routine checking shows that R orders $\mathrm{On} \times \mathrm{On}$. Let y be a non-void subset of $\mathrm{On} \times \mathrm{On}$. By the least ordinal principle y has members $\langle \alpha, \beta \rangle$ with minimal $\max(\alpha, \beta)$; let this minimal $\max(\alpha, \beta)$ be γ. Let $y' = \{\langle \alpha, \beta \rangle \in y \mid \max(\alpha, \beta) = \gamma\}$, and let $\langle \alpha', \beta' \rangle$ be a minimal member of y' by the lexicographic ordering (of $(\gamma \cup \{\gamma\}) \times (\gamma \cup \{\gamma\})$). $\langle \alpha', \beta' \rangle$ is obviously an R-minimal member of y. We still have to show that R is left-narrow. By the definition of R, $R^{-1}[\langle \alpha, \beta \rangle] \subseteq \delta \times \delta$, where δ is the successor of $\max(\alpha, \beta)$; hence, $R^{-1}[\langle \alpha, \beta \rangle]$ is a set. □

3.21 Definition. P will denote the function on $\mathrm{On} \times \mathrm{On}$ which is the isomorphism of $\langle \mathrm{On} \times \mathrm{On}, R \rangle$ on $\langle \mathrm{On}, < \rangle$, where R is the canonical ordering of $\mathrm{On} \times \mathrm{On}$. There exists a unique such isomorphism by 3.20, II.2.18 and II.2.15(ii), since On and $\mathrm{On} \times \mathrm{On}$ are proper classes.

3.22 Proposition (Jourdain 1908). *$P \upharpoonright (\aleph_\alpha \times \aleph_\alpha)$ is a bijection of $\aleph_\alpha \times \aleph_\alpha$ on $\aleph_\alpha$.*

Proof. By induction on α. By the definition of the canonical ordering of $\mathrm{On} \times \mathrm{On}$, $\aleph_\alpha \times \aleph_\alpha$ is an initial segment of $\mathrm{On} \times \mathrm{On}$. Therefore, P maps $\aleph_\alpha \times \aleph_\alpha$ on a section β of On, and we have $\aleph_\alpha \cdot \aleph_\alpha = |\beta|$. We have to prove that $\beta = \aleph_\alpha$. If $\beta < \aleph_\alpha$, then we have $\aleph_\alpha \leqslant \aleph_\alpha \cdot \aleph_\alpha = |\beta|$, but this is a contradiction since $|\beta| < \aleph_\alpha$ because $\beta < \aleph_\alpha$. If $\beta > \aleph_\alpha$, then $\aleph_\alpha \in P[\aleph_\alpha \times \aleph_\alpha] = \beta$; hence, $\aleph_\alpha = P(\gamma, \delta)$ for some $\gamma, \delta < \aleph_\alpha$. Let η be the successor of $\max(\gamma, \delta)$. Since, by 3.10, $\aleph_\alpha$ is a limit ordinal, also $\eta < \aleph_\alpha$. Since $\aleph_\alpha = P(\gamma, \delta)$ and $\gamma, \delta < \eta$, we have, by definition of the canonical ordering $P^{-1}[\aleph_\alpha] \subseteq \eta \times \eta$; hence, $\aleph_\alpha \leqslant |\eta|\,|\eta|$. If η is finite then this cannot hold since by 3.19 $|\eta| \cdot |\eta|$ is finite. If η is infinite, let $|\eta| = \aleph_\xi$. Since $\eta < \aleph_\alpha$ we have $\aleph_\xi < \aleph_\alpha$ and hence, $\xi < \alpha$. By the induction hypothesis $P \upharpoonright (\aleph_\xi \times \aleph_\xi)$ is a bijection of $\aleph_\xi \times \aleph_\xi$ on $\aleph_\xi$; hence, $\aleph_\xi \cdot \aleph_\xi = \aleph_\xi$ and therefore, $\aleph_\alpha \leqslant |\eta| \cdot |\eta| = \aleph_\xi \cdot \aleph_\xi = \aleph_\xi$, contradicting $\aleph_\xi < \aleph_\alpha$.

3.23 Corollary (Hessenberg 1906). $\aleph_\alpha \cdot \aleph_\alpha = \aleph_\alpha$. □

3.24 Corollary. $\aleph_\alpha \cdot \aleph_\beta = \aleph_{\max(\alpha, \beta)} \quad (= \aleph_\alpha + \aleph_\beta)$.

Proof. Assume, without loss of generality $\alpha \leqslant \beta$. We have $\aleph_\beta = 1 \cdot \aleph_\beta \leqslant \aleph_\alpha \cdot \aleph_\beta \leqslant \aleph_\beta \cdot \aleph_\beta = \aleph_\beta$, hence, $\aleph_\alpha \cdot \aleph_\beta = \aleph_\beta = \aleph_{\max(\alpha, \beta)}$. □

3.25 Proposition. *If n is a finite cardinal and $n > 0$, then $n \cdot \aleph_\alpha = \aleph_\alpha$.*

Proof. $\aleph_\alpha = 1 \cdot \aleph_\alpha \leqslant n \cdot \aleph_\alpha \leqslant \aleph_\alpha \cdot \aleph_\alpha = \aleph_\alpha$, hence, $n \cdot \aleph_\alpha = \aleph_\alpha$. □

3.26 Proposition. *If $\mathfrak{a}$ and $\mathfrak{b}$ are cardinals such that $\aleph_\alpha \leqslant \mathfrak{a} \cdot \mathfrak{b}$, then $\aleph_\alpha \leqslant \mathfrak{a}$ or $\aleph_\alpha \leqslant \mathfrak{b}$. As a consequence, if $\mathfrak{a}$ and $\mathfrak{b}$ are Dedekind finite so is $\mathfrak{a} \cdot \mathfrak{b}$.*

Hint of proof. Let $|u|=\mathfrak{a}$, $|v|=\mathfrak{b}$, and $w\subseteq u\times v$ and $|w|=\aleph_\alpha$. Apply 2.37 to the functions

$$1^{\text{st}}\,1w\ (=\{\langle\langle x,y\rangle,x\rangle \mid \langle x,y\rangle\in w\})\quad\text{and}$$
$$2^{\text{nd}}\,1w\ (=\{\langle\langle x,y\rangle,y\rangle \mid \langle x,y\rangle\in w\}),\text{ then apply 3.24 and 3.25.}\quad\square$$

3.27 Lemma. *Let x, y, x', y' be sets such that $x\approx x'$ and $y\approx y'$, then ${}^yx\approx{}^{y'}x'$, where* ${}^yx=\{f|f:y\to x\}$ (see I.6.26).

Hint of Proof. Let g be a bijection of x on x' and let h be a bijection of y on y'. Let F be the function on ${}^yx$ given by $F(f)=gfh^{-1}$ for $f\in{}^yx$; F is a bijection of ${}^yx$ on ${}^{y'}x'$. $\square$

3.28 Definition (Exponentiation of cardinals—Cantor 1895). $\mathfrak{a}^\mathfrak{b}$ is the unique cardinal $\mathfrak{c}$ such that if x and y are sets with $|x|=\mathfrak{a}$ and $|y|=\mathfrak{b}$ then $|{}^yx|=\mathfrak{c}$. This $\mathfrak{c}$ is always unique by 3.27.

3.29 Proposition (Properties of exponentiation—Cantor 1895). (i) $\mathfrak{a}^{\mathfrak{b}+\mathfrak{c}}=\mathfrak{a}^\mathfrak{b}\cdot\mathfrak{a}^\mathfrak{c}$.

(ii) $(\mathfrak{a}^\mathfrak{b})^\mathfrak{c}=\mathfrak{a}^{\mathfrak{b}\cdot\mathfrak{c}}$.

(iii) $(\mathfrak{a}\cdot\mathfrak{b})^\mathfrak{c}=\mathfrak{a}^\mathfrak{c}\cdot\mathfrak{b}^\mathfrak{c}$.

(iv) $0^0=1$.

(v) $0^\mathfrak{a}=0$ *for* $\mathfrak{a}\neq0$.

(vi) $\mathfrak{a}^1=\mathfrak{a}$.

(vii) $1^\mathfrak{a}=1$.

(viii) $\mathfrak{a}^2=\mathfrak{a}\cdot\mathfrak{a}$.

(ix) $\mathfrak{a}\leqslant\mathfrak{c}\wedge\mathfrak{b}\leqslant\mathfrak{d}\to\mathfrak{a}^\mathfrak{b}\leqslant\mathfrak{c}^\mathfrak{d}$.

Hint of proof. (i) Let x, y, z be sets such that y and z are disjoint. Find a natural bijection of ${}^{y\cup z}x$ on ${}^yx\times{}^zx$. (ii) For sets x, y, z find a natural bijection of ${}^z({}^yx)$ on ${}^{y\times z}x$. (viii) Use (i) and (vi). $\square$

3.30 Proposition. *If m and n are finite cardinals then m^n is a finite cardinal. On the finite cardinals the operation of cardinal exponentiation is the ordinary exponentiation of natural numbers, since the former satisfies the recursive definition of the latter, namely $m^0=1$, $m^{n+1}=m^n\cdot m$.*

Hint of proof. To get the first part prove by induction on the finite set b that if a and b are finite sets, then also ${}^ba$ is a finite set. Use the proof of 3.19. $\square$

3.31 Proposition. *For every finite cardinal $n>0$ $\aleph_\alpha^n=\aleph_\alpha$.*

Hint of proof. By induction on n, using $\aleph_\alpha\cdot\aleph_\alpha=\aleph_\alpha$. $\square$

3.32 Proposition. *For every set z, $|\mathbf{P}(z)|=2^{|z|}$.*

Proof. We shall construct a bijection F of $\mathbf{P}(z)$ on ${}^z2$. For every $u \in \mathbf{P}(z)$ we take $F(u)$ to be the function on z such that for every $x \in z$

$$F(u)(x) = \begin{cases} 1 & \text{if } x \in u \\ 0 & \text{if } x \notin u \end{cases}$$

$F(u)$ is called the *characteristic function* of u. F is easily seen to be indeed a bijection as claimed. □

3.33 Corollary. *For every cardinal* $\mathfrak{a}$, $\mathfrak{a} < 2^{\mathfrak{a}}$.

Hint of proof. By Cantor's theorem, 2.13. □

3.34 Corollary. *For every cardinal* $\mathfrak{a}$, $\mathfrak{a}^+ \leqslant 2^{2^{\mathfrak{a}^2}}$.

Hint of proof. Use Hartog's theorem 2.27. □

3.35 Proposition. $n \geqslant 2 \rightarrow n^{\aleph_\alpha} = 2^{\aleph_\alpha}$; $\beta \leqslant \alpha \rightarrow \aleph_\beta^{\aleph_\alpha} = 2^{\aleph_\alpha}$.

Proof. If $n \geqslant 2$ and $\beta \leqslant \alpha$ then

$$\begin{aligned} 2^{\aleph_\alpha} &\leqslant n^{\aleph_\alpha} \leqslant \aleph_\beta^{\aleph_\alpha}, \text{ by the monotonicity of exponentiation;} \\ &\leqslant (2^{\aleph_\alpha})^{\aleph_\alpha}, \text{ since } \aleph_\beta \leqslant \aleph_\alpha \leqslant 2^{\aleph_\alpha} \text{ (by 3.29(ix) and 3.33);} \\ &\leqslant 2^{\aleph_\alpha \cdot \aleph_\alpha} = 2^{\aleph_\alpha}, \text{ by 3.29(ii) and 3.23.} \end{aligned}$$

Hence, all these inequalities are equalities, and our proposition holds. □

3.36 Can We "Compute" $\aleph_\alpha^{\aleph_\beta}$? In the case of addition and multiplication, we were able to determine exactly the values of $\aleph_\alpha + \aleph_\beta$ and $\aleph_\alpha \cdot \aleph_\beta$. As a matter of fact, these turned out to be very simple, and even boring operations. This is very convenient if you need these operations in order to compute the cardinalities of sets with which you deal. If, on the other hand, one expected to obtain operations which give rise to some interesting mathematical theory (such as, for example, the theory of prime integers) then the simplicity of the addition and the multiplication of the alephs is disappointing. The case of exponentiation is completely different; this is where the mystery and the challenge come in. The only nontrivial case where we were able to determine the value was $\aleph_\alpha^n$. As for $\aleph_\beta^{\aleph_\alpha}$, and even $2^{\aleph_0}$, we cannot even prove, from the axioms of ZF, that they are alephs (Cohen 1963). This does not mean that we know very little about $2^{\aleph_\alpha}$. $2^{\aleph_0}$ is, as we shall see, the cardinality of the set of all real numbers, a set with which mathematicians are very familiar. What is not known, on the basis of ZF, is whether the set of all real numbers can be well-ordered, and if it can be well-ordered which aleph is its cardinal. As we shall see in § V.5 the determination of $\aleph_\alpha^{\aleph_\beta}$ is a central point of interest of set theory. This can be viewed as a part of the more general problem of the power set $\mathbf{P}(x)$. Whereas many constructions of set theory yield new sets which are in some good sense completely known once the sets to which the constructions were applied are known, this cannot be said about **P**(x). To know what **P**(x) is we must know how a "general" subset of x is obtained, and this is almost as profound a question as the question of what are the sets in general.

3.37 Exercises (Tarski in 1926—see Lindenbaum–Tarski 1926). (i) If $\mathfrak{a}+\mathfrak{b}=\mathfrak{a}$ we say that $\mathfrak{b}$ is *absorbed* in $\mathfrak{a}$ and write $\mathfrak{b}\ll\mathfrak{a}$. We have

$$\mathfrak{b}\ll\mathfrak{a}\leftrightarrow\mathfrak{b}\cdot\aleph_0\leqslant\mathfrak{a}.$$

therefore a finite $n>0$ is absorbed in $\mathfrak{a}$ iff $\mathfrak{a}\geqslant\aleph_0$, and $\aleph_\alpha$ is absorbed in $\mathfrak{a}$ iff $\mathfrak{a}\geqslant\aleph_\alpha$. If we assume the axiom of choice, which implies that every infinite cardinal is an aleph, then the relation $\ll$ is not an interesting one since then $\mathfrak{b}\ll\mathfrak{a}$ iff $\mathfrak{b}=0$, or $\mathfrak{b},\aleph_0\leqslant\mathfrak{a}$.

(ii) *A partial cancellation law*. We cannot have a cancellation law for addition since $0+\aleph_0=\aleph_0=1+\aleph_0$. What we can still prove is the following: If $\mathfrak{a}+\mathfrak{c}=\mathfrak{b}+\mathfrak{c}$ then $\mathfrak{a}$ and $\mathfrak{b}$ are "equal except for parts absorbed in $\mathfrak{c}$", i.e., there are cardinals $\mathfrak{d}$, $\mathfrak{a}'$, $\mathfrak{b}'$ such that $\mathfrak{a}'$, $\mathfrak{b}'\ll\mathfrak{c}$, $\mathfrak{a}=\mathfrak{d}+\mathfrak{a}'$, and $\mathfrak{b}=\mathfrak{d}+\mathfrak{b}'$. $\mathfrak{d}$ is the "common part of $\mathfrak{a}$ and $\mathfrak{b}$", and $\mathfrak{a}'$, $\mathfrak{b}'$ are the respective parts of $\mathfrak{a}$ and $\mathfrak{b}$ which are absorbed in $\mathfrak{c}$.

(iii) *A cancellation law for Dedekind-finite cardinals*. If $\mathfrak{a}+\mathfrak{c}=\mathfrak{b}+\mathfrak{c}$ and $\mathfrak{c}$ is Dedekind-finite then $\mathfrak{a}=\mathfrak{b}$.

(iv) (Sierpinski 1958). A set A is Dedekind-finite iff for every set B such that $B\approx A$ we have $B\sim A\approx A\sim B$.

(v) (Bernstein 1905). If $\mathfrak{a}+\mathfrak{a}=\mathfrak{a}+\mathfrak{b}$ then $\mathfrak{b}\leqslant\mathfrak{a}$.

Hint. (ii) Let a, b, c be pairwise disjoint sets of the cardinalities $\mathfrak{a}$, $\mathfrak{b}$, $\mathfrak{c}$, respectively. Let f be a bijection of $a\cup c$ on $b\cup c$. Let $d=\{x\in a\mid(\exists n\in\omega)(x\in\mathrm{Dom}(f^n)\wedge f^n(x)\in b)\}$. Let g be a function on d defined by $g(x)=f^n(x)$ for the unique $n<\omega$ such that $x\in\mathrm{Dom}(f^n)\wedge f^n(x)\in b$, then g is an injection of d into b. $|a\sim d|$ is absorbed in $|c|$ since the sets $f^n[a\sim d]$ are pairwise disjoint subsets of c (for different n's) and likewise $|b\sim g[d]|$ is absorbed in $|c|$ since the sets $f^{-n}[b\sim g[d]]$ are pairwise disjoint subsets of c (f^{-n} denotes $(f^{-1})^n$). □

4. The Infinite Arithmetic of the Well Ordered Cardinals

4.1 Terminology. (Indexed families of objects.) When one applies a mathematical operation to finitely or infinitely many objects one often encounters a situation as in the following example. Suppose we want to add the following cardinals 2, 3, 2, 2, 5, 3, 4. We cannot regard the addition as operating on the *set* $\{2, 3, 2, 2, 5, 3, 4\}$ because this is the set $\{2, 3, 4, 5\}$ and this set does not contain the information that 2 should be added three times. We can regard addition as operating on the sequence $\langle 2, 3, 2, 2, 5, 3, 4\rangle$, but by constructing the sequence we impose on the members a certain order which has nothing to do with the operation of addition and which often has nothing to do with whatever led us to this addition. What we really want to apply the addition to is something which is like the set $\{2, 3, 4, 5\}$ except that it contains the number 2 three times, it contains 3 twice and it contains 4 and 5 once apiece. There is no difficulty in getting something which meets our requirements; we can use for this purpose a function f on the set $\{2, 3, 4, 5\}$ into the class of all non-zero cardinals which will count how many times is each member of $\{2, 3, 4, 5\}$ contained. For the particular example we gave above we would take $f(2)=3$, $f(3)=2$, $f(4)=f(5)=1$. Let us call such functions f whose domains are sets and whose values are non-zero cardinals *multiple-membership sets*. The kind of operations which are naturally applied to multiple-membership sets are either the *commutative* operations, where the outcome does not depend at all on the order of the arguments (or of any other structure on them) such as the operations of addition and multiplication of well ordered cardinals, or else the *essentially commutative* operations, where the outcome depends on the order of the arguments only to an extent considered insignificant in the context in which it occurs, such as direct products of structures where changing the order of the arguments transforms the outcome only into something isomorphic to the original outcome.

In spite of the theoretical advantage of dealing with multiple-membership sets we shall only keep them in mind and not deal with them directly. What we shall deal with is illustrated by the following example. For adding the numbers 2, 3, 2, 2, 5, 3, 4 we shall consider a function e on a set u such that $|u|=7$ and such that $2=e(x)$ for three different $x \in u$, $3=e(x)$ for two different $x \in u$ and each of 5 and 4 equals $e(x)$ only for a single $x \in u$. In a general situation we shall use the functional notation of I.6.15(i) and write such a function e as $\langle e_x \mid x \in u\rangle$, where e_x stands for $e(x)$. In those cases, as the example given above, where what we are really interested in are the e_x's and the number of times each one of them occurs as a value of e, rather than the function e itself, we shall refer to the function $\langle e_x \mid x \in u\rangle$ as an *indexed family* and say that x is an *index* of e_x (with respect to e). Unless e is a one–one function an object in $\{e_x \mid x \in u\}$ may have several indices. Thus when we speak about the indexed family $\langle e_x \mid x \in u\rangle$ we really have in mind the multiple-membership set f given by $\mathrm{Dom}(f)=\{e_x \mid x \in u\}$ and for $d \in \mathrm{Dom}(f)$, $f(d)= |\{x | x \in u \wedge e_x = d\}|$. The reasons why we still prefer to speak, if not to think, about the indexed family e rather than the multiple-membership set f, even though e carries along the irrelevant set u of indices and the correspondence between the members of $\{e_x \mid x \in u\}$ and their indices, are the following. In actual situations in mathematics when we apply commutative operations the arguments come often as an indexed family anyway, and converting it into a multiple-membership set requires an additional operation which complicates at least the notation and possibly even the discussion. The second reason is that if d is an object which occurs more than once and if we need therefore in a construction several distinct copies of d, these are somewhat more directly available when we deal with an indexed family rather than a multiple-membership set. We shall also denote indexed families by $\langle \tau(x) \mid x \in u\rangle$, where $\tau(x)$ is a term, according to the notation for functions introduced in I.6.15(i). Since whenever we speak about an indexed family we have in mind the corresponding multiple-membership set, we shall say that $\langle e_x \mid x \in u\rangle$ is an indexed family of objects which have a certain property whenever we want to say that for every $x \in u$ e_x has that property. For example, if we shall say that $\langle e_x \mid x \in u\rangle$ is a family of finite sets we shall mean by it that for every $x \in u$ e_x is a finite set.

Following this line of thought we shall write, e.g., $\langle \mathfrak{e}_x \mid x \in u \rangle$ for a family of cardinals, even though $\mathfrak{e}$ is a function and should therefore be denoted by an italic letter rather than a German one.

The infinite arithmetic of the cardinals is a tool which is most handy only in the presence of the axiom of choice. As long as we do not assume this axiom it is convenient to restrict our discussion to well ordered cardinals. This will still give us the theory of the infinite arithmetic of *all* cardinals once we assume the axiom of choice since then every cardinal is a well ordered cardinal.

4.2 Definition. (General addition of cardinals—Whitehead 1902). Let $\langle \mathfrak{e}_x \mid x \in u \rangle$ be an indexed family of well ordered cardinals, then

$$\Sigma_{x \in u} \mathfrak{e}_x = |\bigcup_{x \in u} \{x\} \times \mathfrak{e}_x| = |\{\langle x, \alpha \rangle \mid x \in u \wedge \alpha \in \mathfrak{e}_x\}|.$$

$\Sigma_{x \in u} \mathfrak{e}_x$ is called the *sum* of the $\mathfrak{e}_x$'s. (When we write $\Sigma_{x \in u} \tau(x)$ where $\tau(x)$ is a term we mean, of course, the sum of the indexed family $\langle \tau(x) \mid x \in u \rangle$).

The reason why we defined general addition for well ordered cardinals only is that for a well ordered cardinal $\mathfrak{e}_x$, $\mathfrak{e}_x$ itself is a set of cardinality $\mathfrak{e}_x$ and we are not faced with the problem of getting a set of cardinality $\mathfrak{e}_x$ in order to "perform" the addition. If $\mathfrak{e}_x$ is not a well ordered cardinal the simplest general way to get a set of cardinality $\mathfrak{e}_x$ is to choose a member of $\mathfrak{e}_x$. To do that, for more than finitely many x's requires the axiom of choice. Moreover, different choices of respective members of the $\mathfrak{e}_x$'s may result in unions of different cardinality, as we shall see in the next paragraph, and thus the sum will not be well defined.

Binary addition of cardinals was defined in 3.2 as the cardinal of the union of *any* two disjoint sets of the respective cardinals. In 4.2 we defined the general addition of cardinals as the cardinal of the union of *particular* disjoint sets of the respective cardinals, as we chose the sets $\{x\} \times \mathfrak{e}_x$ as the sets of the respective cardinals $\mathfrak{e}_x$. We ask now whether the cardinal of a union of *arbitrary* pairwise disjoint sets is the sum of the cardinals of the respective sets. We shall see that the answer is completely positive if we assume the axiom of choice, but the situation is somewhat complicated if we do not. As a matter of fact, we cannot prove in ZF that if $\langle w_x \mid x \in u \rangle$ is such that for each $x \in u$ w_x is well-orderable and, for all $x, y \in u$, $x \neq y \rightarrow w_x \cap w_y = 0$, then

$$(1) \qquad |\bigcup_{x \in u} w_x| = \Sigma_{x \in u} |w_x|$$

(unless we assume the axiom of choice). For example, even while $\Sigma_{x \in \omega} \aleph_0 = \aleph_0$ (by 4.7) we can have, as far as ZF is concerned, an indexed family $\{w_x \mid x \in \omega\}$ of denumerable sets such that $x \neq y \rightarrow w_x \cap w_y = 0$ for all $x, y \in \omega$ and yet $|\bigcup_{x \in \omega} w_x| = 2^{\aleph_0} > \aleph_0$ (Feferman and Levy 1963—see Cohen 1966). To obtain (1) we need the additional assumption (*c*) of the next proposition.

4.3 Proposition. *Let* $\langle w_x \mid x \in u \rangle$ *be such that*

(a) *for all* $x \in u$, w_x *is well-orderable,*

(b) *for all* $x, y \in u$, $x \neq y \rightarrow w_x \cap w_y = 0$, *and*

(c) *the w_x's for $x \in u$ have their cardinals* uniformly, *i.e., there is a function f on u such that for every $x \in u$, $f(x)$ is a bijection of w_x on $|w_x|$.*
Then we have

(1) $$|\bigcup_{x\in u} w_x| = \Sigma_{x\in u}|w_x|.$$

If the axiom of choice holds then (a) *and* (c) *are always true and are, therefore, superfluous.*

Proof. By Definition 4.2 of $\Sigma_{x\in u}|w_x|$, in order to establish (1) we have to present a bijection of $\bigcup_{x\in u} w_x$ on $\bigcup_{x\in u}\{x\}\times|w_x|$. If f is a function as in (c) then such a bijection is the function g on $\bigcup_{x\in u} w_x$ given by $g(t)=\langle x, f(x)(t)\rangle$, where x is that member of u for which $t \in w_x$.

Let h be the function on u defined by $h(x)=$ the set of all bijections of w_x on $|w_x|$. By the axiom of choice (see II.2.25) there is a function f on u such that for every $x \in u$ $f(x)=$ a member of $h(x)$. This is a function f as required by (c). □

We shall now see that if in $\langle \mathfrak{e}_x \mid x \in u\rangle$ all the $\mathfrak{e}_x$'s are well ordered and u is well-orderable then the outcome of the addition is almost as easily "computable" as the outcome of finite addition and multiplication.

4.4 Proposition. *If $\langle \mathfrak{e}_x \mid x \in u\rangle$ is an indexed family of well ordered cardinals and the set u is well-orderable then $\Sigma_{x\in u}\mathfrak{e}_x$ is also a well ordered cardinal. If also $\mathfrak{e}_x > 0$ for every $x \in u$ and either u or one of the $\mathfrak{e}_x$'s is infinite then*

$$\Sigma_{x\in u}\mathfrak{e}_x = \max(|u|, \sup_{x\in u}\mathfrak{e}_x).$$

Proof. Let f be a bijection of u on $|u|$. The function g on $\bigcup_{x\in u}\{x\}\times\mathfrak{e}_x$ defined by $g(x,\alpha)=\langle f(x),\alpha\rangle$ is an injection of $\bigcup_{x\in u}\{x\}\times\mathfrak{e}_x$ into $|u|\times\sup_{x\in u}\mathfrak{e}_x$, hence $\Sigma_{x\in u}\mathfrak{e}_x \leqslant |u|\sup_{x\in u}\mathfrak{e}_x$.

The rest of the proof is left to the reader. □

To deal with the general union of sets which are not necessarily pairwise disjoint we have the following proposition.

4.5 Proposition. *Let $\langle w_x \mid x \in u\rangle$ be an indexed family of sets such that*

(a) *u is well-orderable, and*

(b) *for all $x \in u$ w_x is well-orderable, then by* 4.4, *$\Sigma_{x\in u}|w_x|$ is a well ordered cardinal; let us denote it with $\mathfrak{a}$. If also*

(c) *the w_x's, for $x \in u$, can be injected uniformly into $\mathfrak{a}$ (i.e., there is a function f on u such that, for $x \in u$, $f(x)$ is an injection of w_x into $\mathfrak{a}$) then we have*

$$|\bigcup_{x\in u} w_x| \leqslant \Sigma_{x\in u}|w_x|.$$

Once we assume the axiom of choice the assumptions (a)–(c) *become superfluous.*

Remark. This proposition is very closely related to 4.3. We have to make here one stronger assumption, that u is well-orderable. Other than that the assumptions here are weaker and we get an inequality. An inequality is as useful as an equality, since in practically all cases it is easy to give the best lower estimate for $|\bigcup_{x\in u} w_x|$.

Proof of 4.5. For $x\in u, f(x)$ injects w_x into $\mathfrak{a}$ and thereby well-orders the set w_x (by the relation $\{\langle y,z\rangle | y,z\in w_x \wedge f(x)(y)<f(x)(z)\}$) so that its ordinal is $\leqslant\mathfrak{a}$. Let $j(x)$ be the unique isomorphism of w_x, with this well-ordering, on an ordinal $\leqslant\mathfrak{a}$. Assume first that $\mathfrak{a}$ is finite. $j(x)$ is then a bijection of w_x on a finite ordinal which must be $|w_x|$. We construct now an injection g of $\bigcup_{x\in u} w_x$ into $\bigcup_{x\in u}\{x\}\times|w_x|$ by setting $g(t)=\langle x, j(x)(t)\rangle$, where x is a member of u such that $t\in w_x$ (see II.2.25). This proves that if $\mathfrak{a}$ is finite then $|\bigcup_{x\in u} w_x|\leqslant|\bigcup_{x\in u}\{x\}\times|w_x||=\Sigma_{x\in u}|w_x|$. Let us deal now with the case where $\mathfrak{a}$ is infinite. Let $u'=\{x\in u \mid w_x\neq 0\}$. Since $u'\times\{0\}\subseteq\bigcup_{x\in u}\{x\}\times|w_x|$ we get, by taking the cardinals of both sides, $|u'|\leqslant\mathfrak{a}$. Let k be an injection of u' into $\mathfrak{a}$, and let f be as in (c). We construct an injection g of $\bigcup_{x\in u} w_x$ into $\mathfrak{a}\times\mathfrak{a}$ by setting $g(t)=\langle k(x), f(x)(t)\rangle$ where x is a member of u' such that $t\in w_x$ (there is such an $x\in u'$ since $t\in w_x\rightarrow w_x\neq 0\rightarrow x\in u'$). Thus we have

$$|\textstyle\bigcup_{x\in u} w_x|=|\mathrm{Dom}(g)|=|\mathrm{Rng}(g)|\leqslant|\mathfrak{a}\times\mathfrak{a}|=|\mathfrak{a}|^2=\mathfrak{a}^2=\mathfrak{a}. \quad \square$$

4.6 Proposition. (i) $\Sigma_{x\in\{s,t\}}\mathfrak{e}_x=\mathfrak{e}_s+\mathfrak{e}_t$.

(ii) $\Sigma_{x\in\{s\}}\mathfrak{e}_x=\mathfrak{e}_s$.

(iii) $\Sigma_{x\in 0}\mathfrak{e}_x=0$.

(iv) Commutativity. *Let* $\langle\mathfrak{e}_x \mid x\in u\rangle$ *and* $\langle\mathfrak{e}'_x \mid x\in u'\rangle$ *be indexed families of well ordered cardinals. If there is a bijection g of u on u' such that for every $x\in u$* $\mathfrak{e}'_{g(x)}=\mathfrak{e}_x$ *then* $\Sigma_{x\in u}\mathfrak{e}_x=\Sigma_{x\in u'}\mathfrak{e}'_x$. (*Now we see why we are talking about an* indexed family *of* $\mathfrak{e}_x$*'s rather than about a function* $\mathfrak{e}$; *by the commutativity the sum depends only on the multiple membership set which corresponds to* $\mathfrak{e}$, *provided we assume the axiom of choice*).

(v) Monotonicity. *If* $\langle\mathfrak{e}_x \mid x\in u\rangle$ and $\langle\mathfrak{e}'_x \mid x\in u\rangle$ *are such that for every* $x\in u$ $\mathfrak{e}_x$ *and* $\mathfrak{e}'_x$ *are well ordered cardinals and* $\mathfrak{e}_x\leqslant\mathfrak{e}'_x$, *then* $\Sigma_{x\in u}\mathfrak{e}_x\leqslant\Sigma_{x\in u}\mathfrak{e}'_x$.

(vi) *For every* $y\in u \quad \mathfrak{e}_y\leqslant\Sigma_{x\in u}\mathfrak{e}_x$. $\square$

We could prove now certain versions of the associative law. The full associative and associative–commutative laws can be proved only by means of the axiom of choice. These theorems yield information concerning general addition, but do not have any application in the present book. Therefore they will not be formulated here.

4.7 Proposition (Binary multiplication as multiple addition—Whitehead 1902). *If* $\mathfrak{d}$ *is a well ordered cardinal then* $\Sigma_{x\in u}\mathfrak{d}=|u|\cdot\mathfrak{d}$ ($\Sigma_{x\in u}\mathfrak{d}$ *is the sum of* $\langle\mathfrak{d} \mid x\in u\rangle$). $\square$

4.8 Definition (General Cartesian product—Cantor 1895). Let $\langle w_x \mid x\in u\rangle$ be an indexed family of sets. We define the *Cartesian product* $\bigtimes_{x\in u} w_x$ of this family to be the set $\{f \mid f$ is a function on $u\wedge(\forall x\in u)(f(x)\in w_x)\}$.

4.9 Binary and General Cartesian Product. (While the general operations of set union (I.4.15(i)) and cardinal addition (4.2) are direct generalization of the respective binary operations of union (I.5.16(i)) and addition (3.2), this is not the case with the Cartesian product, for purely formal reasons. If we want to apply the general Cartesian product to the two sets s and t we take for u a set of two distinct objects $\{a, b\}$ and let $w_a=s$, $w_b=t$. Then $\times_{x\in\{a,b\}} w_x$ is the set of all functions $\{\langle a, y\rangle, \langle b, z\rangle\}$, where $y\in s$ and $z\in t$. This set is different from the set $s\times t$ obtained from s and t by the binary Cartesian product (I.6.5). However, since there is a natural bijection of the set $s\times t$ on the set $\times_{x\in\{a,b\}} w_x$ these two sets can be regarded, for all practical purposes, as the same set. As we have already remarked above, the role of the indices in indexed families of sets is rather limited, and the only purpose a and b serve in $\{\langle a, y\rangle, \langle b, z\rangle\}$ is to distinguish between the "a components" and the "b-components" of the members of $\times_{x\in\{a,b\}} w_x$, and this is done in $s\times t$ without the use of indices.

For $\times_{x\in u} w_x$, if for some $x\in u$ $w_x=0$ then obviously $\times_{x\in u} w_x=0$. We shall see, in V.1.5 that the other direction, that if for all $x\in u$ $w_x\neq 0$ then also $\times_{x\in u} w_x\neq 0$, is directly equivalent to the axiom of choice.

4.10 Definition (General multiplication of cardinals—Whitehead 1902). Let $\langle \mathfrak{e}_x \mid x\in u\rangle$ be an indexed family of well ordered cardinals, then $\Pi_{x\in u}\mathfrak{e}_x=|\times_{x\in u}\mathfrak{e}_x|$. $\Pi_{x\in u}\mathfrak{e}_x$ is called the *product* of the $\mathfrak{e}_x$'s. □

Looking at an arbitrary Cartesian product the next proposition will assert that the cardinal of the Cartesian product is the product of the cardinals of the factors to the same extent as the cardinal of a disjoint union of sets in the sum of the cardinals of the participating sets (4.3).

4.11 Proposition. *Let* $\langle w_x \mid x\in u\rangle$ *be such that*

(a) *for all* $x\in u$, w_x *is a well-orderable set, and*

(b) *the* w_x*'s, for* $x\in u$, *have their cardinals uniformly, i.e., there is a function* f *on* u *such that for every* $x\in u$, $f(x)$ *is a bijection of* w_x *on* $|w_x|$. *Then we have*

(2) $$|\times_{x\in u} w_x|=\Pi_{x\in u}|w_x|.$$

If the axiom of choice holds then (a) *and* (b) *hold always and are, therefore, superfluous.* □

4.12 Proposition. (i) $\Pi_{x\in\{s,t\}}\mathfrak{e}_x=\mathfrak{e}_s\cdot\mathfrak{e}_t$.

(ii) $\Pi_{x\in\{s\}}\mathfrak{e}_x=\mathfrak{e}_s$.

(iii) $\Pi_{x\in 0}\mathfrak{e}_x=1$.

(iv) Commutativity. *Let* $\langle \mathfrak{e}_x \mid x\in u\rangle$ *and* $\langle \mathfrak{e}'_x \mid x\in u'\rangle$ *be indexed families of well ordered cardinals. If there is a bijection* g *of* u *on* u' *such that for every* $x\in u$ $\mathfrak{e}'_{g(x)}=\mathfrak{e}_x$, *then* $\Pi_{x\in u}\mathfrak{e}_x=\Pi_{y\in u'}\mathfrak{e}'_y$.

(v) Monotonicity. *If* $\langle \mathfrak{e}_x \mid x\in u\rangle$ *and* $\langle \mathfrak{e}'_x \mid x\in u'\rangle$ *are such that for every* $x\in u$ $\mathfrak{e}_x$ *and* $\mathfrak{e}'_x$ *are well ordered cardinals and* $\mathfrak{e}_x\leqslant\mathfrak{e}'_x$ *then* $\Pi_{x\in u}\mathfrak{e}_x\leqslant\Pi_{x\in u}\mathfrak{e}'_x$.

(vi) Ac Rules of exponentiation. *Let* $\langle \mathfrak{e}_x \mid x \in u \rangle$ *be an indexed family of cardinals. Assuming the axiom of choice we have*

$$\mathfrak{a}^{\Sigma_{x\in u}\mathfrak{e}_x} = \Pi_{x\in u}\mathfrak{a}^{\mathfrak{e}_x} \quad \text{and} \quad (\Pi_{x\in u}\mathfrak{e}_x)^{\mathfrak{b}} = \Pi_{x\in u}\mathfrak{e}_x^{\mathfrak{b}}.$$

(*We have to assume the axiom of choice even if the* $\mathfrak{e}_x$'*s are well ordered cardinals in order that the right-hand sides of the equalities be defined.*)

Hint of proof. (vi) Let $\langle e_x \mid x \in u \rangle$ be an indexed family of pairwise disjoint sets such that for every $x \in u$ $|e_x| = \mathfrak{e}_x$. Prove that the sets ${}^{(\bigcup_{x\in u} e_x)}a$ and $\times_{x\in u} {}^{(e_x)}a$ are equinumerous. Do the same for the sets ${}^{b}(\times_{x\in u} e_x)$ and $\times_{x\in u} {}^{b}e_x$. □

4.13 Proposition. *For an indexed family* $\langle \mathfrak{e}_x \mid x \in u \rangle$ *of well ordered cardinals* $\Pi_{x\in u}\mathfrak{e}_x = 0$ *iff for some* $y \in u$ $\mathfrak{e}_y = 0$. □

4.14 Remark. Notice that since $\Pi_{x\in u}\mathfrak{e}_x$ is a product of actual well ordered cardinals we do not need the axiom of choice to prove 4.13. This is unlike the case of the Cartesian product $\times_{x\in u} w_x$, for which we mentioned in 4.9 that we cannot prove without the axiom of choice that if all the w_x's are non-void then so is $\times_{x\in u} w_x$. If we want to apply the result about cardinal multiplication (4.13) to the case of the Cartesian product, where we assume $w_x \neq 0$ for every $x \in u$, we have to apply 4.11 to get $|\times_{x\in u} w_x| = \Pi_{x\in u}|w_x| > 0$, but to do that we have to prove that the family $\langle w_x \mid x \in u \rangle$ satisfies the hypotheses of 4.11 and this requires, in the general case, the axiom of choice.

4.15 Proposition. (Exponentiation as multiple multiplication). *If* $\mathfrak{d}$ *is a well ordered cardinal, then* $\Pi_{x\in u}\mathfrak{d} = \mathfrak{d}^{|u|}$. □

Unlike what we proved in 4.4 with respect to cardinal addition we cannot prove in ZF that if the $\mathfrak{e}_x$'s are well ordered cardinals and u is well-orderable then $\Pi_{x\in u}\mathfrak{e}_x$ is a well ordered cardinal. For example, by 4.15 $\Pi_{u\in\omega} 2 = 2^{\aleph_0}$, and we have already mentioned that $2^{\aleph_0}$ cannot be shown in ZF to be an aleph. The reason why in the case of addition (4.4) we not only knew that, under the right assumptions, $\Sigma_{x\in u}\mathfrak{e}_x$ is a well ordered cardinal, but we had also a neat formula for "computing" the sum, is that general addition of cardinals is closely related to binary multiplication of cardinals which is a particularly simple operation for alephs; general multiplication of cardinals is closely related to (binary) exponentiation of cardinals, which is an operation on whose outcome we do not know much.

4.16 Exercise (Bernstein 1905, Schönflies 1913). A set of cardinality $\aleph_\alpha$ has exactly $2^{\aleph_\alpha}$ permutations (a *permutation* of a set is a bijection of a set on itself). □

4.17 Proposition. *If* $\langle \mathfrak{e}_x \mid x \in u \rangle$ *is an indexed family of well ordered cardinals such that* $\mathfrak{e}_x \geqslant 2$ *for every* $x \in u$, *then* $\Sigma_{x\in u}\mathfrak{e}_x \leqslant \Pi_{x\in u}\mathfrak{e}_x$.

Hint of proof. See the first part of the proof of Theorem 4.18. □

4.18 Theorem (The Zermelo–König inequality–König 1905, Zermelo 1908). *If $\langle \mathfrak{d}_x \mid x \in u \rangle$ and $\langle \mathfrak{e}_x \mid x \in u \rangle$ are two indexed families of well ordered cardinals with the same set of indices u, such that for every $x \in u$ $\mathfrak{d}_x < \mathfrak{e}_x$, then $\Sigma_{x \in u} \mathfrak{d}_x < \Pi_{x \in u} \mathfrak{e}_x$.*

Remark. Note that the monotonicity of addition, multiplication and exponentiation give us only weak inequalities, i.e., of the $\leqslant$ kind. Our present inequality is a strict inequality, the like of which we do not have many.

Proof of 4.18. The proof we shall give is formally correct also for the case where $u=0$, but it is more convenient to notice that for $u=0$ the inequality we prove is just $0<1$, and deal, for the rest of the proof, with the case where $u \neq 0$.

Consider the sets $p=\bigcup_{x \in u}\{x\} \times \mathfrak{d}_x$ and $q=\bigtimes_{x \in u} \mathfrak{e}_x$; by the definition of addition and multiplication we have $|p|=\Sigma_{x \in u} \mathfrak{d}_x$ and $|q|=\Pi_{x \in u} \mathfrak{e}_x$. We shall first define an injection F of p in q thereby proving $\Sigma_{x \in u} \mathfrak{d}_x \leqslant \Pi_{x \in u} \mathfrak{e}_x$.

If $t \in p$ then there are $x \in u$ and $\alpha < \mathfrak{d}_x$ such that $t=\langle x, \alpha \rangle$. We define $F(t)=F(\langle x, \alpha \rangle)=$ the function f on u such that for $y \in u$

$$f(y)=\begin{cases} \mathfrak{d}_y & \text{if } y \neq x \\ \alpha & \text{if } y=x. \end{cases}$$

Since for every $y \in u$ $\mathfrak{d}_y \in \mathfrak{e}_y$ and since $\alpha \in \mathfrak{d}_x \subseteq \mathfrak{e}_x$ we have $F(t) \in q$. To prove that F is one–one let $t_1, t_2 \in p$, $t_1 \neq t_2$, thus $t_1=\langle x_1, \alpha_1 \rangle$, $t_2=\langle x_2, \alpha_2 \rangle$, and either $x_1 \neq x_2$ or else $x_1=x_2$ and $\alpha_1 \neq \alpha_2$. If $x_1 \neq x_2$ then $F(t_1)(x_1)=\alpha_1<\mathfrak{d}_{x_1}=F(t_2)(x_1)$, and $F(t_1) \neq F(t_2)$; if $x_1=x_2$ and $\alpha_1 \neq \alpha_2$ then $F(t_1)(x_1)=\alpha_1 \neq \alpha_2=F(t_2)(x_1)$, hence, again, $F(t_1) \neq F(t_2)$.

In order to prove the strict inequality we have to show that there is no bijection of p on q; we shall do it by the *diagonal method.* Suppose G is a bijection of p on q. We shall construct a member h of q which will not be in the range of G, getting thereby a contradiction. For a fixed $x \in u$ we aim to set the value $h(x)$ in such a way as to assure that $h \notin G[\{x\} \times \mathfrak{d}_x]$. Doing this for every $x \in u$ we make sure that $h \notin G[p]$, since $p=\bigcup_{x \in u}\{x\} \times \mathfrak{d}_x$. We make sure that $h \notin G[\{x\} \times \mathfrak{d}_x]$ by taking $h(x)$ to be different from all the values $f(x)$ for $f \in G[\{x\} \times \mathfrak{d}_x]$. $h(x)$ can be taken to be such if we prove that $\{f(x) \mid f \in G[\{x\} \times \mathfrak{d}_x]\}$ is of cardinality $\leqslant \mathfrak{d}_x < \mathfrak{e}_x$ and hence a proper subset of $\mathfrak{e}_x$, because then we define $h(x)$ to be the least member of the set $\mathfrak{e}_x \sim \{f(x) \mid f \in G[\{x\} \times \mathfrak{d}_x]\}$. To see that $|\{f(x) \mid f \in G[\{x\} \times \mathfrak{d}_x]\}| \leqslant \mathfrak{d}_x$ we define a function j on $\mathfrak{d}_x$ by $j(\alpha)=G(\langle x, \alpha \rangle)(x)$. We have, by 2.37,

$$\mathfrak{d}_x=|\mathfrak{d}_x|=|\mathrm{Dom}(j)| \geqslant |\mathrm{Rng}(j)|=|\{f(x) | f \in G[\{x\} \times \mathfrak{d}_x]\}|,$$

since $\mathrm{Rng}(j)=\{f(x) \mid f \in G[\{x\} \times \mathfrak{d}_x]\}$. □

For well ordered cardinals $\mathfrak{a}$ the Zermelo–König inequality is a generalization of Cantor's theorem (3.33). To obtain Cantor's theorem for a well ordered cardinal $\mathfrak{a}$ from 4.18 compare the indexed families $\langle 1 \mid x \in \mathfrak{a} \rangle$ and $\langle 2 \mid x \in \mathfrak{a} \rangle$. Since $1<2$ we get by 4.18 $\Sigma_{x \in \mathfrak{a}} 1 < \Pi_{x \in \mathfrak{a}} 2$ i.e., $\mathfrak{a} < 2^{\mathfrak{a}}$ (by 4.7 and 4.15).

4.19 Exercise (Jourdain 1908a). Prove that if λ is a limit ordinal, $\langle \mathfrak{e}_\xi \mid \xi < \lambda \rangle$ is an increasing sequence of cardinals and $\mathfrak{e}_0 \neq 0$ then $\Sigma_{\xi<\lambda}\mathfrak{e}_\xi < \Pi_{\xi<\lambda}\mathfrak{e}_\xi$. $\square$

We conclude this section by computing the cardinals of a few sets.

4.20 Definition. (i) For a class A and an ordinal α, ${}^{\underline{\alpha}}A$ is the set of all functions on ordinals $\beta < \alpha$ into A, i.e., ${}^{\underline{\alpha}}A = \{f \mid (\exists \beta < \alpha) f : \beta \to A\}$ or, informally, ${}^{\underline{\alpha}}A = \bigcup_{\beta<\alpha} {}^{\beta}A$. In particular, ${}^{\underline{\omega}}A$ is the class of all finite sequences of members of A. If A is a set then ${}^{\underline{\alpha}}A$ is a set too.

(ii) For a class A, $[A]^{<\omega}$ is the class of all finite subsets of A.

4.21 Proposition.

(i) $|{}^{\underline{\omega}}\aleph_\alpha| = \aleph_\alpha$.

(ii) $|[\aleph_\alpha]^{<\omega}| = \aleph_\alpha$.

(iii) $|\{f \mid f$ *is a function* $\wedge \operatorname{Dom}(f)$ *is a finite subset of* $\aleph_\alpha \wedge \operatorname{Rng}(f) \subseteq \aleph_\alpha\}| = \aleph_\alpha$.

Proof. (i) If we assume the axiom of choice we apply 4.3 and get, by 3.31, $|{}^{\underline{\omega}}\aleph_\alpha| = |\bigcup_{n<\omega} {}^n\aleph_\alpha| = \Sigma_{n<\omega}|{}^n\aleph_\alpha| = \Sigma_{n<\omega}\aleph_\alpha^n = \aleph_\alpha^0 + \Sigma_{0<n<\omega}\aleph_\alpha^n = 1 + \Sigma_{0<n<\omega}\aleph_\alpha = 1 + \aleph_0 \cdot \aleph_\alpha = 1 + \aleph_\alpha = \aleph_\alpha$. (In taking out the summand $\aleph_\alpha^0$ from the sum $\Sigma_{n<\omega}\aleph_\alpha^n$ we used a trivial instance of the associative law which we don't bother to formulate and prove.) In the absence of the axiom of choice we have to show that condition (c) of 4.3 is satisfied in our application of 4.3, i.e., we have to show that the ${}^n\aleph_\alpha$'s, for $n > 0$, are bijected on $\aleph_\alpha$ uniformly. For this purpose we shall define, by recursion on n, a sequence $\langle p_n \mid 0 < n < \omega \rangle$ such that p_n is a bijection of ${}^n\aleph_\alpha$ on $\aleph_\alpha$. Let p be any bijection of $\aleph_\alpha \times \aleph_\alpha$ on $\aleph_\alpha$. We define p_1 to be the identity function on $\aleph_\alpha$ (we confuse ${}^1\aleph_\alpha$ with $\aleph_\alpha$), and p_{n+1} is the function on ${}^{n+1}\aleph_\alpha$ such that for $s \in {}^{n+1}\aleph_\alpha$ $p_{n+1}(s) = p(p_n(s \upharpoonright n), s(n))$. If we view p_n as a method of coding n-tuples of ordinals $< \aleph_\alpha$ by single ordinals $< \aleph_\alpha$ then p_{n+1} codes an $n+1$-tuple s by the p-code of the pair of ordinals whose first component is the p_n-code of the n-tuple $s \upharpoonright n$ of the first n terms of s and whose second component is the last term of s. It is easily shown, by induction on n, that p_n is indeed a bijection of ${}^n\aleph_\alpha$ on $\aleph_\alpha$.

(ii) Obviously $[\aleph_\alpha]^{<\omega} \supseteq \{\{\beta\} \mid \beta < \aleph_\alpha\}$ and hence $|[\aleph_\alpha]^{<\omega}| \geqslant \aleph_\alpha$. By the Cantor–Bernstein theorem $|[\aleph_\alpha]^{<\omega}| = \aleph_\alpha$ will follow once we prove $|[\aleph_\alpha]^{<\omega}| \leqslant \aleph_\alpha$. We shall prove this inequality by constructing an injection f of $[\aleph_\alpha]^{<\omega}$ into ${}^{\underline{\omega}}\aleph_\alpha$, which is of cardinality $\aleph_\alpha$ by (i). For $u \in [\aleph_\alpha]^{<\omega}$ we set $f(u)$ to be the n-tuple of all members of u in their natural order, where $n = |u|$ (i.e., $f(u)$ is the unique isomorphism of $\langle n, \in \rangle$ on $\langle u, \in \rangle$). f is obviously an injection as claimed.

(iii) Let $a = \{f \mid f$ is a function $\wedge \operatorname{Dom}(f)$ is a finite subset of $\aleph_\alpha \wedge \operatorname{Rng}(f) \subseteq \aleph_\alpha\}$. We define first a function g on $[\aleph_\alpha]^{<\omega}$ as follows: For $u \in [\aleph_\alpha]^{<\omega}$ $g(u) = u \times \{0\}$. g is obviously an injection of $[\aleph_\alpha]^{<\omega}$ into a, hence by (ii) $|a| \geqslant \aleph_\alpha$. On the other hand, $a \subseteq [\aleph_\alpha \times \aleph_\alpha]^{<\omega}$; since $\aleph_\alpha \times \aleph_\alpha \approx \aleph_\alpha$ we have $[\aleph_\alpha \times \aleph_\alpha]^{<\omega} \approx [\aleph_\alpha]^{<\omega}$ and, therefore, by (ii), $|a| \leqslant |[\aleph_\alpha \times \aleph_\alpha]^{<\omega}| = |[\aleph_\alpha]^{<\omega}| = \aleph_\alpha$, and hence $|a| = \aleph_\alpha$. $\square$

4.22 Proposition. $|\{f|f$ *is a function* $\wedge$ Dom(f) *is a finite subset of* $\aleph_\alpha \wedge$ Rng$(f)\subseteq \aleph_\beta\}|=\aleph_{\max(\alpha,\beta)}$.

Hint of proof. Use 4.21(iii). □

The way we formally represented the functions in set theory gives rise to some confusion. For us a function of n variables is a function on ${}^nV$ with $F(x_1,\ldots,x_n)$ standing for $F(\langle x_1,\ldots,x_n\rangle)$. Notice that this makes perfect sense also if $n=0$; then F is a function on ${}^0V=\{0\}$ and $F(\langle x_1,\ldots,x_n\rangle)=F(\langle\rangle)$. Just by looking at a function F we cannot tell whether F is intended to be a (one variable) function whose domain happens to consist of n-tuples or a function of n-variables. This makes a difference in expressions like "the class A is closed under F". If F is regarded as a one variable function then this means $\forall x(x\in A\rightarrow F(x)\in A)$ and if F is regarded as a function of n variables then this means

(3) $$\forall x_1\ldots\forall x_n(x_1,\ldots,x_n\in A\rightarrow F(x_1,\ldots,x_n)\in A)$$

(in the former case we needed $\langle x_1,\ldots,x_n\rangle\in A$ to get $F(x_1,\ldots,x_n)\in A$). Notice that even if $n=1$ the two meanings are not the same. Therefore we shall be careful in the definition to make clear what we mean by A being closed under F. However, throughout the rest of the book we shall make no effort to point out the right interpretation since it is always clear what the right interpretation is.

4.23 Definition. (i) The class A is said to be *closed under* F as a function of n variables ($n<\omega$) if for every finite sequence $\langle s_1,\ldots,s_n\rangle$ if $s_1,\ldots,s_n\in A$ then also $F(s_1,\ldots,s_n)\in A$. For $n=0$ this means just that $F(\,)\in A$ (where $F(\,)=F(\langle\,\rangle)=F(0)$).

(ii) The set b is said to be the *closure* under $F_1,\ldots,F_k$ of the set a, where $F_1,\ldots,F_k$ are functions of $n_1,\ldots,n_k$ variables, respectively, if b is the least set which includes a and is closed under $F_1,\ldots,F_k$.

4.24 Theorem. *Let* $F_1,\ldots,F_k$ *be functions of* $n_1,\ldots,n_k$ *variables, respectively. Every set a has a closure b with respect to* $F_1,\ldots,F_k$. *If* $|a|=\aleph_\alpha$ *then* $|b|=\aleph_\alpha$; *if a is finite then* $|b|\leqslant\aleph_0$.

Proof. We can assume, in order to simplify the proof, that F_1 is the identity function (i.e., $n_1=1$ and $F_1(\langle x\rangle)=x$), since otherwise we can add the identity function to the list $F_1,\ldots,F_k$ as the first member. We define now a function G on ω by recursion as follows.

(4) $$G(0)=a\quad\text{and}\quad G(m+1)=\textstyle\bigcup_{1\leqslant i\leqslant k}F_i[{}^{n_i}G(m)].$$

It follows easily from the axiom of replacement (and the other axioms) that the right-hand side of (4) is indeed a set (which we have to know in order to apply the theorem on definition by recursion). Since F_1 is the identity function we have $G(m+1)\supseteq G(m)$, and hence if $m<l$ then $G(m)\subseteq G(l)$. Let $b=\bigcup_{m<\omega}G(m)$, then, since $G(0)=a$, $b\supseteq a$. We shall now prove that b is closed under F_i, $1\leqslant i\leqslant k$.

Let $x_1,\ldots,x_{n_i}\in b$. Since $b=\bigcup_{m<\omega}G(m)$ let m_j, for each $1\leqslant j\leqslant n_i$, be the least number m such that $x_j\in G(m)$. Let l be the maximal member of the finite set $\{m_1,\ldots,m_{n_i}\}$ then, by the monotonicity of G, $x_1,\ldots,x_{n_i}\in G(l)$. By the definition of $G(l+1)$ we have $F_i(x_1,\ldots,x_{n_i})\in G(l+1)\subseteq b$.

To show that b is the closure of a with respect to $F_1,\ldots,F_k$ we prove that every class S which includes a and which is closed under $F_1,\ldots,F_k$ includes also b. It is very easy to prove by induction on m that $G(m)\subseteq S$, and as a consequence we get $b=\bigcup_{m<\omega}G(m)\subseteq S$.

We still have to compute the cardinality of b. If $|a|=\aleph_\alpha$ then, since $b\supseteq a$, we have $|b|\geqslant\aleph_\alpha$ and it is sufficient to prove $|b|\leqslant\aleph_\alpha$. If a is finite then all we want to prove is $|b|\leqslant\aleph_0$. Thus in either case we assume $|a|\leqslant\aleph_\alpha$ and we have to prove $|b|\leqslant\aleph_\alpha$, where $\alpha=0$ if a is finite. Let us assume for a while the axiom of choice. We prove by induction that $|G(m)|\leqslant\aleph_\alpha$. $|G(0)|=|a|\leqslant\aleph_\alpha$. Assume, as an induction hypothesis, that $|G(m)|\leqslant\aleph_\alpha$. We have

(5) $\quad |G(m+1)|\leqslant\Sigma_{1\leqslant i\leqslant k}|F_i[{}^{n_i}G(m)]|$, by 4.5;

(6) $\quad \Sigma_{1\leqslant i\leqslant k}|F_i[{}^{n_i}G(m)]|\leqslant\Sigma_{1\leqslant i\leqslant k}|{}^{n_i}G(m)|$, by 2.37 and 4.6(v);

(7) $\quad \Sigma_{1\leqslant i\leqslant k}|{}^{n_i}G(m)|=\Sigma_{1\leqslant i\leqslant k}|G(m)|^{n_i}\leqslant\Sigma_{1\leqslant i\leqslant k}\aleph_\alpha^{n_i}\leqslant\Sigma_{1\leqslant i\leqslant k}\aleph_\alpha=k\cdot\aleph_\alpha=\aleph_\alpha$, by 3.29(ix) and 4.6(v).

Combining (5)–(7) we get $|G(m+1)|\leqslant\aleph_\alpha$. $b=\bigcup_{m<\omega}G(m)$ hence, by 4.5, $|b|\leqslant\Sigma_{m<\omega}|G(m)|\leqslant\Sigma_{m<\omega}\aleph_\alpha=\aleph_\alpha\cdot\aleph_0,=\aleph_\alpha$, which is what we had to show.

When we computed the upper estimate of $|b|$ we used the axiom of choice where we applied 4.5 to $b=\bigcup_{m<\omega}G(m)$ without verifying that condition (c) of 4.5 holds, i.e., that the sets $G(m)$ can be injected into $\aleph_\alpha$ uniformly. We used 4.5 also in proving (5), but there we do not need the axiom of choice since the set $\{1,\ldots,k\}$ of indices is finite—see V.1.4. To prove $|b|\leqslant\aleph_\alpha$ without assuming the axiom of choice we shall show that the sets $G(m)$ are injected uniformly into $\aleph_\alpha$ by means of a function H on ω such that, for $m<\omega$, $H(m)$ is an injection of $G(m)$ into $\aleph_\alpha$. Let q be an injection of the set ${}^{\underline{\omega}}\aleph_\alpha$ of all finite sequences members of $\aleph_\alpha$ into $\aleph_\alpha$; such an injection exists by 4.21(i). We define H by recursion. For $H(0)$ we take any injection of $G(0)=a$ into $\aleph_\alpha$. By the induction hypothesis, $H(m)$ is an injection of $G(m)$ into $\aleph_\alpha$. (Don't let this combination of the definition of H by recursion and the proof by induction that $H(m)$ is an injection of $G(m)$ into $\aleph_\alpha$ scare you! What really goes on is that we just define H by recursion, setting $H(m+1)=0$ if $H(m)$ is not an injection of $G(m)$ into $\aleph_\alpha$; then we prove by induction on m that $H(m)$ is an injection of $G(m)$ into $\aleph_\alpha$. From now on we shall use this informal way of simultaneously defining by recursion and proving by induction without drawing the reader's attention to it.) Let

$$w=\bigcup_{1\leqslant i\leqslant k}\{\langle F_i(x_1,\ldots,x_{n_i}),q(\langle i,H(m)(x_1),\ldots,H(m)(x_{n_i})\rangle)\rangle\mid x_1,\ldots x_{n_i}\in G(m)\}.$$

By definition of $G(m+1)$, $\mathrm{Dom}(w)=G(m+1)$; since $\mathrm{Rng}(q)\subseteq\aleph_\alpha$ we have $\mathrm{Rng}(w)\subseteq\aleph_\alpha$; and since $H(m)$ and q are injections w is a one–many relation (i.e., $\langle y,x\rangle\in$

$w \wedge \langle z, x\rangle \in w \rightarrow y=z)$. We set $H(m+1)$ to be the one–one function which is a subset of w and which is obtained from w by keeping for every $y \in \mathrm{Dom}(w)$ only that pair $\langle y, \alpha\rangle$ where α is minimal, i.e., $H(m+1)=\{\langle y, \alpha\rangle \mid \langle y, \alpha\rangle \in w \wedge \forall\delta\,(\delta<\alpha \rightarrow \langle y, \delta\rangle \notin w)\}$. Obviously $\mathrm{Dom}(H(m+1))=\mathrm{Dom}(w)=G(m+1)$, $\mathrm{Rng}(H(m+1)) \subseteq \mathrm{Rng}(w) \subseteq \aleph_\alpha$ and $H(m+1)$ is a one–one function (since w is a one–many relation). □

4.25 Exercise. For a class F denote with F_x the class $F[\{x\}]=\{y \mid \langle x, y\rangle \in F\}$. Let F and N be classes such that for every $\gamma<\aleph_\alpha$ F_γ is a function of $N(\gamma)<\omega$ variables. For every set a such that $|a| \leqslant \aleph_\alpha$ there is a set b of cardinality $\leqslant \aleph_\alpha$ which is the closure of a under the functions F_γ, $\gamma<\aleph_\alpha$, i.e., b is the least set which includes a and which is closed under each F_γ, $\gamma<\aleph_\alpha$. □

4.26 On Closure Under n+1-ary Relations. 4.24 and 4.25 are strengthenings of II.4.29 where we prove the existence, for a given set a, of a set b including it and closed under the functions $F_1, \ldots, F_k$. The present theorems are more general since we consider here also functions of any finite number of variables rather than just of one variable. Also the conclusion here is stronger since we know here also the cardinality of b. It is easy to convert also II.4.33 and II.7.11 which are the generalization of II.4.29 to narrow relations and to general binary relations, to $n+1$-ary relations which are the natural analogues of the functions of n-variables. In order to obtain good results about the cardinality of b in those cases we must assume the axiom of choice, and, if we deal with narrow relations we need also an estimate of how narrow the relations are (for example, we may have for every x $|R[\{x\}]| \leqslant \aleph_\alpha$). To obtain a set b of cardinality $\aleph_\alpha$ which is *nearly* closed under finitely many or even $\leqslant \aleph_\alpha$ many relations which are not necessarily narrow we need no additional assumptions (other than the axiom of choice).

Chapter IV
The Ordinals

In Chapter II we developed the concept of the ordinals to an extent sufficient to define the rank function ρ and the related function $\langle R(\alpha) \mid \alpha \in \mathrm{On} \rangle$ on one hand, and to define the alephs and establish their basic arithmetic on the other hand. The present chapter is devoted to a wider and deeper study of the ordinals.

The first two sections deal with the operations of addition, multiplication and exponentiation of ordinals. These operations were defined and thoroughly studied by Cantor (1883, 1895 and 1897) who proved most of the theorems in these sections. We saw in Chapter III that as far as binary (and finitary) addition and multiplication are concerned the cardinal arithmetic of the alephs is completely trivial, and even the infinitary addition of the alephs is very simple (III.4.4). This is quite different for ordinal addition and multiplication, which are interesting, and by no means trivial, operations. However, the study of the arithmetic of the ordinals does not occupy an important position in set theoretical research and the attraction of this subject in mathematics never came close to that of number theory. Still, basic ordinal arithmetic is very attractive and it is sometimes a useful tool in set theory and in related areas.

In § 3 we start studying the regularity and cofinality properties of the ordinals. These are deeper properties of the ordinals, and they form the basis for a theory of infinitary cardinal arithmetic and cardinal exponentiation. In this section we encounter and study the inaccessible cardinals. This is the first step (to be continued in Chapter IX) of our rather limited foray into the highly interesting and important area of large cardinals. In § 4 we go into the theory of closed unbounded classes and stationary classes. We also apply a part of this theory in order to get more information about the size of the inaccessible cardinals.

While the beginning of this chapter consists mostly of results which are among the earliest ones in set theory, the last theorem was proved only a few years before this book was written.

1. *Ordinal Addition and Multiplication*

1.1 Definition (Order types—Cantor 1895, Scott 1955). Ord will denote a function defined on the class of all ordered sets such that

(a) $\mathrm{Ord}(\langle a, r\rangle) = \mathrm{Ord}(\langle b, s\rangle) \leftrightarrow \langle a, r\rangle \cong \langle b, s\rangle$, and

(b) if $\langle a, r\rangle$ is a well ordered set then $\mathrm{Ord}\,(\langle a, r\rangle)$ is the ordinal of $\langle a, r\rangle$.

The existence of such a function Ord follows from the theorem on the existence of equivalence types (II.7.13—using a modification as in the definition, III.2.2, of the function $|x|$ of cardinality, and the analogues of III.2.3 and III.2.4). The members of the range of Ord will be called *order types*.

1.2 Remark. Because of the clause (b) in the definition of Ord our present definition does not clash with our earlier definition (II.3.24) of Ord as a function on the class of all *well* ordered sets with Ord ($\langle a, r\rangle$) being the ordinal of $\langle a, r\rangle$. The class of all order types includes the class On of ordinals. By III.1.8 and III.1.19 it is exactly the natural numbers which are order types of finite sets.

1.3 Definition (Cardinality of order types). Card is the function on the class of all order types defined by $\mathrm{Card}(\sigma)=|a|$; where $\langle a, r\rangle$ is some ordered set such that $\mathrm{Ord}(\langle a, r\rangle)=\sigma$.

Since ordered sets of the same order type are isomorphic and hence equinumerous, $\mathrm{Card}(\sigma)$ is well defined. Note that for every ordinal α, $\mathrm{Card}(\alpha)=|\alpha|$.

The function Card is different from the cardinality function $|x|$ (i.e., $\langle |x| \mid x$ is a set$\rangle$); as a matter of fact, for Ord as defined in 1.1 we have that for every order type σ which is not an ordinal $|\sigma|>\mathrm{Card}(\sigma)$ (compare III.2.24). Whenever σ occurs in a discussion as an order type we are not interested at all in what kind of a set σ itself is and therefore we shall never care to look at $|\sigma|$; on the other hand $\mathrm{Card}(\sigma)$ is relevant, being the cardinal of the ordered sets of order type σ. Thus there is no danger of confusion if we shall also refer to $\mathrm{Card}(\sigma)$ as the cardinal of σ.

1.4 Exercises. (i) If $\langle a, r\rangle$ is an ordered set, then such is also $\langle a, r^{-1}\rangle$.

(ii) (Cantor 1895). If $\langle a, r\rangle$, $\langle b, s\rangle$ are structures such that $\langle a, r\rangle\cong\langle b, s\rangle$ then also $\langle a, r^{-1}\rangle\cong\langle b, s^{-1}\rangle$. Therefore, one can define a function * on the order types such that for every order type σ, $\sigma^*=\mathrm{Ord}(\langle a, r^{-1}\rangle)$, where $\langle a, r\rangle$ is an ordered set such that $\mathrm{Ord}(\langle a, r\rangle)=\sigma$. Obviously, $\mathrm{Card}(\sigma^*)=\mathrm{Card}(\sigma)$, and $\sigma^{**}=\sigma$.

(iii) For every *ordinal* α, α^* is an ordinal iff α is a finite ordinal. For every finite ordinal n, $n^*=n$.

Hint. (iii) Use III.1.31, III.1.19 and III.1.21. □

1.5 Lemma. *Let $\langle A, R\rangle$, $\langle A', R'\rangle$, $\langle B, S\rangle$, $\langle B', S'\rangle$ be ordered classes such that $\langle A, R\rangle\cong\langle A', R'\rangle$, $\langle B, S\rangle\cong\langle B', S'\rangle$,* $A\cap B=0$ and $A'\cap B'=0$, *then the ordered union $\langle A, R\rangle\oplus\langle B, S\rangle$ is isomorphic to the ordered union $\langle A', R'\rangle\oplus\langle B', S'\rangle$ (see* II.1.14 *for the definition of ordered union*). □

1.6 Definition. (Addition of order types—Cantor 1883, 1895.) Let σ and τ be order types. $\sigma+\tau$ is the unique order type λ such that if $\langle a, r\rangle$ and $\langle b, s\rangle$ are disjoint ordered sets of the respective order types σ and τ then λ is the order type of $\langle a, r\rangle\oplus\langle b, s\rangle$. It is easy to show that for all σ and τ there are indeed such disjoint ordered sets $\langle a, r\rangle$ and $\langle b, s\rangle$ (see III.3.2); the uniqueness of λ follows from 1.5.

Remark. $\sigma\dot{+}\tau$ is to be read "σ plus τ". The dot on top of the addition sign is there in order to distinguish our present addition from cardinal addition (III.3.2). We may drop the dot whenever it is clear from the context that we deal with order type addition; in particular, we shall never use this dot in subscripts.

1.7 Proposition (Cantor 1883, 1895). (i) (Associativity of addition). *For all order types σ, τ, λ, $\sigma\dot{+}(\tau\dot{+}\lambda)=(\sigma\dot{+}\tau)\dot{+}\lambda$.*

(ii) *Order type addition is not commutative.*

(iii) $\sigma\dot{+}0=0\dot{+}\sigma=\sigma$.

Hint of proof. (ii) $1\dot{+}\omega\neq\omega\dot{+}1$, since $1\dot{+}\omega$ is the order type of an ordered set with no last member while $\omega\dot{+}1$ is the order type of a set with a last member. □

1.8 Exercise. For all order types σ and τ, $(\sigma\dot{+}\tau)^*=\tau^*\dot{+}\sigma^*$. □

1.9 Proposition. *For every ordinal α, $\alpha\dot{+}1$ is the successor $\alpha\cup\{\alpha\}$ of α.*

Proof. $\langle\alpha\cup\{\alpha\},<\rangle$ is the ordered union of $\langle\alpha,<\rangle$ and $\langle\{\alpha\},<\rangle$ which are of the respective order types α and 1. □

1.10 The Various Operations of Addition. In the earlier chapters we have already defined two operations of addition. One is the addition of natural numbers defined in II.4.5; the second is the addition of cardinals defined in III.3.2. We have proved in III.3.5 that the addition of cardinals is an extension of the addition of natural numbers. Now we have defined the addition of order types. Since the intersection of the class of all cardinal numbers and the class of all order types is the class $On\cap Cn$ of the well ordered cardinals (III.2.5) it is natural to ask whether the operations of cardinal addition and of order type addition coincide on $On\cap Cn$. First we note that the two operations coincide on the set ω of the finite ordinals. This is a direct consequence of the fact that the operation of order type addition satisfies on the finite ordinals the same recursive equations $m\dot{+}0=m$ and $m\dot{+}(n\cup\{n\})=(m\dot{+}n)\cup\{m\dot{+}n\}$ satisfied by cardinal addition (III.3.5), by 1.7 and 1.9. (Another proof uses 1.11 below together with III.2.17 and III.2.18.) The two operations do not coincide on the alephs, which are the other members of $On\cap Cn$. If γ is an aleph and β a nonzero well ordered cardinal such that $\beta\leqslant\gamma$ then $\gamma\dot{+}\beta>\gamma$, by 1.13 below, while $\gamma+\beta=\gamma$, by III.3.8 and III.3.14. In particular, $\aleph_0+1=\aleph_0$ while $\omega\dot{+}1>\omega$. We shall write also ω_α for $\aleph_\alpha$, mostly when we shall want to perform order-type operations on $\aleph_\alpha$. Thus, e.g., we shall usually write $\aleph_\alpha+1=\aleph_\alpha$ for cardinal addition, but $\omega_\alpha\dot{+}1>\omega_\alpha$ for order-type addition.

1.11 Proposition. *For all order types σ, τ,* $\text{Card}(\sigma\dot{+}\tau)=\text{Card}(\sigma)+\text{Card}(\tau)$. □

Even though we defined addition of general order types we are mainly interested in the ordinals, and the following theorems deal only with them. As we shall see, the properties of ordinal addition and multiplication are very asymmetric with respect to the two sides of the operations $\dot{+}$ and $\bullet$. This asymmetry cannot be traced in the case of ordinal addition to any asymmetry in the operation itself.

Addition of order types is symmetric, as seen by the rule $(\sigma \dotplus \tau)^* = \tau^* \dotplus \sigma^*$ (1.8). The asymmetry of the properties is due to the asymmetry in the definition of well-ordering, since every subset is required to have a least member, but is not required to have a maximal member.

1.12 Proposition. *The sum of two ordinals is an ordinal.*

Hint of proof. Use II.1.14. □

1.13 Proposition. (Strict monotonicity of addition). *For all ordinals* $\alpha, \beta, \gamma, \beta < \gamma \to \alpha \dotplus \beta < \alpha \dotplus \gamma$. *In particular if* $\gamma > 0$ *then* $\alpha < \alpha \dotplus \gamma$.

Proof. Let $\langle a, <_a \rangle$ and $\langle c, <_c \rangle$ be disjoint well ordered sets of the order types α and γ respectively. Since $\beta < \gamma$ there is by II.3.26 a section $\langle c_u, <_c \rangle$ of $\langle c, <_c \rangle$ determined by a member u of c such that $\mathrm{Ord}(\langle c_u, <_c \rangle) = \beta$. $\mathrm{Ord}(\langle a, <_a \rangle \oplus \langle c, <_c \rangle) = \alpha \dotplus \gamma$. $\langle a, <_a \rangle \oplus \langle c_u, <_c \rangle$ is the section of $\langle a, <_a \rangle \oplus \langle c, <_c \rangle$ determined by u and its order type is $\alpha \dotplus \beta$. Hence, by II.3.26, $\alpha \dotplus \beta < \alpha \dotplus \gamma$. □

1.14 Corollary. (Cancellation of the left summand.)

$$\alpha \dotplus \beta = \alpha \dotplus \gamma \to \beta = \gamma, \qquad \alpha \dotplus \beta < \alpha \dotplus \gamma \to \beta < \gamma. \quad \square$$

1.15 Proposition. *If* $\alpha < \gamma$ *then there is a unique ordinal* $\beta > 0$ *such that* $\alpha \dotplus \beta = \gamma$.

Proof. If $\alpha < \gamma$ then α is an initial section of γ and $\alpha \subset \gamma$. Therefore $\langle \gamma, < \rangle$ is the ordered union of $\langle \alpha, < \rangle$ and $\langle \gamma \sim \alpha, < \rangle$. Since $\gamma \sim \alpha \subseteq \gamma$, $\langle \gamma \sim \alpha, < \rangle$ is a well ordered set and $\mathrm{Ord}\,(\langle \gamma \sim \alpha, < \rangle)$ is an ordinal; let us denote it with β. Since $\gamma \sim \alpha \neq 0$, $\beta > 0$. We have $\gamma = \mathrm{Ord}(\langle \gamma, < \rangle) = \mathrm{Ord}(\langle \alpha, < \rangle \oplus \langle \gamma \sim \alpha, < \rangle) = \mathrm{Ord}(\langle \alpha, < \rangle) \dotplus \mathrm{Ord}(\langle \gamma \sim \alpha, < \rangle) = \alpha \dotplus \beta$. The uniqueness of β follows from 1.14. □

1.16 Corollary. $\alpha < \gamma$ *iff there is an ordinal* $\beta > 0$ *such that* $\alpha \dotplus \beta = \gamma$.
$\alpha \leqslant \gamma$ *iff there is an ordinal* $\beta \geqslant 0$ *such that* $\alpha \dotplus \beta = \gamma$. □

1.17 Exercise. Show that it suffices to prove 1.13 for $\beta = 0$ only and then to derive the full 1.13 from 1.15. □

1.18 Proposition (Cantor 1897). *For all ordinals* α, β

$$\beta > 0 \to \alpha \dotplus \beta = \sup{}^{+} \{\alpha \dotplus \gamma \mid \gamma < \beta\},$$

where $\sup^+ a$ *is the least strict upper bound of* a.

Proof. If $\gamma < \beta$ then by 1.13 $\alpha \dotplus \beta > \alpha \dotplus \gamma$, hence $\alpha \dotplus \beta$ is a strict upper bound of $\{\alpha \dotplus \gamma \mid \gamma < \beta\}$. Let ξ be a strict upper bound of $\{\alpha \dotplus \gamma \mid \gamma < \beta\}$. Since $\beta > 0$, $\alpha = \alpha \dotplus 0 \in \{\alpha \dotplus \gamma \mid \gamma < \beta\}$, hence $\xi > \alpha$. By 1.15 there is a δ such that $\xi = \alpha \dotplus \delta$. We have to show that $\delta \geqslant \beta$, since then, by 1.13 $\xi = \alpha \dotplus \delta \geqslant \alpha \dotplus \beta$ and $\alpha \dotplus \beta$ is the *least* strict upper bound of $\{\alpha+\gamma \mid \gamma<\beta\}$. Suppose $\delta<\beta$, then $\xi=\alpha+\delta\in\{\alpha+\gamma \mid \gamma<\beta\}$, but this is a contradiction since ξ is a strict upper bound of the set $\{\alpha+\gamma \mid \gamma<\beta\}$. □

1.19 Recursive Definition of Ordinal Addition (Jacobsthal 1909). $\alpha+0=\alpha$ together with 1.18 are often taken to be a definition by recursion of ordinal addition. If we want to distinguish between the cases of 1.18 where β is a successor cardinal and where β is a limit ordinal we write:

$$\alpha+(\xi+1)=(\alpha+\xi)+1, \quad \text{and for limit numbers } \eta \quad \alpha+\eta=\sup\{\alpha+\xi \mid \xi<\eta\}.$$

Notice that we wrote here sup instead of $\sup^+$; this replacement does not effect the right-hand side since for a limit number η the set $\{\alpha+\xi \mid \xi<\eta\}$ has no greatest member (by II.3.32 and 1.13) and thus, by II.3.31, the sup of it equals the $\sup^+$ of it.

1.20 Proposition. (Weak monotonicity of addition.) *For all ordinals* α, β, γ, $\beta\leqslant\gamma\rightarrow\beta+\alpha\leqslant\gamma+\alpha$; *hence always* $\alpha\leqslant\gamma+\alpha$. *As a consequence* $\gamma+\alpha<\beta+\alpha\rightarrow\gamma<\beta$.

Hint of proof. Use II.2.20. □

1.21 Proposition. *For every finite* n, $n+\omega=\omega$.

Proof. By 1.18 (and 1.10) $n+\omega=\sup^+\{n+m \mid m<\omega\}=\omega$. □

1.22 Exercise. $1+\alpha=\alpha$ iff $\alpha\geqslant\omega$. □

1.23 Asymmetry of Ordinal Addition. We have already mentioned that the properties of ordinal addition are very asymmetric with respect to the two sides of $+$; we shall now go through Propositions 1.13–1.18 checking whether the symmetric analog of each theorem holds.

(i) 1.13. $\alpha+\beta$ is not strictly increasing in α as seen from $0+\omega=\omega=1+\omega$ (by 1.21); this shows that we do not have even $\gamma>0\rightarrow\alpha<\gamma+\alpha$.

(ii) 1.14. Since $\alpha+\beta$ is weakly but not strictly increasing in α we cannot cancel right summands in equalities; in particular $0+\omega=1+\omega$.

(iii) 1.15. The statement that for every $\alpha<\gamma$ the equation $\xi+\alpha=\gamma$ has a solution for ξ is true only for finite γ's. We shall see in 2.20 that for every infinite ordinal γ there are only finitely many α's for which that equation has solutions. For example, we shall now see that $\xi+\alpha=\omega$ has solutions ξ only if $\alpha=\omega$ or $\alpha=0$. If $\xi+\alpha=\omega$ is to have a solution we must have, by 1.20, $\alpha\leqslant\omega$. Thus any such α, other than 0 or ω, is a non-zero finite ordinal and is hence equal to $n+1$ for some finite ordinal n, and $\xi+\alpha=\xi+(n+1)=(\xi+n)+1$. But $\xi+\alpha$, being the successor of the ordinal $\xi+n$ cannot be equal to ω which is a limit ordinal.

(iv) 1.18. Separating the cases where β is a successor ordinal and where β is a limit ordinal, as in 1.19, we notice the following. We do not have always $(\xi+1)+\alpha=(\xi+\alpha)+1$, since $(0+1)+\omega=1+\omega=\omega$, but $(0+\omega)+1=\omega+1>\omega$. We also do not always have for limit numbers η $\eta+\alpha=\sup\{\xi+\alpha \mid \xi<\eta\}$ since $\sup\{n+\omega \mid n<\omega\}=\sup\{\omega\}=\omega$ (by 1.21) but $\omega+\omega>\omega$ (by 1.13).

1.24 Definition (Veblen 1908). A function F is said to be a *normal function* if it is a function on On into On which is strictly monotonic and *continuous*, i.e.,

(a) $\alpha<\beta \rightarrow F(\alpha)<F(\beta)$, and
(b) for every limit ordinal λ $F(\lambda)=\sup\{F(\alpha) \mid \alpha<\lambda\}=\sup^{+}\{F(\alpha) \mid \alpha<\lambda\}$

(by (a) the set $\{F(\alpha) \mid \alpha<\lambda\}$ has no maximal member hence its sup and sup$^+$ are equal). Since (a) already implies that $F(\lambda) \geqslant \sup\{F(\alpha) \mid \alpha<\lambda\}=\sup^{+}\{F(\alpha) \mid \alpha<\lambda\}$ it is sufficient to require in (b) that $F(\lambda) \leqslant \sup\{F(\alpha) \mid \alpha<\lambda\}=\sup^{+}\{F(\alpha) \mid \alpha<\lambda\}$.

1.25 Examples of Normal Functions. One comes often across normal functions when one deals with ordinal numbers. One example of a normal function is ordinal addition regarded as a function of the right-hand summand only (with the left-hand one being fixed); this follows from 1.13 and 1.19. A second example is the following.

1.26 Proposition. *The function* $\aleph$ *is normal.*

Proof. $\aleph$ is strictly monotonic by III.2.33(iii), so we have only to prove the continuity of $\aleph$. Let λ be a limit ordinal; we want to prove $\aleph_\lambda=\sup\{\aleph_\alpha \mid \alpha<\lambda\}$. Denote $\sup\{\aleph_\alpha \mid \alpha<\lambda\}$ by γ. If $|\gamma|<\gamma$ then for some $\beta<\lambda$ $|\gamma|<\aleph_\beta$ hence $\gamma<\aleph_\beta$ (by III.2.22(ii)) and thus $\gamma<\aleph_\beta \leqslant \sup\{\aleph_\alpha \mid \alpha<\lambda\}=\gamma$, which is a contradiction. Therefore $|\gamma|=\gamma$ and γ is a cardinal. Since $\gamma=\sup\{\aleph_\alpha \mid \alpha<\lambda\}=\sup^{+}\{\aleph_\alpha \mid \alpha<\lambda\}$ is the least cardinal greater than all the members of $\{\aleph_\alpha \mid \alpha<\lambda\}$, γ must be $\aleph_\lambda$. □

Normal functions have many interesting properties. For example, it will follow immediately from 4.24 and 4.12 that every normal function has arbitrarily large fixed points. Here we shall mention only a few very simple properties.

1.27 Proposition. *If F is a normal function then for every limit ordinal λ, $F(\lambda)$ is also a limit ordinal.* □

1.28 Proposition. *Let F be a normal function and let w be a non-void set of ordinals; then we have* $\sup\{F(\alpha) \mid \alpha \in w\}=F(\sup w)$.

Proof. Let $\beta=\sup w$. If $\beta \in w$ then, by the monotonicity of F, $\sup\{F(\alpha) \mid \alpha \in w\}= F(\beta)=F(\sup w)$. If $\beta \notin w$ then β is a strict bound of w and hence $\beta=\sup^{+} w$. By the definition of $\sup^{+} w$, w is cofinal in $\sup^{+} w=\beta$ and therefore, by the monotonicity of F,

$$(1) \qquad \sup\{F(\alpha) \mid \alpha \in w\}=\sup\{F(\alpha) \mid \alpha<\beta\}.$$

Since F is continuous the right-hand side of (1) equals $F(\beta)$ and thus we have $\sup\{F(\alpha) \mid \alpha \in w\}=F(\beta)$. □

1.29 Proposition. *If F is a normal function and $\beta \geqslant F(0)$ then there is a maximal α for which $F(\alpha) \leqslant \beta$ (namely, $\alpha=\sup\{\gamma \mid F(\gamma) \leqslant \beta\}$).* □

1.30 Proposition. *If F and G are normal functions so is their composition FG.* □

At this point we return to the arithmetic of the ordinal numbers. The subject of the normal functions will be taken up again in § 4.

1.31 Exercise. Give necessary and sufficient conditions on α and β for $\alpha+\beta$ to be a limit ordinal. Generalize it to order types. □

1.32 Definition. (Multiplication of ordered classes.) Let $\langle A,<_A\rangle$ and $\langle B,<_B\rangle$ be ordered classes; $\langle A,<_A\rangle\otimes\langle B,<_B\rangle$ is defined to be the structure $\langle A\times B, R\rangle$, where R is the *right lexicographic order* of $A\times B$ given by

$$\langle x, y\rangle\, R\, \langle x', y'\rangle \leftrightarrow y<_B y' \vee y=y' \wedge x<_A x'$$

(III.3.11). It is easy to verify that $\langle A\times B, R\rangle$ is an ordered class.

Remark. $\otimes$ is introduced here as a function on things which may not exist, namely ordered pairs of possibly proper classes. Formally $\otimes$ should be taken as a function of the four arguments $A, <_A, B, <_B$ such that $\otimes(A, <_A, B, <_B)$ is the relation R which orders $A\otimes B$.

1.33 Proposition. *If $\langle a,<\rangle$ and $\langle B,<\rangle$ are a well ordered set and a well ordered class, respectively, then $\langle a,<\rangle\otimes\langle B,<\rangle$ is a well ordered class. If also $\langle B,<\rangle$ is a well ordered* set *then so is $\langle a,<\rangle\otimes\langle B,<\rangle$ too.* □

1.34 Lemma. *Let $\langle A, R\rangle, \langle B, S\rangle, \langle A', R'\rangle, \langle B', S'\rangle$ be ordered classes such that $\langle A, R\rangle\cong\langle A', R'\rangle$ and $\langle B, S\rangle\cong\langle B', S'\rangle$, then $\langle A, R\rangle\otimes\langle B, S\rangle\cong\langle A', R'\rangle\otimes\langle B', S'\rangle$.* □

1.35 Definition. (Multiplication of order types—Cantor 1883, 1895). Let σ and τ be order types. $\sigma\bullet\tau$ is the unique order type λ (by 1.34) such that if $\langle a, r\rangle$ and $\langle b, s\rangle$ are ordered sets of the respective order types σ and τ then λ is the order type of $\langle a, r\rangle\otimes\langle b, s\rangle$. $\sigma\bullet\tau$ is to be read: σ *times* τ, or: the *product* of σ and τ. Intuitively, $\sigma\bullet\tau$ is the order obtained by taking "τ copies" of an ordered set of order type σ and placing them one after the other in the order type τ.

A particular equality which we shall need soon is the following.

1.36 Proposition. *$n\bullet\omega=\omega$ for every finite ordinal $n>0$.*

Proof. The function f on $n\times\omega$ given by $f(i,k)=n\cdot k+i$ is easily seen to be an isomorphism of $\langle n,<\rangle\otimes\langle\omega,<\rangle$ on $\langle\omega,<\rangle$. □

1.37 Proposition (Cantor 1883, 1895). (i) (*Associativity of multiplication.*) *For all order types σ, τ, λ, $\sigma\bullet(\tau\bullet\lambda)=(\sigma\bullet\tau)\bullet\lambda$.*

(ii) *Multiplication of order types is not commutative.*

(iii) $\sigma \bullet \tau = 0$ *iff* $\sigma = 0$ *or* $\tau = 0$.

(iv) $\sigma \bullet 1 = 1 \bullet \sigma = \sigma$.

(v) (*The left distributive law.*) *For all order types* σ, τ *and* λ, $\sigma \bullet (\tau \dotplus \lambda) = \sigma \bullet \tau \dotplus \sigma \bullet \lambda$.

(vi) *The right distributive law (for multiplication with respect to addition) fails for the order types.*

(vii) $\sigma \bullet 2 = \sigma \dotplus \sigma$.

Partial proof. (ii) $2 \bullet \omega = \omega$, by 1.36. On the other hand, as can be verified directly or obtained from the left distributive law, we have $\omega \bullet 2 = \omega \dotplus \omega$, which is $> \omega$ by 1.13.

(vi) $(1 \dotplus 1) \bullet \omega = 2 \bullet \omega = \omega < \omega \dotplus \omega = 1 \bullet \omega \dotplus 1 \bullet \omega$ (by 1.36, 1.13, and (iv)). □

1.38 Exercise. For all order types σ, τ, $(\sigma \bullet \tau)^* = \sigma^* \bullet \tau^*$. □

Remark. Notice that while addition of order types is symmetric (though not commutative) the multiplication of order types is not, since the roles of σ and τ in obtaining $\sigma \bullet \tau$ are quite different. This becomes apparent when one compares 1.8, where $*$ inverts the order of addition, with 1.38 where $*$ does not effect the order of multiplication.

1.39 Proposition. *For all order types* σ, τ, $\mathrm{Card}(\sigma \bullet \tau) = \mathrm{Card}(\sigma) \cdot \mathrm{Card}(\tau)$. □

1.40 Proposition. *Order type multiplication coincides with cardinal multiplication on the finite ordinals, but they do not coincide on all ordinals.*

Hint of proof. For the finite ordinals use 1.39, III.3.19, III.2.18 and III.2.17. Alternatively, notice that ordinal and cardinal multiplication satisfy the same recursive equations (see III.3.19). For the second part notice that $\aleph_0 \cdot 2 = \aleph_0$ but $\omega \bullet 2 = \omega \dotplus \omega > \omega$. □

The following propositions establish the properties of ordinals under multiplication. They exhibit the asymmetry which we know already from ordinal addition.

1.41 Proposition. *The product of two ordinals is an ordinal.*

Hint of proof. Use an analogue of III.3.12. □

1.42 Proposition (Strict monotonicity of multiplication). *If* $\alpha > 0$ *and* $\beta < \gamma$ *then* $\alpha \bullet \beta < \alpha \bullet \gamma$. *In particular, if* $\alpha > 0$ *and* $\gamma > 1$ *then* $\alpha \bullet \gamma > \alpha$.

Proof. Since $\beta < \gamma$ there is, by 1.15, an ordinal $\delta > 0$ such that $\beta \dotplus \delta = \gamma$. By the distributive law $\alpha \bullet \gamma = \alpha \bullet (\beta \dotplus \delta) = \alpha \bullet \beta \dotplus \alpha \bullet \delta$. By 1.37(iii) $\alpha > 0$ and $\delta > 0$ imply $\alpha \bullet \delta > 0$ hence $\alpha \bullet \gamma = \alpha \bullet \beta \dotplus \alpha \bullet \delta > \alpha \bullet \beta$, by 1.13. □

1.43 Corollary (Cancellation of a left factor). *If* $\alpha > 0$ *then* $\alpha \bullet \beta = \alpha \bullet \gamma \rightarrow \beta = \gamma$ *and* $\alpha \bullet \beta < \alpha \bullet \gamma \rightarrow \beta < \gamma$. □

1.44 Proposition (Weak monotonicity of multiplication). $\beta \leqslant \gamma \to \beta \cdot \alpha \leqslant \gamma \cdot \alpha$. *As a consequence* $\gamma \geqslant 1 \to \alpha \leqslant \gamma \cdot \alpha$ and $\gamma \cdot \alpha < \beta \cdot \alpha \to \gamma < \beta$.

Hint of proof. Use II.2.20. □

1.45 Theorem (Division with remainder—Cantor 1897). *For all ordinals* α *and* $\beta > 0$ *there are unique ordinals* σ (*which is always* $\leqslant \alpha$) *and* $\rho < \beta$ *such that* $\alpha = \beta \cdot \sigma + \rho$. σ *and* ρ *are called, respectively, the* quotient *and* remainder *of* α *when divided by* β.

Proof. By 1.44 $\alpha \leqslant \beta \cdot \alpha$. If $\alpha = \beta \cdot \alpha$ then the conclusion of the theorem holds with $\sigma = \alpha$ and $\rho = 0$. Otherwise $\alpha < \beta \cdot \alpha$ and then, by II.3.26, the ordered set $\langle \beta, < \rangle \otimes \langle \alpha, < \rangle$ of order type $\beta \cdot \alpha$ has a section p of order type α which is determined by a member $\langle \rho, \sigma \rangle$ of $\beta \times \alpha$ (i.e., $\rho < \beta$ and $\sigma < \alpha$). By the definition of the lexicographic order $p = \{\langle \xi, \eta \rangle \mid \eta < \sigma \wedge \xi < \beta \vee \eta = \sigma \wedge \xi < \rho\}$ and thus p is the ordered union of $\langle \beta, < \rangle \otimes \langle \sigma, < \rangle$ and $\{\langle \xi, \sigma \rangle \mid \xi < \rho\}$ (the latter set being ordered by the order of the ξ's) which are of order types $\beta \cdot \sigma$ and ρ, respectively. Thus $\alpha = \mathrm{Ord}(p) = \beta \cdot \sigma + \rho$.

To prove the uniqueness of σ and ρ assume that $\alpha = \beta \cdot \sigma + \rho = \beta \cdot \sigma' + \rho'$, $\rho, \rho' < \beta$ and $\sigma \leqslant \sigma'$. If $\sigma < \sigma'$ then, by $\rho < \beta$ and $\sigma + 1 \leqslant \sigma'$ we get $\alpha = \beta \cdot \sigma + \rho < \beta \cdot \sigma + \beta = \beta \cdot (\sigma + 1) \leqslant \beta \cdot \sigma' \leqslant \beta \cdot \sigma' + \rho' = \alpha$ and thus $\alpha < \alpha$ which is a contradiction. Therefore, we have $\sigma = \sigma'$ and thus $\beta \cdot \sigma + \rho = \alpha = \beta \cdot \sigma' + \rho' = \beta \cdot \sigma + \rho'$. Cancelling out $\beta \cdot \sigma$ we get $\rho = \rho'$. □

1.46 Exercise. Give an alternative proof of the existence part of 1.45 by proving the existence of ρ and σ by induction on α. □

1.47 Proposition (Right continuity of multiplication—Cantor 1897). *For every limit ordinal* λ, $\sup\{\alpha \cdot \beta \mid \beta < \lambda\} = \alpha \cdot \lambda$. *Therefore, by the right strict monotonicity of multiplication, if* $\alpha > 0$ *then* $\alpha \cdot \gamma$ *is a normal function of the second argument* γ (1.24).

Proof. If $\alpha = 0$ the theorem is trivial, so assume $\alpha > 0$. By the monotonicity of multiplication $\alpha \cdot \lambda$ is an upper bound of $\{\alpha \cdot \beta \mid \beta < \lambda\}$. Suppose δ is an upper bound of $\{\alpha \cdot \beta \mid \beta < \lambda\}$; we have to prove that $\delta \geqslant \alpha \cdot \lambda$. Since $\alpha > 0$ there are, by 1.45, ordinals σ and $\rho < \alpha$ such that $\delta = \alpha \cdot \sigma + \rho$. If $\sigma < \lambda$ then, since λ is a limit number, also $\sigma + 1 < \lambda$ and we have $\delta = \alpha \cdot \sigma + \rho < \alpha \cdot \sigma + \alpha = \alpha \cdot (\sigma + 1)$; thus δ is less than the member $\alpha \cdot (\sigma + 1)$ of $\{\alpha \cdot \beta \mid \beta < \lambda\}$, contradicting our assumption that δ is an upper bound of this set. Now we have $\sigma \geqslant \lambda$, hence $\delta = \alpha \cdot \sigma + \rho \geqslant \alpha \cdot \sigma \geqslant \alpha \cdot \lambda$, and thus $\alpha \cdot \lambda$ is the *least* upper bound of $\{\alpha \cdot \beta \mid \beta < \lambda\}$. □

1.48 Recursive Definition of Ordinal Multiplication (Jacobsthal 1909). The equalities $\alpha \cdot 0 = 0$, $\alpha \cdot (\xi + 1) = \alpha \cdot \xi + \alpha$ and $\alpha \cdot \eta = \sup\{\alpha \cdot \xi \mid \xi < \eta\}$, for limit ordinals η, give a recursive definition of ordinal multiplication. This can serve as an alternative definition of multiplication if one is interested in ordinals only.

1.49 Asymmetry of Ordinal Multiplication. We shall now go through Theorems 1.42–1.47 checking whether the symmetric analog of each one of them holds.

(i) 1.42. $\alpha \cdot \beta$ is not strictly monotonous in α as seen from $1 \cdot \omega = \omega = 2 \cdot \omega$ (1.36); this shows that we do not have even $\alpha > 0 \wedge \gamma > 1 \rightarrow \gamma \cdot \alpha > \alpha$.

(ii) 1.43. Since $\alpha \cdot \beta$ is weakly but not strictly monotonous in α (even for $\beta > 0$) we cannot cancel right factors in equalities; in particular $1 \cdot \omega = 2 \cdot \omega$.

(iii) 1.45. The symmetric analog of 1.45 is: for every α and $\beta > 0$ there are ordinals σ and $\rho < \beta$ such that $\alpha = \sigma \cdot \beta \dotplus \rho$. This fails badly for infinite ordinals α. For example for $\alpha = \omega$ and $1 < \beta < \omega$ there are no such σ and ρ. To see this notice that if $\sigma \geqslant \omega$ then, since $\beta > 1$, $\sigma \cdot \beta \dotplus \rho \geqslant \sigma \cdot \beta > \sigma \geqslant \omega$, hence $\sigma \cdot \beta \dotplus \rho > \omega$, and if $\sigma < \omega$ then since also β, $\rho < \omega$ we have $\sigma \cdot \beta \dotplus \rho < \omega$.

(iv) 1.47. Multiplication is not continuous on the left, since $\sup\{n \cdot \omega \mid n < \omega\} = \sup\{\omega\} = \omega < \omega \cdot \omega (\sup\{n \mid n < \omega\}) \cdot \omega$ (by 1.36 and 1.42).

1.50 Exercise. Give necessary and sufficient conditions on α and β for $\alpha \cdot \beta$ to be a limit ordinal. Generalize what you have proved to general order types. □

1.51 Proposition. *α is a limit ordinal iff it is of the form $\omega \cdot \beta$ for $\beta \neq 0$.*

Hint of proof. Use 1.45 in one direction and 1.50 in the other. □

1.52 Exercise. Prove that one cannot replace $\omega \cdot \beta$ by $\beta \cdot \omega$ in 1.51 by showing, for example, that for $1 < n < \omega$ the limit ordinal $\omega \cdot n$ is not a left multiple of ω. □

1.53 Definition. (General addition of ordinals—Cantor 1897.) Let F be a function on an initial segment A of On into On. We define for $\beta \in A$ $\Sigma_{\alpha<\beta} F(\alpha)$ by recursion on β as follows. $\Sigma_{\alpha<0} F(\alpha) = 0$, $\Sigma_{\alpha<\beta \dotplus 1} F(\alpha) = \Sigma_{\alpha<\beta} F(\alpha) \dotplus F(\beta)$, and for a limit ordinal λ, $\Sigma_{\alpha<\lambda} F(\alpha) = \sup\{\Sigma_{\alpha<\beta} F(\alpha) \mid \beta < \lambda\}$. For a term $\tau(\alpha)$, $\Sigma_{\alpha<\beta}\tau(\alpha)$ denotes $\Sigma_{\alpha<\beta} F(\alpha)$ where F is the function on On given by $F(\alpha) = \tau(\alpha)$. We refer to $\Sigma_{\alpha<\beta}\tau(\alpha)$ as the *sum* of the $\tau(\alpha)$'s for $\alpha < \beta$. For $\delta \geqslant \gamma$, $\Sigma_{\gamma \leqslant \alpha < \delta}\tau(\alpha)$ denotes $\Sigma_{\alpha<\xi}\tau(\gamma + \alpha)$, where ξ is the unique ordinal such that $\gamma + \xi = \delta$. $\Sigma_{\gamma \leqslant \alpha < \delta}\tau(\alpha)$ can, of course, also be defined directly by induction on δ.

Remarks. We use here for general ordinal addition the same symbol we used in III.4.2 to denote general cardinal addition; we shall in the future make clear which operation is intended unless it is anyway clear from the context.

Using the axiom of choice we can generalize 1.53 and define the sum $\Sigma_{x \in a} f(x)$, where $\langle a, < \rangle$ is any ordered set and f is a function on a into the class of order types (the sum depends, of course, also on the ordering $<$ of a).

We formulate here the associative law for the addition of two Σ's. The formulation and proof of the general associative law (for any number of Σ's) is left to the reader.

1.54 Proposition. For all $\beta \leqslant \gamma$ and for every function F on γ into On we have $\Sigma_{\alpha<\beta} F(\alpha) \dotplus \Sigma_{\beta \leqslant \alpha < \gamma} F(\alpha) = \Sigma_{\alpha<\gamma} F(\alpha)$.

Hint of proof. Use induction on γ. □

1.55 Proposition (Multiplication as repeated addition). *For all ordinals β and γ $\beta\boldsymbol{\cdot}\gamma=\Sigma_{\alpha<\gamma}\beta$.*

Hint of proof. Use induction on γ and 1.48. □

Notation. From now on, whenever we shall deal with general sums of ordinals we shall feel free to use informal notation. For example, $\Sigma_{\alpha<\beta+1}F(\alpha)$ may be written as $F(0)\dot{+}F(1)\dot{+}\cdots\dot{+}F(\beta)$; in particular we shall do it often in the case of a finite β. Also $F(0)\dot{+}\cdots\dot{+}F(n-1)$ is understood to be 0 when $n=0$. We shall use the associative law (1.54) without referring to it explicitly in the cases where we add to a sum a single term or a sum on either side of it.

2. *Ordinal Exponentiation*

2.1 How Should We Define Ordinal Exponentiation? As we know from the arithmetic of the natural numbers, if we want to reach large numbers fast we cannot be content with just addition and multiplication and we need the more rapidly increasing operation of exponentiation. A similar situation exists in our case and therefore we aim now at introducing exponentiation of ordinals. Maybe the reader was content with the sizes of ordinals which are available by means of addition and multiplication and feels no urge to look for exponentiation, but this attitude will surely change after we have defined exponentiation and shown the reader what this operation can do for him.

Addition and multiplication of ordinals were introduced as special cases of addition and multiplication of order types, which are natural operations on the order types. We defined addition and multiplication of order types by means of essentially the same constructions we used for the addition and multiplication of cardinals, except that in the case of the order types we had also to supply natural orderings of the sets obtained. This is why we have the theorems $\text{Card}(\sigma\dot{+}\tau)=\text{Card}(\sigma)+\text{Card}(\tau)$ and $\text{Card}(\sigma\boldsymbol{\cdot}\tau)=\text{Card}(\sigma)\cdot\text{Card}(\tau)$ (1.11 and 1.39). We shall now see that unlike the cases of addition and multiplication of order types, we cannot define in ZF a reasonable operation $\sigma^{\cdot\tau}$ of order type exponentiation which satisfies

$$(1)\qquad \text{Card}(\sigma^{\cdot\tau})=(\text{Card}(\sigma))^{\text{Card}(\tau)}.$$

If we take τ to be the order type of the set of real numbers in its natural order, and $\sigma=2$ then (1) would yield $\text{Card}(\sigma^{\cdot\tau})=2^{2^{\aleph_0}}$ (since Card $(\tau)=2^{\aleph_0}$—VII.1.6). Thus we would get in ZF an ordered set of cardinality $2^{2^{\aleph_0}}$, but, as proved by Cohen 1963, one cannot prove in ZF that sets of the cardinality $2^{2^{\aleph_0}}$, such as the set of all sets of real numbers, can be ordered at all. Notice that we do not claim that we refuted (1); we have only shown that (1) cannot be proved in ZF (it is provable in ZFC for an appropriate operation $\sigma^{\cdot\tau}$, as we shall see below).

If we restrict the intended order type exponentiation by requiring the exponent τ to be an ordinal then we can indeed define an operation of exponentiation which satisfies (1) (see 2.2 below), but for no such operation can we prove in ZF that if also σ is an ordinal then $\sigma^{\cdot\tau}$ is an ordinal too. Such a proof would yield, in particular, that $2^{\cdot\omega}$ is an ordinal. By (1) we would have, in ZF, $\mathrm{Card}(2^{\cdot\omega})=2^{\aleph_0}$, and thus every set of cardinality $2^{\aleph_0}$, and in particular the set of all reals, could be well ordered (in the order type $2^{\cdot\omega}$ which is an ordinal). As proved by Cohen 1963 one cannot prove in ZF that the set of all real numbers can be well ordered.

Passing from ZF to the theory ZFC, which contains also the axiom of choice, we are faced with a different situation, since in ZFC every set can be well ordered. It is now easy to define an operation $\sigma^{\cdot\tau}$ which satisfies (1); just take $\sigma^{\cdot\tau}=\mathrm{Card}(\sigma)^{\mathrm{Card}(\tau)}$. However, this cannot be taken seriously as an operation of ordinal exponentiation; it gives us nothing more than cardinal exponentiation since the ordinal which it yields is in no way related to the particular orderings which σ and τ represent. This is also reflected in the fact that the equality $\sigma^{\cdot(\rho+\tau)}=\sigma^{\cdot\rho}\bullet\sigma^{\cdot\tau}$ which we expect to hold for an operation of exponentiation fails here (since now $\omega^{\cdot 2}=\omega$ and $\omega^{\cdot 1}\bullet\omega^{\cdot 1}=\omega\bullet\omega>\omega$). If we look back at ordinal addition and multiplication we can see that an important ingredient in what made those operations natural is the fact that for ordered sets $\langle s,<_s\rangle$ and $\langle t,<_t\rangle$ such that $s\cap t=0$, $\mathrm{Ord}(\langle s,<_s\rangle)=\sigma$ and $\mathrm{Ord}(\langle t,<_t\rangle)=\tau$ the ordering relations on the respective sets $s\cup t$ and $s\times t$ which we used in order to define $\sigma+\tau$ and $\sigma\cdot\tau$ were *uniformly definable* from $\langle s,<_s\rangle$ and $\langle t,<_t\rangle$. This means, in the case of multiplication, for example, that there is a function F, given by a class-term without parameters such that for all ordered sets $\langle s,<_s\rangle,\langle t,<_t\rangle, F(\langle s,<_s\rangle,\langle t,<_t\rangle)$ is the order relation on $s\times t$ which we use for defining $\sigma\bullet\tau$. If we try to obtain order type exponentiation in the same way and also to have (1) we turn, naturally to the set ${}^t s$ and look for a definable function F such that we could prove in ZFC

$$F(\langle s,<_s\rangle,\langle t,<_t\rangle) \text{ is an ordering of } {}^t s,$$

and then define $\sigma^{\cdot\tau}$ by

$$\sigma^{\cdot\tau}=\mathrm{Ord}(\langle {}^t s, F(\langle s,<_s\rangle,\langle t,<_t\rangle)\rangle).$$

If we had such an F then $F(\langle 2,\in\rangle,\langle \mathbb{R},<_{\mathbb{R}}\rangle)$, where $\langle \mathbb{R},<_{\mathbb{R}}\rangle$ is the set of all real numbers ordered by magnitude, were a definable ordering of ${}^{\mathbb{R}}2$, but in ZFC we cannot prove the existence of a definable ordering of the set $\mathbf{P}(\mathbb{R})$, or of the set ${}^{\mathbb{R}}2$ which is equinumerous to it (Feferman 1965—see Levy 1965 and V.3.5).

If we restrict our attention to exponents which are ordinals then we can indeed define an operation $\sigma^{\cdot\alpha}$ by means of such an F even in ZF (see 2.2 below), but we cannot obtain, not even in ZFC, such an operation which would yield ordinals when applied to ordinals. If $2^{\cdot\omega}$ were an ordinal then $F(\langle 2,\in\rangle,\langle\omega,\in\rangle)$ were a definable well-ordering of ${}^{\omega}2$, but, as proved by Feferman 1965, $\mathbf{P}(\omega)$ and ${}^{\omega}2$ cannot be shown in ZFC to possess a definable well-ordering (see V.3.5).

2.2 Exercise (Hausdorff 1904). For an order type σ and an ordinal α define $\sigma^{\cdot\alpha}$ as follows. Let $\langle s,<_s\rangle$ be an ordered set of the order type σ. $\sigma^{\cdot\alpha}$ is taken to be the order type of the set ${}^{\alpha}s$ as ordered by the following left lexicographic ordering $<$. If $f, g \in {}^{\alpha}s$ then $f<g$ if $f \neq g$ and $f(\gamma) <_s g(\gamma)$ for the least ordinal γ such that $f(\gamma) \neq g(\gamma)$.

(i) Prove that this operation of exponentiation is well-defined by showing, first, that the relation $<$ which we defined is indeed an ordering of ${}^{\alpha}s$ and, second, that different choices of the ordered set $\langle s, <_s \rangle$ will not yield different outcomes.

(ii) $\mathrm{Card}(\sigma^{\cdot\alpha}) = \mathrm{Card}(\sigma)^{\mathrm{Card}(\alpha)}$.

(iii) $\sigma^{\cdot(\alpha \dotplus \beta)} = \sigma^{\cdot\beta} \bullet \sigma^{\cdot\alpha}$.

(iv) $(\sigma^{\cdot\alpha})^{\cdot\beta} = \sigma^{\cdot\alpha\bullet\beta}$.

(v) If $\mathrm{Card}(\sigma) \geqslant 2$ and $\alpha \geqslant \omega$ then $\sigma^{\cdot\alpha}$ is not an ordinal. □

As we saw, at great length, we cannot expect to define ordinal exponentiation in a natural way so that for all α and β $\alpha^{\cdot\beta}$ is an ordinal and

(1) $\qquad \mathrm{Card}(\alpha^{\cdot\beta}) = \mathrm{Card}(\alpha)^{\mathrm{Card}(\beta)}$.

Therefore, we abandon (1) and we take up another property of ordinal addition and multiplication to guide us in defining ordinal exponentiation, namely the continuity of the operation $\alpha^{\cdot\beta}$ in the variable β. We shall, however, see in 2.10 that the same operation of ordinal exponentiation can be defined in a way which is closely related to that of defining a natural order relation on the set ${}^{\beta}\alpha$.

2.3 Exercise. Prove that if $\alpha^{\cdot\beta}$ is continuous in β then (1) cannot hold. □

2.4 Definition. (Ordinal exponentiation—Cantor 1897.) $\alpha^{\cdot\beta}$ is defined by recursion on β as follows.

(a) $\alpha^{\cdot 0} = 1$;

(b) $\alpha^{\cdot(\beta+1)} = \alpha^{\cdot\beta} \bullet \alpha$;

(c) for a limit ordinal λ, $\alpha^{\cdot\lambda} = \sup\{\alpha^{\cdot\beta} \mid 0 < \beta < \lambda\}$.

2.5 Proposition (The basic properties of ordinal exponentiation—Cantor 1897).

(i) $\beta > 0 \rightarrow 0^{\cdot\beta} = 0$; $1^{\cdot\beta} = 1$; $\alpha^{\cdot 1} = \alpha$; $\alpha^{\cdot 2} = \alpha \bullet \alpha$.

(ii) $\beta > 0 \rightarrow \alpha^{\cdot\beta} \geqslant \alpha$.

(iii) $\alpha > 0 \wedge \lambda$ *is a limit ordinal* $\rightarrow \alpha^{\cdot\lambda} = \sup\{\alpha^{\cdot\beta} \mid \beta < \lambda\}$.
This differs from 2.4(c) in that we include here in the set whose sup *we take also* $\alpha^{\cdot 0} = 1$.

(iv) $\alpha^{\cdot\beta} = 0 \leftrightarrow \alpha = 0 \wedge \beta > 0$; $\alpha^{\cdot\beta} = 1 \leftrightarrow \beta = 0 \vee \alpha = 1$.

(v) $\alpha > 1 \wedge \beta < \gamma \rightarrow \alpha^{\cdot\beta} < \alpha^{\cdot\gamma}$.

(vi) $\alpha > 1 \wedge \alpha^{\cdot\beta} = \alpha^{\cdot\gamma} \rightarrow \beta = \gamma$; $\alpha \geqslant 1 \wedge \alpha^{\cdot\beta} < \alpha^{\cdot\gamma} \rightarrow \beta < \gamma$.

(vii) *For* $\alpha > 1$ $\quad \alpha^{\cdot\beta}$ *is a normal function of* β.

(viii) $\alpha > 1 \rightarrow \alpha^{\cdot\beta} \geqslant \beta$.

(ix) $\alpha^{\cdot(\beta+\gamma)} = \alpha^{\cdot\beta} \bullet \alpha^{\cdot\gamma}$.

(x) $(\alpha^{\cdot\beta})^{\cdot\gamma} = \alpha^{\cdot\beta\bullet\gamma}$.

Partial proof. (ii). Use induction on β. (iii) and (iv). Use (ii). (v). Prove $\alpha > 1 \wedge \delta > 0 \rightarrow \alpha^{\cdot\beta} < \alpha^{\cdot\beta+\delta}$ by induction on δ, using (iv) to get $\alpha^{\cdot\beta} > 0$. (viii). Use (v) and II.2.13.

(ix). If $\alpha \leqslant 1$ (ix) follows easily from (i). For $\alpha > 1$ we prove (ix) by induction on γ. The cases where $\gamma = 0$ or γ is a successor are easy; we shall deal here with the case where γ is a limit ordinal. We have

$\alpha^{\cdot\beta} \bullet \alpha^{\cdot\gamma} = \alpha^{\cdot\beta} \bullet \sup\{\alpha^{\cdot\delta} \mid \delta < \gamma\}$, by (iii),
$= \sup\{\alpha^{\cdot\beta} \bullet \alpha^{\cdot\delta} \mid \delta < \gamma\}$, by the normality of multiplication (by 1.47 and since $\alpha^{\cdot\beta} > 0$ by (iv)) and 1.28,
$= \sup\{\alpha^{\cdot(\beta \dotplus \delta)} \mid \delta < \gamma\}$, by the induction hypothesis,
$= \alpha^{\cdot\sup\{\beta \dotplus \delta \mid \delta < \gamma\}}$, by the normality of exponentiation (vii) and 1.28,
$= \alpha^{\cdot\beta \dotplus \sup\{\delta \mid \delta < \gamma\}}$, by the normality of addition (1.25) and 1.28,
$= \alpha^{\cdot\beta \dotplus \gamma}$, since γ is a limit ordinal (II.3.33).

(x) If $\beta = 0$ or $\gamma = 0$ or $\alpha \leqslant 1$ (x) follows from (i), so we assume $\beta > 0$, $\gamma > 0$ and $\alpha > 1$, and by (ii) $\alpha^{\cdot\beta} > 1$. We prove now (x) by induction on γ. The case where γ is a successor ordinal is easy, so let us deal with the case where γ is a limit ordinal. We get

$(\alpha^{\cdot\beta})^{\cdot\gamma} = (\alpha^{\cdot\beta})^{\cdot\sup\{\delta \mid \delta < \gamma\}}$, since γ is a limit ordinal (II.3.33),
$= \sup\{(\alpha^{\cdot\beta})^{\cdot\delta} \mid \delta < \gamma\}$, by the normality of exponentiation (vii), since $\alpha^{\cdot\beta} > 1$, and 1.28,
$= \sup\{\alpha^{\cdot(\beta \bullet \delta)} \mid \delta < \gamma\}$, by the induction hypothesis,
$= \alpha^{\cdot\sup\{\beta \bullet \delta \mid \delta < \gamma\}}$, by the normality of exponentiation,
$= \alpha^{\cdot(\beta \bullet \sup\{\delta \mid \delta < \gamma\})}$, by the normality of multiplication (1.47),
$= \alpha^{\cdot(\beta \bullet \gamma)}$, since γ is a limit ordinal (II.3.33). □

2.6 Proposition. *Ordinal exponentiation coincides with cardinal exponentiation, and hence with ordinary exponentiation, on the finite ordinals.*

Proof. Clauses (a) and (b) in the definition 2.4 of ordinal exponentiation are the recursive definition of ordinary exponentiation of the natural numbers, hence ordinal exponentiation coincides on the finite ordinals with the ordinary exponentiation, which, in its turn, coincides there with cardinal exponentiation by III.3.30. □

2.7 Exercise. $\alpha^{\cdot\beta}$ is a limit ordinal iff $\alpha > 1$ and $\beta \geqslant \omega$ or α is a limit ordinal and $\beta > 0$. □

When we come to define the general multiplication of ordinals we run into the same difficulties which occurred when we wanted to define ordinal exponentiation, which is to be expected since exponentiation is a special case of general multiplication. We adopt here essentially the same solution which we adopted for ordinal exponentiation. Therefore, like ordinal exponentiation, and unlike general addition of ordinals, general multiplication of ordinals will not be closely connected with general multiplication of cardinals.

2.8 Definition (General multiplication of ordinals—Hausdorff 1908). Let F be a function on an initial segment A of On. We define for $\beta \subseteq A$ $\Pi_{\alpha<\beta} F(\alpha)$ by recursion on β as follows. $\Pi_{\alpha<0}F(\alpha)=1$, $\Pi_{\alpha<\beta+1}F(\alpha)=(\Pi_{\alpha<\beta}F(\alpha))\bullet F(\beta)$, and for a limit ordinal λ, $\Pi_{\alpha<\lambda}F(\alpha)=\sup\{\Pi_{\alpha<\gamma}F(\alpha)\mid\gamma<\lambda\}$ if for no $\gamma<\lambda$ does $F(\gamma)=0$, and $\Pi_{\alpha<\lambda}F(\alpha)=0$ if there is a $\gamma<\lambda$ such that $F(\gamma)=0$. Notice that we use here, unjustifiably, the same symbol we used for general multiplication of cardinals.

2.9 Exercise (Exponentiation is repeated multiplication). For all β, γ, $\beta^{\cdot\gamma}=\Pi_{\alpha<\gamma}\beta$. □

2.10 Exercise (Hausdorff 1908). (i) Let $\langle s,<_s\rangle$ and $\langle t,<_t\rangle$ be ordered sets such that s has a least member which we shall denote with 0_s. We define exp(s, t) as the subset of ${}^ts$ consisting of all the function $f\in{}^ts$ such that f obtains values other than 0_s only a finite number of times, i.e., $\{x\in t\mid f(x)\neq 0_s\}$ is finite. We define $<$ on the set exp(s, t) as the right lexicographic order on exp(s, t) by setting for $f, g\in\exp(s, t)$, $f<g$ iff $f\neq g$ and for the greatest member y of t for which $f(y)\neq g(y)$ we have $f(y)<_s g(y)$; if $f\neq g$ then there is indeed such a y since the set $\{y\mid f(y)\neq g(y)\}\subseteq\{y\mid f(y)\neq 0_s\}\cup\{y\mid g(y)\neq 0_s\}$ is a finite non-void set, by the definition of exp(s, t), and has therefore a greatest member. Prove that if $\langle s,<_s\rangle\cong\langle s',<_{s'}\rangle$ and $\langle t,<_t\rangle\cong\langle t',<_{t'}\rangle$ then $\langle\exp(s, t),<\rangle\cong\langle\exp(s', t'),<\rangle$.

(ii) Use (i) to define an operation $\sigma^{\cdot\cdot\tau}$ on order types σ, τ such that σ is an order type of an ordered set which has a first member. (We shall write in this case $\sigma\geqslant 1$.)

(iii) For order types σ, τ, ρ such that $\sigma\geqslant 1$ we have $\sigma^{\cdot\cdot(\tau+\rho)}=\sigma^{\cdot\cdot\tau}\bullet\sigma^{\cdot\cdot\rho}$ and $(\sigma^{\cdot\cdot\tau})^{\cdot\cdot\rho}=\sigma^{\cdot\cdot(\tau\bullet\rho)}$ ($\sigma\geqslant 1$ implies $\sigma^{\cdot\cdot\tau}\geqslant 1$).

(iv) If α, β are ordinals such that $\alpha>0$ then $\alpha^{\cdot\cdot\beta}=\alpha^{\cdot\beta}$.

Hint. (iv) Prove it by induction on β, and for a limit ordinal β show first that $\alpha^{\cdot\cdot\beta}$ is an ordinal. □

2.10 gives us an alternative way of defining ordinal exponentiation. This way is more meaningful from the point of view of ordered sets. Ordinal exponentiation differs from cardinal exponentiation in that the latter uses the full set ${}^ts$ while the former uses only its subset exp(s, t). Both approaches are used in algebra for defining general multiplication of algebras (which is a generalization of exponentiation). The direct product of algebras follows the idea of cardinal exponentiation while the weak direct product follows the idea of ordinal exponentiation as given in 2.10. The big difference between the operations of ordinal and cardinal exponentiation is reflected by the next theorem (when contrasted with $2^{\mathfrak{a}}>\mathfrak{a}$).

2.11 Theorem (Schönflies 1913). *If $\beta=0$ or $\alpha\leqslant 1$ or α and β are finite then $\alpha^{\cdot\beta}$ is finite. In all other cases $|\alpha^{\cdot\beta}|=\max(|\alpha|, |\beta|)$.*

Proof. If $\beta=0$ or $\alpha\leqslant 1$ then $\alpha^{\cdot\beta}\leqslant 1$ by 2.4 and 2.5(i). If α and β are finite then $\alpha^{\cdot\beta}$ is finite by 2.6. If $\beta>0$ and $\alpha>1$ then $\alpha^{\cdot\beta}\geqslant\alpha$ and $\alpha^{\cdot\beta}\geqslant\beta$ by 2.5(ii) and 2.5(viii) and hence $|\alpha^{\cdot\beta}|\geqslant|\alpha|, |\beta|$. To complete the proof we shall show that if at least one of $|\alpha|$ and $|\beta|$ is infinite and γ is a cardinal such that $\gamma\geqslant|\alpha|, |\beta|$ then $|\alpha^{\cdot\beta}|\leqslant\gamma$. We shall prove it by induction on β, assuming for a while the axiom of choice.

For $\beta=0$, $\alpha^{\cdot\beta}=1\leqslant\gamma$, since γ is infinite. If $|\alpha^{\cdot\beta}|\leqslant\gamma$ then $|\alpha^{\cdot(\beta+1)}|=|\alpha^{\cdot\beta}\bullet\alpha|=$

$|\alpha^{\cdot\beta}|\cdot|\alpha|\leqslant\gamma$, since $|\alpha^{\cdot\beta}|$, $|\alpha|\leqslant\gamma$. If β is a limit ordinal then $\alpha^{\cdot\beta}=\sup\{\alpha^{\cdot\delta}|\delta<\beta\}=\bigcup\{\alpha^{\cdot\delta}|\delta<\beta\}$ and hence, by III.4.5, III.4.6(v), and III.4.7, $|\alpha^{\cdot\beta}|\leqslant\Sigma_{\delta<\beta}|\alpha^{\cdot\delta}|\leqslant\Sigma_{\delta<\beta}\gamma=|\beta|\cdot\gamma\leqslant\gamma$ (since $|\beta|\leqslant\gamma$). The only place where we used the axiom of choice in the proof of $|\alpha^{\cdot\beta}|\leqslant\gamma$ was where we applied III.4.5 without proving that the ordinals $\alpha^{\cdot\delta}$, for $\delta<\beta$, can be uniformly injected into γ. However, once we are given injections of α, β and $\gamma\times\gamma$ into γ it is easy to define by recursion a function f on β such that for every $\delta<\beta$ $f(\delta)$ is an injection of $\alpha^{\cdot\delta}$ into γ. An alternative proof of the present theorem is given by the next exercise. □

2.12 Exercise. Use the alternative definition of ordinal exponentiation given in 2.10 to prove Theorem 2.11 (for $\alpha\neq 0$).

Hint. Use III.4.21. □

2.13 Corollary. $\omega^{\cdot\omega_\alpha}=\omega_\alpha$.

Hint of proof. Use 2.11 and 2.5(vii). □

2.14 Theorem (Expansion of ordinals as sums of powers of α—Cantor 1897). *Let α be an ordinal >1.*

(i) *For every ordinal β there is a unique finite ordinal k and unique sequences $\gamma_0,\ldots,\gamma_{k-1}$ and $\delta_0,\ldots,\delta_{k-1}$ such that*

$$(1)\qquad \gamma_0>\gamma_1>\cdots>\gamma_{k-1},\qquad 0<\delta_i<\alpha \quad \textit{for}\quad i<k,\quad \textit{and}$$
$$\beta=\alpha^{\cdot\gamma_0}\cdot\delta_0+\alpha^{\cdot\gamma_1}\cdot\delta_1+\cdots+\alpha^{\cdot\gamma_{k-1}}\cdot\delta_{k-1}.$$

(ii) *Let β be as in* (1) *and $\eta>\gamma_0$ then $\alpha^{\cdot\eta}>\beta$.*

(iii) *Let β be as in* (1) *and let $\beta'=\alpha^{\cdot\gamma_0}\cdot\delta'_0+\alpha^{\cdot\gamma'_1}\cdot\delta'_1+\cdots+\alpha^{\cdot\gamma'_{k'-1}}\cdot\delta'_{k'-1}$ where $k'<\omega$, $\gamma'_0>\gamma'_1>\cdots>\gamma'_{k'-1}$ and $0<\delta'_i<\alpha$ for $i<k'$. $\beta<\beta'$* iff

(2) *for some $l\leqslant k, k'$ we have $\gamma_i=\gamma'_i$ and $\delta_i=\delta'_i$ for every $i<l$ and either $l=k<k'$, or else $l<k, k'$, and either $\gamma_l<\gamma'_l$ or $\gamma_l=\gamma'_l$ and $\delta_l<\delta'_l$ (in symbols: $(\exists l\leqslant k,k')[(\forall i<l)(\gamma_i=\gamma'_i\wedge\delta_i=\delta'_i)\wedge[l=k<k'\vee(l<k,k'\wedge(\gamma_l<\gamma'_l\vee(\gamma_l=\gamma'_l\wedge\delta_l<\delta'_l)))]]$).*

In other words, to compare two ordinals expanded in powers of α as in (1) *we compare the terms going from left to right; the smaller ordinal is the one in whose expansion we encounter first a term with a smaller exponent or, in the case of equal exponents, a term with a smaller coefficient δ, or, if neither happens, the ordinal whose expansion runs out first.* (ii) *is a special case of* (iii).

Remark. Notice that the present theorem implies, as a special case, that every finite ordinal can be given in decimal notation (by taking $\alpha=10$).

Proof. (i) We shall now prove, by induction on β, that β can be given by an expansion as in (1). The uniqueness of k and the sequences of the γ's and the δ's will follow from (iii), since if two expansions are different then one of them must be related to the other as in (2) and they yield therefore different ordinals. For $\beta=0$ take $k=0$, the sequences of the δ's and γ's are the null sequences and (1) holds trivially. For $\beta>0$ there is, by the normality of exponentiation (and 1.29) a maximal ordinal γ such that $\alpha^{\cdot\gamma}\leqslant\beta$, hence $\alpha^{\cdot\gamma+1}>\beta$. By the normality of multiplication there is a maximal ordinal δ such that $\alpha^{\cdot\gamma}\boldsymbol{\cdot}\delta\leqslant\beta$, hence $\alpha^{\cdot\gamma}\boldsymbol{\cdot}(\delta\dot{+}1)>\beta$. $\delta\geqslant 1$ since $\alpha^{\cdot\gamma}\boldsymbol{\cdot}1=\alpha^{\cdot\gamma}\leqslant\beta$; $\delta<\alpha$ since if we had $\delta\geqslant\alpha$ we would get $\alpha^{\cdot\gamma}\boldsymbol{\cdot}\delta\geqslant\alpha^{\cdot\gamma}\boldsymbol{\cdot}\alpha=\alpha^{\cdot\gamma+1}>\beta$, contradicting our choice of δ. Let β' be the unique ordinal such that $\alpha^{\cdot\gamma}\boldsymbol{\cdot}\delta\dot{+}\beta'=\beta$. $\beta'<\alpha^{\cdot\gamma}$, since if $\beta'\geqslant\alpha^{\cdot\gamma}$ then $\beta=\alpha^{\cdot\gamma}\boldsymbol{\cdot}\delta\dot{+}\beta'\geqslant\alpha^{\cdot\gamma}\boldsymbol{\cdot}\delta\dot{+}\alpha^{\cdot\gamma}=\alpha^{\cdot\gamma}\boldsymbol{\cdot}(\delta\dot{+}1)$, contradicting $\beta<\alpha^{\cdot\gamma}\boldsymbol{\cdot}(\delta\dot{+}1)$. Since $\beta'<\beta$ there are, by the induction hypothesis, a finite ordinal $k>0$, a sequence $\gamma_1>\gamma_2>\cdots>\gamma_{k-1}$ and a sequence $\delta_1,\ldots,\delta_{k-1}$ of nonzero ordinals $<\alpha$ (we have $k>0$ since we start numbering the sequences with 1) such that

$$\beta'=\alpha^{\cdot\gamma_1}\boldsymbol{\cdot}\delta_1\dot{+}\cdots\dot{+}\alpha^{\cdot\gamma_{k-1}}\boldsymbol{\cdot}\delta_{k-1},\quad\text{and hence}$$
$$\beta=\alpha^{\cdot\gamma}\boldsymbol{\cdot}\delta\dot{+}\beta'=\alpha^{\cdot\gamma}\boldsymbol{\cdot}\delta\dot{+}\alpha^{\cdot\gamma_1}\boldsymbol{\cdot}\delta_1\dot{+}\cdots\dot{+}\alpha^{\cdot\gamma_{k-1}}\boldsymbol{\cdot}\delta_{k-1}.$$

Taking into account what we already know about the γ's and the δ's, this is a representation as required by the theorem once we show that if $k>1$ then $\gamma>\gamma_1$. Obviously $\alpha^{\cdot\gamma_1}\leqslant\beta'$, and since, as we saw above, $\beta'<\alpha^{\cdot\gamma}$ we get $\alpha^{\cdot\gamma_1}<\alpha^{\cdot\gamma}$ and hence $\gamma_1<\gamma$.

(ii) This is proved by induction on $k\geqslant 1$. For $k=1$ we have $\alpha^{\cdot\gamma_0}\cdot\delta_0<\alpha^{\cdot\gamma_0}\cdot\alpha=\alpha^{\cdot\gamma_0+1}\leqslant\alpha^{\cdot\eta}$ (since $\delta_0<\alpha$ and $\eta>\gamma_0$). For $k>1$ we have, by the induction hypothesis, $\alpha^{\cdot\gamma_0}>\alpha^{\gamma_1}\boldsymbol{\cdot}\delta_1\dot{+}\cdots\dot{+}\alpha^{\gamma_{k-1}}\boldsymbol{\cdot}\delta_{k-1}$ (since the sequences $\gamma_1,\ldots,\gamma_{k-1}$ and $\delta_1,\ldots,\delta_{k-1}$ are of length $k-1$) and therefore

$$\beta=\alpha^{\cdot\gamma_0}\boldsymbol{\cdot}\delta_0\dot{+}\alpha^{\cdot\gamma_1}\boldsymbol{\cdot}\delta_1\dot{+}\cdots\dot{+}\alpha^{\cdot\gamma_{k-1}}\boldsymbol{\cdot}\delta_{k-1}<\alpha^{\cdot\gamma_0}\boldsymbol{\cdot}\delta_0\dot{+}\alpha^{\cdot\gamma_0}=\alpha^{\cdot\gamma_0}\boldsymbol{\cdot}(\delta_0\dot{+}1)\leqslant\alpha^{\cdot\gamma_0}\boldsymbol{\cdot}\alpha=\alpha^{\cdot\gamma_0+1}\leqslant\alpha^{\cdot\eta}\ (\text{since }\gamma_0<\eta).$$

(iii) Assume first that (2) holds. Then either the expansions of β and β' have the same terms but that of β runs out before that of β', in which case $\beta<\beta'$ since all the terms are >0, or else

$$\begin{aligned}(3)\qquad &\beta=\alpha^{\cdot\gamma_0}\boldsymbol{\cdot}\delta_0\dot{+}\cdots\dot{+}\alpha^{\cdot\gamma_{l-1}}\boldsymbol{\cdot}\delta_{l-1}\dot{+}\alpha^{\cdot\gamma_l}\boldsymbol{\cdot}\delta_l\dot{+}\cdots\dot{+}\alpha^{\cdot\gamma_{k-1}}\boldsymbol{\cdot}\delta_{k-1}\\ &\beta'=\alpha^{\gamma_0}\boldsymbol{\cdot}\delta_0\dot{+}\cdots\dot{+}\alpha^{\cdot\gamma_{l-1}}\boldsymbol{\cdot}\delta_{l-1}\dot{+}\alpha^{\cdot\gamma_l'}\boldsymbol{\cdot}\delta_l'\dot{+}\cdots\dot{+}\alpha^{\cdot\gamma_{k'-1}'}\boldsymbol{\cdot}\delta_{k'-1}'\end{aligned}$$

where $k-1, k'-1\geqslant l$ and $\gamma_l'>\gamma_l$ or $\gamma_l'=\gamma_l$ and $\delta_l'>\delta_l$. If $\gamma_l'>\gamma_l$ then, by (ii),

$$\alpha^{\cdot\gamma_l'}\boldsymbol{\cdot}\delta_l'\dot{+}\cdots\dot{+}\alpha^{\cdot\gamma_{k'-1}'}\boldsymbol{\cdot}\delta_{k'-1}'\geqslant\alpha^{\cdot\gamma_l'}\boldsymbol{\cdot}\delta_l'\geqslant\alpha^{\cdot\gamma_l'}>\alpha^{\cdot\gamma_l}\boldsymbol{\cdot}\delta_l\dot{+}\cdots\dot{+}\alpha^{\cdot\gamma_{k-1}}\boldsymbol{\cdot}\delta_{k-1}.$$

Adding on the left $\alpha^{\cdot\gamma_0}\boldsymbol{\cdot}\delta_0\dot{+}\cdots\dot{+}\alpha^{\cdot\gamma_{l-1}}\boldsymbol{\cdot}\delta_{l-1}$ to both extreme terms of this inequality we get, by (3), $\beta<\beta'$. If $\gamma_l'=\gamma_l$ and $\delta_l'>\delta_l$ then, again by (ii), we have

$$\begin{aligned}\alpha^{\cdot\gamma_l'}\cdot\delta_l'\dot{+}\cdots\dot{+}\alpha^{\cdot\gamma_{k-1}'}\cdot\delta_{k'-1}'\geqslant\alpha^{\cdot\gamma_l'}\cdot\delta_l'\geqslant\alpha^{\cdot\gamma_l}\cdot(\delta_l\dot{+}1)\\ =\alpha^{\cdot\gamma_l}\cdot\delta_l\dot{+}\alpha^{\gamma_l}>\alpha^{\cdot\gamma_l}\cdot\delta_l\dot{+}\cdots\dot{+}\alpha^{\cdot\gamma_{k-1}}\cdot\delta_{k-1}.\end{aligned}$$

Adding on the left $\alpha^{\cdot\gamma_0}\cdot\delta_0\dot{+}\cdots\dot{+}\alpha^{\cdot\gamma_{l-1}}\cdot\delta_{l-1}$ to both extreme terms of this inequality we get, by (3), $\beta<\beta'$.

Now we assume $\beta<\beta'$ and prove (2). Given respective representations for β and β' (as in (1) and (iii)) they have to be different, since $\beta\neq\beta'$, hence either (2) holds or (2) holds with the roles of β and β' reversed. In the latter case we would have, by the direction of (iii) which we have already proved, $\beta'<\beta$. Since this contradicts $\beta<\beta'$ we conclude that (2) holds. □

As we shall see, the most useful expansion in powers of an ordinal are expansions in powers of ω. This is no surprise since it can be justified on an idealogical level; which ordinal qualifies better than ω as a non arbitrary ordinal >1? We shall now use expansions in powers of ω to further study the ordinal addition and multiplication.

2.15 Definition. Let F be a binary operation on ordinals, i.e. $\mathrm{Dom}(F)=\mathrm{On}\times\mathrm{On}$ and $\mathrm{Rng}(F)\subseteq\mathrm{On}$. α is said to be a *critical point* of F if for all $\beta,\gamma<\alpha$ also $F(\beta,\gamma)<\alpha$.

2.16 Proposition. *The following three conditions on α are equivalent.*

(a) *α is a critical point of ordinal addition.*

(b) *for every $\beta<\alpha$ $\beta\dot{+}\alpha=\alpha$.*

(c) *$\alpha=0$ or $\alpha=\omega^{\cdot\xi}$ for some ordinal ξ.*

Proof in outline. $(c)\to(b)$. First prove by induction on ξ that for all ξ and η if $\xi>\eta$ then $\omega^{\cdot\eta}\dot{+}\omega^{\cdot\xi}=\omega^{\cdot\xi}$. Apply this to prove (b) after you have expanded β in powers of ω. $(b)\to(a)$ follows directly from the monotonicity of addition. $(a)\to(c)$. Expand α in powers of ω and see that unless $\alpha=\omega^{\cdot\xi}$ or $\alpha=0$ then α is the sum of two smaller ordinals. □

2.17 Exercise. For $\alpha>0$, the least critical point of addition greater than α is $\alpha\cdot\omega$. □

2.18 Proposition (Cantor 1883). *The following conditions on α are equivalent.*

(a) *α is a critical point of ordinal multiplication.*

(b) *for every $0<\beta<\alpha$ $\beta\cdot\alpha=\alpha$.*

(c) *$\alpha\leqslant2$ or $\alpha=\omega^{\cdot\omega^{\cdot\xi}}$ for some ordinal ξ.*

Hint of proof. Follow the outline of the proof of 2.16 using the basic facts about expansion in powers of ω (2.14) and about the critical points of addition (2.16), and the facts that $\omega^{\cdot\xi}$ is a limit ordinal for $\xi>0$ (2.7) and that, for $\alpha>2$, if α is a critical point of multiplication it is also a critical point of addition. □

2.19 Exercise. For $\alpha \geqslant 2$, the least critical point of multiplication greater than α is $\alpha^{\cdot\omega}$. □

2.20 Exercise. Every ordinal α has only finitely many right summands, i.e., there are only finitely many ordinals η for which the equation $\alpha = \xi \dotplus \eta$ has a solution ξ.

Hint. Expand α in powers of ω. □

2.21 Definition (Hessenberg 1906). The *natural sum* $\alpha \# \beta$ of the ordinals α and β is defined as follows. Using 2.14 expand α and β in powers of ω to get

$$\alpha = \omega^{\cdot\gamma_0} \bullet m_0 \dotplus \omega^{\cdot\gamma_1} \bullet m_1 \dotplus \cdots \dotplus \omega^{\cdot\gamma_{k-1}} \bullet m_{k-1},$$

$$\beta = \omega^{\cdot\gamma_0} \bullet n_0 \dotplus \omega^{\cdot\gamma_1} \bullet n_1 \dotplus \cdots \dotplus \omega^{\cdot\gamma_{k-1}} \bullet n_{k-1},$$

where $\quad \gamma_0 > \gamma_1 > \cdots > \gamma_{k-1}, \quad$ and $\quad m_i, n_i < \omega, m_i + n_i > 0 \quad$ for $\quad i < k$.

Notice that we have the same exponents of ω in both expansions; this is possible since we allow one of m_i, n_i to be zero. Such expansions are unique by 2.14. We set

$$\alpha \# \beta = \omega^{\cdot\gamma_0} \bullet (m_0 + n_0) \dotplus \omega^{\cdot\gamma_1} \bullet (m_1 + n_1) \dotplus \cdots \dotplus \omega^{\cdot\gamma_{k-1}} \bullet (m_{k-1} + n_{k-1}).$$

2.22 Exercise. (i) (Hessenberg 1906). The operation # is commutative and associative.

(ii) $\alpha_1 \# \alpha_2 \# \cdots \# \alpha_n \leqslant \max(\alpha_1, \ldots, \alpha_n) \bullet (n+1)$.

(iii) For a given ordinal α how many pairs $\langle \xi, \eta \rangle$ satisfy $\xi \# \eta = \alpha$?

(iv) (Hessenberg 1906). Give an alternative proof of $\aleph_\alpha \cdot \aleph_\alpha = \aleph_\alpha$ (III.3.23) as follows. First prove $\aleph_0 \cdot \aleph_\alpha = \aleph_\alpha$ using $\aleph_\alpha + \aleph_\alpha = \aleph_\alpha$. Then use the answer to (iii) to inject $\aleph_\alpha \times \aleph_\alpha$ into $\aleph_0 \times \aleph_\alpha$.

(v) If $\beta < \gamma$ then $\alpha \# \beta < \alpha \# \gamma$.

(vi) Is $\alpha \# \beta$ continuous in one of its variables (while the other one is kept fixed—compare with 1.47)?

(vii) (Carruth 1942). If $A_1, \ldots, A_n$ are sets of ordinals such that $\mathrm{Ord}(A_i) = \alpha_i$ for $i = 1, \ldots, n$ then $\mathrm{Ord}(\bigcup_{1 \leqslant i \leqslant n} A_i) \leqslant \alpha_1 \# \alpha_2 \# \cdots \# \alpha_n$. For all $\alpha_1, \ldots, \alpha_n$ there are sets $A_1, \ldots, A_n$ of ordinals such that Ord $(A_i) = \alpha_i$ for $i = 1, \ldots, n$ and $\mathrm{Ord}(\bigcup_{1 \leqslant i \leqslant n} A_i) = \alpha_1 \# \alpha_2 \# \cdots \# \alpha_n$. This is what makes the natural sum natural; $\alpha_1 \# \alpha_2 \# \cdots \# \alpha_n$ is the maximal ordinal of well ordered sets which can be decomposed into n sets of the respective ordinals $\alpha_1, \ldots, \alpha_n$.

(viii) $\alpha_1 \dotplus \alpha_2 \dotplus \cdots \dotplus \alpha_n \leqslant \alpha_1 \# \alpha_2 \# \cdots \# \alpha_n$.

Hints. (iv) To prove $\aleph_0 \cdot \aleph_\alpha = \aleph_\alpha$ either utilize a bijection of $\aleph_\alpha \times 2$ onto $\aleph_\alpha$, or else use the fact that $\aleph_\alpha$ is a critical point of ordinal addition and apply 2.16 to prove $\omega \bullet \omega_\alpha = \omega_\alpha$. Now obtain an injection F of $\aleph_\alpha \times \aleph_\alpha$ into $\aleph_0 \times \aleph_\alpha$ such that for all $\beta, \gamma < \aleph_\alpha$ $2^{\text{nd}}(F(\beta, \gamma)) = \beta \# \gamma$.

(vii) For $n = 2$ prove by induction on $\langle \alpha_1, \alpha_2 \rangle$ with respect to the canonical ordering of On × On (III.3.20). Use the induction hypothesis and (iii) to prove that the ordinal of every section of $A_1 \cup A_2$ is $< \alpha_1 \# \alpha_2$. □

2.23 Exercise. A *pairing function* on a class A is an injection F of $A \times A$ into A. Pairing functions are useful tools in set theory. The function $\langle x, y\rangle$ (i.e., the function F given by $F(x, y)=\langle x, y\rangle$) is a pairing function on V. In this exercise we shall study only pairing functions on On and therefore refer to them just as *pairing functions.* A pairing function F is said to be *monotonic* if it is monotonic in each one of its arguments, i.e., if $\alpha<\alpha' \rightarrow F(\alpha, \beta)<F(\alpha', \beta)$ and $\beta<\beta' \rightarrow F(\alpha, \beta)<F(\alpha, \beta')$. It is desirable that a monotonic pairing function have as many critical points as possible, since if α is a critical point of the monotonic pairing function F then $F \upharpoonright (\alpha \times \alpha)$ is a monotonic pairing function on α.

(i) If F is a monotonic pairing function and α is a critical point of F then α is 0 or $\omega^{\cdot\xi}$ for some ξ.

(ii) Show that there is a "best" monotonic pairing function, i.e., there is a monotonic pairing function whose critical points are exactly the ordinals 0 and $\omega^{\cdot\xi}$ for all ξ.

(iii) Show that the critical points of the pairing function P of III.3.21 are 0, 1 and the ordinals $\omega^{\cdot\omega^{\cdot\xi}}$ for all ξ.

Hints. (i) Prove that α is a critical point of addition. (ii) Take a pairing function g on ω such that $g(0, 0)=0$. To define $F(\beta, \gamma)$ expand β and γ in powers of ω and use g in an appropriate way. (iii) Use (i) and show that α is a critical point of P iff α is a critical point of multiplication. □

2.24 Proposition. *There is a function H on* On *such that for every infinite ordinal α $H(\alpha)$ is a pairing function on α.*

Hint of proof. For $\alpha=\omega^{\cdot\beta}$, $\beta>0$, use 2.23(ii). For $\alpha=\omega^{\cdot\beta}\cdot n+\gamma$, $\beta>0$, $0\leqslant\gamma<\omega^{\cdot\beta}$, use the fact that $\gamma+n\cdot\omega^{\cdot\beta}=\omega^{\cdot\beta}$. □

2.25Ac Proposition. *Let $\kappa=\omega^{\cdot\xi}$ for some ordinal ξ and let $\langle \mathfrak{a}_\alpha | \alpha<\kappa\rangle$ be a non-decreasing sequence of cardinals with $\mathfrak{a}_0>0$, then $\Pi_{\alpha<\kappa}\mathfrak{a}_\alpha=(\sup_{\alpha<\kappa}\mathfrak{a}_\alpha)^{|\kappa|}$.*

Hint of proof. First prove that $(\Pi_{\alpha<\kappa}\mathfrak{a}_\alpha)^{|\kappa|}=\Pi_{\alpha<\kappa}\mathfrak{a}_\alpha$. By 2.23(ii) there is a monotonic pairing function p on κ. For $\beta, \alpha<\kappa$ define $\mathfrak{b}_{p(\beta,\gamma)}=\mathfrak{a}_\beta$ then $\mathfrak{b}_{p(\beta,\gamma)}\leqslant\mathfrak{a}_{p(\beta,\gamma)}$. Compare $\Pi_{\alpha<\kappa}\mathfrak{b}_\alpha$ with both $\Pi_{\alpha<\kappa}\mathfrak{a}_\alpha$ and $(\Pi_{\alpha<\kappa}\mathfrak{a}_\alpha)^{|\kappa|}$. □

3. *Cofinality and Regular Ordinals*

Given an ordered set $\langle a, <\rangle$ we shall observe the ordinals of those cofinal subsets b of $\langle a, <\rangle$ (II.1.16(i)) which are well ordered by $<$. The following theorem 3.1 will give us much information about the existence of such subsets b. In particular it will establish that if we assume the axiom of choice then every ordered set has such a subset b.

3.1 Theorem (Hausdorff 1908). *Let $\langle a,<\rangle$ be an ordered set, let g be a mapping of an ordinal γ onto a cofinal subset of $\langle a,<\rangle$ (g is not necessarily increasing). And let c be a well-orderable cofinal subset of $\langle a,<\rangle$. There is a cofinal subset b of $\langle c,<\rangle$ (and by II.1.17*(iii) *b is also a cofinal subset of $\langle a,<\rangle$) which is* well *ordered by $<$ and such that* $\mathrm{Ord}(\langle b,<\rangle)\leqslant\gamma$. *In particular, if a is well-orderable and one takes $\gamma=|a|$ and $c=a$ then a has a cofinal subset b well ordered by $<$ such that* $\mathrm{Ord}(\langle b,<\rangle)\leqslant|a|$.

Proof. The idea of the proof is to define an increasing function f on a segment of γ into c such that f is defined as long as possible, using the function g as a pacesetter for f and thus ensuring that the set of values of f will become cofinal in c after at most γ steps.

We define f by recursion (as in II.2.12) by setting $f(\alpha)=$ a strict upper bound of $f[\alpha]$ in $\langle c,<\rangle$ which is also $\geqslant g(\alpha)$, as long as $\alpha<\gamma$ and $f[\alpha]$ is not cofinal in c. Notice that if $f[\alpha]$ is not cofinal in c then it has strict upper bounds in c and some of them must be $\geqslant g(\alpha)$. (The use of the indefinite article in the definition of $f(\alpha)$ is legitimate since c is well-orderable.) $\mathrm{Dom}(f)$ is an initial segment of γ, hence $\mathrm{Dom}(f)$ is an ordinal $\beta\leqslant\gamma$. If $\beta<\gamma$ then the reason why $f(\beta)$ is undefined must be because $f[\beta]$ is cofinal in $\langle c,<\rangle$. If $\beta=\gamma$ then we shall show that $f[\beta]$ is cofinal in $\langle c,<\rangle$ by showing that for every $x\in c$ there is a $y\in\mathrm{Rng}(f)$ such that $y\geqslant x$. Let $x\in c$ then, since $\mathrm{Rng}(g)$ is cofinal in a, there is an ordinal $\delta<\gamma$ such that $g(\delta)\geqslant x$. Since $\mathrm{Dom}(f)=\beta=\gamma$, $\delta\in\mathrm{Dom}(f)$. Set $y=f(\delta)$; by definition of f, $y=f(\delta)\geqslant g(\delta)\geqslant x$. Thus we have seen that, in any case, $f[\beta]$, which is $\mathrm{Rng}(f)$, is a cofinal subset of $\langle c,<\rangle$. For all $\xi<\alpha<\beta$ we have by the definition of $f(\alpha)$, $f(\xi)<f(\alpha)$. Thus f is an isomorphism of β into $\langle c,<\rangle$ and $\mathrm{Rng}(f)$ is a subset of c which is well ordered by $<$ and its ordinal is $\beta\leqslant\gamma$. □

3.2 Remark. One cannot refute in ZF the existence of non-void Dedekind finite ordered sets $\langle a,<\rangle$ with no last member (Cohen 1966). Such ordered sets have no cofinal subsets well-ordered by $<$, for the following reason. Since a is a non-void and has no last member no finite subset of a is cofinal in a; and since a is Dedekind finite a includes no infinite well-orderable set. This shows that the requirement of Theorem 3.1 that c be well-orderable is an essential requirement.

3.3 Definition (Hausdorff 1908). (i) cf is the function defined on the class of all order types σ of well-orderable ordered sets by

$\mathrm{cf}(\sigma)=$ the least ordinal of cofinal subsets b of $\langle a,<\rangle$ which are well-ordered by $<$, where $\langle a,<\rangle$ is an ordered set of order type σ.

cf (σ) does obviously not depend on the choice of the ordered set $\langle a,<\rangle$ (as long as $\langle a,<\rangle$ is of order type σ). We shall also write cf $(\langle a,<\rangle)$ for cf $(\mathrm{Ord}(\langle a,<\rangle))$. $\mathrm{cf}(\sigma)$ and $\mathrm{cf}(\langle a,<\rangle)$ are called the *cofinalities* of σ and $\langle a,<\rangle$, respectively. If a is a set of ordinals and $<$ is the natural ordering of the ordinals we shall also write $\mathrm{cf}(a)$ for $\mathrm{cf}(\langle a,<\rangle)$.

(ii) For every ordinal α, $\mathrm{cf}(\alpha)\leqslant\alpha$. An ordinal α is said to be *regular* if $\mathrm{cf}(\alpha)=\alpha$ and *singular* if $\mathrm{cf}(\alpha)<\alpha$.

3.4 Proposition. (i) *Let $\langle a, <\rangle$ be a well-orderable non-void ordered set. For every ordinal γ we have $\gamma < \mathrm{cf}(\langle a, <\rangle)$ iff*

(1) *for every function f on γ into a, $f[\gamma]$ is a bounded subset of a.*

(ii) *An ordinal α is regular iff*

(2) *for every $\gamma < \alpha$ and every function f on γ into α, $f[\gamma]$ is a bounded subset of α.*

Proof. (i) We shall prove that (1) fails iff $\gamma \geqslant \mathrm{cf}(\langle a, <\rangle)$. If (1) fails then some function f on γ into a is a mapping of γ on a cofinal subset of a. By Theorem 3.1 (for $c = \mathrm{Rng}(f)$) and Definition 3.3 $\mathrm{cf}(\langle a, <\rangle) \leqslant \gamma$, which establishes one direction of (i). By the definition of $\mathrm{cf}(\langle a, <\rangle)$ there is a function g mapping the ordinal $\mathrm{cf}(\langle a, <\rangle)$ on a cofinal subset d of a, and $d \neq 0$ since $a \neq 0$. To prove the other direction of (i) assume that $\gamma \geqslant \mathrm{cf}(\langle a, <\rangle)$. Let u be any member of d, then the function f on γ given by

$$f(\xi) = \begin{cases} g(\xi) & \text{for } \xi < \mathrm{cf}(\langle a, <\rangle) \\ u & \text{for } \mathrm{cf}(\langle a, <\rangle) \leqslant \xi < \gamma \end{cases}$$

maps γ on the cofinal subset d of a, and thus (1) fails.

(ii) If α is regular then $\mathrm{cf}(\alpha) = \alpha$ and (2) follows from (1). If α is singular then $\mathrm{cf}(\alpha) < \alpha$ and if we set $\gamma = \mathrm{cf}(\alpha)$ then (2) fails for this γ by (i). □

3.5 Proposition. *If $\langle a, <\rangle$ is an ordered set such that a is well-orderable, and c is a cofinal subset of $\langle a, <\rangle$, then $\mathrm{cf}(\langle c, <\rangle) = \mathrm{cf}(\langle a, <\rangle)$. Therefore, if d is a subset of a such that $\mathrm{cf}(\langle d, <\rangle) \neq \mathrm{cf}(\langle a, <\rangle)$ then d is a bounded subset of $\langle a, <\rangle$.*

Proof. Every cofinal well-ordered subset of $\langle c, <\rangle$ is a cofinal well-ordered subset of $\langle a, <\rangle$ (by II.1.17(iii)) and therefore $\mathrm{cf}(\langle a, <\rangle) \leqslant \mathrm{cf}(\langle c, <\rangle)$. To prove the opposite inequality let $\mathrm{cf}(\langle a, <\rangle) = \gamma$, then there is a bijection g of γ on a cofinal subset of $\langle a, <\rangle$. Applying Theorem 3.1 we get $\mathrm{cf}(\langle c, <\rangle) \leqslant \gamma = \mathrm{cf}(\langle a, <\rangle)$. □

3.6 Exercise. Prove that if α is a regular ordinal and a is a cofinal subset of α then the order type of a is α. □

3.7 Proposition. *Let $\langle a, <\rangle$ and $\langle b, <\rangle$ be well-orderable ordered sets and let $\langle b, <\rangle$ be without a last member. If there is a monotonic mapping f of a onto a cofinal subset of b then $\mathrm{cf}(\langle a, <\rangle) = \mathrm{cf}(\langle b, <\rangle)$.*

Outline of proof. Let g be the function defined on $\mathrm{Rng}(f)$ by

$$g(x) = \text{a } y \in a \quad \text{such that} \quad f(y) = x.$$

g is an isomorphism of $\mathrm{Rng}(f)$ on $\mathrm{Rng}(g)$. Since $\mathrm{Rng}(g)$ and $\mathrm{Rng}(f)$ are cofinal subsets of a and b, respectively, we have, by 3.5, $\mathrm{cf}(\langle a, <\rangle) = \mathrm{cf}(\langle \mathrm{Rng}(g), <\rangle) = \mathrm{cf}(\langle \mathrm{Rng}(f), <\rangle) = \mathrm{cf}(\langle b, <\rangle)$. □

3.8 Proposition (Hausdorff 1908). (i) *For every order type* σ, $\mathrm{cf}(\sigma)$ *is a regular ordinal.*

(ii) 0 *is the only order type* σ *with* $\mathrm{cf}(\sigma)=0$, *thus* 0 *is regular.*

(iii) *For every ordinal* α, $\mathrm{cf}(\alpha)=1$ *iff* α *is a successor ordinal; hence* 1 *is regular and every other successor ordinal is singular.*

(iv) ω *is a regular ordinal.*

(v) *For every ordinal* α, $\mathrm{cf}(\alpha)\leqslant|\alpha|$; *hence every ordinal which is not a cardinal is singular, or, in other words, every regular ordinal is a cardinal.*

Proof. (i) Let $\mathrm{cf}(\sigma)=\gamma$. By definition of $\mathrm{cf}(\sigma)$ there is an ordered set $\langle a,<\rangle$ of order type σ and a cofinal subset c of $\langle a,<\rangle$ such that $\mathrm{Ord}(\langle c,<\rangle)=\gamma$. By 3.5 $\mathrm{cf}(\gamma)=\mathrm{cf}(\langle c,<\rangle)=\mathrm{cf}(\langle a,<\rangle)=\gamma$, hence γ is regular. (ii) and (iii) are easy. (iv) holds since every cofinal subset of ω is of order type ω (see the proof of III.1.32). (v) Let $\gamma=|\alpha|$ and let g be a mapping of γ on α. By 3.1 (where $c=a=\alpha$ and $<$ is the natural ordering of the ordinals) $\mathrm{cf}(\alpha)\leqslant\gamma=|\alpha|$. $\square$

We have seen till now that only cardinals can be regular ordinals and thus we shall use "*regular cardinal*" as a synonym for "regular ordinal". Among the cardinals 0, 1 and ω are regular while all other finite cardinals are singular. This still leaves open the question which of the $\aleph_\alpha$'s, for $\alpha>0$, are regular. Before we shall further deal with this question we shall present very useful characterizations of the cofinality of cardinals and of the regular cardinals from the point of view of cardinal arithmetic.

3.9 Proposition. *For an aleph* $\mathfrak{a}$ *and a well ordered cardinal* $\mathfrak{c}$ *the following conditions are equivalent.*

(a) $\mathrm{cf}(\mathfrak{a})\leqslant\mathfrak{c}$.

(b) $\mathfrak{a}$ *is the union of* $\leqslant\mathfrak{c}$ *sets each of which is of cardinality* $<\mathfrak{a}$.

If the axiom of choice holds then these conditions are also equivalent to:

(c) $\mathfrak{a}$ *is the sum of* $\leqslant\mathfrak{c}$ *cardinals each of which is* $<\mathfrak{a}$.

Proof. (a)$\to$(b). Assume $\mathrm{cf}(\mathfrak{a})=\mu\leqslant\mathfrak{c}$, and let f be a bijection of μ on a cofinal subset of $\mathfrak{a}$. Since $\mathfrak{a}$ is a limit ordinal $\mathfrak{a}=\bigcup_{\delta<\mu}f(\delta)$. Thus $\mathfrak{a}$ is the union of $\mu\leqslant\mathfrak{c}$ sets each of which is of cardinality $|f(\delta)|\leqslant f(\delta)<\mathfrak{a}$.

(b)$\to$(a). Assume that $\mathfrak{a}=\bigcup_{\alpha<\mu}w_\alpha$, where $\mu\leqslant\mathfrak{c}$ and $0<|w_\alpha|<\mathfrak{a}$ for $\alpha<\mu$. The set $v=\bigcup_{\alpha<\mu}\{\alpha\}\times w_\alpha$ is of cardinality $\geqslant\mathfrak{a}$ (to get an injection f of $\mathfrak{a}$ into v set $f(x)=\langle\alpha,x\rangle$, where α is the least ordinal such that $x\in w_\alpha$); when we order v by the left lexicographic order we get therefore a well ordered set of order type $\geqslant\mathfrak{a}$. The initial segment s of v of order type $\mathfrak{a}$ is either of the form $\bigcup_{\alpha<\eta}\{\alpha\}\times w_\alpha$ for some $\eta\leqslant\mu$ or of the form $\bigcup_{\alpha<\eta}\{\alpha\}\times w_\alpha\cup\{\eta\}\times u$ for some $\eta<\mu$ and $u\subseteq w_\eta$; in either case we can write $s=\bigcup_{\alpha<\theta}\{\alpha\}\times u_\alpha$, where $\theta\leqslant\mu\leqslant\mathfrak{c}$, and for $\alpha<\theta$ $0\neq u_\alpha\subseteq w_\alpha$ and hence $0<|u_\alpha|\leqslant|w_\alpha|<\mathfrak{a}$. We shall now see that θ is a limit ordinal. Assume $\theta=\lambda+1$ then since $\{\lambda\}\times u_\lambda\neq 0$ the set $t=\bigcup_{\alpha<\lambda}\{\alpha\}\times u_\alpha$ is a proper initial segment of s, its order type is $<\mathfrak{a}$ and hence its cardinality is $<\mathfrak{a}$. $s=t\cup(\{\lambda\}\times u_\lambda)$, thus s, which is of cardinality $\mathfrak{a}$, is the union of two sets of cardinalities $<\mathfrak{a}$, contradicting III.3.5, III.3.8 or III.3.14. For each u_α, $\alpha<\theta$, let ξ_α be the least member of u_α. Since θ is a limit ordinal, the set $\{\langle\alpha, \xi_\alpha\rangle \mid \alpha<\theta\}$ is a cofinal subset of s, and since the respective order types of s and $\{\langle\alpha, \xi_\alpha\rangle \mid \alpha<\theta\}$ are $\mathfrak{a}$ and θ we get, by 3.5, $\mathrm{cf}(\mathfrak{a})=\mathrm{cf}(\theta)\leqslant\theta\leqslant\mathfrak{c}$.

(b)$\rightarrow$ (c). Let $\mathfrak{a}=\bigcup_{\alpha<\mu} w_\alpha$ where $\mu\leqslant\mathfrak{c}$ and $|w_\alpha|<\mathfrak{a}$ for $\alpha<\mu$. Without loss of generality we can assume that the w_α's are pairwise disjoint since otherwise we can replace each w_α by $w_\alpha\sim\bigcup_{\xi<\alpha} w_\xi$. Since $\mathfrak{a}=\bigcup_{\alpha<\mu} w_\alpha$ we get, by III.4.3., $\mathfrak{a}=|\mathfrak{a}|=\Sigma_{\alpha<\mu}|w_\alpha|$.

(c)$\rightarrow$ (b). This is left to the reader. □

3.10 Corollary (Hausdorff 1914). *For an aleph* $\mathfrak{a}$ *the following conditions are equivalent.*

(a) $\mathfrak{a}$ *is regular.*

(b) *A set of cardinality* $\mathfrak{a}$ *is not the union of less than* $\mathfrak{a}$ *sets of cardinality* $<\mathfrak{a}$.

If we assume the axiom of choice then also the following conditions are equivalent to (a)–(b).

(c) $\mathfrak{a}$ *is not the sum of* $<\mathfrak{a}$ *smaller cardinals, i.e., if* $|u|<\mathfrak{a}$ *and* $\langle\mathfrak{e}_x|x\in u\rangle$ *is such that for every* $x\in u$ $\mathfrak{e}_x<\mathfrak{a}$ *then* $\Sigma_{x\in u}\mathfrak{e}_x\neq\mathfrak{a}$.

(d) *If* $|u|<\mathfrak{a}$ *and for every* $x\in u$ $\mathfrak{e}_x<\mathfrak{a}$ *then* $\Sigma_{x\in u}\mathfrak{e}_x<\mathfrak{a}$. □

3.11Ac Proposition (Hausdorff 1908). *Every infinite successor cardinal, i.e., every cardinal of the form* $\aleph_{\alpha+1}$, *is regular.*

Proof. Let $\gamma=\mathrm{cf}(\aleph_{\alpha+1})$. If $\aleph_{\alpha+1}$ is singular then $\gamma<\aleph_{\alpha+1}$, hence $|\gamma|\leqslant\aleph_\alpha$. Let g be a bijection of γ on a cofinal subset of $\aleph_{\alpha+1}$. The ordinals in $\mathrm{Rng}(g)$ divide $\aleph_{\alpha+1}$ into $|\gamma|\leqslant\aleph_\alpha$ pieces each of which is of cardinality $\leqslant\aleph_\alpha$, hence $\aleph_{\alpha+1}\leqslant\aleph_\alpha^2=\aleph_\alpha$ which is a contradiction. To give a formal proof we first apply the axiom of choice to define a function f on $\aleph_{\alpha+1}$ by

$$f(\beta)=\text{an injection of } \beta \text{ into } \aleph_\alpha\,;$$

there is such an injection for every $\beta<\aleph_{\alpha+1}$ since $|\beta|\leqslant\aleph_\alpha$. This is an essential use of the axiom of choice in this proof; without the axiom of choice we know that each single ordinal $<\aleph_{\alpha+1}$ can be injected into $\aleph_\alpha$, but we do not know whether this can be done "uniformly" for all ordinals $\beta<\aleph_{\alpha+1}$ by means of a function f as defined here.

Let h be the function on $\aleph_{\alpha+1}$ defined by,

$$h(\delta)=\langle\mu, f(g(\mu))(\delta)\rangle, \text{ where } \mu \text{ is the least ordinal } \beta<\gamma \text{ such that } \delta<g(\beta).$$

It is easily seen that h is an injection of $\aleph_{\alpha+1}$ into $\gamma\times\aleph_\alpha$, which is a contradiction since $|\gamma\times\aleph_\alpha|=|\gamma|\cdot\aleph_\alpha\leqslant\aleph_\alpha^2=\aleph_\alpha$. □

3.12 Exercise. Give a short proof of Proposition 3.11 by showing that $\aleph_{\alpha+1}$ satisfies condition (d) of 3.10. □

3.13 The Axiom of Choice and the Regularity of $\aleph_1$. One cannot prove in ZF (i.e., without the axiom of choice) that $\aleph_1$ is regular (Feferman and Levy 1963—see

Cohen 1966). Since $\aleph_1$ satisfies conditions (c) and (d) of 3.10 this shows that one cannot prove in ZF that conditions (c) and (d) of 3.10 imply regularity. In this connection we have the following open problem, the answer to which is generally conjectured to be negative.

3.14 Open Problem. Can one prove in ZF the existence of a regular ordinal $>\omega$? Or, equivalently, can one prove in ZF the existence of an ordinal α with $\mathrm{cf}(\alpha)>\omega$? While this book was in print this question was given a negative answer by Gitik 1979 (assuming the consistency with ZFC of the existence of a proper class of certain large cardinals).

3.15Ac Exercise. A set x is said to be *hereditarily of cardinality* $<\mathfrak{a}$ if $|x|<\mathfrak{a}$ and every member y of $\mathrm{Tc}(x)$ is of cardinality $<\mathfrak{a}$ (see III.1.34). We denote the class of the well founded sets hereditarily of cardinality $<\mathfrak{a}$ with $\mathrm{HC}_{\mathfrak{a}}$, and we denote the class of the well founded set x such that $|\mathrm{Tc}(x)|<\mathfrak{a}$ with $H_{\mathfrak{a}}$.

(i) For every cardinal $\mathfrak{a}$, $\mathrm{HC}_{\mathfrak{a}}$ and $H_{\mathfrak{a}}$ are transitive classes and $H_{\mathfrak{a}}\subseteq\mathrm{HC}_{\mathfrak{a}}$.

(ii) If $\mathfrak{a}$ is an infinite regular cardinal then $H_{\mathfrak{a}}=\mathrm{HC}_{\mathfrak{a}}$, and if $\mathfrak{a}$ is an infinite singular cardinal then $\mathrm{HC}_{\mathfrak{a}}\subseteq H_{\mathfrak{a}^+}$.

(iii) $H_{\mathfrak{a}}\subseteq R(\mathfrak{a})$, hence if $\mathfrak{a}$ is infinite and regular then $\mathrm{HC}_{\mathfrak{a}}\subseteq R(\mathfrak{a})$ and if $\mathfrak{a}$ is infinite and singular then $\mathrm{HC}_{\mathfrak{a}}\subseteq R(\mathfrak{a}^+)$. Thus $H_{\mathfrak{a}}$ and $\mathrm{HC}_{\mathfrak{a}}$ are sets for every $\mathfrak{a}$.

(iv) For an infinite cardinal $\mathfrak{a}$, $\rho(H_{\mathfrak{a}})=\mathfrak{a}$; for an infinite singular $\mathfrak{a}$, $\rho(\mathrm{HC}_{\mathfrak{a}})=\mathfrak{a}^+$.

(v) For every cardinal $\mathfrak{a}$, $|H_{\mathfrak{a}}|=2^{\underline{\mathfrak{a}}}$ ($2^{\underline{\mathfrak{a}}}$ is defined to be $\sup_{\mathfrak{b}<\mathfrak{a}}2^{\mathfrak{b}}$).

(vi) For every singular infinite cardinal $\mathfrak{a}$, $|\mathrm{HC}_{\mathfrak{a}}|=2^{\mathfrak{a}}$.

Hints. (iii). Use II.6.9(i). (v) Use the fact that, by II.6.16, a well founded set x is uniquely determined by a binary relation on a set of cardinality $|\mathrm{Tc}(x)|$. □

3.16 Exercise (Tarski 1929). (i) For an ordinal $\alpha>0$ let $\mathrm{top}(\alpha)$ be the least non-zero right summand of α, i.e., the least $\eta>0$ such that the equation $\alpha=\xi+\eta$ has a solution. Prove that if the expansion of α in powers of ω, as in 2.14, is $\alpha=\omega^{\cdot\gamma_1}\cdot n_1\dotplus\cdots\dotplus\omega^{\cdot\gamma_k}\cdot n_k$ (where $n_k\neq 0$) then $\mathrm{top}(\alpha)=\omega^{\cdot\gamma_k}$. Give also a simpler proof, using 2.16 rather than 2.14, that $\mathrm{top}(\alpha)$ is a power of ω.

(ii) Prove that for all $\alpha,\beta>0$, α has a cofinal subset of order type β iff $\beta\leqslant\alpha\wedge\mathrm{cf}(\beta)=\mathrm{cf}(\alpha)\wedge\mathrm{top}(\beta)\leqslant\mathrm{top}(\alpha)$.

Hint. (ii) First prove that $\omega^{\cdot\gamma}$ has a cofinal subset of ordinal $\omega^{\cdot\delta}$ iff $\mathrm{cf}(\omega^{\cdot\delta})=\mathrm{cf}(\omega^{\cdot\gamma})$ and $\delta\leqslant\gamma$. To do this represent $\omega^{\cdot\gamma}$ as $\sup_{\lambda<\mathrm{cf}(\omega_\gamma)}g(\lambda)$, where g is an increasing function and for each λ $g(\lambda)$ is a power of ω if γ is a limit ordinal and $g(\lambda)=\omega^{\cdot\gamma-1}\cdot\lambda$ if γ is a successor ordinal. Then show that $\omega^{\cdot\delta}$ can be decomposed so that its successive parts can be injected in the sets $g(\lambda+1)\sim g(\lambda)$ for $\lambda<\mathrm{cf}(\omega^{\cdot\gamma})$. The rest of (ii) is now an easy consequence of the fact that if $\beta<\alpha$ then $\alpha=\beta\dotplus\sigma\dotplus\mathrm{top}(\alpha)$ for some σ. □

3.17 Exercise (Milner and Rado 1965). Let α be an aleph. For every ordinal β such that $\beta<\alpha^+$ there are subsets, $A_n\subseteq\beta$, for $n<\omega$, such that $\mathrm{Ord}(A_n)\leqslant\alpha^{\cdot n}$ and $\beta=\bigcup_{n<\omega}A_n$.

This is, at first sight, a very strange result since by 2.22 all finite unions of A_n's as above have order types less than $\alpha^{\cdot\omega}$, and by 2.11 $|\alpha^{\cdot\omega}|=\alpha$.

Hint. Use induction on β. For a limit ordinal β use the fact that $\mathrm{cf}(\beta)\leqslant\alpha$, decompose β to at most α sets of smaller order type and apply the induction hypothesis to each one of them. An alternative proof which sheds some more light on the matter can be obtained as follows. If σ and τ are order types we say that $\sigma\leqslant\tau$ if there is an ordered set $\langle t,<\rangle$ and a subset s of t such that $\mathrm{Ord}(\langle t,<\rangle)=\tau$ and $\mathrm{Ord}(\langle s,<\rangle)=\sigma$. Define an order $<_l$ on ${}^{\omega}\alpha$ by setting, for $p,q\in{}^{\omega}\alpha$, $p<_l q$ if $p\subseteq q$ or if neither of p and q is included in the other and p precedes q lexicographically. Let $\lambda=\mathrm{Ord}(\langle{}^{\omega}\alpha,<_l\rangle)$, then we have $\lambda\cdot\alpha\leqslant\lambda$. Use this to prove by induction on $\beta<\alpha^+$ that $\beta\leqslant\lambda$. □

The only cardinals which we have not yet determinated whether they are regular or singular are the *limit cardinals*, i.e. the cardinals $\aleph_\alpha$, where α is a limit ordinal. We shall see that the limit cardinals are usually singular; the regular ones, if there are any, are named by the following definition.

3.18 Definition (Hausdorff 1908). An ordinal α is said to be *weakly inaccessible* if (a) $\alpha>\omega$, (b) α is regular, and (c) α is a *limit cardinal*, i.e., $\alpha=\aleph_\lambda$ for some limit ordinal λ.

3.19 Proposition. *If F is a normal function and α is a limit ordinal then* $\mathrm{cf}(F(\alpha))=\mathrm{cf}(\alpha)$, *and hence either $F(\alpha)=\alpha$ or $F(\alpha)$ is a singular ordinal.*

Hint of proof. Use the definition of normality, 1.24, and the fact that $F(\alpha)\geqslant\alpha$, II.2.13. □

3.20 Corollary. (i) *If α is a limit ordinal then* $\mathrm{cf}(\aleph_\alpha)=\mathrm{cf}(\alpha)$ *and either $\aleph_\alpha=\alpha$ or else $\aleph_\alpha$ is singular.*

(ii) *If α is weakly inaccessible then $\aleph_\alpha=\alpha$.*

Hint of proof. Use 1.26. □

3.21 On the Size of the Weakly Inaccessible Cardinals. By 3.20 the least limit cardinal $\aleph_\omega$ is singular because its cofinality ω is $<\aleph_\omega$. When we look at larger limit cardinals, we come first across $\aleph_\alpha$'s with countable α; all these have still the cofinality ω (by 3.20 and 3.8(v)) and are therefore singular. Moreover, it follows from 3.20 that for limit ordinals α $\aleph_\alpha$ is singular as long as $\alpha<\aleph_\omega$ (since $\aleph_\alpha\geqslant\aleph_\omega>\alpha$), and such $\aleph_\alpha$'s are already quite big. By the same token $\aleph_\alpha$ is also singular for every limit ordinal α such that $\aleph_\omega\leqslant\alpha<\aleph_{\aleph_\omega}$. The least ordinal α such $\aleph_\alpha=\alpha$ is already truly huge but it is still singular since it can be shown to be cofinal with ω (see the proof of 4.24) and we have not yet encountered any weakly inaccessible cardinal. As a matter of fact, one cannot prove in ZFC that there are any weakly inaccessible cardinals; we shall explain the reason for it in 3.28. In 4.33 we shall discuss again the question of how big the weakly inaccessible cardinals are, if they exist; the tools which will be available to us then will shed some more light on this question. □

A weakly inaccessible cardinal is a cardinal $\aleph_\gamma$ which is $>\aleph_0$ and cannot be reached from below (i.e., from smaller cardinals) by the operations of passing from a cardinal $\aleph_\beta$ to its successor cardinal $\aleph_{\beta+1}$ (since γ is a limit ordinal) and of general cardinal addition (since by 3.10(d) the sum of $<\aleph_\gamma$ cardinals below $\aleph_\gamma$ does not reach $\aleph_\gamma$). We obtain a concept of an inaccessible cardinal $\aleph_\gamma$ stronger than that of a weakly inaccessible cardinal by replacing the cardinal increasing operation of passing from $\aleph_\beta$ to $\aleph_{\beta+1}$ by the stronger exponential operation of passing from $\aleph_\beta$ to $2^{\aleph_\beta}$. Thus we shall require that if $\beta<\gamma$ then $2^{\aleph_\beta}<\aleph_\gamma$. Such a requirement makes good sense only in the presence of the axiom of choice since without this axiom we do not even know that $2^{\aleph_0}$ is an aleph. Therefore we shall now give a definition which replaces this requirement by another one which makes sense even in the absence of the axiom of choice, and which, in the presence of the axiom of choice is equivalent to this requirement.

3.22 Definition (Tarski in 1930—see Sierpinski and Tarski 1930, Zermelo 1930). An ordinal α is said to be *inaccessible* (or an *inaccessible cardinal*) if

(a) $\alpha>\omega$,

(b) α is regular (and hence α is a cardinal), and

(c) if $x\in R(\alpha)$ then it is not the case that $|x|{*}\geqslant\alpha$, i.e., there is no function with domain x and range α.

Remark. Requirement (c) replaces the requirement that a weakly inaccessible cardinal be a limit cardinal. Requirement (c) introduces an aspect of closure under the exponential operation, since if $x\in R(\alpha)$ also $\mathbf{P}(x)\in R(\alpha)$ (α being a limit ordinal by (a) and (b)) and therefore also $2^{|x|}=|\mathbf{P}(x)|{*}\geqslant\alpha$ fails. In 3.24 it is shown that if the axiom of choice holds then (c) can be replaced by the requirement that for all $\mathfrak{b}<\alpha$ also $2^{\mathfrak{b}}<\alpha$.

3.23 Lemma. *If α is inaccessible, $y\in R(\alpha)$, $x\subseteq R(\alpha)$ and $|x|\leqslant{*}|y|$ then $x\in R(\alpha)$.*

Hint of proof. If $x\notin R(\alpha)$ then the function $\rho\upharpoonright x$ (where ρ is as in II.6.6) maps x on a cofinal subset of α. □

3.24 Theorem. (i) *Every inaccessible cardinal is also a weakly inaccessible cardinal.*

(ii) *Assuming the axiom of choice, part* (c) *of Definition* 3.22 *can be replaced, equivalently, by any one of the following conditions.*

(d) *If $\mathfrak{b}<\alpha$ then also $2^{\mathfrak{b}}<\alpha$.*

(e) *If $x\in R(\alpha)$ then $|x|<\alpha$.*

(f) *If $\beta<\alpha$ then $|R(\beta)|<\alpha$.*

(g) $|R(\alpha)|=\alpha$.

(iii) *If the* generalized continuum hypothesis *holds, i.e., if $2^{\aleph_\gamma}=\aleph_{\gamma+1}$ for all γ, then the inaccessible cardinals are the same as the weakly inaccessible cardinals.*

Proof. (i) All we have to show here is that α is a limit cardinal. If α were a successor cardinal let β be the cardinal predecessor of α. Let $t=\{r \mid r$ is a reflexive well-ordering of a subset of $\beta\}$, where by r being a reflexive well-ordering of w we mean that $r = q \cup \{\langle x, x\rangle \mid x \in w\}$, where q is a well-ordering of w and $q \subseteq w \times w$ (i.e., r is to q like $\leqslant$ to $<$). Let s be the set of equivalence classes of t under the relation

$$\{\langle r, r'\rangle \mid \langle \mathrm{Dom}(r), r\rangle \cong \langle \mathrm{Dom}(r'), r'\rangle\}, \text{ i.e.,}$$
$$s = \{\{r' \mid r' \in t \wedge \langle \mathrm{Dom}(r'), r'\rangle \cong \langle \mathrm{Dom}(r), r\rangle\} \mid r \in t\}.$$

By the proof of Hartogs' theorem (III.2.25) $|s|$ is a well ordered cardinal greater than β, hence $|s| \geqslant \alpha$. Once we prove that $s \in R(\alpha)$ this will contradict (c) since $|s| \geqslant \alpha$ obviously implies $|s| * \geqslant \alpha$. By II.6.8 $\rho(\beta) = \beta$, hence $\beta \subseteq R(\beta)$. Since β is a limit ordinal also $\beta \times \beta \subseteq R(\beta)$. $s \in \mathbf{P}(\mathbf{P}(\mathbf{P}(\beta \times \beta)))$ hence $s \in R(\beta + 3)$. Since α is a limit ordinal (being an aleph) and $\beta < \alpha$ also $\beta + 3 < \alpha$, hence $s \in R(\beta + 3) \subseteq R(\alpha)$ which is what was left to prove.

(ii). (c) $\leftrightarrow$ (e). By the axiom of choice the relations $*\geqslant$ and $\geqslant$ coincide (III.2.39(iii)) and all cardinals are well ordered cardinals, hence $|x| * \geqslant \alpha$ fails just in case that $|x| < \alpha$, which proves the equivalence of (c) and (e).

(e) $\rightarrow$ (f). If $\beta < \alpha$ then $R(\beta) \in R(\alpha)$ (by II.6.12(v)) hence $|R(\beta)| < \alpha$.

(f) $\rightarrow$ (e). For every x, $x \subseteq R(\rho(x))$ hence $|x| \leqslant |R(\rho(x))|$. If $x \in R(\alpha)$ then $\rho(x) < \alpha$ and by (f) $|x| \leqslant |R(\rho(x))| < \alpha$.

(f) $\rightarrow$ (g). Since α is a limit ordinal we have, by II.6.12(viii), $R(\alpha) = \bigcup_{\beta < \alpha} R(\beta)$. Since by (f) $\beta < \alpha \rightarrow |R(\beta)| < \alpha$ we get, by the axiom of choice and III.4.5, $|R(\alpha)| \leqslant \Sigma_{\beta < \alpha} |R(\beta)| \leqslant \Sigma_{\beta < \alpha} \alpha = \alpha \cdot \alpha = \alpha$. We have shown $|R(\alpha)| \leqslant \alpha$; $\alpha \leqslant |R(\alpha)|$ is a direct consequence of $\alpha \subseteq R(\alpha)$ (II.6.8).

(g) $\rightarrow$ (f). If $\beta < \alpha$ then, by II.6.12(v) $\mathbf{P}(R(\beta)) \subseteq R(\alpha)$, hence, by (g), $|R(\beta)| < |\mathbf{P}(R(\beta))| \leqslant |R(\alpha)| = \alpha$.

(d) $\rightarrow$ (f). We shall prove, by induction on β that $\beta < \alpha \rightarrow |R(\beta)| < \alpha$. For $\beta = 0$ $|R(0)| = 0 < \omega < \alpha$. If $\beta = \gamma + 1$ then by the induction hypothesis $|R(\gamma)| < \alpha$, and since, by II.6.12(vii) $R(\beta) = \mathbf{P}(R(\gamma))$ we have by (d) $|R(\beta)| = 2^{|R(\gamma)|} < \alpha$. If β is a limit ordinal then $R(\beta) = \bigcup_{\gamma < \beta} R(\gamma)$, and hence, by III.4.5 $|R(\beta)| \leqslant \Sigma_{\gamma < \beta} |R(\gamma)|$. Since α is regular the sum of $< \alpha$ cardinals each of which is $< \alpha$ is $< \alpha$ (by 3.10(d)), hence we have $|R(\beta)| \leqslant \Sigma_{\gamma < \beta} |R(\gamma)| < \alpha$.

(f) $\rightarrow$ (d). Since $\mathfrak{b} < \alpha$ and since α is a limit ordinal $\mathfrak{b} \dotplus 1 < \alpha$. By II.6.8 and II.6.12(vii) $\mathbf{P}(\mathfrak{b}) \subseteq R(\mathfrak{b} \dotplus 1)$. Hence, by (f) and $\mathfrak{b} \dotplus 1 < \alpha$, $2^{\mathfrak{b}} = |\mathbf{P}(\mathfrak{b})| \leqslant |R(\mathfrak{b} \dotplus 1)| < \alpha$.

(iii) Once the generalized continuum hypothesis is assumed then (d) is obviously equivalent to the requirement that α is a limit cardinal. $\square$

How big is the first inaccessible cardinal, if there is one? We can immediately say that, by 3.24(i) it is at least as big as the first weakly inaccessible ordinal; and if the generalized continuum hypothesis holds the two are the same. Assuming the axiom of choice we can say about the size of the inaccessible cardinals all what we said above about the size of the weakly inaccessible cardinals replacing the operation $\mathfrak{a}^+$ by the operation $2^{\mathfrak{a}}$.

3.25Ac Definition (The function $\beth$). $\beth$ (beth) is the function on On into Cn defined by recursion as follows. $\beth_0 = \aleph_0$, $\beth_{\alpha+1} = 2^{\beth_\alpha}$, and for a limit number γ, $\beth_\gamma = \sup_{\alpha < \gamma} \beth_\alpha$. Thus $\beth$ is a normal function.

The Ac after the number 3.25 of the definition means that our definition makes sense only if we assume the axiom of choice.

3.26Ac Proposition. (i) *For all* α, $\beth_\alpha \geqslant \aleph_\alpha$, *and if the generalized continuum hypothesis holds then* $\beth_\alpha = \aleph_\alpha$.

(ii) *For every* α, $|R(\omega)+\alpha| = \beth_\alpha$, *and hence, for* $\alpha \geqslant \omega^{\cdot 2}$, $|R(\alpha)| = \beth_\alpha$.

Hint of proof. (i) Use induction on α and Cantor's theorem. (ii) Use induction on α and 2.17. □

3.27 On the Size of the Inaccessible Cardinals. Whatever we said in 3.21 about how big the weakly inaccessible cardinals are applies to the inaccessible cardinals with $\aleph$ replaced by $\beth$. Thus no cardinal $\leqslant \beth_\omega$ is inaccessible, and for all countable α's no cardinal $\leqslant \beth_\alpha$ is inaccessible. It will even follow from the proof of 4.24 that even the least ordinal α such that $\alpha = \beth_\alpha$ is of cofinality ω and hence not inaccessible. More on this question will be said in 4.33.

Now we arrive at the question of comparing the inaccessible cardinals with the weakly inaccessible cardinals as to size. Clearly, the least inaccessible cardinal is not less than the least weakly inaccessible cardinal, but to obtain more information we must be able to compare the operation $2^{\mathfrak{a}}$ with the operation $\mathfrak{a}^+$. We already encountered this problem in III.3.36 where we asked how to "compute" $\aleph_\alpha^{\aleph_\beta}$. We shall present in §V.5 most of what we can prove in ZFC concerning how to "compute" the $2^{\aleph_\alpha}$'s in terms of the $\aleph_\alpha$'s, which is not much. This means that the axioms of ZFC still leave open many options concerning the $2^{\aleph_\alpha}$'s. As to the relationship of the inaccessible cardinals to the weakly inaccessible cardinals the open options include, under certain natural assumptions, the possibility that all weakly inaccessible cardinals are inaccessible (this follows from the generalized continuum hypothesis—3.24(iii)), and also (as follows from the results of Cohen 1963) the possibility that $2^{\aleph_0}$ is already weakly inaccessible and even that there are $2^{\aleph_0}$ weakly inaccessible cardinals below $2^{\aleph_0}$, while $2^{\aleph_0}$ is just the first small step in the long way towards the first inaccessible cardinal (if there is such at all).

3.28 Do Inaccessible Cardinals Exist? Let us now attack the question of whether one can prove in ZFC the existence of inaccessible cardinals. One can actually show that the existence of such ordinals cannot be proved in ZFC. A rigorous proof of that, while still very simple, requires some metamathematical concepts and methods which are not handled in this book. Therefore, we shall only sketch here the idea of that proof (Kuratowski 1925). Suppose there is a proof in ZFC that there is an inaccessible cardinal. Let θ be the least inaccessible cardinal. Let us observe $R(\theta)$ and see that it is closed under all the set-constructing operations introduced by the axioms of ZFC. The axiom of union introduces the operation $\bigcup x$ of union; as easily seen if $x \in R(\theta)$ also $\bigcup x \in R(\theta)$. Similarly $R(\omega) \in R(\theta)$ and if $x \in R(\theta)$ also $\mathbf{P}(x) \in R(\theta)$, which takes care of the axioms of infinity and power set. As for the axiom of replacement, we have, by Proposition 3.23 that for every function f, if $\mathrm{Dom}(f) \in R(\theta)$ and $\mathrm{Rng}(f) \subseteq R(\theta)$ then $\mathrm{Rng}(f) \in R(\theta)$, and this embodies the operation of the axiom of replacement in a very strong way. The axiom of choice is no problem either since, if we assume it, then every member x of $R(\theta)$ has a well-ordering which is a member of $R(\theta)$. In $R(\theta)$ there are no inaccessible cardinals since θ is the least inaccessible cardinal. We have shown that $R(\theta)$ is a model for all the axioms of ZFC; since we assumed that one can prove in ZFC the existence of an inaccessible cardinal $R(\theta)$ must contain an inaccessible cardinal contradicting what we said above. Thus we have obtained a contradiction in ZFC, while it is our basic assumption for all metamathematical results mentioned in this book that ZF and ZFC are consistent.

We outlined a proof that one cannot prove in ZFC the existence of inaccessible cardinals. What about the existence of weakly inaccessible cardinals? Here the answer is, as we shall see, essentially the same. Suppose one could prove in ZFC the existence of weakly inaccessible cardinals. As shown by Gödel 1938 ZFC remains consistent when we add to it the generalized continuum hypothesis as an additional axiom; let us call the resulting axiom system ZFC$^+$. By our assumption we can prove in ZFC$^+$, which is an extension of ZFC, the existence of weakly inaccessible cardinals; let θ be the least such cardinal. By 3.24(iii) θ is inaccessible. As we saw above this leads to a contradiction. By the choice of θ there are no weakly inaccessible cardinals below it, but, on the other hand, since all the axioms of ZFC$^+$ hold in $R(\theta)$ we can prove that there is a weakly inaccessible cardinal in $R(\theta)$. Thus ZFC$^+$ is inconsistent, which, as we know, is not the case.

How should our attitude towards the inaccessible cardinals be effected by the fact that their existence is unprovable in ZFC? This should not have a great effect since we have no reason to believe that ZFC is exactly the right axiom system for set theory. The axioms of ZFC were taken up because they seemed to be "true" basic facts about the universe of sets, but we never assumed that they include all the "true" facts. Thus what we saw above does not answer the question of whether we should assume the existence of inaccessible cardinals, as distinguished from the question whether their existence is provable in ZFC.

4. *Closed Unbounded Classes and Stationary Classes*

4.1 Definition. Let X be a non-void set.

(i) (Cartan 1937). A *filter* on X is a non-void subset F of $\mathbf{P}(X)$ which satisfies the following conditions for all $A, B \subseteq X$.

(a) If $A \in F$ and $B \supseteq A$ then also $B \in F$,

(b) If $A, B \in F$ then also $A \cap B \in F$,

(c) $0 \notin F$.

(ii) (Stone 1934). The concept of an *ideal* on X is the concept dual to that of a filter on X, i.e., we obtain the definition of an ideal by replacing in (a)–(c) above the symbols $\supseteq$, $\cap$ and 0 by $\subseteq$, $\cup$ and X, respectively. I is an ideal on X if $0 \neq I \subseteq \mathbf{P}(X)$ and I satisfies, for all $A, B \subseteq X$,

(a′) If $A \in I$ and $B \subseteq A$ then also $B \in I$,

(b′) If $A, B \in I$ then also $A \cup B \in I$,

(c′) $X \notin I$.

4.2 On Filters and Ideals. If F is a filter on X then clearly $X \in F$ (by $F \neq 0$ and (a) of 4.1(i)). Also, because of (a) we can replace (c), equivalently, by the requirement $F \neq \mathbf{P}(X)$. (b) implies that F is also closed under any finite intersection, i.e., if G is a non-void finite subset of F then $\bigcap G \in F$ (this is proved by induction on the finite set G—III.1.6). The dual facts about an ideal I are the following: $0 \in I$; by (a′), (c′) is equivalent to $I \neq \mathbf{P}(X)$; and (b′) implies that if G is a non-void finite subset of I then $\bigcup G \in I$.

4.3 Exercise. Let X be a non-void set, let $F \subseteq \mathbf{P}(X)$ and let I be the set of the complements of the members of F, i.e., $I = \{X \sim A \mid A \in F\}$. I is an ideal iff F is a filter. □

4.4 Examples of Filters and Ideals. In the following examples we denote the filters with F and the ideals with I.

(i) X is a non-void set and $F = \{X\}$.

(ii) $|X| \geqslant \aleph_\alpha$ and $I = \{A \subseteq X \mid |A| < \aleph_\alpha\}$.

(iii) $\langle X, < \rangle$ is a non-void ordered set and I is the set of all strictly bounded subsets of X.

(iv) Let $0 \neq E \subseteq X$ and $F = \{A \subseteq X \mid A \supseteq E\}$. This F is called the *principal filter generated by* E. Dually, for $E \subset X$ $\mathbf{P}(E)$ is the *principal ideal generated by E*.

(v) Let $\langle X, \mu \rangle$ be a complete measure space (VII.3.12), and let I be the set of all subsets of X of measure 0.

(vi) Let $0 \neq G \subseteq \mathbf{P}(X)$ be such that $0 \notin G$ and for all $A, B \in G$ there is a $C \in G$ such that $C \subseteq A \cap B$. Then $F = \{D \subseteq X \mid (\exists A \in G)\ (D \supseteq A)\}$ is the least filter on X which includes G. F is called the filter *generated* by G, and G is called a *base* for F.

4.5 Definition. A filter F on X is said to be $\mathfrak{a}$-*complete* if F is closed under the intersection of $< \mathfrak{a}$ of its members, i.e., if for every $G \subseteq F$, if $|G| < \mathfrak{a}$ then also $\bigcap G \in F$. An ideal I is said to be $\mathfrak{a}$-complete if I is closed under the union of $< \mathfrak{a}$ of its members. (Obviously, F is an $\mathfrak{a}$-complete filter on X iff $\{X \sim B \mid B \in F\}$ is an $\mathfrak{a}$-complete ideal on X.)

It rubs one the wrong way to call a filter which is closed under countable intersections $\aleph_1$-complete rather than $\aleph_0$-complete. However, had we decided to use the adjective $\mathfrak{a}$-complete for filters F closed under the intersection of $\leqslant \mathfrak{a}$ of their members we would not have a convenient way of referring to filters closed under the intersection of $< \aleph_\alpha$ members, where α is a limit ordinal or 0. The present choice of terminology seems to be the lesser evil. Following established terminology we shall also refer to the $\aleph_1$-complete ideals as σ-*ideals*.

4.6Ac Examples and Exercises. (i) Prove that the ideal I of 4.4(ii) is $\mathrm{cf}(\aleph_\alpha)$-complete but not $\mathrm{cf}(\aleph_\alpha)^+$-complete.

(ii) Prove that if $\langle X, < \rangle$ has no last member then the ideal I of 4.4(iii) is $\mathrm{cf}(\langle X, < \rangle)$-complete, but not $\mathrm{cf}(\langle X, < \rangle)^+$-complete.

(iii) Prove that an ideal I is $\mathfrak{a}$-complete for every cardinal $\mathfrak{a}$ iff I is a principal ideal.

(iv) If an ideal I is such that for every subset J of I of cardinality $< \mathfrak{a}$ which consists of pairwise disjoint sets $\bigcup J \in I$ then I is $\mathfrak{a}$-complete.

Hint. (iv) Use the identity $\bigcup_{\sigma<\lambda} A_\sigma = \bigcup_{\sigma<\lambda}(A_\sigma \sim \bigcup_{\tau<\sigma} A_\tau)$. □

4.7 Members of an Ideal as Insignificant Sets. Given a set X one often considers an ideal I over X when one chooses to regard the members of I as "infinitely small" or "insignificant" subsets of X in the context handled. These insignificant sets are such that the union of finitely many insignificant sets is still insignificant, and if the ideal is $\mathfrak{a}$-complete then even the union of any number $<\mathfrak{a}$ of insignificant sets is insignificant. Viewing the members of a filter as the complements of the members of an ideal (4.3) we can regard the members of a filter F on X as those subsets of X which contain almost all of X (except for an insignificant part).

According to the point of view which we have just presented, if a subset B of X is obtained from a subset A of X by first substracting from A a set of the ideal and then adding a set in the ideal then the changes we made in A in order to obtain B are insignificant and B is almost equal to A. This leads to the following definition.

4.8 Definition. (i) For sets A, B we denote with $A \vartriangle B$ the *symmetric difference* of A and B, i.e., the set $(A \sim B) \cup (B \sim A)$.

(ii) Let I be an ideal on X. We write, for $A, B \subseteq X$, $A =_{\mathrm{mod}\, I} B$ (*A equals B modulo I*, or modulo a member of I) if $A \vartriangle B \in I$.

4.9 Proposition. *Let I be an ideal on a set X. The relation $=_{\mathrm{mod}\, I}$ on $\mathbf{P}(X)$ is an equivalence relation on $\mathbf{P}(X)$ (i.e., $=_{\mathrm{mod}\, I}$ is reflexive on $\mathbf{P}(X)$, symmetric and transitive). Moreover, $=_{\mathrm{mod}\, I}$ is a congruence relation with respect to the operations of set difference, binary union and binary intersection, i.e., if $A_1 =_{\mathrm{mod}\, I} A_2$ and $B_1 =_{\mathrm{mod}\, I} B_2$ then*

$$A_1 \sim B_1 =_{\mathrm{mod}\, I} A_2 \sim B_2, \qquad A_1 \cup B_1 =_{\mathrm{mod}\, I} A_2 \cup B_2, \qquad A_1 \cap B_1 =_{\mathrm{mod}\, I} A_2 \cap B_2.$$

If I is an $\mathfrak{a}$-complete ideal then $=_{\mathrm{mod}\, I}$ is also a congruence relation with respect to union and intersection of $<\mathfrak{a}$ sets, i.e., if $|u|<\mathfrak{a}$ and $\langle P_x \mid x \in u \rangle$, $\langle Q_x \mid x \in u \rangle$ are such that for every $x \in u$ $P_x =_{\mathrm{mod}\, I} Q_x$ then also $\bigcup_{x\in u} P_x =_{\mathrm{mod}\, I} \bigcup_{x \in u} Q_x$ and $\bigcap_{x\in u} P_x =_{\mathrm{mod}\, I} \bigcap_{x\in u} Q_x$.

Hint of proof. To show $A_1 \sim B_1 =_{\mathrm{mod}\, I} A_2 \sim B_2$ use

$$(A_1 \sim B_1) \vartriangle (A_2 \sim B_2) \subseteq (A_1 \vartriangle A_2) \cup (B_1 \vartriangle B_2).$$

For the other parts use $A_1 \cup B_1 = X \sim [(X \sim A_1) \sim B_1]$, $A_1 \cap B_1 = X \sim [(X \sim A_1) \cup (X \sim B_1)]$, $\bigcup_{x\in u} P_x \vartriangle \bigcup_{x\in u} Q_x \subseteq \bigcup_{x\in u}(P_x \vartriangle Q_x)$. □

4.10 Conventions. For the rest of the present section Ω will denote either an ordinal with $\mathrm{cf}(\Omega) > \omega$ or the class On.

We want to give a unified treatment to the case where Ω is an ordinal and to the case where Ω is On. Since we shall mostly speak about Ω as if it is an ordinal we have to explain what will be the meaning of what we say if Ω is On. Let $\Omega = \mathrm{On}$; then $\gamma < \Omega$ means just that $\gamma \in \mathrm{On}$; also, by a sequence $\langle A_\alpha \mid \alpha < \gamma \rangle$ of subclasses of Ω, with $\gamma \leqslant \Omega$, we mean a class $A \subseteq \gamma \times \Omega$ such that for all $\alpha < \gamma$ A_α denotes the class $\{\xi \mid \langle \alpha, \xi \rangle \in A\}$, which can be viewed as the α-th component of A.

A subclass A of Ω is said to be *bounded in* Ω if some $\alpha < \Omega$ is a strict upper bound of A (i.e., $A \subseteq \alpha$).

If Ω is an ordinal we denote by Θ the cofinality of Ω, i.e., $\Theta = \mathrm{cf}(\Omega)$. By 3.4 Θ is characterized by

(1) For every $\gamma < \Theta$ and every function F on γ into Ω, $F[\gamma]$ is a bounded subclass of Ω, while there is a function F on Θ into Ω such that $F[\Theta]$ is a cofinal subclass of Ω.

For the case where $\Omega = \mathrm{On}$ we set $\Theta = \mathrm{cf}(\mathrm{On}) = \mathrm{On}$. (1) holds also for this case since for $\gamma < \mathrm{On}$ $F[\gamma]$ is a set, by the axiom of replacement, and hence it is strictly bounded by $\sup^+ F[\gamma]$. Thus we can also say that On is regular since in both cases where Ω is a regular ordinal and where Ω is On, and in no other case, we have (see 3.4(ii))

(2) If Ω is regular then for every $\gamma < \Omega$ and every function F on γ into Ω, $F[\gamma]$ is a bounded subset of Ω.

Since we assumed that Ω is either an ordinal with $\mathrm{cf}(\Omega) > \omega$ or Ω is the class On we have in either case

(3) $\Theta > \omega$.

4.11 Definition. (i) An ordinal ξ is said to be a *limit point* of a class A of ordinals if ξ is a limit ordinal such that arbitrarily large ordinals below ξ belong to A, i.e., $(\forall \alpha < \xi) \exists \beta (\alpha < \beta < \xi \wedge \beta \in A)$ or, equivalently, $\xi = \sup(\xi \cap A)$.

(ii) A subclass A of Ω is said to be *closed* (with respect to Ω) if it contains all its limit points which are $< \Omega$. This concept of a closed class is exactly the same as the topological concept of a closed class in the order topology on Ω (VI.1.6).

(iii) By a *normal function* (for Ω) we mean a function F on some $\gamma \leqslant \Omega$ which is increasing and continuous, i.e., for all $\alpha, \beta < \gamma$ $\alpha < \beta \rightarrow F(\alpha) < F(\beta)$, and for every limit ordinal $\alpha < \gamma$ $F(\alpha) = \sup_{\beta < \alpha} F(\beta)$. (This is like the definition of a normal function in 1.24 except that the domain of F does not have to be all of On.)

(iv) For every class A of ordinals the *enumerating function* of A is the function F on an ordinal or on On such that $F(\alpha)$ is the α-th member of A (where the least member of A is called the 0-th member of A). F is clearly the unique isomorphism of On or an ordinal on A (II.3.23). Notice that every increasing function F on $\gamma \leqslant \mathrm{On}$ into On is the enumerating function of its range.

4.12 Proposition (Veblen 1908). (i) *Let A be a cofinal subclass of Ω and let F be its enumerating function, then F is normal iff A is closed.*

(ii) *Let F be an increasing function on Ω into Ω.* Rng(F) *is closed iff F is a normal function.* □

4.13 Proposition. (i) *Let A be a class of ordinals, and let F be its enumerating function. ξ is a limit point of A iff there is a non-void subset b of A without a largest member such that* $\sup b=\xi$, *iff for some limit ordinal* $\eta\leqslant\mathrm{Dom}(F)$ $\xi=\sup_{\alpha<\eta}F(\alpha)$.

(ii) *If A is a closed subclass of Ω and $b\subseteq A$ is non-void and bounded in Ω then* $\sup b\in A$.

(iii) *If A is an unbounded subclass of Ω then the class B of all limit points of A which are $<\Omega$ is a closed unbounded subclass of Ω.*

Partial proof. (iii) First we shall show that B is unbounded. Let $\alpha<\Omega$; we shall find a $\beta\in B$ such that $\beta>\alpha$. Since A is an unbounded subclass of Ω and $\alpha<\Omega$ also $A\sim\alpha$ is an unbounded subclass of Ω. Let F be the enumerating function of $A\sim\alpha$ and let $\gamma=\mathrm{Dom}(F)$. By 4.10(1) $\gamma\geqslant\Theta$, hence, since $\Theta>\omega$, $\gamma>\omega$. $\{F(n)\mid n<\omega\}$ is a subset of A bounded by $F(\omega)<\Omega$, hence $\sup_{n<\omega}F(n)<\Omega$ and by (i) $\sup_{n<\omega}F(n)$ is a limit point of A. Thus $\sup_{n<\omega}F(n)\in B$ and obviously $\sup_{n<\omega}F(n)>F(0)\geqslant\alpha$.

To show that B is closed let $\beta<\Omega$ be a limit point of B. We shall prove that β is also a limit point of A and hence $\beta\in B$. We have to show that for every $\alpha<\beta$ there is a $\xi\in A$ such that $\alpha<\xi<\beta$. Let $\alpha<\beta$; since β is a limit point of B there is an $\eta\in B$ such that $\alpha<\eta<\beta$. Since $\eta\in B$, η is a limit point of A and hence there is a $\xi\in A$ such that $\alpha<\xi<\eta$. We have therefore for this ξ, $\alpha<\xi<\eta<\beta$, which is what we had to show. □

4.14 Proposition. *The intersection of any family of closed subclasses of Ω is a closed subclass of Ω. (This is actually true for closed subclasses in any topological space.)*

Proof. Let $\langle A_t\mid t\in T\rangle$ be a family of closed subclasses of Ω and let α be a limit point of $\bigcap_{t\in T}A_t$ which is $<\Omega$. For each $s\in T$, $A_s\supseteq\bigcap_{t\in T}A_t$ and hence α is also a limit point of A_s. Since A_s is a closed subclass of Ω we have $\alpha\in A_s$, for each $s\in T$. Therefore $\alpha\in\bigcap_{t\in T}A_t$, and thus we have shown that $\bigcap_{t\in T}A_t$ is a closed subclass of Ω. □

4.15 Theorem (Veblen 1908, Bachmann 1955). *For $\mathfrak{a}$ such that $0<\mathfrak{a}<\Theta=\mathrm{cf}(\Omega)$ the intersection of $\mathfrak{a}$ closed unbounded subclasses of Ω is a closed unbounded subclass of Ω.*

Proof. Let $\langle A_\xi\mid\xi<\mathfrak{a}\rangle$ be a sequence of closed unbounded subclasses of Ω, where $0<\mathfrak{a}<\Theta$; as a consequence of 4.14 it suffices to prove that $\bigcap_{\xi<\mathfrak{a}}A_\xi$ is an unbounded subclass of Ω. The idea of the proof is as follows. For every $\alpha<\Omega$ we look for an ordinal η such that $\alpha<\eta<\Omega$ and η is a limit point of each one of the classes A_ξ. Since each class A_ξ is closed η belongs to each A_ξ and hence also to their intersection. Such an η is obtained by constructing an increasing sequence f which runs through all the A_ξ's ω times and by choosing η to be the least upper bound of its terms. We shall define f on $\mathfrak{a}\cdot\omega$ by recursion but let us verify first that $\mathfrak{a}\cdot\omega<\Theta$. If $\Theta=\mathrm{On}$ this is obvious. If Θ is an ordinal it must be regular and hence a cardinal, and since $\mathfrak{a}$, $\omega<\Theta$ (by 4.10(3)) also $\mathfrak{a}\cdot\omega<\Theta$ (by 1.39). To define f we set

$$f(0)=\text{the least member of } A_0 \text{ which is } >\alpha,$$

and for all $n<\omega$ and $\xi<\mathfrak{a}$ (other than $n=\xi=0$)

$$f(\mathfrak{a}\cdot n\dotplus\xi)=\text{the least member of } A_\xi \text{ which is greater than all the members of } f[\mathfrak{a}\cdot n\dotplus\xi].$$

Since $\mathfrak{a}$, n, $\xi<\Theta$ we have $\mathfrak{a}\cdot n\dotplus\xi<\Theta$ (by 1.11 and 1.39) and hence, by 4.10(1), $f[\mathfrak{a}\cdot n\dotplus\xi]$ is a bounded subset of Ω. A_ξ, being unbounded, does therefore have a member greater than all the ordinals in $f[\mathfrak{a}\cdot n\dotplus\xi]$ and thus $f(\mathfrak{a}\cdot n\dotplus\xi)$ is defined. We take now $\eta=\sup f[\mathfrak{a}\cdot\omega]$. It follows directly from $\mathfrak{a}\cdot\omega<\Theta$ and 4.10(1) that $\eta<\Omega$. We still have to prove that for every $\xi<\mathfrak{a}$ η is a limit point of A_ξ. The set $\{\mathfrak{a}\cdot n\dotplus\xi \mid n<\omega\}$ (where $\mathfrak{a}$ and ξ are held fixed) is a cofinal subset of $\mathfrak{a}\cdot\omega$ (by 1.45) and hence, since f is an increasing sequence, the set $f[\{\mathfrak{a}\cdot n\dotplus\xi \mid n<\omega\}]$ is a cofinal subset of $f[\mathfrak{a}\cdot\omega]$ and we have $\sup f[\{\mathfrak{a}\cdot n\dotplus\xi \mid n<\omega\}]=\sup f[\mathfrak{a}\cdot\omega]=\eta$. By the definition of f the set $f[\{\mathfrak{a}\cdot n\dotplus\xi \mid n<\omega\}]$ is a subset of A_ξ without a largest member, hence, by 4.13(i), its least upper bound η is a limit point of A_ξ, which is what was left to prove. □

4.16 Exercises. (i) Show that our assumption in 4.10 that $\Theta=\mathrm{cf}(\Omega)>\omega$ is essential for Proposition 4.15 by proving that whenever $\mathrm{cf}(\Omega)=\omega$ there are two disjoint closed unbounded subsets of Ω. Can you also obtain such subsets of cardinality Ω?

(ii) Show that 4.15 fails for $\mathfrak{a}=\Theta$. □

4.17Ac The Filter Generated by the Closed Unbounded Sets. It is a consequence of 4.15 that if Ω is an ordinal then the set of the closed unbounded subsets of Ω is a base for a Θ-complete filter, which is the filter of all subsets of Ω which include a closed unbounded set.

4.18 Definition. Let $\langle A_\alpha \mid \alpha<\Omega\rangle$ be a sequence of subclasses of Ω. The *diagonal intersection* $D_{\alpha<\Omega}A_\alpha$ of this sequence is defined to be the class

$$B=\{\beta<\Omega \mid (\forall\alpha<\beta)\,(\beta\in A_\alpha)\}=\{\beta<\Omega \mid \beta\in\textstyle\bigcap_{\alpha<\beta}A_\alpha\}.$$

4.19 Proposition. *If $\gamma<\Omega$ then $D_{\alpha<\Omega}A_\alpha\subseteq\bigcap_{\alpha<\gamma}A_\alpha\cup\gamma$. Thus the diagonal intersection $D_{\alpha<\Omega}A_\alpha$ is included in the intersection of the first $\gamma<\Omega$ A_α's modulo a bounded subclass of Ω.* □

4.20 Theorem. *If $\langle A_\alpha \mid \alpha<\Omega\rangle$ is a sequence of closed unbounded subclasses of Ω, and*

(4) *for all $\beta<\Omega$ $\bigcap_{\alpha<\beta}A_\alpha$ is unbounded in Ω,*

then also the diagonal intersection $B=D_{\alpha<\Omega}A_\alpha$ is a closed unbounded subclass of Ω. If Ω is regular then assumption (4) is superfluous (since it follows from 4.15). If the *sequence* $\langle A_\alpha \mid \alpha<\Omega\rangle$ *is descending* (*i.e.*, $\alpha<\beta\to A_\beta\subseteq A_\alpha$) *then* (*4*) *holds* (*since* $\bigcap_{\alpha<\beta}A_\alpha\supseteq A_\beta$).

Proof. First we prove that B is closed. Let b be any bounded subset of B without a greatest member and let $\beta=\sup b$; by 4.13(i) it suffices to prove that $\beta\in B$. Let $\gamma<\beta$. For every $\delta\in b\subseteq B$ such that $\gamma<\delta<\beta$ we have, by the definition of B, $\delta\in\bigcap_{\alpha<\delta}A_\alpha\subseteq A_\gamma$, hence $b\sim(\gamma\cup\{\gamma\})\subseteq A_\gamma$. β is a limit ordinal (since b has no greatest member) and $\gamma<\beta$ hence $\sup(b\sim(\gamma\cup\{\gamma\}))=\sup b=\beta$. Since A_γ is closed and $b\sim(\gamma\cup\{\gamma\})$ is a subset of A_γ without a greatest member, its least upper bound β is in A_γ. Thus we have shown that $\beta\in A_\gamma$ for every $\gamma<\beta$, i.e., $\beta\in B$.

Now we shall prove that B is unbounded by showing that for every $\lambda<\Omega$ B has a member $\eta\geqslant\lambda$. We define an ascending sequence $\langle\delta_n\mid n<\omega\rangle$ of ordinals $<\Omega$ as follows. $\delta_0=\lambda$, $\delta_{n+1}=$ the least member of $\bigcap_{\alpha<\delta_n}A_\alpha$ which is greater than δ_n; there is such an ordinal by (4). Let $\eta=\sup_{n<\omega}\delta_n$. $\eta<\Omega$ by 4.10(1) since $\Theta>\omega$; also $\eta\geqslant\delta_0=\lambda$; we shall now prove that $\eta\in B$. Let $n<\omega$; for every $m>n$ in ω we have, by the definition of δ_m, $\delta_m\in\bigcap_{\alpha<\delta_{m-1}}A_\alpha\subseteq\bigcap_{\alpha<\delta_n}A_\alpha$ hence $\{\delta_m\mid n<m<\omega\}\subseteq\bigcap_{\alpha<\delta_n}A_\alpha$. $\bigcap_{\alpha<\delta_n}A_\alpha$ is a closed set, by 4.14, and $\{\delta_m\mid n<m<\omega\}$ is a subset of it without a greatest member, hence $\eta=\sup_{n<m<\omega}\delta_m\in\bigcap_{\alpha<\delta_n}A_\alpha$. Since this holds for every $n<\omega$ we get $\eta\in\bigcap_{n<\omega}\bigcap_{\alpha<\delta_n}A_\alpha=\bigcap_{\alpha<\eta}A_\alpha$ (since $\eta=\sup_{n<\omega}\delta_n$), i.e., $\eta\in B$. □

4.21 Exercise. (i) Show that 4.20 fails to hold for every ordinal Ω such that $\mathrm{cf}(\Omega)=\omega$.

(ii) Show that assumption (4) is essential for 4.20 for every singular ordinal Ω. □

4.22Ac Ω-Completeness and Normality. By 4.20, the filter generated by the closed unbounded subsets of a regular ordinal $\Omega>\omega$ is closed also under diagonal intersection of any of its members. A filter closed under diagonal intersection of its members is called a *normal filter*. The property of normality can be viewed as being stronger than Ω-completeness, since by 4.19 every normal filter F on Ω which contains the complements of all bounded subsets of Ω is Ω-complete. Since the filter generated by the closed unbounded sets contains the complements of all bounded sets, Theorem 4.20, which asserts the normality of this filter, is stronger than Theorem 4.15, which asserts its Ω-completeness, as far as regular Ω's are concerned. □

We want to have a method which, for a given subclass A of Ω, yields "large" members of A. One particular case where such a method is helpful is when we discuss, as in 3.21 the size of the least weakly inaccessible cardinal. In 3.21 we were looking for "large" members of the class of all alephs. One large member of that class which we found was the least α such that $\alpha=\aleph_\alpha$; we mentioned that even this large aleph is still less than the least weakly inaccessible cardinal. If we shall have a method for getting yet larger alephs and if we shall be able to prove that those alephs, too, are less than the least inaccessible cardinal, then we shall possess additional information on the size of that cardinal. Just as we considered in 3.21

the α's such that $\alpha = \aleph_\alpha$ we shall now consider, for an arbitrary subclass A of Ω, all those α's in A which are the α-th members of A.

4.23 Definition. Let A be a subclass of Ω. A' will denote the subclass $\{\alpha \mid \alpha$ is the α-th member of $A\}$ of A. A' is thus the class of the fixed points of the enumerating function F of A (i.e., $A' = \{\alpha \mid F(\alpha) = \alpha\}$).

4.24 Proposition (Veblen 1908). *If Ω is regular and A is a closed unbounded subclass of Ω then so is also A'.*

Hint of proof. Let F be the enumerating function of A. For $\beta < \Omega$, $\sup_{n<\omega} F^n(\beta)$ is a fixed point of F, by 4.12(i) and 1.28 (it is in fact the least fixed point of F which is $\geqslant \beta$). □

If the class A is sparse (i.e., the ordinals in A are far apart) and $0 \notin A$ then even the least member of A' must be a considerably large member of A (it is $\sup_{n<\omega} F^n(0)$, where F is the function enumerating A). A' is still sparser than A, since A' is a subclass of A which omits many members of A. To get still sparser classes, with larger fixed points of their enumerating function, we iterate the operation leading from A to A'. This will result in a descending sequence of classes, and the natural step to take at limit ordinal stages is to take the intersection of all classes obtained earlier.

4.25 Definition (Veblen 1908). Let A be a closed unbounded subclass of a regular Ω. For $\alpha < \Omega$ we define $A^{(\alpha)}$ as follows. $A^{(0)} = A$, $A^{(\alpha+1)} = (A^{(\alpha)})'$ and for a limit ordinal α, $A^{(\alpha)} = \bigcap_{\beta<\alpha} A^{(\beta)}$. For every $\alpha < \Omega$, $A^{(\alpha)}$ is a closed unbounded subclass of Ω by 4.24 and 4.15.

4.26 On the Validity of Definition 4.25. In the case where $\Omega = \mathrm{On}$ the class A of 4.25 is a proper class and the definition in 4.25 is not a valid definition by recursion (since the theorem on definition by recursion—II.2.9—allows only to define set valued functions but not "class valued functions"). This can be overcome as follows. Suppose that $A^{(\alpha)}$ satisfies the recursive clauses of 4.25. For an ordinal $\beta < \Omega$ we set $W_\beta(\alpha) = A^{(\alpha)} \cap \beta$. The reader is invited to define $W_\beta(\alpha)$ from $\langle W_\beta(\gamma) \mid \gamma < \alpha \rangle$, (without using the sequence $\langle A^{(\alpha)} \mid \alpha \in \mathrm{On} \rangle$) and use this to define by recursion the function $\langle W_\beta(\alpha) \mid \alpha < \Omega \rangle$ (where β and $W_\beta(0) = A \cap \beta$ are parameters). $A^{(\alpha)}$ can now be defined by $A^{(\alpha)} = \bigcup_{\beta<\Omega} W_\beta(\alpha)$.

This way of handling the "function" $\langle A^{(\alpha)} \mid \alpha \in \mathrm{On} \rangle$ will be formulated in 4.27 below as a general proposition on definition by recursion of "class valued functions" $\langle A_\alpha \mid \alpha \in \mathrm{On} \rangle$, from which we can, of course, derive the existence of the sequence $\langle A^{(\alpha)} \mid \alpha \in \mathrm{On} \rangle$ of 4.25.

4.27 Exercise (Definition by recursion of class valued functions). Let $\tau(\alpha, X)$ be a class term and let G be a function on $\mathrm{On} \times V$ such that for all α and y $G(\alpha, y) \subseteq \alpha \times V$. We also assume that for every $X \subseteq \alpha \times V$ the answer to the question whether

$y \in \tau(\alpha, X)$ depends only on $X \cap G(\alpha, y)$, i.e., we can prove that

$$y \in \tau(\alpha, X) \leftrightarrow y \in \tau(\alpha, X \cap G(\alpha, y)).$$

Then there is a unique class $A \subseteq \mathrm{On} \times V$ such that for all α $A_\alpha = \tau(\alpha, A \upharpoonright \alpha)$, where $A_\alpha = \{z \mid \langle \alpha, z \rangle \in A\} = A[\{\alpha\}]$. Notice that τ is the recursion term, i.e., the term which gives the dependence of A_α on "$\langle A_\beta \mid \beta < \alpha \rangle$" and $G(\alpha, y)$ is the function that tells us how far we have to look into each A_β, $\beta < \alpha$, to determine whether $y \in A_\alpha$. In A_β we have to look at the set $A_\beta \cap G(\alpha, y)[\{\beta\}]$; this may be a very large set compared to y or to the transitive closure $\mathrm{Tc}(y)$ of y, but it is still a set. □

4.28 Exercise. Let A be a closed unbounded subclass of Ω. If $0 \notin A$ then for every $\alpha < \Omega$ the least member of $A^{(\alpha)}$ is $\geqslant \omega^{\cdot \alpha}$. Show that with $0 \notin A$ as the only assumption this is the best possible result. □

We started with a closed unbounded subclass A of Ω and define the $A^{(\alpha)}$'s as a way of getting sparser and sparser classes. A still sparser class is obtained by taking their diagonal intersection $B = D_{\alpha < \Omega} A^{(\alpha)}$. We can now start the same process all over again with B, looking at the classes $B^{(\alpha)}$ and $D_{\alpha < \Omega} B^{(\alpha)}$. The natural thing to do now is, at least in the case where Ω is an ordinal, to look at the set of all sets which can be obtained from A by the operations $'$ (of 4.23), intersection of $< \Omega$ sets, and diagonal intersection of Ω sets. This leads us to discuss the closure of a set under infinitary operations; till now we have only discussed, in II.4.25–4.33 and in III.4.23–4.26, only closures of sets under finitary operations and relations.

4.29 Proposition. *Let $\mathfrak{d}$ be a regular cardinal and let $F_1, \ldots, F_k, k < \omega$ be functions regarded as functions on sequences of length $< \mathfrak{d}$, not necessarily defined for all such sequences. For every set a there is a set b which is the* closure *of a under $F_1, \ldots, F_k$, i.e., $b \supseteq a$ and for every sequence x of length $< \mathfrak{d}$ of members of b, if $F_i(x)$ is defined then $F_i(x) \in b$, $1 \leqslant i \leqslant k$, and b is the least such set.*

Proof. We define the sequence $\langle c_\alpha \mid \alpha < \mathfrak{d} \rangle$ by recursion as follows. $c_0 = a$, $c_{\alpha+1} = c_\alpha \cup \bigcup_{1 \leqslant i \leqslant k} F_i[{}^{\mathfrak{d}}c_\alpha]$ (where ${}^{\mathfrak{d}}c_\alpha$, defined in III.4.20 is the set of all sequences of length $< \mathfrak{d}$ of members of c_α), and for a limit ordinal α, $c_\alpha = \bigcup_{\beta < \alpha} c_\beta$. We set $b = \bigcup_{\alpha < \mathfrak{d}} c_\alpha$. Let x be a sequence of members of b of length $\gamma < \mathfrak{d}$. For each $\alpha < \gamma$ let $G(\alpha)$ be the least $\beta < \mathfrak{d}$ such that $x_\alpha \in c_\beta$. Since $\mathfrak{d}$ is regular and $\gamma < \mathfrak{d}$, $G[\gamma]$ is bounded by some $\delta < \mathfrak{d}$ (by 3.4(ii)). Thus for every $\alpha < \gamma$ $x_\alpha \in c_{G(\alpha)} \subseteq c_\delta$, hence $x \in {}^{\mathfrak{d}}c_\delta$. By the definition of $c_{\delta+1}$, if $F_i(x)$ is defined then $F_i(x) \in c_{\delta+1} \subseteq b$.

If b$'$ is another set such that $b' \supseteq a$ and b' is closed under $F_1, \ldots, F_k$ then, as easily seen, $c_\alpha \subseteq b'$ for all $\alpha < \mathfrak{d}$, and hence $b \subseteq b'$. □

4.30 Exercise. Let us weaken the assumption of 4.29 about $\mathfrak{d}$ and assume only that $\mathfrak{d}$ is an infinite cardinal. Let $\mathfrak{e}$ be the least regular cardinal $\geqslant \mathfrak{d}$ (if there is such; we know how to prove its existence only from the axiom of choice, see 3.11). Prove that the closure b of a is obtained from a in $\mathfrak{e}$ steps, as given in 4.29, and give an example where all the $\mathfrak{e}$ steps are necessary. Prove that if there is no regular cardinal

$\geqslant \mathfrak{d}$ then a class B which is the closure of a under $F_1, \ldots, F_k$ is obtained by On-many steps. Give an example where B is a proper class. □

4.31 Obtaining Sparse Closed Unbounded Sets. Returning to the question of obtaining from a closed and unbounded subset A of the regular ordinal Ω sparser and sparser closed unbounded subsets, the thing to do is to look for such sets in the closure W of $\{A\}$ under the operations $'$ (of 4.23), intersection of $<\Omega$ sets and diagonal intersection of Ω sets. All the members of W are closed unbounded subsets of Ω. If we assume the axiom of choice we can actually extend the sequence $\langle A^{(\alpha)} \mid \alpha<\Omega\rangle$ all the way up to Ω^+, staying all the time within W. Define a function G on $\Omega^+ \sim \Omega$ by: $G(\alpha)=$a cofinal subset of α of order type $\mathrm{cf}(\alpha)\leqslant\Omega$. Now define $A^{(\alpha)}$, for $\alpha\in\Omega^+\sim\Omega$ by recursion as follows.

$$A^{(\alpha)}=\begin{cases}(A^{(\alpha-1)})' & \text{if } \alpha \text{ is a successor ordinal,}\\ \bigcap_{\beta\in G(\alpha)} A^{(\beta)} & \text{if } \alpha \text{ is a limit ordinal and } \mathrm{cf}(\alpha)<\Omega,\\ D_{\beta<\Omega} A^{(f(\beta))}, \text{ where } f \text{ is the unique isomorphism of } \Omega \text{ on } G(\alpha) & \text{if } \mathrm{cf}(\alpha)=\Omega.\end{cases}$$

The values of $A^{(\alpha)}$, for $\alpha\geqslant\Omega$, depend of course on our choice of the function G.

4.32 Proposition. *Let A be a closed unbounded subclass of Ω and let $\gamma>\omega$ be a regular ordinal in A. Then one of the following two cases holds.*

(a) *γ is the 0-th, or the $\alpha+1$-th member of A for some $\alpha<\Omega$.*

(b) *γ is the γ-th member of A.*

In the latter case γ belongs also to every class obtained from A by the operations $'$ (of 4.23), *intersection of $<\gamma$ classes and diagonal intersection.*

Remarks. For γ as in (b), classes obtained from A by the above operations and intersection of γ (rather than $<\gamma$) classes need no more to contain γ; for example, if $0\notin A$ then $A^{(\gamma+1)}$ does not contain γ since its least member is, by 4.28, $\omega^{\cdot\gamma+1}>\omega^{\cdot\gamma}\geqslant\gamma$.

The second part of 4.32 says that $\gamma\in B$ for every subclass B of Ω which is obtained from A by certain operations, to which we shall now refer as the admissible operations. If Ω is an ordinal then the statement "B is obtained from A by the admissible operations" means that B is a member of the closure of $\{A\}$ under those operations (4.29). If Ω is On then it seems impossible to write a formal statement which captures the full content of the second part of 4.32 for this case. It will, however, become obvious from the proof of 4.32 that the second part of 4.32 applies to any subclass B of On which is obtained from A by an explicit construction involving only the operations mentioned there.

Proof of 4.32. Let F be the function enumerating A. If (a) of 4.32 does not hold, then $\gamma=F(\alpha)$ for some limit ordinal α. Since F is normal (by 4.12(i)) and γ is regular we have, by 3.19, $\alpha=\gamma$ and thus $F(\gamma)=\gamma$ and (b) holds.

To prove the last part of the proposition we notice first that if we apply the operations mentioned there, which we call admissible, to closed subclasses of Ω we get again closed subclasses of Ω (this follows immediately from 4.14 and the first parts of the proofs of 4.24 and 4.20). We assume now that (b) holds, i.e. that γ is the γ-th member of A. Let F be the enumerating function of A, then $F(\gamma)=\gamma$. Since A is closed F is a normal function (4.12(i)) and therefore $\gamma=F(\gamma)=\sup F[\gamma]=\sup A\cap\gamma$ and hence $A\cap\gamma$ is an unbounded subset of γ. Since A is a closed subset of Ω, $A\cap\gamma$ is a closed unbounded subset of γ. To prove what we need it suffices to establish

(5) if we apply any of the admissible operations to subclasses B of Ω such that $B\cap\gamma$ is a closed unbounded subset of γ then also the subclass of Ω obtained by the operation has this property.

Once we prove (5) it is clear that every class C obtained from A by the admissible operations is such that $C\cap\gamma$ is a closed unbounded subset of γ. Formally this is shown as follows. Let U be the class of all closed subclasses B of Ω such that $B\cap\gamma$ is a closed unbounded subset of γ. By what we mentioned concerning closed classes and by (5) U is closed under the admissible operations. By what we proved about A $A\in U$. Let W be the closure of $\{A\}$ under the admissible operations. Since W is the *least* class containing A which is closed under the admissible operations we get $W\subseteq U$. Therefore every $C\in W$ is a member of U, i.e., C is closed and $C\cap\gamma$ is a closed unbounded subset of γ. Since C is closed and $\sup C\cap\gamma=\gamma<\Omega$ we get $\gamma\in C$ which is what is claimed by the last part of the proposition.

We still have to prove (5). Let B be as in (5). We shall first show that

(6) $$B'\cap\gamma=(B\cap\gamma)'.$$

If (6) holds, the right-hand side of (6) is a closed unbounded subset of γ by 4.24 (applied to the regular ordinal γ instead of Ω), hence $B'\cap\gamma$ is a closed unbounded subset of γ, which is one of the three components of (5) which we have to prove. To prove (6), let F be the enumerating function of B, then for some ordinal δ $B\cap\gamma=F[\delta]$ (take for δ the least ordinal such that $F(\delta)\geqslant\gamma$ or $F(\delta)$ is undefined). Since $F[\delta]=B\cap\gamma$ is a cofinal subset of γ, δ cannot be $<\gamma$ (by 3.4(ii)), hence $\delta\geqslant\gamma$. δ cannot be $>\gamma$ since then $F(\gamma)$ were defined and since F is normal $F(\gamma)\geqslant\gamma$, contradicting $F(\gamma)\in F[\delta]=B\cap\gamma$. Thus $\delta=\gamma$ and $B\cap\gamma=F[\gamma]$ and $F\restriction\gamma$ is the enumerating function of $B\cap\gamma$ (by 4.11(iv)). We get

$$(B\cap\gamma)'=\{\alpha<\gamma\mid(F\restriction\gamma)(\alpha)=\alpha\}=\{\alpha<\Omega\mid F(\alpha)=\alpha\}\cap\gamma=B'\cap\gamma,$$

and (6) holds.

It follows immediately from 4.15 that the intersection of $<\gamma$ classes as in (5) is again as in (5). Let us finally deal with diagonal intersection. Let $\langle B_\alpha\mid\alpha<\Omega\rangle$ be a sequence of subclasses of Ω as in (5). $(D_{\alpha<\Omega}B_\alpha)\cap\gamma=\{\alpha<\Omega\mid\alpha\in\bigcap_{\beta<\alpha}B_\beta\}\cap\gamma=\{\alpha<\gamma\mid\alpha\in\bigcap_{\beta<\alpha}(B_\beta\cap\gamma)\}=D_{\alpha<\gamma}(B_\alpha\cap\gamma)$. By Theorem 4.20, applied to γ instead of Ω, $D_{\alpha<\gamma}(B_\alpha\cap\gamma)$ is a closed unbounded subset of γ, hence also $(D_{\alpha<\Omega}B_\alpha)\cap\gamma$ is a closed unbounded subset of γ, thereby establishing (5). □

4.33 More on the Size of the Weakly Inaccessible Cardinals. As we mentioned when we started to describe the process of getting sparser and sparser closed unbounded classes, a major aim of that was to be able to shed some more light on the question of the size of the weakly inaccessible cardinals. For this purpose we shall observe the status of the weakly inaccessible cardinals in the class A of all alephs. Since $\aleph$ is the enumerating function of A, and since by definition a weakly inaccessible cardinal is a regular cardinal $\aleph_\gamma$ where γ is a limit ordinal, we get that case (b) of 4.32 holds here, and we have $\aleph_\gamma = \gamma$, which we know already from 3.20(ii). Moreover, by the last part of 4.32, $\gamma \in A^{(2)}$ and then γ is the γ-th ordinal α such that $\aleph_\alpha = \alpha$, and so on. Since γ does not get omitted by diagonal intersection γ is even the γ-th ordinal α such that $\alpha \in A^{(\beta)}$ for all $\beta < \alpha$, and so on. This gives us very high lower estimates on the size of the least weakly inaccessible cardinal. Already the least α such that $\aleph_\alpha = \alpha$ is a huge cardinal, not to speak of the least β such that β is the β-th cardinal α such that $\aleph_\alpha = \alpha$ and so on. Whatever we said just now on the status of the least weakly inaccessible cardinal in the class of all alephs applies equally well to the status of the least inaccessible cardinal in the class $\{\beth_\alpha \mid \alpha \in \mathrm{On}\}$ of all beths (see 3.25 and 3.27). □

We have already mentioned, in 4.17, that if Ω is an ordinal then the closed unbounded subsets of Ω generate a filter on Ω which is, if one assumes the axiom of choice, Θ-complete. The complements of the sets in this filter, namely the subset of Ω which are disjoint from at least one closed unbounded subset of Ω, constitute an ideal, which, if one assumes the axiom of choice, is Θ-complete. As we mentioned in 4.7, members of an ideal can be regarded as insignificant sets.

4.34 Definition. A is called an *insignificant* subclass of Ω if A is disjoint from some closed unbounded subclass of Ω.

4.35 Proposition. *In the case where Ω is an ordinal, the insignificant subsets of Ω constitute an ideal which is, if one assumes the axiom of choice, Θ-complete. Every bounded subset of Ω is insignificant, but there are also unbounded insignificant subsets of Ω, such as $\{\alpha+1 \mid \alpha < \Omega\}$.* □

4.36 Definition (Bloch 1953). A subclass of Ω which is not insignificant is called *stationary*. (The reason for the choice of this name is given later). I.e., $A \subseteq \Omega$ is a stationary subclass of Ω if $A \cap B \neq 0$ for every closed unbounded subclass B of Ω.

4.37 Remarks. In the case where $\Omega = \mathrm{On}$ the definition of "A is stationary" involves a universal class quantifier "for all B". Therefore in this case the statement "A is stationary" can be validly asserted only if A is given by a class term τ and then the statement is asserted by the schema "$\tau \subseteq \Omega$ and if B is a closed unbounded subclass of Ω then $B \cap \tau \neq 0$", where B varies over all class terms. The definition of "A is insignificant" involves in the case $\Omega = \mathrm{On}$ an existential class quantifier "there exists B" and is to be interpreted, as mentioned in I.4.8, not as a statement of the extended language, but as a promise to provide a class term $\tau(y_1, \ldots, y_n)$ and a proof of $\exists y_1 \exists y_2 \ldots \exists y_n (A \subseteq \Omega \wedge \tau(y_1, \ldots, y_n)$ is a closed unbounded subclass of $\Omega \wedge A \cap \tau(y_1, \ldots, y_n) = 0$). Throughout the rest of the section we shall disregard the case $\Omega = \mathrm{On}$, letting the interested reader provide the interpretation for the case $\Omega = \mathrm{On}$ of what we say.

The situation with the closed unbounded, insignificant and stationary subclasses of Ω can be made more familiar to some of the readers by observing its analogy to the situation of the measurable subsets of the unit interval $[0, 1]$ on the real line. Classes which include closed unbounded classes correspond to sets of measure one (with Θ corresponding to $\aleph_1$ since the intersection of $\aleph_0$ sets of measure one is again a set of measure one), insignificant classes correspond to sets of measure zero, and stationary classes correspond to sets of positive measure.

4.38 Proposition. (i) *Every subclass of Ω which includes a stationary subclass of Ω is stationary. Every closed unbounded subclass of Ω is a stationary subclass of Ω.*

(ii) *If A is a closed unbounded subclass of Ω and B is a stationary subclass of Ω then also $A \cap B$ is a stationary subclass of Ω.*

(iii) *For every regular limit ordinal $\gamma < \Theta$, $\{\alpha < \Omega \mid \mathrm{cf}(\alpha) = \gamma\}$ is a stationary subclass of Ω.*

Hint of proof. (iii) If C is a closed unbounded subclass of Ω and F is its enumerating function then $F(\gamma) \in C \cap \{\alpha < \Omega \mid \mathrm{cf}(\alpha) = \gamma\}$. □

4.39 Definition (Bloch 1953). Let F be a function on a subclass of Ω into Ω. We call F *regressive* if for every $\alpha \in \mathrm{Dom}(F) \sim \{0\}$ we have $F(\alpha) < \alpha$.

4.40 Theorem (Fodor 1956). *Let A be a subset of Ω. The following conditions* (a), (b) *on A are equivalent.*

(a) *A is stationary.*

(b) *For every regressive function F on A there is an $\eta < \Omega$ such that $F^{-1}[\eta]$ is an unbounded subclass of Ω, i.e., values $< \eta$ are obtained by the function F for arbitrarily large $\alpha < \Omega$.*

If the axiom of choice holds then (a) *and* (b) *are equivalent also to:*

(c) *For every regressive function F on A there is an $\eta < \Omega$ such that $F^{-1}[\eta]$ is a stationary subset of Ω.*

Proof. (c) → (b). This is trivial since every stationary class is unbounded, by 4.35.

(b) → (a). We shall show that if A is not stationary then there is a regressive function on A such that for every $\eta < \Omega$ $F^{-1}[\eta]$ is a bounded subclass of Ω. Assume that A is an insignificant subclass of Ω, then there is a closed unbounded subclass B of Ω disjoint from A. Let β be the least member of B. Define F on A by

$$F(\alpha) = \begin{cases} \text{the greatest ordinal } \gamma < \alpha \text{ in } B & \text{if } \alpha > \beta, \\ 0 & \text{if } \alpha \leqslant \beta. \end{cases} \tag{7}$$

To see that for $\alpha > \beta$ there is an ordinal γ as in (7) take $\gamma = \sup(B \cap \alpha)$; clearly $\gamma \leqslant \alpha$. Since in this case $B \cap \alpha$ is a non-void bounded subclass of the closed class B we have $\gamma = \sup(B \cap \alpha) \in B$. $\alpha \in A$, $y \in B$ and $A \cap B = 0$, therefore $\gamma \neq \alpha$ and hence $\gamma < \alpha$ and

γ is the greatest ordinal $<\alpha$ in B. Since $\gamma<\alpha$ we have $F(\alpha)<\alpha$ and F is a regressive function. To see that (b) fails we notice that from the definition of F it follows that for every $\eta<\Omega$ $F^{-1}[\eta]\subseteq\gamma<\Omega$, where γ is the least member of the unbounded subclass B of Ω which is $\geqslant\eta$.

(a)$\to$ (b), and if we assume the axiom of choice then also (a)$\to$ (c). Let A be a stationary class and let F be a regressive function on A. If (c) fails to hold then for every $\eta<\Omega$ the class $F^{-1}[\eta]$ is not stationary and hence there is a closed unbounded class disjoint from it. Using our assumption that Ω is an ordinal and that the axiom of choice holds we define $\langle C_\eta \mid \eta<\Omega\rangle$ by

(8) $\qquad C_\eta=$ a closed unbounded subset of Ω disjoint from $F^{-1}[\eta]$.

If we make the stronger assumption that (b) fails, i.e., that for every η $F^{-1}[\eta]$ is bounded then we define $\langle C_\eta \mid \eta<\Omega\rangle$ by

(9) $\qquad C_\eta=\Omega\sim\sup{}^{+}F^{-1}[\eta],$

and for this definition we do not have to assume that Ω is an ordinal or that the axiom of choice holds and we have anyway that C_η is a closed unbounded subclass of Ω disjoint from $F^{-1}[\eta]$. Let $D=D_{\eta<\Omega}C_\eta$; by 4.20 D is a closed unbounded subclass of Ω once we show that the sequence $\langle C_\eta \mid \eta<\Omega\rangle$ is descending. If $\langle C_\eta \mid \eta<\Omega\rangle$ is defined by (9) it is obviously descending, but definition (8) requires some tinkering if we want $\langle C_\eta \mid \eta<\Omega\rangle$ to be descending. Let $P\subseteq\Omega$ be a cofinal subset of Ω of order type Θ. We replace (8) by

(10) $\qquad$ for $\xi\in P$ $E_\xi=$ a closed unbounded subset of Ω disjoint from $F^{-1}[\xi]$, and for $\eta<\Omega$
$C_\eta=\bigcap_{\xi\in P\cap(\lambda+1)}E_\xi$, where λ is the least member of P which is $\geqslant\eta$.

Θ is a regular infinite ordinal and is hence a cardinal and a limit ordinal. Since P is of order type Θ and $\lambda\in P$, $|P\cap(\lambda+1)|<\Theta$ and therefore C_η is a closed unbounded subclass of Ω, being the intersection of $<\Theta$ such classes (4.15). Since, by (10), $C_\eta\subseteq E_\lambda$, where $\lambda\in P$ and $\lambda\geqslant\eta$, we get $C_\eta\cap F^{-1}[\lambda]=0$ and hence $C_\eta\cap F^{-1}[\eta]=0$. $\langle C_\eta \mid \eta<\Omega\rangle$ is obviously descending.

Now we have established that $D=D_{\eta<\Omega}C_\eta$ is a closed unbounded subclass of Ω. Since A is stationary we get, by 4.38(ii), that $A\cap D$ is stationary; let $\beta\in A\cap D\sim\{0\}$. Since $\beta\in A$, $F(\beta)$ is defined. Since $\beta\in D$, $\beta\in C_\eta$ for every $\eta<\beta$, hence $\beta\notin F^{-1}[\eta]$ for every $\eta<\beta$ (because $C_\eta\cap F^{-1}[\eta]=0$), i.e., $F(\beta)\neq\eta$ for every $\eta<\beta$ and thus $F(\beta)\geqslant\beta$. However, $F(\beta)\geqslant\beta$ contradicts the regressiveness of F. □

For a regular Ω we can strengthen Theorem 4.40 as follows.

4.41 Theorem (Alexandroff and Urysohn 1929, Fodor 1956). *Let Ω be regular and let A be a stationary subclass of Ω. For every regressive function F on A there is a $\xi<\Omega$ such that $F^{-1}[\{\xi\}]$ is an unbounded subclass of Ω, i.e., the value ξ is obtained by the function F for arbitrarily large $\alpha<\Omega$. If the axiom of choice holds then, under the same assumptions, there is a $\xi<\Omega$ such that $F^{-1}[\{\xi\}]$ is a stationary subset of Ω.*

Remark. This is where the term stationary comes from. A subclass of Ω is stationary if every regressive function on it is stationary in the sense that it obtains a certain value again and again for an unbounded number of times.

Proof. By Theorem 4.40 there is an $\eta<\Omega$ such that $F^{-1}[\eta]$ is an unbounded subclass of Ω. We shall see that for some $\xi<\eta$ $F^{-1}[\{\xi\}]$ is an unbounded subclass of Ω. Suppose that for every $\xi<\eta$ $F^{-1}[\{\xi\}]$ is bounded in Ω; then if we define G on η by $G(\xi)=\sup^{+} F^{-1}[\{\xi\}]$ we have $G(\xi)<\Omega$ for $\xi<\eta$. By 3.4(ii) $G[\eta]$ is a bounded subclass of Ω; let $\gamma<\Omega$ be a bound of $G[\eta]$. $F^{-1}[\eta]=\bigcup_{\xi<\eta}F^{-1}[\{\xi\}]\subseteq \bigcup_{\xi<\eta}G(\xi)\subseteq\gamma<\Omega$, contradicting what we got that $F^{-1}[\eta]$ is an unbounded subclass of Ω.

If the axiom of choice holds then, by Theorem 4.40, there is an $\eta<\Omega$ such that $F^{-1}[\eta]$ is a stationary subset of Ω. For some $\xi<\eta$ $F^{-1}[\{\xi\}]$ is a stationary subset of Ω, since if no such set $F^{-1}[\{\xi\}]$ were stationary then, by 4.35, also $F[\eta]=\bigcup_{\xi<\eta}F^{-1}[\{\xi\}]$ were not stationary. □

4.42 Proposition. *Let C be a closed unbounded subclass of Ω, let F be the enumerating function of C and let $\Lambda=\mathrm{Dom}(F)$ $(\Theta\leqslant\Lambda\leqslant\Omega)$.*

(i) *For every subclass A of Λ, A is a closed unbounded subclass of Λ iff $F[A]$ is a closed unbounded subclass of Ω.*

(ii) *For every subclass B of Ω, B is a stationary subclass of Ω iff $F^{-1}[B]$ is a stationary subclass of Λ.* □

4.43 Proposition. (i) *If A is a subclass of Ω then the subclass C of Ω obtained from A by adding to A all its limit points which are $<\Omega$ (i.e., $C=\{\sup b \mid b$ is a subset of A bounded in $\Omega\}$) is the least closed subclass of Ω which includes A, and is therefore called the* closure *of A (in Ω).*

(ii) *For every unbounded subclass A of Ω the order type of the closure C of A equals the order type of A.*

(iii) *Ω has a closed unbounded subclass of order type Θ.*

Hint of proof. (ii) Use the isomorphism F of C into A given by $F(\alpha)=$ the least member of A greater than α. □

4.44 Exercise. (i) Give a direct proof of Theorem 4.41, i.e., without using Theorem 4.40.

(ii) Use Theorem 4.41 to prove that in Theorem 4.40 (a) implies (b) and (c).

Hint. Use 4.43(iii) and 4.42(ii). □

4.45 On Obtaining Small Stationary Classes. The only subclasses of Ω which we know till now to be stationary are the classes which include a closed unbounded subclass of Ω. As the reader can easily verify, a stationary subclass A of Ω does not include a closed unbounded subclass of Ω iff also $\Omega\sim A$ is stationary. Thus our quest for stationary subclasses of Ω which do not include closed unbounded classes leads us to look for a decomposition $\Omega=A\cup B$ of Ω into two disjoint stationary subclasses. Such a decomposition is easily obtained if Θ is greater than the least uncountable regular ordinal γ ($\gamma=\omega_1$ if the axiom of choice holds); in this case we define $A=\{\alpha<\Omega \mid \mathrm{cf}(\alpha)\leqslant\omega\}$, $B=\Omega\sim A$. In order to prove that both A and B are stationary we must show that for every closed unbounded class C, $C\cap A\neq 0$ and $C\cap B\neq 0$. Let F be the enumerating function of such a class C. Since F is normal we get, by 3.19, $\mathrm{cf}(F(\omega))=\omega$, hence $F(\omega)\in C\cap A$, and $\mathrm{cf}(F(\gamma))=\gamma>\omega$, hence $F(\gamma)\in C\sim A=C\cap B$. The proof of the same result for the case where Θ is the least uncountable regular cardinal requires the axiom of choice and is considerably more difficult. Theorem 4.48 below will cover this case. As is customary with mathematicians we generalize the question about the decomposition of Ω into two disjoint stationary subclasses in two steps as follows. First we ask whether Ω is the union of Ω pairwise disjoint stationary subclasses, and then we ask whether every stationary subclass of Ω is the union of Ω pairwise disjoint stationary classes. Theorem 4.48 gives a positive answer for the case where Ω is a regular ordinal, under the assumption that the axiom of choice holds.

4.46 Definition. If $A\subseteq\Omega$ and $\alpha<\Omega$ then α is said to be a *stationary point of A* if $\mathrm{cf}(\alpha)>\omega$ and $A\cap\alpha$ is a stationary subset of α.

4.47 Proposition. *Let A be a stationary subset of Ω. The class of the members of A which are not stationary points of A is a stationary subclass of Ω.*

Outline of proof. If the conclusion of the proposition fails to hold then there is a closed unbounded subclass K of Ω such that every member of $K\cap A$ is a stationary point of A. Let L be the class of all limit points of K, $L\subseteq K$. Let β be the least member of $L\cap A$. $\mathrm{cf}(\beta)>\omega$, and since $K\cap\beta$ is unbounded in β so is $L\cap\beta$; hence $(L\cap\beta)\cap(A\cap\beta)\neq 0$, which contradicts our choice of β. □

4.48Ac Theorem (Bloch 1953, Fodor 1966, Solovay 1971). *If Ω is a regular ordinal then every stationary subset A of Ω is the union of Ω pairwise disjoint stationary subsets of Ω.*

Outline of proof. Given a subset B of Ω we shall say that for *almost all* $\alpha\in B$ $\Phi(\alpha)$ holds, if the set $\{\alpha\in B \mid \Phi(\alpha) \text{ fails}\}$ is insignificant. Given a function G on B into Ω we say that G is *almost bounded* if there is a $\gamma<\Omega$ such that $G(\alpha)\leqslant\gamma$ for almost every $\alpha\in B$; such a γ is called an *almost bound* of G.

(i) If on some stationary subset B of A there is a regressive function F which is not almost bounded then B, and hence also A, is the union of Ω pairwise disjoint stationary sets.

To prove (i) use 4.41 to prove that there are arbitrarily large λ's such that for a stationary subset D of B $F[D]=\{\lambda\}$.

By (i) it suffices to prove now that on some stationary subset B of A there is a regressive function F which is not almost bounded; we shall prove it by deriving a contradiction from the assumption that every regressive function on a stationary subset B of A is almost bounded.

(ii) For every $\lambda<\Omega$ the set $\{\alpha\in A \mid \mathrm{cf}(\alpha)\leqslant\lambda\}$ is insignificant.

To prove (ii) set $B = \{\alpha \in A \mid \mathrm{cf}(\alpha) \leqslant \lambda\}$. For $\alpha \in B \sim \{0\}$ let f_α be a mapping of λ on a cofinal subset of α. For $\mu < \lambda$ define g_μ on $B \sim \{0\}$ by $g_\mu(\alpha) = f_\alpha(\mu)$. Assume that B is stationary; let $\beta_\mu < \Omega$ be an almost bound of g_μ on B and let $\beta = \sup_{\mu < \lambda} \beta_\mu$. $\beta < \Omega$ is therefore greater than almost all the members of B, which is a contradiction.

Now we proceed to the proof of the contradiction. Let $D = \{\alpha \in A \mid \mathrm{cf}(\alpha) > \omega \wedge \alpha$ is not a stationary point of $A\}$; by 4.47 and (ii) D is a stationary subset of Ω. For $\gamma \in D$ there is a closed unbounded subset K_γ of γ such that $K_\gamma \cap A = 0$. Let f_γ be the normal function on an ordinal $\leqslant \gamma$ which enumerates K_γ, and let g_ξ, for $\xi < \Omega$, be the function on D defined by

$$g_\xi(\gamma) = \begin{cases} f_\gamma(\xi) & \text{if} \quad \xi \in \mathrm{Dom}(f_\gamma) \\ 0 & \text{if} \quad \xi \notin \mathrm{Dom}(f_\gamma). \end{cases}$$

Each g_ξ is regressive on D, therefore let δ_ξ be the least almost bound of g_ξ. $\langle \delta_\xi \mid \xi < \Omega \rangle$ is a normal function on Ω, and hence, by 4.24, $\{\xi < \Omega \mid \delta_\xi = \xi\rangle$ is a closed unbounded subset of Ω. Since A is stationary there is an $\eta \in A \cap \{\xi < \Omega \mid \delta_\xi = \xi\}$. Since $\delta_\eta = \eta$ we have for almost all $\gamma \in D$ $f_\gamma(\eta) = \eta$ or $\eta \notin \mathrm{Dom}(f_\gamma)$. By (ii) there is such a $\gamma \in D$ with $\mathrm{cf}(\gamma) > \eta$, hence $\eta \in \mathrm{Dom}(f_\gamma)$ and therefore $f_\gamma(\eta) = \eta$. Thus $\eta \in \mathrm{Rng}(f_\gamma) = K_\gamma$, contradicting $\eta \in A$. □

4.49Ac Exercise. To how many pairwise disjoint stationary subsets can one decompose a stationary subset of a singular Ω? □

Chapter V
The Axiom of Choice and Some of its Consequences

The axiom of choice turned out to be a most important and interesting axiom of set theory. While we succeeded to develop a considerable part of set theory without using the axiom of choice this axiom is essential for most of the advanced quantitative and combinatorial theory in set theory, as well as for many applications in various mathematical fields.

In § 1 the axiom of choice is introduced and discussed, and several statements of set theory and of cardinal arithmetic are proved to be equivalent to it. In § 2 we look at several weaker versions of the axiom of choice and we partially establish the relations between them. In § 3 we handle the concept of a definable set and we discuss the existence of definable well-orderings and orderings of various familiar sets. In § 4 we discuss the existence of a global choice function and of a well-ordering of the whole universe. We study there an extention ZFGC of ZFC introduced so as to guarantee their existence. In § 5 we go into a detailed study of the exponential functions $\langle 2^{\mathfrak{a}} \mid \mathfrak{a} \in \mathrm{Cn} \rangle$ and $\langle \mathfrak{a}^{\mathfrak{b}} \mid \mathfrak{a}, \mathfrak{b} \in \mathrm{Cn} \rangle$. We discuss also assumptions which simplify these functions, such as the generalized continuum hypothesis. We conclude § 5 by studying sets of almost disjoint sets.

Even though we introduce the real numbers only in Chapter VII we shall use them, and their well known properties, already in the present chapter.

1. *The Axiom of Choice and Equivalent Statements*

1.1 The Axiom of Choice (Beppo Levi 1902; Erhard Schmidt in 1904—see Zermelo 1904). For every set a there exists a function f on a such that for every non-void $x \in a$, $f(x) \in x$.

1.2 Definition. If F is a function on a class A such that for every $x \in A$, $x \neq 0 \rightarrow F(x) \in x$ then F is said to be a *choice function* on A (since F chooses the member $F(x)$ out of all the members of x). By the axiom of choice there is a choice function on every set.

1.3 Why is the Axiom of Choice Needed? As was shown by Paul Cohen 1963, the axiom of choice is unprovable in ZF, but it is still worthwhile to look at possible ways which one might have taken in trying to prove the axiom of choice in ZF. The principal tool for proving the existence of sets in ZF is the use of instances of the axiom schema of comprehension I.2.3. The version of this axiom schema which is suitable for defining functions is the rule of explicit definition of functions (I.6.15) which, for our present purposes, can be formulated as follows. "Given a set term $\tau(x)$, for every set a there is a function f on a such that for every $x \in a$, $f(x) = \tau(x)$". In order to use this rule to prove the existence of a choice function on a set a, one must find a term $\tau(x)$ which will single out one member of x for every $x \in a$; for a general set a we see no way of getting such a term $\tau(x)$ (and indeed there is no such term). This argumentation does not rule out the possibility that we could prove the existence of a choice function f on a in an indirect way, and this possibility makes the question of the independence of the axiom of choice such a deep question. On those sets a for which we can provide a term $\tau(x)$ as above we obtain a choice function f without using the axiom of choice. For example, if $\bigcup a$ is well-orderable, then let $<$ be a well-ordering of $\bigcup a$ and we can define a choice function on x by setting

$$f(x) = \begin{cases} \text{the least member of } x \text{, with respect to } < & \text{if } x \neq 0, \\ 0 & \text{if } x = 0; \end{cases}$$

the existence of f is provable in ZF (by the rule of explicit definition of functions—I.6.15).

What leads mathematicians to adopt the axiom of choice as an axiom of set theory, in addition to the opportunistic reason that it enables them to prove many theorems, is the following consideration. The basic idea behind the axiom of comprehension is that every collection of given objects should be a set (or at least a class). The only collections which we can handle easily are those we called in §I.2 specifiable, namely those collections which can be described as the collection of all x's such that $\phi(x)$, where $\phi(x)$ is a formula of the (basic) language. However, there is little reason to assume that only the specifiable collections should be admitted as sets. One may choose to introduce also other collections as sets, such as the collection f obtained by the mental process of simultaneously choosing for every non-void $x \in a$ a member y of x and putting all these pairs $\langle x, y \rangle$ in f (together with the pair $\langle 0, 0 \rangle$, if $0 \in a$); the f thus obtained is a choice function on a.

The well-ordering theorem, which we have already seen to be a very useful theorem, is shown in 1.7 below to be an equivalent form of the axiom of choice. However, it is difficult to give a direct intuitive justification of the well-ordering theorem. Such a justification for the existence of a well-ordering of a set a would involve a process of choosing members of a one after the other till a well-ordering of a is obtained. Such a process is much less intuitive than that which leads to the acceptance of the axiom of choice, since it involves carrying out many choices one after the other for a well ordered sequence of steps. This is by far less simple than the simultaneous choice we use for the axiom of choice which can be carried out "in a single step". It is interesting to notice that the intuitive construction leading to the axiom of choice can actually be carried out in ZF in the case where a is a finite set as we shall now see.

1.4 Proposition (Whitehead and Russell 1912). *On every finite set a there is a choice function f.*

Proof. By induction on the finite set a (III.1.6). For $a=0$ $f=0$ is a choice function. For a finite a, let g be a choice function on a and let $z \notin a$. If $z \neq 0$ then z has a member y, set $f=g \cup \{\langle z, y\rangle\}$; if $z=0$ set $f=g \cup \{\langle 0,0\rangle\}$. In either case f is a choice function on $a \cup \{z\}$. □

1.5 Proposition (Russell 1906). *The axiom of choice is equivalent to each one of the following statements.*

(i) *For every indexed family $\langle w_x \mid x \in c\rangle$ of non-void sets the Cartesian product $\bigtimes_{x \in c} w_x$ is non-void.*

(ii) *For every set b of pairwise disjoint non-void sets (i.e., $x, y \in b \wedge x \neq y \rightarrow x \cap y=0$) there is a set u which has exactly one member in common with each member b. (Why is the requirement that the members of b be pairwise disjoint necessary?)*

Proof. (i) Given $\langle w_x \mid x \in c\rangle$, let $a=\{w_x \mid x \in c\}$ and let f be a choice function on a then $\langle f(w_x) \mid x \in c\rangle \in \bigtimes_{x \in c} w_x$. In the other direction, if a is given and $0 \notin a$ then every member of $\bigtimes_{x \in a} x$ is a choice function on a.

(ii) If a is a set such that $0 \notin a$ then the set $b=\{\{x\} \times x \mid x \in a\}$ is a set of non-void pairwise disjoint sets, and if $u \subseteq \bigcup b$ is such that for every member y of b $|u \cap y|=1$ then u is a choice function on a. The other direction is easy. □

We shall now establish the equivalence of several statements of set theory with the axiom of choice. Out of those statements the well-ordering theorem and Zorn's lemma are among the most central consequences of the axiom of choice, being the forms of the axiom of choice used in almost all applications of the axiom. In fact, whenever we proved till now theorems by means of the axiom of choice all we did was to use the well-ordering theorem. The other statements to be handled here are easy consequences of the well-ordering theorem considered mainly because they are equivalent to the axiom of choice. For a very comprehensive treatment of statements equivalent to the axiom of choice, taken from various areas of mathematics, see Rubin and Rubin 1963. There is a twofold aim in studying a large variety of statements equivalent to the axiom of choice. First, it shows the usefulness of the axiom of choice in many areas of mathematics; second, these equivalences establish the "stability" of the axiom of choice, i.e., they show that the axiom of choice is not a random product of the historical development of mathematics but a fundamental principle which is not likely to be replaced by some other axiom which is not equivalent to it.

1.6Ac The Well-Ordering Theorem (Conjectured by Cantor 1883; proved by Zermelo 1904). *Every set is well-orderable.*

Proof. Let a be a set, then, by the axiom of choice, there is a choice function f on the power set $\mathbf{P}(a)$ of a. We define now by recursion a function H on an initial segment of On into a as follows.

$$H(\alpha)=f(a\sim H[\alpha]) \quad \text{as long as} \quad a\sim H[\alpha]\neq 0.$$

This is a definition by recursion as long as some condition is met, established in II.2.12. Obviously, $H(\alpha)\in a$ for every $\alpha\in \mathrm{Dom}(H)$. Let us show that H is one–one. For $\alpha, \beta\in \mathrm{Dom}(H)$, $\alpha<\beta$, we have $a\sim H[\beta]\neq 0$ and hence $H(\beta)=f(a\sim H[\beta])\in a\sim H[\beta]$, i.e., $H(\beta)\notin H[\beta]$, while $H(\alpha)\in H[\beta]$ and thus $H(\alpha)\neq H(\beta)$. $\mathrm{Dom}(H)$ cannot be all of On since in such a case H would be an injection of the proper class On into the set a; therefore, $\mathrm{Dom}(H)$ is an ordinal γ. (In fact, since H is an injection of γ in a we have $\gamma<|a|^{+}$.) $H(\gamma)$ is not defined therefore the condition $a\sim H[\beta]\neq 0$ does no longer hold for $\beta=\gamma$, thus $a\sim H[\gamma]=0$ and since $\mathrm{Rng}(H)\subseteq a$ we get $\mathrm{Rng}(H)=a$. H is a bijection of γ on a and hence it induces on a a well-ordering of order type γ (II.1.5(iv)). □

1.7 Corollary. *The well-ordering theorem is equivalent to the axiom of choice.*

Proof. By 1.6 the axiom of choice implies the well-ordering theorem; we have to prove the other direction. Let a be a set, then, by the well-ordering theorem there is a well-ordering $<$ of $\bigcup a$. By means of $<$ we can easily define a choice function on a (as done in 1.3). □

1.8 Proposition (Hartogs 1915). *The statement "Any two cardinals $\mathfrak{a}$ and $\mathfrak{b}$ are comparable, i.e., $\mathfrak{a}\leqslant\mathfrak{b}$ or $\mathfrak{b}\leqslant\mathfrak{a}$" (or, "the partial ordering of the cardinals is a total ordering") is equivalent to the well-ordering theorem, and hence also to the axiom of choice.*

Proof. By the well-ordering theorem every set is well-orderable and hence every cardinal is an ordinal; thus the class of all cardinals is a subclass of On and, by III.2.20, the natural partial ordering $<$ of the cardinals is the same as their ordering as ordinals, which is a total ordering.

Let us assume now the comparability of the cardinals and prove the well ordering theorem. The comparability of cardinals is obviously equivalent to the comparability of sets, i.e., that for all sets a and b, $a\preccurlyeq b$ or $b\preccurlyeq a$. Let a be any set; by Hartogs' theorem (III.2.25) there is an ordinal α such that $\alpha\not\preccurlyeq a$, hence, by the comparability of sets $a\preccurlyeq\alpha$. Thus there exists an injection h of a into α, and h^{-1} induces a well ordering of a (II.1.5(iv)). □

A very useful consequence of the axiom of choice, which we shall prove to be equivalent to the axiom of choice, is the following theorem, which is referred to as a lemma for reasons of tradition.

1.9Ac Zorn's Lemma (Hausdorff 1914). *Let $\langle u, <\rangle$ be a partially ordered set such that*

(1) *every subset v of u which is totally ordered by $<$ has an upper bound (in u)*

then u has a maximal member z with respect to the partial ordering $<$, i.e., there is no $x\in u$ such that $x>z$.

Remark. Notice that (1) is required to hold also in the case where v is the null-set; in this case (1) just means that u is non-void. Be careful not to overlook the case $v=0$ when applying Zorn's lemma.

Proof. Let f be a choice function on the power set $\mathbf{P}(u)$ of u. We define by recursion a function H on an initial segment of On into u as follows.

$H(\alpha)=f$ (the set of strict upper bounds of $H[\alpha]$),
as long as $H[\alpha]$ has at least one strict upper bound.

If $\alpha, \beta \in \mathrm{Dom}(H)$ and $\alpha<\beta$ then, by the definition of H, $H(\beta)$ is a strict upper bound of the set $H[\beta]$ which contains $H(\alpha)$, and hence $H(\alpha)<H(\beta)$ and H is an increasing function. Since H is an injection into the set u, $\mathrm{Dom}(H)$ cannot be On and it is therefore an ordinal γ. By the definition of H, $H[\gamma]$ has no strict upper bound in u. However, $H[\gamma]$ is totally ordered by $<$ (since H is increasing—II.1.13 and II.1.11(i)) and hence, by (1), $H[\gamma]$ has an upper bound z. z is a maximal member of u, since if there were a member y of u such that $y>z$ then y would be a *strict* upper bound of $H[\gamma]$ which, we know, is impossible. □

1.10 Theorem (Zorn 1935). *Zorn's lemma is equivalent to the axiom of choice.*

Proof. By 1.9 the axiom of choice implies Zorn's lemma; let us now use Zorn's lemma to prove the axiom of choice. Let a be any set; we shall show that there is a choice function on a. Let u be the set of all choice functions on subsets of a. We shall now apply Zorn's lemma to the partially ordered set $\langle u, \subset\rangle$, where $\subset$ denotes the relation of proper inclusion. Let v be a subset of u which is totally ordered by $\subset$, i.e., for all $g, h \in v$ we have $g \subseteq h$ or $h \subseteq g$, which says that g and h are compatible. By I.6.30(ii) $f=\bigcup v$ is a function. $\mathrm{Dom}(f)=\bigcup\{\mathrm{Dom}(g) \mid g \in v\} \subseteq a$. For $x \in \mathrm{Dom}(f)$ we have $f(x)=g(x)$ for some $g \in v$; since such a g is a choice function we have, if $x \neq 0$, $f(x)=g(x) \in x$. We have seen that f is a choice function on a subset of a, therefore $f \in u$ and $f=\bigcup v$ is obviously an upper bound of v in u. Thus condition (1) of 1.9 is fulfilled and by Zorn's lemma u has a maximal member h. We claim that $\mathrm{Dom}(h)=a$, and thus h is a choice function on a, which is what we set out to get. If $\mathrm{Dom}(h) \subset a$ let $x \in a \sim \mathrm{Dom}(f)$. If $x \neq 0$ let y be a member of x, and if $x=0$ let $y=0$. Since h is a choice function on a subset of a so is also, obviously, $h \cup\{\langle x, y\rangle\}$ which properly includes h, contradicting the maximality of h. □

1.11 Exercises. Many of the consequences of the axiom of choice follow in a very neat way from Zorn's lemma. Prove the following statements directly from it.

(i) The well-ordering theorem.

(ii) (Zorn 1944.) The comparability of sets, i.e., for any two sets a and b $a \preccurlyeq b$ or $b \preccurlyeq a$.

In the presence of the axiom of choice every infinite cardinal is an aleph and therefore, by III.3.13 and III.3.23, for all infinite cardinals $\mathfrak{a}$, $\mathfrak{a}+\mathfrak{a}=\mathfrak{a}$ and $\mathfrak{a} \cdot \mathfrak{a}=\mathfrak{a}$. These equalities can be proved from Zorn's lemma without any use of the theory of well ordered sets, as we shall now see.

(iii) For every infinite set a, $a \approx a \times 2$ (use also (ii)).

(iv) (Zorn 1944) For every infinite set a, $a \approx a \times a$ (use also (ii) and (iii)).

(v) (de Bruijn and Erdös 1951). We say that g is a *graph* on a set a if $g \subseteq \{\{x, y\} \mid x, y \in a \wedge x \neq y\}$. If $\{x, y\} \in g$ we say that x and y are *joined* by the *edge* $\{x, y\}$ of g. For a subset b of a a *coloring of b with n colors consistent with g* is a function f on b into n such that for all $x, y \in b$ $\{x, y\} \in g \rightarrow f(x) \neq f(y)$. (The colors are $0, 1, \ldots, n-1$. If $f(x)=i$ then we say that x is coloured with the color i. The requirement is that two points which are joined by an edge should not be colored with the same color). Prove that, for $n<\omega$, if every finite subset of a has a coloring with n colors consistent with g then a itself has such a coloring.

Hints. (i) To well-order a set a, take for u of 1.9 the set of all well ordered sets $\langle b, r\rangle$ where $b \subseteq a$ and $r \subseteq b \times b$, and take for $<$ the relation given by: $\langle b, r\rangle < \langle c, s\rangle$ if $\langle b, r\rangle$ is a section of $\langle c, s\rangle$.

(iii) Take for u in 1.9 the set of all bijections of subsets b of a on $b \times 2$, and take for $<$ proper inclusion. Let f be a maximal member of u, then $a \sim \mathrm{Dom}(f)$ is finite since otherwise $a \sim \mathrm{Dom}(f)$ includes a denumerable set and f is not maximal. Finally prove $|\mathrm{Dom}(f)| = |a|$.

(iv) Using Zorn's lemma get a maximal function f such that f is a bijection of b on $b \times b$, where $b \subseteq a$. If $b \preccurlyeq a \sim b$ take a subset c of $a \sim b$ such that $c \approx b$ and use (iii) to extend f to a bijection of $b \cup c$ on $(b \cup c) \times (b \cup c)$. If $a \sim b \preccurlyeq b$ then, by (iii), $b \approx a$.

(v) Consider the finitely extendible colorings consistent with g on subsets b of a, i.e., colorings f of b such that for every finite subset c of a f can be extended to a coloring of $b \cup c$ consistent with g. □

1.12 Zorn's Lemma Eliminates Recursion. In applications Zorn's lemma often replaces the use of definition by recursion of a function H where one proceeds as long as some condition is met (II.2.12) in situations where at the recursion step one may have to choose $H(\alpha)$ as an arbitrary member of some set. This is exactly the kind of recursion we used to prove Zorn's lemma itself in 1.9. In that proof the recursion step consisted of setting $H(\alpha) = f$(the set of strict upper bounds of $H[\alpha]$), where f was a choice function obtained by the axiom of choice; as far as the proof is concerned all we needed was that $H(\alpha)$ be a strict upper bound of $H[\alpha]$. To see an example of how the use of Zorn's lemma replaces recursion as described above let us consider 1.11(ii). We can prove that for any two sets a, b $a \preccurlyeq b$ or $b \preccurlyeq a$ as follows. We define a function H on an initial segment of On by setting

$$H(\alpha) = \text{some } \langle x, y\rangle \text{ such that } x \in a \sim \mathrm{Dom}(H[\alpha]) \text{ and } y \in b \sim \mathrm{Rng}(H[\alpha])$$
$$\text{as long as both } a \sim \mathrm{Dom}(H[\alpha]) \text{ and } b \sim \mathrm{Rng}(H[\alpha]) \text{ are non-void.}$$

Such an $H(\alpha)$ can be specified completely by a choice function on $\mathbf{P}(a \cup b)$ or by a well ordering of $a \cup b$. As easily seen, one gets that $\mathrm{Dom}(H)$ is an ordinal γ and $\mathrm{Rng}(H)$ is an injection of a into b or a bijection of a subset of a onto b. If one compares this proof to the proof of the same result by Zorn's lemma one sees that the two proofs contain the same ingredients, except that the use of Zorn's lemma replaces the definition by recursion.

1.13 Lemma. *Let $\mathfrak{a}$ be a cardinal and let $\mathfrak{b}$ be a well ordered cardinal. If $\mathfrak{a}\cdot\mathfrak{b}\leqslant\mathfrak{a}+\mathfrak{b}$ then $\mathfrak{a}$ and $\mathfrak{b}$ are comparable (i.e., $\mathfrak{a}\leqslant\mathfrak{b}$ or $\mathfrak{b}\leqslant\mathfrak{a}$).*

Proof. Let a be a set of cardinality $\mathfrak{a}$ disjoint from $\mathfrak{b}$, then since $\mathfrak{a}\cdot\mathfrak{b}\leqslant\mathfrak{a}+\mathfrak{b}$ there is an injection h of $a\times\mathfrak{b}$ into $a\cup\mathfrak{b}$. We distinguish now two cases.

Case 1. There is $x\in a$ such that $h[\{x\}\times\mathfrak{b}]\subseteq a$. Then, since h is an injection we get $|h[\{x\}\times\mathfrak{b}]|=|\{x\}\times\mathfrak{b}|=\mathfrak{b}$, and hence $\mathfrak{b}\leqslant|a|=\mathfrak{a}$.

Case 2. If Case 1 does not hold then for every $x\in a$ there is a $\beta\in\mathfrak{b}$ such that $h(x,\beta)\notin a$, i.e., $h(x,\beta)\in\mathfrak{b}$. Define now the function f on a by $f(x)=$ the least β such that $h(x,\beta)\in\mathfrak{b}$. The function g on a given by $g(x)=h(x,f(x))$ is clearly an injection of a into $\mathfrak{b}$ and hence $\mathfrak{a}=|a|\leqslant\mathfrak{b}$. □

1.14 Proposition (Tarski 1924). *The axiom of choice is equivalent to the statement "For every infinite cardinal $\mathfrak{a}$, $\mathfrak{a}^2=\mathfrak{a}$."*

Proof. If the axiom of choice holds then, by the well-ordering theorem, every cardinal is a well ordered cardinal and hence every infinite cardinal is an aleph. Since, by III.3.23 $\aleph_\alpha^2=\aleph_\alpha$ for every α, we have $\mathfrak{a}^2=\mathfrak{a}$ for every infinite $\mathfrak{a}$. (For an alternative proof see 1.11(iv)). Assume now that $\mathfrak{a}^2=\mathfrak{a}$ for every infinite cardinal $\mathfrak{a}$. We shall now prove the well-ordering theorem by showing that every infinite cardinal is a well ordered cardinal. Let $\mathfrak{b}$ be an infinite cardinal, then such is also $\mathfrak{b}+\mathfrak{b}^+$. Since $\mathfrak{a}=\mathfrak{a}^2$ for every infinite $\mathfrak{a}$ we get

$$(2)\qquad \mathfrak{b}+\mathfrak{b}^+=(\mathfrak{b}+\mathfrak{b}^+)^2=\mathfrak{b}^2+2\cdot\mathfrak{b}\cdot\mathfrak{b}^++(\mathfrak{b}^+)^2.$$

Since $2\cdot\mathfrak{b}\cdot\mathfrak{b}^+$ is a term in the right hand side of (2) we get $\mathfrak{b}\cdot\mathfrak{b}^+\leqslant\mathfrak{b}+\mathfrak{b}^+$. Since $\mathfrak{b}^+$ is a well ordered cardinal we get, by Lemma 1.13, that $\mathfrak{b}$ and $\mathfrak{b}^+$ are comparable. By the definition of $\mathfrak{b}^+$ (see III.2.26) $\mathfrak{b}^+\nleqslant\mathfrak{b}$, hence $\mathfrak{b}\leqslant\mathfrak{b}^+$. Since $\mathfrak{b}^+$ is a well ordered cardinal so is also $\mathfrak{b}$ (III.2.16). □

Until now all the equivalences with the axiom of choice were proved in a "local" way, i.e., all sets used in the proofs were related to the sets with which we started. Therefore these proofs did not require the use of the axiom of foundation. The equivalence which we handle next is different in this respect since it makes an essential use of the axiom of foundation.

1.15 Proposition (H. Rubin 1960). *The axiom of choice is equivalent to the statement "The power set of every ordinal is well-orderable".*

Proof. If we assume the axiom of choice then, by the well-ordering theorem, every set is well-orderable and in particular every power set of an ordinal.

Let us assume now that the power set of every ordinal is well-orderable, and prove that every set a is well-orderable. Since by the axiom of foundation every set a is a subset of $R(\alpha)$ for some α (II.7.5) it is sufficient to prove that every $R(\alpha)$

can be well ordered. First let us make a simple remark. Let $\langle b, <\rangle$ be any well ordered set, then there is an isomorphism f of $\langle b, <\rangle$ into $\langle \beta, <\rangle$ for some ordinal β. By our hypothesis there is a relation r which well-orders $\mathbf{P}(\beta)$. Also $\mathbf{P}(b)$ is now well ordered by the relation $\{\langle u, v\rangle \mid u, v \subseteq b \wedge f[u]\, r\, f[v]\}$ which we shall call the well-ordering induced on $\mathbf{P}(b)$ by f and r. (This concept is very closely related but not quite identical with the concept of an induced relation of II.1.5(iv)). Thus we have seen that the power set of a well-orderable set is well-orderable.

Now let us try to prove by induction on α that $R(\alpha)$ is well-orderable. $R(0)=0$ is obviously well-orderable. Assuming that $R(\alpha)$ is well-orderable we have, since $R(\alpha+1)=\mathbf{P}(R(\alpha))$ (II.6.12(vii)), that also $R(\alpha+1)$ is well-orderable. If α is a limit ordinal then $R(\alpha)=\bigcup_{\beta<\alpha}R(\beta)$ (II.6.12(viii)), and by the induction hypothesis each $R(\beta)$, for $\beta<\alpha$, is well-orderable, but this does not suffice to well-order $R(\alpha)$. For this purpose we must have a *uniform* well-ordering of all $R(\beta)$'s, i.e. a function s on α such that for every $\beta<\alpha$ $s(\beta)$ is a well-ordering of $R(\beta)$. If there is such an s, the relation

$$(3) \qquad \{\langle u, v\rangle \mid u, v \in R(\alpha) \wedge (\rho(v)<\rho(v) \vee \rho(u)=\rho(v) \wedge u\, s(\rho(u)+1)v)\}$$

is clearly a well-ordering of $R(\alpha)$. In order to get such a uniform well-ordering s we have to start all over again and define s on α. Let us take a sufficiently large ordinal δ; it will be only at the end of the proof that we shall see how large we have to choose δ. We pick also a particular well-ordering r of $\mathbf{P}(\delta)$. We define s by induction as follows. $s(0)=0$, this is obviously a well-ordering of $R(0)$. Given $s(\gamma)$, for $\gamma<\alpha$, let ξ be the order type of $\langle R(\gamma), s(\gamma)\rangle$ and let f be the unique isomorphism of $\langle R(\gamma), s(\gamma)\rangle$ on $\langle \xi, <\rangle$, which is, if $\xi \leqslant \delta$, also an isomorphism into $\langle \delta, <\rangle$. We set $s(\gamma+1)$ to be the well-ordering of $R(\gamma+1)=\mathbf{P}(R(\gamma))$ induced by f and the well-ordering r of $\mathbf{P}(\delta)$. If γ is a limit ordinal then $s(\gamma)$ is defined as in (3) above, namely

$$s(\gamma)=\{\langle u, v\rangle \mid u, v \in R(\gamma) \wedge (\rho(u)<\rho(v) \vee \rho(u)=\rho(v) \wedge \langle u, v\rangle \in s(\rho(u)+1))\}$$

For the induction step to work we must have $\mathrm{Ord}(\langle R(\gamma), s(\gamma)\rangle)<\delta$ for every $\gamma<\alpha$; this can be secured by choosing $\delta=(\sup_{\gamma<\alpha}|R(\gamma)|)^{+}$. Notice that $|R(\gamma)|$ is a well ordered cardinal by the induction hypothesis that for all $\gamma<\alpha$, $R(\gamma)$ is well-orderable. □

1.16 Exercise (Katuzi Ono). Prove that the following statement is equivalent to the axiom of choice. For every set b there is a set $u \subseteq \bigcup b$ such that u has at most one member in common with each member of b, and such that no $v \subseteq \bigcup b$ which properly includes u has this property. □

1.17 Exercises (Tarski 1924). Prove each of the following statements to be equivalent to the axiom of choice.

(a) For all infinite cardinals $\mathfrak{a}$, $\mathfrak{b}$, $\mathfrak{a}+\mathfrak{b}=\mathfrak{a}\cdot\mathfrak{b}$.

(b) Every infinite cardinal $\mathfrak{a}$ is a square (i.e., there is a $\mathfrak{b}$ such that $\mathfrak{a}=\mathfrak{b}^2$).

(c) For all infinite cardinals $\mathfrak{a}$, $\mathfrak{b}$, if $\mathfrak{a}^2=\mathfrak{b}^2$ then $\mathfrak{a}=\mathfrak{b}$.

(d) For all cardinals $\mathfrak{a}$, $\mathfrak{b}$, $\mathfrak{c}$, $\mathfrak{d}$, if $\mathfrak{a}<\mathfrak{b}$ and $\mathfrak{c}<\mathfrak{d}$ then $\mathfrak{a}+\mathfrak{c}<\mathfrak{b}+\mathfrak{d}$.
(e) For all cardinals $\mathfrak{a}$, $\mathfrak{b}$, $\mathfrak{c}$, $\mathfrak{d}$, if $\mathfrak{a}<\mathfrak{b}$ and $\mathfrak{c}<\mathfrak{d}$ then $\mathfrak{a}\cdot\mathfrak{c}<\mathfrak{b}\cdot\mathfrak{d}$.
(f) For all cardinals $\mathfrak{a}$, $\mathfrak{b}$, $\mathfrak{c}$, if $\mathfrak{a}+\mathfrak{c}<\mathfrak{b}+\mathfrak{c}$ then $\mathfrak{a}<\mathfrak{b}$.
(g) For all cardinals $\mathfrak{a}$, $\mathfrak{b}$, $\mathfrak{c}$, if $\mathfrak{a}\cdot\mathfrak{c}<\mathfrak{b}\cdot\mathfrak{c}$ then $\mathfrak{a}<\mathfrak{b}$.

Hints. (b) Given a cardinal $\mathfrak{e}$ take $\mathfrak{a}$ such that $\mathfrak{a}^2=\mathfrak{e}+\mathfrak{e}^+$. Prove $\mathfrak{e}^+\leqslant\mathfrak{a}$ and let $\mathfrak{a}=\mathfrak{e}^++\mathfrak{c}$. Use $\mathfrak{a}^2=\mathfrak{e}+\mathfrak{e}^+$ to prove $\mathfrak{e}^+\cdot\mathfrak{c}\leqslant\mathfrak{e}+\mathfrak{e}^+$; hence $\mathfrak{c}\leqslant\mathfrak{e}^+$ and $\mathfrak{e}$ is well ordered. (c) Given $\mathfrak{e}$, take $\mathfrak{a}\geqslant\mathfrak{e}$ such that $\mathfrak{a}^2=\mathfrak{a}$. Show that $(\mathfrak{a}+\mathfrak{a}^+)^2=(\mathfrak{a}\cdot\mathfrak{a}^+)^2$, hence $\mathfrak{a}+\mathfrak{a}^+=\mathfrak{a}\cdot\mathfrak{a}^+$. Thus $\mathfrak{a}$ and $\mathfrak{e}$ are well ordered. (d) Given $\mathfrak{e}$, take $\mathfrak{a}\geqslant\mathfrak{e}$ such that $\mathfrak{a}+\mathfrak{a}=\mathfrak{a}$. If $\mathfrak{a}$ is not well ordered get a contradiction from $\mathfrak{a}<\mathfrak{a}+\mathfrak{a}^+$ and $\mathfrak{a}^+<\mathfrak{a}+\mathfrak{a}^+$. (f) If $\mathfrak{e}$ is not well ordered then $\mathfrak{e}^++\mathfrak{e}^+<\mathfrak{e}+\mathfrak{e}^+$. □

1.18 Exercises (Tarski 1954a). (i) Prove that for every cardinal $\mathfrak{a}$ there is a cardinal $\mathfrak{b}>\mathfrak{a}$ such that for no cardinal $\mathfrak{c}$ does $\mathfrak{a}<\mathfrak{c}<\mathfrak{b}$.

(ii) Prove that the following statement is equivalent to the axiom of choice: For every cardinal $\mathfrak{a}$ there is a cardinal $\mathfrak{b}>\mathfrak{a}$ such that for every cardinal $\mathfrak{c}>\mathfrak{a}$ we have $\mathfrak{c}\geqslant\mathfrak{b}$.

Hints. (i) Take $\mathfrak{b}=\mathfrak{a}+\mathfrak{a}^+$. (ii) Use (i) and prove, by contradiction, that the condition of 1.14 holds. □

2. Some Weaker Versions of the Axiom of Choice

As we have already seen in § 1, and as we shall continue to see, the axiom of choice is a very strong axiom indeed. It is interesting to discuss various weaker versions of this axiom. Candidates for such weaker versions are statements Φ which are theorems of ZFC. Not all the theorems of ZFC are equally interesting from this point of view. One would tend to pay attention to Φ if Φ can replace the full axiom of choice in some body of applications, or if Φ is stable in the sense that many different natural statements turn out to be equivalent to Φ. The first question we ask about Φ is whether it is already a theorem of ZF, in which case Φ does not interest us in this context. The next question we ask is whether Φ is equivalent to the axiom of choice. (A positive answer to this question implies a negative answer to the first question.) If the answer to both these questions is negative then Φ can be taken as a weaker version of the axiom of choice and we can ask how it relates to the other weaker versions of the axiom; i.e., given another weaker version Ψ of the axiom of choice we can ask whether Φ implies Ψ in ZF and whether Ψ implies Φ in ZF. Why are these questions interesting? First, in proving a theorem in set theory it is often interesting to see which part of the full power of the axiom of choice is needed to obtain it, since applying a scalpel is more elegant than applying an axe. Second, one is sometimes led by other considerations to study an extension of ZF which does not contain the full axiom of choice but contains some weaker versions of that axiom; it is helpful if one knows beforehand what can be proved from those weaker versions. On the negative side, proving that various weaker versions of the axiom of choice do not imply other weaker versions of this axiom has created an interesting body of proofs in the metamathematics of set theory. In the present section we take a very small sample of the weaker versions of the axiom of choice consisting of three such statements; a fourth one, the Boolean prime ideal theorem, will be discussed in § VIII.2.

Throughout the rest of the present section "implies" will mean "implies in ZF".

2.1 The Axiom of Choice for Denumerable Sets—AC_ω. There is a choice function on every denumerable set a.

The existence of choice functions on *finite* sets a is a theorem of ZF (1.4). AC_ω goes just one step further, as it deals with a denumerable a. This step is sufficient to make AC_ω unprovable in ZF; AC_ω is not provable in ZF even for the special cases where a is a set of (infinite) sets of real numbers (Cohen 1963) or a set of pairs of sets of real numbers (Feferman 1965). Notice that if a is a set of finite sets of real numbers then we can obtain in ZF a choice function f on a by defining, for all $x \in a, f(x) =$ the least member of x in the natural ordering of the real numbers. While AC_ω is unprovable in ZF it is not equivalent to the axiom of choice; we shall mention below that it does not imply, in particular, a certain consequence DC_ω of the axiom of choice.

2.2 Proposition. *Assume the axiom of choice for denumerable sets* AC_ω, *then:*

(i) *The union of* $\aleph_0$ *countable sets is countable.*

(ii) ω_1 *is a regular ordinal.*

(iii) (Whitehead and Russell 1912.) *Every infinite set includes a denumerable set, or, equivalently, every Dedekind finite set is finite.*

Proof. (i) Let $\langle w_x \mid x \in u \rangle$ be such that $|u| = \aleph_0$ and $|w_x| \leqslant \aleph_0$ for $x \in u$. Let $\mathfrak{a} = \Sigma_{x \in u} |w_x|$; by III.4.6(v) $\mathfrak{a} = \Sigma_{x \in u} |w_x| \leqslant \Sigma_{x \in u} \aleph_0 = |u| \cdot \aleph_0 = \aleph_0 \cdot \aleph_0 = \aleph_0$. Let G be the function on u given by $G(x) =$ the set of all injections of w_x into $\mathfrak{a}$. By the definition of $\mathfrak{a}$ $\mathfrak{a} \geqslant |w_x|$ hence $G(x) \neq 0$ for $x \in u$. By AC_ω (and 1.4) there is a choice function F on the countable set $\{G(x) \mid x \in u\}$. The function $\langle F(G(x)) \mid x \in u \rangle$ provides a uniform injection of all w_x's into $\mathfrak{a}$. Therefore, by III.4.5, $|\bigcup_{x \in u} w_x| \leqslant \Sigma_{x \in u} |w_x| = \mathfrak{a} \leqslant \aleph_0$.

(ii) We shall prove (ii) from AC_ω by proving that (i) implies (ii) without any use of AC_ω. Since ω_1 is a limit ordinal $\mathrm{cf}(\omega_1)$ must be an infinite cardinal $\leqslant \omega_1$ (by IV.3.8). Let us assume that ω_1 is singular and obtain a contradiction. If ω_1 is singular then $\mathrm{cf}(\omega_1) < \omega_1$ hence $\mathrm{cf}(\omega_1) = \omega$. Then ω_1 has a cofinal subset b of order type ω. Since ω_1 is a limit ordinal we have, as easily seen, $\bigcup b = \bigcup \omega_1 = \omega_1$. $\bigcup b$ is a countable union of countable sets, hence, by (i), $\bigcup b = \omega_1$ is countable, which is a contradiction.

(iii) Let a be an infinite set. Let Q be a function on ω such that for $n < \omega$ $Q(n)$ is the set of all injections of n into a. Since a is infinite $Q(n) \neq 0$ (by III.1.27). By AC_ω there is a choice function F on the denumerable set $\mathrm{Rng}(Q)$. Let G be the function on ω such that $G(n) = F(Q(n))$ for $n < \omega$, then $G(n)$ is an injection of n into a. Writing the sequences $G(0), G(1), \ldots$ one after the other and erasing all occurrences of each member u of a other than its first occurrence yields an infinite sequence H of length ω of different members of a. H is indeed infinite since for every n the sequence $G(n)$ consists of n different terms. The formal definition of the sequence H is left to the reader. $\mathrm{Rng}(H)$ is a denumerable subset of a. □

2.3 The Axiom of ω Dependent Choices—DC_ω (Bernays 1942). Let a be a set, let $u \in a$ and let R be a relation. If for every $x \in a$ there is a $y \in a$ such that xRy, then there is a sequence $\langle z_n \mid n<\omega\rangle$ of members of a such that $z_0 = u$ and for all $n<\omega$ $z_n R z_{n+1}$.

2.4 Proposition (Bernays 1942). (i) *The axiom* DC_ω *of* ω *dependent choices is a theorem of* ZFC.

(ii) *The axiom* DC_ω *of* ω *dependent choices implies the axiom of choice* AC_ω *for denumerable sets.*

Proof. (i) Given a, u and R as in 2.3 we define the sequence $\langle z_n \mid n<\omega\rangle$ by recursion as follows $z_0 = u$, $z_{n+1} =$ a member of a such that $z_n R z_{n+1}$; by our assumption in 2.3 there is such a z_{n+1}.

(ii) Let $\{b_n \mid n<\omega\}$ be a denumerable set, with $b_m \neq b_n$ for $m \neq n$. Let $a = {}^{\omega}(\bigcup_{n<\omega} b_n)$, i.e., a is the set of all finite sequences of members of $\bigcup_{n<\omega} b_n$, and

$$R = \{\langle x, y\rangle \mid \exists n(x \in a \wedge \mathrm{Dom}(x) = n \wedge \mathrm{Dom}(y) = n+1 \wedge x \subseteq y \wedge (b_n \neq 0 \rightarrow y(n) \in b_n)\}.$$

By DC_ω there is a sequence $\langle z_n \mid n<\omega\rangle$ such that $z_0 = 0$ and for all $n<\omega$ $z_n R z_{n+1}$. One can easily prove by induction that $\mathrm{Dom}(z_n) = n$ for all $n<\omega$. Since $z_n R z_{n+1}$ we get that if $b_n \neq 0$ then $z_{n+1}(n) \in b_n$. Therefore if we define $f = \{\langle b_n, z_{n+1}(n)\rangle \mid n<\omega\}$ then f is a choice function on $\{b_n \mid n<\omega\}$. □

2.5 Exercise. Let a be a set of real numbers and let t be an accumulation point of a (i.e., $(\forall \varepsilon>0)(\exists z \in a)(-\varepsilon < t - z < \varepsilon)$).

(i) Prove, using AC_ω, that there is a sequence $\langle x_n \mid n<\omega\rangle$ of members of a such that $\lim_{n\to\infty} x_n = t$.

(ii) Prove the same result, for the case where a is a set of rational numbers, without using any weak form of the axiom of choice (use VII.1.4). □

2.6 Proposition. DC_ω *is equivalent to each of the following statements*

(a) *Let a be a non-void set and R a relation such that* $(\forall x \in a)(\exists y \in a)xRy$, *then there is a sequence* $\langle z_n \mid n<\omega\rangle$ *of members of a such that* $z_n R z_{n+1}$ *for all* $n<\omega$. (*This is like* DC_ω *except that the first member u of* $\langle z_n \mid n<\omega\rangle$ *is not predetermined.*)

(b) *If s is a relation which is not well founded then there is a sequence* $\langle z_n \mid n<\omega\rangle$ *such that* $z_{n+1} s z_n$ *for all* $n<\omega$.

Proof. $\mathrm{DC}_\omega \rightarrow$ (a) is trivial.

(a) $\rightarrow \mathrm{DC}_\omega$. Let a, R and u be as in DC_ω. Let $R^* = \{\langle x, y\rangle \mid$ there is an R-chain from x to $y\}$ (II.4.14), and let $a' = a \cap R^*[\{u\}]$. We shall show below that a' satisfies the assumptions of (a) for a and therefore, by (a), there is a sequence $\langle z'_n \mid n<\omega\rangle$ of members of $a' \subseteq a$ such that $z'_n R z'_{n+1}$ for $n < \omega$. Since $z'_0 \in a' \subseteq R^*[\{u\}]$ there is a finite sequence $\langle w_n \mid n \leqslant k \rangle$, $k > 0$, which is an R-chain leading from $u = w_0$ to $z'_0 = w_k$. Combining the sequences w and z' to a sequence $\langle z_n \mid n < \omega \rangle$ by setting $z_n = w_n$ for $n < k$ and $z_{k+m} = z'_m$ for $m < \omega$ we get $z_0 = w_0 = u$ and $z_n R z_{n+1}$ for all $n < \omega$, which is the conclusion of DC_ω. In order to complete the proof we still have to show that a' satisfies the assumptions of (a) for a. By our assumptions there is a $v \in a$ such that uRv, hence $v \in a'$ and $a' \neq 0$. If $x \in a' \subseteq a$ then there is a $y \in a$ such that xRy. Since $x \in a' \subseteq R^*[\{u\}]$ we have uR^*x, and since also xRy we get uR^*y (by II.4.20), i.e., also $y \in a'$. Thus we have $a' \neq 0 \wedge (\forall x \in a')(\exists y \in a')xRy$, which is the hypothesis of (a).

(a) → (b). Let s be a relation which is not well founded. Then, by the definition of well-foundedness there is a non-void set a which has no s-minimal member, i.e., $(\forall x \in a)(\exists y \in a)ysx$. We apply (a) to a and $R = s^{-1}$ and get a sequence $\langle z_n \mid n < \omega \rangle$ as required by (b).

(b) → (a). Let R and a be as in (a), then $s = R^{-1} \mid a$ is not well founded, since the non-void set a has no R^{-1}-minimal member. By (b) there is a sequence $\langle z_n \mid n < \omega \rangle$ such that for all $n < \omega$ $z_{n+1} s z_n$, hence $z_n \in a$ and $z_n R z_{n+1}$, which establishes (a). □

The second part of II.5.3(ii) asserts that, assuming the axiom of choice, the condition "every non-void set has an R-minimal member" in the definition of well-foundedness can be equivalently replaced by the condition "there is no sequence x of length ω such that $x_{n+1} R x_n$ for all $n < \omega$". By 2.6 it suffices to assume DC_ω, rather than the full axiom of choice, in order to obtain this equivalence.

2.7 The Relation of AC_ω to DC_ω. We have already shown that the axiom of choice implies DC_ω. DC_ω does not imply the axiom of choice; we shall mention in 2.10 below that DC_ω does not even imply a certain weaker version AC_{ω_1} of the axiom of choice. As to the relationship of DC_ω to AC_ω, by 2.4(ii) DC_ω implies AC_ω, but the reverse implication does not hold (Jensen 1966). This is no wonder once we see what each of AC_ω and DC_ω asserts. We shall use here the following version of AC_ω whose equivalence with 2.1 is obvious: For every sequence $\langle w_n \mid n < \omega \rangle$ of non-void sets there is a sequence $\langle z_n \mid n < \omega \rangle$ such that for all $n < \omega$ $z_n \in w_n$. To compare this version of AC_ω with DC_ω we observe from which set we have to choose z_n in each case. In the case of AC_ω we choose z_n from the set w_n which does not depend on any component of $\langle z_m \mid m < n \rangle$; in the case of DC_ω we choose z_n, for $n > 0$, from the set $\{x \mid z_{n-1} R x\}$ which depends on z_{n-1}. This is why DC_ω is called the axiom of (ω) dependent choices.

2.8 The Axiom of Choice for Well-Orderable Sets—AC_{wo}. There is a choice function on every well-orderable set.

2.9 Proposition (Jensen 1967). AC_{wo} *implies* DC_ω.

Proof. We shall prove version 2.6(b) of DC_ω by showing that if s is a relation for which there is no sequence $\langle z_n \mid n < \omega \rangle$ such that $z_{n+1} s z_n$ for all $n < \omega$, then s is a well-founded relation. Let a be the field of s, then $s \subseteq a \times a$. A natural way of proving that s is well founded is to obtain a function λ on a into On such that for all $x, y \in a$ if xsy then $\lambda(x) < \lambda(y)$ (II.5.14(ii)). What we know right away is that if b is a *well-orderable* subset of a then, by II.5.3(ii), the relation $s \mid b$ is well-founded since there is no sequence $\langle z_n \mid n < \omega \rangle$ such that for all $n < \omega$ $z_n \in b \wedge z_{n+1}\, s z_n$. Therefore on b we have the function $\langle \rho_{s|b}(x) \mid x \in b \rangle$ where $\rho_{s|b}(x)$ is the rank of x with respect to the well-founded relation $s \mid b$ (II.5.12). We set for $x \in b$ $\tau(b, x) = \rho_{s|b}(x)$, and we define the function λ on a as follows.

(9) for $x \in a$ $\lambda(x) = \sup\{\tau(b, x) \mid b \text{ is a well-orderable subset of } a \text{ and } x \in b\}$.

We can now prove that for $x, y \in a$, if xsy then $\lambda(x) < \lambda(y)$, which is all we need to prove our proposition. First we prove that the sup in (9) is really attained, i.e., for every $x \in a$ there is a well-orderable subset c of a such that $x \in c$ and $\lambda(x) = \tau(c, x)$. If $\lambda(x) = 0$ then $\lambda(x) = \tau(\{x\}, x)$; if $\lambda(x)$ is a successor ordinal then the sup in (9) is attained, by II.3.31; so we are left to deal with the case where $\lambda(x)$ is a limit ordinal. Set $\alpha = \lambda(x)$, then, by the definition of α, for every $\beta < \alpha$ there is a well-orderable subset b of a such that $x \in b$ and $\tau(b, x) \geqslant \beta$. Let G be the function on α defined by

(10) $G(\beta) = \{f \mid f$ is an injection of an ordinal into a such that $x \in \text{Rng}(f)$ and $\tau(\text{Rng}(f), x) \geqslant \beta\}$.

(The right-hand side of (10) is a set since it is a subclass of $\mathbf{P}(a^+ \times a)$). By what we have just said, $G(\beta) \neq 0$ for $\beta < \alpha$. Applying AC_{wo}, let F be a choice function on the well-orderable set $\{G(\beta) \mid \beta < \alpha\}$ and let H be the function on α given by $H(\beta) = F(G(\beta))$. $H(\beta)$ is an injection of an ordinal into a such that $x \in \text{Rng}(H(\beta))$ and $\tau(\text{Rng}(H(\beta)), x) \geqslant \beta$. Since the sets $\text{Rng}(H(\beta))$, $\beta < \alpha$, are uniformly well-ordered their union $c = \bigcup_{\beta < \alpha} \text{Rng}(H(\beta))$ is a well-orderable subset of a (by the proof of III.4.5). Also $x \in \text{Rng}(H(0)) \subseteq c$. Since $c \supseteq \text{Rng}(H(\beta))$ for every $\beta < \alpha$ we have $s \mid c \supseteq s \mid \text{Rng}(H(\beta))$ and therefore, by II.5.14(iv),

$$\tau(c, x) = \rho_{s|c}(x) \geqslant \rho_{s|\text{Rng}(H(\beta))}(x) = \tau(\text{Rng}\, H(\beta), x) \geqslant \beta.$$

Since α is a limit ordinal we have $\tau(c, x) \geqslant \alpha$. Since also $\tau(c, x) \leqslant \lambda(x) = \alpha$, by the definition of $\lambda(x)$, we get $\tau(c, x) = \lambda(x)$, which is what we wanted to show.

Let $x, y \in a$ and xsy, then, as we have shown, there is a well-orderable subset c of a such that $x \in c$ and $\lambda(x) = \tau(c, x)$. Since c is well-orderable so is also $c \cup \{y\}$ and we have, by II.5.14(i) and II.5.14(iv)),

$$\lambda(y) \geqslant \tau(c \cup \{y\}, y) = \rho_{s|(c \cup \{y\})}(y) > \rho_{s|(c \cup \{y\})}(x) \geqslant \rho_{s|c}(x) = \lambda(x).$$

Thus we have shown $xsy \rightarrow \lambda(x) < \lambda(y)$ which, as we have mentioned above, completes our proof. □

2.10 On the Strength of DC_ω **and** AC_{wo}. DC_ω does not imply AC_{wo}, it does not even imply the statement AC_{ω_1} which asserts that on every set a of cardinality $\aleph_1$

there is a choice function (Mostowski 1948, Jech 1966). Also does not imply the axiom of choice; it does not even imply the statement DC_{ω_1} which permits ω_1 choices such that each choice depends on all the earlier ones (Levy 1964, Pincus 1972).

2.11 Exercise. Formulate DC_{ω_1} (of 2.10). □

3. Definable Sets

3.1 On the Existence of Certain Orders and Well-Orderings. There are sets which can be proved in ZF to be well-orderable, such as the set ω of all natural numbers, or the set of all rational numbers. Other sets, such as the set $\mathbb{R}$ of all real numbers, or the set $\mathbf{P}(\omega)$ cannot be shown in ZF to be well-orderable (Cohen 1963). Of course we cannot prove in ZF that there is any set which is not well-orderable, since such a proof would amount to a refutation of the axiom of choice in ZF and this cannot be done (if ZF is consistent—see the introduction).

Sets like the set $\mathbb{R}$ of all real numbers and the set $\mathbf{P}(\omega)$ can be proved in ZF to be *orderable*; on $\mathbb{R}$ we have the ordering by magnitude, and on $\mathbf{P}(\omega)$ we have the *lexicographic ordering* (i.e., the ordering induced by the lexicographic ordering of the characteristic functions) which is

$$\{\langle u, v\rangle \mid (\exists n \in \omega)\,(n \notin u \wedge n \in v \wedge u \cap n = v \cap n)\}.$$

Yet sets like $\mathbf{P}(\mathbf{P}(\omega))$, or $\mathbf{P}(\mathbb{R})$, or the set ${}^{\mathbb{R}}\mathbb{R}$ of all real functions cannot even be shown in ZF to be *orderable* (Cohen 1963). The axiom of choice solves the question of the existence of orderings and well-orderings of the various sets by decree; the well-ordering theorem, which is a version of the axiom of choice, decrees that every set has a well-ordering. However, once we start asking questions about the nature of these well-orderings we are immediately faced with the fact that the well-ordering theorem asserts nothing, at least not directly, concerning these well-orderings in addition to their mere existence. Our enquiry concerning these well-orderings leads us to the concepts of a definable class and a definable set.

3.2 Definition. A *definable class* is a class given by a class term τ with no free variables.

Examples of definable classes are $V(=\{x \mid x=x\})$, $0(=\{x \mid x \neq x\})$, $\{0\}$ $(=\{x \mid x=0\})$, Wf, On, Cn, On$\cap$Cn, the cardinality function $|x|$ and the ordinal addition function $+$.

When we assert "for every definable class X, $\Phi(X)$" we mean the schema $\Phi(\tau)$ where τ varies over all class terms without free variables. The statement "there exists a definable class X such that $\Phi(X)$" does not represent a particular statement of set theory; when asserted as a theorem it is to be understood as a promise to provide a term τ with no free variables and to prove $\Phi(\tau)$. This is similar to our interpretation of "there exists a class X such that $\Phi(X)$" in I.4.8.

A *definable set* is a definable class which is a set.

Examples of definable sets are 0, $\{0\}$, ω, $\mathbf{P}(\omega)$, ${}^{\omega}\omega$, ω_1, the set of all prime natural numbers, the function $\langle n^2 \mid n<\omega \rangle$ and the set of all well-orderings of the set $\mathbb{R}$ of all real numbers.

When we say "for every definable set x, $\Phi(x)$" we mean the schema $\forall x(x=\tau \rightarrow \Phi(x))$, where τ varies over all class terms without free variables. The statement "there exists a definable set x such that $\Phi(x)$" does not represent a particular statement of set theory; when asserted as a theorem it is to be understood as a promise to provide a class term τ without free variables and to prove "τ is a set and $\Phi(\tau)$".

Statements of the form "there exists a definable set x such that . . ." may occur in somewhat more complicated situations than the one just mentioned, but the reader should have no trouble in interpreting them correctly. For example, we may assert a theorem of the form "If there is a definable set x such that $\Phi(x)$ then there is a definable set y such that $\Psi(y)$"; this is to be understood as a promise to provide for each class term τ without free variables a class term τ' without free variables and to prove the statement "If τ is a set and $\Phi(\tau)$ then τ' is a set and $\Psi(\tau')$."

3.3 Definable Sets. Definable sets can be given by set terms of the extended language which have no explicit or implicit free set or class variables, as indicated by the examples 0, $\{0\}$, ω, $\mathbf{P}(\omega)$, ${}^{\omega}\omega$ and ω_1 mentioned above. If σ is such a term then the class term $\{x \mid x \in \sigma\}$ is a class term with no free variables and it yields a definable set since $\exists y(y=\{x \mid x \in \sigma\})$ is obviously provable.

If F is a definable function (i.e., a definable class which is a function) and σ is a set term with no free variables then $F(\sigma)$ is a definable set. To see this, let F be given by the class term τ without free variables, then $F(\sigma)$ is given by the class term $\{x \mid \exists y(\langle \sigma, y \rangle \in \tau \wedge x \in y)\}$.

3.4 On Asserting the Existence of a Non-Definable Class. We cannot name an example of a non-definable class since once it can be named it is definable. Moreover, there is no way to give some meaning to the statement "there is a non-definable class". Of course once we have a structure $\langle a, r \rangle$ which is a model of the theory ZF then we can recognize among the subsets of a those which are *classes* of the model $\langle a, r \rangle$ and those that are *definable classes* of the model $\langle a, r \rangle$. Therefore it makes perfect sense to ask whether the model $\langle a, r \rangle$ of ZF has classes which are not definable classes. However, as the subject of models of ZF is not handled in the present book, so we shall not go into the matter of definable and undefinable classes in such models.

In the language of set theory we can make an assertion which will imply the existence of non-definable classes (and sets) in the following way. For some formula $\Phi(X)$ and term $\tau(y_1, \ldots, y_n)$, with no free variables other than indicated, we can assert the statement $\exists y_1 \ldots \exists y_n \Phi(\tau(y_1 \ldots y_n))$ together with the schema $\neg \Phi(\tau')$, where τ' varies over all terms with no free variables. Informally we have in this case the statements that there is a class X such that $\Phi(X)$, but there is no such definable class X.

3.5 On the Existence of Definable Orders and Well-Orderings. Let us return to the orders and well-orderings we discussed in 3.1 and look at them from the point of view of definability. The well-ordering of ω is clearly definable by the term $\{\langle x, y\rangle \mid x \in y \wedge y \in \omega\}$. In ZFC the axiom of choice implies the existence of well-orderings of $\mathbb{R}$ and $\mathbf{P}(\omega)$, but one cannot prove in ZFC that $\mathbb{R}$ or $\mathbf{P}(\omega)$ have *definable* well-orderings (Feferman 1965). The ordering of $\mathbb{R}$ by magnitude and the lexicographic ordering of $\mathbf{P}(\omega)$ are definable sets; the latter is given by the term $\{\langle u, v\rangle \mid u, v \subseteq \omega \wedge (\exists n \in \omega)(n \notin u \wedge n \in v \wedge n \cap u = n \cap v)\}$. As we have already mentioned in 3.1, we cannot prove in ZF that the sets $\mathbf{P}(\mathbf{P}(\omega))$, $\mathbf{P}(\mathbb{R})$ and ${}^{\mathbb{R}}\mathbb{R}$, have an ordering; in ZFC all these sets are of course even well-orderable, but one cannot prove in ZFC that any of these sets posesses even a definable order (Feferman 1965).

3.6 Exercise. We denote with $\mathbf{P}(\omega)/[\omega]^{<\omega}$ the set of the subsets of ω modulo finite sets, i.e., the set $\{\{y \subseteq \omega \mid y \triangle x \text{ is finite}\} \mid x \subseteq \omega\}$, where $y \triangle x$ is the symmetric difference of y and x, defined in IV.4.8(i). The operation $\langle \omega \sim x \mid x \subseteq \omega\rangle$ on $\mathbf{P}(\omega)$ "induces" on $\mathbf{P}(\omega)/[\omega]^{<\omega}$ the operation *minus* defined by

$$- = \langle \{\omega \sim x \mid x \in t\} \mid t \in \mathbf{P}(\omega)/[\omega]^{<\omega}\rangle.$$

It is easily seen that the operation $-$ is a permutation of $\mathbf{P}(\omega)/[\omega]^{<\omega}$ of order 2, i.e., the composition $--$ of $-$ with itself is the identity function $\langle x \mid x \in \mathbf{P}(\omega)/[\omega]^{<\omega}\rangle$ on $\mathbf{P}(\omega)/[\omega]^{<\omega}$.

Let us consider the following statements

(a) x is a well ordering of $\mathbf{P}(\omega)$,
(b) y is an ordering of $\mathbf{P}(\mathbf{P}(\omega))$,
(c) z is an ordering of $\mathbf{P}(\omega)/[\omega]^{<\omega}$,
(d) u is a choice function on the set $\{\{t, -t\} \mid t \in \mathbf{P}(\omega)/[\omega]^{<\omega}\}$.

Prove that there are *definable* functions f, g, h such that if x satisfies (a) then $y = f(x)$ satisfies (b), if y satisfies (b) then $z = g(y)$ satisfies (c), and if z satisfies (c) then $u = h(z)$ satisfies (d).

Since f, g and h are definable and since the value $j(t)$ of a definable function j at a definable argument t is definable, this proves also that if there is a definable x which satisfies (a) then there are also definable y, z and u which satisfy (b), (c) and (d).

It has been shown that in ZF one cannot prove that there is a u which satisfies (d) (Feferman 1965) and, therefore, one cannot prove in ZF that there are x, y, z which satisfy (a), (b) or (c). It has also been shown that even in ZFC one cannot prove that there is a *definable* u which satisfies (d) (Feferman 1965) and hence one cannot prove in ZFC that there are definable x, y and z which satisfy (a), (b) or (c). □

3.7 Exercise. The results which we have just now discussed apply also to sets which are even dearer to most mathematicians than the sets mentioned in (a)–(d) of 3.6.

(i) First let us consider the following statement.

(a′) x is a well-ordering of the set $\mathbb{R}$ of all real numbers.

Prove that there is a definable bijection of $\mathbf{P}(\omega)$ on $\mathbb{R}$. (Notice that the bijection given by the Cantor–Bernstein theorem is obtained from the given two injections via a definable function.) Hence, there is an x which satisfies (a) of 3.6 iff there is an x which satisfies (a′), and there is a definable x which satisfies (a) iff there is a definable x which satisfies (a′).

By 3.6, one cannot prove in ZF the existence of a well-ordering of $\mathbb{R}$, and one cannot prove in ZFC the existence of a definable well-ordering of $\mathbb{R}$.

(ii) Now let us consider the following statements.

(b′) y is an order on $\mathbf{P}(\mathbb{R})$.
(b″) y is an order on the set ${}^{\mathbb{R}}\mathbb{R}$ of all real functions.

Prove that there are definable bijections between any two of the three sets $\mathbf{P}(\mathbf{P}(\omega))$, $\mathbf{P}(\mathbb{R})$ and ${}^{\mathbb{R}}\mathbb{R}$ and hence the existence of a solution, or a definable solution, to (b) of 3.6 is equivalent to the existence of a solution of the same kind to (b′) and to (b″).

By 3.6, one cannot prove in ZF the existence of a solution y to (b′) or to (b″), and one cannot prove in ZFC the existence of a definable solution y to (b′) or to (b″).

(iii) Let us denote by $\mathbb{R}/\mathbb{Q}$ the set of all real numbers modulo the rational numbers, i.e., $\mathbb{R}/\mathbb{Q}=\{\{x+r \mid r\in\mathbb{Q}\} \mid x\in\mathbb{R}\}$, where $\mathbb{Q}$ is the set of all rational numbers. We define an operation $-$ on $\mathbb{R}/\mathbb{Q}$ by defining, for $t\in\mathbb{R}/\mathbb{Q}$, $-t=\{-x \mid x\in t\}$. The operation $-$ is easily seen to be a permutation of $\mathbb{R}/\mathbb{Q}$ of order 2. This is, in fact, the natural minus operation of the quotient group of the additive group of $\mathbb{R}$ by the additive group of $\mathbb{Q}$. We consider now the following statement.

(c′) z is an order on $\mathbb{R}/\mathbb{Q}$.

Prove that there is a definable bijection of $\mathbb{R}/\mathbb{Q}$ on $\mathbf{P}(\omega)/[\omega]^{<\omega}$. (Here you may have to use the fact that every real number can be represented in an almost unique way as $\Sigma_{n=1}^{\infty} i_n/n!$, where i_n is an integer, and for $n\geqslant 2$ $0\leqslant i_n<n$.) Therefore the existence of a solution, or a definable solution, to (c) of 3.6 is equivalent to the existence of a solution of the same kind to (c′).

By 3.6, one cannot prove in ZF the existence of a z as in (c′), and one cannot prove in ZFC the existence of a definable z as in (c′).

(iv) Finally we consider the following statement.

(d′) u is a choice function on $\{\{t, -t\} \mid t\in\mathbb{R}/\mathbb{Q}\}$.

Prove that (d) of 3.6 has a solution, or a definable solution, iff (d′) has a solution of the same kind, by providing appropriate definable injections of $\mathbb{R}/\mathbb{Q}$ into $\mathbf{P}(\omega)/[\omega]^{<\omega}$ and of $\mathbf{P}(\omega)/[\omega]^{<\omega}$ into $\mathbb{R}/\mathbb{Q}$.

By 3.6, one cannot prove in ZF the existence of a u as in (d′), and one cannot prove in ZFC the existence of a definable u as in (d′). □

3.8 On the Non-Definability of Sets Which Exist by the Axiom of Choice. We have by now seen several examples of statements $\Phi(x)$ for which one can prove in ZF the existence of a definable x such that $\Phi(x)$. We saw also examples of statements $\Psi(x)$ for which one can prove in ZFC, but not in ZF, the existence of sets x such that $\Psi(x)$, and yet one cannot prove in ZFC the existence of definable x's such that $\Psi(x)$. This points to close relationship, for a given statement $\Gamma(x)$, between the provability of $\exists x\Gamma(x)$ in ZF and the existence (in ZF) of a definable x such that $\Gamma(x)$ on one hand, and between the non-provability of $\exists x\Gamma(x)$ in ZF and the non-provability in ZFC of the existence of a definable x such that $\Gamma(x)$ on the other hand. The reader who is now ready to jump to far-reaching conclusions is invited to consider the following examples. Let $\Gamma(x)$ be the statement "x is a well-ordering of $\mathbb{R}$ if $\mathbb{R}$ is well-orderable and x is an ordering of $\mathbb{R}$ if $\mathbb{R}$ is not well-orderable". Obviously we can prove $\exists x\Gamma(x)$ in ZF (i.e., without using the axiom of choice), but we cannot prove even in ZFC the existence of a definable x such that $\Gamma(x)$, because such a proof would obviously establish in ZFC the existence of a definable well-ordering of $\mathbb{R}$, contradicting what we saw in 3.7. It is at least as easy to make up an example of a statement $\Gamma(x)$ such that $\exists x\Gamma(x)$ is not provable in ZF, but in ZFC one can prove the existence of a definable x such that $\Gamma(x)$ (have you found such a $\Gamma(x)$?). If the reader will claim that the counterexamples we mentioned here are unnatural he will not be totally wrong, but he will probably be at a loss suggesting a clear-cut distinction between natural and unnatural statements $\Gamma(x)$.

3.9 On the Existence of a Definable Global Choice Function. In 3.6 and 3.7 we saw that one cannot prove in ZFC the existence of definable members of certain definable non-void classes such as the set of all well-orderings of the set $\mathbb{R}$ of all real numbers. Let us ask whether a different situation may exist in some set theories T obtained from ZFC by addition of extra axioms. The best possible situation, from the point of view of definability, which may exist is that in such a T there is a definable *global choice function* C, i.e., a choice function C on V. In that case all the questions asked in 3.6 and 3.7 concerning the existence of several kinds of definable sets have a positive answer. For example, $C(\{x \mid x \text{ is a well-ordering of } \mathbb{R}\})$ is a definable well-ordering of the real numbers. In general, if a is a definable non-void set then $C(a)$ is a definable member of a (by 3.3) and if A is a definable non-void class then $C(\{x \mid x \text{ is a member of } A \text{ of least rank}\})$ is a definable member of A.

How far is the requirement that T have a definable global choice function removed from the down-to-earth requirement that there be a definable well-ordering of the real numbers, a definable ordering of the real functions, etc.? Not much indeed; if we want every definable set to have a definable member in T we have to make sure that a definable global choice function exists for T. (This can be easily shown by means of the concept of ordinal-definability mentioned below.)

It is worth noticing that, as follows easily from the proof of 4.1 below, there is in T a definable global choice function C iff there is in T a definable well-ordering of the whole universe V.

If we try to formulate directly the requirement that every non-void definable set have a definable member, or that there is a definable global choice function, we realize soon that this is impossible. However, in a rather roundabout way we can obtain a statement which says just that. This is the *axiom of ordinal-definability* which asserts that every set is ordinal-definable (Gödel in 1946—see Gödel 1965). An exact formulation of that axiom is beyond the scope of this book; it asserts, roughly, that every set x is definable from some ordinal number α (this does not mean that x is definable since α is not necessarily definable). The axiom of ordinal-definability is a consequence of the much stronger axiom of constructibility (§ IX.1).

4. Set Theory with Global Choice

4.1 Proposition.

(1) *There is a global choice function C*

iff

(2) *there is a well-ordering S of V.*

Proof. Given a well-ordering S of V we define C by $C(0)=0$ and for $x\neq 0$

$$C(x) = \text{the least member of } x \text{ with respect to } S.$$

Given a global choice function C we define S as follows. First we define a function H on V as follows. In the proof of the well-ordering theorem, 1.6, one starts with a choice function on the power set $\mathbf{P}(x)$ of a set x, called there a, and one constructs a well-ordering of x; we shall take $H(x)$ to be the well-ordering of x obtained in that way from the choice function $C \upharpoonright \mathbf{P}(x)$ on $\mathbf{P}(x)$. We set now

$$S=\{\langle x,y\rangle \mid \rho(x)<\rho(y) \vee \rho(x)=\rho(y) \wedge xH(R(\rho(x)+1))y\}.$$

We prove that S is a well-ordering as follows. If $0\neq u\subseteq S$ let α be the least ordinal such that $u\cap R(\alpha)\neq 0$, then the least member of $u\cap R(\alpha)$ with respect to $H(R(\alpha))$ is the least member of u with respect to S. S is left-narrow since for every x $S^{-1}[\{x\}]\subseteq R(\rho(x)+1)$. Notice that in this direction of the proof we make an essential use of the axiom of foundation. □

4.2 How to Obtain Global Choice. (1) implies directly the axiom of choice in ZF, also (2) implies directly the well-ordering theorem. Thus (1) can be viewed as a strong version of the axiom of choice, and (2) can be viewed as a strong version of the well-ordering theorem. We have already seen that the existence of a *definable* global choice function C and hence a *definable* well-ordering S of V cannot be proved in ZFC, moreover, it has also been shown (by Easton 1964) that one cannot even prove (1) in ZFC (even if no definability is required) and hence, by 4.1, one also cannot prove (2) in ZFC.

(1) and (2) are convenient versions of the axiom of choice to have around. Since they are unprovable in ZFC we ask how we have to strengthen ZFC in order to get them. One way is the following. We can actually assert (1) indirectly but exactly by a weaker version of the axiom of ordinal-definability (3.9), namely, by the axiom asserting that there exists an x such that every set is ordinal-definable from x. However, the advantage in (1) or (2) is just a technical advantage of convenience; to add to ZFC an axiom which puts some real restriction on the universe of sets is too high a price to pay for it (unless one has other good reasons for adopting that axiom, or a stronger one).

The second way to obtain (1) and (2) is to pass to the theory ZFGC described below.

4.3 The Set Theory ZFGC. The *basic language of* ZFGC is the basic language of ZF enriched by a unary function symbol σ which we call the *selector*. Notice that σ is not a class symbol; its linguistic status is like the status of the unary $-$ (minus) in arithmetic. σ occurs in the expressions of the language of ZFGC as $\sigma(\tau)$, where τ is any term. The axioms of ZFGC are all the axioms of ZF together with the following ones.

(a) The Axiom of Global Choice. $x \neq 0 \rightarrow \sigma(x) \in x$.

(b) All the instances of the axiom schema of replacement (I.5.6)

$$\forall u \forall v \forall w(\psi(u, v) \wedge \psi(u, w) \rightarrow v = w) \rightarrow \forall z \exists y \forall v(v \in y \leftrightarrow (\exists u \in z)\psi(u, v))$$

where ψ is now any formula of the basic language of ZFGC.

4.4 The Role of the New Axioms of ZFGC. Let us investigate now the theory ZFGC and, in particular, the role of its additional axioms (a) and (b). The role of (a) is clear; (a) implies that the function C which we define by $C = \langle \sigma(x) \mid x \in V \rangle$ is indeed a global choice function; but it is even unnecessary to define C since, by (a), σ itself acts as a global choice function. To clarify the role of (b) let us first prove the axiom of choice in ZFGC. For a given set a let us take for the formula $\psi(u, v)$ in the axiom schema of replacement

$$\forall u \forall v \forall w(\psi(u, v) \wedge \psi(u, w) \rightarrow v = w) \rightarrow \forall a \exists f \forall v(v \in f \leftrightarrow (\exists u \in a)\psi(u, v)) \tag{3}$$

the formula $v = \langle u, \sigma(u) \rangle$ then the set f obtained by (3) is a function on a such that for every $x \in a$ $f(x) = \sigma(x)$ and hence, by (a), f is a choice function on a. Notice that in this proof of the axiom of choice we used not only (a) but also (b). The use of (b) occurred when we took for $\psi(u, v)$ in (3) the formula $v = \langle u, \sigma(u) \rangle$ in which the symbol σ occurs. In fact, had we added to ZF only the axiom (a) we could not prove in the theory obtained the axiom of choice or any other statement in the language of ZF which is not already provable in ZF (Hilbert and Bernays 1939, § 1). Thus the addition of (a) alone would not increase the proving power of ZF and thus would serve no purpose.

4.5 The Theory ZFGC **With Classes.** The axioms of the extended theory of ZF, i.e., ZF with classes, are listed in I.4.3 and I.6.20. As long as we pass from ZF to a new theory ZF in the language of ZF by adding new axioms to ZF the conservation theorem I.4.6 tells us what we have to add to the axioms of the extended theory in order to obtain the extended theory of ZF. If a single axiom is added to ZF then the same axiom is also added to the axioms of the extended theory; if an axiom schema of a certain simple kind is added to ZF then a certain corresponding statement involving a class variable is added to the extended theory— see I.4.7. However, in passing from ZF to ZFGC we did not stay within the language of ZF since we added the new function symbol σ; therefore we have to go into the question of setting up a class version of ZFGC, which we shall call *the extended theory of* ZFGC.

The language of the extended theory of ZFGC is the same as that of the extended theory of ZF, which is described in I.4.1, except for the following differences. First, the language of the extended theory of ZFGC contains the function symbol σ which is applied only to set variables (and set terms). Second, class terms are now expressions $\{x \mid \Phi(x)\}$, where $\Phi(x)$ is any formula of the language of ZFGC, i.e., $\Phi(x)$ may contain also the symbol σ. Notice that this is not merely a small change in the grammar of the language; this changes the concept of a class. In (the extended theory of) ZF a class was understood to be a *specifiable* collection of sets (see I.2.2 and I.§ 3), i.e. a collection A such that for some formula $\phi(x)$ of the basic language of ZF A consists exactly of those x's for which $\phi(x)$ holds (for some fixed values of the parameters of $\phi(x)$). In admitting as class terms of the extended theory of ZFGC all the expressions $\{x \mid \Phi(x)\}$, where Φ is any formula of ZFGC, we adopt for ZFGC a wider concept of class; a class of ZFGC is a collection *A specifiable in the language of* ZFGC, i.e., a collection of sets such that for some formula $\phi(x)$ of the basic language of ZFGC A consists exactly of those x's for which $\phi(x)$ holds. (We have class terms $\{x \mid \Phi(x)\}$ for every formula $\Phi(x)$ of the *extended* language of ZFGC, but by the eliminability theorem of ZFGC, which is like I.4.5 and which we shall discuss below, if $\Phi'(x)$ is any formula without class variables obtained from $\Phi(x)$ by substituting class terms for the class variables of $\Phi(x)$, if any, then we have $\{x \mid \Phi'(x)\} = \{x \mid \phi(x)\}$ for some formula $\phi(x)$ of the *basic* language of ZFGC.) The concept of a class of ZFGC is indeed wider than that of ZF since there are classes of ZFGC, such as $\langle \sigma(x) \mid x \in V \rangle$, which cannot be proved in ZFGC to be equal to any class of ZF (see Exercise 4.7 below). Why do we adopt a wider concept of class for ZFGC? Because the natural concept of class for any theory is obtained by taking the collections of the objects of that theory which are specifiable in the language of that theory; this approach allows one to admit as classes all the $\{x \mid \Phi(x)\}$'s without having to check $\Phi(x)$ carefully as to which symbols are contained in it.

The extended theory of ZFGC has the same rule of inference I.4.2 as that of ZF, and its axioms are the axioms I.4.3 and I.6.20 of the extended theory of ZF, where in the axiom schema $y \in \{x \mid \Phi(x)\} \leftrightarrow \Phi(y)$ of I.4.3 Φ ranges now over all formulas of the extended language of ZFGC, together with the axiom $x \neq 0 \rightarrow \sigma(x) \in x$ (4.3(a)) of global choice.

Following the proof of the eliminability theorem I.4.5 of ZF one can easily prove such a theorem also for ZFGC, i.e., that every formula Φ of the extended language of ZFGC without class variables is equivalent (by means of the axioms of I.4.3) to a formula ϕ of the basic language of ZFGC. As in the case of ZF, in order to justify our calling this theory with classes the extended theory of ZFGC we have also to prove that every statement ϕ of the basic language of ZFGC is provable in the extended theory of ZFGC iff it is provable in ZFGC. The proof of this is essentially the same as the corresponding proof for ZF. We shall only pay special attention to two points in it. First, when we proceed as in I.6.19 to show that, essentially, the substitution instances of the class version

(4) If F is a function then $F[z]$ is a set

of the axiom of replacement yield exactly the axiom schema of replacement.

(5) $\forall u \forall v \forall w(\phi(u, v) \wedge \phi(u, w) \rightarrow v=w) \rightarrow \forall z \exists y \forall v(v \in y \leftrightarrow (\exists u \in z)\phi(u, v))$,

then, for ZFGC, these substitution instances include also all the instances of (5) where ϕ contains the symbol σ, and this where the extension 4.3(b) of the axiom schema of replacement to the basic language of ZFGC comes in again. Notice that 4.3(b) has not been assumed as an axiom of the extended theory of ZFGC; it comes in via the axiom schema $y \in \{x \mid \Phi(x)\} \leftrightarrow \Phi(y)$ of I.4.3, in which Φ may now contain also σ, and the class version (4) of the axiom of replacement. The second point which we want to mention is that we retain the axiom of foundation

(6) $A \neq 0 \rightarrow (\exists x \in A) x \cap A = 0$

even though in ZFGC it stands for a schema which contains more instances than the corresponding schema of ZF. This makes no difference since, as we saw in II.7.4, all the instances of (6) in ZF and ZFGC are anyway consequences of the single instance $a \neq 0 \rightarrow (\exists x \in a)(x \cap a = 0)$ of (6) in ZF.

4.6 Another Formal Rendering of ZFGC **With Classes.** As we have already mentioned earlier, the function $C = \langle \sigma(x) \mid x \in V \rangle$ is a global choice function. The extended theory of ZFGC can also be formulated, equivalently, as follows. We add to the extended theory of ZF a class constant C together with the single axiom

(7) C is a global choice function.

In this theory σ can be *defined* by $\sigma(x) = C(x)$. As in our earlier version of ZFGC, the schema $y \in \{x \mid \Phi(x)\} \leftrightarrow \Phi(x)$ is now extended also to formulas which contain the class constant C.

4.7 Exercise. Show that there is no formula $\phi(x, y, z_1, \ldots, z_n)$ of the basic language of ZF such that

(8) $\exists z_1 \ldots \exists z_n (y = \sigma(x) \leftrightarrow \phi(x, y, z_1, \ldots, z_n))$

is provable in ZFGC. (I.e., even in ZFGC we cannot prove that the global choice function is a particular function defined in the language of ZF from a parameter.) The reason for this is, roughly, that the additional axioms of ZFGC are far from specifying the function σ, thus if one function will do for σ many others will do too, hence it is impossible to prove that σ is one particular function.

Hint. Reduce first the general case to the case where $n=1$. For this case and for a given formula $\phi(x, y, z)$ define by the diagonal method a set term σ such that if ϕ satisfies $\exists z \forall x \exists t((x \neq 0 \rightarrow t \in x) \wedge \forall y(y=t \leftrightarrow \phi(x, y, z)))$ then σ satisfies the axioms of ZFGC but does not satisfy (8). □

4.8 ZFGC **is a Conservative Extension of** ZFC. Finally let us ask how ZFGC is related to ZF and ZFC. As we have seen, the axiom of choice is provable in ZFGC so it is better to compare ZFGC with ZFC. Obviously, every theorem of ZFC is also a theorem of ZFGC. Going in the other direction, theorems of ZFGC which contain an occurrence of the symbol σ cannot be theorems of ZFC; however every theorem of ZFGC which contains no occurrence of σ is already a theorem of ZFC (see Felgner 1971* or Gaifman 1975). Thus the theory ZFGC is a *conservative extension* of ZFC in the sense that it adds no new theorems in the language of ZFC. This makes ZFGC a very good extension of ZFC; on one hand ZFGC has global choice, which ZFC does not have, and on the other hand ZFGC does not commit us to any new assumptions concerning sets. If we wanted to follow the first way mentioned in 4.2 and obtain a global choice function without changing the language of ZFC, we would have to admit an axiom which asserts, at least, that there is an x such that every set is ordinal-definable from x and this is a new assumption about the universe of sets which calls for justification. In contrast, since ZFGC is a conservative extension of ZFC the passage from ZFC to ZFGC does not require any set theoretical commitment and is of the same kind as our passage from ZF with sets only to ZF with classes.

4.9 Exercise. Prove in ZFGC with classes that a class A is proper iff there is a bijection F of A on the universal class V. (*Hint:* Use 4.1.) Explain what the existential class quantifiers which occur implicitly in this statement mean in ZFGC. □

5. *Cardinal Exponentiation*

We have discussed cardinal exponentiation already in III.3.36. We saw that in order to obtain any results beyond the most elementary ones we must assume the axiom of choice, without which we cannot even prove that $\mathfrak{a}^{\aleph_\beta}$ is an aleph for $\mathfrak{a} \geq 2$. Moreover, we have mentioned also that even with the axiom of choice we cannot even come close to "computing" $\mathfrak{a}^{\aleph_\beta}$. For example, for all we know $2^{\aleph_0}$ may be $\aleph_1$, it may be a weakly inaccessible cardinal (in which case $2^{\aleph_0} = \aleph_\alpha$ for a very big α) and it may even be bigger than many weakly inaccessible cardinals (IV.3.27). Here we shall carry out a further examination of cardinal exponentiation. We shall see how to compute some values of $\mathfrak{a}^{\aleph_0}$ from other values of this function without and with simplifying assumptions. For the reason just mentioned we shall assume the axiom of choice throughout the present section. We start with a discussion of the function $\langle 2^{\mathfrak{a}} \mid \mathfrak{a} \in \mathrm{Cn} \rangle$.

5.1 Definition. For every well-ordered cardinal $\mathfrak{a}$ we denote with $\mathfrak{a}^{+\lambda}$ the λ-th cardinal following $\mathfrak{a}$, i.e., if $\mathfrak{a}=\aleph_\alpha$ then $\mathfrak{a}^{+\lambda}=\aleph_{\alpha+\lambda}$.

5.2Ac Theorem. *The function* $\langle 2^{\mathfrak{a}} \mid \mathfrak{a} \in \mathrm{Cn} \rangle$ *satisfies the following conditions for all infinite cardinals* $\mathfrak{a}$

(i) *If* $\mathfrak{a}<\mathfrak{b}$ *then* $2^{\mathfrak{a}} \leqslant 2^{\mathfrak{b}}$.

(ii) $\mathrm{cf}(2^{\mathfrak{a}})>\mathfrak{a}$, *and hence* $2^{\mathfrak{a}}>\mathfrak{a}$.

(iii) (Bukovšky 1965, Hechler 1973). *If* $\mathfrak{a}$ *is a singular cardinal and there is a* $\mathfrak{b}<\mathfrak{a}$ *such that for every cardinal* $\mathfrak{c}$ *such that* $\mathfrak{b}<\mathfrak{c}<\mathfrak{a}$ *we have* $2^{\mathfrak{c}}=2^{\mathfrak{b}}$ *then also* $2^{\mathfrak{a}}=2^{\mathfrak{b}}$ *(i.e., if* $2^{\mathfrak{c}}$ *has a fixed value for* $\mathfrak{c}<\mathfrak{a}$ *from a certain point on then it has the same value at* $\mathfrak{a}$ *too).*

(iv) (Silver 1975). *If* $\mathfrak{a}$ *is a singular cardinal with* $\mathrm{cf}(\mathfrak{a})>\omega$ *and* $\mu<\mathrm{cf}(\mathfrak{a})$ *is such that the set* $\{\mathfrak{b}<\mathfrak{a} \mid 2^{\mathfrak{b}} \leqslant \mathfrak{b}^{+\mu}\}$ *is a stationary subset of* $\mathfrak{a}$ *then also* $2^{\mathfrak{a}} \leqslant \mathfrak{a}^{+\mu}$. *(In particular, if* $\mathfrak{a}>\mathrm{cf}(\mathfrak{a})>\omega$ *and* $\{\mathfrak{b}<\mathfrak{a} \mid 2^{\mathfrak{b}}=\mathfrak{b}^{+}\}$ *is a stationary subset of* $\mathfrak{a}$ *then also* $2^{\mathfrak{a}}=\mathfrak{a}^{+}$.)

(v) (Galvin and Hajnal 1975). *If* $\kappa>\omega$ *is a regular ordinal and for all* $\alpha<\kappa$ $2^{\mathfrak{c}^{+\alpha}}<\mathfrak{c}^{+(2^\kappa)^+}$ *then also* $2^{\mathfrak{c}^{+\kappa}}<\mathfrak{c}^{+(2^\kappa)^+}$.

5.3Ac The Behavior of the Power Function $2^{\mathfrak{a}}$. Before discussing Theorem 5.2 from a global point of view let us throw some light on parts (iv) and (v) of it. In part (iv) $\mathfrak{a}$ is a singular cardinal and hence, by IV.3.11, $\mathfrak{a}$ is a limit cardinal. Since $\mathfrak{a}>\omega$ and the class of all alephs is closed (by IV.1.26 and IV.4.12), the set of all alephs below $\mathfrak{a}$ is a closed unbounded subset of $\mathfrak{a}$. Thus the set of ordinals $\beta<\mathfrak{a}$ which are not alephs is an insignificant subset of $\mathfrak{a}$ and almost all the ordinals below $\mathfrak{a}$ are alephs, and if for "sufficiently many" such $\mathfrak{b}<\mathfrak{a}$ $2^{\mathfrak{b}} \leqslant \mathfrak{b}^{+\mu}$ then also $2^{\mathfrak{a}} \leqslant \mathfrak{a}^{+\mu}$.

In 5.2(v) we may have $\mathfrak{c}^{+\kappa}=\kappa$ (just in case $\mathfrak{c}<\kappa$ and κ is weakly inaccessible), but in this case 5.2(v) is trivial; in all other cases we have $\mathfrak{c}^{+\kappa}>\kappa=\mathrm{cf}(\mathfrak{c}^{+\kappa})$ and thus $\mathfrak{c}^{+\kappa}$, for which we estimate $2^{\mathfrak{c}^{+\kappa}}$ is a singular cardinal, and 2^κ, which we use to obtain a strict upper bound for $2^{\mathfrak{c}^{+\kappa}}$, is the value of the power function at the cardinal κ which is less than $\mathfrak{c}^{+\kappa}$. The assumption of 5.2(v), that $2^{\mathfrak{c}^{+\alpha}}<\mathfrak{c}^{+(2^\kappa)^+}$ for $\alpha<\kappa$, holds in particular when $\mathfrak{c}^{+\kappa}$ is a *strong limit cardinal* (i.e., for $\mathfrak{b}<\mathfrak{c}^{+\kappa}$, $2^{\mathfrak{b}}<\mathfrak{c}^{+\kappa}$). The instance of 5.2(v) with the least $\mathfrak{c}$ and κ is the following one, obtained by taking $\mathfrak{c}=\aleph_0$ and $\kappa=\omega_1$. If for every $\alpha<\omega_1$ $2^{\aleph_\alpha}<\aleph_{(2^{\aleph_1})^+}$ then $2^{\aleph_{\omega_1}}<\aleph_{(2^{\aleph_1})^+}$ and, in particular, if $\aleph_{\omega_1}$ is a strong limit cardinal then $2^{\aleph_{\omega_1}}<\aleph_{(2^{\aleph_1})^+}$.

It has been shown by Easton 1964 that in ZFC (i) and (ii) are the only rules which govern the behavior of the function $\langle 2^{\mathfrak{a}} \mid \mathfrak{a} \in \mathrm{Cn} \rangle$ at regular cardinals $\mathfrak{a}$. That is, for every function F on the class of all regular alephs into Cn which is given by an appropriate definition and which obeys (i) and (ii) for all regular cardinals (i.e., $\mathfrak{a}<\mathfrak{b} \rightarrow F(\mathfrak{a}) \leqslant F(\mathfrak{b})$ and $\mathrm{cf}(F(\mathfrak{a}))>\mathfrak{a}$) it is consistent with the axioms of ZFC that $2^{\mathfrak{a}}=F(\mathfrak{a})$ for all regular cardinals $\mathfrak{a}$. Examples of stements about the function $\langle 2^{\mathfrak{a}} \mid \mathfrak{a} \in \mathrm{Cn} \rangle$ which are consistent with ZFC are: $2^{\aleph_\alpha}=\aleph_{\alpha+13}$ for all regular $\aleph_\alpha$;

$2^{\aleph_0}=\aleph_{243}$, $2^{\aleph_1}=\aleph_{368}$ and $2^{\aleph_\alpha}=\aleph_{(\aleph_\alpha)^+}$ for all regular $\aleph_\alpha>\aleph_1$; $2^{\aleph_n}=\aleph_{\omega+n}$ for $n<\omega$, $2^{\aleph_\alpha}=\aleph_{\alpha\cdot\alpha+13}$ for all even $\alpha\geqslant\omega$ (i.e., for all $\alpha\geqslant\omega$ such that $\alpha=2\cdot\beta$ for some β) and $2^{\aleph_\alpha}=\aleph_{\alpha\cdot\alpha+47}$ for all odd $\alpha\geqslant\omega$. In order to explain what we mean by requiring that F be given by an "appropriate definition" let us point out what kind of definition we want to avoid. We clearly have to avoid definitions such as $F(\mathfrak{a})=(2^{\mathfrak{a}})^+$ since it makes use of the very function it comes to determine. (Indeed, this particular function F satisfies (i) and (ii) of 5.2 but for it $F(\mathfrak{a})=2^{\mathfrak{a}}$ cannot hold for any $\mathfrak{a}$.) Thus we mean by "an appropriate definition" any definition into which nothing like the function $\langle 2^{\mathfrak{a}}\mid\mathfrak{a}\in\mathrm{Cn}\rangle$ does enter, directly or indirectly. There is no clear way of faithfully translating this vague requirement on the definition of F to a strict mathematical requirement. However, our freedom of defining F definitely includes any definition which uses only the concepts of an ordinal, a cardinal and $<$, constants for the finite ordinals and ω, and the operations of ordinal addition, multiplication and exponentiation, cardinal addition and multiplication and the functions $\langle\aleph_\alpha\mid\alpha\in\mathrm{On}\rangle$ and $\langle\mathrm{cf}(\alpha)\mid\alpha\in\mathrm{On}\rangle$ (but not the more general concept of a set).

Let us see how a simple property of the function $\langle 2^{\mathfrak{a}}\mid\mathfrak{a}\in\mathrm{Cn}\sim\omega\rangle$ which is not one of 5.2(i)–(v) follows easily from 5.2(i)–(v) and the properties of addition and multiplication. To prove $2^{\mathfrak{a}+\mathfrak{b}}=2^{\mathfrak{a}}\cdot 2^{\mathfrak{b}}$ for infinite $\mathfrak{a}$ and $\mathfrak{b}$ we proceed as follows

$$\begin{aligned}2^{\mathfrak{a}+\mathfrak{b}}&=2^{\max(\mathfrak{a},\mathfrak{b})}, \text{ by the properties of addition,}\\&=\max(2^{\mathfrak{a}},2^{\mathfrak{b}}), \text{ by 5.2(i),}\\&=2^{\mathfrak{a}}\cdot 2^{\mathfrak{b}}, \text{ by the properties of multiplication.}\end{aligned}$$

Unlike the case of the regular cardinals, Theorem 5.2 does not give us a complete answer as to which behavior of the function $\langle 2^{\mathfrak{a}}\mid\mathfrak{a}\in\mathrm{Cn}\rangle$ is permissible on the singular cardinals. To mention the simplest example, Theorem 5.2 tells us nothing about $2^{\aleph_\omega}$. If we assume that $2^{\aleph_n}<\aleph_\omega$ for all $n<\omega$ then all what is known about $2^{\aleph_\omega}$ is that it can be $\aleph_{\omega+1}$ (Gödel 1938, Easton 1964) or $\aleph_{\omega+k}$, for any $1<k<\omega$ or $\aleph_{\omega+\omega+1}$ (Magidor 1977). Theorem 5.2 does not even contain all what has been proved about the values of $2^{\mathfrak{a}}$ for singular cardinals $\mathfrak{a}$; the results obtained following the breakthrough of Silver 1975 beyond 5.2(iv) and 5.2(v) are rather fragmentary and it would be tedious to go through all of them. A few additional results are given in Galvin and Hajnal 1975.

We shall now prove parts (ii) and (iii) of 5.2; part (iv) is Corollary 5.7 below and part (v) is Theorem IX.3.33.

Proof of 5.2(ii)–(iii). (ii) Assume that $\lambda=\mathrm{cf}(2^{\mathfrak{a}})\leqslant\mathfrak{a}$, then, by IV.3.9, there is a sequence $\langle\mathfrak{b}_\alpha\mid\alpha<\lambda\rangle$ of cardinals $<2^{\mathfrak{a}}$ such that $\Sigma_{\alpha<\lambda}\mathfrak{b}_\alpha=2^{\mathfrak{a}}$. Since for every $\alpha<\lambda$ $\mathfrak{b}_\alpha<2^{\mathfrak{a}}$ we get, by the Zermelo–König inequality (III.4.18),

(1) $\quad\Sigma_{\alpha<\lambda}\mathfrak{b}_\alpha<\Pi_{\alpha<\lambda}2^{\mathfrak{a}}$.

The left-hand side of (1) is $2^{\mathfrak{a}}$ and the right-hand side of (1) is, by III.4.15, $(2^{\mathfrak{a}})^{|\lambda|}=2^{\mathfrak{a}\cdot|\lambda|}=2^{\mathfrak{a}}$ (since we assumed that $\lambda\leqslant\mathfrak{a}$). Thus (1) asserts that $2^{\mathfrak{a}}<2^{\mathfrak{a}}$, which is a contradiction.

(iii) Since $\mathfrak{a}$ is singular there is, by IV.3.9, a $\lambda<\mathfrak{a}$ and a sequence $\langle \mathfrak{c}_\alpha \mid \alpha<\lambda\rangle$ of cardinals $<\mathfrak{a}$ such that $\Sigma_{\alpha<\lambda}\,\mathfrak{c}_\alpha=\mathfrak{a}$. We have, by III.4.12(vi), III.4.12(v) and III.4.15,

$$(2) \qquad 2^{\mathfrak{a}}=2^{\Sigma_{\alpha<\lambda}\mathfrak{c}_\alpha}=\Pi_{\alpha<\lambda}2^{\mathfrak{c}_\alpha}\leqslant\Pi_{\alpha<\lambda}2^{\mathfrak{b}}=(2^{\mathfrak{b}})^{|\lambda|}=2^{\mathfrak{b}\cdot|\lambda|},$$

since, by our assumption, $2^{\mathfrak{c}}\leqslant 2^{\mathfrak{b}}$ for every $\mathfrak{c}<\mathfrak{a}$. Since $\lambda<\mathfrak{a}$ also $|\lambda|<\mathfrak{a}$ and $\mathfrak{b}\cdot|\lambda|<\mathfrak{a}$, therefore, by our assumption the right-hand side $2^{\mathfrak{b}\cdot|\lambda|}$ of (2) equals $2^{\mathfrak{b}}$. Thus (2) yields $2^{\mathfrak{a}}\leqslant 2^{\mathfrak{b}}$, hence $2^{\mathfrak{a}}=2^{\mathfrak{b}}$. □

5.4 Discussion. We can formulate Theorem 5.2(iv) as follows. If $\mathfrak{a}>\mathrm{cf}(\mathfrak{a})>\omega$, $\mu<\mathrm{cf}(\mathfrak{a})$ and for a stationary subset $Q\subseteq\mathfrak{a}$ of β's we have $|\{x\cap\beta \mid x\in\mathbf{P}(\mathfrak{a})\}|=|\mathbf{P}(\beta)|\leqslant|\beta|^{+\mu}$, then $|\mathbf{P}(\mathfrak{a})|\leqslant\mathfrak{a}^{+\mu}$. A natural generalization of this is to replace $\mathbf{P}(\mathfrak{a})$ by a general subset T of $\mathbf{P}(\mathfrak{a})$ and to ask what we can tell about the cardinal of T if we know the cardinals of the sets $\{x\cap\beta \mid x\in T\}$ for all $\beta<\mathfrak{a}$. Such questions turn out to be very interesting and some of them are known not to have an answer in ZFC, even if one assumes the generalized continuum hypothesis (i.e., $2^{\aleph_\alpha}=\aleph_{\alpha+1}$ for all α). An example of such a question is the following. Suppose $T\subseteq\mathbf{P}(\omega_1)$ is such that for all $\beta<\omega_1$ $|\{x\cap\beta \mid x\in T\}|\leqslant\aleph_0$; does such a T have to be of cardinality $\aleph_1$ or can it be also of cardinality $\aleph_2$? The assertion that there exists such a T of cardinality $\aleph_2$ is known as *Kurepa's hypothesis*. Kurepa's hypothesis is known to be neither provable nor refutable in ZFC even if we add the generalized continuum hypothesis to the axiom of ZFC. Here we shall handle only the case which is a direct generalization of Theorem 5.2(iv); we shall return to other cases in § IX.2.

5.5Ac Definition. For all cardinals $\mathfrak{a}$, $\mathfrak{b}$ we define $\mathfrak{a}^{\underline{\mathfrak{b}}}=\sup_{\mathfrak{d}<\mathfrak{b}}\mathfrak{a}^{\mathfrak{d}}$ and $\underline{\mathfrak{a}}^{\mathfrak{b}}=\sup_{\mathfrak{c}<\mathfrak{a}}\mathfrak{c}^{\mathfrak{b}}$. (Recall that we have already encountered $\mathfrak{a}^{\underline{\mathfrak{b}}}$ in IV.3.15 where we asserted that for every infinite cardinal $\mathfrak{a}$ $|H_{\mathfrak{a}}|=2^{\underline{\mathfrak{a}}}$.)

5.6Ac Theorem. (Baumgartner and Prikry 1976.) *Let* $\mathfrak{a}$ *be a singular cardinal such that* $\mathrm{cf}(\mathfrak{a})>\omega$ *and let* T *be a subset of* $\mathbf{P}(\mathfrak{a})$. *Denote* $\mathrm{cf}(\mathfrak{a})$ *by* λ. *If* $\mu<\lambda$ *is such that the set* $\{\mathfrak{b}<\mathfrak{a} \mid |\{x\cap\mathfrak{b} \mid x\in T\}|\leqslant\mathfrak{b}^{+\mu}\}$ *is a stationary subset of* $\mathfrak{a}$ *then* $|T|\leqslant(\underline{\mathfrak{a}}^{\lambda})^{+\mu}$.

Proof. We prove the theorem by induction on μ. Before dealing separately with the three induction cases we shall carry out some preparations which we need for all of them. Let $P=\{\mathfrak{b}<\mathfrak{a} \mid |\{x\cap\mathfrak{b} \mid x\in T\}|\leqslant\mathfrak{b}^{+\mu}\}$; by the hypothesis of the theorem P is a stationary subset of $\mathfrak{a}$. Since $\mathrm{cf}(\mathfrak{a})=\lambda$, $\mathfrak{a}$ has a closed unbounded subset W of cardinality λ (by IV.4.43(iii)). Since $\mathfrak{a}>\omega$ also $\mathfrak{a}\sim\omega$ is a closed unbounded subset of $\mathfrak{a}$. Let $S=P\cap W\cap(a\sim\omega)$, then S is a stationary subset of $\mathfrak{a}$ by IV.4.38(ii) and S is a set of alephs of cardinality λ. The fact that $|S|=\lambda$ is crucial for the proof of the theorem since we shall be able to reduce questions about what happens for the whole of $\mathfrak{a}$ to similar questions concerning S, and S is a set of smaller cardinality than $\mathfrak{a}$. For a regular $\mathfrak{a}$ we cannot get such an S with $|S|<\mathfrak{a}$ and therefore our theorem does not hold for a regular $\mathfrak{a}$. For every $\mathfrak{b}\in S\subseteq P$ $|\{x\cap\mathfrak{b} \mid x\in T\}|\leqslant\mathfrak{b}^{+\mu}$, and hence there is an injection $f_{\mathfrak{b}}$ of $\{x\cap\mathfrak{b} \mid x\in T\}$ into $\mathfrak{b}^{+\mu}$. For every $x\in T$ we define a function g_x on S as follows:

(3) For $\mathfrak{b} \in S$, $g_x(\mathfrak{b}) = f_{\mathfrak{b}}(x \cap \mathfrak{b}) < \mathfrak{b}^{+\mu}$.

Let us start the induction by proving the case where $\mu = 0$. For $x \in T$ we have, by (3), $g_x(\mathfrak{b}) < \mathfrak{b}$ for every $\mathfrak{b} \in S$, thus g_x is a regressive function on S. By Theorem IV.4.40 g_x is strictly bounded by some ordinal $\gamma_x < \mathfrak{a}$ on a subset Q_x of S which is stationary in $\mathfrak{a}$. Let J be the relation

$$\{\langle\langle Q, \gamma, h\rangle, x\rangle \mid x \in T,\ Q \text{ is a subset of } S \text{ stationary in } \mathfrak{a},\ h \in {}^{Q}\gamma, \text{ and } h = g_x \upharpoonright Q\}.$$

For every $x \in T$ we have, by the way we obtained Q_x and γ_x, $\langle\langle Q_x, \gamma_x, g_x \upharpoonright Q_x\rangle, x\rangle \in J$, hence $\mathrm{Rng}(J) = T$. For $\langle\langle Q, \gamma, h\rangle, x\rangle \in J$, Q is stationary in $\mathfrak{a}$ and hence unbounded in $\mathfrak{a}$ and therefore

$$x = \bigcup_{\mathfrak{b} \in Q}(x \cap \mathfrak{b}) = \bigcup_{\mathfrak{b} \in Q} f_{\mathfrak{b}}^{-1}(f_{\mathfrak{b}}(x \cap \mathfrak{b})) = \bigcup_{\mathfrak{b} \in Q} f_{\mathfrak{b}}^{-1}(g_x(\mathfrak{b})) = \bigcup_{\mathfrak{b} \in Q} f_{\mathfrak{b}}^{-1}(h(\mathfrak{b}));$$

thus x is uniquely determined by Q and h and hence J is a function. As a consequence $|T| = |\mathrm{Rng}(J)| \leqslant |\mathrm{Dom}(J)|$; thus it suffices to prove $|\mathrm{Dom}(J)| \leqslant \underset{\sim}{\mathfrak{a}}^{\lambda}$.

$$\mathrm{Dom}(J) \subseteq \{\langle Q, \gamma, h\rangle \mid Q \subseteq S, \gamma < \mathfrak{a}, h \in {}^{Q}\gamma\} = \bigcup_{Q \in \mathbf{P}(S)} \bigcup_{\gamma < \mathfrak{a}} \{\langle Q, \gamma, h\rangle \mid h \in {}^{Q}\gamma\}$$

Therefore, by III.4.5,

$$|\mathrm{Dom}(J)| \leqslant \Sigma_{Q \in \mathbf{P}(S)} \Sigma_{\gamma < \mathfrak{a}} |{}^{Q}\gamma| \leqslant \Sigma_{Q \in \mathbf{P}(S)} \Sigma_{\gamma < \mathfrak{a}} \underset{\sim}{\mathfrak{a}}^{\lambda} = 2^{\lambda} \cdot \dot{\mathfrak{a}} \cdot \underset{\sim}{\mathfrak{a}}^{\lambda}.$$

As easily seen $\mathfrak{a} \leqslant \underset{\sim}{\mathfrak{a}}^{\lambda}$ and obviously $2^{\lambda} \leqslant \underset{\sim}{\mathfrak{a}}^{\lambda}$ hence $|\mathrm{Dom}(J)| \leqslant \underset{\sim}{\mathfrak{a}}^{\lambda}$, which is what we had to prove.

Now we shall carry out the induction step for a successor ordinal μ. Let $\mu = \nu + 1$ and we assume that the theorem holds for ν. First we prove:

(4) For every $x \in T$ the set $R_x = \{y \in T \mid \{\mathfrak{b} \in S \mid g_y(\mathfrak{b}) \leqslant g_x(\mathfrak{b})\}$ is a stationary subset of $\mathfrak{a}\}$ is of cardinality $\leqslant (\underset{\sim}{\mathfrak{a}}^{\lambda})^{+\nu}$.

For a subset Q of S stationary in $\mathfrak{a}$ let $T_Q = \{y \in T \mid (\forall \mathfrak{b} \in Q) g_y(\mathfrak{b}) \leqslant g_x(\mathfrak{b})\}$, then $R_x = \bigcup_{Q \text{ is a subset of } S \text{ stationary in } \mathfrak{a}} T_Q$. For every $\mathfrak{b} \in Q$

$$\{y \cap \mathfrak{b} \mid y \in T_Q\} \subseteq \{y \cap \mathfrak{b} \mid y \in T \wedge g_y(\mathfrak{b}) \leqslant g_x(\mathfrak{b})\} = \{y \cap \mathfrak{b} \mid y \in T \wedge f_{\mathfrak{b}}(y \cap \mathfrak{b}) \leqslant g_x(\mathfrak{b})\} \subseteq f_{\mathfrak{b}}^{-1}[g_x(\mathfrak{b}) + 1].$$

Every $\mathfrak{b} \in Q \subseteq S$ is an aleph and hence also $\mathfrak{b}^{+\mu}$ is an aleph. Using this and (3) we get, since $f_{\mathfrak{b}}$ is an injection,

$$|\{y \cap \mathfrak{b} \mid y \in T_Q\}| \leqslant |f_{\mathfrak{b}}^{-1}[g_x(b) + 1]| \leqslant |g_x(b) + 1| < \mathfrak{b}^{+\mu} = \mathfrak{b}^{+(\nu+1)}.$$

Therefore T_Q is a subset of $\mathbf{P}(\mathfrak{a})$ such that for every $\mathfrak{b} \in Q$ $|\{y \cap \mathfrak{b} \mid y \in T_Q\}| \leqslant b^{+\nu}$. By the induction hypothesis on μ the theorem holds for ν and we have $|T_Q| \leqslant (\underset{\sim}{\mathfrak{a}}^{\lambda})^{+\nu}$.

Therefore, by III.4.5,

$$|R_x| \leqslant \Sigma_{Q \text{ is a subset of } S \text{ stationary in } \mathfrak{a}} |T_Q| \leqslant \Sigma_{Q \in \mathbf{P}(S)} (\underset{\sim}{\mathfrak{a}}^{\lambda})^{+\nu} = 2^{\lambda} \cdot (\underset{\sim}{\mathfrak{a}}^{\lambda})^{+\nu} = (\underset{\sim}{\mathfrak{a}}^{\lambda})^{+\nu}$$

which establishes (4). Now we claim:

(5) There is a set $W \subseteq T$ with $|W| \leqslant (\underset{\sim}{\mathfrak{a}}^{\lambda})^{+\mu}$ such that $\bigcup_{x \in W} R_x = T$

(5) implies $|T| \leqslant \Sigma_{x \in W} |R_x| \leqslant \Sigma_{x \in W} (\underset{\sim}{\mathfrak{a}}^{\lambda})^{+\nu} \leqslant (\underset{\sim}{\mathfrak{a}}^{\lambda})^{+\mu} \cdot (\underset{\sim}{\mathfrak{a}}^{\lambda})^{+\nu} = (\underset{\sim}{\mathfrak{a}}^{\lambda})^{+\mu}$, which is what we had to prove. We establish (5) by contradiction as follows. Suppose that (5) fails to hold, then we define a sequence $\langle x_\beta \mid \beta \leqslant (\underset{\sim}{\mathfrak{a}}^{\lambda})^{+\mu} \rangle$ of members of T by choosing x_β to be a member of T not in $\bigcup_{\gamma<\beta} R_{x_\gamma}$; there is a member of T which is not in $\bigcup_{\gamma<\beta} R_{x_\gamma}$ since otherwise the set $W = \{x_\gamma \mid \gamma < \beta\}$ would be as required by (5). Denote $(\underset{\sim}{\mathfrak{a}}^{\lambda})^{+\mu}$ with δ. By the definition of the sequence $\langle x_\beta \mid \beta \leqslant \delta \rangle$ we have $x_\delta \notin R_{x_\beta}$ for $\beta < \delta$. Hence, by (4), the set $\{\mathfrak{b} \in S \mid g_{x_\delta}(\mathfrak{b}) \leqslant g_{x_\beta}(\mathfrak{b})\}$ is not a stationary subset of S, and its complement with respect to S, which is $\{\mathfrak{b} \in S \mid g_{x_\beta}(\mathfrak{b}) < g_{x_\delta}(\mathfrak{b})\}$, is a stationary subset of $\mathfrak{a}$. This implies, by the definition of R_{x_δ}, $x_\beta \in R_{x_\delta}$. Now we have $\{x_\beta \mid \beta < \delta\} \subseteq R_{x_\delta}$. For every $x \in T$ we have, obviously $x \in R_x$, and since, for $\beta < \delta$, $x_\beta \notin \bigcup_{\gamma<\beta} R_{x_\gamma}$, we have also $x_\beta \neq x_\gamma$ for $\gamma < \beta$. Thus $|R_{x_\delta}| \geqslant |\{x_\beta \mid \beta < \delta\}| = \delta = (\underset{\sim}{\mathfrak{a}}^{\lambda})^{+\mu}$, contradicting (4) which asserts that for every $x \in T$ $|R_x| \leqslant (\underset{\sim}{\mathfrak{a}}^{\lambda})^{+\nu} < (\underset{\sim}{\mathfrak{a}}^{\lambda})^{+\mu}$.

We finally come to deal with the case where μ is a limit ordinal. For a fixed $x \in T$ and $\nu < \mu$ let $P_\nu = \{\mathfrak{b} \in S \mid g_x(\mathfrak{b}) < \mathfrak{b}^{+\nu}\}$. Since μ is a limit ordinal we have, by (3), $S = \bigcup_{\nu<\mu} P_\nu$. Since S is a stationary subset of $\mathfrak{a}$ and $\mu < \lambda$ we get, by IV.4.35, that for some $\nu < \mu$ P_ν is a stationary subset of $\mathfrak{a}$. Thus we have shown that for every $x \in T$ there is a $\nu < \mu$ such that the set $\{\mathfrak{b} \in S \mid g_x(\mathfrak{b}) < \mathfrak{b}^{+\nu}\}$ is stationary. In other words, if we denote with $T_{Q,\nu}$ the set $\{x \in T \mid (\forall \mathfrak{b} \in Q)\ (g_x(\mathfrak{b}) < \mathfrak{b}^{+\nu})\}$ then $T = \bigcup_{\nu<\mu,\, Q \text{ is a subset of } S \text{ stationary in } \mathfrak{a}} T_{Q,\nu}$. For every $\mathfrak{b} \in Q$ and $x \in T_{Q,\nu}$ we have $f_{\mathfrak{b}}(x \cap \mathfrak{b}) = g_x(\mathfrak{b}) < \mathfrak{b}^{+\nu}$, hence $\{x \cap \mathfrak{b} \mid x \in T_{Q,\nu}\} \subseteq f_{\mathfrak{b}}^{-1}[\mathfrak{b}^{+\nu}]$; and since $f_{\mathfrak{b}}$ is an injection $|\{x \cap \mathfrak{b} \mid x \in T_{Q,\nu}\}| \leqslant \mathfrak{b}^{+\nu}$ for every $\mathfrak{b} \in Q$. Thus for a stationary Q the induction hypothesis on μ yields $|T_{Q,\nu}| \leqslant (\underset{\sim}{\mathfrak{a}}^{\lambda})^{+\nu} < (\underset{\sim}{\mathfrak{a}}^{\lambda})^{+\mu}$. Since

$$T = \bigcup_{\nu<\mu,\, Q \text{ is a subset of } S \text{ stationary in } \mathfrak{a}} T_{Q,\nu}$$

we get

$$|T| \leqslant \Sigma_{\nu<\mu} \Sigma_{Q \text{ is a subset of } S \text{ stationary in } \mathfrak{a}} (\underset{\sim}{\mathfrak{a}}^{\lambda})^{+\mu} \leqslant |\mu| \cdot 2^{\lambda} \cdot (\underset{\sim}{\mathfrak{a}}^{\lambda})^{+\mu} = (\underset{\sim}{\mathfrak{a}}^{\lambda})^{+\mu},$$

which is what we had to prove. □

5.7Ac Corollary (Silver 1975). *If* $\mathfrak{a}$ *is a singular cardinal with* $\mathrm{cf}(\mathfrak{a}) > \omega$ *and* $\mu < \mathrm{cf}(\mathfrak{a})$ *is such that the set* $\{\mathfrak{b} < \mathfrak{a} \mid 2^{\mathfrak{b}} \leqslant \mathfrak{b}^{+\mu}\}$ *is a stationary subset of* $\mathfrak{a}$ *then also* $2^{\mathfrak{a}} \leqslant \mathfrak{a}^{+\mu}$. (*This is Theorem* 5.2(iv).)

Proof. By our hypothesis the set $T = \mathbf{P}(\mathfrak{a})$ satisfies the hypothesis of Theorem 5.6 (since for $\mathfrak{b} < \mathfrak{a}$ $\{x \cap \mathfrak{b} \mid x \in \mathbf{P}(\mathfrak{a})\} = \mathbf{P}(\mathfrak{b})$) hence we get, by 5.6, $2^{\mathfrak{a}} = |\mathbf{P}(\mathfrak{a})| \leqslant (\underset{\sim}{\mathfrak{a}}^{\lambda})^{+\mu}$, where $\lambda = \mathrm{cf}(\mathfrak{a})$. Once we prove $\underset{\sim}{\mathfrak{a}}^{\lambda} = \mathfrak{a}$ we get $2^{\mathfrak{a}} \leqslant \mathfrak{a}^{+\mu}$, which is what we have to prove.

Let us notice that for $\mathfrak{b}<\mathfrak{a}$ and $\nu<\lambda$ we have $\mathfrak{b}^{+\nu}<\mathfrak{a}$. This is shown by induction on ν using the facts that if ν is a successor ordinal then $\mathfrak{b}^{+\nu}\neq\mathfrak{a}$ since $\mathfrak{b}^{+\nu}$ is regular whereas $\mathfrak{a}$ is singular, and if ν is a limit ordinal then $\mathrm{cf}(\mathfrak{b}^{+\nu})=\mathrm{cf}(\nu)\leqslant\nu<\lambda=\mathrm{cf}(\mathfrak{a})$ and hence, again, $\mathfrak{b}^{+\nu}\neq\mathfrak{a}$. Now let us estimate $\mathfrak{c}^{\lambda}$ for $\mathfrak{c}<\mathfrak{a}$. Let $\mathfrak{b}$ be a member of the unbounded subset $\{\mathfrak{b}<\mathfrak{a}\mid 2^{\mathfrak{b}}\leqslant\mathfrak{b}^{+\mu}\}$ of $\mathfrak{a}$ which is both $\geqslant\mathfrak{c}$ and $\geqslant\lambda$. We have $\mathfrak{c}^{\lambda}\leqslant\mathfrak{b}^{\mathfrak{b}}=2^{\mathfrak{b}}\leqslant\mathfrak{b}^{+\mu}$, and since we have shown that for $\nu<\lambda$ $\mathfrak{b}^{+\nu}<\mathfrak{a}$ we get $\mathfrak{c}^{\lambda}<\mathfrak{a}$. As a consequence $\underline{\mathfrak{a}}^{\lambda}=\sup_{\mathfrak{c}<\mathfrak{a}}\mathfrak{c}^{\lambda}=\mathfrak{a}$ (since $\mathfrak{a}$ is a limit cardinal), which completes our proof. □

5.8Ac Does the Power Function Determine the Cardinal Exponentiation? Till now we dealt only with the properties of the *power function* $\langle 2^{\mathfrak{a}}\mid\mathfrak{a}\in\mathrm{Cn}\rangle$, but we want to explore cardinal exponentiation in general, where the base is any cardinal, not just 2. We have already mentioned, in 5.3, that the theory ZFC leaves even the power function $\langle 2^{\mathfrak{a}}\mid\mathfrak{a}\in\mathrm{Cn}\rangle$ much leeway; what we ask now is if once the power function is fixed whether this determines completely cardinal exponentiation. The answer to this question is negative. ZFC is consistent with the statement

$$(6)\qquad 2^{\aleph_0}=\aleph_1,\ 2^{\aleph_n}=\aleph_{\omega+2}\ \text{for } n\leqslant\omega \text{ and } 2^{\mathfrak{a}}=\mathfrak{a}^{+}\ \text{for } \mathfrak{a}>\aleph_{\omega},$$

together with each one, separately, of $\aleph_{\omega}^{\aleph_0}=\aleph_{\omega+1}$ and $\aleph_{\omega}^{\aleph_0}=\aleph_{\omega+2}$ (Magidor 1977, strengthening a result of Jech 1974, and assuming that the existence of a supercompact cardinal is consistent with ZFC). Thus the axioms of ZFC together with (6), which determines completely the power function, do not suffice to determine the value of $\aleph_{\omega}^{\aleph_0}$. There is, however, another function of one cardinal variable which does determine cardinal exponentiation completely; this is the gimel function discussed below.

5.9 Definition (The gimel function). $\gimel$ (gimel) is the function on the alephs defined by $\gimel(\mathfrak{a})=\mathfrak{a}^{\mathrm{cf}(\mathfrak{a})}$.

5.10Ac Proposition. *For all alephs* $\mathfrak{a}$ *and* $\mathfrak{b}$ $\mathrm{cf}(\mathfrak{a}^{\mathfrak{b}})>\mathfrak{b}$. *In particular* $\mathrm{cf}(\gimel(\mathfrak{a}))=\mathrm{cf}(\mathfrak{a}^{\mathrm{cf}(\mathfrak{a})})>\mathrm{cf}(\mathfrak{a})$ *and hence* $\gimel(\mathfrak{a})>\mathfrak{a}$.

Hint of proof. Use the Zermelo–König inequality, III.4.18, in a way similar to that used in the proof of 5.2(ii). □

5.11Ac How Does $\gimel(\mathfrak{a})$ Compare With $2^{\mathfrak{a}}$? For every regular aleph $\mathfrak{a}$ we have $\gimel(\mathfrak{a})=\mathfrak{a}^{\mathfrak{a}}=2^{\mathfrak{a}}$, and for every singular aleph we have $\mathfrak{a}<\gimel(\mathfrak{a})\leqslant 2^{\mathfrak{a}}$. If $2^{\mathfrak{a}}=\mathfrak{a}^{+}$ for every aleph $\mathfrak{a}$ then we have $\gimel(\mathfrak{a})=2^{\mathfrak{a}}=\mathfrak{a}^{+}$ for every aleph $\mathfrak{a}$, but it is also consistent with ZFC that $\gimel(\mathfrak{a})<2^{\mathfrak{a}}$ for some aleph $\mathfrak{a}$ (as will be shown in 5.14).

5.12Ac Theorem (Bukovsky 1965). (i) *For all singular alephs* $\mathfrak{a}$, $2^{\mathfrak{a}}$ *obtains the least value it can obtain in view of the values of* $2^{\mathfrak{b}}$ *for* $\mathfrak{b}<\mathfrak{a}$, *i.e.*, $2^{\mathfrak{a}}=2^{\underline{\mathfrak{a}}}$, *unless this clashes with the condition* $\mathrm{cf}(2^{\mathfrak{a}})>\mathfrak{a}$ (5.2(ii)), *in which case* $2^{\mathfrak{a}}=\gimel(2^{\underline{\mathfrak{a}}})$.

(ii) *For all* $\mathfrak{a}>2$ *and* $\mathfrak{b}\geqslant\aleph_0$, $\mathfrak{a}^{\mathfrak{b}}$ *obtains the least value it can obtain in view of the values of* $\mathfrak{c}^{\mathfrak{b}}$ *for* $\mathfrak{c}<\mathfrak{a}$ *and in view of the inequality* $\mathfrak{a}^{\mathfrak{b}}\geqslant\mathfrak{a}$, *i.e., for a successor cardinal*

$\mathfrak{a}=\mathfrak{c}^+$ $\mathfrak{a}^{\mathfrak{b}}=\max(\mathfrak{c}^{\mathfrak{b}}, \mathfrak{a})$ *(this is the Hausdorff recursion formula—Hausdorff 1904), and for a limit cardinal* $\mathfrak{a}$, $\mathfrak{a}^{\mathfrak{b}}=\underset{\smile}{\mathfrak{a}}^{\mathfrak{b}}$ *(Tarski 1925), unless this violates the inequality* $\mathrm{cf}(\mathfrak{a}^{\mathfrak{b}})>\mathfrak{b}$ (5.10) *in which case* $\mathfrak{a}^{\mathfrak{b}}=\gimel(\underset{\smile}{\mathfrak{a}}^{\mathfrak{b}})$.

Remarks. In part (ii), the value $\max(\mathfrak{c}^{\mathfrak{b}}, \mathfrak{c}^+)$ for $(\mathfrak{c}^+)^{\mathfrak{b}}$ never violates the inequality $\mathrm{cf}((\mathfrak{c}^+)^{\mathfrak{b}})>\mathfrak{b}$ since if $\mathfrak{c}^+\leqslant\mathfrak{c}^{\mathfrak{b}}$ then $\max(\mathfrak{c}^{\mathfrak{b}}, \mathfrak{c}^+)=\mathfrak{c}^{\mathfrak{b}}$ hence $\mathrm{cf}(\max(\mathfrak{c}^{\mathfrak{b}}, \mathfrak{c}^+))=\mathrm{cf}(\mathfrak{c}^{\mathfrak{b}})>\mathfrak{b}$ (by 5.10) and if $\mathfrak{c}^+>\mathfrak{c}^{\mathfrak{b}}$ then we have $\mathfrak{c}^+>\mathfrak{c}^{\mathfrak{b}}\geqslant 2^{\mathfrak{b}}\geqslant 2^{\aleph_0}$; therefore $\mathfrak{c}\geqslant\mathfrak{b}$ (since $\mathfrak{c}^+\geqslant 2^{\mathfrak{b}}\geqslant\mathfrak{b}^+$), $\mathfrak{c}$ is an aleph and $\mathfrak{c}^+$ is regular (by IV.3.11), hence $\mathrm{cf}(\max(\mathfrak{c}^{\mathfrak{b}}, \mathfrak{c}^+))=\mathrm{cf}(\mathfrak{c}^+)=\mathfrak{c}^+>\mathfrak{c}\geqslant\mathfrak{b}$.

If $\mathfrak{a}\leqslant 2^{\mathfrak{b}}$ then $\mathfrak{a}^{\mathfrak{b}}=2^{\mathfrak{b}}$ since $\mathfrak{a}^{\mathfrak{b}}\leqslant(2^{\mathfrak{b}})^{\mathfrak{b}}=2^{\mathfrak{b}}$; prove it also from 5.12(ii)!

Proof. (i) We distinguish two cases. *Case a:* There is a $\mathfrak{b}<\mathfrak{a}$ such that for all $\mathfrak{c}$ $\mathfrak{b}<\mathfrak{c}<\mathfrak{a}\rightarrow 2^{\mathfrak{c}}=2^{\mathfrak{b}}$. In this case, by 5.2(iii), $2^{\mathfrak{a}}=2^{\mathfrak{b}}=2^{\underset{\smile}{\mathfrak{a}}}$. *Case b:* For every $\mathfrak{b}<\mathfrak{a}$ there is a $\mathfrak{c}$ such that $\mathfrak{b}<\mathfrak{c}<\mathfrak{a}$ and $2^{\mathfrak{c}}>2^{\mathfrak{b}}$. In this case the set $\{2^{\mathfrak{c}}\mid\mathfrak{c}<\mathfrak{a}\}$ is clearly a cofinal subset of $2^{\underset{\smile}{\mathfrak{a}}}$ and hence, by IV.3.7, $\mathrm{cf}(2^{\underset{\smile}{\mathfrak{a}}})=\mathrm{cf}(\mathfrak{a})\leqslant\mathfrak{a}$. Therefore, by 5.2(ii), we cannot have $2^{\mathfrak{a}}=2^{\underset{\smile}{\mathfrak{a}}}$; we shall see that

$$(7)\qquad 2^{\mathfrak{a}}=(2^{\underset{\smile}{\mathfrak{a}}})^{\mathrm{cf}(\mathfrak{a})}=(2^{\underset{\smile}{\mathfrak{a}}})^{\mathrm{cf}(2^{\underset{\smile}{\mathfrak{a}}})}=\gimel(2^{\underset{\smile}{\mathfrak{a}}}).$$

Since $\mathrm{cf}(2^{\underset{\smile}{\mathfrak{a}}})=\mathrm{cf}(\mathfrak{a})$ we need only to prove the first equality of (7). As $2^{\underset{\smile}{\mathfrak{a}}}\leqslant 2^{\mathfrak{a}}$, we have

$$(2^{\underset{\smile}{\mathfrak{a}}})^{\mathrm{cf}(\mathfrak{a})}\leqslant(2^{\mathfrak{a}})^{\mathrm{cf}(\mathfrak{a})}=2^{\mathfrak{a}\cdot\mathrm{cf}(\mathfrak{a})}=2^{\mathfrak{a}}.$$

To prove the opposite inequality, let $\mathrm{cf}(\mathfrak{a})=\lambda$ then, by IV.3.9, $\mathfrak{a}=\Sigma_{\alpha<\lambda}\mathfrak{b}_\alpha$, where $\mathfrak{b}_\alpha<\mathfrak{a}$ for $\alpha<\lambda$. Now, by III.4.12(vi) and III.4.12(v),

$$2^{\mathfrak{a}}=2^{\Sigma_{\alpha<\lambda}\mathfrak{b}_\alpha}=\Pi_{\alpha<\lambda}2^{\mathfrak{b}_\alpha}\leqslant\Pi_{\alpha<\lambda}2^{\underset{\smile}{\mathfrak{a}}}=(2^{\underset{\smile}{\mathfrak{a}}})^{|\lambda|}=(2^{\underset{\smile}{\mathfrak{a}}})^{\mathrm{cf}(\mathfrak{a})},$$

which is what had to be shown.

(ii) We distinguish the following three cases *a–c*.
Case a: $\mathrm{cf}(\mathfrak{a})>\mathfrak{b}$, and this includes the case where $\mathfrak{a}$ is a successor cardinal $>\mathfrak{b}$. The range of every function from $\mathfrak{b}$ into $\mathfrak{a}$ is bounded in $\mathfrak{a}$ (by IV.3.4(i)) hence ${}^{\mathfrak{b}}\mathfrak{a}=\bigcup_{\gamma<\mathfrak{a}}{}^{\mathfrak{b}}\gamma$, and therefore, by III.4.5, III.4.6(v) and III.4.7,

$$(8)\qquad \mathfrak{a}^{\mathfrak{b}}\leqslant\Sigma_{\gamma<\mathfrak{a}}|\gamma|^{\mathfrak{b}}\leqslant\Sigma_{\gamma<\mathfrak{a}}\underset{\smile}{\mathfrak{a}}^{\mathfrak{b}}=\mathfrak{a}\cdot\underset{\smile}{\mathfrak{a}}^{\mathfrak{b}}=\max(\mathfrak{a}, \underset{\smile}{\mathfrak{a}}^{\mathfrak{b}}).$$

If $\mathfrak{a}=\mathfrak{c}^+$ then $\underset{\smile}{\mathfrak{a}}^{\mathfrak{b}}=\mathfrak{c}^{\mathfrak{b}}$ and by (8) $\mathfrak{a}^{\mathfrak{b}}\leqslant\max(\mathfrak{a}, \mathfrak{c}^{\mathfrak{b}})\leqslant\mathfrak{a}^{\mathfrak{b}}$, hence $\mathfrak{a}^{\mathfrak{b}}=\max(\mathfrak{a}, \mathfrak{c}^{\mathfrak{b}})$. If $\mathfrak{a}$ is a limit cardinal then, as easily seen, $\underset{\smile}{\mathfrak{a}}^{\mathfrak{b}}\geqslant\mathfrak{a}$ and by (8) $\mathfrak{a}^{\mathfrak{b}}\leqslant\max(\mathfrak{a}, \underset{\smile}{\mathfrak{a}}^{\mathfrak{b}})\leqslant\underset{\smile}{\mathfrak{a}}^{\mathfrak{b}}\leqslant\mathfrak{a}^{\mathfrak{b}}$, hence $\mathfrak{a}^{\mathfrak{b}}=\underset{\smile}{\mathfrak{a}}^{\mathfrak{b}}$.

Case b: $\mathrm{cf}(\mathfrak{a})\leqslant\mathfrak{b}$ and there is a $\mathfrak{c}<\mathfrak{a}$ such that for all cardinals $\mathfrak{d}$, $\mathfrak{c}<\mathfrak{d}<\mathfrak{a}\rightarrow\mathfrak{d}^{\mathfrak{b}}=\mathfrak{c}^{\mathfrak{b}}$. If $\mathfrak{a}$ is a successor cardinal then either $\mathfrak{a}$ is finite and hence $\mathfrak{a}<\mathfrak{b}$ or else $\mathfrak{a}$ is regular and $\mathfrak{a}=\mathrm{cf}(\mathfrak{a})\leqslant\mathfrak{b}$. In either case $\mathfrak{a}^{\mathfrak{b}}\leqslant\mathfrak{b}^{\mathfrak{b}}=2^{\mathfrak{b}}\leqslant\underset{\smile}{\mathfrak{a}}^{\mathfrak{b}}\leqslant\mathfrak{a}^{\mathfrak{b}}$, hence $\mathfrak{a}^{\mathfrak{b}}=\underset{\smile}{\mathfrak{a}}^{\mathfrak{b}}$. If $\mathfrak{a}$ is a limit cardinal then we shall show, in analogy to 5.2(iii), $\mathfrak{a}^{\mathfrak{b}}=\mathfrak{c}^{\mathfrak{b}}=\underset{\smile}{\mathfrak{a}}^{\mathfrak{b}}$. To prove $\mathfrak{a}^{\mathfrak{b}}\leqslant\mathfrak{c}^{\mathfrak{b}}$, let $\lambda=\mathrm{cf}(\mathfrak{a})$ and let, by IV.3.9, $\mathfrak{a}=\Sigma_{\alpha<\lambda}\mathfrak{d}_\alpha$, where $\mathfrak{d}_\alpha<\mathfrak{a}$ for $\alpha<\lambda$. We can assume, without loss of generality that $\mathfrak{d}_\alpha\geqslant 2$ for every $\alpha<\lambda$ (since $|\lambda|=\mathrm{cf}(\mathfrak{a})\leqslant\mathfrak{a}$). By III.4.17 we have $\mathfrak{a}=\Sigma_{\alpha<\lambda}\mathfrak{d}_\alpha\leqslant\Pi_{\alpha<\lambda}\mathfrak{d}_\alpha$ and hence, by III.4.12(vi), by the assumption of Case *b*, and by III.4.15,

$$\mathfrak{a}^{\mathfrak{b}}\leqslant(\Pi_{\alpha<\lambda}\mathfrak{d}_\alpha)^{\mathfrak{b}}=\Pi_{\alpha<\lambda}\mathfrak{d}_\alpha^{\mathfrak{b}}\leqslant\Pi_{\alpha<\lambda}\mathfrak{c}^{\mathfrak{b}}=\mathfrak{c}^{\mathfrak{b}\cdot|\lambda|}=\mathfrak{c}^{\mathfrak{b}},$$

since $|\lambda|=\mathrm{cf}(\mathfrak{a})\leqslant\mathfrak{b}$. Thus we have $\mathfrak{a}^{\mathfrak{b}}\leqslant\mathfrak{c}^{\mathfrak{b}}\leqslant\underset{\smile}{\mathfrak{a}}^{\mathfrak{b}}\leqslant\mathfrak{a}^{\mathfrak{b}}$.

Case c: $\mathrm{cf}(\mathfrak{a})\leqslant\mathfrak{b}$ and for every $\mathfrak{c}<\mathfrak{a}$ there is a $\mathfrak{d}$ such that $\mathfrak{c}<\mathfrak{d}<\mathfrak{a}$ and $\mathfrak{d}^{\mathfrak{b}}>\mathfrak{c}^{\mathfrak{b}}$. Then the set $\{\mathfrak{c}^{\mathfrak{b}}\mid\mathfrak{c}<\mathfrak{a}\}$ is a cofinal subset of $\underset{\smile}{\mathfrak{a}}^{\mathfrak{b}}$ and hence, by IV.3.7, $\mathrm{cf}(\underset{\smile}{\mathfrak{a}}^{\mathfrak{b}})=\mathrm{cf}(\mathfrak{a})\leqslant\mathfrak{b}$ and by 5.10 we cannot have $\mathfrak{a}^{\mathfrak{b}}=\underset{\smile}{\mathfrak{a}}^{\mathfrak{b}}$. We shall now show that we have, in the present case,

$$(9)\qquad \mathfrak{a}^{\mathfrak{b}}=(\underset{\smile}{\mathfrak{a}}^{\mathfrak{b}})^{\mathrm{cf}(\mathfrak{a})}=(\underset{\smile}{\mathfrak{a}}^{\mathfrak{b}})^{\mathrm{cf}(\underset{\smile}{\mathfrak{a}}^{\mathfrak{b}})}=\gimel(\underset{\smile}{\mathfrak{a}}^{\mathfrak{b}}).$$

Clearly $\underset{\smile}{\mathfrak{a}}^{\mathfrak{b}}\leqslant\mathfrak{a}^{\mathfrak{b}}$, hence $(\underset{\smile}{\mathfrak{a}}^{\mathfrak{b}})^{\mathrm{cf}(\mathfrak{a})}\leqslant\mathfrak{a}^{\mathfrak{b}\cdot\mathrm{cf}(\mathfrak{a})}=\mathfrak{a}^{\mathfrak{b}}$, since $\mathrm{cf}(\mathfrak{a})\leqslant\mathfrak{b}$; thus in order to establish (9) we have only to prove $\mathfrak{a}^{\mathfrak{b}}\leqslant(\underset{\smile}{\mathfrak{a}}^{\mathfrak{b}})^{\mathrm{cf}(\mathfrak{a})}$. As in Case *b* let $\mathfrak{a}=\Sigma_{\alpha<\lambda}\mathfrak{d}_\alpha$ where $\lambda=\mathrm{cf}(\mathfrak{a})$ and $2\leqslant\mathfrak{d}_\alpha<\mathfrak{a}$ for $\alpha<\lambda$. We get

$$\mathfrak{a}^{\mathfrak{b}}=(\Sigma_{\alpha<\lambda}\mathfrak{d}_\alpha)^{\mathfrak{b}}\leqslant(\Pi_{\alpha<\lambda}\mathfrak{d}_\alpha)^{\mathfrak{b}}=\Pi_{\alpha<\lambda}\mathfrak{d}_\alpha^{\mathfrak{b}}\quad\Pi_{\alpha<\lambda}\underset{\smile}{\mathfrak{a}}^{\mathfrak{b}}=(\underset{\smile}{\mathfrak{a}}^{\mathfrak{b}})^{|\lambda|}=(\underset{\smile}{\mathfrak{a}}^{\mathfrak{b}})^{\mathrm{cf}(\mathfrak{a})}.\qquad\square$$

5.13Ac Computing Cardinal Exponentiation from the Gimel Function (Bukovsky 1965). As a consequence of Theorem 5.12 the values of $\mathfrak{a}^{\mathfrak{b}}$ for all cardinals $\mathfrak{a}\geqslant 2$ and $\mathfrak{b}\geqslant\aleph_0$ are completely determined by the gimel function provided we are given the order of the cardinal numbers and the cofinality function cf and we can use recursion as we shall now see. Let us also remark that in all other cases of $\mathfrak{a}^{\mathfrak{b}}$, i.e. where $\mathfrak{a}\leqslant 1$ or $\mathfrak{b}<\aleph_0$ the value of $\mathfrak{a}^{\mathfrak{b}}$ is determined even without any use of gimel function. (In the case of finite $\mathfrak{a}$ and $\mathfrak{b}$ we have to use recursion.)

First let us show that the function $\langle 2^{\mathfrak{a}}\mid\mathfrak{a}\geqslant\aleph_0\rangle$ is determined by the gimel and cofinality functions. If $\mathfrak{a}$ is regular then $2^{\mathfrak{a}}=\gimel(\mathfrak{a})$. If $\mathfrak{a}$ is singular then $2^{\mathfrak{a}}$ is determined by recursion as follows. If $\mathrm{cf}(2^{\underset{\smile}{\mathfrak{a}}})>\mathrm{cf}(\mathfrak{a})$ then $2^{\mathfrak{a}}=2^{\underset{\smile}{\mathfrak{a}}}$ and if $\mathrm{cf}(2^{\underset{\smile}{\mathfrak{a}}})\leqslant\mathrm{cf}(\mathfrak{a})$ then $2^{\mathfrak{a}}=\gimel(2^{\underset{\smile}{\mathfrak{a}}})$; thus $2^{\mathfrak{a}}$ is determined from $2^{\underset{\smile}{\mathfrak{a}}}=\sup_{\mathfrak{b}<\mathfrak{a}}2^{\mathfrak{b}}$ via the cf and $\gimel$ functions. Now let us determine $\mathfrak{a}^{\mathfrak{b}}$ for $\mathfrak{a}>2$, $\mathfrak{b}\geqslant\aleph_0$ by recursion on $\mathfrak{a}$. If $\mathfrak{a}$ is a successor cardinal $\mathfrak{c}^+$ then $\mathfrak{a}^{\mathfrak{b}}=\max(\mathfrak{c}^{\mathfrak{b}},\mathfrak{a})$ and if $\mathfrak{a}$ is a limit cardinal then if $\mathrm{cf}(\mathfrak{a}^{\underset{\smile}{\mathfrak{b}}})>\mathrm{cf}(\mathfrak{b})$ then $\mathfrak{a}^{\mathfrak{b}}=\underset{\smile}{\mathfrak{a}}^{\mathfrak{b}}$ and if $\mathrm{cf}\,(\underset{\smile}{\mathfrak{a}}^{\mathfrak{b}})\leqslant\mathrm{cf}(\mathfrak{b})$ then $\mathfrak{a}^{\mathfrak{b}}=\gimel(\underset{\smile}{\mathfrak{a}}^{\mathfrak{b}})$; thus $\mathfrak{a}^{\mathfrak{b}}$ is determined from $\underset{\smile}{\mathfrak{a}}^{\mathfrak{b}}=\sup_{\mathfrak{c}<\mathfrak{a}}\mathfrak{c}^{\mathfrak{b}}$ via the cf and $\gimel$ functions.

Even though 5.12 covers all cases of $\mathfrak{a}^{\mathfrak{b}}$ with $\mathfrak{a}\geqslant 2$ and $\mathfrak{b}\geqslant\aleph_0$ there is a simple case for which the relatively heavy proof of 5.12 is unnecessary. If $2\leqslant\mathfrak{a}\leqslant\mathfrak{b}$ or even $2\leqslant\mathfrak{a}\leqslant 2^{\mathfrak{b}}$ then $\mathfrak{a}^{\mathfrak{b}}=2^{\mathfrak{b}}$ since $2^{\mathfrak{b}}\leqslant\mathfrak{a}^{\mathfrak{b}}\leqslant(2^{\mathfrak{b}})^{\mathfrak{b}}=2^{\mathfrak{b}}$.

5.14Ac Exercise. As we have mentioned in 5.3, the only laws the power function has to observe on the regular alephs are 5.2(i) and 5.2(ii) (provided this freedom is not abused). Therefore, it is consistent with the axioms of ZFC that for all $n<\omega$, $2^{\aleph_n}=\aleph_{\omega+n+1}$. Assuming this, prove that $\aleph_\omega^{\aleph_0}<2^{\aleph_\omega}$, providing thereby an example where $\gimel(\mathfrak{a})<2^{\mathfrak{a}}$. Use this also to show that the function $\gimel(\mathfrak{a})$ is not necessarily a monotonic function of $\mathfrak{a}$. $\square$

5.15Ac Exercises. (i) Give an example where IV.2.25 fails to hold for a limit ordinal κ which is not of the form $\omega^{\cdot\xi}$.

(ii) Let κ be a limit ordinal and let $\langle \mathfrak{a}_\alpha \mid \alpha < \kappa \rangle$ be a non-decreasing sequence of cardinals with $\mathfrak{a}_0 \geqslant 2$. Prove that $\mathrm{cf}(\Pi_{\alpha<\kappa}\mathfrak{a}_\alpha) > \mathrm{top}(\kappa) \geqslant \mathrm{cf}(\kappa)$ (see IV.3.16 for the definition of $\mathrm{top}(\kappa)$).

Hints. (i) Take $\kappa = \lambda + \mu$, where λ and μ are regular infinite ordinals and $\mu < \lambda$. Take for all $\mathfrak{a}_{\lambda+\nu}$, $\nu < \mu$, a cardinal $\mathfrak{a}$ such that $\mathfrak{a} \geqslant \Pi_{\alpha<\lambda}\mathfrak{a}_\alpha$, $\mathfrak{a}^\mu = \mathfrak{a}$ and $\mathfrak{a}^\lambda > \mathfrak{a}$. Use 5.10 to obtain such an $\mathfrak{a}$.

(ii) Use IV.2.14, IV.2.25 and 5.10. □

5.16 The Generalized Continuum Hypothesis. The statement $2^{\aleph_0} = \aleph_1$, announced as a conjecture by Cantor 1878, is known as the *continuum hypothesis*. The question whether $2^{\aleph_0}$ is $\aleph_1$ or not was considered by Hilbert to be so important that he made this question the first in his famous list of open mathematical problems (Hilbert 1900). Cantor 1883 conjectured also, implicitly, that $2^{\aleph_1} = \aleph_2$. The more general statement, formulated first by Hausdorff 1908, that for all alephs $\mathfrak{a}$ $2^{\mathfrak{a}} = \mathfrak{a}^+$, is known as the *generalized continuum hypothesis*. This statement can also be formulated as "$2^{\aleph_\alpha} = \aleph_{\alpha+1}$, for all ordinals α". Gödel 1938 proved, on one hand, that the generalized continuum hypothesis is consistent with ZFC (see § IX.1) and Cohen 1963 proved, on the other hand, that even the continuum hypothesis $2^{\aleph_0} = \aleph_1$ is unprovable in ZFC. (We have already encountered, in 5.3, Easton's generalization of the unprovability of $2^{\aleph_0} = \aleph_1$.) Assuming the generalized continuum hypothesis leads to an extreme simplification of cardinal exponentiation (see 5.17 and 5.18 below) and is altogether a strong set theoretical assumption. Mathematicians do not tend to assume it as an additional axiom of set theory mostly since they cannot convince themselves that this statement is "true" as many of them have done for the axioms of ZFC including the axiom of choice. However, a mathematician trying to prove a theorem will usually regard a proof of the theorem from the generalized continuum hypothesis as a partial success. Several important theorems have been proved by means of the continuum hypothesis or the generalized continuum hypothesis quite a while before they were proved in ZFC (while others turned out to be unprovable in ZFC). In § VIII.4 we discuss Martin's axiom, which is an interesting weakening of the continuum hypothesis.

5.17Ac Proposition. If the generalized continuum hypothesis holds then for all $\mathfrak{a} \geqslant 2$ and for all alephs $\mathfrak{b}$

$$\mathfrak{a}^{\mathfrak{b}} = \begin{cases} \mathfrak{a} & \text{if } \mathfrak{b} < \mathrm{cf}(\mathfrak{a}) \\ \mathfrak{a}^+ & \text{if } \mathrm{cf}(\mathfrak{a}) \leqslant \mathfrak{b} \leqslant \mathfrak{a} \\ \mathfrak{b}^+ & \text{if } \mathfrak{b} \geqslant \mathfrak{a}. \end{cases}$$

Hint of proof. Use the generalized continuum hypothesis and 5.10 to prove that for infinite $\mathfrak{a}$ $\mathfrak{a}^{\mathrm{cf}(\mathfrak{a})} = \mathfrak{a}^+$. Then use 5.12 to compute $\mathfrak{a}^{\mathfrak{b}}$. A direct proof of the proposition is equally easy and involves, in addition to giving trivial bounds, only (8) of 5.12. □

5.18Ac Exercises. (i) Assuming the generalized continuum hypothesis, compute, for $\mathfrak{a} \geqslant 2$ and $\mathfrak{b} \geqslant \aleph_0$, the values of $\mathfrak{a}^{\underline{\mathfrak{b}}}$ and $\underline{\mathfrak{a}}^{\mathfrak{b}}$ (as we did in 5.16 for $\mathfrak{a}^{\mathfrak{b}}$).

(ii) Let κ be a limit ordinal, let $\langle \mathfrak{a}_\lambda \mid \lambda < \kappa \rangle$ be an increasing sequence of non-zero cardinals and let $\mathfrak{a} = \sup\{\mathfrak{a}_\lambda \mid \lambda < \kappa\}$. Assuming the generalized continuum hypothesis prove that $\Pi_{\lambda<\kappa} \mathfrak{a}_\lambda = \mathfrak{a}^+$. (*Hint:* Use III.4.19.) □

The relationship between the generalized continuum hypothesis and the axiom of choice has been given considerable attention. As we have already seen in 5.3 the axiom of choice does not imply even the continuum hypothesis; the question is whether the generalized continuum hypothesis implies the axiom of choice. The answer is essentially positive, but the handling of it depends on the way in which the generalized continuum hypothesis is formulated. In the form "for all ordinals α $2^{\aleph_\alpha} = \aleph_{\alpha+1}$" the generalized continuum hypothesis implies the axiom of choice with the aid of the axiom of foundation; this form implies that the power set of a well-orderable set is well-orderable, which is equivalent to the axiom of choice by 1.15. Another formulation of the generalized continuum hypothesis is "for every infinite cardinal $\mathfrak{a}$, $2^{\mathfrak{a}}$ is an immediate successor of $\mathfrak{a}$ in the sense that there is no cardinal $\mathfrak{b}$ such that $\mathfrak{a} < \mathfrak{b} < 2^{\mathfrak{a}}$". Lindenbaum and Tarski 1926 asserted that this form of the generalized continuum hypothesis implies the axiom of choice; this implication is an immediate consequence of Theorem 5.21 below (and 1.7).

5.19 Proposition (Specker 1954). *If* $\mathfrak{a} \geqslant 5$ *then* $2^{\mathfrak{a}} \nleqslant \mathfrak{a}^2$.

Proof. Let A be a set with $|A| \geqslant 5$; we assume that there is an injection G of $\mathbf{P}(A)$ into $A \times A$ and shall get a contradiction by constructing by recursion an injection F of On into the set A. For $F(0), \ldots, F(4)$ we choose five arbitrary members of A. We define now $F(n)$ for every finite $n \geqslant 5$. Let $<_l$ be any well-ordering of the denumerable set of all finite subsets of ω. Since $2^n > n^2$ for every finite $n \geqslant 5$ (as easily seen by induction on n) we have $|\mathbf{P}(F[n])| > |F[n] \times F[n]|$, hence there is a $u \subseteq F[n]$ such that $G(u) \notin F[n] \times F[n]$. Among those u's let v be the one with the least set $F^{-1}[u]$ (in the well-ordering $<_l$ of the finite subsets of ω). We have $G(v) = \langle v_1, v_2 \rangle \notin F[n] \times F[n]$. We set $F(n) = v_1$ if $v_1 \notin F[n]$ and $F(n) = v_2$ if $v_1 \in F[n]$. In either case we have $F(n) \in A \sim F[n]$. For $\alpha \geqslant \omega$ we define $F(\alpha)$ as follows. By IV.2.24 there is a function H on On such that for every infinite ordinal α $H(\alpha)$ is a bijection of $\alpha \times \alpha$ on α. $H(\alpha)$ induces a bijection j of $F[\alpha] \times F[\alpha]$ on $F[\alpha]$, i.e., for $\beta, \gamma < \alpha$ $j(F(\beta), F(\gamma)) = F(H(\alpha)(\beta, \gamma))$. Let $g = (jG) \upharpoonright \mathbf{P}(F[\alpha])$ then $\mathrm{Dom}(g) \subseteq \mathbf{P}(F[\alpha])$ and g is an injection into $F[\alpha]$. Let h be the mapping of $F[\alpha]$ into $\mathbf{P}(F[\alpha])$ given by $h(x) = g^{-1}(x)$ if $x \in \mathrm{Rng}(g)$ and $h(x) = 0$ otherwise. As in the proof of Cantor's theorem $|\mathbf{P}(z)| > |z|$ (III.2.13) the subset $t = \{x \in F[\alpha] \mid x \notin h(x)\}$ of $F[\alpha]$ is not in the range of h. By the definition of h $t \notin \mathrm{Dom}(g)$, i.e., $G(t) \notin \mathrm{Dom}(j) = F[\alpha] \times F[\alpha]$. Let $G(t) = \langle v_1, v_2 \rangle \in A \times A$, we set, again, $F(\alpha) = v_1$ if $v_1 \notin F[\alpha]$ and $F(\alpha) = v_2$ if $v_1 \in F[\alpha]$. In either case $F(\alpha) \in A \sim F[\alpha]$, which completes our construction and our proof that F is an injection of On into A. □

5.20 Corollary. *For every infinite cardinal* $\mathfrak{a}$ *and every finite cardinal* n $n\mathfrak{a} < 2^{\mathfrak{a}}$.

Proof. Since $\mathfrak{a}$ is infinite and n is finite there is an infinite cardinal $\mathfrak{b}$ such that $\mathfrak{a}=\mathfrak{b}+n$. We have

$$n\mathfrak{a}=n(\mathfrak{b}+n)=n\mathfrak{b}+n^2\leqslant(2^n-1)\mathfrak{b}+n^2\leqslant(2^n-1)\mathfrak{b}+\mathfrak{b}=2^n\mathfrak{b}\leqslant 2^n\cdot 2^{\mathfrak{b}}=2^{n+\mathfrak{b}}=2^{\mathfrak{a}}.$$

We cannot have $n\mathfrak{a}=2^{\mathfrak{a}}$ since this would imply $2^{\mathfrak{a}}=n\mathfrak{a}\leqslant\mathfrak{a}^2$, contradicting 5.19. □

5.21 Theorem (Specker 1954). *If* $\mathfrak{a}$ *is a cardinal such that*

(1) $2^{\mathfrak{a}}$ *is an immediate successor of* $\mathfrak{a}$, *and*

(2) $2^{2^{\mathfrak{a}}}$ *is an immediate successor of* $2^{\mathfrak{a}}$

(*i.e., there is no cardinal* $\mathfrak{b}$ *such that* $\mathfrak{a}<\mathfrak{b}<2^{\mathfrak{a}}$ *or* $2^{\mathfrak{a}}<\mathfrak{b}<2^{2^{\mathfrak{a}}}$) *then* $2^{\mathfrak{a}}$, *and hence also* $\mathfrak{a}$, *is a well-ordered cardinal.*

Proof. We may assume that $\mathfrak{a}$ is infinite since if $\mathfrak{a}$ is finite $2^{\mathfrak{a}}$ is anyway well-ordered. By 5.20 $\mathfrak{a}\leqslant\mathfrak{a}+1\leqslant 2\mathfrak{a}<2^{\mathfrak{a}}$ hence, by (1), $\mathfrak{a}=\mathfrak{a}+1=2\mathfrak{a}$. Now $\mathfrak{a}\leqslant\mathfrak{a}^2\leqslant(2^{\mathfrak{a}})^2=2^{2\mathfrak{a}}=2^{\mathfrak{a}}$ and since, by 5.19, $\mathfrak{a}^2\neq 2^{\mathfrak{a}}$ we get, by (1) $\mathfrak{a}^2=\mathfrak{a}$. By III.2.27 $\mathfrak{a}^+\leqslant 2^{2^{\mathfrak{a}^2}}=2^{2^{\mathfrak{a}}}$, where $\mathfrak{a}^+$ is the least aleph not $\leqslant\mathfrak{a}$. We have

(3) $2^{\mathfrak{a}}\leqslant 2^{\mathfrak{a}}+\mathfrak{a}^+\leqslant 2^{\mathfrak{a}}\cdot\mathfrak{a}^+\leqslant 2^{2^{\mathfrak{a}}}\cdot 2^{2^{\mathfrak{a}}}\leqslant 2^{2^{(\mathfrak{a}+1)}}=2^{2^{\mathfrak{a}}}.$

Let us consider the inequality.

(4) $2^{\mathfrak{a}}+\mathfrak{a}^+\leqslant 2^{\mathfrak{a}}\cdot\mathfrak{a}^+$

in (3). If the two sides of (4) are equal then, by 1.13 $2^{\mathfrak{a}}$ and $\mathfrak{a}^+$ are comparable. Thus either $2^{\mathfrak{a}}\leqslant\mathfrak{a}^+$ and then $2^{\mathfrak{a}}$ is a well-ordered cardinal, which is what we have to prove, or else $\mathfrak{a}^+\leqslant 2^{\mathfrak{a}}$. If a strict inequality holds in (4) then by (2) all other inequalities in (3) are equalities and we have $2^{\mathfrak{a}}=2^{\mathfrak{a}}+\mathfrak{a}^+$ hence, again $\mathfrak{a}^+\leqslant 2^{\mathfrak{a}}$. Thus we have $\mathfrak{a}<\mathfrak{a}+\mathfrak{a}^+\leqslant\mathfrak{a}\cdot\mathfrak{a}^+\leqslant 2^{\mathfrak{a}}\cdot 2^{\mathfrak{a}}=2^{2\mathfrak{a}}=2^{\mathfrak{a}}$ hence, by (1) $\mathfrak{a}+\mathfrak{a}^+=\mathfrak{a}\cdot\mathfrak{a}^+$. By 1.13 $\mathfrak{a}$ and $\mathfrak{a}^+$ are comparable and since $\mathfrak{a}^+\nleqslant\mathfrak{a}$ we have $\mathfrak{a}<\mathfrak{a}^+\leqslant 2^{\mathfrak{a}}$, thus by (1), $\mathfrak{a}^+=2^{\mathfrak{a}}$ and $2^{\mathfrak{a}}$ is well-ordered. □

5.22Ac The Gimel Hypothesis (Solovay 1974). The *gimel hypothesis* is the statement that for every singular aleph $\mathfrak{a}$ $\gimel(\mathfrak{a})=\max(2^{\mathrm{cf}(\mathfrak{a})},\mathfrak{a}^+)$.

By $\gimel(\mathfrak{a})=\mathfrak{a}^{\mathrm{cf}(\mathfrak{a})}\geqslant 2^{\mathrm{cf}(\mathfrak{a})}$ and $\gimel(\mathfrak{a})>\mathfrak{a}$ (5.10) we can prove in ZFC that $\gimel(\mathfrak{a})\geqslant\max(2^{\mathrm{cf}(\mathfrak{a})},\mathfrak{a}^+)$, thus the gimel hypothesis asserts that for singular alephs $\mathfrak{a}$ $\gimel(\mathfrak{a})$ obtains the least value permitted by the axioms of ZFC (with respect to given values of $\gimel(\mathfrak{b})=2^{\mathfrak{b}}$ for the regular cardinals $\mathfrak{b}$).

For any "given" function F on the class of the regular alephs into Cn which satisfies $\mathfrak{a}<\mathfrak{b}\rightarrow F(\mathfrak{a})\leqslant F(\mathfrak{b})$ and $\mathrm{cf}(F(\mathfrak{a}))>\mathfrak{a}$ (5.2(i) and 5.2(ii)) it is consistent with ZFC and with the gimel hypothesis that $2^{\mathfrak{a}}=F(\mathfrak{a})$ for all regular alephs $\mathfrak{a}$ (this follows from the proof of Easton 1964—compare with 5.3). This means that the gimel hypothesis neither says nor implies anything at all concerning the values of $2^{\mathfrak{a}}=\gimel(\mathfrak{a})$ for regular alephs $\mathfrak{a}$. Once these values are given the gimel hypothesis determines completely all values $\gimel(\mathfrak{a})$ for the singular alephs $\mathfrak{a}$ and hence, by 5.13, all the values of $\mathfrak{a}^{\mathfrak{b}}$.

Let us compare the gimel hypothesis with the generalized continuum hypothesis. The gimel hypothesis is a consequence of the generalized continuum hypothesis (why?). Both are assumptions which simplify cardinal exponentiation; the generalized continuum hypothesis is an extreme simplification of the functions $2^{\mathfrak{a}}$ and $\mathfrak{a}^{\mathfrak{b}}$, the gimel hypothesis leads also to a considerable simplification of cardinal exponentiation (see 5.23 below) but here we have to justify why we chose to present an hypothesis which simplifies $\gimel(\mathfrak{a})$ for singular $\mathfrak{a}$'s but leaves complete freedom to $\gimel(\mathfrak{a})=2^{\mathfrak{a}}$ for regular alephs $\mathfrak{a}$. The distinction between regular and singular alephs is natural here; this is evident already from 5.2 and 5.3, where the value of $2^{\mathfrak{a}}$ for a regular $\mathfrak{a}$ depends on the values of $2^{\mathfrak{b}}$ for $\mathfrak{b}<\mathfrak{a}$ only in the trivial sense that $2^{\mathfrak{a}}$ has to be $\geqslant 2^{\mathfrak{b}}$, but the value of $2^{\mathfrak{a}}$ for a singular $\mathfrak{a}$ depends on the values of $2^{\mathfrak{b}}$ for $\mathfrak{b}<\mathfrak{a}$ in a much stronger way. This argument, however, does not suffice to explain why we came across the gimel hypothesis at all. It turns out that the gimel hypothesis is a consequence of certain natural statements of set theory which do not seem to be related to it at all.

Jensen has shown (see Devlin and Jensen 1976) that a certain statement "for some subset x of ω, $x^{\#}$ does not exist" implies the gimel hypothesis. It is beyond the scope of the present book to explain what "$x^{\#}$ exists" means (see Solovay 1967). Let us only remark that the contrapositive of Jensen's result is that the negation of the gimel hypothesis implies "for all $x\subseteq\omega$, $x^{\#}$ exists", and this is a very strong set-theoretical hypothesis. Another assumption which leads to the gimel hypothesis is the assumption of the existence of a strongly compact cardinal. (The strongly compact cardinals are a special kind of measurable cardinals—IX.4.11.) Solovay 1974 proved that if there exists a strongly compact cardinal $\mathfrak{e}$ then the gimel hypothesis holds above $\mathfrak{e}$, i.e., for all $\mathfrak{a}>\mathfrak{e}$ $\gimel(\mathfrak{a})=\max(2^{\mathrm{cf}(\mathfrak{a})}, \mathfrak{a}^{+})$.

The gimel hypothesis is consistent with ZFC since it is a consequence of the generalized continuum hypothesis which is consistent with ZFC. It is not a theorem of ZFC (if the existence of a supercompact cardinal is consistent with ZFC—Solovay 1974) since by the example of Magidor mentioned in 5.8 we may have $\gimel(\aleph_{\omega})=\aleph_{\omega+2}$ while $\max(2^{\mathrm{cf}(\aleph_{\omega})}, \aleph_{\omega}^{+})=\max(2^{\aleph_0}, \aleph_{\omega+1})=\aleph_{\omega+1}$.

5.23Ac Proposition. *If the gimel hypothesis holds then for every singular* $\mathfrak{a}$

$$2^{\mathfrak{a}}=\begin{cases}2^{\underline{\mathfrak{a}}} & \text{if } \mathrm{cf}(2^{\underline{\mathfrak{a}}})>\mathrm{cf}(\mathfrak{a})\\ (2^{\underline{\mathfrak{a}}})^{+} & \text{if } \mathrm{cf}(2^{\underline{\mathfrak{a}}})\leqslant\mathrm{cf}(\mathfrak{a}),\end{cases}$$

and for all $\mathfrak{a}>2$ *and* $\mathfrak{b}\geqslant\aleph_0$

$$\mathfrak{a}^{\mathfrak{b}}=\begin{cases}2^{\mathfrak{b}} & \text{if } \mathfrak{a}\leqslant 2^{\mathfrak{b}}\\ \mathfrak{a} & \text{if } \mathfrak{a}>2^{\mathfrak{b}} \text{ and } \mathrm{cf}(\mathfrak{a})>\mathfrak{b}\\ \mathfrak{a}^{+} & \text{if } \mathfrak{a}>2^{\mathfrak{b}} \text{ and } \mathrm{cf}(\mathfrak{a})\leqslant\mathfrak{b}. \quad\square\end{cases}$$

5.24Ac Exercise. The gimel hypothesis is equivalent to each one of the following statements: (a) For all singular $\mathfrak{a}$, if $2^{\mathrm{cf}(\mathfrak{a})}<\mathfrak{a}$ then $\gimel(\mathfrak{a})=\mathfrak{a}^+$. (b) For all regular alephs $\mathfrak{b}$ and $\mathfrak{c}$ if $2^{\mathfrak{c}}<\mathfrak{b}$ then $\mathfrak{b}^{\mathfrak{c}}=\mathfrak{b}$. □

5.25 Sets of Almost Disjoint Sets (Sierpinski 1928). After we have investigated the cardinality $2^{\mathfrak{a}}$ of the power set of a set A with $|A|=\mathfrak{a}$ we conclude this section by studying the cardinality of some special subsets of $\mathbf{P}(A)$, namely those which consist of almost disjoint sets. A subset P of the power set $\mathbf{P}(A)$ of A is called a *partition* of A if $\bigcup P=A$ and the members of P are pairwise disjoint. Given a set A we run often into questions of obtaining partitions P of A with desirable properties. One conceivably desirable property of a partition P is that it has as many parts (i.e., members) as possible while having each part as big as possible. We dealt at length with such a question in IV.4.48 where we showed how to partition a stationary subset of Ω to Ω many stationary sets. If we approach partitions from the point of view of cardinality alone the answers are trivial; a set A of cardinality $\mathfrak{a}$ can be partitioned into $\mathfrak{a}$ sets of cardinality $\mathfrak{a}$ each, and this can obviously not be improved. The situation changes when we drop the requirement of disjointness of the members of P and require only that different members of P be *almost disjoint* i.e., for all $B, C\in P$, if $B\neq C$ then $B\cap C$ is of "small" cardinality. To be more specific, a general question which we ask in this direction is the following. Let $\mathfrak{a}\geqslant\mathfrak{b}\geqslant\mathfrak{c}\geqslant\aleph_0$ and let A be a set of cardinality $\mathfrak{a}$. How big can $|P|$ get under the following conditions: P is a set of subsets of A; for each $B\in P$ $|B|=\mathfrak{b}$; and for all $B, C\in P$ $|B\cap C|<\mathfrak{c}$. (Notice that we do not have to require that $\bigcup P=A$ since any such P can be easily changed to a set Q satisfying the same conditions and such that $|Q|\geqslant|P|$ and $\bigcup Q=A$ by either adding members to P or by extending a single member of P.) The place to look for such big $|P|$'s is the range extending from $\mathfrak{a}$ to $2^{\mathfrak{a}}$, since we can, trivially, obtain such sets P of cardinality $\mathfrak{a}$ and since any such set P is at most of the cardinality $2^{\mathfrak{a}}$, being a subset of $\mathbf{P}(A)$.

We shall handle first the case where $\mathfrak{a}=\mathfrak{b}=\mathfrak{c}=\aleph_0$, being a simple case which illustrates the main method used in the general case.

5.26 Proposition. *For a denumerable set A, there is a set P of cardinality $2^{\aleph_0}$ of almost disjoint denumerable subsets of A, i.e., for all $B, C\in P$, if $B\neq C$ then $B\cap C$ is finite.* (*This gives us the largest possible $|P|$ since $|\mathbf{P}(A)|=2^{\aleph_0}$.*)

Proof. Since it does not matter which denumerable set A we take let us chose for A the set ${}^{\circledast}\omega$ of all finite sequences of finite ordinals, which is denumerable by III.4.21. Let F be the function on ${}^{\omega}\omega$ given by $F(g)=\{g\restriction n\mid n\in\omega\}$, for $g\in{}^{\omega}\omega$. As easily seen, F is an injection of ${}^{\omega}\omega$ into $\mathbf{P}(A)$. Put $P=\mathrm{Rng}(F)$, then clearly $P\subseteq\mathbf{P}(A)$, $|B|=\aleph_0$ for $B\in P$, and $|P|=2^{\aleph_0}$. If $B, C\in P$ and $B\neq C$ then there are $g, h\in{}^{\omega}\omega$, $g\neq h$, such that $B=F(g)$ and $C=F(h)$. Since $g\neq h$ there is a $k\in\omega$ such that $g(k)\neq h(k)$. Let $s\in B\cap C=F(g)\cap F(h)$, then s is a member of ${}^{\circledast}\omega$ of length $n\in\omega$, hence $s=g\restriction n=h\restriction n$. Since $g(k)\neq h(k)$ we have $n\leqslant k$ and thus $s=g\restriction n\in\{g\restriction m\mid m\leqslant k\}$. We have shown that $B\cap C\subseteq\{g\restriction m\mid m\leqslant k\}$ and hence $B\cap C$ is a finite set. □

5.27 Definition. For $\mathfrak{a} \geqslant \mathfrak{b} \geqslant \mathfrak{c} \geqslant \aleph_0$, we say that a set P is an $\mathfrak{a}, \mathfrak{b}, \mathfrak{c}$-*set* if for some set A of cardinality $\mathfrak{a}$ $P \subseteq \mathbf{P}(A)$, and for all $B, C \in P$ $|B| = \mathfrak{b}$ and if $B \neq C$ then $|B \cap C| < \mathfrak{c}$. As we mentioned in 5.25 there is always an $\mathfrak{a}, \mathfrak{b}, \mathfrak{c}$-set of cardinality $\mathfrak{a}$, and every $\mathfrak{a}, \mathfrak{b}, \mathfrak{c}$-set is of cardinality $\leqslant 2^{\mathfrak{a}}$. In 5.26 we proved the existence of an $\aleph_0, \aleph_0, \aleph_0$-set of cardinality $2^{\aleph_0}$.

5.28Ac Proposition (Sierpinski 1928, Tarski 1928). *If* $\mathfrak{a} \geqslant \aleph_0$, *for some* $\mathfrak{d} > 1$ $\mathfrak{e}$ *is the least aleph such that* $\mathfrak{d}^{\mathfrak{e}} > \mathfrak{a}$, $\mathfrak{b} \leqslant \mathfrak{e}$ *and* $\mathrm{cf}(\mathfrak{b}) = \mathrm{cf}(\mathfrak{e})$ *then there is an* $\mathfrak{a}, \mathfrak{b}, \mathfrak{b}$-set *of cardinality* $\mathfrak{d}^{\mathfrak{e}} > \mathfrak{a}$.

Proof. Since $\mathfrak{d} \geqslant 2$, $\mathfrak{d}^{\mathfrak{a}} \geqslant 2^{\mathfrak{a}} > \mathfrak{a}$, hence $\mathfrak{e} \leqslant \mathfrak{a}$. Since $\mathfrak{e}$ is the least cardinal such that $\mathfrak{d}^{\mathfrak{e}} > \mathfrak{a}$ we have $|{}^{\underline{\mathfrak{e}}}\mathfrak{d}| \leqslant \mathfrak{a}$ where ${}^{\mathfrak{e}}\mathfrak{d}$ is the set of all functions into $\mathfrak{d}$ whose domains are ordinals $< \mathfrak{e}$. Since $\mathfrak{b} \leqslant \mathfrak{e}$ and $\mathrm{cf}(\mathfrak{b}) = \mathrm{cf}(\mathfrak{e})$, $\mathfrak{e}$ has a cofinal subset B of the order type $\mathfrak{b}$ (by IV.3.16). The function F on ${}^{\mathfrak{e}}\mathfrak{d}$ given by $F(h) = \{h \upharpoonright \lambda \mid \lambda \in B\}$, for $h \in {}^{\mathfrak{e}}\mathfrak{d}$, is an injection of ${}^{\mathfrak{e}}\mathfrak{d}$ in $\mathbf{P}({}^{\underline{\mathfrak{e}}}\mathfrak{d})$. It is easily seen that the range of F is an $|{}^{\underline{\mathfrak{e}}}\mathfrak{d}|, \mathfrak{b}, \mathfrak{b}$-set of cardinality $\mathfrak{d}^{\mathfrak{e}} > \mathfrak{a}$, and since $\mathfrak{a} \geqslant |{}^{\underline{\mathfrak{e}}}\mathfrak{d}|$ it is also an $\mathfrak{a}, \mathfrak{b}, \mathfrak{b}$-set. □

5.29Ac Corollary (Tarski 1929). (i) *For every aleph* $\mathfrak{a}$ *there is an* $\mathfrak{e} \leqslant \mathrm{cf}(\mathfrak{a})$ *such that there is an* $\mathfrak{a}, \mathfrak{e}, \mathfrak{e}$-*set of cardinality* $> \mathfrak{a}$.

(ii) *For every aleph* $\mathfrak{a}$ *such that* $2^{\underline{\mathfrak{a}}} = \mathfrak{a}$ (*and this is always the case if the generalized continuum hypothesis holds*) *and for every* $\mathfrak{b} \leqslant \mathfrak{a}$ *such that* $\mathrm{cf}(\mathfrak{b}) = \mathrm{cf}(\mathfrak{a})$ *there is an* $\mathfrak{a}, \mathfrak{b}, \mathfrak{b}$-*set of cardinality* $> \mathfrak{a}$.

Proof. (i) Since $\mathfrak{a}^{\mathrm{cf}(\mathfrak{a})} > \mathfrak{a}$ (5.10) the least cardinal $\mathfrak{e}$ such that $\mathfrak{a}^{\mathfrak{e}} > \mathfrak{a}$ is $\leqslant \mathrm{cf}(\mathfrak{a})$ and by 5.28 there is an $\mathfrak{a}, \mathfrak{e}, \mathfrak{e}$-set of cardinality $> \mathfrak{a}$.

(ii) Since $2^{\underline{\mathfrak{a}}} = \mathfrak{a}$, $\mathfrak{a}$ is the least cardinal $\mathfrak{e}$ such that $2^{\mathfrak{e}} > \mathfrak{a}$, hence, by 5.28, there is an $\mathfrak{a}, \mathfrak{b}, \mathfrak{b}$-set of cardinality $> \mathfrak{a}$. □

Assuming the generalized continuum hypothesis, or even only $2^{\mathfrak{a}} = \mathfrak{a}$, 5.29(ii) is the strongest possible result concerning the existence of $\mathfrak{a}, \mathfrak{b}, \mathfrak{c}$-sets; we shall see in the next two propositions that for alephs $\mathfrak{a} \geqslant \mathfrak{b} \geqslant \mathfrak{c}$ if $2^{\underline{\mathfrak{a}}} = \mathfrak{a}$ then the conditions $\mathrm{cf}(\mathfrak{b}) = \mathrm{cf}(\mathfrak{a})$ and $\mathfrak{c} = \mathfrak{b}$ are not only sufficient but also necessary for the existence of an $\mathfrak{a}, \mathfrak{b}, \mathfrak{c}$-set of cardinality $> \mathfrak{a}$.

5.30Ac Proposition (Tarski 1929). *If* $\mathfrak{a}$ *is an aleph*, $\mathfrak{b} < \mathfrak{a}$, $\mathrm{cf}(\mathfrak{b}) \neq \mathrm{cf}(\mathfrak{a})$ *then for every* $\mathfrak{a}, \mathfrak{b}, \mathfrak{b}$-*set* P, $|P| \leqslant \max(\mathfrak{a}, \underline{\mathfrak{a}}^{\mathfrak{b}}) \leqslant \max(\mathfrak{a}, 2^{\underline{\mathfrak{a}}})$. *Thus if also* $2^{\underline{\mathfrak{a}}} = \mathfrak{a}$ *then* $|P| \leqslant \mathfrak{a}$.

Outline of proof. Take $\bigcup P = \mathfrak{a}$ and define F on P by setting $F(B) =$ the set of the first $\mathfrak{b}$ members of B, for $B \in P$. F is one–one, and since $\mathrm{cf}(\mathfrak{b}) \neq \mathrm{cf}(\mathfrak{a})$ F is an injection of P into the set $\bigcup_{\lambda < \mathfrak{a}} {}^{\mathfrak{b}}\lambda$, whose cardinality is $\max(\mathfrak{a}, \underline{\mathfrak{a}}^{\mathfrak{b}})$. □

5.31Ac Proposition (Tarski 1929). *If* $\mathfrak{a} \geqslant \mathfrak{b} > \mathfrak{c}$ *then every* $\mathfrak{a}, \mathfrak{b}, \mathfrak{c}$-*set is of cardinality* $\leqslant \max(\mathfrak{a}, 2^{\underline{\mathfrak{a}}})$.

Proof. There is always a cardinal $\mathfrak{d}$ such that $\mathfrak{c} \leqslant \mathfrak{d} \leqslant \mathfrak{b}$, $\mathfrak{d} < \mathfrak{a}$ and $\mathrm{cf}(\mathfrak{d}) \neq \mathrm{cf}(\mathfrak{a})$, for the following reasons. If $\mathrm{cf}(\mathfrak{c}) \neq \mathrm{cf}(\mathfrak{a})$ we take $\mathfrak{d} = \mathfrak{c} < \mathfrak{b} \leqslant \mathfrak{a}$. If $\mathrm{cf}(\mathfrak{c}) = \mathrm{cf}(\mathfrak{a})$ we take $\mathfrak{d} = \mathfrak{c}^+$ and then $\mathrm{cf}(\mathfrak{d}) = \mathfrak{c}^+ > \mathrm{cf}(\mathfrak{c}) = \mathrm{cf}(\mathfrak{a})$; also, since $\mathfrak{c} < \mathfrak{a}$ $\mathfrak{d} = \mathfrak{c}^+ \leqslant \mathfrak{a}$, and since $\mathrm{cf}(\mathfrak{d}) \neq \mathrm{cf}(\mathfrak{a})$ we have $\mathfrak{d} < \mathfrak{a}$. By 5.30 every $\mathfrak{a}$, $\mathfrak{d}$, $\mathfrak{d}$-set Q is of cardinality $\leqslant \max(\mathfrak{a}, 2^{\mathfrak{a}})$. Let P be an $\mathfrak{a}$, $\mathfrak{b}$, $\mathfrak{c}$-set and assume $\bigcup P \subseteq \mathfrak{a}$. We define a function F on P by $F(B) =$ the set of the first $\mathfrak{d}$ members of B, for $B \in P$. F is a one–one mapping of P on an $\mathfrak{a}$, $\mathfrak{d}$, $\mathfrak{c}$-set Q, which is also an $\mathfrak{a}$, $\mathfrak{d}$, $\mathfrak{d}$-set, and hence $|P| = |Q| \leqslant \max(\mathfrak{a}, 2^{\mathfrak{a}})$. □

Part B

Applications and Advanced Topics

Chapter VI

A Review of Point Set Topology

Point set topology is a field of mathematics which depends almost entirely on the relatively elementary parts of set theory. When Cantor studied in the 1870's trigonometric series he was naturally led to the concepts of point set topology; these he could not handle without developing set theory. Thus point set topology led to Cantors discovery of set theory.

Point set topology is now a mathematical field on its own and therefore we do not intend to present a full account even of its basic theory; there are many texts available on this subject. However, we shall delve in the next chapter quite a bit into descriptive set theory and we need for that a fair amount of point set topology; the present chapter contains all the results of point set topology needed there. Almost all these results will be presented without proof. The few proofs and hints we shall give are those which are of special interest for set theory and which are particularly related to material which will be presented later on. The results of point set topology presented here do not constitute a mere list of theorems; they are given in a logical order and the reader who has already learned some point set topology should have no difficulty reconstructing the proofs. The reader who knows no point set topology is advised to study a textbook on the subject before attempting the next two chapters, but if he prefers not to do it he can continue with Chapter VIII without difficulty.

1. *Basic Concepts*

1.1 Definition (Hausdorff 1914). A *topology* O on $X \neq 0$ is a subset of $\mathbf{P}(X)$ which contains X and is such that O is closed under arbitrary unions and under intersections of any two of its members (and hence also under the intersection of any finite number of its members), i.e., if $Q \subseteq O$ then $\bigcup Q \in O$ and if $A, B \in O$ also $A \cap B \in O$. If O is a topology on $X \neq 0$ then $\langle X, O \rangle$ is said to be a *topological space*, or just a *space*. The members of O are called *open sets* (of the topology O), the complements with respect to X of the open sets (i.e., the sets $X \sim A$, where $A \in O$) are called *closed sets* (of O). Thus the complement of a closed set is an open set (and vice versa). The members of X are called *points* (of $\langle X, O \rangle$).

1.2 The Null Set and the Whole Space. Since the null set 0 is a subset of O we have $0=\bigcup 0 \in O$ and thus 0 is an open set. Since $0, X \in O$ their complements $X, 0$ are also closed sets.

1.3 The Discrete Topology. For $X \neq 0$ $\mathbf{P}(X)$ is a topology on X; it is called the *discrete* topology. Prove that a topology $O \subseteq \mathbf{P}(X)$ is discrete iff all subsets of X are both open and closed, and also if for every $x \in X$ $\{x\}$ is an open set. □

In order to give also some non-trivial examples we shall first present a general method for constructing topologies.

1.4 Definition. (i) Let $X \neq 0$ and $Q \subseteq \mathbf{P}(X)$. We say that Q is a *base for a topology* (on X) if $\bigcup Q = X$ and for all $A, B \in Q$ $A \cap B$ is the union of members of Q (i.e., $A \cap B = \bigcup\{C \in Q \mid C \subseteq A \cap B\}$). Proposition 1.5 below asserts that if Q is a base for a topology on X then $O = \{\bigcup P \mid P \subseteq Q\}$ is a topology on X; obviously $O \supseteq Q$. This topology O is said to be *the topology generated by* Q (since it is the least topology which includes Q); Q is said to be *a base* for the topology O; and the members of Q are called *basic open sets*. (A space has usually many bases, so when we speak of basic open sets we have a particular base in mind).

(ii) If $P \subseteq \mathbf{P}(X)$ is such that $\bigcup P = X$ then the set of all intersections of finitely many members of P, i.e., the set $\{\bigcap S \mid 0 \neq S \subseteq P \text{ and } S \text{ is finite}\}$ is obviously a base for a topology O on X. O is said to be the topology generated by P, being the least topology which includes P. P is said to be a *sub-base* of O.

1.5 Proposition. *Let $X \neq 0$ and let Q be a subset of* $\mathbf{P}(X)$.

(i) *Q is a base for a topology iff $\bigcup Q = X$ and for all $A, B \in Q$ and for all $x \in A \cap B$ there is a $C \in Q$ such that $x \in C \subseteq A \cap B$.*

(ii) *Q is a base for a topology iff the set $O = \{\bigcup P \mid P \subseteq Q\}$ is a topology on X.*

(iii) *If Q is a base for a topology O then $A \subseteq X$ is open iff for every $x \in A$ there is a $B \in Q$ such that $x \in B \subseteq A$.* □

1.6 The Order Topology. Let $\langle X, < \rangle$ be an ordered set. An *open interval* of X is a set which is of the form $(u, v) = \{z \in X \mid u < z < v\}$, where $u < v$. An *open half-line* of X is a set of the form $(-\infty, v) = \{z \in X \mid z < v\}$ or $(u, \infty) = \{z \in X \mid u < z\}$. The set which consists of the open intervals of X, the open half-lines of X and X itself is a base for a topology on X which is called the *order topology* on X (with respect to $<$).

Prove that for a member x of X which is neither the first nor the last member of X, $\{x\}$ is an open set in the order topology iff x has both a predecessor and a successor in X. Give a necessary and sufficient condition on the order of X for the order topology to be the discrete topology. Prove that if Ω is an ordinal then the closed subsets of Ω as defined in IV.4.11(ii) are exactly the closed subsets of Ω in the order topology of Ω. [Show that if C is a subset of Ω which is closed in the sense of IV.4.11, $\alpha < \Omega$ and $\alpha \notin C$ then α belongs to some open interval of Ω disjoint from C.]

1.7 The Tree Topology. Let $\langle X, <\rangle$ be a partially ordered set. For $x \in X$ we write $X_{(x)}$ for $\{y \in X \mid y \geqslant x\}$. The set $\{X_{(x)} \mid x \in X\}$ is a base for a topology on X which we shall call the *tree topology* on X. Prove that a subset A of X is open iff A is a *terminal segment* of X, i.e., if for all $x, y \in X$ if $x \in A$ and $y > x$ then also $y \in A$. Prove that $A \subseteq X$ is a closed subset of X iff A is an initial segment of X. When is the tree topology identical with the discrete topology on X? If $\langle X, <\rangle$ is an ordered set then the order topology and the tree topology are different in practically all cases; which is the exceptional case where they are identical?

1.8 Sequence Spaces. Let $H = \langle H_i \mid i < \omega\rangle$ be a sequence of sets such that $\times_{i<\omega} H_i$ is non-void. (Assuming the axiom of choice this means that $H_i \neq 0$ for all $i < \omega$). By a *node* we mean a finite sequence $\langle a_i \mid i < n\rangle$ such that $a_i \in H_i$ for $i < n$. By an H-sequence we mean a sequence $f \in \times_{i<\omega} H_i$. For a node $a = \langle a_i \mid i < n\rangle$ we set $B_a = \{f \in \times_{i<\omega} H_i \mid f_i = a_i$ for all $i < n\}$. The set $\{B_a \mid a$ is a node$\}$ is a basis for a topology O on the set $\times_{i<\omega} H_i$, which will be called the *natural topology* of the H-sequences. This space $\langle X, O\rangle$ is said to be a *sequence space*. When is the natural topology identical with the discrete topology?

1.9 Metric Spaces. Let X be a non-void set and let d be a function on $X \times X$ into the set of non-negative real numbers. d is called a *metric*, or a *distance function*, on X, and $\langle X, d\rangle$ is called a *metric space* if d satisfies, for all $x, y, z \in X$ the following (a)–(c).

(a) $d(x, y) = 0 \leftrightarrow x = y$.
(b) Symmetry: $d(x, y) = d(y, x)$.
(c) The triangle inequality: $d(x, z) \leqslant d(x, y) + d(y, z)$.

$d(x, y)$ is called the *distance* between x and y. For $z \in X$ and a positive real number r the set $S(z, r) = \{x \in X \mid d(z, x) < r\}$ is called the *open sphere* with *center* z and *radius* r. Prove, using 1.5(i), that the open spheres are a basis for a topology on X. This topology is called the *topology associated with the metric d*. The most natural example of a metric is the distance function of the n-dimensional Euclidean space ${}^n\mathbb{R}$ given by $d(x, y) = (\Sigma_{i=1}^n (x_i - y_i)^2)^{1/2}$. This function obviously satisfies (a) and (b), and it requires some elementary algebra to prove that it also satisfies (c).

Another example of a metric space is as follows. Let X be any non-void set. Define d on $X \times X$ by $d(x, y) = 1$ if $x \neq y$ and $d(x, y) = 0$ if $x = y$. Prove that d is a metric and that the associated topology is discrete.

1.10 Proposition. Let $\langle X, d\rangle$ *be a metric space.*

(i) *The* closed sphere $S^c(z, r) = \{x \in X \mid d(z, x) \leqslant r\}$ *is a closed set.*

(ii) *A subset A of X is open iff for every $z \in A$ there is a positive real r such that* $S(z, r) = \{x \in X \mid d(z, x) < r\} \subseteq A$. □

1.11 Exercise. Let $H = \langle H_n \mid n < \omega\rangle$ be such that $X = \times_{n<\omega} H_n \neq 0$. Define d on $X \times X$ as follows. For $f, g \in X$ $d(f, g) = \Sigma_{f(n) \neq g(n)} 2^{-n}$ (i.e., $d(f, g)$ is the sum of the 2^{-n}'s for all n's such that $f(n) \neq g(n)$). Prove that d is a metric. Use 1.5(iii) to prove that the topology associated with d is the natural topology of the H-sequences (1.8). □

1.12 Topology on a Proper Class. Sometimes we want to define a topology on a proper class X. For example, we may want to use the order topology on On. In this case O would be a collection of classes, some of which are proper classes, and therefore O is not even a class and we have to describe it indirectly. The neatest way to do it is to provide a class B of subsets of X which is a base (or a sub-base) for a topology on X. A subclass A of X is said to be open if for every $x \in A$ there is a $P \in B$ such that $x \in P \subseteq A$. For example, the order topology on On can be given by the basis $\mathrm{On} \cup \{(\alpha, \beta) \mid \alpha < \beta\}$ (where $(\alpha, \beta) = \beta \sim (\alpha+1)$).

1.13 Proposition. *The intersection of any number of closed sets is a closed set, i.e., if S is a non-void set of closed then $\bigcap S$ is a closed set too. The union of a finite number of closed sets is a closed set.* □

1.14 Definition. In a metric space $\langle X, d\rangle$, a set A can be said to contain all points sufficiently close to a point $z \in A$ if for some $r > 0$ the open sphere $S(z, r)$ is included in A. By 1.10(ii) A contains all points sufficiently close to z iff there is an open set D such that $z \in D \subseteq A$. Therefore we define for a general topological space $\langle X, O\rangle$ that A is a *neighborhood* of z if for some open set D $z \in D \subseteq A$.

1.15 Proposition. (i) *The set of all neighborhoods of a point is a filter.*

(ii) *A set A is open iff it is a neighbourhood of each one of its points.*

(iii) *A set B is closed iff every point z which is not in B has a neighborhood disjoint from B.* □

1.16 Definition. For a topological space $\langle X, O\rangle$, the *interior* $\mathrm{Int}(A)$ of a subset A of X is the union of all open subsets of A. The *closure* $\mathrm{Cl}(A)$ of a subset A of X is the intersection of all closed sets which include A.

1.17 Proposition. *Let $\langle X, O\rangle$ be a topological space and $A \subseteq X$.*

(i) $\mathrm{Int}(A) \subseteq A$, $\mathrm{Int}(A)$ *is open, and* $\mathrm{Int}(A)$ *is the maximal open set included in* A.

(ii) $\mathrm{Cl}(A) \supseteq A$, $\mathrm{Cl}(A)$ *is closed, and* $\mathrm{Cl}(A)$ *is the least closed set which includes* A.

(iii) $\mathrm{Int}(A) = A$ *iff A is open, hence* $\mathrm{Int}(\mathrm{Int}(A)) = \mathrm{Int}(A)$.

(iv) $\mathrm{Cl}(A) = A$ *iff A is closed, hence* $\mathrm{Cl}(\mathrm{Cl}(A)) = \mathrm{Cl}(A)$.

(v) *If A is open and $A \cap B = 0$ then also $A \cap \mathrm{Cl}(B) = 0$.*

(vi) $\mathrm{Int}(A) = X \sim \mathrm{Cl}(X \sim A)$, $\mathrm{Cl}(A) = X \sim \mathrm{Int}(X \sim A)$.

(vii) *If $A \subseteq B$ then* $\mathrm{Int}(A) \subseteq \mathrm{Int}(B)$, $\mathrm{Cl}(A) \subseteq \mathrm{Cl}(B)$. □

1.18 Exercises. (i) In a metric space $\langle X, d\rangle$ is it always the case that the closure of the open sphere $S(z, r) = \{x \in X \mid d(z, x) < r\}$ is the closed sphere $S^c(z, r) = \{x \in X \mid d(z, x) \leq r\}$?

(ii) Look at the tree topology from the point of view of a special kind of a snail who sits on a point in the partially ordered set X and who can only climb up (i.e., he can only go from $y \in X$ to any $z \in X$ such that $z > y$). For such a snail $\mathrm{Cl}(A)$ is the set of all points in X from which he can go to some point in A; $\mathrm{Int}(A)$ is the set of all points of A from which he can never get out of A. Characterize in the same manner also the sets $\mathrm{Int}(\mathrm{Cl}(A))$ and $\mathrm{Cl}(\mathrm{Int}(A))$. □

1.19 Definition. Let $\langle X, O\rangle$ be a topological space and let $0 \neq Y \subseteq X$. It is outright trivial to see that $\{A \cap Y \mid A \in O\}$ is a topology on Y; it is called the topology on Y *induced* by O. A space $\langle Y, O'\rangle$, where $Y \subseteq X$ and O' is the topology induced on Y by O, is called a *subspace* of $\langle X, O\rangle$.

1.20 Exercises. (i) Let $<$ be a partial ordering of a non-void set X, and let O be the tree topology on X. Y is partially ordered by $<|Y$ and this partial ordering yields a tree topology on Y. Is this tree topology on Y the same as the topology induced on Y by the tree topology of X?

(ii) Let $\langle X, O\rangle$ be a topological space, and let $\langle Y, O'\rangle$ be a subspace of $\langle X, O\rangle$. Prove that a subset A of Y is closed in $\langle Y, O'\rangle$ iff there is a set B closed in $\langle X, O\rangle$ such that $A = B \cap Y$. □

1.21 Proposition. (i) *Let $\langle X, <\rangle$ be an ordered set and let $Y \subseteq X$ be a segment of it, i.e., for all $a, b, c \in X$, if $a, c \in Y$ and $a < b < c$ then also $b \in Y$. The topology induced on Y by the order topology on X is the same as the order topology on Y. (Is this true also for a general subset Y of X?)*

(ii) *Let $\langle X, d\rangle$ be a metric space, and let $0 \neq Y \subseteq X$. $\langle Y, d \upharpoonright (Y \times Y)\rangle$ is also a metric space. The topology induced on Y by the topology associated with the metric on X is the same as the topology associated with the metric on Y.* □

1.22 Definition. Let $\langle X, O\rangle$ and $\langle X', O'\rangle$ be topological spaces. A bijection F of X on X' is said to be a *homeomorphism* of the two spaces if $O' = \{F[A] \mid A \in O\}$, and if there is a homeomorphism of $\langle X, O\rangle$ on $\langle X', O'\rangle$ then the two spaces are said to be *homeomorphic*. The homeomorphism of topological spaces is clearly a particular instance of the isomorphism of general structures. Two homeomorphic topological spaces differ only in their underlying set, but they obviously share all topological properties. For example, if $\langle X, O\rangle$ is discrete, or Hausdorff, or compact, or connected (these terms will be defined in § 2) and $\langle X, O\rangle$ is homeomorphic to $\langle X', O'\rangle$ then also $\langle X', O'\rangle$ is discrete, or Hausdorff, or compact, or connected, respectively.

1.23 Definition. Let $\langle X, O\rangle$ and $\langle Y, O'\rangle$ be topological spaces, let $z \in X$ and let F be a mapping of X into Y. We say that F is *continuous at* z if for x's which are sufficiently close to z $F(x)$ is sufficiently close to $F(z)$, i.e., for every neighborhood B of $F(z)$ (in O') there is a neighborhood A of z (in O) such that $F[A] \subseteq B$ (or, equivalently, for every neighborhood B of $F(z)$ $F^{-1}[B]$ is a neighborhood of z). F is said to be *continuous* if it is continuous at every point $z \in X$.

1.24 Proposition. *Let $\langle X, O\rangle$ and $\langle Y, O'\rangle$ be topological spaces and let F be a mapping from X to Y. F is continuous iff*

(1) *for every open set B (in O') $F^{-1}[B]$ is open (in O), and also iff*

(2) *for every closed set C $F^{-1}[C]$ is closed.* □

1.25 Proposition. *Let $\langle X, O\rangle$ and $\langle Y, O'\rangle$ be topological spaces, let $Z \subseteq Y$, let O'' be the topology on Z induced by O', and let F be a mapping of X into Z. F is continuous at a point t of X as a function into $\langle Y, O'\rangle$ iff it is continuous as t as a function into $\langle Z, O''\rangle$. As a consequence, F itself is a continuous function as a function into $\langle Y, O'\rangle$ iff it is a continuous function as a function into $\langle Z, O''\rangle$.* □

Remark. Notice that when we defined continuity in 1.23 both the domain $\langle X, O\rangle$ and the "target" $\langle Y, O'\rangle$ figured in the definition. Thus if we have, for example, a function on the real line into a circle in the Euclidean plane the function might be continuous as a function into the Euclidean plane but not as a function into the circle or vice versa. Proposition 1.25 tells us that this is not the case; and that in order to know whether F is continuous at z, or whether F is continuous, all we have to know about the target space $\langle Y, O'\rangle$ is what is the topology it induces on $\mathrm{Rng}(F)$ (which is the least set Y into which F maps).

1.26 Remark. We deal often with a situation where we have two topological spaces $\langle X, O\rangle$ and $\langle Y, O'\rangle$ and a function F such that $\mathrm{Dom}(F) \subseteq X$ (and not necessarily $\mathrm{Dom}(F) = X$ as in Definition 1.23) and $\mathrm{Rng}(F) \subseteq Y$. In such a case, when we say that F is *continuous at a point* z of $\mathrm{Dom}(F)$, or that F is a *continuous function*, we consider F as a function from $\langle \mathrm{Dom}(F), O''\rangle$ into $\langle Y, O'\rangle$ where O'' is the topology on $\mathrm{Dom}(F)$ induced by O.

1.27 Proposition. *Let F be a function on a space $\langle X, O\rangle$ into a space $\langle Y, O'\rangle$, and let $\langle Z, O''\rangle$ be a subspace of $\langle X, O\rangle$. If F is continuous at a point $x \in Z$ then also $F \restriction Z$ is continuous at x. If F is continuous then so is also $F \restriction Z$.* □

1.28 Proposition. *If F and G are functions such that F is continuous at x and G is continuous at $F(x)$ then GF is continuous at x. As a consequence, if F and G are continuous functions then GF is a continuous function too.* □

1.29 Proposition. *Let $\langle X, O\rangle$ and $\langle Y, O'\rangle$ be topological spaces and let W be a set of functions on subsets of X into Y such that for every $F \in W$ $\mathrm{Dom}(F)$ is disjoint from the closure of the union of the domains of the other functions in W (i.e., $\mathrm{Dom}(F) \cap \mathrm{Cl}(\bigcup\{\mathrm{Dom}(G) \mid G \in W \wedge G \neq F\}) = 0$). If $F \in W$ and F is continuous at $x \in \mathrm{Dom}(F)$ then also $\bigcup W$ is continuous at x. As a consequence, if every $F \in W$ is continuous so is $\bigcup W$.* □

1.30 Proposition. *Let $\langle X, O\rangle$ and $\langle Y, O'\rangle$ be topological spaces. A bijection F of X on Y is a homeomorphism iff both F and F^{-1} are continuous functions.* □

1.31 Exercise. Let Ω be an ordinal or On and let F be a monotonic function on Ω into On. F is a continuous function with respect to the order topologies on Ω and On iff for every limit ordinal $\gamma < \Omega$ $F(\gamma) = \sup_{\beta<\gamma} F(\beta)$. □

Remark. This justifies our giving the name continuity in IV.1.24 to the requirement that $F(\gamma) = \sup_{\beta<\gamma} F(\beta)$ for limit ordinals γ.

2. Useful Properties and Operations

2.1 Definition. A topological space $\langle X, O\rangle$ is called a *Hausdorff space* if any two different points in X have disjoint neighborhoods (or, equivalently, if for all $u, v \in X$, if $u \neq v$ then there are open sets A, B such that $A \cap B = 0$, $u \in A$ and $v \in B$).

2.2 Examples of Hausdorff Spaces. The discrete topology and the order topology are always Hausdorff. The topology of a metric space is always Hausdorff (because of requirement (a) of 1.9). The tree topology on any partially ordered set is not Hausdorff, unless it is discrete. A subspace of a Hausdorff space is always Hausdorff.

2.3 Proposition. *In a Hausdorff space $\langle X, O\rangle$, for every $z \in X$ $\{z\}$ is a closed set.* □

2.4 Definition. (i) Let $\langle X, O\rangle$ be a topological space, and $A \subseteq X$. A subset Q of $\mathbf{P}(X)$ is called an *open cover* of A if $Q \subseteq O$ and $\bigcup Q \supseteq A$.

(ii) A set A in a topological space $\langle X, O\rangle$ is said to be a *compact set* if every open cover Q of A has a finite subset S which is also a cover of A (i.e., $\bigcup S \supseteq A$).

(iii) A topological space $\langle X, O\rangle$ is said to be a *compact space* if X itself is a compact set in $\langle X, O\rangle$, i.e., if every open cover Q of X has a finite subset S which covers X.

Remark. The well-known Heine–Borel theorem asserts that every closed interval of the real line is a compact set (in the order topology). This turned out to be a very important property of topological spaces and this is what led to Definition 2.4.

2.5 Proposition. *A space $\langle X, O\rangle$ is compact iff for every set W of closed subsets of X, if the intersection $\bigcap S$ of every finite subset S of W is non-void then also $\bigcap W$ is non-void.* □

2.6 Examples of Compact and Non-Compact Sets. It is well known that every closed and bounded subset of the n-dimensional Euclidean space is compact. No unbounded set in a metric space is compact (A is *unbounded* if the set of reals $\{d(x, y) \mid x, y \in A\}$ is unbounded).

Additional examples of compact sets are all finite sets, and all subsets of the trivial spaces $\langle X, \{0, X\}\rangle$. No infinite subset of a discrete space is compact.

2.7 Exercise. Prove that a closed set C in a compact space is compact. □

2.8 Proposition. *Let $\langle X, O\rangle$ be a topological space and let $A \subseteq X$. A is a compact set in $\langle X, O\rangle$ iff the subspace $\langle A, O'\rangle$ of $\langle X, O\rangle$ is a compact space (where O' is the topology on A induced by O). Thus the question whether A is a compact set or not depends only on the topology induced on A by O and not on the way the topology O behaves outside A.* □

2.9 Proposition. *A compact set in a Hausdorff space is always closed. (This is a converse of* 2.7). □

2.10 Proposition. *Let F be a continuous function on a space $\langle X, O\rangle$ into a space $\langle Y, O'\rangle$. If a subset A of X is compact so is its image $F[A]$.* □

2.11 Corollary. (i) *Let F be a continuous function on a compact space $\langle X, O\rangle$ into a Hausdorff space $\langle Y, O'\rangle$, then for every closed set A in $\langle X, O\rangle$, $F[A]$ is closed too.*

(ii) *Let F be a continuous bijection of a compact space onto a Hausdorff space then F is a homeomorphism.*

Hint of proof. (i) Use 2.7, 1.27, 2.10 and 2.9. (ii) Use 1.30 and show that F^{-1} is continuous by 1.24 and (i). □

2.12 Definition. A topological space is said to be *regular* if it is a Hausdorff space and for every closed set C in the space and for every point x which is not in C there are disjoint open set A, B such that $x \in A$ and $C \subseteq B$.

2.13 Proposition. *A Hausdorff space $\langle X, O\rangle$ is regular iff for every open set A and $x \in A$ there is an open set B such that $x \in B$ and* $\mathrm{Cl}(B) \subseteq A$. □

2.14 Proposition. (i) *Every metric space is regular.*

(ii) *Every compact Hausdorff space is regular.*

Hint of proof. (ii) If C is closed and $x \notin C$ let Q be the set of all open sets which are disjoint from some neighborhood of x. Q is a cover of C. □

2.15 Definition (Products and powers of topological spaces). Let $\langle\langle X_r, O_r\rangle \mid r \in w\rangle$ be an indexed family of topological spaces.

(i) The *projection* p_s of $\times_{r\in w} X_r$ on the s-th factor X_s is the function on $\times_{r\in w} X_r$ given by $p_s(f) = f(s)$.

(ii) Assuming that $\times_{r\in w} X_r \neq 0$ (which is always the case if we assume the axiom of choice—V.1.5(i)), the *product* of the spaces $\langle\langle X_r, O_r\rangle \mid r \in w\rangle$ is the topological space $\langle \times_{r\in w} X_r, O'\rangle$ where O' is the topology generated by the sub-base consisting of all the sets $p_s^{-1}[A]$, where $s \in w$ and A is an open set in $\langle X_s, O_s\rangle$. (The set $p_s^{-1}[A]$ can also be given as $\times_{r\in w} Y_r$, where $Y_r = X_r$ for $r \neq s$ and $Y_s = A$).

(iii) For a set w, the *w-th power* of a space $\langle X, O\rangle$ is the product of the indexed family $\langle\langle X, O\rangle \mid r \in w\rangle$, i.e., it is the product of "w many copies" of $\langle X, O\rangle$.

2.16 The Product of Topological Spaces. It follows directly from the definition of the product of topological spaces (and 1.24) that each projection is a continuous function on $\times_{r\in w} X_r$ onto X_s, and that the product topology O' is the least topology on $\times_{r\in w} X_r$ such that all the projections p_s are continuous, i.e., O' is included in every topology O'' on $\times_{r\in w} X_r$ such that each projection p_s is a continuous function from $\langle \times_{r\in w} X_r, O''\rangle$ onto $\langle X_s, O_s\rangle$.

In 2.15(ii) we defined the product topology by means of a sub-base. The corresponding base (1.4(ii)) is the set of all sets $\bigcap_{r\in u} p_r^{-1}[A_r]$, where u is a finite subset of w and for every $r\in u$ A_r is an open set in $\langle X_r, O_r\rangle$. The basic open sets can also be represented as $\times_{r\in w} Y_r$, where $Y_r\in O_r$ for every $r\in w$, and $Y_r=X_r$ for all $r\in w$ except finitely many. If the number of the factors is finite, i.e., if w is finite, then the base of the product topology is the set $\{\times_{r\in w} Y_r \mid (\forall r\in w)(Y_r\in O_r)\}$. The usual topology of the n-dimensional Euclidean space (which can be taken to be the topology associated with the Euclidean metric $(\Sigma_{i=1}^n x_i^2)^{1/2}$ on it) is identical with the product topology of n copies of the real line $\mathbb{R}$ (with the order topology).

The reader can easily verify that if we replace each factor $\langle X_r, O_r\rangle$ in the product $\langle \times_{r\in w} X_r, O''\rangle$ by a homeomorphic space $\langle X_r', O_r'\rangle$, then the product $\langle \times_{r\in w} X_r', O'''\rangle$ thus obtained is homeomorphic to $\langle \times_{r\in w} X_r, O''\rangle$. This proof uses the axiom of choice unless w is finite or the homeomorphism of the $\langle X_r, O_r\rangle$'s on the $\langle X_r', O_r'\rangle$'s is uniform.

2.17 Exercises. (i) Let $\langle\langle X_r, O_r\rangle \mid r\in w\rangle$ be an indexed family of topological spaces, let O' be the product topology of those spaces, and let $\langle\langle Y_r, O_r''\rangle \mid r\in w\rangle$ be an indexed family such that $0\neq Y_r\subseteq X_r$ and O_r'' is the topology induced on Y_r by O_r, for $r\in w$. Prove that the product topology of $\langle\langle Y_r, O_r''\rangle \mid r\in w\rangle$ is the same as the topology induced by O' on the subset $\times_{r\in w} Y_r$ of $\times_{r\in w} X_r$.

(ii) Let f be a bijection of a set v on a set w and let $\langle\langle X_r, O_r\rangle \mid r\in v\rangle$ and $\langle\langle Y_s, O_s'\rangle \mid s\in w\rangle$ be indexed families of spaces such that for all $r\in v$ $\langle X_r, O_r\rangle = \langle Y_{f(r)}, O_{f(r)}'\rangle$, (i.e., the two families differ only in the identity of their indices), then the respective products of the two indexed families are homeomorphic.

(iii) The product of finitely many discrete spaces is a discrete space, but the product of infinitely many spaces each of which has at least two points is not discrete.

(iv) The natural topology of the H-sequences, given in 1.8, is the product of the discrete topologies on the factors H_i, $i\in\omega$. □

2.18 Definition. For a cardinal $\mathfrak{a}$ we refer to any w-th power of $\langle X, O\rangle$ with $|w|=\mathfrak{a}$ as an *$\mathfrak{a}$-power* of $\langle X, O\rangle$. By 2.17(ii) all the $\mathfrak{a}$-powers of $\langle X, O\rangle$ are homeomorphic to each other and thus topologically indistinguishable. Notice that we speak of the $\mathfrak{a}$-power and not the $\mathfrak{a}$-th power when we use $\mathfrak{a}$ as a cardinal and not as whatever set $\mathfrak{a}$ happens to be. If $\mathfrak{a}$ is a well-ordered cardinal then the $\mathfrak{a}$-th power of $\langle X, O\rangle$ is an $\mathfrak{a}$-power of $\langle X, O\rangle$ and can therefore be taken to be the standard $\mathfrak{a}$-power of $\langle X, O\rangle$.

2.19 Proposition (The general associativity of the product of topological spaces). (i) *Let $\langle w_t \mid t\in q\rangle$ be an indexed family of sets, and for $r\in w_t$ let $\langle X_{t,r}, O_{t,r}\rangle$ be a topological space. Let O' be the product topology, on $\times_{t\in q\wedge r\in w_t} X_{t,r}$, of all spaces $\langle X_{t,r}, O_{t,r}\rangle$. For $t\in q$ let O_t'' be the product topology on $\times_{r\in w_t} X_{t,r}$,* i.e., *$O_t''$ is the product of all spaces $\langle X_{t,r}, O_{t,r}\rangle$ with a fixed t. Then the product of the spaces $\langle \times_{r\in w_t} X_{t,r}, O_t''\rangle$, for $t\in q$, is homeomorphic to the space $\langle \times_{t\in q\wedge r\in w_t} X_{t,r}, O'\rangle$.*

(ii) *A $\mathfrak{b}$-power of an $\mathfrak{a}$-power of a space $\langle X, O\rangle$ is an $\mathfrak{a}\cdot\mathfrak{b}$-power of $\langle X, O\rangle$.* □

2.20 Proposition. *The product of a family of topological spaces is Hausdorff iff each one of the factors is Hausdorff.* □

2.21 The Product of Compact Spaces. If the product $\langle \times_{r\in w} X_r, O' \rangle$ of topological spaces $\langle X_r, O_r \rangle$, $r \in w$, is compact so is each one of the factors, being the image of the product under the corresponding projection function, which is continuous (2.16 and 2.10). The other direction, that if all the factors are compact then also the product is compact, is the important Tychonoff product theorem. It will be proved now for a product of finitely many spaces. The proof of the general case, where the number of factors may be infinite, requires the axiom of choice and will be given in VIII.2.20.

2.22 Proposition (The Tychonoff product theorem for finite products—Tychonoff 1935). *If $\langle \langle X_i, O_i \rangle \mid i<n \rangle$ is a family of n compact topological spaces, $2 \leqslant n < \omega$, then its product is compact.*

Proof. We shall prove below the proposition for the case $n=2$; all other cases follow easily from it by induction as follows. For $n \geqslant 2$ consider the product $\langle \times_{i<n+1} X_i, O'_{n+1} \rangle$ of the compact factors $\langle X_i, O_i \rangle$, $i<n+1$. By the associativity of the product operation (2.19) $\langle \times_{i<n+1} X_i, O'_{n+1} \rangle$ is homeomorphic to the product of $\langle \times_{i<n} X_i, O'_n \rangle$ and $\langle X_n, O_n \rangle$ (to apply 2.19 one has to make the trivial observations that the product of a single space is homeomorphic to the space itself). By the induction hypothesis the space $\langle \times_{i<n} X_i, O'_n \rangle$ is compact; by the case $n=2$ its product with the compact space $\langle X_n, O_n \rangle$ is compact; hence $\langle \times_{i<n+1} X_i, O'_{n+1} \rangle$, which is homeomorphic to this product, is compact too.

Before we deal with the case $n=2$ let us observe that when we come to prove that a set A in some space $\langle X, O \rangle$ is compact it is sufficient to consider open covers of A which consist of basic open sets only. To see this, let Q be an open cover of A, and let P be the set of all basic open sets which are subsets of some member of Q. Since every open set is the union of its basic open subsets $\bigcup P = \bigcup Q \supseteq A$, and $\bigcup P$ is a cover of A. If the compactness requirement holds for the cover P which consists of basic open sets, then P has a finite subset T which is a cover of A. For each $W \in T \subseteq P$ let D_W be a member of Q such that $W \subseteq D_W$, and let $S = \{D_W \mid W \in T\} \subseteq Q$. $\bigcup S \supseteq \bigcup T \supseteq A$, thus $S \subseteq Q$ is a finite cover of A.

For the case $n=2$, let $\langle X_0, O_0 \rangle$ and $\langle X_1, O_1 \rangle$ be two compact spaces and let Q be a cover of their product $X_0 \times X_1$ which consists of basic open sets. Let P be the set of all open subsets A of X_0 such that the subset $A \times X_1$ of $X_0 \times X_1$ is covered by a finite subset of Q. We shall now prove that P is an open cover of X_0. Let $x \in X_0$, we shall prove that there is an $A \in P$ which contains x. The subset $\{x\} \times X_1$ of $X_0 \times X_1$ is of course covered by Q and hence also by $Q' = \{W \in Q \mid W \cap (\{x\} \times X_1) \neq 0\}$. Every basic open set W is of the form $B_0 \times B_1$ where $B_0 \in O_0$ and $B_1 \in O_1$; B_0 and B_1 are the projections of W and can be written as $p_0[W]$ and $p_1[W]$ (see 2.15(i)) Since Q' covers $\{x\} \times X_1$ the set $\{p_1[W] \mid W \in Q'\}$ is an open cover of X_1. Since $\langle X_1, O_1 \rangle$ is compact this set has a finite subset R which is an open cover of X_1. Let Q'' be a finite subset of Q' such that $\{p_1[W] \mid W \in Q''\} = R$, and let $A = \bigcap_{W \in Q''} p_0[W]$. A is open, being the intersection of finitely many open sets. For $W \in Q'' \subseteq Q'$ $x \in p_0[W]$, by the definition of Q', hence $x \in A$. $\bigcup Q'' = \bigcup_{W \in Q''}(p_0[W] \times p_1[W]) \supseteq \bigcup_{W \in Q''}(A \times p_1[W]) = A \times (\bigcup_{W \in Q''} p_1[W]) = A \times \bigcup R = A \times X_1$, since R is a cover of X_1. Thus $Q'' \subseteq Q$ is a cover of $A \times X_1$ and $A \in P$. Thus we have shown that P is an open cover of X_0. Since $\langle X_0, O_0 \rangle$ is compact there is a finite subset T of P which covers X_0. For each $A \in T \subseteq P$ there is a finite subset Q_A of Q which covers $A \times X_1$ (by the definition of P). Let $S = \bigcup \{Q_A \mid A \in T\}$. S is clearly a finite subset of Q. $\bigcup S = \bigcup_{A \in T}(\bigcup Q_A) \supseteq \bigcup_{A \in T}(A \times X_1) = (\bigcup_{A \in T} A) \times X_1 = X_0 \times X_1$, thus S is a finite subset of Q which covers $X_0 \times X_1$. □

2.23 Definition. (i) In a topological space a set D is said to be *dense* in a set A if $A \subseteq \mathrm{Cl}(D)$ or, equivalently if $E \cap D$ is non-void for every open set E such that $E \cap A$ is non-void. (Prove this equivalence and show also that it suffices to require $E \cap A \neq 0 \rightarrow E \cap D \neq 0$ for *basic* open sets E).

(ii) A topological space $\langle X, O \rangle$ is said to be *separable* if there is a countable set D dense in the whole space X.

2.24 Examples of Separable Spaces. If the topological space $\langle X, O \rangle$ is countable (i.e., $|X| \leqslant \aleph_0$) then it is trivially separable. An example of a separable uncountable space is the space of the real numbers with the usual topology, since the countable set of all rational numbers is dense in it. Also every finite-dimensional Euclidean space is separable since the countable set of the points with only rational co-ordinates is dense in it.

2.25 Proposition. *If the metric space $\langle X, d \rangle$ is separable then its topology has a countable base. Conversely, if the topology of $\langle X, d \rangle$ has a countable base then, assuming the axiom of choice, $\langle X, d \rangle$ is separable.*

Hint of proof. If $D \subseteq X$ is a countable set dense in X then $\{S(x, \frac{1}{n}) \mid x \in D \wedge n \in \omega\}$ is a countable base. If $\{B_i \mid i \in \omega\}$ is a base and $y_i \in B_i$ then $\{y_i \mid i \in \omega\}$ is dense in X. □

2.26Ac Proposition. *A product of countably many separable spaces is separable. The axiom of choice is not needed for the cases of a product of finitely many separable spaces and of a countable power of a separable space.*

Hint of proof. Let $\langle\langle X_r, O_r \rangle \mid r \in w\rangle$ be a family of separable spaces with $|w| \leqslant \aleph_0$. Let F be a function on w such that for $r \in w$ $F(r)$ is a mapping of ω on a dense subset of X_r. In general one uses the axiom of choice to get such an F, but if w is finite or all spaces $\langle X_r, O_r \rangle$ are identical then one can obtain such an F without using that axiom. Let g be a point in $\times_{r \in w} X_r$. The set $\{h \in \times_{r \in w} X_r \mid$ there is a finite $u \subseteq w$ such that for $r \in u$ $h(r) \in \mathrm{Rng}(F(r))$ and for $r \in w \sim u$ $h(r) = g(r)\}$ is a countable set dense in the product space. □

2.27 Definition. (i) A set A in a topological space is said to be *clopen* if it is both closed and open.

(ii) A topological space $\langle X, O\rangle$ is said to be *connected* if it has no clopen sets beside 0 and X, or, equivalently, if X is not the union of two disjoint non-void open sets.

2.28 Proposition. *A product of topological spaces is connected iff each one of its factors is.* □

2.29 Proposition. *Let $\langle X,<\rangle$ be an ordered set. X is connected in the order topology iff $\langle X,<\rangle$ is* (a) *dense, i.e., for all $a, c \in X$, if $a<c$ then there is a $b \in X$ such that $a<b<c$, and* (b) gapless, *i.e., every proper initial segment (and hence every bounded subset) of X has a least upper bound.* □

2.30 Definition. (i) A sequence $\langle x_n \mid n<\omega\rangle$ of points in a metric space is said to *converge* to a point y if for every positive real number ε there is an $n_0 \in \omega$ such that $d(y, x_n)<\varepsilon$ for every $n \geqslant n_0$. If the sequence $\langle x_n \mid n<\omega\rangle$ converges to y then y is said to be the *limit* of the sequence.

(ii) A sequence $\langle x_n \mid n<\omega\rangle$ in a metric space is called a *Cauchy sequence* if for every positive real number ε there is an $n_0 \in \omega$ such that for all $n, m \geqslant n_0$ $d(x_n, x_m)<\varepsilon$.

(iii) A metric space is said to be *complete* if every Cauchy sequence in it converges to some point in it.

2.31 Proposition. *Let $\langle X, d\rangle$ be a metric space.*

(i) *Every sequence has at most one limit point.*

(ii) *Every sequence which has a limit is a Cauchy sequence.*

(iii) *If A is a closed set in $\langle X, d\rangle$ and $\langle x_n \mid n<\omega\rangle$ is a converging sequence of members of A then its limit is also in A.*

(iv) *If $\langle X, d\rangle$ is a complete metric space and A is a non-void closed set in it then also $\langle A, d \mid (A\times A)\rangle$ is a complete metric space.* □

2.32 Examples of Complete Metric Spaces. By the well known Cauchy convergence theorem the n-dimensional Euclidean spaces are complete metric spaces. Also the metric spaces where $x \neq y \rightarrow d(x, y)=1$ are clearly complete.

2.33 Proposition. *Let $X = \times_{n<\omega} H_n \neq 0$ and let d be the function on $X\times X$ given by $d(f, g)=\Sigma_{f(n)\neq g(n)} 2^{-n}$. $\langle X, d\rangle$, which is a metric space by* 1.11, *is a complete metric space.*

Hint of proof. Let $\langle f_\kappa \mid \kappa<\omega\rangle$ be a Cauchy sequence, then for every $n \in \omega$ there is an k_0 such that $d(f_k, f_l)<2^{-n}$ for all $k, l \geqslant k_0$. By the definition of d, for all $k, l \geqslant k_0$ we have $f_k(m)=f_l(m)$ for every $m \leqslant n$. Hence for every $n \in \omega$ the sequence $\langle f_k(n) \mid k \in \omega\rangle$ obtains a fixed value $g(n)$ from a certain value of k on. The sequence $\langle f_k \mid k \in \omega\rangle$ converges to $g(=\langle g(n) \mid n \in \omega\rangle)$. □

2.34 Definition. A set A in a metric space $\langle X, d\rangle$ is said to be *bounded* if the set $\{d(x, y) \mid x, y \in A\}$ is a bounded set of real numbers. The *diameter* of a non-void bounded set A is defined to be $\sup\{d(x, y) \mid x, y \in A\}$.

2.35 Exercises. (i) In a metric space the open and closed spheres of radius r are bounded sets of diameter $\leqslant 2r$.

(ii) If a set A in a metric space is bounded so is also $\mathrm{Cl}(A)$, and the diameter of $\mathrm{Cl}(A)$ equals the diameter of A.

(iii) If B is a base of a metric space $\langle X, d \rangle$, ε is a positive real number and B_ε is the set of the members of B whose diameter is $<\varepsilon$ then also B_ε is a base of $\langle X, d \rangle$. □

2.36 Proposition. *In a metric space let $\langle A_n \mid n \in \omega \rangle$ be a descending sequence of sets (i.e., $A_{n+1} \subseteq A_n$ for $n \in \omega$) such that some A_m is bounded (and hence also every A_n with $n>m$ is bounded) and the sequence $\langle d_n \mid n<\omega \rangle$ of their respective diameters converges to 0, then $\bigcap_{n \in \omega} A_n$ contains at most one point. If, in addition, the space is separable and complete and each A_n is a non-void closed set then $\bigcap_{n \in \omega} A_n$ contains exactly one point. If we assume the axiom of choice then the conclusion holds even without requiring the space to be separable.*

Hint of proof. If the space is separable and complete let $\{y_i \mid i<\omega\}$ be a set dense in the space. Let z_n be the first member of $\langle y_i \mid i<\omega \rangle$ such that $d(z_n, x)<\frac{1}{n}$ for some $x \in A_n$. The sequence $\langle z_n \mid n<\omega \rangle$ is a Cauchy sequence and it converges to a point z. Prove that for every n $z \in A_n$. If we assume the axiom of choice we take z_n to be any member of A_n. □

2.37 A Characterization of Completeness. If we assume the axiom of choice then, by 2.36, a metric space is complete iff every intersection $\bigcap_{n \in \omega} A_n$ of a descending sequence of closed sets with diameters converging to zero is non-void, as easily seen. In the absence of the axiom of choice this property seems to be a better definition of completeness than the one we gave in 2.30(iii). This equivalence immediately implies that, assuming the axiom of choice, every compact metric space is complete (by 2.5).

2.38 Proposition. *Let $\mathbb{R}'$ be the set of all non-negative real numbers, and let f be a function on $\mathbb{R}' \times \mathbb{R}'$ into $\mathbb{R}'$ which satisfies, for all $a, a', b, b' \in \mathbb{R}'$,*

(a) $f(a,b)=0$ *iff* $a=0$ *and* $b=0$,

(b) $f(a,0)=a$ *and* $f(0,b)=b$,

(c) *f is monotonic in both arguments, i.e., if $a \leqslant a'$ and $b \leqslant b'$ then $f(a,b) \leqslant f(a',b')$,*

(d) $f(a+a', b+b') \leqslant f(a,b)+f(a',b')$.

Examples of such functions are $\max(a,b)$, $(a^2+b^2)^{1/2}$ *and* $a+b$. *Let $\langle X, d \rangle$ and $\langle X', d' \rangle$ be metric spaces. Let d'' be the function defined on $(X \times X') \times (X \times X')$ by*

$$d''(\langle x, x' \rangle, \langle y, y' \rangle) = f(d(x,y), d'(x',y')).$$

Under these assumptions we have:

(i) *$\langle X \times X', d'' \rangle$ is a metric space.*

(ii) *The topology associated with the metric space $\langle X \times X', d'' \rangle$ is the product of the topologies associated with $\langle X, d \rangle$ and $\langle X', d' \rangle$.*

(iii) *If* $\langle X, d\rangle$ *and* $\langle X', d'\rangle$ *are complete metric spaces then so is also* $\langle X\times X', d''\rangle$.

Hint of proof. Use the inequalities $a, b \leqslant f(a, b) \leqslant a+b$. □

3. Category, Baire and Borel Sets

3.1 Definition (Baire 1899). (i) A set A in a topological space is said to be *nowhere dense* if $\mathrm{Int}(\mathrm{Cl}(A))=0$.

(ii) A set A in a topological space is said to be *meager* (or *of the first category*) if it is the union of at most $\aleph_0$ nowhere dense sets.

3.2 Proposition. *A set A is nowhere dense iff*

(1) *every non-void open set B has a non-void open subset C disjoint from A.* □

3.3 Proposition. (i) *The set N of nowhere dense sets in a topological space* $\langle X, O\rangle$ *is an ideal on X.*

(ii) *If X itself is not meager in the space* $\langle X, O\rangle$ *(and by the Baire category theorem,* 3.6, *this is the case with many spaces) then the set M of all meager subsets of X is an ideal on X. If the axiom of choice holds then M is a* σ*-ideal on X (i.e., an* $\aleph_1$*-complete ideal on X*—IV.4.5). □

3.4 Definition. The *boundary* $\mathrm{Bd}(A)$ of a set A in a topological space is defined by $\mathrm{Bd}(A)=\mathrm{Cl}(A)\sim\mathrm{Int}(A)$; it is always a closed set since $\mathrm{Bd}(A)=\mathrm{Cl}(A)\cap(X\sim\mathrm{Int}(A))$.

3.5 Proposition. *The boundaries of the open sets and of the closed sets are nowhere dense.* □

3.6Ac The Baire Category Theorem (Baire 1899). In a complete metric space no non-void open set is meager. If the space is also separable we do not have to use the axiom of choice in the proof.

Proof. Let A be a non-void open set and let B be meager; we shall prove that A is not included in B. By our choice of B, $B=\bigcup_{i\in\omega}C_i$, where each C_i is a nowhere dense set. We shall now define a descending sequence $\langle D_i \mid i\in\omega\rangle$ of closed sets such that $D_0\subseteq A$, for all $i\in\omega$ $D_{i+1}\cap C_i=0$, and if d_i is the diameter of D_i then $\lim_{i\to\infty}d_i=0$. Once we have such a sequence $\langle D_i \mid i\in\omega\rangle$ we get, by 2.36 (and the axiom of choice), $\bigcap_{i\in\omega}D_i=\{y\}$ for some $y\in X$. Since $D_0\subseteq A$ we have $y\in A$; since $D_{i+1}\cap C_i=0$ we get $y\notin C_i$ for every $i\in\omega$. Thus $y\notin\bigcup_{i\in\omega}C_i=B$, hence $y\in A\sim B$ and A is not a subset of B.

We still have to define the sequence $\langle D_i \mid i\in\omega\rangle$ and prove its properties. We shall choose the D_i's to be closed spheres; by 1.10(i) they are non-void closed sets. Since A is open it is a union of open spheres, and since $A\neq 0$ it includes an open sphere $S(z,r)$. Set $D_0=S^c(z,\frac{r}{2})\subseteq S(z,r)\subseteq A$. For $i\in\omega$ we assume that D_i is a closed sphere $S^c(t,p)$ and proceed to construct D_{i+1}. Since C_i is nowhere dense the open set $S(t,p)$ has, by 3.2, a non-void open subset E disjoint from C_i. Since E is non-void and open it includes an open sphere $S(y,q)$. We take now $D_{i+1}=S^c(y,w)$ where $\omega=\min(\frac{q}{2},2^{-i})$. We have $D_{i+1}=S^c(y,w)\subseteq S(y,q)\subseteq E\subseteq S(t,p)\subseteq S^c(t,p)=D_i$ and since $E\cap C_i=0$ also $D_{i+1}\cap C_i=0$. By 2.35(i) the diameter d_{i+1} of D_{i+1} is $\leqslant 2\cdot 2^{-i}=2^{-(i-1)}$, hence $\lim_{i\to\infty}d_i=0$, which terminates our proof.

If the space is separable then we do not need the axiom of choice in our application of 2.36. We can also eliminate the use of the indefinite article (see II.2.22) in our construction of the closed spheres D_i as follows. Let $\{y_i \mid i\in\omega\}$ be a set dense in the space. If D_i is $S^c(t,p)$ then let y be the first member of the sequence $\langle y_i \mid i\in\omega\rangle$ which is a member of the open set $S(t,p)\sim C_i$ and let $q=\frac{1}{m}$ for the least m such that $S(y,\frac{1}{m})\subseteq S(t,p)\sim C_i$. Now we proceed to define D_{i+1} from $S(y,q)$ as we did above. □

3.7 The Use of the Diagonal Method. The method of proof we used in 3.6 is the diagonal method known to us already from the proof of the Zermelo-König inequality (III.4.18). Our choice of the closed sphere D_{i+1} guarantees that the point y given by $\{y\}=\bigcap_{i\in\omega}D_i$ will not belong to C_i. Since we choose D_{i+1} in that way for every $i\in\omega$ we get $y\notin\bigcup_{i\in\omega}C_i=B$.

We shall now see what the Baire category theorem tells us about the number of points in a complete metric space. If the space has no isolated points (i.e., the space has no points z such that $\{z\}$ is an open set; and the metric spaces which we usually deal with are such) then every singleton $\{z\}$ is nowhere dense (since $\{z\}$ is closed) and hence every countable subset of X is meager. Thus the Baire category theorem implies that in such spaces every open set is uncountable. Indeed, the diagonal method used in the proof of the Baire category theorem is typical for proofs of uncountability.

3.8Ac Proposition (The Baire category theorem for compact spaces). *In a compact Hausdorff space no non-void open set is meager.*

Hint of proof. Proceed as in the proof of 3.6, and use 2.14 and 2.13 to choose D_i to be a closed set with non-void interior such that $D_i\cap C_i=0$, $D_i\subseteq \mathrm{Int}(D_{i-1})$. □

3.9 Definition (Baire 1899, Lebesgue 1905). A *Baire set* in a topological space is a set A which is equal to some open set B modulo a meager set, i.e., the symmetric difference $A\vartriangle B$ is meager. In other words, if M is the σ-ideal of all meager sets then A is a Baire set if $A=_{\mathrm{mod}\,M}B$ for some open set B.

3.10 Proposition. (i) *Every meager set is a Baire set.*

(ii) *If A is a Baire set so is its complement $X\sim A$.*

(iii)Ac *If each A_i, for $i\in\omega$, is a Baire set so are $\bigcup_{i\in\omega}A_i$ and $\bigcap_{i\in\omega}A_i$.* □

3.11 Definition (Kuratowski 1922). A *regular open set* is a set A such that $A=\mathrm{Int}(\mathrm{Cl}(A))$; such a set is obviously open.

Remark. A regular open set can be described as an open set without "cracks". For example, if A is the Euclidean plane without the x-axis then A is an open set, but A is not a regular open set since $\mathrm{Int}(\mathrm{Cl}(A))=\mathrm{Cl}(A)$ is the whole Euclidean plane; by applying the closure we fill in the crack which is in this case the x-axis. On the other hand the subset $\{\langle x, y\rangle \mid x\cdot y>0\}$ of the Euclidean plane is an open set with no cracks and it is also a regular open set.

3.12 Proposition. (i) *For every set A,* $\mathrm{Int}(\mathrm{Cl}(A))$ *is a regular open set. If A is an open set then* $\mathrm{Int}(\mathrm{Cl}(A))$ *is the least regular open set which includes A.*

(ii) *The intersection of two regular open sets is a regular open set.* □

3.13 Theorem. *Every Baire set A is equal, modulo a meager set, to a regular open set G; if in the space no non-void open set is meager (which is the conclusion of the Baire category theorems* 3.6 *and* 3.8) *then G is unique.* □

3.14 Definition. A *field of sets* is a set F of subsets of a non-void set X such that

(a) $0\in F$,

(b) if $A\in F$ then also $X\sim A\in F$, and

(c) if $A, B\in F$ then also $A\cup B\in F$.

Such an F is also said to be a *field of subsets of X*.

If F is a field of subsets of X then, by (a) and (b) also $X=X\sim 0\in F$, and since $A\cap B=X\sim((X\sim A)\cup(X\sim B))$ (b) and (c) imply that if $A, B\in F$ also $A\cap B\in F$.

A field F of subsets of X is said to be a *σ-field* if it satisfies also

(d) if $A_n\in F$ for every $n\in\omega$ then also $\bigcup_{n\in\omega}A_n\in F$.

If F is a σ-field of subsets of X then, since $\bigcap_{n\in\omega}A_n=X\sim\bigcup_{n\in\omega}(X\sim A_n)$, (b) and (d) imply that if $A_n\in F$ for every $n\in\omega$ then also $\bigcap_{n\in\omega}A_n\in F$.

3.15Ac Examples of Fields of Sets. For an arbitrary set X, the sets $\{0, X\}$ and $\mathbf{P}(X)$ are σ-fields of subsets of X, and also the set of all subsets of X which are countable or have countable complements is such. For a topological space $\langle X, O\rangle$ the set of all Baire sets of the space is a σ-field of subsets of X (by 3.9 and 3.10). If X is a set which includes a denumerable subset then the set F of all subsets of X which are finite or have a finite complement is a field of subsets of X but is not a σ-field.

3.16 Definition (Lebesgue 1905). Let $\langle X, O\rangle$ be a topological space. We want to define the Borel sets of this space as the sets obtained from the open sets of the space by any number, finite or infinite, or application of the operations of complementation, union of countably many sets and intersection of countably many sets. We shall give at this point a somewhat different definition of the Borel sets and we shall establish shortly that this definition captures correctly our intuitive concept described above.

A subset A of X is a *Borel set* of $\langle X, O\rangle$ if A is a member of every σ-field of subsets of X which contains all open sets.

3.17 Proposition. *In a topological space $\langle X, O\rangle$ let B be the set of all Borel sets.*

(i) *B is a σ-field of subsets of X which contains all open sets, and is the least such σ-field.*

(ii) *B contains also all the closed sets, and is the least σ-field of subsets of X containing them.*

(iii) *Every Borel set is a Baire set.* □

3.18 Definition (The hierarchy of Borel sets—Lebesgue 1905, Hausdorff 1919). Let $\langle X, O\rangle$ be a topological space. We define the subsets Σ_α and Π_α of $\mathbf{P}(X)$, for $0<\alpha\in \mathrm{On}$, as follows: $\Sigma_1=O$ is the set of all open sets. Π_1 is the set of all closed sets. For $\alpha>1$

$$\Sigma_\alpha = \text{the set of all unions of countable subsets of } \bigcup_{\beta<\alpha}\Pi_\beta,$$

$$\Pi_\alpha = \text{the set of all intersections of countable subsets of } \bigcup_{\beta<\alpha}\Sigma_\beta.$$

The members of Σ_α and Π_α are called Σ_α-sets and Π_α-sets, respectively. Π_2 and Σ_2 are often referred to as G_δ and F_σ.

3.19 Proposition. (i) *For every $\alpha>0$, Π_α is the set of all complements of the members of Σ_α.*

(ii) *If $0<\beta<\alpha$ then $\Sigma_\beta\subseteq\Pi_\alpha$ and $\Pi_\beta\subseteq\Sigma_\alpha$.*

(iii) *If $\alpha>2$ and $0<\beta<\alpha$ then $\Sigma_\beta\subseteq\Sigma_\alpha$ and $\Pi_\beta\subseteq\Pi_\alpha$; if $\langle X, O\rangle$ is the topology associated with a metric on X then also $\Sigma_1\subseteq\Sigma_2$ and $\Pi_1\subseteq\Pi_2$.*

(iv) *$\bigcup_{0<\alpha\in\mathrm{On}}\Sigma_\alpha=\bigcup_{0<\alpha\in\mathrm{On}}\Pi_\alpha$ is the set of all Borel sets.*

(v) *If there is a regular ordinal $>\omega$ let Ω be the least such; $\Omega=\omega_1$ if the axiom of choice holds. Then for every $\gamma>\Omega$ $\Sigma_\gamma=\Sigma_\Omega=\bigcup_{0<\alpha<\Omega}\Sigma_\alpha$ and $\Pi_\gamma=\Pi_\Omega=\bigcup_{0<\alpha<\Omega}\Pi_\alpha$, and thus each of $\bigcup_{0<\alpha<\Omega}\Sigma_\alpha=\bigcup_{0<\alpha<\Omega}\Pi_\alpha$ is the set of all Borel sets.* □

Remark. By 3.19(iv) the Borel sets are indeed the sets obtained from the open sets by any number of application of the operations of complementation, union of countably many sets, and intersection of countably many sets. By 3.19(v) each Borel set can be obtained by less than Ω applications of those operations (where $\Omega=\omega_1$ if the axiom of choice holds).

3.20 Proposition. *Let f be a continuous function on $\langle X, O\rangle$* into *$\langle X', O'\rangle$. Let A be a subset of X' which is in Σ_α* or in *Π_α, then $f^{-1}[A]$ is also in Σ_α or in Π_α, respectively.* □

3.21 Definition. A set B is said to be a *Baire cover* of a set A if B is a Baire set which includes A and every Baire set C which includes A includes also B modulo a meager set, i.e., $B\sim C$ is meager.

3.22Ac Theorem. *Every set has a Baire cover.*

Hint of proof. For a proof of this theorem see, e.g., Kuratowski 1966, § 11.IV.Cor. 1. We shall use this theorem in Chapter VII only for spaces with a countable base, so let us hint at a proof of this case. Let Q be a countable base of the space $\langle X, O\rangle$. Choose the cover B of A to be $(X\sim\bigcup\{E \mid E\in Q \text{ and } E\cap A \text{ is meager}\})\cup A$. □

Chapter VII
The Real Spaces

In Part A we developed a fair amount of set theory with essentially little reference to the real numbers. We did refer to the real numbers in a few places where we wanted to apply the methods of set theory in a context particularly familiar to most readers. For example, in V.3.5 we mentioned that in ZFC one cannot prove the existence of a definable well-ordering of the real numbers. However, from the point of view of set theory it would be sufficient to consider the fact that in ZFC one cannot prove the existence of a definable well-ordering of $\mathbf{P}(\omega)$, and we could thus completely avoid the subject of the real numbers. In the present chapter we shall devote some time to a study of the real numbers.

A large part of what we shall do in the present chapter is part of *descriptive set theory*. This field can be roughly described as the study of sets of real numbers from the point of view of classifying these sets to simpler and more complicated sets according to several criteria. Unlike Part A, which is self sufficient, the present part has a large intersection with several mathematical disciplines such as point set topology, measure theory and recursive function theory. We present here some basic descriptive set theory as developed during the first quarter of the present century. The newer theory of recursive functions leads to the same subject from a somewhat different point of view and enabled mathematicians to make better distinctions and gain deeper insight. However, the classical theory retains its value because of its more general approach.

From the point of view of descriptive set theory the space of the real numbers has three faces: it is the space of real numbers, it is the Cantor space ${}^{\omega}2$ and is the Baire space ${}^{\omega}\omega$. These are three non-homeomorphic spaces, but from the point of view of the present chapter the difference between them is negligible, as will be established in § 3. Therefore we refer to these spaces as *the real spaces*. It is very convenient that the space of real numbers occurs in various forms since we can always handle that form which is most convenient for the matter at hand. Many of the common properties of the real spaces are shared also by the other members of the wider family of the separable complete metric spaces; in § 2 we shall establish some of the common properties of such spaces.

1. The Real Numbers

In this section we shall define the integers, the rational numbers and the real numbers as certain sets. For example, we shall define the integer -1 as the set $\langle 0, 1\rangle$. This does not mean that the integer -1 is really $\langle 0, 1\rangle$, just as an ordered pair $\langle x, y\rangle$ is not really the set $\{\{x\}, \{x, y\}\}$ and a function is not really a set of ordered pairs. Mathematical objects are characterized by their essential properties not by the "substance they are made of". Thus what the integers really are, taken together with their arithmetical operations, is an integral domain which has certain algebraic properties. Those properties are such that they determine this ring up to isomorphism, i.e., any two integral domains with those properties are isomorphic. What is important to know, and this is the purpose of the definitions which we shall give, is that the "ideal" sets of integers, rationals and reals, and the arithmetical operations and relations on them, can be obtained as sets of set theory in some way, and once they are obtained and their essential properties are proved we do not care how we obtained them.

As the reader is assumed to be familiar with the integers, the rational numbers and the real numbers, we shall only mention, but not prove, the basic arithmetical facts about them.

1.1 Definition (The integers). For every natural number $n>0$ we denote with $\dot{-}n$ the pair $\langle 0, n\rangle$. Such a pair is not an ordinal (why?). We denote with $\mathbb{Z}$ the set $\omega\cup\{\dot{-}n \mid 0<n\in\omega\}$ and call it the set of the *integers*. We extend the operations of addition and multiplication which we have on the natural numbers to all integers as follows. For $m, n, r\in\omega$ $n, r\neq 0$ we set

$$m+(\dot{-}n)=(\dot{-}n)+m=\begin{cases} m-n & \text{if } m\geqslant n \\ \dot{-}(n-m) & \text{if } m<n, \end{cases}$$

$$(\dot{-}n)+(\dot{-}r)=\dot{-}(n+r), \qquad 0\cdot(\dot{-}n)=(\dot{-}n)\cdot 0=0,$$
$$n\cdot(\dot{-}r)=(\dot{-}r)\cdot n=\dot{-}(n\cdot r), \qquad (\dot{-}n)\cdot(\dot{-}r)=n\cdot r.$$

We define also a unary operation $-$ on the integers by

$$-0=0, \qquad -n=\dot{-}n, \qquad -(\dot{-}n)=n.$$

It is easily seen that the structure $\langle\mathbb{Z}, +, \cdot, -, 0, 1\rangle$ is an *integral domain*, i.e., it satisfies the following axioms:

Group A (Axioms of addition):

A1. (Commutativity). $x+y=y+x$,

A2. (Associativity). $(x+y)+z=x+(y+z)$,

A3. (Zero element). $x+0=x$,

A4. (Additive inverse). $x+(-x)=0$.

Group B (Axioms of multiplication):

B1. (Commutativity). $x \cdot y = y \cdot x$, B2. (Associativity). $(x \cdot y) \cdot z = x \cdot (y \cdot z)$,
B3. (Unit element). $x \cdot 1 = x$, B4. (No zero divisors). $x \cdot y = 0 \rightarrow x = 0 \vee y = 0$.

Group C (The distributive law): $x \cdot (y+z) = x \cdot y + x \cdot z$.

We extend the order relation $<$ on the natural numbers by setting

$$\begin{aligned} < = &\{\langle m, n\rangle \mid m, n \in \omega \wedge m<n\} \cup \{\langle \dot{-}m, n\rangle \mid m, n \in \omega \wedge m \neq 0\} \\ &\cup \{\langle \dot{-}m, \dot{-}n\rangle \mid m, n \in \omega \sim \{0\} \wedge m>n\}. \end{aligned}$$

The structure $\langle \mathbb{Z}, +, \cdot, -, 0, 1, < \rangle$ is now easily seen to be an *ordered integral domain*, i.e., it satisfies, in addition to the axioms of groups A–C above, also the following.

Group D (Axioms of order):

D1. $<$ is an order relation (on $\mathbb{Z}$), D2. $x<y \rightarrow x+z<y+z$,
D3. $x<y \wedge z>0 \rightarrow x \cdot z < y \cdot z$.

We shall use the symbol $\mathbb{Z}$ ambiguously for the set of all integers and for the ordered integral domain $\langle \mathbb{Z}, +, \cdot, -, 0, 1, < \rangle$. We shall assume without proof all elementary algebraic facts about $\mathbb{Z}$ which we may need.

1.2 Proposition. *The set $\mathbb{Z}$ of all integers is denumerable.* □

1.3 Definition (The rational numbers). Let

$$\mathbb{Q} = \mathbb{Z} \cup \{\langle k, l\rangle \mid k, l \in \mathbb{Z} \wedge k \neq 0 \wedge l>1 \wedge k \text{ and } l \text{ are relatively prime}\};$$

as easily seen, a pair $\langle k, l\rangle$ with $k \neq 0$ cannot be a member of $\mathbb{Z}$. The members of $\mathbb{Q}$ are called *rational numbers*. We define k/l for any two integers k, l such that $l \neq 0$ as follows. If l divides k (in the integral domain $\mathbb{Z}$) then k/l is the integer which is the quotient of this division; if l does not divide k then $k/l = \langle k', l'\rangle$ where k' and l' are the unique integers such that $l'>0$, k' and l' are relatively prime and $k' \cdot l = l' \cdot k$. (k' and l' are obtained from k and l by dividing k and l by their greatest common divisor, and by changing the signs of k and l if $l<0$). It is obvious that if $\langle k, l\rangle \in \mathbb{Q} \sim \mathbb{Z}$ then $k/l = \langle k, l\rangle$.

We extend the operations $+$, $\cdot$ and $-$ from $\mathbb{Z}$ to $\mathbb{Q}$ by setting for all $m \in \mathbb{Z}$ and $\langle k, l\rangle, \langle k', l'\rangle \in \mathbb{Q}$, $k, k' \neq 0$,

$$m + \langle k, l\rangle = \langle k, l\rangle + m = \frac{m \cdot l + k}{l}, \qquad \langle k, l\rangle + \langle k', l'\rangle = \frac{k \cdot l' + k' \cdot l}{l \cdot l'},$$

$$m \cdot \langle k, l\rangle = \langle k, l\rangle \cdot m = \frac{m \cdot k}{l}, \qquad \langle k, l\rangle \cdot \langle k', l'\rangle = \frac{k \cdot k'}{l \cdot l'},$$

$$-\langle k, l\rangle = \langle -k, l\rangle.$$

It is easy to prove that $\langle \mathbb{Q}, +, \cdot, -, 0, 1\rangle$ is a *field*, i.e., it satisfies the axioms of groups A–C of 1.1 with the addition of

B5. (Inverse element). For all $x \neq 0$ there is a y such that $x \cdot y = 1$.

We extend the order relation $<$ on $\mathbb{Z}$ to $\mathbb{Q}$ by

$$\begin{aligned} < = &\{\langle m, n\rangle \mid m, n \in \mathbb{Z} \wedge m < n\} \\ &\cup \{\langle m, \langle k, l\rangle\rangle \mid m \in \mathbb{Z} \wedge \langle k, l\rangle \in \mathbb{Q} \sim \mathbb{Z} \wedge m \cdot l < k\} \\ &\cup \{\langle\langle k, l\rangle, m\rangle \mid m \in \mathbb{Z} \wedge \langle k, l\rangle \in \mathbb{Q} \sim \mathbb{Z} \wedge k < m \cdot l\} \\ &\cup \{\langle\langle k, l\rangle, \langle k', l'\rangle\rangle \mid \langle k, l\rangle, \langle k', l'\rangle \in \mathbb{Q} \sim \mathbb{Z} \wedge k \cdot l' < k' \cdot l\}. \end{aligned}$$

It is easily seen that $\langle \mathbb{Q}, +, \cdot, -, 0, 1, <\rangle$ is an ordered field, i.e., satisfies the axioms of group D of 1.1 in addition to the field axioms. $\langle \mathbb{Q}, <\rangle$ is a dense ordered set, since for $x, y \in \mathbb{Q}$, if $x < y$ then $x < (x+y)/2 < y$, and $\mathbb{Q}$ has no first or last member since for every $x \in \mathbb{Q}$ $x-1 < x$ and $x+1 > x$.

We shall use the symbol $\mathbb{Q}$ ambiguously for the set of all rational numbers and for the ordered field $\langle \mathbb{Q}, +, \cdot, -, 0, 1, <\rangle$.

1.4 Proposition. *The set $\mathbb{Q}$ of all rational numbers is denumerable.*

Hint of proof. Use 1.2 and $\aleph_0 \cdot \aleph_0 = \aleph_0$. □

1.5 Definition (The real numbers). An *irrational number* is a non-void proper initial segment d of $\mathbb{Q}$ (i.e., $0 \subset d \subset \mathbb{Q}$ and $y < x \in d \rightarrow y \in d$) such that d has no greatest member and $\mathbb{Q} \sim d$ has no least member. An irrational number cannot be a rational number since every member of $\mathbb{Q}$ is a finite set while every irrational number is an infinite set (by III.1.11). An object x is said to be a *real number* if it is a rational number or an irrational number. We denote with $\mathbb{R}$ the set of all real numbers ($\mathbb{R} \subseteq \mathbb{Q} \cup \mathbf{P}(\mathbb{Q})$), which is also called *the real line*.

A real number x is said to be *positive* if $x \in \mathbb{Q}$ and $x > 0$ or if $x \subseteq \mathbb{Q}$ and $0 \in x$. We extend the operations $+$, $-$ and $\cdot$ to the real numbers as follows. For $d, e \in \mathbb{R} \sim \mathbb{Q}$ and $z \in \mathbb{Q}$ we set:

$$z + d = d + z = \{z + x \mid x \in d\},$$

$$d + e = \begin{cases} \{x + y \mid x \in d \wedge y \in e\} & \text{if } \mathbb{Q} \sim \{x+y \mid x \in d \wedge y \in e\} \text{ has no least member} \\ u & \text{if } u \text{ is the least member of } \mathbb{Q} \sim \{x+y \mid x \in d \wedge y \in e\}, \end{cases}$$

$$-d = \{-x \mid x \in \mathbb{Q} \sim d\}.$$

Notice that every non-positive irrational number is of the form $-d$ for some positive irrational number d.

$$d \cdot 0 = 0 \cdot d = 0.$$

For positive $z \in \mathbb{Q}$, and positive $d, e \in \mathbb{R} \sim \mathbb{Q}$ we set: $z \cdot d = d \cdot z = \{x \cdot z \mid x \in d\}$,

$d \cdot e = \{x \cdot y \mid x, y > 0 \wedge x \in d \wedge y \in e\} \cup \{x \in \mathbb{Q} \mid x \leqslant 0\}$, if $\mathbb{Q} \sim \{x \cdot y \mid x, y > 0 \wedge x \in d \wedge y \in e\}$ has no least member, and if $\mathbb{Q} \sim \{x \cdot y \mid x, y > 0 \wedge x \in d \wedge y \in e\}$ has a least member then $d \cdot e$ is defined to be that least member.

For positive $a, b \in \mathbb{R}$ not both in $\mathbb{Q}$ we set: $(-a) \cdot b = a \cdot (-b) = -(a \cdot b)$, $(-a) \cdot (-b) = a \cdot b$.

It is not difficult to prove that the structure $\langle \mathbb{R}, +, \cdot, -, 0, 1 \rangle$ is a field (see 1.3). We write $x - y$ for $x + (-y)$, which is the unique solution u of the equation $u + y = x$. For $z \neq 0$ we denote by z^{-1} the unique u such that $z \cdot u = 1$ and we write y/z for $y \cdot z^{-1}$ which is the unique solution u of the equation $z \cdot u = y$. The properties of all these operations on the reals are assumed to be well known.

We extend the relation $<$ on $\mathbb{Q}$ to the whole of $\mathbb{R}$ by setting

$$\begin{aligned} < = \{\langle x, y \rangle \mid & (x, y \in \mathbb{Q} \wedge x < y) \vee (x \in \mathbb{Q} \wedge y \in \mathbb{R} \sim \mathbb{Q} \wedge x \in y) \\ & \vee (x \in \mathbb{R} \sim \mathbb{Q} \wedge y \in \mathbb{Q} \wedge y \notin x) \vee (x, y \in \mathbb{R} \sim \mathbb{Q} \wedge x \subset y)\}. \end{aligned}$$

$\mathbb{Q}$ is dense in $\mathbb{R}$ with respect to $<$, i.e., for all $x, y \in \mathbb{R}$ if $x < y$ then there is a $z \in \mathbb{Q}$ such that $x < z < y$. Every bounded subset A of $\mathbb{R}$ has a least upper bound b (i.e., $(\forall x \in A)(x \leqslant b) \wedge (\forall c \in \mathbb{R})[(\forall x \in A)(x \leqslant c) \rightarrow b \leqslant c]$). The structure $\langle \mathbb{R}, +, \cdot, -, 0, 1, < \rangle$ is an ordered field.

We shall use the symbol $\mathbb{R}$ ambiguously for the set of all real numbers and for the ordered field $\langle \mathbb{R}, +, \cdot, -, 0, 1, < \rangle$. Whenever we shall mention *the* order of $\mathbb{R}$ we mean, of course, the relation $<$ defined here. We shall denote with $\mathbb{I}$ the closed unit interval of $\mathbb{R}$, i.e., $\mathbb{I} = \{x \in \mathbb{R} \mid 0 \leqslant x \leqslant 1\}$.

1.6 Theorem (Cantor 1895). $|\mathbb{R}| = |\mathbb{I}| = 2^{\aleph_0}$.

Proof. By our definition of $\mathbb{R}$ we have $\mathbb{R} \subseteq \mathbb{Q} \cup \mathbf{P}(\mathbb{Q})$, hence $|\mathbb{R}| \leqslant |\mathbf{P}(\mathbb{Q})| + |\mathbb{Q}| = 2^{\aleph_0} + \aleph_0 \leqslant 2^{\aleph_0} + 2^{\aleph_0} = 2 \cdot 2^{\aleph_0} = 2^{\aleph_0 + 1} = 2^{\aleph_0}$, since, by 1.4, $|\mathbb{Q}| = \aleph_0$.

Since $\mathbb{I} \subseteq \mathbb{R}$ our theorem is proved once we show that $2^{\aleph_0} \leqslant |\mathbb{I}|$, since then we shall have $2^{\aleph_0} \leqslant |\mathbb{I}| \leqslant |\mathbb{R}| \leqslant 2^{\aleph_0}$. We get an injection F of ${}^{\omega}2$ into $\mathbb{I}$ by setting, for every $x \in {}^{\omega}2$, $F(x) = \Sigma_{n \in \omega} 2 \cdot x_n \cdot 3^{-(n+1)}$. We leave the proof that F is an injection to the diligent reader. The lazy reader can rely on the fact that the inequality $2^{\aleph_0} \leqslant |\mathbb{I}|$ is an immediate consequence of 2.15 below. We shall also construct in 3.7 below an injection of the set ${}^{\omega}\omega$ (which has the cardinality $\aleph_0^{\aleph_0} = 2^{\aleph_0}$) into $\mathbb{I}$. □

1.7 Basic Topological Properties of $\mathbb{R}$ and $\mathbb{I}$. A *closed interval* (of $\mathbb{R}$) is a set of the form $\{x \in \mathbb{R} \mid a \leqslant x \leqslant b\}$, where $a, b \in \mathbb{R}$, $a < b$; we denote it with $[a, b]$. $[0, 1]$ was denoted above by $\mathbb{I}$. $\langle \mathbb{I}, < \rangle$ is isomorphic to every closed interval $\langle [a, b], < \rangle$ by means of the linear function $F(x) = a + (b - a) \cdot x$. The open unit interval $(0, 1)$ is isomorphic to every open interval (a, b) by the same function F (restricted to $(0, 1)$). Also $(0, 1)$ is isomorphic to the open half-lines (a, ∞) and $(-\infty, b)$ by the respective functions given by $F(x)=a+x/(1-x)$ and $G(x)=b+1-1/x$. Finally $(0, 1)$ is order-isomorphic to $\mathbb{R}$ itself via the function $H(x)=\tan\ \pi\cdot(x-\frac{1}{2})$.

The standard topology of $\mathbb{R}$ is the order topology obtained from the natural ordering $<$ of $\mathbb{R}$. On every interval and half-line we have the topology induced by the topology of $\mathbb{R}$ which is, by VI.1.21(i), also the order topology on that set. Since $\mathbb{R}$ is order-isomorphic to every open interval and every open half-line, $\mathbb{R}$ is also homeomorphic to each such set (in their order topologies). For the same reason also every closed interval is homeomorphic to $\mathbb{I}$.

Henceforth $\mathbb{R}$ and $\mathbb{I}$ will also denote the spaces which consist of these sets and their order topologies. These two spaces are the most interesting topological spaces related directly to the real line.

Since the topology on $\mathbb{R}$ and $\mathbb{I}$ is an order topology $\mathbb{R}$ and $\mathbb{I}$ are Hausdorff spaces (VI.2.2). $\mathbb{R}$ and $\mathbb{I}$ are dense ordered sets (since between any two real numbers there is a real number), and they are also gapless (since every bounded subset of $\mathbb{R}$ and $\mathbb{I}$ has a least upper bound), hence, by VI.2.29, $\mathbb{R}$ and $\mathbb{I}$ are connected spaces. Every open interval of $\mathbb{R}$ contains a rational number and hence, the sets $\mathbb{Q}$ and $\mathbb{Q}\cap\mathbb{I}$ are dense in $\mathbb{R}$ and $\mathbb{I}$, respectively, and thus $\mathbb{R}$ and $\mathbb{I}$ are separable spaces (VI.2.23).

On the real line we have the metric given by $d(a, b)=$ the absolute value of $a-b$; it is easy to verify that d is indeed a metric. Let us prove now that the topology associated with this metric is the same as the order topology. Every open sphere is of the form $S(x, r)=(x-r, x+r)$ and this is an open set also in the order topology. Going in the other direction, a basic open set of the order topology is of one of the forms (a, b), $(-\infty, b)$, (a, ∞) and $\mathbb{R}$. $(a, b)=S((a+b)/2, (b-a)/2)$ is an open sphere; (a, ∞), $(-\infty, b)$ and $\mathbb{R}$ are unions of open spheres since $(a, \infty)=\bigcup_{n\in\omega}(a+n, a+n+2)$, $(-\infty, b)=\bigcup_{n\in\omega}(b-n-2, b-n)$ and $\mathbb{R}=\bigcup_{n\in\mathbb{Z}}(n, n+2)$. Thus the two topologies are identical. By the well known Cauchy convergence theorem the metric space $\langle\mathbb{R}, d\rangle$ is a complete metric space. Since $\mathbb{I}$ is a closed subset of $\mathbb{R}$ also $\langle\mathbb{I}, d\upharpoonright(\mathbb{I}\times\mathbb{I})\rangle$ is a complete metric space, by VI.2.31(iv) (this can also easily be verified directly).

$\mathbb{R}$ is not compact since the cover $\{(n, n+2) \mid n\in\mathbb{Z}\}$ of $\mathbb{R}$ has no finite subset which is a cover. $\mathbb{I}$, on the other hand, is compact by the well-known Heine–Borel theorem.

1.8 Basic Topological Properties of the Euclidean Spaces. For every ordinal $1<n\leqslant\omega$ we have a topology on ${}^n\mathbb{R}$ which is the topology of the n-th power of $\mathbb{R}$. For $n<\omega$ ${}^n\mathbb{R}$ is called the *n-dimensional Euclidean space*, and ${}^\omega\mathbb{R}$ is called the *Fréchet space*. By VI.2.17(i) the topology induced on ${}^n\mathbb{I}$ by the topology of ${}^n\mathbb{R}$ is identical with the topology of the n-th power of $\mathbb{I}$. ${}^n\mathbb{I}$ is called the *n-dimensional closed cube* and ${}^\omega\mathbb{I}$ is called the *Hilbert cube*. Being products of countably many of the separable connected and Hausdorff spaces $\mathbb{R}$ and $\mathbb{I}$, also ${}^n\mathbb{R}$ and ${}^n\mathbb{I}$, for $1\leqslant n\leqslant\omega$, are separable connected Hausdorff spaces (by VI.2.20, VI.2.26 and VI.2.28). ${}^n\mathbb{R}$ is obviously not compact (by VI.2.21), but ${}^n\mathbb{I}$ is compact, by the Tychonoff product theorem (VI.2.22 and VIII.2.20), being the product of compact spaces (The axiom of choice is used only in the proof of the compactness of ${}^\omega\mathbb{I}$).

For $n \in \omega$ we have on ${}^n\mathbb{R}$ a metric given by $d(x, y) = (\Sigma_{i<n}(x_i - y_i)^2)^{1/2}$, for $x, y \in {}^n\mathbb{R}$. The topology associated with this metric is easily seen to be the same as the product topology. It is also easily seen that ${}^n\mathbb{R}$ is a complete metric space.

2. *The Separable Complete Metric Spaces*

In the first section of this chapter we started studying the topology of the real line $\mathbb{R}$ and the Euclidean spaces ${}^n\mathbb{R}$. These spaces are of major importance in mathematics. However, in set theory and in recursion theory one mostly considers the Baire space ${}^\omega\omega$ and the Cantor space ${}^\omega 2$, since these spaces are, in many respects, similar to $\mathbb{R}$ and yet they have a considerably simpler structure. We shall now study all these spaces, but we shall not deal in much detail with the Euclidean spaces ${}^n\mathbb{R}$ with $n > 1$.

2.1 The Cantor and Baire Spaces. The *Cantor space* is the space whose set is ${}^\omega 2$ and whose topology is the topology of the ω-th power of the space 2, where 2 is endowed with the discrete topology. We shall use ${}^\omega 2$ ambiguously for the set ${}^\omega 2$ and for the Cantor space. As mentioned in VI.2.17(iv) the topology of the Cantor space is generated by the base consisting of the sets $B_s = \{f \in {}^\omega 2 \mid f \supseteq s\}$ where $s \in {}^{\underline{\omega}}2$ is a finite sequence of 0's and 1's.

The *Baire space* is the space whose set is ${}^\omega\omega$ and whose topology is the topology of the ω-th power of ω, where ω is endowed with the discrete topology. We shall use ${}^\omega\omega$ ambiguously for the set ${}^\omega\omega$ and for the Baire space. The topology of the Baire space is generated by the base consisting of the sets $B_s = \{f \in {}^\omega\omega \mid f \supseteq s\}$ where $s \in {}^{\underline{\omega}}\omega$ is a finite sequence of natural numbers.

2.2 Basic Topological Properties of the Cantor and Baire Spaces. Both ${}^\omega 2$ and ${}^\omega\omega$ are Hausdorff spaces, being products of discrete spaces (by VI.2.20). Since 2 and ω are trivially separable also ${}^\omega 2$ and ${}^\omega\omega$ are separable (by VI.2.26). Neither ${}^\omega 2$ nor ${}^\omega\omega$ are connected since their factors 2 and ω are not connected (VI.2.28). ${}^\omega\omega$ is not compact (by VI.2.21) since its factors ω, being infinite discrete spaces, are not compact. On the other hand, since the finite space 2 is compact, so is, by the Tychonoff product theorem (VIII.2.20) also ${}^\omega 2$ (see also Exercise 2.3 below). Both ${}^\omega 2$ and ${}^\omega\omega$ can be given the metric defined on the sequences of length ω in VI.1.11 by $d(f, g) = \Sigma_{f(n) \neq g(n)} 2^{-n}$; the topologies associated with this metric are the product topologies we use on ${}^\omega 2$ and ${}^\omega\omega$ (by VI.1.11 and VI.2.17(iv)). With respect to this metric both spaces are complete metric spaces (by VI.2.33), and hence both satisfy the Baire category theorem (VI.3.6). It is obvious that ${}^\omega 2$ is a subspace of ${}^\omega\omega$ as a metric space, hence, by VI.1.21(ii), ${}^\omega 2$ is a subspace of ${}^\omega\omega$ as a topological space. Since ${}^\omega 2$ is the ω-th power of 2 every finite or denumerable power of ${}^\omega 2$ is an ω-power of 2 and is therefore homeomorphic to ${}^\omega 2$ (by VI.2.19(ii) and VI.2.17(ii)). Similarly, every finite or denumerable power of ${}^\omega\omega$ is homeomorphic to ${}^\omega\omega$. Therefore the Cantor and Baire spaces are "dimensionless", unlike the real line $\mathbb{R}$.

2.3 Exercise. Prove directly that the space ${}^{\omega}2$ is compact. Do not use the axiom of choice. □

2.4 The Non-Homeomorphism of the Spaces. The spaces ${}^{\omega}2$ and ${}^{\omega}\omega$ are not homeomorphic to one another since ${}^{\omega}2$ is compact and ${}^{\omega}\omega$ is not. Neither of these spaces is homeomorphic to any ${}^{n}\mathbb{R}$ or ${}^{n}\mathbb{I}$, $1 \leqslant n \leqslant \omega$, since ${}^{n}\mathbb{R}$ and ${}^{n}\mathbb{I}$ are connected while ${}^{\omega}2$ and ${}^{\omega}\omega$ are not. Another way of showing that ${}^{\omega}2$ and ${}^{\omega}\omega$ are not homeomorphic to ${}^{n}\mathbb{I}$ or ${}^{n}\mathbb{R}$ for $1 \leqslant n < \omega$ is to use the fact that while ${}^{\omega}2 \times {}^{\omega}2 \cong {}^{\omega}2$ and ${}^{\omega}\omega \times {}^{\omega}\omega \cong {}^{\omega}\omega$ (where $\cong$ denotes homeomorphism) we have that ${}^{n}\mathbb{R} \times {}^{n}\mathbb{R} \cong {}^{2n}\mathbb{R}$ is not homeomorphic to ${}^{n}\mathbb{R}$ (this is not shown here), and the same holds for ${}^{n}\mathbb{I}$ too.

2.5 Polish Spaces. The spaces which we have studied till now, namely the Euclidean spaces, the Cantor space and the Baire space, have many properties in common. We shall study a family of topological spaces which includes all these spaces, namely the family of all separable complete metric spaces, which we shall call, henceforth, *Polish spaces*. Sometimes we shall add also the requirement that the space be perfect, i.e., that no singleton $\{x\}$ be an open set. Studying the family of all Polish spaces is only a limited enlargement of our scope, since every such space is homeomorphic to a Π_2-subset (i.e., G_δ-subset) of the Hilbert cube ${}^{\omega}\mathbb{I}$. (For the proof see Kuratowski 1966, § 35.III). By VI.2.25 all separable metric spaces have a countable base.

2.6 Proposition. *If* $\langle X, O\rangle$ *is a Hausdorff space with a countable base then* $|X| \leqslant 2^{\aleph_0}$.

Hint of proof. Let B be a countable base of $\langle X, O\rangle$. The function F on X given by $F(x) = \{Y \in B \mid x \in Y\}$ injects X into $\mathbf{P}(B)$. □

2.7 The Cardinality of a Polish Space. By 2.6 (and VI.2.25) all separable metric spaces have at most $2^{\aleph_0}$ points. We shall see in 2.14 below that if the space is complete (i.e., Polish) and has no isolated points (i.e., no points x such that $\{x\}$ is open) then it has exactly $2^{\aleph_0}$ points. Polish spaces which have isolated points may also be countable (e.g., the closed subset $\{0\} \cup \{\frac{1}{n} \mid n \in \omega\}$ of $\mathbb{I}$ is a countable complete metric space by VI.2.31(iv)), but we shall see in 2.22 that if a Polish space is uncountable it must be, again, of the cardinality $2^{\aleph_0}$.

2.8 Definition. (i) Let A be a set in a topological space. A point $x \in A$ is said to be an *isolated point of* A if x has a neighbourhood C such that $C \cap A = \{x\}$.

(ii) (Cantor 1883) A set A in a topological space is called *perfect* if $A \neq 0$ and for every $x \in A$ $x \in \mathrm{Cl}(A \sim \{x\})$ or, in other words, A has no isolated points. The space $\langle X, O\rangle$ is called perfect if X itself is a perfect set in itself, i.e., if there is no $x \in X$ such that $\{x\}$ is open.

(iii) A set A in a topological space $\langle X, O\rangle$ is said to be a *Cantor set* if it is homeomorphic, as a subspace of $\langle X, O\rangle$, to the Cantor space ${}^{\omega}2$. Since ${}^{\omega}2$ is a compact space every Cantor set is a compact set (by VI.2.10 or by VI.2.8) and in a Hausdorff space every Cantor set is closed (by VI.2.9).

2.9 Proposition. *Let $\langle X, O\rangle$ be a topological space and let $\langle A, O'\rangle$ be a subspace of it. $\langle A, O'\rangle$ is perfect iff A is a perfect set in $\langle X, O\rangle$. As a consequence, every Cantor set in $\langle X, O\rangle$ is a perfect set.* □

2.10 Proposition. *If A is a perfect set in a topological space, so is also* $\mathrm{Cl}(A)$. □

2.11 Examples of Perfect and Non-Perfect Spaces. Every point in a discrete space is isolated and hence no subset of a discrete space is perfect. If B is a base of a non-perfect space then B must contain a singleton $\{x\}$ (since in a non-perfect space there is a singleton $\{x\}$ which is open, and it is a union of members of B). Therefore all Euclidean spaces, as well as ${}^{\omega}2$ and ${}^{\omega}\omega$, are perfect spaces.

2.12 The Operation (A) (Souslin 1917). Our present aim is to prove that every perfect complete metric space has at least $2^{\aleph_0}$ points, but we want to obtain a method of proof with a wider range of applicability. What we are looking for is a good method of obtaining mappings of the set ${}^{\omega}2$ into a given metric space $\langle X, d\rangle$; such a method for mapping the set ${}^{\omega}e$, where e is any well-orderable set, is being given now.

Let e be a well-orderable set topologized by the discrete topology. ${}^{\omega}e$ will denote the topological space which is the ω-th power of e. Our aim is to construct a mapping F of the set ${}^{\omega}e$ into a given complete metric space $\langle X, d\rangle$. The arguments of F are infinite sequences $t=\langle t_i \mid i\in\omega\rangle\in{}^{\omega}e$. Let us think of t as not completely given at any moment, but as a sequence that is given to us one term at a time. At each stage we know only a finite part $t\upharpoonright n$ of t; accordingly we cannot tell the exact value of $F(t)$ at this stage, we can only give a subset $A_{t\upharpoonright n}$ of X which is known to contain $F(t)$ on the basis of the information contained in $t\upharpoonright n$ (i.e., $A_{t\upharpoonright n}\supseteq\{F(t')\mid t'\in{}^{\omega}e\wedge t'\supseteq t\upharpoonright n\}$). Therefore we shall assume that we are given a mapping A of the set ${}^{\underline{\omega}}e$ of all finite sequences of the members of e into $\mathbf{P}(X)$, where for each $s\in{}^{\underline{\omega}}e$ we understand A_s as telling us that $F(t)\in A_s$ for every $t\in{}^{\omega}e$ such that $t\supseteq s$. Clearly, the more we know of t we should know more of $F(t)$, i.e., the set of the possible values of $F(t)$ should become smaller. Accordingly we expect to have, for all $r, s\in{}^{\underline{\omega}}e$, if $r\subseteq s$ then $A_s\subseteq A_r$. To get this it clearly suffices to require

(1) for all $s\in{}^{\underline{\omega}}e$ and $w\in e$, $A_{s\frown\langle w\rangle}\subseteq A_s$.

(Where if s is $\langle s_0,\ldots,s_{n-1}\rangle$ $s\frown\langle w\rangle$ is $\langle s_0,\ldots,s_{n-1},w\rangle$—II.4.12).

We want now to make sure that for $t\in{}^{\omega}e$ we have $\bigcap_{n\in\omega}A_{t\upharpoonright n}\neq 0$ since $F(t)$ is supposed to belong to this intersection. Since the space $\langle X, d\rangle$ is complete we can use Proposition VI.2.36 and make sure that $\bigcap_{n\in\omega}A_{t\upharpoonright n}\neq 0$ for every $t\in{}^{\omega}e$ by adding to (1) the requirements

(2) for every $s\in{}^{\underline{\omega}}e$ A_s is a non-void closed set, and

(3) for every $t\in{}^{\omega}e$ $\lim_{n\to\infty}(\text{diameter of } A_{t\upharpoonright n})=0$

(i.e., $A_{t\upharpoonright n}$ is bounded from some value n_0 of n and on, and the sequence of the diameters of $A_{t\restriction n}$, for $n \geqslant n_0$, converges to 0). Moreover, by VI.2.36 $\bigcap_{n\in\omega} A_{t\restriction n}$ contains exactly one member. Thus once requirements (1)–(3) are satisfied we can define F from A by setting, for $t \in {}^{\omega}e$,

$$F(t) = \text{the single member of } \bigcap_{n\in\omega} A_{t\restriction n}.$$

We shall now prove that the function F thus defined is continuous. It clearly suffices to prove that for $t \in {}^{\omega}e$ and a positive real ε t has a neighbourhood D such that $F[D] \subseteq S(F(t), \varepsilon)$. For such t and ε there is, by (3), an $n \in \omega$ such that the diameter of $A_{t\restriction n}$ is $< \varepsilon$. Recalling our notation of writing B_s for the basic open set $\{u \in {}^{\omega}e \mid u \supseteq s\}$, where $s \in {}^{\underline{\omega}}e$, we have here

$$F[B_{t\restriction n}] = \{F(t') \mid t' \in {}^{\omega}e \wedge t' \supseteq t\restriction n\} = \bigcup\{\bigcap_{m\in\omega} A_{t'\restriction m} \mid t' \supseteq t\restriction n\} \subseteq$$

$$\bigcup\{A_{t\restriction n}\} = A_{t\restriction n} \subseteq S(F(t), \varepsilon), \text{ which is what we had to prove.}$$

For F to be one–one we need

(4) for all $t, t' \in {}^{\omega}e$, if $t \neq t'$ then there is an $n < \omega$ such that $A_{t\restriction n} \cap A_{t'\restriction n} = 0$.

(What happens if (4) does not hold for some particular t and t'?)

(4) is clearly sufficient to establish that F is an injection.

If we have, in addition to (1)–(3) also

(5) $A_0 = X$ and for every $s \in {}^{\underline{\omega}}e$ $A_s = \bigcup_{w\in e} A_{s^\frown\langle w\rangle}$,

then we shall see that F is a surjection on X. Let $x \in X$; we define now by induction on n a sequence $\langle s(n) \mid n \in \omega\rangle$ such that for $n \in \omega$ $s(n) \in {}^{n}e$, $s(n) \subseteq s(n+1)$ and $x \in A_{s(n)}$. We set first $s(0) = 0$ and get $x \in A_{s(0)} = A_0 = X$. Given $s(n)$ such that $x \in A_{s(n)}$ we notice that since, by (5), $A_{s(n)} = \bigcup_{w\in e} A_{s(n)^\frown\langle w\rangle}$ there is a $w \in e$ such that $x \in A_{s(n)^\frown\langle w\rangle}$; let $s(n+1)$ be $s(n)^\frown\langle w\rangle$ for such a w (this definition is alright since e is well-orderable—II.2.25). Take $t = \bigcup_{n\in\omega} s(n)$, then for all $n \in \omega$ $t\restriction n = s(n)$. Thus $x \in \bigcap_{n\in\omega} A_{s(n)} = \bigcap_{n\in\omega} A_{t\restriction n} = \{F(t)\}$ and $x = F(t)$. Summing up what we have shown, we can write:

2.13 Theorem. *Let $\langle X, d\rangle$ be a complete metric space. Let e be a well-orderable set and A a mapping of the set ${}^{\underline{\omega}}e$ of all finite sequences of members of e into* $\mathbf{P}(X)$ *such that*

(1) *for all $s \in {}^{\underline{\omega}}e$ and $w \in e$ $A_{s^\frown\langle w\rangle} \subseteq A_s$,*

(2) *for every $s \in {}^{\underline{\omega}}e$ A_s is a non-void closed set,*

(3) *for every $t \in {}^{\omega}e$ $\lim_{n\to\infty}$(diameter of $A_{t\restriction n}$) $= 0$*

(i.e., $A_{t\restriction n}$ is bounded from some value n_0 of n and on and the sequence of the diameters of $A_{t\restriction n}$, for $n \geqslant n_0$, converges to zero).

Then there is a unique function F on ${}^{\omega}e$ into X such that for every $t \in {}^{\omega}e$ $\bigcap_{n\in\omega} A_{t\restriction n} = \{F(t)\}$. F is a continuous function on ${}^{\omega}e$ (where e is taken to have the discrete topology).

If A satisfies also

(4) *for all $t, t' \in {}^{\omega}e$, if $t \neq t'$ then there is an $n \in \omega$ such that $A_{t\restriction n} \cap A_{t'\restriction n} = 0$,*

then F is one–one.

If A satisfies, in addition to (1)–(3) *also*

(5) *$A_0 = X$ and for all $s \in {}^{\underline{\omega}}e$ $\bigcup_{w\in e} A_{s\frown\langle w\rangle} = A_s$,*

then $\mathrm{Rng}(F) = X$. □

2.14Ac Theorem. *Every perfect complete metric space $\langle X, d\rangle$ includes a Cantor set. (The use of the axiom of choice is superfluous if $\langle X, d\rangle$ is separable). This implies $|X| \geqslant 2^{\aleph_0}$. As a consequence, if $\langle X, d\rangle$ is a perfect Polish space then, by 2.7, $|X| = 2^{\aleph_0}$.*

Proof. In 2.13 we take the set 2 for e, and for every sequence $s \in {}^{\underline{\omega}}2$ of length n we shall take A_s to be a closed sphere of radius $\leqslant 2^{-n}$. We define A_0 as some closed sphere of radius 1. Given $s \in {}^{\underline{\omega}}2$ of length n and $A_s = S^c(u, p)$ we have, by the induction hypothesis, $0 < p \leqslant 2^{-n}$. Since the space is perfect u is not an isolated point and hence $S(u, p)$ contains at least two different points u' and u''. Choose $q > 0$ such that $q \leqslant 2^{-(n+1)}$ and $q \leqslant \frac{1}{4}d(u', u'')$, then $S^c(u', q) \cap S^c(u'', q) = 0$. Since $S(u, p)$ is an open set we can also choose q small enough so that $S^c(u', q), S^c(u'', q) \subseteq S(u, p)$. We set $A_{s\frown\langle 0\rangle} = S^c(u', q)$, $A_{s\frown\langle 1\rangle} = S^c(u'', q)$. It is now easy to see that (1)–(4) hold ((4) follows from $A_{s\frown\langle 0\rangle} \cap A_{s\frown\langle 1\rangle} = 0$), hence F is a continuous injection of ${}^{\omega}2$ into X. Since ${}^{\omega}2$ is a compact space this injection is, by VI.2.11(ii) and VI.1.25 a homeomorphism onto a subspace of X.

If $\langle X, d\rangle$ is separable and D is a fixed countable set dense in X then all the sets A_s can be chosen from the countable set $\{S^c(u, p) \mid u \in D \wedge p \in \mathbb{Q} \wedge p > 0\}$. Such a choice does not require the axiom of choice. □

2.15Ac Corollary. *Every perfect closed set in a complete metric space includes a Cantor set and is, therefore, of cardinality $\geqslant 2^{\aleph_0}$. The use of the axiom of choice is superfluous if the space is separable.*

Hint of proof. Use VI.2.31(iv), 2.9 and VI.2.25. □

From now on we shall discuss only Polish spaces, since these are the spaces we are interested in, even though some of what we shall say generalizes immediately to all complete metric spaces. Our first step will be to determine the cardinals of the open sets in the Polish spaces.

2.16 Corollary. *In a perfect Polish space $\langle X, d\rangle$ every non-void open set A includes a Cantor set and is, hence, of the cardinality $2^{\aleph_0}$.*

Proof. Since A is open and non-void A includes an open sphere $S(x,p)$. $S(x,\frac{p}{2})$, being an open set in a perfect space, is easily seen to be perfect. By 2.10 also $\mathrm{Cl}(S(x,\frac{p}{2}))$ is perfect, and by 2.15 $\mathrm{Cl}(S(x,\frac{p}{2}))$ includes a Cantor set C. We have $C\subseteq\mathrm{Cl}(S(x,\frac{p}{2}))\subseteq S^c(x,\frac{p}{2})\subseteq S(x,p)\subseteq A$; thus A includes a Cantor set. □

While all open sets in perfect separable metric spaces are of one of the cardinals 0 and $2^{\aleph_0}$ we have a wider choice of cardinals for the closed sets. In a Hausdorff space every finite set is closed. There are closed denumerable sets such as $\{0\}\cup\{\frac{1}{n}\mid n\in\omega\}$ in $\mathbb{R}$. A more general fact is the following.

2.17Ac Exercise. Prove that every metric space $\langle X, d\rangle$ with $|X|\geqslant\aleph_0$ includes a closed denumerable set.

Hint. Assume first that the space has a point x which is not isolated. □

As we have seen, closed sets in Polish spaces can have any cardinal $\leqslant\aleph_0$, as well as $2^{\aleph_0}$ (which is the cardinal of the space itself, if the space is perfect). If we assume the continuum hypothesis $2^{\aleph_0}=\aleph_1$ then, by what we saw, closed sets can have every cardinal $\leqslant 2^{\aleph_0}$. However, if we do not assume the continuum hypothesis it makes good sense to ask whether there are also closed sets C such that $\aleph_0<|C|<2^{\aleph_0}$. We shall prove a negative answer which will apply not only to closed sets, but also to all Borel sets and to the analytic sets as well.

2.18 Definition. Let A be a set in a topological space $\langle X, O\rangle$. A point $x\in X$ (not necessarily in A) is called a *condensation point* of A if for every neighbourhood W of x $W\cap A$ is uncountable.

2.19Ac Proposition. *Let $\langle X, O\rangle$ be a topological space with a countable base. Let A be an uncountable set in $\langle X, O\rangle$ and let D be the set of the condensation points of A. Then we have:*

(i) $D\subseteq\mathrm{Cl}(A)$.
(ii) *D is a closed set.*
(iii) *$A\sim D$ is countable (i.e., all the members of A, except countably many, are in D).*
(iv) *$D\cap A$ is a perfect set.*

Partial proof. (iii) Let B be a countable base for the topology. By the definition of D there is for every $x\in X\sim D$ a neighbourhood W_x such that $W_x\cap A$ is countable. Since B is a base we can choose W_x to be in B. Let $E=\bigcup_{x\in X\sim D}W_x\supseteq X\sim D$. Intersecting with A we get

(6) $$E\cap A=\bigcup_{x\in X\sim D}(W_x\cap A)\supseteq A\sim D.$$

The union in (6) is a countable union since the set B, which contains all the W_x's, is countable. Each term $W_x \cap A$ of the union is countable by our choice of W_x. Thus $E \cap A$ is seen in (6) to be a countable union of countable sets and hence $E \cap A$ is countable and, by (6), also $A \sim D$ is countable.

(iv) Let $y \in D$; we shall show that for every neighbourhood W of y $|W \cap (D \cap A)| > 1$. Since $y \in D$ $|W \cap A| > \aleph_0$, but since $W \cap A = (W \cap (A \sim D)) \cup (W \cap (D \cap A))$, and $|A \sim D| \leqslant \aleph_0$ by (iii), we get $|W \cap (D \cap A)| > \aleph_0$. □

2.20 Theorem (Cantor 1883, Bendixson 1883). *Let $\langle X, O \rangle$ be a topological space with a countable base. Every uncountable closed set A in $\langle X, O \rangle$ is the union of a perfect closed set and a countable set.*

Proof. We shall give here a proof which makes use of the axiom of choice by using 2.19; a proof which does not use the axiom of choice is outlined in the next exercise 2.21.

Let D be the set of all condensation points of A. By 2.19(i) $D \subseteq A$, since A is closed. By 2.19(ii) and 2.19(iv) D is a perfect closed set, and by 2.19(iii) $A \sim D$ is countable. $A = D \cup (A \sim D)$. □

2.21 Exercise. Prove the Cantor–Bendixson theorem 2.20 without using the axiom of choice.

Hint of proof. Define C_α for $\alpha < \omega_1$ by $C_0 = A$, $C_{\alpha+1} =$ the set of all non-isolated points of C_α, and for a limit ordinal λ $C_\lambda = \bigcap_{\alpha<\lambda} C_\alpha$. Using the countable base define a well-ordering of order type $\leqslant \omega$ of the set $\bigcup_{\alpha<\omega_1} C_\alpha \sim C_{\alpha+1}$. Use this to show that for some $\beta < \omega_1$ $C_{\beta+1} = C_\beta$. Show that C_β is a perfect closed set and $A \sim C_\beta$ is countable. □

2.22 The Cardinality of a Closed Set. It is now a trivial consequence of the Cantor–Bendixson theorem (2.20), together with 2.7 and 2.15, that in a Polish space every closed set is either countable or else is of the cardinality $2^{\aleph_0}$. In 2.21 we outlined how to prove this fact without using the axiom of choice, but the proof we gave in 2.20 uses this axiom. We chose to do it this way because the main lemma we used, 2.19, will also be useful later. From now on we shall use the axiom of choice in the proofs of most propositions and theorems in the present section. As a matter of fact, as far as the Baire space, the Cantor space and the Euclidean spaces are concerned the general result that every uncountable analytic set is of cardinality $2^{\aleph_0}$ can be proved without using the axiom of choice. This is usually proved from the point of view of the recursive hierarchy (in particular for the Baire space). A general proposition which will enable us to show that many sets include a Cantor set is the following.

2.23Ac Proposition. *For every continuous function G on a Polish space $\langle X, d \rangle$ into a Hausdorff space, if* Rng(G) *is uncountable then there is a Cantor set $W \subseteq X$ such that $G \upharpoonright W$ is a homeomorphism (of the appropriate subspaces) and $G[W]$ is a Cantor set too.*

Proof. By the axiom of choice there is a subset C of X such that $G[C]=\mathrm{Rng}(G)$ and $G \upharpoonright C$ is a one–one function (III.2.37). Since $\mathrm{Rng}(G)$ is uncountable C is uncountable too. By 2.19(iv) C has a perfect subset D. We shall use the operation (A) of Theorem 2.13 as follows. Let $x_0 \in D$ and set $A_0 = S^c(x_0, 1)$. We shall define, for $s \in {}^{\underline{\omega}}2$, A_s by recursion so that for a sequence s of length n A_s is a closed sphere $S^c(x, r)$ in $\langle X, d\rangle$ with $x \in D$ and $r \leqslant 2^{-n}$. Let $A_s = S^c(x, r)$ be as just mentioned. Since $x \in D$ and D is perfect $S(x, r) \cap D$ must contain at least two distinct points $p \neq q$. Since G is one–one on C $G(p) \neq G(q)$. Since $G(p)$ and $G(q)$ are distinct points in a Hausdorff space they have respective disjoint open neighbourhoods P and Q. By the continuity of G there is an $\varepsilon > 0$ such that $G[S^c(p, \varepsilon)] \subseteq P$ and $G[S^c(q, \varepsilon)] \subseteq Q$. Since $P \cap Q = 0$ also $S^c(p, \varepsilon) \cap S^c(q, \varepsilon) = 0$. We can, of course, require that $\varepsilon \leqslant 2^{-(n+1)}$; also, since $p, q \in S(x, r)$ we can always choose ε small enough so that $S^c(p, \varepsilon)$, $S^c(q, \varepsilon) \subseteq S(x, r)$. We set $A_{s^\frown\langle 0\rangle} = S^c(p, \varepsilon)$, $A_{s^\frown\langle 1\rangle} = S^c(q, \varepsilon)$. The function $\langle A_s \mid s \in {}^{\underline{\omega}}2\rangle$ thus defined satisfies conditions (1)–(4) of 2.13 and therefore there is a continuous injection F on ${}^\omega 2$ into X given by $\{F(t)\} = \bigcap_{n\in\omega} A_{t\upharpoonright n}$. Since ${}^\omega 2$ is a compact set the injection F is a homeomorphism (by VI.2.11(ii)); we denote its range $F[{}^\omega 2]$ by W.

Let us prove now that G is one–one on $W = F[{}^\omega 2]$. Let $r, t \in {}^\omega 2$, $r \neq t$, and let n be the least such that $r(n) \neq t(n)$. Without loss of generality we can assume that $r(n) = 0$ and $t(n) = 1$. Let us denote $r\upharpoonright n = t\upharpoonright n$ with s. By the definition of $A_{s^\frown\langle 0\rangle}$ and $A_{s^\frown\langle 1\rangle}$ we have $G[A_{s^\frown\langle 0\rangle}] \cap G[A_{s^\frown\langle 1\rangle}] = 0$ and since $\{F(r)\} = \bigcap_{n\in\omega} A_{r\upharpoonright n} \subseteq A_{s^\frown\langle 0\rangle}$ and $\{F(t)\} = \bigcap_{n\in\omega} A_{t\upharpoonright n} \subseteq A_{s^\frown\langle 1\rangle}$ we have $G(F(r)) \neq G(F(t))$. Thus $G\upharpoonright W$ is one–one. Since W is compact, being homeomorphic to ${}^\omega 2$, the injection $G\upharpoonright W$, which is continuous (by VI.1.27) is a homeomorphism (by VI.2.11(ii)). □

2.23 tells us that a continuous image of a Polish space in a Hausdorff space is either countable or else includes a Cantor set. Our immediate aim is to prove that every Borel set in a Polish space is such a continuous image, and we shall do it in two steps.

2.24 Theorem. *Every Polish space $\langle X, d\rangle$ is a continuous image of the Baire space ${}^\omega\omega$.*

Proof. We use Theorem 2.13 on the (A) operation, with $e = \omega$. We set $A_0 = X$ and we define the non-void closed sets A_s, for $s \in {}^{\underline{\omega}}\omega$, by recursion as follows. Let A_s be given, for a sequence s of length n. By VI.2.25 $\langle X, d\rangle$ has a countable base B. Let B' be the set of those members of B whose diameter is $\leqslant 2^{-(n+1)}$, then by VI.2.35(iii) B' is also a base and hence $\bigcup B' = X$. Since $A_s \neq 0$ the set $\{W \in B' \mid W \cap A_s \neq 0\}$ is a non-void countable set, let us write it as $\{W_k \mid k \in \omega\}$ where $\langle W_k \mid k \in \omega\rangle$ may contain repetitions. Clearly $\langle \mathrm{Cl}(W_k) \cap A_s \mid k \in \omega\rangle$ is a sequence of non-void closed sets. Since $\bigcup B' = X$ we have $\bigcup_{k\in\omega} \mathrm{Cl}(W_k) \supseteq \bigcup_{k\in\omega} W_k \supseteq A_s$ and hence $\bigcup_{k\in\omega} \mathrm{Cl}(W_k) \cap A_s = A_s$, i.e., the sequence $\langle \mathrm{Cl}(W_k) \cap A_s \mid k \in \omega\rangle$ covers A_s. We set now $A_{s^\frown\langle k\rangle} = \mathrm{Cl}(W_k) \cap A_s$. By VI.2.35(ii)

$$\begin{aligned}&\text{the diameter of } A_{s^\frown\langle k\rangle} \leqslant \text{the diameter of } \mathrm{Cl}(W_k)\\ &= \text{the diameter of } W_k \leqslant 2^{-(n+1)}.\end{aligned}$$

Thus requirements (1)–(3), (5) of Theorem 2.13 are satisfied and the function F on ${}^\omega\omega$ given by $\{F(t)\} = \bigcap_{n\in\omega} A_{t\upharpoonright n}$ is a continuous surjection of ${}^\omega\omega$ on X. □

2.25 Exercise. (i) Every compact Polish space is a continuous image of the Cantor space.

(ii) Let us call a topological space 0-*dimensional* if the clopen sets of the space are a base for its topology. The Cantor and Baire spaces are 0-dimensional.

(iii)Ac Every 0-dimensional compact Polish space is homeomorphic to the Cantor space.

Hints. (i) Proceed as in the proof of 2.24 but take $e=2$. Since the space is compact finitely many $\mathrm{Cl}(W_k)$'s cover A_s, say $\mathrm{Cl}(W_0),\ldots,\mathrm{Cl}(W_{l-1})$. For the least m such that $2^m\geqslant l$ take $\{A_{s\frown s'}\mid s'\in{}^m2\}=\{\mathrm{Cl}(W_i)\cap A_s\mid i<l\}$.

(iii) Let $\{B_i\mid i\in\omega\}$ be a base of clopen sets. Proceed as in (i) but see to it that for every $j\in\omega$ there be an $n\in\omega$ such that for every $s\in{}^n2$ and $i<j$ $A_s\subseteq B_i$ or $A_s\cap B_i=0$. □

2.26Ac Theorem. *In a Polish space $\langle X,d\rangle$ every non-void Borel set C is a continuous image of the Baire space.*

Proof. If the Borel set C is closed then, by VI.2.31(iv), the subspace $\langle C, d\restriction(C\times C)\rangle$ of $\langle X,d\rangle$ is a complete metric space with a countable base and hence, by 2.24, C is a continuous image of the Baire space.

In a metric space the open sets are countable intersections of the closed sets (VI.3.19(iii)) and all Borel sets are obtained from the open sets and the closed sets by repeated applications of countable union and intersection (VI.3.19), hence all Borel sets are obtained from the closed sets by repeated application of countable union and intersection. Therefore our theorem will follow once we prove that if the sets C_n, $n\in\omega$, are continuous images of the Baire space so are also $\bigcup_{n\in\omega}C_n$ and $\bigcap_{n\in\omega}C_n$ (provided that $\bigcap_{n\in\omega}C_n\neq 0$). Let $\langle f_n\mid n\in\omega\rangle$ be such that f_n is a continuous surjection of the Baire space ${}^\omega\omega$ on C_n. Let us denote with h the mapping of ${}^\omega\omega$ on ${}^\omega\omega$ given by $h(t)=\langle t_{i+1}\mid i\in\omega\rangle=\langle t_1,t_2,\ldots\rangle$ (i.e., $h(t)$ is t with its first term deleted). h is obviously continuous; and if we write $B_{\langle k\rangle}$ for the basic open set $\{t\in{}^\omega\omega\mid t_0=k\}$ we have, for every $k\in\omega$, $h[B_{\langle k\rangle}]={}^\omega\omega$. We define a function g on ${}^\omega\omega$ by $g(t)=f_{t_0}(h(t))$. Then

$$\begin{aligned}\mathrm{Rng}(g)&=\textstyle\bigcup_{n\in\omega}g[B_{\langle n\rangle}]=\bigcup_{n\in\omega}(f_nh)[B_{\langle n\rangle}]\\&=\textstyle\bigcup_{n\in\omega}f_n[h[B_{\langle n\rangle}]]=\bigcup_{n\in\omega}f_n[{}^\omega\omega]=\bigcup_{n\in\omega}C_n.\end{aligned}$$

On each set $B_{\langle n\rangle}$ the function $g\restriction B_{\langle n\rangle}=(f_nh)\restriction B_{\langle n\rangle}$ is continuous (by VI.1.28 and VI.1.27). Since each $B_{\langle n\rangle}$ is a clopen set also $g=\bigcup_{n\in\omega}g\restriction B_{\langle n\rangle}$ is continuous by VI.1.29.

To deal with $\bigcap_{n\in\omega}C_n$ let us assume $\bigcap_{n\in\omega}C_n\neq 0$ and define a mapping h of $\bigtimes_{n\in\omega}{}^\omega\omega$ onto $\bigtimes_{n\in\omega}C_n$ by $h(s)=\langle f_n(s_n)\mid n\in\omega\rangle$, where $s\in\bigtimes_{n\in\omega}{}^\omega\omega$ (thus each s_n is in ${}^\omega\omega$). It is easy to verify that since each f_n is a continuous function so is h. Let j be an homeomorphism of ${}^\omega\omega$ on $\bigtimes_{n\in\omega}{}^\omega\omega$ (such a j exists by VI.2.19(ii)), then hj is a continuous surjection of ${}^\omega\omega$ on $\bigtimes_{n\in\omega}C_n$. Let $W=\{r\in{}^\omega X\mid(\forall n\in\omega)(r_n=r_0)\}$. Since each factor X of ${}^\omega X$ is a Hausdorff space it is easy to show that W is a closed set. It is also obvious that $W \cap \times_{n\in\omega} C_n$ consists exactly of the sequences $\langle u \mid n \in \omega \rangle$ where $u \in \bigcap_{n\in\omega} C_n$. The projection function p_0 on ${}^\omega X$, given by $p_0(r) = r_0$, is clearly continuous. Since $W \cap \times_{n\in\omega} C_n = \{\langle u \mid n \in \omega \rangle \mid u \in \bigcap_{n\in\omega} C_n\}$ we have $p_0[W \cap \times_{n\in\omega} C_n] = \bigcap_{n\in\omega} C_n$. Since hj is continuous and W is closed also $(hj)^{-1}[W] = (hj)^{-1}[W \cap \times_{n\in\omega} C_n]$ is a closed subset of ${}^\omega\omega$ (VI.1.24) and $p_0 hj$ is a continuous function mapping $(hj)^{-1}[W]$ onto $\bigcap_{n\in\omega} C_n$. Since $\bigcap_{n\in\omega} C_n \neq 0$ also $(hj)^{-1}[W] \neq 0$. As a consequence $(hj)^{-1}[W]$ is a non-void closed subset of ${}^\omega\omega$ and, by the first part of the present proof, there is a continuous mapping g of ${}^\omega\omega$ onto $(hj)^{-1}[W]$. $p_0 hjg$ is thus a continuous function mapping ${}^\omega\omega$ onto $\bigcap_{n\in\omega} C_n$. □

2.27Ac Theorem (Alexandrov 1916, Hausdorff 1916). *In a Polish space every uncountable Borel set includes a Cantor set, and is hence, of cardinality* $2^{\aleph_0}$.

Proof. This follows immediately from 2.26 and 2.23. □

Looking back at Proposition 2.23 and asking which is the widest family of subsets of the Polish space $\langle X, d\rangle$ for which that proposition will yield a result like Theorem 2.27 we arrive at the following concept.

2.28 Definition (Souslin 1917). A subset A of a separable complete metric space $\langle X, d\rangle$ is called *analytic* if it is the null-set or a continuous image of the Baire space (or, for that matter, of any Polish space $\langle Y, d'\rangle$, since any such Y is a continuous image of ${}^\omega\omega$).

This definition of the concept of an analytic set turns out to be a good definition only if the axiom of choice is assumed. We shall now study briefly the analytic sets.

2.29Ac Corollaries (Souslin 1917). *In a Polish space we have*:

(i) *Every Borel set is analytic.*

(ii) *A continuous image of an analytic set is analytic.*

(iii) *The union and intersection of countably many analytic sets are analytic.*

(iv) *Every uncountable analytic set includes a Cantor set, and is therefore of cardinality* $2^{\aleph_0}$.

Hint of proof. (i) Use 2.26. (ii) Use VI.1.28. (iii) See the proof of 2.26. (iv) Use 2.23. □

2.30Ac Proposition. *Let $\langle X, d\rangle$ be a Polish space. For a subset A of X the following conditions are equivalent.*

(a) *A is analytic.*

(b) *A is the projection on X of a closed subset of ${}^\omega\omega \times X$.*

(c) *A is a continuous image of a Borel subset of X.*

(d) *A is the projection on X of a Borel subset of $X \times X$.*

(e) *A is the image of the Baire space under a continuous mapping into X obtained by the operation (A) of 2.12 from a family $\{A_s \mid s \in {}^{\underline{\omega}}\omega\}$ of closed sets which satisfies (1)–(3) of 2.13.*

Hint of proof. To show that each one of (b)–(e) implies that A is analytic all we need are VI.2.39(iii), VI.2.26 and 2.29(ii).

(a)$\to$(b) Let f be a continuous surjection of ${}^{\omega}\omega$ on A then the graph $\{\langle x, f(x)\rangle \mid x \in {}^{\omega}\omega\}$ of f is a closed subset of ${}^{\omega}\omega \times X$.

(a)$\to$(c) If A is countable this is trivial. If A is uncountable then by 2.27 X includes a Cantor set C. By removing $\aleph_0$ points from the closed set C one gets a set B homeomorphic to the Baire space (see 3.4 and 3.3 below) which is a G_δ (Π_2) set.

(a)$\to$(d) Use (a)$\to$(c) and the idea of the proof of (a)$\to$(b).

(a)$\to$(e) Let F be a continuous surjection of ${}^{\omega}\omega$ on A. For $s \in {}^{\underline{\omega}}\omega$ set $A_s = \mathrm{Cl}(F[B_s])$, where $B_s = \{t \in {}^{\omega}\omega \mid t \supseteq s\}$. Use the continuity of F and VI.2.35(ii) to prove that for $t \in {}^{\omega}\omega$ the diameter of $A_{t \upharpoonright n}$ converges to 0. □

2.31Ac Proposition. *Let $\langle X, d\rangle$ be a Polish space, let W be a σ-field of subsets of X which contains all closed sets and let I be a σ-ideal on X which is included in W. We assume:*

(7) *Every subset A of X has a* minimal cover *in W modulo I, i.e., there is a $B \in W$ such that $B \supseteq A$ and for every $C \in W$ such that $C \supseteq A$ C includes B modulo a set in I, i.e., $B \sim C \in I$.*

Then every analytic set is in W.

2.32Ac Corollary (Lusin and Sierpinski 1923). *In a Polish space every analytic set is a Baire set.*

Proof of 2.32. Apply 2.31 to the σ-field W of all Baire sets, (VI.3.9) and to the σ-ideal of all meager sets (VI.3.3). (7) holds in this case by VI.3.22. □

Proof of 2.31. By 2.30 an analytic set can be given as $F[{}^{\omega}\omega]$ where for $t \in {}^{\omega}\omega$ $\{F(t)\} = \bigcap_{n\in\omega} A_{t\upharpoonright n}$ and $\langle A_s \mid s \in {}^{\underline{\omega}}\omega\rangle$ is a family of closed sets which satisfies (1)–(3) of 2.13. For $s \in {}^{\underline{\omega}}\omega$ let $B_s = \{t \in {}^{\omega}\omega \mid t \supseteq s\}$ and let C_s be a minimal cover of $F[B_s]$. For $t \in B_s$ $\{F(t)\} = \bigcap_{n\in\omega} A_{t\upharpoonright n} \subseteq A_s$, hence $F[B_s] \subseteq A_s$ and since A_s is closed we can assume that $C_s \subseteq A_s$ (otherwise we can replace C_s by $C_s \cap A_s$ which is also a minimal cover of $F[B_s]$ in W). We have, for $t \in {}^{\omega}\omega$, $\{F(t)\} = \bigcap_{n\in\omega} A_{t\upharpoonright n} \supseteq \bigcap_{n\in\omega} C_{t\upharpoonright n} \supseteq \bigcap_{n\in\omega} F[B_{t\upharpoonright n}] \supseteq \{F(t)\}$ hence $F(t) = \bigcap_{n\in\omega} C_{t\upharpoonright n}$. We claim that

$$C_0 \sim F[{}^{\omega}\omega] \subseteq \bigcup\{C_s \sim \bigcup_{k\in\omega} C_{s\frown\langle k\rangle} \mid s \in {}^{\underline{\omega}}\omega\}. \tag{8}$$

If (8) holds then, since $\bigcup_{k\in\omega} C_{s\frown\langle k\rangle} \supseteq \bigcup_{k\in\omega} F[B_{s\frown\langle k\rangle}] = F[B_s]$, $\bigcup_{k\in\omega} C_{s\frown\langle k\rangle}$ is a cover of $F[B_s]$ in W while C_s is a minimal cover of $F[B_s]$ in W, hence $C_s \sim \bigcup_{k\in\omega} C_{s\frown\langle k\rangle} \in I$. Thus the right-hand side of (8) is the union of $\aleph_0$ sets in I hence the right-hand side and the left-hand side of (8) are both in I, hence $F[{}^{\omega}\omega] = C_0 \sim (C_0 \sim F[{}^{\omega}\omega]) \in W$, which is what we had to prove. To prove (8) let $y \in C_0$. If for every $s \in {}^{\underline{\omega}}\omega$ such that $y \in C_s$ there is a $k \in \omega$ such that $y \in C_{s\frown\langle k\rangle}$ then we define $t \in {}^{\omega}\omega$ by recursion so that $y \in C_{t\upharpoonright n}$ for every n as follows. We set $t(n)$ to be a

$k\in\omega$ such that $y\in C_{t\restriction n^\frown\langle k\rangle}$. Thus $y\in\bigcap_{n\in\omega}C_{t\restriction n}=\{F(t)\}\subseteq F[{}^{\omega}\omega]$. Therefore, if $y\in C_0\sim F[{}^{\omega}\omega]$ there is an $s\in{}^{\underset{\smile}{\omega}}\omega$ such that $y\in C_s$ and there is no $k\in\omega$ such that $y\in C_{s^\frown\langle k\rangle}$, i.e., y belongs to the right-hand side of (8). □

2.33Ac Theorem (Separation—Lusin 1927). *If P and Q are disjoint analytic sets in a Polish space $\langle X, d\rangle$ then there is a Borel set E which separates them in the sense that $P\subseteq E$ and $Q\cap E=0$ (i.e., $Q\subseteq X\sim E$).*

2.34Ac Corollary (Souslin 1917). *In a Polish space $\langle X, d\rangle$ if A is a set such that both A and its complement $X\sim A$ are analytic then A is a Borel set.* □

2.35Ac Lemma. *If $\langle P_i\mid i\in\omega\rangle$ and $\langle Q_j\mid j\in\omega\rangle$ are families of sets in a topological space $\langle X, O\rangle$ such that for all $i,j\in\omega$ P_i is separated from Q_j by a Borel set then $\bigcup_{i\in\omega}P_i$ is separated from $\bigcup_{j\in\omega}Q_j$ by a Borel set.*

Proof of 2.35. If $E_{i,j}$ is a Borel set such that $P_i\subseteq E_{i,j}$ and $E_{i,j}\cap Q_j=0$ then $\bigcup_{i\in\omega}P_i\subseteq\bigcup_{i\in\omega}\bigcap_{j\in\omega}E_{i,j}$ and $(\bigcup_{i\in\omega}\bigcap_{j\in\omega}E_{i,j})\cap\bigcup_{j\in\omega}Q_j=0$. Thus the Borel set $\bigcup_{i\in\omega}\bigcap_{j\in\omega}E_{i,j}$ separates $\bigcup_{i\in\omega}P_i$ from $\bigcup_{j\in\omega}Q_j$. □

Proof of 2.33. Let $P=F[{}^{\omega}\omega]$ and $Q=G[{}^{\omega}\omega]$, where F and G are continuous mappings of ${}^{\omega}\omega$ into X. For all $s\in{}^{\underset{\smile}{\omega}}\omega$ let $B_s=\{t\in{}^{\omega}\omega\mid t\supseteq s\}$, then $F[B_s]=\bigcup_{i\in\omega}F[B_{s^\frown\langle i\rangle}]$ and $G[B_s]=\bigcup_{j\in\omega}G[B_{s^\frown\langle j\rangle}]$. Therefore, by lemma 2.35, if for $s,s'\in{}^{\underset{\smile}{\omega}}\omega$ $F[B_s]$ and $G[B_{s'}]$ are not separated by any Borel set then for some $i,j\in\omega$ $F[B_{s^\frown\langle i\rangle}]$ and $G[B_{s'^\frown\langle j\rangle}]$ are not separated by any Borel set. Assume now, in order to get a contradiction that P and Q (i.e., $F[{}^{\omega}\omega]=F[B_0]$ and $G[{}^{\omega}\omega]=G[B_0]$) are not separated by any Borel set. We define by recursion $t,t'\in{}^{\omega}\omega$ such that $F[B_{t\restriction n}]$ and $G[B_{t'\restriction n}]$ are not separated by any Borel set. Since $F[B_{t\restriction n}]$ and $G[B_{t'\restriction n}]$ are not separated by any Borel set, we know, as we mentioned above, that there are $i,j\in\omega$ such that $F[B_{t\restriction n^\frown\langle i\rangle}]$ and $G[B_{t'\restriction n^\frown\langle j\rangle}]$ are not separated by any Borel set; take $t(n)$ and $t'(n)$ to be such i and j respectively, then $F[B_{t\restriction(n+1)}]$ and $G[B_{t'\restriction(n+1)}]$ are not separated by any Borel set. $F(t)\in F[{}^{\omega}\omega]=P$, $G(t')\in G[{}^{\omega}\omega]=Q$ and $P\cap Q=0$ hence $F(t)\neq G(t')$. Let D and E be open neighborhoods of $F(t)$ and $G(t')$, respectively, such that $D\cap E=0$. By the continuity of F and G, t and t' have respective neighborhoods D' and E' such that $F[D']\subseteq D$ and $G[E']\subseteq E$. Inside D' and E' we pick basic open sets $B_{t\restriction k}$ and $B_{t'\restriction l}$ which contain t and t' respectively. Let $n=\max(k,l)$ then $B_{t\restriction n}\subseteq B_{t\restriction k}\subseteq D'$, $B_{t'\restriction n}\subseteq B_{t'\restriction l}\subseteq E'$, hence $F[B_{t\restriction n}]\subseteq F[D']\subseteq D$, $G[B_{t'\restriction n}]\subseteq G(E')\subseteq E$ and since $E\cap D=0$ $G[B_{t'\restriction n}]\cap D=0$. Thus the open set D separates $F[B_{t\restriction n}]$ from $G[B_{t'\restriction n}]$, which is a contradiction since we proved that $F[B_{t\restriction n}]$ and $G[B_{t'\restriction n}]$ are not separated by any Borel set. □

2.36 Definition (Lusin 1925). Let $J\subseteq\mathbf{P}(X)$. A subset Y of ${}^{\omega}\omega\times X$ is said to be *universal* for the members of J if J equals the set of all fibers of Y "parallel to the X-axis", i.e., if $J=\{A\subseteq X\mid(\exists t\in{}^{\omega}\omega)((\{t\}\times X)\cap Y=\{t\}\times A)\}$.

2.37 Exercise. (i) If $\langle X, O\rangle$ is a topological space with a countable base then there is an open subset D of ${}^{\omega}\omega\times X$ which is universal for all open subsets of X and there is a closed subset E of ${}^{\omega}\omega \times X$ which is universal for all closed subsets of X.

(ii) If $\langle X, d\rangle$ is a Polish space then there is an analytic subset of ${}^{\omega}\omega \times X$ which is universal for all analytic subsets of X.

Hint. (i) If $B_0 = 0$ and $B_1, B_2, \ldots$ are the basic open sets of $\langle X, O\rangle$ take $D = \bigcup\{\{t\} \times \bigcup_{n \in \omega} B_{t(n)} \mid t \in {}^{\omega}\omega\}$ and $E = {}^{\omega}\omega \times X \sim D$.

(ii) Let D be a closed subset of ${}^{\omega}\omega \times ({}^{\omega}\omega \times X)$ which is universal for all closed subsets of ${}^{\omega}\omega \times X$. Let F be the projection of ${}^{\omega}\omega \times ({}^{\omega}\omega \times X)$ onto ${}^{\omega}\omega \times X$ given by $F(\langle x, \langle y, z\rangle\rangle) = \langle x, z\rangle$, then $F[D]$ is an analytic subset of ${}^{\omega}\omega \times X$ which, by 2.30, is universal for all analytic subsets of X. □

2.38 Exercise (Souslin 1917). There is an analytic subset A of ${}^{\omega}\omega$ whose complement is not analytic, and hence A is not a Borel set.

Hint. Let D be an analytic subset of ${}^{\omega}\omega \times {}^{\omega}\omega$ which is universal for the analytic subsets of ${}^{\omega}\omega$. $\{t \in {}^{\omega}\omega \mid \langle t, t\rangle \in D\}$ is analytic, but its complement is not, as follows from a diagonal argument. □

2.39Ac Exercise. If $\langle X, O\rangle$ is a topological space with a countable base and with $|X| \geqslant 2^{\aleph_0}$ then X has an uncountable subset Y which includes no closed set of cardinality $2^{\aleph_0}$, and hence if $\langle X, O\rangle$ is a Hausdorff space Y includes no Cantor set, and if $\langle X, d\rangle$ is a complete metric space Y includes no perfect closed set. Notice that Y is not analytic.

Hint. Prove that there are at most $2^{\aleph_0}$ open sets and at most $2^{\aleph_0}$ closed sets. Let $\langle C_\alpha \mid \alpha < 2^{\aleph_0}\rangle$ be a sequence of all closed sets of cardinality $\geqslant 2^{\aleph_0}$. Define by recursion $x_\alpha, y_\alpha \in C_\alpha \sim \{x_\beta, y_\beta \mid \beta < \alpha\}$, $x_\alpha \neq y_\alpha$, then take $Y = \{y_\alpha \mid \alpha < 2^{\aleph_0}\}$. □

3. The Close Relationship Between the Real Numbers, the Cantor Space and the Baire Space

We saw in § 2 that all Polish spaces have interesting common set theoretical and topological properties. The spaces we are really interested in are the real numbers, the Baire space and the Cantor space; and we shall now set out to establish stronger relationships between them which are nearly homeomorphisms.

3.1 How to map ${}^{\omega}\omega$ into $\mathbb{I}$. Let us try to construct a mapping j of the Baire space ${}^{\omega}\omega$ into the unit real interval $\mathbb{I}$ such that j will nearly be a homeomorphism. One way to try to do it is as follows. Given $t \in {}^{\omega}\omega$ we first look at t_0. If t_0 is 0 we put $j(t)$ in the interval $[0, \frac{1}{2}]$, if t_0 is 1 we put $j(t)$ in the interval $[\frac{1}{2}, \frac{3}{4}]$, if $t_0 = 2$ we put $j(t)$ in the interval $[\frac{3}{4}, \frac{7}{8}]$ and in general if $t_0 = n$ we put $j(t)$ in the interval $[1-2^{-n}, 1-2^{-(n+1)}]$. Having looked at t_0 we look now also at t_1 and continue specifying the place of $j(t)$ by continuing essentially in the same way. Thus, for example, if $t_0 = 1$ and $t_1 = 0$

we put $j(t)$ in the first half of $[\frac{1}{2}, \frac{3}{4}]$, i.e., $j(t) \in [\frac{1}{2}, \frac{5}{8}]$, if $t_0 = 1$ and $t_1 = 1$ then $j(t) \in [\frac{5}{8}, \frac{11}{16}]$, and so on (see Figure 4). We can carry out this idea fully, by what essentially amounts to an application

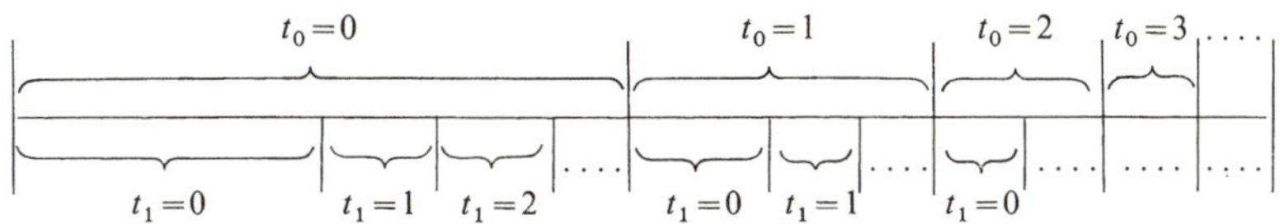

Fig. 4. The location of $j(t)$ according to the value of t, for the temporary j.

of the operation (A) of 2.12 (the formal details of such an application will be described below). For example, we get $j(\langle 1, 0, 0, 0, \ldots \rangle) = \frac{1}{2}$. The function j which we obtain this way is continuous but is not an homeomorphism on its range since the set $B_{\langle 0 \rangle} = \{t \in {}^{\omega}\omega \mid t_0 = 0\}$ is a closed (and open) set in the Baire space whereas $j[B_{\langle 0 \rangle}]$ is not a closed set in Rng(j) for the following reason, $j[B_{\langle 0 \rangle}]$ has members arbitrarily close (on the left) to $\frac{1}{2}$ but $\frac{1}{2}$ itself is $j(\langle 1, 0, 0, 0, \ldots \rangle)$ and is not in $j[B_{\langle 0 \rangle}]$. Analyzing what happens we see that the left endpoints of the intervals of $\mathbb{I}$ we consider can be approached from the left in $\mathbb{I}$ while their sources in ${}^{\omega}\omega$ (such as $\langle 1, 0, 0, \ldots \rangle$) cannot be approached from the "left" in ${}^{\omega}\omega$. This prevents j from being an homeomorphism. This can be cured by changing j so that those endpoints are dropped from the range of j. To achieve this we alternatingly reverse the direction of choosing the different intervals according to the value of t_n. I.e., while proceeding with t_0 as above, when we come to t_1 we

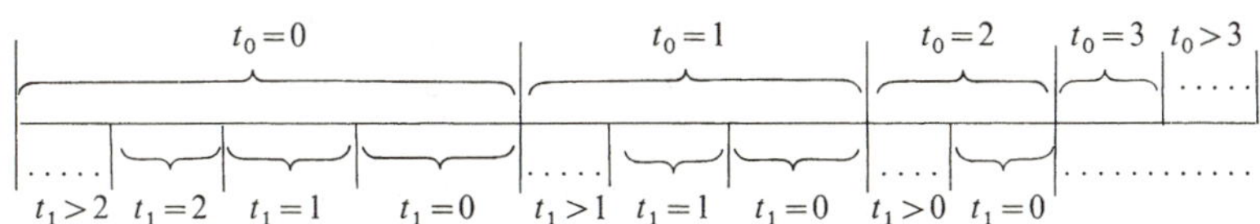

Fig. 5. The location of $j(t)$ according to the value of t for the final j.

start choosing intervals from right to left. Thus if $t_0 = 1$ and $t_1 = 0$ we shall put $j(t)$ in the *right* half of $[\frac{1}{2}, \frac{3}{4}]$, i.e., in $[\frac{5}{8}, \frac{3}{4}]$; if $t_0 = 1$ and $t_1 = 1$ we put $j(t)$ in $[\frac{9}{16}, \frac{5}{8}]$, and so on as in Fig. 5.

We shall now proceed to define j formally, but since the binary notation for the reals clearly affords the most convenient way of handling j we shall define j as a mapping from ${}^{\omega}\omega$ into ${}^{\omega}2$, combining it later with the binary representation of the real numbers to obtain a mapping of ${}^{\omega}\omega$ into $\mathbb{I}$.

3.2 Definition (The standard injection j of ${}^{\omega}\omega$ into ${}^{\omega}2$). For $n, k \in \omega$ let a_n^k be an $n+1$-tuple of n 1's and a single 0 if k is even and an $n+1$-tuple of n 0's and a single

1 if k is odd, i.e.,

$$a_n^k = \begin{cases} \langle \overbrace{1, 1, \ldots, 1, 1}^{n}, 0\rangle & \text{if } k \text{ is even,} \\ \langle \overbrace{0, 0, \ldots, 0, 0}^{n}, 1\rangle & \text{if } k \text{ is odd.} \end{cases}$$

If $w = \langle w_i \mid i < \omega\rangle$ is a sequence of finite sequences then we denote by Concat(w) the concatenation of w, i.e., the sequence obtained by putting the w_i's one after the other. Formally, $v = \text{Concat}(w)$ iff $\text{Length}(v) = \omega$ and $v_m = (w_k)_l$, where $m = (\Sigma_{i=o}^{k-1} \text{Length}(w_i)) + l$ and $l \in \text{Length}(w_k)$. Finally we define, for $t \in {}^\omega\omega$, $j(t) = \text{Concat}(\langle a_{t(k)}^k \mid k \in \omega\rangle)$. j is called the *standard mapping* of ${}^\omega\omega$ into ${}^\omega 2$.

3.3 Definition. We call a sequence t of length ω *stagnant* if all its terms are the same from some place onwards, i.e., if there is an $m \in \omega$ such that $t_n = t_m$ for all $n > m$. There are obviously exactly $\aleph_0$ stagnant sequences in ${}^\omega 2$.

3.4 Proposition. *j is a homeomorphism of the Baire space on the subspace of all non-stagnant members of the Cantor space.*

Proof. The proof is left to the reader, except that we shall show how one finds $t = j^{-1}(s)$ for a non-stagnant $s \in {}^\omega 2$. One looks for the first 0 in s. Let $n \geqslant 0$ be the number of 1's which precede this 0; we set $t_0 = n$. We look now for the first 1 to the right of this 0, and let m be the number of 0's which precede it and come after the 0 which we dealt with before then $t_1 = m$, and so on. For example $j^{-1}(\langle 0, 0, 0, 1, 0, 1, 1, 1, 0, 0, 1, 1, 0, \ldots\rangle) = \langle 0, 2, 0, 0, 2, 1, 1, \ldots\rangle$. Since s is non-stagnant we shall always go on finding such 0's and 1's. □

3.5 Theorem (Binary expansion of the real numbers). *For every $t \in {}^\omega 2$ the sum $\Sigma_{n \in \omega} t_n 2^{-(n+1)}$ converges and $0 \leqslant \Sigma_{n<\omega} t_n 2^{-(n+1)} \leqslant 1$. For every real $0 \leqslant x \leqslant 1$ there is a $t \in {}^\omega 2$ such that $x = \Sigma_{n<\omega} t_n 2^{-(n+1)}$. For $s, t \in {}^\omega 2$ $\Sigma_{n<\omega} s_n 2^{-(n+1)} < \Sigma_{n<\omega} t_n 2^{-(n+1)}$ iff for some $m < \omega$ we have $s_n = t_n$ for all $n < m$, $s_m = 0$, $t_m = 1$ and for some $n > m$ either $s_n = 0$ or $t_n = 1$. For $s, t \in {}^\omega 2$ $\Sigma_{n<\omega} s_n 2^{-(n+1)} = \Sigma_{n<\omega} t_n 2^{-(n+1)}$ iff $s = t$ or else for some $m \in \omega$ $s_n = t_n$ for all $n < m$, $s_m \neq t_m$ (i.e., one of them is 0 and the other one is 1) and for all $n > m$ $s_n = t_m$ and $t_n = s_m$. Thus for a given $0 \leqslant x \leqslant 1$ the $t \in {}^\omega 2$ such that $x = \Sigma_{n<\omega} t_n 2^{-(n+1)}$ is unique except for the case where $x > 0$ and x is a finite binary fraction, i.e., where for some finite sequence $\langle r_0, \ldots, r_{m-1}\rangle$ in ${}^{\omega}2$ $x = \Sigma_{n<m} r_n 2^{-(n+1)}$. In this case, since $x > 0$, we can choose m so as to obtain $r_{m-1} = 1$ and then $x = \Sigma_{n<\omega} t_n 2^{-(n+1)}$ for exactly the following two sequences t, $\langle r_0, r_1, \ldots, r_{m-2}, 1, 0, 0, \ldots\rangle$ and $\langle r_0 . r_1, \ldots, r_{m-2}, 0, 1, 1, \ldots\rangle$. There are, obviously, only $\aleph_0$ finite binary fractions.* □

3.6 Definition. Let k be the function on ${}^\omega 2$ given by $k(t) = \Sigma_{n<\omega} t_n 2^{-(n+1)}$. k is called the *standard surjection* of ${}^\omega 2$ on $\mathbb{I}$.

3.7 Proposition. *k is a continuous surjection of ${}^{\omega}2$ on $\mathbb{I}$. If W is the set of non-stagnant members of ${}^{\omega}2$ then $k \upharpoonright W$ is a homeomorphism of W onto the set of all reals in $\mathbb{I}$ which are not finite binary fractions. As a consequence (of 3.4) kj is a homeomorphism of ${}^{\omega}\omega$ on the set of all real numbers in $\mathbb{I}$ which are not finite binary fractions.*

Proof. To prove the continuity let $k(t)=x$ and let U be an open neighborhood of x, then there is an $m \in \omega$ such that $[x-2^{-m}, x+2^{-m}] \subseteq U$. For a finite sequence r of 0's and 1's let $B_r = \{s \in {}^{\omega}2 \mid s \supseteq r\}$. We shall prove that $k[B_{t \upharpoonright m}] \subseteq [x-2^{-m}, x+2^{-m}] \subseteq U$, thereby establishing that k is continuous at t. We have

(1) $$x = \Sigma_{n<\omega} t_n 2^{-(n+1)} = \Sigma_{n<m} t_n 2^{-(n+1)} + \Sigma_{m \leqslant n<\omega} t_n 2^{-(n+1)}.$$

Let us write

(2) $$y = \Sigma_{n<m} t_n 2^{-(n+1)}$$

and then, by (1), $x \geqslant y$ and also

$$x = y + \Sigma_{m \leqslant n<\omega} t_n 2^{-(n+1)} \leqslant y + \Sigma_{m \leqslant n<\omega} 2^{-(n+1)} = y + 2^{-m},$$

hence $y \leqslant x \leqslant y+2^{-m}$. If s is any member of $B_{t \upharpoonright m}$ then $s_n = t_n$ for $n<m$ and hence we have, in the same way as above, $y \leqslant k(s) \leqslant y+2^{-m}$ and therefore $|k(s)-x| \leqslant 2^{-m}$ and $k(s) \in [x-2^{-m}, x+2^{-m}]$.

Let W be the set of non-stagnant members of ${}^{\omega}2$. To prove that $k \upharpoonright W$ is a homeomorphism we must prove that $k^{-1} \upharpoonright k[W]$ is continuous. Let $x=k(t)$, $t \in W$, and let U be an open neighborhood of t, then for some $m<\omega$ $B_{t \upharpoonright m} \subseteq U$. Since $t \in W$ t is not stagnant, hence the sequence $t_m, t_{m+1}, \ldots$ contains both 0's and 1's, and there is a $p>m+1$ such that $t_p \neq t_m$. As can be easily seen, we have, for y as in (2), $y+2^{-(p+1)} \leqslant x \leqslant (y+2^{-m})-2^{-(p+1)}$. Therefore, for every $z \in (x-2^{-(p+1)}, x+2^{-(p+1)})$ we have $y<z<y+2^{-m}$, and hence $(y, y+2^{-m})$ is a neighborhood of x. By 3.5, since $y=k(\langle t_0, \ldots, t_{m-1}, 0, 0, \ldots \rangle)$ and $y+2^{-m}=k(\langle t_0, \ldots, t_{m-1}, 1, 1, 1, \ldots \rangle)$ we get for every $z \in (y, y+2^{-m})$ $k^{-1}(z) \in B_{t \upharpoonright m} \subseteq U$, thus $k^{-1}[(y, y+2^{-m})] \subseteq B_{t \upharpoonright m}$, which proves the continuity of k^{-1} on $k[W]$. □

3.8 The Standard Mappings. By 3.4 and 3.7 we have standard homeomorphisms of the Baire space on a subspace of the Cantor space which contains all members of the Cantor space except denumerably many, and of that subset of the Cantor space on a subset of $\mathbb{I}$ which contains all members of $\mathbb{I}$ except denumerably many. Since in each case the exceptional set is denumerable and hence Σ_2 (i.e., F_σ) and meager these homeomorphisms are a very good tool in carrying over results from one of the spaces $\mathbb{I}$, $\mathbb{R}$, ${}^{\omega}\omega$, ${}^{\omega}2$ to the other ones. We shall demonstrate several examples along this line, but before doing that let us notice one difference between the way we defined the standard mapping of the Baire space into $\mathbb{I}$ and the way this is usually done in the literature. Here the Baire space is mapped by kj on the set of all members of $\mathbb{I}$ which are not finite binary fractions, as described towards the end of 3.1. In the literature one usually maps the Baire space into $\mathbb{I}$ by the mapping

$$f(t)=\cfrac{1}{t_0+\cfrac{1}{t_1+\cdots}}$$

defined on the space ${}^{\omega}(\omega\sim\{0\})$, and this mapping is a homeomorphism on the set of all irrational numbers in $\mathbb{I}$. We choose to use the mapping kj we gave above for two reasons. First, the way we chose gives a clearer idea of how the mapping works and why such a particular mapping was chosen. Second, our treatment enables us to avoid continued fractions, a subject which is not related to what we do here.

As we saw in VI.3.20 if f is a continuous function from X to Y and A is a Σ_α (or Π_α) subset of Y then $f^{-1}[A]$ is a Σ_α (or Π_α, respectively) subset of X. Also, by 2.29(ii), if C is an analytic subset of X then $f[C]$ is an analytic subset of Y. Combining these with the specific properties of our mappings we get:

3.9 Proposition. *Let A, B, C be subsets of the Baire space, the Cantor space and the unit real interval $\mathbb{I}$, respectively. The following holds, for the standard mappings j of ${}^{\omega}\omega$ into ${}^{\omega}2$ and k of ${}^{\omega}2$ onto $\mathbb{I}$.*

(i) *B is a Σ_α-set, or a Π_α-set, for $\alpha\geqslant 3$, iff $j^{-1}[B]$ is such. A is a Π_2-set iff $j[A]$ is such.*

(ii) *B is an analytic set iff $j^{-1}[B]$ is such.*

(iii) *B is a Π_α-set, or a Σ_α-set, for $\alpha\geqslant 3$, iff $k[B]$ is such. If B is a closed set or a Σ_2-set then also $k[B]$ is such. If $k[B]$ is a Π_2-set then so is also B.*

(iv) *B is an analytic set iff $k[B]$ is such.*

(v) *C is a Π_α-set, or a Σ_α-set, for $\alpha\geqslant 3$, iff $(kj)^{-1}[C]$ is such. A is a Π_2-set iff $kj[A]$ is such.*

(vi) *C is an analytic set iff $(kj)^{-1}[C]$ is such.*

Remark. We mentioned here only the strongest facts we know, other than mere direct consequences of the continuity of j and k. For example, in (i) we did not mention that A is a Σ_α-set, or a Π_α-set, for $\alpha\geqslant 3$, iff $j[A]$ is such, since this follows directly from what is written in (i) by substituting $j[A]$ for B.

Hint of proof. Use the following facts: Every countable set is a Σ_2-set, and is, a fortiori, analytic. A closed set in a compact space is compact and so is its continuous image. The union and intersection of two analytic sets are analytic (by 2.29, but in the case of a union or intersection of two sets we do not need the axiom of choice). In a space $\langle X, O\rangle$ if D is a Π_2-set and E is a Π_2-set of the subspace $\langle D, O'\rangle$ of $\langle X, O\rangle$ then E is a Π_2-set also of $\langle X, O\rangle$. □

3.10 Proposition. *Let A, B, C be subsets of the Baire space, the Cantor space and the closed unit interval $\mathbb{I}$, respectively.*

(i) *$j[A]$ is nowhere dense iff A is.*

(ii) *$k[B]$ is nowhere dense iff B is.*

(iii) *$j^{-1}[B]$ is meager iff B is.*

(iv) *$k[B]$ is meager iff B is.*
(v) *$j^{-1}[B]$ is a Baire set iff B is.*
(vi) *$k[B]$ is a Baire set iff B is.*

Hint of proof. (v) Use 3.9(i) and VI.3.17(iii). □

3.11 The Relationship Between $\mathbb{R}$ and $\mathbb{I}$. As mentioned in 1.7 $\mathbb{R}$ is order-isomorphic to the open unit interval (0, 1) and thus $\mathbb{R}$ differs topologically from $\mathbb{I}$ only in that $\mathbb{I}$ has two endpoints. Therefore if f is any order-isomorphism of $\mathbb{R}$ on (0, 1) we can easily prove the following statements. For every subset C of $\mathbb{I}$, if $\alpha \geqslant 1$ then $C \in \Sigma_\alpha$ iff $f^{-1}[C] \in \Sigma_\alpha$ and for $\alpha > 1$ $C \in \Pi_\alpha$ iff $f^{-1}[C] \in \Pi_\alpha$. Each of the following properties is possessed by C iff it is possessed by $f^{-1}[C]$: being analytic, being nowhere dense, being meager and being a Baire set. □

Another important feature of the spaces which we deal with is the measure on these spaces.

3.12 Definition (measure). (i) Let X be a set and F a σ-field of subsets of X (VI.3.14). We shall refer to the members of F as *measurable sets.* μ is said to be a *measure* on F (or, if F is understood, on X) if μ is a function on F into the set which consists of all non-negative real numbers and of a single additional member which we denote by ∞ (and which we take to be greater than every real number), and μ satisfies:

(3) $\mu(0)=0$, and

(4) for every sequence $\langle A_n \mid n \in \omega \rangle$ of pairwise disjoint members of F we have $\mu(\bigcup_{n\in\omega} A_n) = \Sigma_{n\in\omega}\mu(A_n)$, where $\Sigma_{n\in\omega}\mu(A_n)$ is ∞ if for some n $\mu(A_n)=\infty$ or if $\Sigma_{n\in\omega}\mu(A_n)$ is a diverging series (see IX.4.1).

(ii) A measure μ on X is said to be a *probability measure* if $\mu(X)=1$. μ is said to be a *complete measure* if for every set A such that $\mu(A)=0$ every subset B of A is measurable.

3.13 Proposition (i) *If $A_0, \ldots, A_{n-1}$ are pairwise disjoint measurable sets then $\mu(\bigcup_{i<n} A_i) = \Sigma_{i<n}\mu(A_i)$.*

(ii) *If A, B are measurable sets and $B \subseteq A$ then $\mu(B) \leqslant \mu(A)$ and if $\mu(A) < \infty$ then $\mu(A \sim B) = \mu(A) - \mu(B)$.*

(iii) *If $\langle A_i \mid i \in \omega \rangle$ is a sequence of measurable sets, not necessarily pairwise disjoint, then $\mu(\bigcup_{i\in\omega} A_i) \leqslant \Sigma_{i\in\omega}\mu(A_i)$* and *$\mu(\bigcup_{i<n} A_i) \leqslant \Sigma_{i<n}\mu(A_i)$ for $n<\omega$.*

Hint of proof. (i) Choose in (4) $A_i = 0$ for $i \geqslant n$. (ii) Apply (i) to B and $A \sim B$. (iii) Consider the sequence $\langle A_i \sim \bigcup_{j<i} A_j \mid i \in \omega \rangle$ of pairwise disjoint sets and use $\mu(A_i \sim \bigcup_{j<i} A_j) \leqslant \mu(A_i)$ (by (ii)). □

3.14Ac Proposition. *Let F be a σ-field of subsets of X and let μ be a measure on F. Let G be the set of subsets of X given by $G = \{A \cup N \mid A \in F \wedge (\exists Z \in F)(\mu(Z)=0$*

$\wedge N \subseteq Z)\}$, *then G is a σ-field of subsets of X which includes F. Let*

$$\bar{\mu} = \{\langle A \cup N, s\rangle \mid A \in F \wedge \mu(A) = s \wedge (\exists Z \in F)(\mu(Z) = 0 \wedge N \subseteq Z)\},$$

then $\bar{\mu}$ is the unique measure on G which extends μ. $\bar{\mu}$ is a complete measure and it is called the completion *of μ.*

Proof. See, e.g., Halmos 1950, Theorem 13B. □

3.15Ac The Lebesgue Measure on ℝ and 𝕀. There is a unique measure μ defined on all Borel subsets of ℝ such that

(5) for every open interval (a, b), with $a < b$, $\mu((a, b)) = b - a$.

μ, restricted to the Borel subsets of 𝕀, is also the unique measure such that (5) holds for open intervals included in the unit interval. All this is a well-known theorem of measure theory (see, e.g., Halmos 1950, Theorem 13A). The completion of this measure is also denoted with μ and is called the *Lebesgue measure*. The subsets A of ℝ (and 𝕀) for which the Lebesgue measure $\mu(A)$ is defined are called *Lebesgue measurable sets* or, in brief, *measurable sets*. It is easily seen that if we start with the completion of the measure μ on the Borel subsets of ℝ and restrict it to 𝕀 we obtain the measure which is the completion of the measure on the Borel subsets of 𝕀. The Lebesgue measure on 𝕀 is a complete probability measure.

3.16Ac The Natural Measure on the Cantor Space $^{\omega}2$. This measure is obtained as follows. We consider a sequence of 0's and 1's as the outcome of ω tosses of a coin where 0 denotes heads and 1 tails. For a subset A of $^{\omega}2$ $\mu(A)$ is the probability that ω tosses of an honest coin will yield a sequence which belongs to A. μ is formally obtained as follows. For a finite sequence s of 0's and 1's let $B_s = \{t \in {}^{\omega}2 \mid t \supseteq s\}$. By a theorem of measure theory (Halmos 1950, Theorem 38B) there is a unique measure μ defined on all Borel subsets of the Cantor space such that for $s \in {}^{n}2$ $\mu(B_s) = 2^{-n}$. This is in accordance with the fact that the probability of getting the sequence $s \in {}^{n}2$ in n tosses is 2^{-n}. We denote the completion of μ also by μ and call it *the measure* on the Cantor space, it is clearly a complete probability measure. The subsets of $^{\omega}2$ on which this measure is defined will be called *measurable* subsets of $^{\omega}2$.

3.17Ac The Measure on the Baire Space $^{\omega}\omega$. Let us define first a probability measure π on ω. A natural measure is given by $\pi(\{i\}) = 2^{-(i+1)}$ and, as a consequence, for every set $E \subseteq \omega$ $\pi(E) = \Sigma_{i \in E} 2^{-(i+1)}$. The product measure μ on the set of all Borel subsets of $^{\omega}\omega$ is the unique measure μ such that for every $s \in {}^{n}\omega$ $\mu(B_s) = \Pi_{i<n} 2^{-(s(i)+1)}$, where $B_s = \{t \in {}^{\omega}\omega \mid t \supseteq s\}$ (Halmos 1950, Theorem 38B). We denote also the completion of μ by μ and call it *the measure* on the Baire space; it is clearly a complete probability measure. The subsets of $^{\omega}\omega$ for which this measure is defined will be called the *measurable* subsets of $^{\omega}\omega$.

3.18Ac Proposition. *Let j be the standard injection of $^{\omega}\omega$ into $^{\omega}2$.*

(i) *For every subset A of $^{\omega}\omega$, A is measurable iff $j[A]$ is, and if both are measurable then $\mu(j[A])=\mu(A)$.*

(ii) *For every subset B of $^{\omega}2$, B is measurable iff $j^{-1}[B]$ is, and if both are measurable then $\mu(j^{-1}[B])=\mu(B)$.*

Proof. (i) By 3.9(i) $j[A]$ is a Borel set iff A is such. We shall now prove that $\mu(j[A])=\mu(A)$ for every Borel set A and we shall first deal with sets A of the form $B_s=\{t\in{}^{\omega}\omega \mid t\supseteq s\}$, where $s\in{}^{n}\omega$, $n\in\omega$. By our definition 3.2 of the injection j if $t\in B_s$ then $j(t)$ is a sequence which starts with the concatenation of the finite sequences $a^{o}_{s(0)},\ldots,a^{n-1}_{s(n-1)}$. Let us denote the concatenation of these finite sequences by r then $j[B_s]\subseteq B_r$, where $B_r=\{t\in{}^{\omega}2 \mid t\supseteq r\}$. It is easily seen that every non-stagnant member of B_r is an image under j of some member of B_s. Since there are only $\aleph_0$ stagnant members of $^{\omega}2$ we have $|B_r\sim j[B_s]|\leqslant\aleph_0$. A subset of $^{\omega}2$ consisting of a single point t has measure 0 (since $\{t\}\subseteq B_{t\restriction n}$ and hence $\mu(\{t\})\leqslant\mu(B_{t\restriction n})=2^{-n}$ for every n) and therefore, by 3.12(4), every countable subset of $^{\omega}2$ has measure 0. Thus by $j[B_s]=B_r\sim(B_r\sim j[B_s])$ we have, by 3.13(ii),

$$\mu(j[B_s])=\mu(B_r)-\mu(B_r\sim j[B_s])=\mu(B_r)-0=\mu(B_r). \tag{9}$$

Let m be the length of r. Since r is the concatenation of $a^{o}_{s(0)},\ldots,a^{n-1}_{s(n-1)}$ and the length of $a^{i}_{s(i)}$ is $s(i)+1$ we have $m=\Sigma_{i<n}(s(i)+1)$. By the definition of the measure on $^{\omega}2$ we have $\mu(B_r)=2^{-m}$. By the definition of the measure on $^{\omega}\omega$ we have $\mu(B_s)=\Pi_{i<n}2^{-(s(i)+1)}=2^{-\Sigma_{i<n}(s(i)+1)}=2^{-m}$. Thus we have, by (9), $\mu(j[B_s])=\mu(B_r)=2^{-m}=\mu(B_s)$, and this establishes the equality $\mu(j[A])=\mu(A)$ for every subset A of $^{\omega}\omega$ of the form B_s.

Let $\bar{\mu}$ be the function defined on all Borel subsets of $^{\omega}\omega$ by $\bar{\mu}(A)=\mu(j[A])$. $\bar{\mu}$ is easily seen to be a measure. For every sequence $s\in{}^{n}\omega$ we have $\bar{\mu}(B_s)=\mu(j[B_s])=\mu(B_s)=\Pi_{i<n}2^{-(s(i)+1)}$, by what we showed above concerning the sets B_s. In 3.17 we mentioned that μ is the unique measure on the σ-field of all Borel subsets of $^{\omega}\omega$ such that for $s\in{}^{n}\omega$ $\mu(B_s)=\Pi_{i<n}2^{-(s(i)+1)}$, hence for every Borel subset A of $^{\omega}\omega$ we have $\bar{\mu}(A)=\mu(A)$, i.e., $\mu(j[A])=\mu(A)$.

We still have to prove that for every subset A of $^{\omega}\omega$ A is measurable iff $j[A]$ is, and if both are measurable then $\mu(j[A])=\mu(A)$. Let A be a measurable subset of $^{\omega}\omega$. Then there are Borel subsets C, Z of $^{\omega}\omega$ and a subset N of Z such that $A=C\cup N$, $N\subseteq Z$ and $\mu(Z)=0$. Since C and Z are Borel sets also $j[C]$ and $j[Z]$ are such and we have, by what we have already shown, $\mu(j[C])=\mu(C)$, $\mu(j[Z])=\mu(Z)=0$. Since $j[A]=j[C]\cup j[N]$ where $j[C]$ is a Borel set and $j[N]$ is a subset of the Borel set $j[Z]$ of measure 0, $j[A]$ is measurable and $\mu(j[A])=\mu(j[C])=\mu(C)=\mu(A)$.

The only thing left to prove now is that if for a subset A of $^{\omega}\omega$ $j[A]$ is measurable then A is measurable too. Let $j[A]$ be measurable then there are Borel sets P, Q and a set $T\subseteq Q$ such that $j[A]=P\cup T$ and $\mu(Q)=0$. Since j is an injection we have $A=j^{-1}[P]\cup j^{-1}[T]$ and $j^{-1}[T]\subseteq j^{-1}[Q]$. Since P and Q are Borel sets so are, by 3.9(i), $j^{-1}[P]$ and $j^{-1}[Q]$. By what we have already shown, $\mu(j^{-1}[Q])=\mu(j[j^{-1}[Q]])\leqslant\mu(Q)=0$. Thus $A=j^{-1}[P]\cup j^{-1}[T]$ is the union of Borel set

$j^{-1}[P]$ and a subset $j^{-1}[T]$ of a Borel set $j^{-1}[Q]$ of measure 0 and hence A is measurable.

(ii) This is an easy consequence of (i) and the fact that ${}^{\omega}2 \sim j[{}^{\omega}\omega]$ is a denumerable set and hence $\mu({}^{\omega}2 \sim j[{}^{\omega}\omega]) = 0$. □

3.19Ac Proposition. *Let k be the standard surjection of ${}^{\omega}2$ on $\mathbb{I}$.*

(i) *For every subset B of ${}^{\omega}2$, B is measurable iff $k[B]$ is. If both are measurable then $\mu(k[B]) = \mu(B)$.*

(ii *For every subset C of $\mathbb{I}$, C is measurable iff $k^{-1}[C]$ is. If both are measurable then $\mu(k^{-1}[C]) = \mu(C)$.*

Proof. (i) By 3.9(iii) B is a Borel set iff $k[B]$ is. Let us prove now that $\mu(k[B]) = \mu(B)$ for every Borel set B, and first we shall do it for the sets B_s, where $s \in {}^{n}2$, $n \in \omega$. By the definition of μ on ${}^{\omega}2$ $\mu(B_s) = 2^{-n}$. By the binary expansion theorem 3.5 $k[B_s] = [\Sigma_{i<n} s(i) \cdot 2^{-(i+1)}, \Sigma_{i<n} s(i) \cdot 2^{-(i+1)} + 2^{-n}]$ and thus $\mu(k[B_s]) = 2^{-n} = \mu(B_s)$. Having proved $\mu(k[B]) = \mu(B)$ for the sets B_s let us turn to all Borel sets B. Let $\bar{\mu}$ be the function defined on all Borel subsets B of ${}^{\omega}2$ by $\bar{\mu}(B) = \mu(k[B])$. We shall first see that $\bar{\mu}$ is a measure. The values of $\bar{\mu}$ are real numbers $\leqslant 1$. $\bar{\mu}(0) = \mu(k[0]) = \mu(0) = 0$. Let $\langle D_i \mid i \in \omega \rangle$ be a sequence of pairwise disjoint Borel subsets of ${}^{\omega}2$. Let Z be the denumerable set of all finite binary fractions in $[0, 1]$, then $\langle k[D_i] \sim Z \mid i \in \omega \rangle$ is a sequence of pairwise disjoint Borel subsets of $\mathbb{I}$. Since $\mu(Z) = 0$ we have

$$\bar{\mu}(\bigcup_{i \in \omega} D_i) = \mu(k[\bigcup_{i \in \omega} D_i]) = \mu(k[\bigcup_{i \in \omega} D_i] \sim Z)$$
$$= \mu(\bigcup_{i \in \omega}(k[D_i] \sim Z)) = \Sigma_{i \in \omega} \mu(k[D_i] \sim Z) = \Sigma_{i \in \omega} \mu(k[D_i]) = \Sigma_{i \in \omega} \bar{\mu}(D_i).$$

Thus we have seen that $\bar{\mu}$ is a measure on the Borel subsets of ${}^{\omega}2$. By what we have seen above, for every $s \in {}^{n}2$ $\bar{\mu}(B_s) = \mu(k[B_s]) = \mu(B_s) = 2^{-n}$. In 3.16 we mentioned that the measure μ on the Borel subsets of ${}^{\omega}2$ is the unique measure μ for which $\mu(B_s) = 2^{-n}$ for every $s \in {}^{n}2$, hence $\bar{\mu}(B) = \mu(B)$ for every Borel subset B of ${}^{\omega}2$.

The reader who went through the proof of 3.18 will have no trouble completing the proof of (i) and proving (ii). □

3.20 Definition. Let l be the bijection of the real line $\mathbb{R}$ on the open unit interval $(0, 1)$ which, for $n \geqslant 1$, linearly maps the closed interval $[n-1, n]$ on the closed interval $[1-2^{-n}, 1-2^{-(n+1)}]$ and the closed interval $[-n, -n+1]$ on the closed interval $[2^{-(n+1)}, 2^{-n}]$, i.e., l is defined on $\mathbb{R}$ by

$$l(x) = \begin{cases} 1-2^{-(n+1)}(n+1-x) & \text{for} \quad n-1 \leqslant x \leqslant n, \ n \geqslant 1 \\ 2^{-(n+1)}(x+n+1) & \text{for} \quad -n \leqslant x \leqslant -n+1, \ n \geqslant 1. \end{cases}$$

3.21Ac Relating the Borel Measures on $\mathbb{R}$ and $\mathbb{I}$. As mentioned in 3.11, for every subset D of $\mathbb{I}$ D is a Borel subset of $\mathbb{I}$ iff $l^{-1}[D]$ is a Borel subset of $\mathbb{R}$. Since for $n \geqslant 1$ the derivative of l on $[n-1, n]$ is $2^{-(n+1)}$ we have, for every Borel subset C of $[n-1, n]$, $\mu(l[C]) = 2^{-(n+1)}\mu(C)$, as follows from an easy and well-known theorem on the Lebesgue measure. Similarly, if C is a Borel subset of $[-n, -n+1]$ then

$\mu(l[C])=2^{-(n+1)}\mu(C)$. As a consequence, if C is a Borel set of $\mathbb{R}$ then

$$\mu(l[C])=\Sigma_{n=1}^{\infty}2^{-(n+1)}\mu(C\cap[n-1,n])+\Sigma_{n=1}^{\infty}2^{-(n+1)}\mu(C\cap[-n,-n+1]).$$

(The endpoints of the intervals are taken into account twice in the right-hand side but this does not matter since the measure of a countable set is 0). Thus if $\mu(C)=0$ then also $\mu(l[C])=0$. The same proof works also in the other direction, namely, if $\mu(l[C])=0$ also $\mu(C)=0$.

3.22Ac Proposition. *For every subset D of $\mathbb{I}$, D is Lebesgue measurable iff $l^{-1}[D]$ is, and $\mu(D)=0$ iff $\mu(l^{-1}[D])=0$.*

Hint of proof. Use 3.21 and the definition in 3.14 of the completion of a measure. □

3.23 On Measure-Related Properties of the Real Spaces. As a consequence of 3.18, 3.19 and 3.21 we can easily transfer results about measure from any of the spaces ${}^{\omega}\omega$, ${}^{\omega}2$ and $\mathbb{I}$ to any other one of these spaces. Using also 3.9 and 3.10 we can transfer also results which connect topological and measure theoretic properties of these spaces. As a consequence of 3.22 and 3.11 the same applies to the space $\mathbb{R}$ too, except that the value of the measure μ, or even its finiteness, is not carried over.

3.24Ac Proposition (Lusin 1917). *Every analytic set of reals, as well as every analytic subset of ${}^{\omega}2$ and ${}^{\omega}\omega$, is measurable.*

Proof. It is a well-known theorem of measure theory that every set A of reals has a measurable cover B which is a G_δ (i.e., Π_2) set, where by "B is a measurable cover of A" we mean that $B\supseteq A$, B is measurable, and for every $C\supseteq A$ which is measurable C includes B up to a set of measure zero (i.e., $\mu(B\sim C)=0$). Therefore, by 2.31 every analytic subset of $\mathbb{R}$ (and $\mathbb{I}$) is measurable. If D is an analytic subset of ${}^{\omega}2$ or ${}^{\omega}\omega$ then by 2.29 its image under the appropriate standard mapping into $\mathbb{I}$ is also analytic and hence measurable. Therefore by 3.18(i) and 3.19(i), also D itself is measurable. □

Once we know that every subset of ${}^{\omega}2$ and ${}^{\omega}\omega$ has a measurable cover, which is easy to prove, we can use 2.31 to prove this proposition directly for ${}^{\omega}2$ and ${}^{\omega}\omega$. However, the use of 3.18 and 3.19 enabled us to prove this proposition for ${}^{\omega}2$ and ${}^{\omega}\omega$ without any detailed knowledge of the properties of the measure on those spaces.

Chapter VIII

Boolean Algebras

Boolean algebra is not a part of set theory proper, but it is an off and on companion to set theory and it comes close to set theory at many points. Originally Boolean algebra was a precursor of set theory as the discoveries of Boole preceded those of Cantor. The part of Boolean algebra developed by Boole provided only trivial information about sets, yet it was a first step toward the mathematical investigation of sets. At this point the ways of set theory and Boolean algebra parted to a considerable extent. Set theory studied properties of sets which cannot be naturally defined in terms of the Boolean structure; Boolean algebra turned mostly to the study of Boolean algebras in which the interesting operations cannot be conveniently described as natural operations on sets. Boolean algebra used quite a bit of set theory, but in this it differed possibly in degree, but not in principle, from other branches of mathematics. The first non-trivial feedback from Boolean algebra to set theory came about when it turned out that the Boolean prime ideal theorem (2.12) is not just another consequence of the axiom of choice but is a very interesting weaker version of the axiom of choice, as we shall see in § 2. Later, methods related to the prime ideal theorem and to the Boolean representation theory played an important role in the study of large cardinal numbers such as the weakly compact cardinals and the measurable cardinals discussed in the next chapter. Finally, complete Boolean algebras turned out to play a major role in Cohen's concept of forcing. Here the contact was two-way in a strong sense; while the method of forcing can be viewed as an application of Boolean algebra to set theory, this method, which originated in set theory, led to a deeper understanding of several aspects of Boolean algebra. Most of § 4 is devoted to the study of Martin's hypothesis and related concepts, all of which came about as a result of this new close contact between set theory and Boolean algebra brought about by Cohen's forcing.

This chapter is by no means a complete or an unbiased survey of Boolean algebra. Its contents were chosen mostly on the ground of how they relate to set theory.

1. The Basic Theory

In this section we shall see how we arrive at Boolean algebra once we study certain aspects of sets. We shall present the basic concepts of Boolean algebra, including the axioms and their immediate consequences. Then we shall define the infinitary operations Σ and Π and establish their basic properties. Finally we shall say a few words about the concept of a subalgebra.

1.1 Fields of Sets. Let us recall, from VI.3.14, that a field of subsets of a non-void set X is a set $F \subseteq \mathbf{P}(X)$ such that $0 \in F$, if $A \in F$ then also $X \sim A \in F$, and if $A, B \in F$ then also $A \cup B \in F$ (and, as a consequence, also $A \cap B \in F$). We have already encountered several examples of fields of sets such as the set of all Borel sets and the set of all Baire sets in a topological space (VI.3.17 and VI.3.10) and the set of all measurable sets in a measure space (VII.3.12). These fields of sets are even closed under countable union (and intersection) and thus they are σ-fields of sets. Additional examples of fields of subsets of a non-void set X are the sets $\{A \subseteq X \mid |A| < \aleph_\alpha$ or $|X \sim A| < \aleph_\alpha\}$ (for $\alpha = 0$ this is the set of all finite and cofinite subsets of X), the set of all clopen (closed and open) sets of a topological space $\langle X, O\rangle$ and the set $F_\sigma \cap G_\delta (= \Pi_2 \cap \Sigma_2)$ of a topological space. Two trivial and extreme examples of fields of subsets of X are $\mathbf{P}(X)$ and $\{0, X\}$.

We shall view a field F of subsets of X as a structure which consists of the set F, the three operations $\langle A \cup B \mid \langle A, B\rangle \in F \times F\rangle$, $\langle A \cap B \mid \langle A, B\rangle \in F \times F\rangle$, $\langle X \sim A \mid A \in F\rangle$ and the members 0 and X of F. We shall use for these operations and members the symbols $+$, $\cdot$, $^-$, 0 and 1, respectively; thus we shall also write $A+B$ for $A \cup B$, $A \cdot B$ for $A \cap B$, $\bar{A}$ for $X \sim A$, and 1 for X (while we have in mind a particular X).

1.2 Boolean Algebras. In the present context we mean by an *equation* any expression of the form $\Gamma = \Delta$ where Γ and Δ are expressions built up from variables and 0, 1 by the use of the operation symbols $+$, $\cdot$ and $^-$ (and parentheses). We shall say that an equation $\Gamma = \Delta$ *holds* in a structure $\langle B, +, \cdot, ^-, 0, 1\rangle$, where B is a set, $+$ and $\cdot$ are binary operations, $^-$ is a unary operation, and 0, 1 are members of B, if for every assignment of values in B to the variables in Γ and Δ Γ and Δ get always equal values. For example, the equation $x+y=y+x$ is said to hold in $\langle B, +, \cdot, ^-, 0, 1\rangle$ if for all $a, b \in B$ $a+b=b+a$, i.e., if $+$ is a commutative operation.

We generalize now the concrete concept of a field of sets to the abstract concept of a Boolean algebra. We shall describe now what we intend by a Boolean algebra; the formal definition of a Boolean algebra is given in 1.3 below. For our present purposes a *structure* is a set B with binary operations $+$ and $\cdot$ on B, a unary operation $^-$ on B and distinct members 0, 1 of B (i.e., an ordered sextuple $\langle B, +, \cdot, ^-, 0, 1\rangle$). We refer to $+$ and $\cdot$ as *addition* and *multiplication*, respectively, and to $^-$ as *complementation*, and thus $\bar{x}$ is said to be the *complement* of x. A Boolean algebra is a structure $\langle B, +, \cdot, ^-, 0, 1\rangle$ such that every equation $\Gamma = \Delta$ which holds in every field of sets holds also in $\langle B, +, \cdot, ^-, 0, 1\rangle$. Thus, trivially, every field of sets is a Boolean algebra. This definition of a Boolean algebra, which is a good one from a conceptual point of view, is rather inconvenient from a practical point of view. Luckily, there are several short lists of equations which hold in all fields of sets such that if $\langle B, +, \cdot, ^-, 0, 1\rangle$ is any structure in which every equation of such a list holds then all the equations which hold in every field of sets hold in that structure. This enables us to take up any one of those lists to serve as a simple definition of the concept of a Boolean algebra; and this is indeed what we do in Definition 1.3 where the equations of such a list are taken to be the axioms of Boolean algebra. What we just claimed about such a list is proved in 2.14 below.

1.3 Definition. A *Boolean algebra* is a set B with binary operations $+$ and $\cdot$ on B, a unary operation $^-$ on B, and distinct members 0, 1 of B such that for all $a, b, c \in B$,

1a) $a+a=a$	1b) $a\cdot a=a$
2a) $a+b=b+a$	2b) $a\cdot b=b\cdot a$
3a) $(a+b)+c=a+(b+c)$	3b) $(a\cdot b)\cdot c=a\cdot(b\cdot c)$
4a) $a\cdot(b+c)=a\cdot b+a\cdot c$	4b) $a+b\cdot c=(a+b)\cdot(a+c)$
5a) $a+\bar{a}=1$	5b) $a\cdot\bar{a}=0$
6a) $a+0=a$	6b) $a\cdot 1=a$.

Requirements 1a–6b are called the axioms of Boolean algebra. As these requirements are obviously satisfied by every field of sets, every field of sets is a Boolean algebra.

Most of the time we shall refer to the Boolean algebra $\langle B, +, \cdot, ^-, 0, 1\rangle$ as the Boolean algebra B, without mentioning explicitly the operations and distinguished members, which we shall have in mind implicitly.

1.4 Proposition. *In every Boolean algebra B we have, for all $a \in B$, $a+1=1$ and $a\cdot 0=0$.*

Proof. Read each of the following two columns downward.

$a+1=a+(a+\bar{a})$	by 5a,	$a\cdot 0=a\cdot(a\cdot\bar{a})$	by 5b,
$=(a+a)+\bar{a}$	by 3a,	$=(a\cdot a)\cdot\bar{a}$	by 3b,
$=a+\bar{a}$	by 1a,	$=a\cdot\bar{a}$	by 1b,
$=1$	by 5a.	$=0$	by 5b. □

1.5 Duality. Looking at the axioms of Boolean algebra in 1.3 we see that if in any of the axioms we replace each one of the symbols $+$ and $\cdot$ by the other one and each one of the symbols 0 and 1 by the other one we obtain the other axiom on the same line. Let Φ be a statement about members of a Boolean algebra which uses only the symbols $+, \cdot, ^-, 0$ and 1 in addition to variables and equality. The statement Ψ obtained from Φ by replacing $+, \cdot, 0$ and 1 by $\cdot, +, 1$ and 0, respectively, is called the *dual* of Φ. For example, the dual of $a+1=1$ is $a\cdot 0=0$. Evidently, any two axioms on the same line are duals of each other. It is also obvious that for every statement Φ as above the dual of the dual of Φ is Φ itself. If Φ is a theorem then its dual Ψ is a theorem too since we can prove Ψ by going through the proof of Φ and replacing each statement which occurs in it by its dual. What we obtain this way is indeed a proof since the duals of the axioms are axioms. This is illustrated in our proof of 1.4 where two dual statements are proved in two columns and the proof of $a \cdot 0 = 0$ is indeed obtained from the proof of $a + 1 = 1$ by replacing each equation in the proof of $a + 1 = 1$ by its dual. From now on, once we have proved a statement Φ we shall assert its dual Ψ without a proof. For a concept of Boolean algebra defined by means of $+, \cdot, ^-, 0$ and 1 there is a *dual concept* defined by the same definition with $+, \cdot, 0$ and 1 replaced by $\cdot$, $+$, 1 and 0, respectively. If a statement Φ contains in addition to $+, \cdot, ^-, 0$ and 1 also concepts defined by means of these concepts then the dual of Φ is taken to be the statement Ψ obtained from Φ by replacing $+, \cdot, 0$ and 1 by $\cdot$, $+$, 1 and 0, respectively, and by replacing all the defined concepts by their duals. We shall soon see examples of such dual defined concepts. Also for this wider notion of a dual it is still the case that if Φ is a theorem and Ψ is the dual of Φ then Ψ is a theorem too. We do not prove this fact here since any such proof would require some concepts of logic or model theory. No proof of the general principle of duality is really needed for our purposes since the reader can easily convince himself of the truth of what we claim for every theorem Φ given below by actually writing down a proof of its dual Ψ by making the appropriate replacements in the proof of Φ.

1.6 Proposition. *In a Boolean algebra we have,*

(i) $a + a \cdot b = a$,

(ii) $a \cdot (a + b) = a$,

(iii) $a + b = b \leftrightarrow a \cdot b = a$.

Proof. (i) $a + a \cdot b = a \cdot 1 + a \cdot b = a \cdot (1 + b) = a \cdot 1 = a$, by the axioms and 1.4. (ii) is the dual of (i). (iii) If $a + b = b$ then $a \cdot b = a \cdot (a + b) = a \cdot a + a \cdot b = a + a \cdot b = a$, by the axioms, 1.4 and (i). If $a \cdot b = a$ then $a + b = a \cdot b + b = b + b \cdot a = b$, by (i). □

1.7 Definition (partial order—Boole 1847). In a field of sets we can define the inclusion relation $\subseteq$ from the union operation by defining $a \subseteq b$ to hold iff $a \cup b = b$, and from the intersection operation by setting $a \subseteq b$ iff $a \cap b = a$. We use this to define on every Boolean algebra B a relation $\leqslant$ which is like $\subseteq$, and which is obviously identical with $\subseteq$ if B is a field of sets, as follows. For $a, b \in B$ we define $a \leqslant b$ if $a + b = b$. As usual we define $a \geqslant b$ if $b \leqslant a$, $a < b$ if $a \leqslant b$ and $a \neq b$, $a > b$ if $b < a$. By 1.6(iii) we have

(1) $$a \leqslant b \leftrightarrow a + b = b \leftrightarrow a \cdot b = a.$$

The dual of $a \leqslant b$ is obtained by taking the dual of its defining statement $a + b = b$, which is $a \cdot b = b$. By (1) it is equivalent to $a \geqslant b$. Hence the dual of $a < b$ is $a > b$.

1.8 Proposition. (i) *In a Boolean algebra B $\leqslant$ is a partial order in the $\leqslant$ sense, i.e., for all $a, b, c \in B$ $a \leqslant a$, $a \leqslant b \wedge b \leqslant a \rightarrow a = b$,* and *$a \leqslant b \wedge b \leqslant c \rightarrow a \leqslant c$.*

For all $a, b, c \in B$, where B is a Boolean algebra, we have:

(ii) $0 \leqslant a \leqslant 1$.

(iii) $a \cdot b \leqslant a, b, \quad a+b \geqslant a, b$.

(iv) *If* $c \leqslant a$ *and* $c \leqslant b$ *then* $c \leqslant a \cdot b$; *thus, by* (iii), $a \cdot b$ *is the greatest lower bound of* a *and* b. *Dually, if* $c \geqslant a$ *and* $c \geqslant b$ *then* $c \geqslant a+b$ *and thus* $a+b$ *is the least upper bound of* a *and* b.

(v) *If* $a \leqslant b$ *then* $a \cdot c \leqslant b \cdot c$ *and* $a+c \leqslant b+c$.

(vi) $a+b=1$ *iff* $a \geqslant \bar{b}$; $a \cdot b=0$ *iff* $a \leqslant \bar{b}$.

Partial proof. (i) We shall prove only $a \leqslant b \wedge b \leqslant c \rightarrow a \leqslant c$ since the rest is trivial. Assuming $a \leqslant b$ and $b \leqslant c$ we have $a \cdot b=a$ and $b \cdot c=b$. Therefore $a \cdot c=(a \cdot b) \cdot c= a \cdot (b \cdot c)=a \cdot b=a$, thus $a \leqslant c$.

(iv) If $c \leqslant a$ and $c \leqslant b$ then $c \cdot a=c$ and $c \cdot b=c$ hence $c \cdot (a \cdot b)=(c \cdot a) \cdot b= c \cdot b=c$, i.e., $c \leqslant a \cdot b$.

(v) If $a \leqslant b$ then $a \cdot b=a$ hence $(a \cdot b) \cdot c=a \cdot c$. By the associativity and commutativity of multiplication and by what we already know we have $(a \cdot c) \cdot (b \cdot c)= (a \cdot b) \cdot (c \cdot c)=a \cdot c$, i.e., $a \cdot c \leqslant b \cdot c$.

(vi) Assume $1=a+b$, then $\bar{b}=1 \cdot \bar{b}=(a+b) \cdot \bar{b}=a \cdot \bar{b}+b \cdot \bar{b}=a \cdot \bar{b}+0=a \cdot \bar{b}$, i.e., $\bar{b} \leqslant a$. □

1.9 Corollaries. *In a Boolean algebra we have, for all* a *and* b:

(i) $a+b=1$ and $a \cdot b=0$ iff $b=\bar{a}$.

(ii) $\bar{0}=1, \ \bar{1}=0$.

(iii) $\bar{\bar{a}}=a$.

(iv) $\overline{a+b}=\bar{a} \cdot \bar{b}, \ \overline{a \cdot b}=\bar{a}+\bar{b}$.

(v) $a \leqslant b$ iff $\bar{b} \leqslant \bar{a}$. □

1.10 Definition (Schröder 1891). A member a of a Boolean algebra is called an *atom* if $a>0$ and there is no $b<a$ except for 0. In general we can say that if $0<b<a$ then b is a *proper part* of a, so an atom is a non-zero member which has no proper parts (that is what a physical atom was thought to be when it was given this name). A Boolean algebra B is said to be *atomic* if every non-zero member of B is $\geqslant$ some atom. B is said to be *atomless* if it contains no atoms.

1.11 Examples and Exercises. (i) In a field of sets every singleton is an atom and every finite set includes an atom.

(ii) In the field of all subsets of a set the singletons are the only atoms. Such a field of sets is an atomic Boolean algebra.

(iii) Provide an example of a field of set with an atom which is not a singleton.

(iv) If in a Boolean algebra B $b \neq 0$ is such that $\{c \in B \mid c \leqslant b\}$ is finite then there is an atom $a \leqslant b$. In particular, every finite Boolean algebra is atomic.

(v) A topological space is called *totally disconnected* if its topology has a base which consists entirely of clopen sets. Examples of such spaces are the discrete spaces, ${}^{\omega}2$ and ${}^{\omega}\omega$. The field of all clopen sets of a totally disconnected perfect Hausdorff space is atomless.

(vi) Give an example of a Boolean algebra which is neither atomic nor atomless.

(vii) Prove that in a Boolean algebra B a is an atom iff for all $b, c \in B$ if $a \leqslant b+c$ then $a \leqslant b$ or $a \leqslant c$. □

1.12 Definition. On every Boolean algebra B we define the following operations.

(i) The *difference* of a and b: $a-b=a \cdot \bar{b}$.

(ii) The *symmetric difference* of a and b: $a \vartriangle b=(a-b)+(b-a)$.

1.13 Proposition. *In a Boolean algebra B the following statements hold for all $a, b, c, a_1, b_1, a_2, b_2 \in B$.*

(i) $a-b=0$ *iff* $a \leqslant b$.

(ii) $a-c \leqslant (a-b)+(b-c)$.

(iii) $(a+b)-c=(a-c)+(b-c)$.

(iv) $a \vartriangle b=0$ *iff* $a=b$.

(v) $a \vartriangle b+b \vartriangle c \geqslant a \vartriangle c$.

(vi) $\bar{a} \vartriangle \bar{b}=a \vartriangle b$.

(vii) $(a_1+b_1) \vartriangle (a_2+b_2) \leqslant (a_1 \vartriangle a_2)+(b_1 \vartriangle b_2)$.

(viii) $a_1 \cdot b_1 \vartriangle a_2 \cdot b_2 \leqslant (a_1 \vartriangle a_2)+(b_1 \vartriangle b_2)$. □

1.14 Definition (Peirce 1880). Let $\langle B, < \rangle$ be a poset and let $A \subseteq B$. A member d of B is said to be an *upper bound* of A if for every $a \in A$ $d \geqslant a$. d is a *least upper bound* of A if d is an upper bound of A and for every upper bound e of A $d \leqslant e$. If both d and e are least upper bounds of A then we have $d \leqslant e$ and $e \leqslant d$ and hence $d=e$. Thus A has at most one least upper bound and we can speak of *the* least upper bound of A, if it exists, and we denote it by $\Sigma^B A$ or by ΣA. When we say "ΣA exists" or "ΣA is defined" we mean that the subset A of B has a least upper bound. Now let us define the concepts dual to those already defined. d is said to be a *lower bound* of A if $d \leqslant a$ for every $a \in A$. d is the *greatest lower bound* of A if d is a lower bound of A and for every lower bound e of A $d \geqslant e$. The greatest lower bound of A is denoted by $\Pi^B A$ or by ΠA, if it exists.

1.15 Proposition. *In a poset $\langle B, < \rangle$ if $A, C \subseteq B$ and ΣA, ΣC, $\Sigma\{\Sigma A, \Sigma C\}$ exist then also $\Sigma(A \cup C)$ exists and $\Sigma(A \cup C)=\Sigma\{\Sigma A, \Sigma C\}$. The same holds also for Π instead of Σ.* □

1.16 Σ and Π in Boolean Algebras. In a general poset even a set of two members does not necessarily have a least upper bound or a greatest lower bound; for example, in the poset $\langle \{p, q\}, 0 \rangle$, where $p \neq q$, $\Sigma\{p, q\}$ and $\Pi\{p, q\}$ do not exist. In a Boolean algebra, however, by 1.8(iii) and 1.8(iv) $\Sigma\{a, b\}$ and $\Pi\{a, b\}$ always exist and we have $\Sigma\{a, b\}=a+b$ and $\Pi\{a, b\}=a \cdot b$. Thus, for Boolean algebras the operations Σ and Π can be viewed as extending the operations $+$ and $\cdot$ to arbitrary subsets of the Boolean algebra and we shall often refer to ΣA and ΠA as the *sum* and *product*, respectively, of A. In the case of a Boolean algebra we can formulate 1.15 as asserting that if ΣA and ΣC exist then $\Sigma(A \cup C)=\Sigma A+\Sigma C$, and if

ΠA and ΠC exist then $\Pi(A\cup C)=\Pi A\cdot\Pi C$. If C consists of a single member c then trivially $\Sigma C=\Pi C=c$ and we get that

(2) if ΣA exists then $\Sigma(A\cup\{c\})=(\Sigma A)+c$, and

(3) if ΠA exists then $\Pi(A\cup\{c\})=(\Pi A)\cdot c$.

Since, as easily seen, $\Sigma 0=0$ and $\Pi 0=1$ (where 0 in the left-hand sides is the null-set), we can use (2) and (3) to prove by induction on finite sets (III.1.6) that in a Boolean algebra ΣA and ΠA exist for every finite set A. As follows easily from (2) and (3), if $A=\{a_0,\ldots,a_{n-1}\}$ then $\Sigma A=a_0+a_1+\cdots+a_{n-1}$ and $\Pi A=a_0\cdot a_1\cdot\cdots\cdot a_{n-1}$ (because of the associativity of $+$ and $\cdot$ it does not matter where we put the parentheses in the right-hand sides). We shall see below that for infinite subsets A of a Boolean algebra ΣA and ΠA do not always exist.

1.17 Proposition. *For every subset A of a Boolean algebra B we have*

$$\overline{\Sigma A}=\Pi\{\bar{a}\mid a\in A\} \quad\text{and}\quad \overline{\Pi A}=\Sigma\{\bar{a}\mid a\in A\}$$

in the sense that if one side of an equation is defined so is the other and they are equal.

Proof. Assume that ΣA is defined and denote it with d. Then $d\geqslant a$ for every $a\in A$. By 1.9(v) $\bar{d}\leqslant\bar{a}$ for every $a\in A$ hence $\bar{d}$ is a lower bound of $\{\bar{a}\mid a\in A\}$. If e is a lower bound of $\{\bar{a}\mid a\in A\}$ then, by 1.9(v) and 1.9(iii), $\bar{e}\geqslant a$ for every $a\in A$, and hence $\bar{e}\geqslant\Sigma A=d$. Again by 1.9(v) and 1.9(iii) $e\leqslant\bar{d}$ and therefore $\bar{d}$ is the greatest lower bound of $\{\bar{a}\mid a\in A\}$. A dual proof shows that if ΠA is defined then $\overline{\Pi A}$ is the least upper bound of $\{\bar{a}\mid a\in A\}$.

We have already shown that if ΠA is defined then $\Sigma\{\bar{a}\mid a\in A\}$ is defined and $\Sigma\{\bar{a}\mid a\in A\}=\overline{\Pi A}$. Substituting $\{\bar{a}\mid a\in A\}$ for A we get that if $\Pi\{\bar{a}\mid a\in A\}$ is defined then $\Sigma\{\bar{b}\mid b\in\{\bar{a}\mid a\in A\}\}=\overline{\Sigma\{\bar{\bar{a}}\mid a\in A\}}=\overline{\Sigma A}$ is defined and $\overline{\Sigma A}=\Pi\{\bar{a}\mid a\in A\}$. Again a dual proof shows that if $\Sigma\{\bar{a}\mid a\in A\}$ is defined then ΠA is defined and $\overline{\Pi A}=\Sigma\{\bar{a}\mid a\in A\}$. □

1.18 Proposition (A distributive law). *If A and C are subsets of a Boolean algebra B and ΣA, ΣC exist then $\Sigma\{a\cdot c\mid a\in A\wedge c\in C\}=\Sigma A\cdot\Sigma C$.*

Proof. By 1.8(v) $\Sigma A\cdot\Sigma C$ is an upper bound of the set $\{a\cdot c\mid a\in A\wedge c\in C\}$. To prove that $\Sigma A\cdot\Sigma C$ is the least upper bound of this set let d be any upper bound of this set; we have to prove $d\geqslant\Sigma A\cdot\Sigma C$. Suppose $d\ngeqslant\Sigma A\cdot\Sigma C$ then by 1.13(i) $\Sigma A\cdot\Sigma C-d\neq 0$, i.e., $\Sigma A\cdot\Sigma C\cdot\bar{d}\neq 0$. If for every $a\in A$ $a\cdot\Sigma C\cdot\bar{d}=0$ then by 1.8(vi) $\overline{\Sigma C\cdot\bar{d}}$ is an upper bound of A hence $\Sigma A\leqslant\overline{\Sigma C\cdot\bar{d}}$ and therefore by 1.8(vi) $\Sigma A\cdot\Sigma C\cdot\bar{d}=0$, contradicting $\Sigma A\cdot\Sigma C\cdot\bar{d}\neq 0$. Thus for some $a_0\in A$ $a_0\cdot\Sigma C\cdot\bar{d}\neq 0$. An entirely similar argument shows that for some $c_0\in C$ $a_0\cdot c_0\cdot\bar{d}\neq 0$. Therefore, by 1.8(vi) $d\ngeqslant a_0\cdot c_0$, contradicting our assumption that d is an upper bound of the set $\{a\cdot c\mid a\in A\wedge c\in C\}$. □

1.19 Corollary. *If ΣA is defined then $\Sigma A \cdot c = \Sigma\{a \cdot c \mid a \in A\}$.* □

1.20 Exercises. (i) Formulate the duals of 1.18 and 1.19.

(ii) A Boolean algebra B is atomic iff for every $b \in B$ $b = \Sigma\{a \in B \mid a$ is an atom and $a \leqslant b\}$. □

1.21 Definition. A subset C of a Boolean algebra B is called a *subalgebra* of B if C is non-void and closed under the operations $+$, $\cdot$ and $^{-}$. Since a subalgebra C is non-void and $a + \bar{a} = 1$ and $a \cdot \bar{a} = 0$ we get $0, 1 \in C$. Since all the axioms of Boolean algebra are equations they hold also in each subalgebra of a Boolean algebra and therefore every subalgebra of a Boolean algebra is also a Boolean algebra. Notice that $\{0, 1\}$ is a subalgebra of every Boolean algebra. Notice also that in order to prove that a subset C of B is a subalgebra it suffices to establish that it is closed under $^{-}$ and under one of the operations $+$ and $\cdot$; by 1.9 we have $a + b = \overline{\bar{a} \cdot \bar{b}}$ and $a \cdot b = \overline{\bar{a} + \bar{b}}$, thus in either case C is closed under both $+$ and $\cdot$.

Given a Boolean algebra B and a subset A of B, the closure of $A \cup \{0, 1\}$ under the operations $+$, $\cdot$ and $^{-}$ is obviously a subalgebra of B; we call it the subalgebra of B *generated by* A.

If C is a subalgebra of a Boolean algebra B then we also say that B is an *extension* of C.

1.22 Proposition. *For a Boolean algebra B, a subalgebra of B generated by a finite set A is finite.*

Proof. We shall prove the proposition by induction on the finite set A (III.1.6). If $A = 0$ then the set $A \cup \{0, 1\} = \{0, 1\}$ is closed under $+$, $\cdot$ and $^{-}$ and therefore it is the subalgebra generated by 0 and it is finite. Assume now that A is a finite set which generates a finite subalgebra C of B and let $z \in B$; we shall prove that the set $E = \{c \cdot z + d \cdot \bar{z} \mid c, d \in C\}$, which is obviously finite (since it has at most $|C|^2$ members), is the subalgebra generated by $A \cup \{z\}$. Every member of E is obtained from the members of $A \cup \{z\}$ by the operations $+$, $\cdot$, $^{-}$ since for a general member $c \cdot z + d \cdot \bar{z}$ of E c and d, being members of C, are obtained from the members of A by these operations. To prove that E is the subalgebra generated by $A \cup \{z\}$ we still have to show that every member obtained from $A \cup \{z\}$ by the operations $+$, $\cdot$, $^{-}$ is in E. This follows immediately once we establish that $A \cup \{z\} \subseteq E$ and that E is closed under $+$ and $^{-}$. $E \supseteq C \supseteq A$ since for $c \in C$ $c = c \cdot z + c \cdot \bar{z}$. $z \in E$ since $z = 1 \cdot z + 0 \cdot \bar{z}$. If $e = c \cdot z + d \cdot \bar{z}$ and $e' = c' \cdot z + d' \cdot \bar{z}$ are members of E then $e + e' = (c + c') \cdot z + (d + d') \cdot \bar{z} \in E$ and $\bar{e} = \bar{c} \cdot z + \bar{d} \cdot \bar{z} \in E$. □

2. *Prime Ideals and Representation*

We shall start this section with a study of the concept of a homomorphism. This will lead us to a study of the ideals and the filters, and in particular the prime ideals and the ultrafilters. The existence of the latter ideals and filters will be established by means of the prime ideal theorem. The Stone representation theorem will establish that every Boolean algebra is isomorphic to a field of sets of a particular kind. Then we shall have a closer look at the axiomatic status of the Boolean prime ideal theorem and the Stone representation theorem with respect to the axiom of choice. As part of this study we shall also prove and discuss Tychonoff's theorem about the product of compact topological spaces.

As is the case with other algebras, the concept of a homomorphism and related concepts play an important role in the study of Boolean algebra. In our case a homomorphism is a mapping of one Boolean algebra into another which preserves the operations of the Boolean algebra, but possibly fuses different members.

2.1 Definition. A mapping H of a Boolean algebra $\langle B, +, \cdot, {}^-, 0, 1\rangle$ into a Boolean algebra $\langle B', +', \cdot', {}^{-\prime}, 0', 1'\rangle$ is called a *homomorphism* if for all $a, b \in B$ we have $H(a+b)=H(a)+'H(b)$, $H(a\cdot b)=H(a)\cdot' H(b)$ and $H(\bar{a})=\overline{H(a)}'$. The subset $\{a \in B \mid H(a)=0'\}$ of B is called the *kernel* of H. If a homomorphism H is one–one it is called an *isomorphism*. If there is a homomorphism of one Boolean algebra onto another one we say that the second Boolean algebra is a *homomorphic image* of the first. If there is an isomorphism of one Boolean algebra onto another one we say that the two Boolean algebras are *isomorphic*.

This is the only occasion where we use different symbols to denote corresponding operations in different Boolean algebras. From now on we shall use $+$ to denote addition in all Boolean algebras, and similarly for $\cdot, {}^-, 0$ and 1. Thus we shall write the requirements on a homomorphism as $H(a+b)=H(a)+H(b)$, $H(a\cdot b)=H(a)\cdot H(b)$ and $H(\bar{a})=\overline{H(a)}$. It will always be clear which operation is denoted by each symbol.

2.2 Proposition. (i) *If H is a mapping of a Boolean algebra B into a Boolean algebra B' such that $H(\bar{a})=\overline{H(a)}$ and one of the conditions $H(a+b)=H(a)+H(b)$ and $H(a\cdot b)=H(a)\cdot H(b)$ holds for all $a, b \in B$ then also the other condition holds and H is a homomorphism.*

If H is a homomorphism then:

(ii) $H(0)=0$.

(iii) $H(1)=1$.

(iv) *If $a \leqslant b$ then $H(a) \leqslant H(b)$.*

(v) *$H(a)=H(b)$ iff $a \vartriangle b$ is in the kernel of H.*

Hint of proof. (i) Use 1.9(iv). (ii) $0=a\cdot\bar{a}$ hence $H(0)=H(a)\cdot H(\bar{a})=H(a)\cdot\overline{H(a)}=0$. (v) Since $d\vartriangle b=a\cdot\bar{b}+\bar{a}\cdot b$ $H(a\vartriangle b)=H(a)\cdot\overline{H(b)}+\overline{H(a)}\cdot H(b)=H(a)\vartriangle H(b)$. By 1.13(iv) $H(a)=H(b)$ iff $H(a)\vartriangle H(b)=0$, iff $H(a\vartriangle b)=0$, i.e., iff $a\vartriangle b$ is in the kernel of H. □

2.3 Proposition. *If H is a bijection of a Boolean algebra B onto a Boolean algebra B' such that $a \leqslant b \leftrightarrow H(a) \leqslant H(b)$ then H is an isomorphism.*

Proof. This follows from the fact that the operations $+, \cdot, {}^-$ are definable from $\leqslant$ in a Boolean algebra; $a+b$ is the least upper bound of a and b, $a\cdot b$ is the greatest lower bound of a and b, and $\bar{a}$ is the unique member b of B such that $a+b$ is the maximal member of B and $a \cdot b$ is the least member of B. For example, since $a+b \geqslant a, b$ we get $H(a+b) \geqslant H(a), H(b)$; if $d \in B'$ is any upper bound of $H(a)$ and $H(b)$ then, since H is onto B' $d=H(c)$ for some $c \in B$; since $H(c)=d \geqslant H(a)$, $H(b)$ we get $c \geqslant a, b$ hence $c \geqslant a+b$ and $d=H(c) \geqslant H(a+b)$. Thus we have shown that $H(a+b)$ is the least upper bound of $H(a)$ and $H(b)$, hence $H(a+b)=H(a)+H(b)$. □

2.4 Definition (Stone 1934). Let B be a Boolean algebra.

(i) An *ideal* in B is a non-void subset I of B such that (a) $a \in I \wedge b \leqslant a \rightarrow b \in I$, (b) $a, b \in I \rightarrow a+b \in I$, and (c) $1 \notin I$.

(ii) An ideal I in B is said to be *prime* if for all $a \in B$ $a \in I$ or $\bar{a} \in I$.

(iii) A *filter* in B is a non-void subset F of B such that (a) $a \in F \wedge b \geqslant a \rightarrow b \in F$, (b) $a, b \in F \rightarrow a \cdot b \in F$, and (c) $0 \notin F$.

(iv) A filter F in B is said to be an *ultrafilter* if for all $a \in B$ $a \in F$ or $\bar{a} \in F$.

(v) A *principal ideal* is a set of the form $\{a \in B \mid a \leqslant b\}$, where b is a member of B other than 1. A principal ideal is obviously an ideal.

(vi) A *principal filter* is a set of the form $\{a \in B \mid a \geqslant b\}$, where $b \neq 0$.

2.5 Ideals and Filters. The concepts of an ideal and a filter in a Boolean algebra are direct generalizations of the same concepts for the field of all subsets of a set X, which we introduced in IV.4.1. Following the remark in IV.4.7, if we regard the relation $\leqslant$ on the Boolean algebra as representing inequality of magnitude then the members of an ideal I can be considered to be very small objects and the members of a filter F can be regarded as objects almost equal to the maximal object 1. The concepts of an ideal and a filter are dual concepts, and such are also the concepts of a prime ideal and an ultrafilter.

2.6 Proposition. (i) *For every ideal I in a Boolean algebra B $0 \in I$ and for all $a \in B$ at most one of a and $\bar{a}$ is in I. Thus if I is a prime ideal then exactly one of a and $\bar{a}$ is in I.*

(ii) *An ideal I in a Boolean algebra B is prime iff for all $a, b \in B$ if $a \cdot b \in I$ then $a \in I$ or $b \in I$. Thus for a prime ideal I we have $a \cdot b \in I \leftrightarrow a \in I \vee b \in I$.*

(iii) *For every filter F in a Boolean algebra B $1 \in F$ and for all $a \in B$ at most one of a and $\bar{a}$ is in F. Thus if F is an ultrafilter then exactly one of a and $\bar{a}$ is in F.*

(iv) *A filter F in a Boolean algebra B is an ultrafilter iff for all $a, b \in B$ if $a+b \in F$ then $a \in F$ or $b \in F$. Thus for an ultrafilter F we have $a+b \in F \leftrightarrow a \in F \vee b \in F$.*

(v) *A subset I of B is an ideal iff the set $\{\bar{a} \mid a \in I\}$ is a filter; I is a prime ideal iff the set $\{\bar{a} \mid a \in I\}$ is an ultrafilter.*

(vi) *A principal filter $F=\{a \in B \mid a \geqslant b\}$ is an ultrafilter iff b is an atom.*

Partial proof. (i) Since $I \neq 0$ let $a \in I$; since $0 \leqslant a$ also $0 \in I$. If $a, \bar{a} \in I$ also $1=a+\bar{a} \in I$, but $1 \notin I$.

(ii) If I is prime and $a \cdot b \in I$, assume $a \notin I$ and $b \notin I$. Then $\bar{a}, \bar{b} \in I$ and hence $\overline{a \cdot b}=\bar{a}+\bar{b} \in I$, contradicting $a \cdot b \in I$ (by (i)). Therefore $a \in I$ or $b \in I$. If I is now any ideal such that $a \cdot b \in I \rightarrow a \in I \vee b \in I$ then because of $a \cdot \bar{a}=0 \in I$ we get $a \in I$ or $\bar{a} \in I$, i.e., I is prime. □

2.7 Proposition. *Let A be a subset of a Boolean algebra.*

(i) *If for no finite subset C of A does $\Sigma C=1$ then the set $D=\{d\in B \mid$ for some finite subset C of A $d\leqslant\Sigma C\}$ is the least ideal which includes A; D is called* the ideal generated by A.

(ii) *If for no finite subset C of A does $\Pi C=0$ then the set $D=\{d\in B \mid$ for some finite subset C of A $d\geqslant\Pi C\}$ is the least filter which includes A; D is called* the filter generated by A.

Proof. (i) Assume that for no finite $C\subseteq A$ does $\Sigma C=1$, and let us prove first that D is an ideal. It is obvious that if $d\in D$ and $e\leqslant d$ then $e\in D$. Now let d, $e\in D$ then there are finite subsets C and C' of A such that $d\leqslant\Sigma C$ and $e\leqslant\Sigma C'$; by 1.15 $d+e\leqslant\Sigma(C\cup C')\in D$. If $1\in D$ then $1\leqslant\Sigma C$ for some finite subset C of A; but then $\Sigma C=1$, contradicting our assumption. Having thus proved that D is an ideal we still have to prove that D is included in every ideal I which includes A, but from the definition of D and 1.16 it follows immediately that every member of D is also a member of I. □

2.8 Proposition. *In a Boolean algebra an ideal J is prime iff it is maximal, i.e., iff there is no ideal I which properly includes J. A filter is an ultrafilter iff it is maximal.*

Proof. If J is prime then for every $a\in B$ J contains a or $\bar{a}$. Thus any subset of B which properly includes J must contain both a and $\bar{a}$ for some $a\in B$; by 2.6(i) such a set is not an ideal.

Now let us show that if J is not prime then it is not maximal. If J is not prime then for some $a\in B$ neither a nor $\bar{a}$ is in J. It is easily seen that the set $\{b+j \mid b\leqslant a\wedge j\in J\}$ is an ideal, let us only check that it does not contain 1. Suppose $b+j=1$, where $b\leqslant a$ and $j\in J$, then, by 1.9(v) and 1.8(vi), $\bar{a}\leqslant\bar{b}\leqslant j$, hence $\bar{a}\in J$, contradicting our choice of a. The ideal $\{b+j \mid b\leqslant a\wedge j\in J\}$ properly includes J since it also contains a and $a\notin J$, thus J is not maximal. □

2.9 Proposition. (i) *Let H be a homomorphism of a Boolean algebra B into a Boolean algebra B'. The kernel of H is an ideal in B, and the set $\{a\in B \mid H(a)=1\}$ is a filter in B.*

(ii) *For H as in* (i) $\mathrm{Rng}(H)=\{0,1\}$ *iff the kernel of H is a prime ideal, iff the set $\{a\in B \mid H(a)=1\}$ is an ultrafilter.*

(iii) *If I is a prime ideal in a Boolean algebra then the function H on B given by $H(a)=0$ if $a\in I$ and $H(a)=1$ if $a\notin I$ is a homomorphism of B onto the Boolean algebra $\{0,1\}$.* □

2.10 Proposition. *Let G and H be homomorphisms of a Boolean algebra B onto Boolean algebras B' and B'' with the same kernel. Then there is an isomorphism J of B' onto B'' such that $H=JG$. In this case we say that the diagram in Fig. 6 is commutative in the sense that going from B to B'' via B' by G and J takes us from a member of B to the same member of B'' to which H takes us directly).*

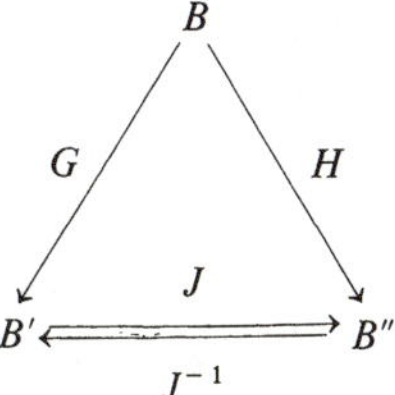

Fig. 6

Proof. Define $J=\{\langle G(a), H(a)\rangle \mid a\in B\}$. Obviously $\mathrm{Dom}(J)=B'$ and $\mathrm{Rng}(J)=B''$. To prove that J is a bijection we have to prove that for $a, b\in B$ $G(a)=G(b)$ iff $H(a)=H(b)$. By 2.2(v) $G(a)=G(b)$ iff $a \vartriangle b$ is in the kernel of G which is also the kernel of H, and this holds, again by 2.2(v), iff $H(a)=H(b)$. To prove that J is an isomorphism let us do only the case $J(d+e)=J(d)+J(e)$. For $d, e\in B'$ $d=G(a)$, $e=G(b)$ for some $a, b\in B$, then

$$J(d+e)=J(G(a)+G(b))=J(G(a+b))=H(a+b)=H(a)+H(b)=J(d)+J(e). \quad \square$$

We have just seen that the kernel of a homomorphism H determines the homomorphic image up to isomorphism, therefore we can claim that we know all homomorphic images of B if we know all the kernels of the homomorphisms of B. We know that every kernel of a homomorphism of B is an ideal in B; 2.11 will establish that every ideal in B is a kernel of a homomorphism of B.

2.11 Proposition (Stone 1934). *For every ideal I in a Boolean algebra B there is a homomorphism H of B onto a Boolean algebra B' such that I is the kernel of H. In the proof we shall construct a particular Boolean algebra B' as required; this B' is called the* quotient algebra *of B over I. Notice that this will be our first encounter with Boolean algebras which are not fields of sets (except for the general treatment of all Boolean algebras).*

Proof. It is clear from 2.2(v) that in order to obtain such a homomorphism we have to identify members a, b of B such that $a \vartriangle b\in I$. Formally this is done as follows. For $a, b\in B$ we write $a\equiv_I b$ if $a \vartriangle b\in I$. We shall now see that $\equiv_I$ is an equivalence relation. $a\equiv_I a$ since $a \vartriangle a=0\in I$. If $a\equiv_I b$ then $b\equiv_I a$ since $\vartriangle$ is a symmetric operation. If $a\equiv_I b$ and $b\equiv_I c$ then $a \vartriangle b\in I$ and $b \vartriangle c\in I$; by 1.13(v) $a \vartriangle c\leqslant a \vartriangle b + b \vartriangle c\in I$, hence $a\equiv_I c$. We take for the quotient algebra B' the set $\{\{b\in B \mid b\equiv_I a\} \mid a\in B\}$ of the equivalence classes of $\equiv_I$. For $s, t\in B'$, $a_1, a_2\in s$ and $b_1, b_2\in t$ we have $a_1+b_1\equiv_I a_2+b_2$, by 1.13(vii), $a_1\cdot b_1\equiv_I a_2\cdot b_2$, by 1.13(viii), and $\bar{a}_1\equiv_I \bar{a}_2$, by 1.13(vi). Therefore we can define for $s, t\in B'$ $s+t, s\cdot t$ and $\bar{s}$ as the equivalence classes of $a+b$, $a\cdot b$ and $\bar{a}$, respectively, where a and b are any members of s and t, respectively. By what we just mentioned our choice of a and b does not effect the outcome of $s+t$, $s\cdot t$ and $\bar{s}$. Let H be the function on B given by $H(a)=\{b\in B \mid b\equiv_I a\}$; obviously $\mathrm{Rng}(H)=B'$. By our definition of the operations in B' we have, for all $a, b\in B$, $H(a)+H(b)=H(a+b)$, $H(a)\cdot H(b)=H(a\cdot b)$, $\overline{H(a)}=H(\bar{a})$, thus H is a homomorphism provided B' is a Boolean algebra. Take in B' $0=H(0)$, $1=H(1)$. $H(0)\neq H(1)$ since $0\in H(0)$ while $0\notin H(1)$ as $0 \vartriangle 1=1\notin I$. Since H commutes with the operations $+$, and $^{-}$ and takes 0 and 1 to 0 and 1 every equality which holds in B holds also in B'. Since all the axioms of Boolean algebra are equalities they all hold in B' too and thus B' is indeed a Boolean algebra. To give a particular example of this general principle let us consider the equality $a\cdot\bar{a}=0$. For every $s\in B'$ there is an $a\in B$ such that $H(a)=s$. We have

$$s\cdot\bar{s}=H(a)\cdot\overline{H(a)}=H(a)\cdot H(\bar{a})=H(a\cdot\bar{a})=H(0)=0,$$

by the established properties of H, by the fact that $a\cdot\bar{a}=0$, and by the definition of 0 of B'. Thus $s\cdot\bar{s}=0$ holds also in B'. Finally, the kernel of H is $\{a\in B \mid H(a)=0\}=\{a\in B \mid H(a)=H(0)\}=\{a\in B \mid a\equiv_I 0\}=I$. □

2.12Ac The Boolean Prime Ideal Theorem (Ulam 1929, Tarski 1930). *Let B be a Boolean algebra. The following three conditions, which are obviously equivalent, hold for B.*

(a) *For every ideal K in B there is a prime ideal J in B which includes K.*

(b) *For every ideal K in B there is an homomorphism H of B on* $\{0, 1\}$ *such that the kernel of H includes K.*

(c) *For every filter F in B there is an ultrafilter U which includes F.*

Proof. Let T be the set of all ideals in B which include K. T is partially ordered by the proper inclusion relation $\subset$. We shall show that T satisfies the requirement of Zorn's lemma (V.1.9), i.e., every subset S of T which is totally ordered by proper inclusion has an upper bound in T. If $S=0$ then K itself is an upper bound of T. If $S\neq 0$ then, as we shall see, $\bigcup S\in T$ and thus $\bigcup S$ is an upper bound of S in T. We shall now see that $\bigcup S$ is an ideal. If $a\leqslant b\in \bigcup S$ then for some $I\in S$ $b\in I$, hence $a\in I$ and $a\in\bigcup S$. If $a, b\in\bigcup S$ then for some $I_1, I_2\in S$ $a\in I_1$ and $b\in I_2$. Since S is totally ordered by $\subset$ one of I_1, I_2 is included in the other, say $I_1\subseteq I_2$. Thus $a, b\in I_2$, hence $a+b\in I_2\subseteq\bigcup S$. Obviously $1\notin\bigcup S$ since 1 belongs to no member of S. Having shown that $\bigcup S$ is an ideal we notice also that since $S\neq 0$ and K is a subset of every member of S we have $\bigcup S\supseteq K$, thus $\bigcup S\in T$. By the conclusion of Zorn's lemma T has a maximal member J. J is obviously a maximal ideal, hence by 2.8 it is a prime ideal. Since $J\in T$ $J\supseteq K$. □

2.13Ac Corollary (Stone 1934). *For all members a, b of a Boolean algebra B the following are equivalent.*

(a) $a\nleqslant b$.

(b) *There exists a prime ideal I in B such that* $b\in I$ *but* $a\notin I$.

(c) *There is a homomorphism H of B onto* $\{0, 1\}$ *such that* $H(a)=1$ *and* $H(b)=0$.

(d) *There exists an ultrafilter U in B such that* $a\in U$ *but* $b\notin U$.

Hint of proof. If $a\nleqslant b$ apply 2.12 to $K=\{c\in B \mid c\leqslant b+\bar{a}\}$. □

2.14Ac Corollary. *An equation which holds in a single Boolean algebra holds in every Boolean algebra.*

Proof. Let $\Gamma=\Delta$ be an equation which holds in a Boolean algebra B but does not hold in a Boolean algebra C. Let us denote the variables which occur in Γ and Δ with $x_1, \ldots, x_n$ and write $\Gamma=\Delta$ as $\Gamma(x_1, \ldots x_n)=\Delta(x_1, \ldots, x_n)$. Since this equation does not hold in C there are members $a_1, \ldots, a_n$ of C such that $\Gamma(a_1, \ldots, a_n) \neq \Delta(a_1, \ldots, a_n)$. Denote $\Gamma(a_1, \ldots, a_n)$ with b and $\Delta(a_1, \ldots, a_n)$ with c, then $b \neq c$. Since we cannot have both $b \leqslant c$ and $c \leqslant b$ we can assume, without loss of generality, that $b \nleqslant c$. By the Boolean prime ideal theorem (Corollary 2.13) there is a homomorphism H of C onto the subalgebra $\{0, 1\}$ of B such that $H(b)=1$ and $H(c)=0$. Since $\Gamma(a_1, \ldots, a_n)$ and $\Delta(a_1, \ldots, a_n)$ consist of applications of $+$, $\cdot$ and $^{-}$ to $a_1, \ldots, a_n$ we get, by the basic properties of homomorphisms,

$$H(\Gamma(a_1, \ldots, a_n))=\Gamma(H(a_1), \ldots, H(a_n))$$
$$H(\Delta(a_1, \ldots, a_n))=\Delta(H(a_1), \ldots, H(a_n)).$$

Therefore we have

$$\Gamma(H(a_1), \ldots, H(a_n))=H(\Gamma(a_1, \ldots, a_n))=H(b)=1 \neq$$
$$\neq 0=H(c)=H(\Delta(a_1, \ldots, a_n))=\Delta(H(a_1), \ldots, H(a_n)).$$

Thus the equation $\Gamma(x_1, \ldots, x_n)=\Delta(x_1, \ldots, x_n)$ does not hold in B since it fails for the members $H(a_1), \ldots, H(a_n)$ of B. □

2.15Ac Remark. As we explained in detail in 1.2 we intended the Boolean algebras to be sets B with operations $+$, $\cdot$ and $^{-}$ and with distinguished members 0 and 1 such that all the equations which hold in every field of sets hold in B. However, in the formal definition 1.3 of the concept of a Boolean algebra we used only a small number of those equations. Now we know that the apparently weaker requirements of 1.3 are sufficient to establish our intended stronger requirements since we have proved that it is not only the case that every equation which holds in every field of sets holds also in every Boolean algebra, but even every equation which holds in a single field of sets holds in every Boolean algebra. A remarkable consequence of this state of affairs is that it turns out to be very easy to check whether an equation holds in all fields of sets or, equivalently, in all Boolean algebras; it suffices to check whether the equation holds in the Boolean algebra $\{0, 1\}$.

The next theorem will show that the concept of a Boolean algebra is even closer to that of a field of sets than what we know it now to be.

2.16Ac The Stone Representation Theorem (Stone 1934). *Every Boolean algebra B is isomorphic to a field of sets. This field of sets can be taken to be the field of clopen sets in a totally disconnected compact Hausdorff space, which is called the* Stone space *of B.*

Proof. The only way in which we shall use the axiom of choice in this proof is by applying the Boolean prime ideal theorem.

First let us deal with the first part of the theorem. If the Boolean algebra B is atomic then B is very easily seen to be isomorphic to a field of subsets of the set of all atoms of B, by the isomorphism H given by $H(b)=\{a\in B \mid a \text{ is an atom} \wedge a\leqslant b\}$. However, we saw already that not all Boolean algebras are atomic, therefore we look for some objects similar to atoms to serve as the members of the sets in the field of sets. It turns out that a good choice for such objects are the ultrafilters of B. By 2.6(vi) there is a natural correspondence between the principal ultrafilters and the atoms, thus we can regard the set of the ultrafilters of B as an extension of the set of the atoms of B. For $b\in B$ and an atom a of B $a\leqslant b$ iff b is in the principal ultrafilter $\{c\in B \mid c\geqslant a\}$ which corresponds to a, thus for a general ultrafilter F let us regard it as representing something like an atom which is $\leqslant b$ if $b\in F$. A typical property of an atom a is that whenever $a\leqslant b+c$ then $a\leqslant b$ or $a\leqslant c$ (1.11(vii)); this property of the atoms extends to all ultrafilters F since whenever $b+c\in F$ then $b\in F$ or $c\in F$ (by 2.6(iv)). Thus we take X to be the set of all ultrafilters of B. We define H on B by taking $H(b)$ to be the set of all ultrafilters of B which contain b. By Corollary 2.13 of the Boolean prime ideal theorem H is an injection of B into $\mathbf{P}(X)$. We shall now prove that H is indeed an isomorphism of $\langle B, +, \cdot, {}^{-}, 0, 1\rangle$ onto $\langle \mathrm{Rng}(H), \cup, \cap, \langle X\sim A \mid A\in \mathrm{Rng}(H)\rangle, 0, X\rangle$, and this will also establish that $\mathrm{Rng}(H)$ is closed under the operations $\cap$ and $\langle X\sim A \mid A\subseteq X\rangle$ and hence $\mathrm{Rng}(H)$ is a field of subsets of X. $H(a\cdot b)=H(a)\cap H(b)$ holds for all $a, b\in B$ since for every filter F $a\cdot b\in F$ iff both a and b are in F. $H(\bar{a})=X\sim H(a)$ follows from the fact that every ultrafilter contains exactly one of a and $\bar{a}$ (2.6(iii)) and hence the sets $H(a)$ and $H(\bar{a})$ are complementary. Thus the first part of the theorem is proved.

We turn X into a topological space, called the *Stone space of B*, by taking the members of $\mathrm{Rng}(H)$ to be the basic open sets of the topology, except for $H(0)$ which is the null-set. $\mathrm{Rng}(H)\sim\{0\}$ is indeed a base for a topology since, as we have shown, $H(a)\cap H(b)=H(a\cdot b)$. Since for every $a\in B$ $H(a)=X\sim H(\bar{a})$ every such set is also closed and hence it is clopen. Thus the space is totally disconnected, i.e., the non-void clopen sets of the space are a base of the topology. To see that the topology is Hausdorff take $F, G\in X$, $F\neq G$; we can assume, without loss of generality, that there is an $a\in F$ such that $a\notin G$. We have $F\in H(a)$ and $G\in X\sim H(a)= H(\bar{a})$, and thus $H(a)$ and $H(\bar{a})$ are disjoint open neighbourhoods of F and G.

To prove the compactness of the Stone space it suffices, as mentioned in the proof of VI.2.22, to prove that every cover T of X which consists of basic open sets has a finite subset S which is a cover of X. Let $T^*=\{a\in B \mid H(a)\in T\}$. We shall prove that there is a finite subset $A=\{a_1, \ldots, a_n\}$ of T^* such that $\Sigma A=1$. In this case, since H is an isomorphism, $H(a_1)\cup\cdots\cup H(a_n)=H(a_1+\cdots+a_n)=H(1)=X$ and thus $\{H(a_1), \ldots, H(a_n)\}$ is a finite subset of T which is a cover of X. Let us assume now, in order to obtain a contradiction, that there is no finite subset A of T^* such that $\Sigma A=1$. Then, by 2.7(i), T^* generates an ideal I in B. Let J be a prime ideal which includes I and let $F=\{\bar{a} \mid a\in J\}$. By 2.6(v) F is an ultrafilter in B, thus $F\in X$. Since T is a cover of X $F\in H(a)$ for some $H(a)\in T$ and by the definition of H $a\in F$. On the other hand, since $H(a)\in T$, and by the definition of T^*, I, J and F, $a\in T^*\subseteq I\subseteq J$, hence $\bar{a}\in F$, which contradicts $a\in F$.

Finally, let us prove that every clopen set in the Stone space is in $\mathrm{Rng}(H)$. First we see that in every compact space every clopen set W is the union of finitely many basic open sets. W, being open, is a union of basic open sets. Since W is also closed it is compact, by VI.2.7, and therefore it is already the union of finitely many of the basic open sets. In the Stone space the union of finitely many basic open sets is again in $\mathrm{Rng}(H)$ since $H(a_1)\cup\cdots\cup H(a_n)=H(\Sigma\{a_1,\ldots,a_n\})$. Thus in it every clopen set is in $\mathrm{Rng}(H)$. □

Remark. The Stone representation theorem tells us that when we generalized the concept of a field of sets to that of a Boolean algebra we stayed closer to home than we expected; our generalization was the absolutely minimal generalization one could carry out when passing to abstract structures, namely extending the class of structures with which we started to the class of all structures isomorphic to them. Since every Boolean algebra is isomorphic to a field of sets we can always deal with the isomorphic field of sets instead of dealing with a Boolean algebra, so why do we need Boolean algebras at all? The answer is that one encounters also under natural circumstances Boolean algebras which are not fields of sets; passing to the isomorphic fields of sets involves a representation which may have nothing to do with the problem at hand and the extra effort needed to go through it is thus entirely wasted.

2.17 Exercise. Let B be a Boolean algebra, and let $\langle X, O\rangle$ be its Stone space.

(i) $U\in X$ is a principal ultrafilter in B iff U is an isolated point in X.

(ii) B is atomless iff $\langle X, O\rangle$ is a perfect space.

(iii) B is atomic iff the set of isolated points is dense in X. □

2.18 The Boolean Prime Ideal Theorem as a Weaker Version of the Axiom of Choice. Section V.2 was devoted to the study of weaker versions of the axiom of choice. The Boolean prime ideal theorem, which we proved by means of the axiom of choice, is another weaker version of this axiom. First let us mention that the Boolean prime ideal theorem cannot be proved without the axiom of choice (Anne Davis in 1948—see Tarski 1954, Los and Ryll-Nardzewski 1954) and it does not imply the axiom of choice (Halpern 1964, Halpern and Levy 1971). One criterion for a good axiom, which we applied to the axiom of choice itself in the discussion leading to V.1.6, is the stability criterion. This criterion is fulfilled by an axiom if many, apparently unrelated, statements taken from different areas of mathematics turn out to be equivalent to the axiom. The Boolean prime ideal theorem cannot compete with the axiom of choice on this ground, but it is still a quite "stable" axiom, as we shall see.

2.19 Theorem (Stone 1936). *The following statements are equivalent in* ZF (*i.e., even in the absence of the axiom of choice*).

(a) *In every Boolean algebra B there is a prime ideal* (*or, equivalently, an ultrafilter*).

(b) *The Boolean prime ideal theorem* (2.12).

(c) *The Stone representation theorem* (2.16).

Proof. (a)→(b). Let B be a Boolean algebra and let K be an ideal in B. By 2.11 there is a homomorphism H of B onto the quotient algebra B' of B over K, and K

is the kernel of H. By (a) there is a prime ideal I in B'. Let $J=H^{-1}[I]$. Obviously $J\supseteq K$ since $K=H^{-1}[\{0\}]$ and $0\in I$. We shall establish (b) by showing that J is a prime ideal. If $a\leqslant b$ and $b\in J$ then $H(b)\in I$ and by 2.2(iv) $H(a)\leqslant H(b)\in I$ hence $H(a)\in I$ and therefore $a\in J$. If $a, b\in J$ then $H(a)$, $H(b)\in I$ hence $H(a+b)=H(a)+H(b)\in I$ and $a+b\in J$. Finally, since I is a prime ideal exactly one of $H(a)$ and $\overline{H(a)}=H(\bar{a})$ is in I, hence exactly one of a and $\bar{a}$ is in J.

(b)$\rightarrow$(c). This was shown in the proof of the Stone representation theorem 2.16.

(c)$\rightarrow$(a). Let B be a Boolean algebra. By (c) there is an isomorphism H of B on a field W of subsets of a non-void set X; let u be a member of X. The subset $J=\{A\in W \mid u\notin A\}$ of W is obviously a prime ideal in W and hence $H^{-1}[J]$ is a prime ideal in B. □

2.20Ac The Tychonoff Product Theorem (Tychonoff 1935). *The product of any number of compact topological spaces is a compact space.*

Proof. Let $\langle\langle X_i, O_i\rangle \mid i\in I\rangle$ be a family of compact topological spaces. To prove that the product $\langle \times_{i\in I} X_i, O\rangle$ of these spaces is compact we shall show that if C is a set of closed subsets of $\times_{i\in I}X_i$ such that for every non-void finite subset D of C $\bigcap D\neq 0$ then also $\bigcap C\neq 0$ (VI.2.5). Since the intersection of any finite subset of C is non-void C generates a filter F in the field of all subsets of $\times_{i\in I}X_i$ (2.7(ii)); by the Boolean prime ideal theorem there is an ultrafilter U such that $\mathbf{P}(\times_{i\in I}X_i)\supseteq U\supseteq F$. For $i\in I$ let p_i be the function which projects $\times_{i\in I}X_i$ on X_i (VI.2.15(i)), and let $C_i=\{\mathrm{Cl}(p_i[A]) \mid A\in U\}\subseteq\mathbf{P}(X_i)$. Let us show that the intersection of any non-void finite subset of C_i is non-void. Let $\{E_1, \ldots, E_n\}$ be a finite subset of C_i, and for $1\leqslant j\leqslant n$ let $A_j\in U$ be a set such that $E_j=\mathrm{Cl}(p_i[A_j])$. Since U is an ultrafilter we have $\bigcap_{j=1}^n A_j\neq 0$. Let s be any member of $\bigcap_{j=1}^n A_j$, then, for $1\leqslant j\leqslant n$, $p_i(s)\in p_i[A_j]\subseteq\mathrm{Cl}(p_i[A_j])=E_j$ hence $p_i(s)\in\bigcap_{j=1}^n E_j$ and $\bigcap_{j=1}^n E_j\neq 0$. Since $\langle X_i, O_i\rangle$ is a compact space and C_i is a set of closed sets in it such that the intersection of any non-void finite subset of it is non-void we have, by VI.2.5, $\bigcap C_i\neq 0$. Using the axiom of choice let us take a $t\in\times_{i\in I}X_i$ such that $t_i\in\bigcap C_i$ for all $i\in I$. We claim that $t\in\bigcap C$ and thus $\bigcap C\neq 0$ which is what we have to prove. Assume $t\notin\bigcap C$ then $t\notin A$ for some $A\in C$. Since A is closed t has a basic open neighborhood disjoint from A. A basic open neighborhood is of the form $\bigcap_{i\in J}p_i^{-1}[Y_i]$, where $Y_i\in O_i$ and J is a finite subset of I. Since $t\in\bigcap_{i\in J}p_i^{-1}[Y_i]$ we have $t_i\in Y_i$ for $i\in J$. Since $\bigcap_{i\in J}p_i^{-1}[Y_i]\cap A=0$ and $A\in C\subseteq U$ we have $\bigcap_{i\in J}p_i^{-1}[Y_i]\notin U$. Since U is closed under binary, and hence finite, intersection there is a $j\in J$ such that $p_j^{-1}[Y_j]\notin U$, hence the complement $p_j^{-1}[X_j\sim Y_j]$ of $p_j^{-1}[Y_j]$ is in U. By the definition of C_j the closure of the projection p_j of $p_j^{-1}[X_j\sim Y_j]$ is in C_j, and since $X_j\sim Y_j$ is a closed set $\mathrm{Cl}(p_j[p_j^{-1}[X_j\sim Y_j]])=X_j\sim Y_j\in C_j$. Since $t_j\in Y_j$ we get $t_j\notin\bigcap C_j$, contradicting our choice of t.

The theorem is now proved, but let us still see that if all the spaces $\langle X_i, O_i\rangle$ are Hausdorff spaces then our use of the axiom of choice to choose t can be dispensed with. We shall prove that in this case $\bigcap C_i$ consists of a single point only and therefore we can define t by taking t_i to be the only member of $\bigcap C_i$. Assume that $\bigcap C_i$ has two distinct members z and w. Since $\langle X_i, O_i\rangle$ is a Hausdorff space z and w

have disjoint neighborhoods Z and W. The sets $p_i^{-1}[Z]$ and $p_i^{-1}[W]$ are disjoint, hence at least one of them is not in U; without loss of generality assume $p_i^{-1}[Z] \notin U$, hence $p_i^{-1}[X_i \sim Z] \in U$. Since $X_i \sim Z$ is closed we get, by the definition of C_i, $X_i \sim Z \in C_i$. Since $z \in Z$ we have $z \notin X_i \sim Z$, contradicting $z \in \bigcap C_i$. □

2.21 Theorem. *The Boolean prime ideal theorem is equivalent in* ZF *to each one of the following statements.*

(a) (Tarski 1954). *In the field of all subsets of a non-void set X every ideal is included in a prime ideal.*

(b) (Łoś and Ryll-Nardzewski 1954, Rubin and Scott 1954). *The Tychonoff product theorem for Hausdorff spaces (in the version that if the product is non-void then it is compact; also the stronger version which asserts also that the product is non-void is equivalent to the Boolean prime ideal theorem—see* 2.24(i)).

(c) *Every power of the two-point discrete space is compact.*

Proof. The Boolean prime ideal theorem obviously implies (a). (a) implies (b) since in our proof of the Tychonoff product theorem the only place we need to use the axiom of choice in the case of Hausdorff spaces is when we extend the filter F in the field of all subsets of $\times_{i \in I} X_i$ to an ultrafilter U; to do this (a) clearly suffices. (b) implies (c) since the two-point discrete space is a compact Hausdorff space.

Let us prove now that (c) implies that in every Boolean algebra B there is a prime ideal, which, by 2.19, is equivalent to the Boolean prime ideal theorem. Given a Boolean algebra B we consider the space ${}^B2$ with the topology which is the B-th power of the discrete topology on $2=\{0, 1\}$. The members of ${}^B2$ are mappings of B into the set $\{0, 1\}$, which can be viewed as a Boolean algebra. What we are looking for is a member of ${}^B2$ which is a homomorphism, since its kernel will be a prime ideal in B by 2.9(ii). We shall now list the requirements which $t \in {}^B2$ has to satisfy in order to be a homomorphism. We shall see that the set of all members of ${}^B2$ which satisfy such a single requirement is a closed subset W of ${}^B2$, and that for every finite number of such requirements there is a member of ${}^B2$ which satisfies all of them, and hence the intersection of those finitely many W's is non-void. Since ${}^B2$ is compact there is a t in the intersection of all the W's and such a t is a homomorphism as required.

For $u \in B$ and $i<2$ let $Q_{u,i}=\{t \in {}^B2 \mid t(u)=i\}$; $Q_{u,i}$ is obviously clopen. For $u, v \in B$ the following sets are clopen in ${}^B2$:

$$W_{u,v}^{+}=\{t \in {}^B2 \mid t(u)+t(v)=t(u+v)\}=(Q_{u,1} \cup Q_{v,1}) \cap Q_{u+v,1} \cup Q_{u,0} \cap Q_{v,0} \cap Q_{u+v,0},$$

$$W_{u}^{-}=\{t \in {}^B2 \mid t(\bar{u})=\overline{t(u)}\}=Q_{u,0} \cap Q_{\bar{u},1} \cup Q_{u,1} \cap Q_{\bar{u},0}.$$

Let $Z=\{W_{u,v}^{+} \mid u, v \in B\} \cup \{W_{u}^{-} \mid u \in B\}$. All we need to finish the proof is to show that $\bigcap Z$ is non-void, since every member of $\bigcap Z$ is obviously a homomorphism of B on $\{0, 1\}$. Since all the members of Z are closed and the space ${}^B2$ is compact it suffices to prove that the intersection of every finite subset of Z is non-void. Given a finite subset Z_0 of Z there is clearly a finite subset C of B such that $Z_0 \subseteq \{W_{u,v}^{+} \mid u, v \in C\} \cup \{W_{u}^{-} \mid u \in C\}$. We can assume that C is a subalgebra of B since otherwise we could replace C by the finite subalgebra of B generated by it (by 1.22). C is atomic, by 1.11(iv), hence, by 2.6(vi), there is an ultrafilter U in C, and thus there is a homomorphism s of C on $\{0, 1\}$. Define t on B by $t(a)=s(a)$ for $a \in C$ and $t(a)=0$ for $a \in B \sim C$. Clearly

$$t \in \bigcap(\{W^{+}_{u,v} \mid u, v \in C\} \cup \{W^{-}_{u} \mid u \in C\}) \subseteq \bigcap Z_0,$$

hence $\bigcap Z_0 \neq 0$, which is what was left to prove. □

2.22 Exercises. Use the method we used to prove that if for a Boolean algebra B ${}^{B}2$ is a compact space then B has a prime ideal to obtain the following results.

(i) Give a direct proof that if ${}^{B}2$ is a compact space then every ideal I in B can be extended to a prime ideal.

(ii) Prove from the Boolean prime ideal theorem, in the version (c) of 2.21 that every space ${}^{X}2$ is compact, each one of the following theorems.

(a) The ordering theorem: Every set can be ordered.

(b) (Szpilrajn 1930, Łoś and Ryll-Nardzewski 1951). The order-extension theorem: Every partial order on a set can be extended to a total order.

(c) (Mycielski 1961) The n-coloring theorem of V.1.11(v). (This theorem is equivalent to the Boolean prime ideal theorem—Läuchli 1971). □

2.23 Proposition (Kelley 1950). *The Tychonoff product theorem for arbitrary spaces is equivalent to the axiom of choice.*

Hint of proof. To obtain a choice function on a set W with $0 \notin W$ take a $p \notin \bigcup W$ and for every $A \in W$ topologize $A \cup \{p\}$ by a topology O_A such that $\langle A \cup \{p\}, O_A \rangle$ is a compact space and A is a closed set in it. Prove now that in the space $\times_{A \in W}(A \cup \{p\})$ $\times_{A \in W} A = \bigcap_{A \in W} p_A^{-1}[A]$ is non-void (where p_A is the projection on the A-axis). □

2.24 Exercises. (i) Prove, from the Tychonoff product theorem for Hausdorff spaces, that the product of compact Hausdorff spaces is non-void. Therefore the Boolean prime ideal theorem is also equivalent to the Tychonoff product theorem for Hausdorff spaces in the version that the product of compact Hausdorff spaces is non-void and compact.

(ii) Prove that the Boolean prime ideal theorem implies the following strengthening of it: For every set W of pairs $\langle B, F \rangle$ such that B is a Boolean algebra and F is a filter in it there is a function G on W such that for all $\langle B, F \rangle \in W$ $G(B, F)$ is an ultrafilter in B which includes F.

Hint. (i) Follow the proof of 2.23. (ii) Take the product of the Stone spaces of the Boolean algebras B. Notice that the set of ultrafilters in B which include F is a closed set in the Stone space of B. □

2.25 Proposition. *A Boolean algebra B is finite iff in B 1 is the sum of a finite set of atoms.*

Outline of proof. If B is finite let A be the set of all atoms in B; if $\Sigma A<1$ then, since by 1.11(iv) B is atomic, there is an atom $a \leqslant 1-\Sigma A$, which is a contradiction. If $1=\Sigma C$, where C is a finite set of atoms, then the function H on B defined by $H(b)=\{a \in C \mid a \leqslant b\}$ is an injection of B into $\mathbf{P}(C)$. □

2.26 Proposition. *If the Boolean prime ideal theorem holds then in every infinite Boolean algebra there is a non-principal prime ideal.*

Hint of proof. Apply the prime ideal theorem to the ideal generated by the atoms of the Boolean algebras. □

2.27 Exercise. The Stone space of a Boolean algebra B is discrete iff B is finite.

Hint. Use 2.26 and 2.17. □

3. Complete Boolean Algebras

This section is to a large extent a deeper study of the operations Σ and Π. We study in particular complete Boolean algebras, i.e., Boolean algebras in which these operations are always defined. We shall see how to extend arbitrary Boolean algebras, and even partially ordered sets, to complete Boolean algebras. Toward the end of the section we deal with the countable antichain condition for Boolean algebras, posets and topological spaces.

3.1 Definition. A Boolean algebra is said to be a *complete Boolean algebra* if for every subset A of B ΣA exists.

3.2 Proposition. *A Boolean algebra B is complete iff for every subset A of B ΠA exists. Thus the concept of a complete Boolean algebra is self dual.*

Hint of proof. Use 1.17. □

3.3 Examples of Complete Boolean Algebras. For every non-void set X the field of all subsets of X is a complete Boolean algebra where $\Sigma A=\bigcup A$ and $\Pi A=\bigcap A$. (Are there any other fields of sets which are complete Boolean algebras?) Every finite Boolean algebra is complete by 1.16. Now we shall mention some less trivial examples of complete Boolean algebras. Let $\langle X, \mu\rangle$ be a *σ-finite* measure space, i.e., a measure space in which X is the union of $\aleph_0$ measurable sets each of which is of finite measure. The Boolean algebra of all measurable sets in this space modulo sets of measure zero (i.e., the quotient algebra of the field of all measurable sets over the ideal of all sets of measure zero) is a complete Boolean algebra, by 3.35(i) below. An analogous example is the Boolean algebra of all Baire sets modulo meager sets in a non-meager topological space with a countable base (3.35(ii)). A very important extension of the latter example is the following.

3.4 Definition and Proposition. Let $\langle X, O\rangle$ be a topological space and let B be the set of all regular open sets in this space. We recall from VI.3.11 that a regular open set is an open set A such that $A=\text{Int}(\text{Cl}(A))$. For regular open sets A, C we define: $A\cdot C=A\cap C$, which is a regular open set by VI.3.12(ii); $A+C=\text{Int}(\text{Cl}(A\cup C))$, which is, by VI.3.12(i), the least regular open set which includes both A and C; and $\bar{A}=\text{Int}(X\sim A)$, which is the maximal regular open set disjoint from A. Once we take the null-set for 0 and X for 1 $\langle B, +, \cdot, {}^{-}, 0, 1\rangle$ satisfies all the axioms of Boolean algebra. The $\leqslant$ relation of Boolean algebra coincides for this Boolean algebra with the inclusion relation $\subseteq$. B is a complete Boolean algebra; for all non-void $W\subseteq B$ $\Sigma W=\text{Int}(\text{Cl}(\bigcup W))$ and $\Pi W=\text{Int}(\text{Cl}(\bigcap W))$. If no non-void open set is meager in $\langle X, O\rangle$ (this is the conclusion of the Baire category theorem—VI.3.6 and VI.3.8) then B is isomorphic to the Boolean algebra of the Baire sets of $\langle X, O\rangle$ modulo the meager sets.

Remark. The Boolean algebra B of all regular open sets in a topological space $\langle X, O\rangle$ is in general not a field of sets. For B to be a field of sets we must have for every $A\in B$ $\text{Int}(X\sim A)=\bar{A}=X\sim A$, i.e., $X\sim A$ is open and therefore A is clopen. We shall see in 3.7(i) below that this property is equivalent to the space being extremally disconnected. In those spaces every regular open set is clopen, hence the Boolean algebra of all regular open sets of such a space coincides with the field of clopen sets of the space.

Hint of proof. It follows easily from the basic properties of $+$, $\cdot$ and ${}^{-}$, and from additional simple considerations, that B is a complete Boolean algebra. The isomorphism H of B onto the Boolean algebra of the Baire sets modulo the meager sets is defined by setting for $A\in B$

$$H(A)=\{C \mid C \text{ is a Baire set in } \langle X, O\rangle \text{ and } C \vartriangle A \text{ is meager}\}$$

(use VI.3.13). Use VI.3.5 to show that when applied to members of B the operations $+$, $\cdot$ and ${}^{-}$ of B have values which differ from the respective values of $\cup$, $\cap$ and $\langle X\sim A \mid A\subseteq X\rangle$ only by a nowhere dense set. Use VI.3.13 to prove that H is a bijection. □

3.5 Definition. A topological space $\langle X, O\rangle$ is said to be *extremally disconnected* if the closure of every open set in it is open, or, equivalently, if the interior of every closed set in it is closed.

3.6 Extremally Disconnected Spaces. Every discrete space is obviously extremally disconnected. Spaces which are not discrete but are extremally disconnected are the Stone spaces of infinite complete Boolean algebras (see 3.15(ii) below and 2.27). If a topological space is regular (VI.2.12) and extremally disconnected then it is totally disconnected, as easily seen too. On the other hand, there are totally disconnected spaces which are not extremally disconnected, such as the Cantor and Baire spaces. For example, let A be the set of all infinite sequences of 0's and 1's in the Cantor space where the first 1 is immediately followed by a 0, and let t be the infinite sequence which consists only of zeros. As easily seen A is open, $\mathrm{Cl}(A) = A \cup \{t\}$ and $A \cup \{t\}$ is not open. Moreover, as we shall establish in 3.15(ii), the Stone space of any Boolean algebra which is not complete is not extremally disconnected while all Stone spaces are totally disconnected.

3.7 Exercise. (i) A topological space $\langle X, O\rangle$ is extremally disconnected iff every regular open set in $\langle X, O\rangle$ is clopen.

(ii) If a topological space $\langle X, O\rangle$ is extremally disconnected then

(1) The field B of all clopen sets in $\langle X, O\rangle$ is a complete Boolean algebra.

If $\langle X, O\rangle$ is a totally disconnected topological space which satisfies (1) then $\langle X, O\rangle$ is extremally disconnected.

Remark. (1) does not assert that if A is a family of clopen sets in $\langle X, O\rangle$ then $\bigcup A$ is clopen, it only asserts that A has a least upper bound among the clopen sets.

Hint. For the first part of (ii) use (i). For the second part of (ii) let C be a closed set; we have $\mathrm{Int}(C) = \bigcup\{A \in B \mid A \subseteq C\}$; prove that $\mathrm{Int}(C) = \Sigma^B\{A \in B \mid A \subseteq C\}$. □

3.8 Exercise. Let B be a Boolean algebra and let C be a subalgebra of B. For every finite subset $A = \{a_1, \ldots, a_n\}$ of C $a_1 + \cdots + a_n$ is a member of C and by 1.16 it is the least upper bound of $\{a_1, \ldots, a_n\}$ both in B and in C. The situation is completely different for infinite subsets A of C. Let us denote with $\Sigma^B A$ the least upper bound of A in B and with $\Sigma^C A$ the least upper bound of A in C.

(i) Prove that if both $\Sigma^B A$ and $\Sigma^C A$ exist then $\Sigma^C A \geqslant \Sigma^B A$.

We shall now see that (i) is the only connection there is between $\Sigma^C A$ and $\Sigma^B A$, i.e., there is no connection at all between the existence of one of $\Sigma^C A$ and $\Sigma^B A$ and the existence of the other, and if both exist they do not have to be equal. In order to prove this claim let B be the field of all subsets of ω, let C be the field of all finite subsets of ω and their complements, and let D be the field of all finite subsets of ω which consist of even numbers only and all the complements of those sets. Clearly, C is a subalgebra of B and D is a subalgebra of C and B. Let A be the set of all finite subsets of ω which consist of even numbers and let E be the set of all finite subsets of ω which consist of numbers divisible by 4.

(ii) Prove that $\Sigma^B A = \{2 \cdot n \mid n \in \omega\}$, $\Sigma^C A$ does not exist, $\Sigma^D A = \omega$, and $\Sigma^C E$ and $\Sigma^D E$ do not exist.

Whatever was said here about Σ applies equally well to Π, as can be easily seen by applying 1.17. □

3.9 Definition. A Boolean algebra C which is an extension of a Boolean algebra B is said to be a *faithful extension* of B if for every subset A of B if $\Sigma^B A$ exists then also $\Sigma^C A$ exists and $\Sigma^C A = \Sigma^B A$.

3.10 Proposition. *Let B be a subalgebra of the Boolean algebra C. C is a faithful extension of B iff for every subset A of B if $\Pi^B A$ exists then also $\Pi^C A$ exists and $\Pi^C A = \Pi^B A$. Thus the concept of a faithful extension is self dual.* □

3.11 Definition. A Boolean algebra C is said to be a *completion* of the Boolean algebra B if C is a complete Boolean algebra, C is an extension of B and

(2) for every $y \in C$ there is a subset A of B such that $y = \Sigma^C A$.

(2) is a minimality requirement since it asserts that from the point of view of the Σ operation in C no member of C is superfluous if we want C to include B. The strong connection between a Boolean algebra and its completion is manifest in the fact which will be established in 3.21 that the completion of a Boolean algebra is unique up to isomorphism in a very strong sense.

By 1.17, (2) is equivalent to the statement obtained from it by replacing Σ with Π.

3.12 Proposition. *Let C be a complete Boolean algebra and let B be a subalgebra of C. C is a completion of B iff B is* dense *in C, i.e., iff for every member $y \neq 0$ of C there is a member $z \neq 0$ of B such that $z \leqslant y$.*

Hint of proof. Suppose that B is dense in C, $y \in C$ but $\Sigma\{z \in B \mid z \leqslant y\} \neq y$. Then since $y - \Sigma\{z \in B \mid z \leqslant y\} \neq 0$ we get a contradiction by applying the density of B in C. □

3.13 Theorem (MacNeille 1937). *Every Boolean algebra has a completion.*

Proof. We shall present here a simple proof of this theorem which uses the axiom of choice via the Stone representation theorem. A proof of a more general theorem which does not use the axiom of choice is given in 3.18(i) below.

By 2.16 a Boolean algebra B is isomorphic to the field B' of the clopen sets in the Stone space $\langle X, O\rangle$ of B. Let C be the complete Boolean algebra of all regular open sets of $\langle X, O\rangle$. B' is obviously a subset of C; to prove that B' is a subalgebra of C we have to show that for the members of B' the operations of B' and those of C coincide. $Y \cdot Z$ is $Y \cap Z$ in both algebras. $Y + Z$ is $Y \cup Z$ in B' and $\mathrm{Int}(\mathrm{Cl}(Y \cup Z))$ in C, but since Y and Z are clopen sets so is also $Y \cup Z$ and hence $\mathrm{Int}(\mathrm{Cl}(Y \cup Z)) = Y \cup Z$. $\bar{Y}$ is $X \sim Y$ in B' and $\mathrm{Int}(X \sim Y)$ in C, but $X \sim Y$ is clopen hence $X \sim Y = \mathrm{Int}(X \sim Y)$. Finally, to prove that C is a completion of B it suffices to show, by 3.11, that B' is dense in C, but this follows immediately from the fact that the clopen sets are a base for the topology of every Stone space.

We have shown that C is a completion of B'. Since B' is isomorphic to B one can use C to construct a completion of B by copying the operations of C on an appropriate set which includes B. □

3.14 Proposition. *A completion C of a Boolean algebra B is a faithful extension of B.*

Proof. Let $A \subseteq B$ and assume that $\Sigma^B A$ exists; let us denote it with y. We have to prove that $\Sigma^C A = y$. y is obviously an upper bound of A also in C. To show that y is the least upper bound of A in C assume that for some upper bound $u \in C$ of A $u \ngeqslant y$. We have $y - u \neq 0$. By 3.12 B is dense in C hence there is a $0 \neq z \in B$ such that

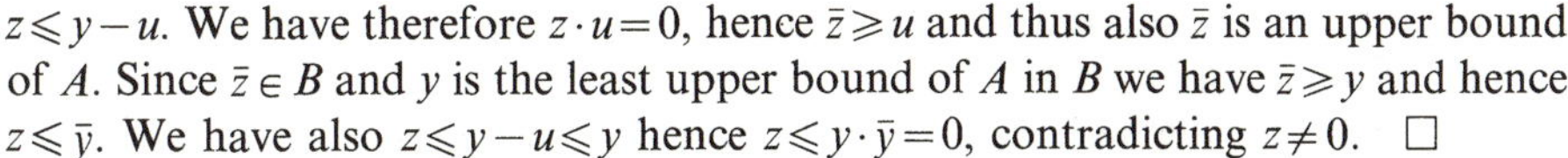

$z \leqslant y-u$. We have therefore $z \cdot u=0$, hence $\bar{z} \geqslant u$ and thus also $\bar{z}$ is an upper bound of A. Since $\bar{z} \in B$ and y is the least upper bound of A in B we have $\bar{z} \geqslant y$ and hence $z \leqslant \bar{y}$. We have also $z \leqslant y-u \leqslant y$ hence $z \leqslant y \cdot \bar{y}=0$, contradicting $z \neq 0$. □

3.15 Corollaries. (i) *A complete Boolean algebra is the one and only completion of itself. This is a particular case of the uniqueness of the completion of a Boolean algebra* (3.21).

(ii) *A Boolean algebra is complete iff its Stone space is extremally disconnected.*

Hint of proof. (i) Let C be a completion of the complete Boolean algebra B. For $y \in C$ we have, for some $A \subseteq B$, $y=\Sigma^C A=\Sigma^B A \in B$, by 3.11 and 3.14. (ii) Use 3.7. □

3.16 Toward a Complete Boolean Algebra of Information. We shall now apply to partially ordered sets $\langle P,<\rangle$ the idea of extending a Boolean algebra to a complete Boolean algebra which is its completion. As we do not care whether the complete Boolean algebra extends $\langle P,<\rangle$ or whether it extends a poset isomorphic to $\langle P,<\rangle$, what we are after is an *embedding* H of $\langle P,<\rangle$ into a complete Boolean algebra. An embedding H of one structure into another is another name for an isomorphism of one structure into another. Without explaining yet what we exactly mean here by an isomorphism, it turns out that if we want the embedding to have the same good properties as the natural embedding of a Boolean algebra into its completion we have, in the case of a general poset $\langle P,<\rangle$, to give up the one-one-ness of H and agree to have an H which will be something like a homomorphism. Under a certain natural assumption on $\langle P,<\rangle$ the embedding will turn out to be an injection after all.

When we say that H is something like a homomorphism of $\langle P,<\rangle$ into the complete Boolean algebra C we mean that H takes the relation $\preccurlyeq$ of $\langle P,<\rangle$ over to a similar relation in C. By $x \preccurlyeq y$ we mean, of course, that $x<y$ or $x=y$. Our first impulse is to require the $\preccurlyeq$ relation of the poset to go over to the $\leqslant$ relation of the Boolean algebra C. However, it turns out that often it is more natural to have the relation of the poset go over to the "opposite" relation $\geqslant$ of the Boolean algebra. The reason for this is that the members of the poset often denote information, with $p \succcurlyeq q$ if p represents all the information represented by q and possibly some more, while in the Boolean algebra C we identify p with the set of all objects for which the information of p is true. Therefore, if p represents all the information in q and possibly some more then the set of all objects for which p is true is a subset of the set of all objects for which q is true and we expect $p \leqslant q$ to hold (while in the poset we have $p \succcurlyeq q$). Let us illustrate these ideas in a concrete case. Let C be the field of all subsets of the Baire space ${}^{\omega}\omega$. Let P be the set $\{p \mid p$ is a function, $\mathrm{Dom}(p)$ is a finite subset of ω, and $\mathrm{Rng}(p) \subseteq \omega\}$. A member p of P is understood to be supplying information on some of the values of a member f of ${}^{\omega}\omega$. For example, $\{\langle 3,15\rangle, \langle 8,0\rangle\}$ asserts that for an $f \in {}^{\omega}\omega$ $f(3)=15$ and $f(8)=0$. Thus the set R_p of all the members of ${}^{\omega}\omega$ for which p is true is the set $\{f \in {}^{\omega}\omega \mid f \supseteq p\}$. If we want $p \preccurlyeq q$ to mean that q contains all the information in p and possibly some more we define $p \preccurlyeq q$ to hold if q is a function which extends p, i.e., if $p \subseteq q$. We have, ob-

viously, $p \leqslant q$ iff $R_q \subseteq R_p$. If we identify p with R_p for every $p \in P$ and if we take $\geqslant$ to denote the relation of inclusion among subsets of ${}^{\omega}\omega$ we get $p \leqslant q$ iff $p \geqslant q$.

In order to be able to look at the members of P both as pieces of information and as the sets of objects for which that information is true we decide that for any given poset $\langle P, <\rangle$, $p \leqslant q$ will be understood as asserting that q contains all the information in p and possibly some more, and $q \leqslant p$ will be, by definition, $p \leqslant q$ and will be understood to assert that the set of all objects for which q is true is a subset of the set of all objects for which p is true.

3.17 Definition. Most of what has been said in 3.16 is informal guidance on how to look at the various concepts with which we deal here. The only formal fact established there is that in every given poset $\langle P, <\rangle$ $p \leqslant q$ is defined to be $p \geqslant q$, and therefore also $p < q \leftrightarrow p > q$.

Following in our terminology the intuitive image presented in 3.16 we shall say, for $p, q \in P$, that q *extends* p if $p \leqslant q$ (or, equivalently, if $q \leqslant p$). We shall say that $p, q \in P$ are *compatible* if there is an $r \in P$ which extends both p and q.

In a Boolean algebra B we have $0 \leqslant p$ for every p, therefore 0 is to be viewed as containing all the information contained in all the members of B. Thus, unless we agree to be in the trivial situation where conflicting information does not exist, we have to consider 0 as containing conflicting information or, in other words, inconsistent information. When we deal with posets $\langle P, <\rangle$ we consider all members of P as representing consistent information, therefore when we deal with a Boolean algebra B the poset we consider for our present purposes is $\langle B \sim \{0\}, >\rangle$, and thereby we consider each non-zero member of B as representing consistent information. Accordingly, for $p, q \in B \sim \{0\}$ we say that p and q are compatible if there is an $r \in B \sim \{0\}$ such that $r \leqslant p, q$, i.e., if $p \cdot q \neq 0$.

3.18 Theorem. (i) *For every poset $\langle P, <\rangle$ there is a complete Boolean algebra C and a mapping F of P into $C \sim \{0\}$ such that the following conditions are met.*

(a) *If $p, q \in P$ and $p \leqslant q$ then $F(p) \leqslant F(q)$.*

(b) *If $p, q \in P$ are incompatible in P then $F(p)$ and $F(q)$ are incompatible in C. (The opposite direction follows from* (a)).

(c) *For all $0 \neq a \in C$ there is a $p \in P$ such that $F(p) \leqslant a$.*

(ii) *All the mappings F as in* (i) *are equivalent in the sense that if F and F' are such mappings of P into respective complete Boolean algebras C and C' then there is an isomorphism H of C on C' such that $HF = F'$, i.e., the diagram in Fig. 7 is commutative.*

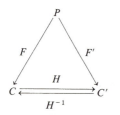

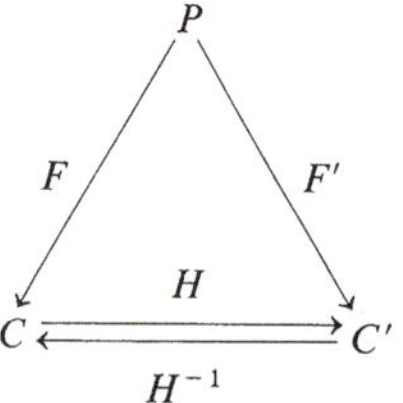

Fig. 7

Remark. Part (i) of this theorem is a generalization of Theorem 3.13 which asserts that every Boolean algebra has a completion, in the following sense. In the particular case where B is a Boolean algebra and $P=B\sim\{0\}$ and F is as claimed by (i) then F is an isomorphism of B into C (when F is extended to all of B by setting $F(0)=0$, by 3.20 below and 2.3), and the range of F is dense in C by (c). Identifying B with the range of F we obtain that C is the completion of B. For a general poset P we shall see in the example of 3.19 that F is not an injection. 3.20 will give a necessary and sufficient condition for F to be an isomorphism of $\langle P,\leqslant\rangle$ into $\langle C,\leqslant\rangle$. If this condition is met then C can be regarded as a completion of P to a complete Boolean algebra. Part (ii) of the present theorem implies that this completion is unique in a very strong sense (see 3.21 below where this is made explicit for the case where P is a Boolean algebra.

Proof. (i) For $p\in P$ let $O_p=\{q\in P\mid q\leqslant p\}$. For all $p,\ q\in P$ $O_p\cap O_q=\bigcup_{r\leqslant p,q}O_r$. Therefore the set $\{O_p\mid p\in P\}$ is a base for a topology on P (see VI.1.7). Let C be the complete Boolean algebra of all regular open sets in this space and let $F(p)=\mathrm{Int}(\mathrm{Cl}(O_p))$. (a) holds trivially. $F(p)$ is the set of all $q\in P$ such that every extension of q is compatible with p (see VI.1.18(ii)). To prove (b) assume that $F(p)$ and $F(q)$ are compatible and let $r\in F(p)\cap F(q)=F(p)\cdot F(q)$. Thus every extension of r is compatible with both p and q. In particular, there is an $s\in P$ such that $s\geqslant r,p$, and since s is an extension of r there is a $t\in P$ such that $t\geqslant s,q$. Thus we have $t\geqslant s\geqslant p$ and $t\geqslant q$, hence p and q are compatible. (c) holds trivially since for $a\in C$ if $a\neq 0$ then there is a $p\in a$ and we have $F(p)\leqslant a$.

(ii) We define mappings $H:C\to C'$ and $H':C'\to C$ by setting

$$H(a)=\Sigma\{F'(p)\mid p\in P\wedge F(p)\leqslant a\},\text{ and}$$
$$(3)\qquad H'(b)=\Sigma\{F(p)\mid p\in P\wedge F'(p)\leqslant b\}.$$

We shall prove below that for all $a\in C$ $H'H(a)=a$. For reasons of symmetry this will also establish that for all $b\in C'$ $HH'(b)=b$. Thus HH' and $H'H$ are identity functions and therefore H is a bijection of C on C'. If $a,c\in C$ and $a\leqslant c$ then, by (3), $H(a)\leqslant H(c)$ follows from the obvious inclusion

$$\{F'(p)\mid p\in P\wedge F(p)\leqslant a\}\subseteq\{F'(p)\mid p\in P\wedge F(p)\leqslant c\}.$$

Therefore, by 2.3, H is an isomorphism of C on C'.

To prove $H'H(a)=a$ we show first that $H'H(a)\geqslant a$. Suppose $H'H(a)\ngeqslant a$ then, by (c), there is a $q\in P$ such that $F(q)\leqslant a\sim H'H(a)$. Since $F(q)\leqslant a$ (3) implies $F'(q)\leqslant H(a)$ and hence also $F(q)\leqslant H'H(a)$, contradicting our choice of q (since $0\notin\mathrm{Rng}(F)$). To prove now $a\geqslant H'H(a)$ it suffices, by (3), to prove that $a\geqslant F(p)$ for every p such that $F'(p)\leqslant H(a)$. Suppose $a\ngeqslant F(p)$ for some such p, then, by (c), there is an $r\in P$ such that $F(r)\leqslant F(p)\sim a$. Since $F(r)\leqslant F(p)$ (b) implies that r is compatible with p. Therefore we can choose r so that we have already $r\leqslant p$. Therefore we have, by (a), $F'(r)\leqslant F'(p)\leqslant H(a)$. By (3)

$$0\neq F'(r)=F'(r)\cdot H(a)=F'(r)\cdot\Sigma\{F'(q)\mid q\in P\wedge F(q)\leqslant a\},$$

hence $F'(r)\cdot F'(q)\neq 0$ for some $q\in P$ such that $F(q)\leqslant a$. By (b) r and q are compatible; let $s\leqslant r, q$. We have $s\leqslant q$ hence $F(s)\leqslant F(q)\leqslant a$. On the other hand, since $s\leqslant r$ $F(s)\leqslant F(r)\leqslant F(p)\sim a$, hence $F(s)\leqslant a\cdot(F(p)\sim a)=0$, contradicting $0\notin \text{Rng}(F)$.

Having established that H is an isomorphism of C on C' we still have to prove that $F'=HF$. For $q\in P$ we have, by (3), $H(F(q))\geqslant F'(q)$. For reasons of symmetry we have also $H'(F'(q))\geqslant F(q)$ and since H' is an isomorphism this implies

$$(4)\qquad F(q)=H'(H(F(q)))\geqslant H'(F'(q))\geqslant F(q).$$

All the inequalities in (4) are now equalities, hence $H'(H(F(q)))=H'(F'(q))$ and therefore $H(F(q))=F'(q)$. Thus we have proved $HF=F'$, which is what was left to prove. □

3.19 Proposition. *Let F be a mapping of a poset $\langle P,<\rangle$ into a complete Boolean algebra as in* 3.18(i). *For $p, q\in P$ the following conditions are equivalent.*

(d) $F(p)\leqslant F(q)$.

(e) *Every $r\leqslant p$ is compatible with q.*

(f) *Every s compatible with p is also compatible with q.*

An example. Suppose $\langle P,<\rangle$ is a set of six members as shown in Figure 8(a) and $x<y$ (i.e., $y<x$!) if one can go upwards from x to y along one or more of the connecting segments (thus $a<d$, $a<e$ but neither $d<e$ nor $e<f$). According to the present lemma

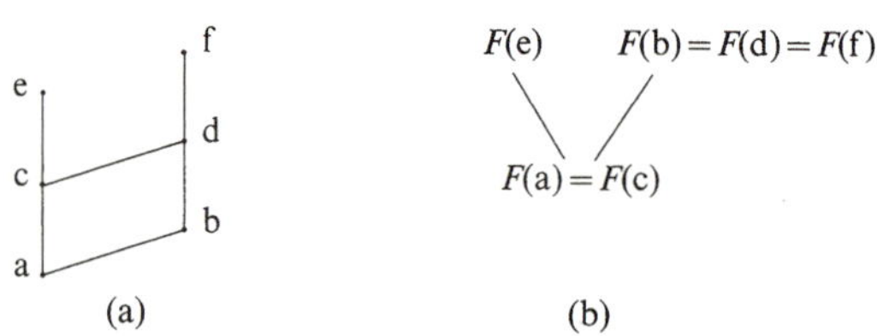

Fig. 8

the values of F on P are as given in Fig. 8(b), as can be easily verified. The reason for the values of F being what they are can be roughly described as follows. From a and c both "branches" leading to e and f can be reached by going upwards, from b, d and f only the f-branch can be reached, and from e only the e-branch is reached.

Proof. (d) → (f). Assume that $F(p)\leqslant F(q)$ and that s is compatible with p. Then there is an $r\in P$ such that $r\leqslant s, p$. By (a) of 3.18 $F(s)\cdot F(p)\geqslant F(r)>0$, and since $F(p)\leqslant F(q)$ also $F(s)\cdot F(q)>0$. By (b) of 3.18 s and q are compatible.

(f) → (e). This is immediate.

(e) → (d). Assume that (d) fails, i.e., $F(p)\nleqslant F(q)$ then, by (c) of 3.18, there is an $s\in P$ such that $F(s)\leqslant F(p)\sim F(q)$, hence $F(s)\leqslant F(p)$, $F(s)\cdot F(q)=0$. Since $F(s)\leqslant F(p)$ s and p are compatible (by (b) of 3.18), let $r\leqslant s, p$. By (a) of 3.18 $F(r)\leqslant F(s)$ hence $F(r)\cdot F(q)\leqslant F(s)\cdot F(q)=0$, and by (a) of 3.18 r and q are incompatible, even though $r\leqslant p$. Thus (e) fails too. □

3.20 Corollary. *Let $\langle P,<\rangle$ be a poset and let F be a mapping of P into a complete Boolean algebra as in* 3.18(i). *We have*

(g) *For all $p, q \in P$ $p \leqslant q \leftrightarrow F(p) \leqslant F(q)$, and hence F is an injection,*

iff

(h) *for all $p, q \in P$, if $p \nleqslant q$ then there is an $r \leqslant p$ incompatible with q.*

If $P = B \sim \{0\}$ where B is a Boolean algebra, then (h) *always holds and therefore F is an isomorphism of B into C (after it is extended by setting $F(0)=0$).* □

3.21 Corollary. *Let B be a Boolean algebra and let C_1, C_2 be two completions of B, then C_1 and C_2 are isomorphic. Moreover, there is an isomorphism H of C_1 on C_2 such that H is the identity on B.*

Hint of proof. The identity map $F = \langle x \mid x \in B \sim \{0\}\rangle$ of $B \sim \{0\}$ into C_1 and C_2 satisfies requirements (a)–(c) of 3.18(i) (by 3.12). Hence 3.21 follows from 3.18(ii). □

3.22 Exercises. (i) Let C be a complete Boolean algebra and let B be a subalgebra of C which is dense in C. Let H be a homomorphism of C into some Boolean algebra D such that $H \upharpoonright B$ is an isomorphism, then H itself is an isomorphism.

(ii)Ac (Sikorski 1948). Let B be a subalgebra of a Boolean algebra C and let H be a homomorphism of B into a complete Boolean algebra D. Then H can be extended to a homomorphism H' of C into D. (This is a strengthening of the Boolean prime ideal theorem 2.12(b), where we take B of 2.12(b) for our C, $K \cup \{\bar{x} \mid x \in K\}$ for our B and $\{0, 1\}$ for our D).

(iii) The requirement that D be complete is essential in (ii).

Hints. (ii) Use Zorn's lemma. Show that for every homomorphism such as H if $B \subset C$ then H can be extended to a homomorphism H' of the subalgebra $\{b \cdot u + c \cdot \bar{u} \mid b, c \in B\}$, where $u \in C \sim B$, by setting

$$H'(b \cdot u + c \cdot \bar{u}) = H(b) \cdot v + H(c) \cdot \bar{v}, \text{ where } v = \Sigma\{H(x) \mid x \in B \wedge x \leqslant u\}.$$

(iii) To get a counterexample choose for B and D an incomplete Boolean algebra, for C the completion of B and for H the identity function on B. □

3.23 Definition. A poset $\langle P, <\rangle$ is said to satisfy the *countable antichain condition* (c.a.c.) if

(j) every subset Q of P which consists of pairwise incompatible members is countable.

A Boolean algebra B is said to satisfy the c.a.c. if (j) holds for $P = B \sim \{0\}$ (i.e., where p and q are considered incompatible if $p \cdot q = 0$). A topological space $\langle X, O\rangle$ is said to satisfy the c.a.c. if the poset $\langle O \sim \{0\}, \supset\rangle$ satisfies the c.a.c., i.e., if every set of pairwise disjoint non-void open sets is countable.

3.24Ac Proposition. (i) *We say that the set Q is* dense *in the poset $\langle P,<\rangle$ if for every $p\in P$ there is a $q\in Q$ such that $p\leqslant q$. Let Q be dense in $\langle P,<\rangle$, then $\langle Q,<\rangle$ satisfies the* c.a.c. *iff $\langle P,<\rangle$ does.*

(ii) *A Boolean algebra satisfies the* c.a.c. *iff its Stone space satisfies the* c.a.c.

(iii) *A topological space satisfies the* c.a.c. *iff the complete Boolean algebra of its regular open sets satisfies the* c.a.c.

(iv) *Let B be a base for a topological space $\langle X, O\rangle$. The space $\langle X, O\rangle$ satisfies the* c.a.c. *iff the poset $\langle B\sim\{0\},\supset\rangle$ does.* □

3.25 Examples Where the Countable Antichain Condition Holds. Every countable poset satisfies the c.a.c. By 3.24(iv) every topological space with a countable base satisfies the c.a.c. We shall prove below that there are topological spaces and Boolean algebras of arbitrarily large cardinality which satisfy the c.a.c. For this purpose we need the following lemma.

3.26 Lemma. *Let X be a well-orderable set* (*or class*), *and let W be a finite set. The poset*

$$\langle\{p\mid p:\mathrm{Dom}(p)\rightarrow W\wedge \mathrm{Dom}(p)\in[X]^{<\omega}\},\subset\rangle \text{ satisfies the c.a.c.}$$

Proof. Let us denote this poset with $\langle P,\subset\rangle$. Since X is well-orderable it is easily seen that P is well-orderable too. Let Q be any subset of P. We shall construct a countable subset T of Q such that every member q of Q is compatible with some member t of T. If Q is a set of pairwise incompatible members of P then the only member of Q compatible with q is q itself, hence $T=Q$; therefore Q is countable, which is what we set out to prove.

If $Q=0$ we take $T=0$. If $Q\neq 0$ we define first a sequence $\langle T_n\mid n<\omega\rangle$ of finite subsets of Q which is increasing in the sense that $T_{n+1}\supseteq T_n$ for all $n\in\omega$. We set $T_0=0$. Given T_n let $Y_n=\bigcup\{\mathrm{Dom}(p)\mid p\in T_n\}$. Since T_n is a finite set $\{\mathrm{Dom}(p)\mid p\in T_n\}$ is a finite set of finite sets, hence also Y_n is finite. Let F_n be the set of all functions from subsets of Y_n into W; F_n is clearly finite. Let H_n be the function on F_n defined by

> $H_n(p)=$ a member q of Q such that q is compatible with p, if there is such, or any member of Q otherwise.

We set $T_{n+1}=T_n\cup\mathrm{Rng}(H_n)$; since $\mathrm{Dom}(H_n)=F_n$ is finite also T_{n+1} is finite. Let $T=\bigcup_{n<\omega}T_n$. Since $T\subseteq Q\subseteq P$ is well-orderable $T=\bigcup_{n<\omega}T_n$ is countable (by III.4.5). Now let us take a $q\in Q$ and prove that T has a member t compatible with q. Let $Y=\bigcup\{\mathrm{Dom}(p)\mid p\in T\}=\bigcup_{n<\omega}\bigcup\{\mathrm{Dom}(p)\mid p\in T_n\}=\bigcup_{n<\omega}Y_n$. The function g on $\mathrm{Dom}(q)\cap Y$ given by $g(x)=$ the least n such that $x\in Y_n$ has a finite domain and therefore must have a bounded range (by III.1.32 or III.1.11); let $m<\omega$ be an upper bound of $\mathrm{Rng}(g)$. Since the sequence $\langle Y_n\mid n<\omega\rangle$ is increasing we have $\mathrm{Dom}(q)\cap Y=\mathrm{Dom}(q)\cap Y_m$. By the definition of F_n $q\restriction Y_m\in F_m$. Let $t=H_n(q\restriction Y_m)$; since $q\in Q$ is compatible with $q\restriction Y_m$ we know that t is compatible with $q\restriction Y_m$, since $H_n(q\restriction Y_m)$ is defined to be a member of Q compatible with $q\restriction Y_m$ if there is such. By the definition of T_{m+1} $t \in T_{m+1}$, hence $\mathrm{Dom}(t) \subseteq Y_{m+1} \subseteq Y$. Thus $\mathrm{Dom}(t) \cap \mathrm{Dom}(q) \subseteq Y \cap \mathrm{Dom}(q) = Y_m \cap \mathrm{Dom}(q)$, hence $q \upharpoonright \mathrm{Dom}(t) \subseteq q \upharpoonright Y_m$ and since t is compatible with $q \upharpoonright Y_m$ t is compatible, a fortiori, with $q \upharpoonright \mathrm{Dom}(t)$, i.e., t is compatible with q. □

3.27Ac Proposition. *Let $\langle X, O \rangle$ be a topological space with a base of cardinality $\mathfrak{a}$. $\langle X, O \rangle$ has at most $2^{\mathfrak{a}}$ open sets. If the* c.a.c. *holds in $\langle X, O \rangle$ then $\langle X, O \rangle$ has at most $\mathfrak{a}^{\aleph_0}+1$ regular open sets.*

Proof. Let B be a base of $\langle X, O \rangle$ such that $|B| = \mathfrak{a}$. It is obvious that $|O| \leqslant 2^{\mathfrak{a}}$. We define a mapping H of ${}^{\aleph_0}B$ into the set of all regular open sets of $\langle X, O \rangle$ by setting, for $W = \langle W_i \mid i < \omega \rangle \in {}^{\aleph_0}B$, $H(W) = \mathrm{Int}(\mathrm{Cl}(\bigcup_{i<\omega} W_i))$; once we prove that $\mathrm{Rng}(H)$ contains all non-void regular open sets of the space we know that there are at most $|{}^{\aleph_0}B| = \mathfrak{a}^{\aleph_0}$ non-void regular open sets in this space (III.2.37). Let A be a non-void regular open set in this space. Let S be the set of all subsets C of the base B such that C consists of pairwise disjoint sets and $\bigcup C \subseteq A$. Obviously $0 \in S$. The poset $\langle S, \subset \rangle$ is easily seen to satisfy the condition of Zorn's lemma ((1) of V.1.9), hence S has a maximal member C^* with respect to $\subset$. Since A is non-void $C^* \neq 0$. Since $\langle X, O \rangle$ satisfies the c.a.c. C^* is countable, hence for some member W of ${}^{\aleph_0}B$ $C^* = \mathrm{Rng}(W)$. We shall now prove that $A = \mathrm{Int}(\mathrm{Cl}(\bigcup_{i<\omega} W_i)) = H(W)$, establishing thereby that $\mathrm{Rng}(H)$ contains all the non-void regular open sets. Since $C^* \in S$ $\bigcup_{i<\omega} W_i = \bigcup C^* \subseteq A$, hence $H(W) = \mathrm{Int}(\mathrm{Cl}(\bigcup_{i<\omega} W_i)) \subseteq \mathrm{Int}(\mathrm{Cl}(A)) = \mathrm{A}$. Assume that $\mathrm{Int}(\mathrm{Cl}(\bigcup_{i<\omega} W_i)) = H(W) \subset A$ then $A \nsubseteq \mathrm{Cl}(\bigcup_{i<\omega} W_i)$ and $A \sim \mathrm{Cl}(\bigcup_{i<\omega} W_i)$ is a non-void open subset of A. Therefore there is a non-void member E of B such that $E \subseteq A \sim \mathrm{Cl}(\bigcup_{i<\omega} W_i)$. We have, obviously, $E \cap W_i = 0$ for every $i < \omega$, hence $\{E\} \cup \{W_i \mid i < \omega\} = \{E\} \cup C^*$ is a member of S which properly includes C^*, contradicting the maximality of C^*. □

3.28 Proposition. *For a well ordered cardinal $\mathfrak{a}$ the topological space ${}^{\mathfrak{a}}2$, which is the $\mathfrak{a}$-th power of the discrete two point space 2, has $2^{\mathfrak{a}}$ open sets and it satisfies the countable antichain condition. If the axiom of choice holds and $\mathfrak{a}$ is infinite then the complete Boolean algebra of all regular open sets in ${}^{\mathfrak{a}}2$ has $\mathfrak{a}^{\aleph_0}$ regular open sets and it satisfies the* c.a.c.

Outline of proof. ${}^{\mathfrak{a}}2$ has at least $2^{\mathfrak{a}}$ closed sets since it has that many points. ${}^{\mathfrak{a}}2$ has at most $2^{\mathfrak{a}}$ open sets since its natural base is of cardinality $\mathfrak{a}$ (by III.4.21(iii)). To show that the topological space ${}^{\mathfrak{a}}2$ satisfies the c.a.c. proceed as follows. Let P be the set of all functions on finite subsets of $\mathfrak{a}$ into 2. For $p \in P$ let $B_p = \{s \in {}^{\mathfrak{a}}2 \mid s \supseteq p\}$. Use Lemma 3.26 with respect to $\langle P, \subset \rangle$ to show that the poset $\langle \{B_p \mid p \in P\}, \supset \rangle$ satisfies the c.a.c., and then apply 3.24(iv).

To prove that ${}^{\mathfrak{a}}2$ has at least $\mathfrak{a}^{\aleph_0}$ regular open sets one constructs an injection $A = \langle A_g \mid g \text{ is an injection of } \omega \text{ into } \mathfrak{a} \rangle$ into the set of all regular open sets as follows. For an injection g of ω into $\mathfrak{a}$ set $A_g = \{f \in {}^{\mathfrak{a}}2 \mid f(g(n)) = 1$ for some $n < \omega$ and the least such n is odd$\}$. Prove that A_g is a regular open set (notice that $\mathrm{Cl}(A_g) = A_g \cup \{f \in {}^{\mathfrak{a}}2 \mid \text{for all } n < \omega\ f(g(n)) = 0\}$). The function A is an injection since one can find out from A_g what g is (prove it!). □

3.29 Definition. An ideal I in a Boolean algebra B is said to be $\mathfrak{a}$-*saturated* if every subset C of B of cardinality $\mathfrak{a}$ which consists of pairwise incompatible members contains a member of I.

3.30 Examples of $\mathfrak{a}$-Saturated Ideals. No ideal is 0-saturated or 1-saturated. An ideal is 2-saturated iff it is prime. If an ideal I is $\mathfrak{a}$-saturated and $\mathfrak{a} \leqslant \mathfrak{b}$ then it is also $\mathfrak{b}$-saturated. Find a 3-saturated ideal which is not 2-saturated in the field of all subsets of a set X with $|X| \geqslant 2$! The next two propositions establish two examples of $\aleph_1$-saturated ideals.

3.31Ac Proposition. *In a topological space with a countable base, if the space itself is not meager then the ideal of all meager sets is $\aleph_1$-saturated in the field of all Baire sets of the space.*

Outline of proof. Let $\langle A_\alpha \mid \alpha < \aleph_1 \rangle$ be a sequence of $\aleph_1$ non-meager pairwise disjoint Baire sets. For $\alpha < \aleph_1$ let B_α be a basic open set such that $B_\alpha \sim A_\alpha$ is meager, and B_α is non-meager. Since there are only $\aleph_0$ basic open sets we get $B_\alpha = B_\beta$ for some $\alpha \neq \beta$, which contradicts $A_\alpha \cap A_\beta = 0$. □

3.32Ac Proposition. *If a measure space is of finite measure or is the union of $\aleph_0$ measurable sets of finite measure then the ideal of sets of measure 0 is an $\aleph_1$-saturated ideal in the σ-field of measurable sets.*

Outline of proof. Let $\langle X, \mu \rangle$ be a measure space and let $X = \bigcup_{n<\omega} X_n$, where $\mu(X_n) < \infty$ for $n < \omega$. Let $\langle A_\alpha \mid \alpha < \omega_1 \rangle$ be a sequence of $\aleph_1$ pairwise disjoint measurable sets of positive measure. For each $\alpha < \omega_1$ there is an n such that $\mu(A_\alpha \cap X_n) > 0$ and there is an m such that $\mu(A_\alpha \cap X_n) \geqslant \frac{1}{m}$. Let g and h be functions on ω_1 into ω such that $\mu(A_\alpha \cap X_{g(\alpha)}) \geqslant (1/h(\alpha))$. There must be therefore a pair $\langle n, m \rangle$ such that $\{\alpha \mid g(\alpha) = n \wedge h(\alpha) = m\}$ is infinite. X_n has therefore infinitely many pairwise disjoint measurable subsets of measure $\geqslant \frac{1}{m}$, contradicting $\mu(X_n) < \infty$. □

3.33 Definition. (i) A Boolean algebra B is said to be $\mathfrak{a}$-*complete* if for every subset A of B with $|A| < \mathfrak{a}$ ΣA exists. By 1.17, this condition is equivalent to the condition that for every subset A of B with $|A| < \mathfrak{a}$ ΠA exists. The $\aleph_1$-complete Boolean algebras are usually called *countably complete* since a Boolean algebra is countably complete iff ΣA exists for every countable subset A of B. Every σ-field of sets is obviously a countably complete Boolean algebra.

(ii) An ideal I in a Boolean algebra is said to be $\mathfrak{a}$-*complete* if for every subset A of I with $|A| < \mathfrak{a}$ ΣA exists and is a member of I. We say that I is *countably complete* if I is $\aleph_1$-complete. This is an extension of the definition in IV.4.5 of the same concept for the field of all subsets of a set.

Every ideal is $\aleph_0$-complete. Examples of $\aleph_1$-complete ideals are the ideal of all meager sets in a non-meager topological space and the ideal of all sets of measure 0 in a measure space.

3.34Ac Theorem (Smith and Tarski 1957). *If B is a countably complete Boolean algebra and I is a countably complete $\aleph_1$-saturated ideal in B then the quotient algebra B/I is complete.*

Proof. We have to prove that for every $A \subseteq B/I$ ΣA exists; without loss of generality we can assume that $0 \notin A$. The members of $A \subseteq B/I$ are equivalence classes of members of B. Let D be the set of all members of $B \sim I$ which are $\leqslant$ some member of an equivalence class in A. Let Q be the set of all subsets of D which consist of pairwise incompatible members. $\langle Q, \subset \rangle$ is a poset which satisfies the condition of Zorn's lemma (V.1.9) hence Q has a maximal member E. E consists of pairwise incompatible members and $E \subseteq D \subseteq B \sim I$, hence $|E| \leqslant \aleph_0$ since I is $\aleph_1$-saturated. ΣE exists since B is countably complete; let $w = \Sigma E$. Let H be the natural homomorphism of B on B/I defined by $H(y) = \{x \in B \mid x \vartriangle y \in I\}$. We shall prove that $H(w) = \Sigma A$.

First let us show that $H(w)$ is an upper bound of A. Suppose $y \in A$ and $y \nleqslant H(w)$ then $y - H(w) \neq 0$. Let u be a member of y then $H(u-w) = H(u) - H(w) = y - H(w) \neq 0$, hence $u - w \notin I$. Since $u - w \leqslant u \in y \in A$ we have, by the definition of D, $u - w \in D$. Since w is an upper bound of E and $u - w$ is incompatible with w, $u - w$ is also incompatible with every member of E. Thus $E \cup \{u - w\}$ is a member of Q which properly includes E, contradicting the maximality of E.

Now let us prove that $H(w)$ is the least upper bound of A. Let z be an upper bound of A, let $t \in z$, and let u be any member of a member of y of A. Since $z \geqslant y$ we have $H(u-t) = H(u) - H(t) = y - z = 0$, hence $u - t \in I$. If $v \in E$ then $v \leqslant u$ for some u which is a member of a member of A, hence $v - t \leqslant u - t \in I$ and therefore $v - t \in I$. $\{v - t \mid v \in E\}$ is a countable subset of I, and since I is countably complete $\Sigma\{v - t \mid v \in E\} \in I$. Let us denote $\Sigma\{v - t \mid v \in E\}$ with s. For any $v \in E$ $v \leqslant t + (v - t) \leqslant t + s$, hence $t + s$ is an upper bound of E. Since $w = \Sigma E$ we have $t + s \geqslant w$, hence $z = H(t) = H(t) + 0 = H(t) + H(s) = H(t+s) \geqslant H(w)$, which ends the proof that $H(w)$ is the least upper bound of A. □

Now we shall use Theorem 3.34 to obtain complete Boolean algebras.

3.35Ac Corollaries. (i) *In every measure space $\langle X, \mu \rangle$ where $\mu(X) < \infty$, or where X is the union of $\aleph_0$ sets of finite measure, the Boolean algebra of the measurable sets modulo sets of measure* 0 *is a complete Boolean algebra.*

(ii) *In a non-meager topological space with a countable base the Boolean algebra of the Baire sets modulo the meager sets is a complete Boolean algebra.* □

Remark. If a topological space is such that no non-void open set in it is meager then, as shown in 3.4, the Boolean algebra of the Baire sets modulo the meager sets is isomorphic in a very natural way to the Boolean algebra of all regular open sets, and is therefore complete. In 3.35(ii) we replaced the requirement that no non-void open set be meager by the requirement that the space be non-meager and have a countable base.

4. *Martin's Axiom*

We start this section by asking to what extent are the homomorphisms of a Boolean algebra also homomorphisms with respect to the infinitary operations Σ and Π. The study of this question will lead us to a theorem which asserts that for every Boolean algebra there is a homomorphism of that Boolean algebra onto the Boolean algebra $\{0, 1\}$ which is also a homomorphism with respect to Σ and Π for a given countable number of applications of Σ and Π. When we try to extend this theorem to more than a countable number of applications of Σ and Π we run into statements which are neither provable nor refutable in the set theory ZFC. From those statements as assumptions we derive several interesting consequences.

4.1 Completeness of Homomorphisms and Filters. We defined in 2.1 a homomorphism to be a mapping of one Boolean algebra B into another Boolean algebra C which commutes with the operations $+$, $\cdot$ and $^-$ of the Boolean algebras in the sense that for $a, b \in B$ when we apply one of the operations $+, \cdot, ^-$ of B to a, b and then apply H to the result we obtain the same member of C obtained by first applying H to a and b and then applying the corresponding operation of C to $H(a)$ and $H(b)$. It is now natural to ask whether, and to what extent, does H commute with the infinitary operations Σ and Π. We shall now see a simple example where H does not commute with Σ. Let I be a non-principal prime ideal in a Boolean algebra B, and let H be the natural homomorphism of B onto B/I. It is obvious that $\Sigma I = 1$; however $\Sigma\{H(i) \mid i \in I\} = \Sigma\{0\} = 0 \neq 1 = H(1) = H(\Sigma I)$. Since a homomorphism H and the ideal which is the kernel of H determine each other, the properties of the homomorphism related to commuting with Σ and Π are reflected in the properties of its kernel, as we shall now see.

4.2 Definition. A homomorphism of a Boolean algebra B into a Boolean algebra C is said to be $\mathfrak{a}$-*complete* if for every subset A of B with $|A| < \mathfrak{a}$ if ΣA exists then $\Sigma\{H(b) \mid b \in A\}$ exists and equals $H(\Sigma A)$.

4.3 Exercise. For a homomorphism H of a Boolean algebra B into a Boolean algebra C the following four conditions are equivalent.

(a) H is $\mathfrak{a}$-complete.

(b) For every subset A of B with $|A| < \mathfrak{a}$ if ΠA exists then $\Pi\{H(b) \mid b \in A\}$ exists and equals $H(\Pi A)$.

(c) For every subset A of B with $|A| < \mathfrak{a}$ if $\Sigma A = 1$ then $\Sigma\{H(b) \mid b \in A\} = 1$.

(d) For every subset A of B with $|A| < \mathfrak{a}$ if $\Pi A = 0$ then $\Pi\{H(b) \mid b \in A\} = 0$. □

4.4 Proposition. (i) *If H is an $\mathfrak{a}$-complete homomorphism then its kernel I is $\mathfrak{a}$-complete in the weaker sense that if $A \subseteq I$, $|A| < \mathfrak{a}$ and ΣA exists then $\Sigma A \in I$.*

(ii) *Let H be a homomorphism of a Boolean algebra B* onto *a Boolean algebra C. If the kernel I of H is $\mathfrak{a}$-complete in the sense that if $A \subseteq I$ and $|A| < \mathfrak{a}$ then A has an upper bound in I (this sense is stronger than that in* (i) *but weaker than that of* 3.33(ii)), *then if $E \subseteq B$, $|E| < \mathfrak{a}$ and ΣE exists then $\Sigma\{H(b) \mid b \in E\}$ exists and equals $H(\Sigma E)$.*

Proof. (i) If $A \subseteq I$, $|A| < \mathfrak{a}$ and ΣA is defined then, by the $\mathfrak{a}$-completeness of H, $H(\Sigma A) = \Sigma\{H(b) \mid b \in A\} = \Sigma\{0\} = 0$; hence $\Sigma A \in I$.

(ii) Let $E \subseteq B$, $|E| < \mathfrak{a}$ and $\Sigma E = c$, then $H(c)$ is obviously an upper bound of $\{H(b) \mid b \in E\}$ (by 2.2(iv)). We have to prove that $H(c)$ is the least upper bound of $\{H(b) \mid b \in E\}$ by showing that if $z \geqslant H(b)$ for every $b \in E$ then $z \geqslant H(c)$. Since H is onto C then for some $d \in B$ $z = H(d)$. For $b \in E$, $H(c)$, $H(d) \geqslant H(b)$ hence $H(b - c \cdot d) = H(b) - H(c) \cdot H(d) = 0$ and therefore $b - c \cdot d \in I$. The set $\{b - c \cdot d \mid b \in E\}$ is a subset of I of cardinality $|E| < \mathfrak{a}$ and hence, by our assumption on I, that set has an upper bound i in I. As a consequence $c \cdot d + i$ is an upper bound of E. Therefore $c \cdot d + i \geqslant \Sigma E = c$. Hence $H(c) \cdot H(d) = H(c) \cdot H(d) + 0 = H(c \cdot d) + H(i) = H(c \cdot d + i) \geqslant H(c)$ and therefore $z = H(d) \geqslant H(c)$, which is what was left to prove. □

4.5 Exercise. Show that in 4.4(ii) one can replace the requirement that H is onto C by the weaker requirement that $\mathrm{Rng}(H)$ is dense in C (3.12) and still prove the proposition. □

The next theorem will establish the existence of homomorphisms of a Boolean algebra on the Boolean algebra $\{0, 1\}$ which have the completeness property to a certain extent.

4.6Ac Theorem (Rasiowa and Sikorski 1950). *Let B be a Boolean algebra, let J and K be countable sets, let $\{S_j \mid j \in J\}$ and $\{T_k \mid k \in K\}$ be families of subsets of B such that ΣS_j and ΠT_k exist for $j \in J$ and $k \in K$. There is a homomorphism H of B on $\{0, 1\}$ which commutes with the operations ΣS_j and ΠT_k, i.e., $\Sigma\{H(a) \mid a \in S_j\} = H(\Sigma S_j)$ and $\Pi\{H(a) \mid a \in T_k\} = H(\Pi T_k)$, for $j \in J$ and $k \in K$. Equivalently, there is an ultrafilter F such that for $j \in J$ if $\Sigma S_j \in F$ then $S_j \cap F \neq 0$ and for $k \in K$ if $T_k \subseteq F$ then $\Pi T_k \in F$.*

Proof. We shall first prove the theorem for the case where $J = \omega$, $K = 0$ and $\Sigma S_j = 1$ for every $j \in J$, and then we shall reduce the general case of the theorem to this case. For the particular case we have just mentioned we define $\langle a_j \mid j \in \omega \rangle$ by recursion so that for $j \in \omega$ $a_j \in S_j$ and $\Pi_{i \leqslant j} a_i \neq 0$. Given $j \in \omega$ we have $\Pi_{i<j} a_i \neq 0$ by the induction hypothesis (if $j > 0$; if $j = 0$ $\Pi_{i<j} a_i = \Pi 0 = 1 \neq 0$). Since $\Sigma S_j = 1$ there must be a member b of S_j such that $b \cdot \Pi_{i<j} a_i \neq 0$ (otherwise $\Sigma S_j \leqslant \overline{\Pi_{i<j} a_i} < 1$); let a_j be such a b. Thus every finite product of the a_j's is non-zero, hence $\{a_j \mid j \in \omega\}$ generates a filter G (2.7(ii)). We take F to be any ultrafilter which includes G (2.12). We have $S_j \cap F \neq 0$ for every $j \in J$, hence the theorem holds for this case.

In order to deal with a more general case we give up the assumption that $\Sigma S_j = 1$. For $j \in J$ let $\Sigma S_j = s_j$ and $S'_j = \{\bar{s}_j + a \mid a \in S_j\}$. As easily seen $\Sigma S'_j = \bar{s}_j + \Sigma S_j = \bar{s}_j + s_j = 1$. Applying to the S'_j the case of the theorem which we have already proved we get an ultrafilter F such that $F \cap S'_j \neq 0$ for $j \in J$, i.e., for some $a \in S_j$ $\bar{s}_j + a \in F$. If $s_j = \Sigma S_j \in F$ then $\bar{s}_j \notin F$ and since $\bar{s}_j + a \in F$ we have $a \in F$ and hence $S_j \cap F \neq 0$, which concludes the proof of the present case.

When we come to deal with the general case we notice first that since J and K are just countable sets of indices we can assume, without loss of generality, that $J \cap K = 0$ and $J \cup K = \omega$. If our original J and K were finite but not both void we still can make $J \cup K$ denumerable by repeating one of the S_j's or the T_k's; if $J=K=0$ our theorem is just the Boolean prime ideal theorem. For $k \in K$ let $S_k = \{\bar{a} \mid a \in T_k\}$; since ΠT_k exists also ΣS_k exists and $\Sigma S_k = \overline{\Pi T_k}$. By the case which we have already proved there is an ultrafilter F such that for $j \in J \cup K$ if $\Sigma S_j \in F$ then $S_j \cap F \neq 0$. For $k \in K$, if $T_k \subseteq F$ then since $S_k = \{\bar{a} \mid a \in T_k\}$ we have $S_k \cap F = 0$ and hence, by the property of this F, $\Sigma S_k \notin F$. Since $\Sigma S_k = \overline{\Pi T_k}$ we have $\Pi T_k \in F$, which concludes our proof. □

Let us analyze the first part of the proof of 4.6 which contains the construction which is the key step of the proof. We want to obtain an ultrafilter F which contains at least one member out of each S_j. We cannot construct such an F in a single step since by putting in F one member out of each one of $S_0, S_1, S_2, \ldots$ we may have put in F incompatible members. Therefore we go about satisfying the requirements $S_j \cap F \neq 0$ one after the other, noticing that since at each step only finitely many members have been put into F it is possible to put in F a member of another S_j. It is clear that this argument will not work for an uncountable set J since once we put $\aleph_0$ members $a_0, a_1, \ldots$ into F we may come across an S_j such that each one of the complements of its members is $\geqslant$ than one of $a_0, a_1, \ldots$ and we can put no member of S_j in F. We shall now investigate the situation where the theorem holds also for uncountable J. Such a situation is interesting since, as we shall see, if it occurs then one can use the theorem for uncountable J to carry out beyond ω steps constructions like that of 4.6 which we cannot carry out without it.

4.7 Definition. For a Boolean algebra B and a cardinal $\mathfrak{a}$ $A_{\mathfrak{a}}(B)$ stands for the statement

$A_{\mathfrak{a}}(B)$ For every set J such that $|J| \leqslant \mathfrak{a}$ and for every family $\langle S_j \mid j \in J \rangle$ of subsets of B such that $\Sigma S_j = 1$ for $j \in J$ there exists a filter F in B such that $F \cap S_j \neq 0$ for all $j \in J$.

Notice that F is not required to be an ultrafilter, but F can be extended to an ultrafilter and every ultrafilter U which includes F satisfies $U \cap S_j \neq 0$ for all $j \in J$ just as F does.

4.8 Exercise. Prove that if $A_{\mathfrak{a}}(B)$ holds, if J, K are sets such that $|J|$, $|K| \leqslant \mathfrak{a}$, and if $\langle S_j \mid j \in J \rangle$ and $\langle T_k \mid k \in K \rangle$ are families of subsets of B such that ΣS_j and ΠT_k exist for $j \in J$ and $k \in K$ then there is an ultrafilter F in B such that for all $j \in J$ if $\Sigma S_j \in F$ then $S_j \cap F \neq 0$ and for all $k \in K$ if $T_k \subseteq F$ then $\Pi T_k \in F$. □

4.9 Proposition (An example of a failure of $A_{\mathfrak{a}}(B)$). *$A_{2^{\aleph_0}}(B)$ fails in every Boolean algebra B in which there is a sequence $\langle a_i \mid i < \omega \rangle$ of members such that for all $w \in {}^{\omega}2$ $\Pi_{i \in \omega} a_i^{(w_i)} = 0$, where $a^{(0)} = a$ and $a^{(1)} = \bar{a}$.*

Proof. Take in 4.8 $J=0$, $K={}^{\omega}2$ and $T_w = \{a_i^{(w_i)} \mid i < \omega\}$ for $w \in K$. Let F be any ultrafilter in B; define $v \in {}^{\omega}2$ by $v_i = 0$ if $a_i \in F$ and $v_i = 1$ if $a_i \notin F$, then for all i $a_i^{(v_i)} \in F$ hence $T_v \subseteq F$, but by the assumption of the proposition $\Pi T_v = \Pi_{i<\omega} a_i^{(v_i)} = 0 \notin F$, hence $A_{2^{\aleph_0}}(B)$ fails. □

The Boolean algebra B of all regular open sets in the Cantor space is a Boolean algebra as in 4.9. If we set for $i<\omega$ $a_i=\{f\in{}^{\omega}2\mid f(i)=0\}$ then a_i is a clopen set and is therefore in B. It is easily seen that for $w\in{}^{\omega}2$ $\bigcap_{i<\omega}a_i^{(w_i)}=\{w\}$ hence $\Pi_{i<\omega}a_i^{(w_i)}=\mathrm{Int}(\mathrm{Cl}(\{w\}))=\mathrm{Int}(\{w\})=0$.

4.10 Proposition (A second example of a failure of $A_{\mathfrak{a}}(B)$). *$A_{\mathfrak{a}}(B)$ fails in every Boolean algebra B for which there are alephs $\mathfrak{c}<\mathfrak{d}\leqslant\mathfrak{a}$ and a rectangle $\langle b_{\alpha,\beta}\mid\alpha<\mathfrak{c}\wedge\beta<\mathfrak{d}\rangle$ of members of B such that:*

(a) *For all $\beta<\mathfrak{d}$ $\Sigma_{\alpha<\mathfrak{c}}b_{\alpha,\beta}=1$ (the sum of each column is 1).*

(b) *For all $\alpha<\mathfrak{c}$, if $\beta,\gamma<\mathfrak{d}$ and $\beta\neq\gamma$ then $b_{\alpha,\beta}\cdot b_{\alpha,\gamma}=0$.*

(The terms in different places in each row are pairwise incompatible).

Proof. If $A_{\mathfrak{a}}(B)$ holds then, by (a), we can apply it to the family $\langle\langle b_{\alpha,\beta}\mid\alpha<\mathfrak{c}\rangle\mid\beta<\mathfrak{d}\rangle$ of the columns of the rectangle to obtain an ultrafilter F which contains at least one $b_{\alpha,\beta}$ from every column. Since the number $\mathfrak{c}$ of rows is less than the number $\mathfrak{d}$ of columns there must be at least one row containing members of F in two different places, which contradicts (b). □

To obtain a Boolean algebra B as in 4.10 consider the space ${}^{\omega}\omega_1$ with the topology which is the ω-th power of the discrete topology on ω_1, and take B to be the Boolean algebra of all regular open sets in ω_1. Take $\mathfrak{c}=\aleph_0$, $\mathfrak{d}=\aleph_1$ and for $\alpha<\omega$, $\beta<\omega_1$ $b_{\alpha,\beta}=\{f\in{}^{\omega}\omega_1\mid f(\alpha)=\beta\}$. $b_{\alpha,\beta}$ is clearly a clopen set and is therefore in B. It is also obvious that for $\gamma\neq\beta$ $b_{\alpha,\beta}\cdot b_{\alpha,\gamma}=b_{\alpha,\beta}\cap b_{\alpha,\gamma}=0$. $\Sigma_{\alpha<\omega}b_{\alpha,\beta}=\mathrm{Int}(\mathrm{Cl}(\bigcup_{\alpha<\omega}b_{\alpha,\beta}))$; hence it suffices to prove that $\bigcup_{\alpha<\omega}b_{\alpha,\beta}$ is dense in ${}^{\omega}\omega_1$ in order to obtain $\Sigma_{\alpha<\omega}b_{\alpha,\beta}=1$. A basic open set of ${}^{\omega}\omega_1$ is of the form $\{f\in{}^{\omega}\omega_1\mid f(\alpha_i)=\beta_i$ for $i=1,\ldots,n\}$ for some $n<\omega$, $\alpha_1,\ldots,\alpha_n<\omega$, $\beta_1,\ldots,\beta_n<\omega_1$. If $\alpha\neq\alpha_1,\ldots,\alpha_n$ then $b_{\alpha,\beta}$, and hence also $\bigcup_{\alpha<\omega}b_{\alpha,\beta}$, has a non-void intersection with this basic open set, which establishes the density of $\bigcup_{\alpha<\omega}b_{\alpha,\beta}$. □

Everybody who is familiar with the technique of forcing in set theory can easily come up with more counterexamples to $A_{\mathfrak{a}}(B)$ but the ones in 4.9 and 4.10 are sufficient to serve our purpose. Ideally one would like to know exactly for which $\mathfrak{a}$ and B can one prove that $A_{\mathfrak{a}}(B)$ fails. Then we could investigate the hypothesis that $A_{\mathfrak{a}}(B)$ holds in all other cases. No such information is in sight and the only cases for which we know that $A_{\mathfrak{a}}(B)$ is not refutable are embodied in Martin's axiom of 4.11.

4.11 Martin's Axiom (Martin and Rowbottom—see Martin and Solovay 1970). Let $A_{\mathfrak{a}}$ stand for the statement "$A_{\mathfrak{a}}(B)$ holds for every Boolean algebra B which satisfies the countable antichain condition". As we saw in 4.9 $A_{\mathfrak{a}}$ can be refuted for $\mathfrak{a}\geqslant 2^{\aleph_0}$. (The Boolean algebra B of the regular open sets in the Cantor space, for which $A_{\mathfrak{a}}(B)$ fails, satisfies the c.a.c. since the space has a countable base—3.24(iv).) We shall call by the name of *Martin's axiom* the statement that $A_{\mathfrak{a}}$ holds for every $\mathfrak{a}<2^{\aleph_0}$.

$A_{\aleph_0}$ is a restricted version of Theorem 4.6. Therefore Martin's axiom is a consequence of the continuum hypothesis, and as such it is obviously consistent with the axioms of ZFC. What is interesting and by no means trivial is that Martin's axiom essentially adds no information on the value of $2^{\aleph_0}$ (Solovay and Tennenbaum 1971) and is therefore consistent also with any one of the statements $2^{\aleph_0}=\aleph_2$, $2^{\aleph_0}=\aleph_3$, etc. How does Martin's axiom avoid the two examples 4.9 and 4.10 where $A_{\mathfrak{a}}(B)$ fails? The two restrictions we imposed on Martin's axiom, namely that $\mathfrak{a}$ be $<2^{\aleph_0}$ and that B satisfy the c.a.c. were expressly introduced so that Martin's hypothesis stays clear of the two counterexamples (even though, for all we know, weaker restrictions might have been sufficient). The restriction that $\mathfrak{a}<2^{\aleph_0}$ rules out 4.9 where $\mathfrak{a}=2^{\aleph_0}$. The restriction that B satisfy the c.a.c. rules out 4.10 as follows. If $\langle b_{\alpha,\beta} \mid \alpha<\mathfrak{c} \wedge \beta<\mathfrak{d}\rangle$ is a rectangle as in 4.10 then, since terms in different places in the same row are incompatible, the c.a.c. implies that there are at most $\aleph_0$ non-zero terms in each row. Since there are $\mathfrak{c}$ rows there are at most $\mathfrak{c}$ non-zero terms in the whole rectangle. Since the rectangle has $\mathfrak{d}$ columns and $\mathfrak{d}>\mathfrak{c}$ there must be a column consisting of zeros only, contradicting the assumption that the sum of every column is 1.

Since Martin's axiom is a consequence of the continuum hypothesis it can be viewed as a weak version of the continuum hypothesis which, instead of outright denying the existence of cardinals between $\aleph_0$ and $2^{\aleph_0}$, asserts that if there are such cardinals they behave like $\aleph_0$ to a certain extent. We shall indeed see below how Martin's axiom serves to replace $\aleph_0$ by any cardinal $<2^{\aleph_0}$ in several important theorems. Even though Martin's axiom is weaker than the continuum hypothesis it is known not to be a theorem of set theory; even its case $A_{\aleph_1}$ (which is the weakest case, except for $A_{\aleph_0}$ which we know is provable) is unprovable in ZFC. One way to prove the independence of Martin's axiom is to use the fact that $\aleph_1<2^{\aleph_0}<2^{\aleph_1}$ is consistent with ZFC (V.5.3) while Martin's axiom implies $\aleph_1<2^{\aleph_0}\rightarrow 2^{\aleph_1}=2^{\aleph_0}$ (4.22).

Martin's axiom is not an axiom in the same sense that the axioms of ZFC are axioms. While the axioms of ZFC are assumptions about the sets which mathematicians usually accept either as being "true" or as constituting a reasonable basis for set theory, Martin's axiom is an assumption which nobody seems to be taking seriously as expressing a "true" or acceptable statement about the state of affairs in set theory. Martin's axiom should have been called Martin's hypothesis, but the present name is unfortunately already in general use. Martin's axiom is important because it has many consequences in mathematics, either by itself or together with the additional hypothesis $2^{\aleph_0}>\aleph_1$ (some of those consequences are presented in the present section and in IX.2.41 and IX.2.58). Since Martin's axiom is consistent with ZFC, even together with the additional hypothesis $2^{\aleph_0}>\aleph_1$, whenever we prove a statement Φ from Martin's axiom together with $2^{\aleph_0}>\aleph_1$ we know that Φ is not refutable in ZFC. When this happens we can sometimes also show that Φ is not provable in ZFC (as we shall do in the cases of Souslin's hypothesis of IX.2.38 and the coloring theorem of IX.2.56). In other cases we can use the proof of Φ from these additional assumptions to help us find a proof of Φ in ZFC. This help can range anywhere from moral support to the situation where the proof of Φ from these additional assumptions is an integral part of the proof of Φ in ZFC.

In order to facilitate the application of Martin's axiom we shall formulate it also in terms of posets.

4.12 Definition. A poset $\langle P,\prec\rangle$ is said to be a *net* if for all $p, q \in P$ there is an $r \in P$ such that $r \geqslant p$ and $r \geqslant q$; we shall say in this case that p and q have a *common extension* r. A *subnet* of a poset $\langle P,\prec\rangle$ is a subset of P which is a net under the partial ordering $\prec$. Recall our convention of 3.16 that we write $x \prec y$ for $x \succ y$.

4.13 Exercises. (i) Let $\langle P,\prec\rangle$ be a poset and let $D \subseteq P$. Prove that D is dense in P in the sense of 3.24(i) iff D is dense in the tree topology on P as defined in VI.1.7.

(ii) Let $\langle P,\prec\rangle$ be a non-void poset and let $\langle D_n \mid n<\omega\rangle$ be a sequence of dense subsets of P. There is a subnet of P which has at least one member in common with each D_n.

Hint. (ii) Choose $p_n \in D_n$ such that $p_n \geqslant p_{n-1}$. □

4.14 Lemma. *Let $\langle P,\prec\rangle$ be a poset and H a mapping of P into the set of non-zero members of a Boolean algebra B as in* 3.18, *i.e.*,

(a) *if $p, q \in P$ and $p \leqslant q$ then $H(p) \leqslant H(q)$,*

(b) *if $p, q \in P$ are incompatible so are also $H(p)$ and $H(q)$, and*

(c) $\mathrm{Rng}(H)$ *is dense in B.*

Let us denote with $A'_{\mathfrak{a}}(P)$ the following statement.

$A'_{\mathfrak{a}}(P)$ If $|J| = \leqslant \mathfrak{a}$ and $\langle D_j \mid j \in J\rangle$ is a family of dense subsets of P then P has a subset G such that $G \cap D_j \neq 0$ for every $j \in J$.

If $A_{\mathfrak{a}}(B)$ holds and $|P| \leqslant \mathfrak{a}$ then $A'_{\mathfrak{a}}(P)$ holds.

Proof. Assume $A_{\mathfrak{a}}(B)$. Let P and $\langle D_j \mid j \in J\rangle$ be as in $A'_{\mathfrak{a}}(P)$ and $|P| \leqslant \mathfrak{a}$. First let us see that $\Sigma H[D_j]=1$ for every $j \in J$. If 1 is not the least upper bound of $H[D_j]$ then $H[D_j]$ has also another upper bound u and $\bar{u} \neq 0$. By the density of $\mathrm{Rng}(H)$ there is a $p \in P$ with $H(p) \leqslant \bar{u}$. Since D_j is dense in P there is a $q \in D_j$ with $q \leqslant p$. $H(q) \leqslant H(p) \leqslant \bar{u}$; but since u is an upper bound of $H[D_j]$ we have also $H(q) \leqslant u$, contradicting $H(q) \neq 0$, and thus $\Sigma H[D_j]=1$. For $p, q \in P$ let

$$Q_{p,q} = \{\overline{H(p) \cdot H(q)}\} \cup \{H(r) \mid r \leqslant p, q\}.$$

To prove $\Sigma Q_{p,q}=1$ assume, in order to obtain a contradiction, that $Q_{p,q}$ has also an upper bound u other than 1. By the density of $\mathrm{Rng}(H)$ there is an $s \in P$ such that $H(s) \leqslant \bar{u}$. If s is incompatible with p then by (b) also $H(s) \leqslant \overline{H(p)} \leqslant \overline{H(p) \cdot H(q)} \leqslant u$, contradicting $H(s) \neq 0$. Thus s is compatible with p and there is a $t \in P$ such that $t \leqslant s, p$. If t is incompatible with q then $H(t) \leqslant \overline{H(q)} \leqslant \overline{H(p) \cdot H(q)} \leqslant u$ contradicting $H(t) \leqslant H(s) \leqslant \bar{u}$. Thus there is an $r \in P$ such that $r \leqslant t, q$. On one hand $r \leqslant t \leqslant p$ and $r \leqslant q$ and hence $H(r) \in Q_{p,q}$ and $H(r) \leqslant u$; on the other hand $r \leqslant t \leqslant s$ and hence $H(r) \leqslant H(s) \leqslant \bar{u}$; this contradicts $H(r) \neq 0$.

Let us apply $A_{\mathfrak{a}}(B)$ to the family $\{H[D_j] \mid j \in J\} \cup \{Q_{p,q} \mid p, q \in P\}$. Since $|J|$, $|P| \leqslant \mathfrak{a}$ the cardinality of this family is $\leqslant \mathfrak{a}$, hence there is a filter F such that $F \cap H[D_j] \neq 0$ for $j \in J$ and $F \cap Q_{p,q} \neq 0$ for $p, q \in P$. Let $G = H^{-1}[F]$. Since $F \cap H[D_j] \neq 0$ we have obviously $G \cap D_j \neq 0$. To prove that G is a net let $p, q \in G$, then by definition of G $H(p), H(q) \in F$, hence $H(p)\cdot H(q) \in F$ and $\overline{H(p)\cdot H(q)} \notin F$. Therefore every member of $Q_{p,q} \cap F$ must be, by the definition of $Q_{p,q}$, an $H(r)$ such that $r \leqslant p, q$. Therefore $r \in G$, which establishes that G is a net. □

4.15Ac Proposition (Martin and Solovay 1970). *The following statements are equivalent, for all alephs* $\mathfrak{a}$.

(a) $A_{\mathfrak{a}}(B)$ *holds for every Boolean algebra* B *which satisfies the* c.a.c.

(b) $A_{\mathfrak{a}}(B)$ *holds for every complete Boolean algebra which satisfies the* c.a.c.

(c) $A'_{\mathfrak{a}}(P)$ *holds for every poset* P *with* $|P| \leqslant \mathfrak{a}$ *which satisfies the* c.a.c.

(d) $A'_{\mathfrak{a}}(P)$ *holds for every poset which satisfies the* c.a.c.

We shall denote statement (d) *with* $A'_{\mathfrak{a}}$.

Proof. (a) → (b) is trivial.

(b) → (c). Given a poset $\langle P, < \rangle$ with $|P| \leqslant \mathfrak{a}$ which satisfies the c.a.c. let B be the complete Boolean algebra which is the "completion" of P as in 3.18, and let H be the appropriate mapping of P into B. By 3.24(i) B satisfies the c.a.c., hence $A_{\mathfrak{a}}(B)$ holds. By 4.14 $A'_{\mathfrak{a}}(P)$ holds.

(c) → (d). Let $\langle P, < \rangle$ be a poset which satisfies the c.a.c. and for $|J| \leqslant \mathfrak{a}$ let $\langle D_j \mid j \in J \rangle$ be a family of dense subsets of P. For $j \in J$ let E_j be a function on P into D_j such that for every $p \in P$ $E_j(p) \in D_j$ and $E_j(p) \geqslant p$. Let $p_0 \in P$ and let P' be the closure of $\{p_0\}$ under the functions $E_j, j \in J$. Since $|J| \leqslant \mathfrak{a}$ also $|P'| \leqslant \mathfrak{a}$ (by III.4.25). Let $D'_j = D_j \cap P'$ for $j \in J$. For $p \in P'$ $p \leqslant E_j(p) \in D_j \cap P' = D'_j$ hence D'_j is dense in P'. By (c) there is a net G in P' such that $G \cap D'_j \neq 0$ for $j \in J$. This G is also a net as required for P since $G \cap D_j \supseteq G \cap D'_j \neq 0$ for every j.

(d) → (a). Let B be a Boolean algebra which satisfies the c.a.c. and let $\langle S_j \mid j \in J \rangle$ be a family of subsets of B such that $\Sigma S_j = 1$ for $j \in J$. Let $\langle P, < \rangle = \langle B \sim \{0\}, > \rangle$ and $D_j = \{a \in P \mid a \leqslant b$ for some $b \in S_j\}$. Let us see that D_j is dense in P. If for $c \in P$ there is no $a \in D_j$ such that $a \leqslant c$ then $c \cdot b = 0$ for every $b \in S_j$ hence $\bar{c}$ is an upper bound of S_j contradicting $\Sigma S_j = 1$. Since P satisfies the c.a.c. there is, by (d), a subnet G of P such that $G \cap D_j \neq 0$ for every $j \in J$. Since G is a net, if for some $n < \omega$ $a_1, \ldots, a_n \in G$ then there is a $b \in G$ such that $a_1, \ldots, a_n \leqslant b$, hence $0 < b \leqslant \Pi\{a_1, \ldots, a_n\}$. Therefore G generates the filter $F = \{a \in B \mid a \geqslant b$ for some $b \in G\}$ in B. Let $b \in G \cap D_j$ then by the definition of D_j there is an $a \in S_j$ such that $b \leqslant a$, but then, by the definition of F, $a \in F$ and thus (a) holds. □

4.16 Exercises. (i) Give a direct proof of $A'_{\aleph_0}$.

(ii) Show that $A'_{2^{\aleph_0}}$ fails for the poset $\langle {}^{\omega}2, \subset \rangle$. □

4.17 Definition. Let $f, g \in {}^{\omega}\omega$. We say that g *majorizes* f if $g(n) > f(n)$ for all n from some n_0 upwards.

4.18 Proposition (Martin and Solovay 1970). *If* $A'_{\mathfrak{a}}$ *holds for an aleph* $\mathfrak{a}$ *then for every subset* T *of* ${}^{\omega}\omega$ *of cardinality* $\leqslant \mathfrak{a}$ *there is a* $g \in {}^{\omega}\omega$ *which majorizes each member of* T.

Proof. Let $P = \{\langle u, t \rangle \mid u \in {}^{\underline{\omega}}\omega \wedge t \subseteq T \wedge |t| < \aleph_0\}$. The member $\langle u, t \rangle$ of P is to be understood to assert the following information, which may be true or false. $\langle u, t\rangle$ asserts that u is the beginning of a g as in the proposition and for every n greater than the length of u and for every $f \in t$ $g(n) > f(n)$. Accordingly we define $\langle u, t\rangle \leqslant \langle u', t'\rangle$ if $u \subseteq u'$, $t \subseteq t'$ and for all n such that Length$(u) \leqslant n <$ Length(u') and $f \in t$ $u'(n) > f(n)$. The last condition in this definition of $<$ is included to make $\langle u', t'\rangle$ consistent with $\langle u, t\rangle$; since $\langle u, t\rangle$ asserts that for $f \in t$ and $n \geqslant$ Length(u) $g(n) > f(n)$ and $\langle u', t'\rangle$ asserts that u' is the beginning of g we must have $u'(n) > f(n)$ for the two pieces of information to be consistent. $\langle P, <\rangle$ is clearly a poset; we shall now see that it satisfies the c.a.c. If $\langle u, t\rangle$ and $\langle u', t'\rangle$ are incompatible then $u \neq u'$, since if $u = u'$ then $\langle u, t \cup t'\rangle$ is a common extension of $\langle u, t\rangle$ and $\langle u', t'\rangle$. Since there are only $\aleph_0$ u's in ${}^{\omega}\omega$, every set of pairwise incompatible members of P is countable.

For every $f \in T$ let $D_f = \{\langle u, t\rangle \in P \mid f \in t\}$. D_f is dense in P since for all $\langle u, t\rangle \in P$ $\langle u, t\rangle \leqslant \langle u, t \cup \{f\}\rangle \in D_f$. For every $n \in \omega$ let $D_n = \{\langle u, t\rangle \in P \mid \text{Length}(u) \geqslant n\}$. Also D_n is dense in P since if $\langle u, t\rangle \in P$ and Length$(u) \geqslant n$ then $\langle u, t\rangle \in D_n$, and if Length$(u) < n$ let $u' \in {}^{n}\omega$ be such that $u' \supseteq u$ and $u'(i) = \max\{f(i) \mid f \in t\} + 1$ for Length$(u) \leqslant i < n$, and then clearly $\langle u, t\rangle \leqslant \langle u', t\rangle \in D_n$.

Since $|T \cup \omega| = T + \aleph_0 \leqslant \mathfrak{a} + \aleph_0 = \mathfrak{a}$ there is, by $A'_{\mathfrak{a}}$, a subnet G of P such that $G \cap D_f \neq 0$ and $G \cap D_n \neq 0$ for each $f \in T$ and $n \in \omega$. If $\langle u, t\rangle$, $\langle u', t'\rangle \in G$ then for some $\langle u'', t''\rangle$ in G $\langle u, t\rangle$, $\langle u', t'\rangle \leqslant \langle u'', t''\rangle$ hence u, $u' \subseteq u''$ and thus u, u' are compatible. Therefore $g = \bigcup\{u \in {}^{\omega}\omega \mid \text{for some } t \subseteq T, \langle u, t\rangle \in G\}$ is a function into ω and Length$(g) = \omega$ or Length$(g) \in \omega$. Since $G \cap D_n \neq 0$ Length(g) must be at least n, for every n, hence Length$(g) = \omega$. Let us prove now that g majorizes each $f \in T$. For $f \in T$ $G \cap D_f \neq 0$, let $\langle u, t\rangle \in G \cap D_f$ and let Length$(u) = n_0$. For any $n \geqslant n_0$ let $\langle u', t'\rangle \in G \cap D_{n+1}$. Since G is a net G contains a member $\langle u'', t''\rangle$ extending both $\langle u, t\rangle$ and $\langle u', t'\rangle$, hence Length$(u'') \geqslant$ Length$(u') > n$. Since $\langle u'', t''\rangle \geqslant \langle u, t\rangle$ we have, by the definition of $\geqslant$, $u''(n) > f(n)$. Since $\langle u'', t''\rangle \in G$ we have $g \supseteq u''$ hence $g(n) = u''(n) > f(n)$, which is what we were out to prove. □

Proposition 4.18 and its proof show how $A_{\mathfrak{a}}$ can be used to extend standard diagonalization methods used for $\aleph_0$ up to the cardinal $\mathfrak{a}$. For $\mathfrak{a} = \aleph_0$ 4.18 can be easily proved by a straightforward diagonalization argument; if $T = \{f_n \mid n < \omega\}$ we take $g(n) = \max\{f_i(n) \mid i \leqslant n\} + 1$. In Proposition 4.18 we are able to satisfy $\mathfrak{a}$ requirements on g, rather than only $\aleph_0$, by means of $A_{\mathfrak{a}}$. □

Given subsets A, B of $P(\omega)$ we look for a $w \subseteq \omega$ such that $|w \cap a| < \aleph_0$ for every $a \in A$ but $|w \cap b| = \aleph_0$ for $b \in B$. It is useful to have such a w since this w can be taken to be a "code" of the set $C = \{c \subseteq \omega \mid |w \cap c| < \aleph_0\}$ which separates A and B in the sense that it includes A and is disjoint from B. If such a w exists and $b \in B$ then $|w \cap b| = \aleph_0$ while every finite union of members of A contains only finitely many members of w, hence $|b \sim \bigcup A_0| = \aleph_0$ for every finite subset A_0 of A. If $|A \cup B| \leqslant \mathfrak{a}$ and we assume $A'_{\mathfrak{a}}$ then this necessary condition for the existence of w is also sufficient, as we shall now see.

4.19 Definition. We say that the set D *almost covers* the set E if $|E \sim \bigcup D| < \aleph_0$, i.e., the members of D contain all the members of E except finitely many.

4.20 Theorem (Martin and Solovay 1970). *If $A'_{\mathfrak{a}}$ holds and $A, B \subseteq P(\omega)$ are such that $|A|, |B| \leqslant \mathfrak{a}$ and no finite subset of A almost covers a member of B then there is a $w \subseteq \omega$ such that $|w \cap a| < \aleph_0$ for every $a \in A$ and $|w \cap b| = \aleph_0$ for every $b \in B$.*

Proof. Let $P = [\omega]^{<\omega} \times [A]^{<\omega}$. The member $\langle u, t\rangle$ of P is understood to assert that $u \subseteq w$ (where w is the subset of ω we want to construct) and w contains no members of $\bigcup t$ in addition to those contained in u. Accordingly we define $\langle u, t\rangle \leqslant \langle u', t'\rangle$ if $u' \supseteq u$, $t' \supseteq t$ and $u' \cap \bigcup t = u \cap \bigcup t$. The last condition is essential to prevent a clash between what $\langle u, t\rangle$ asserts and what $\langle u', t'\rangle$ asserts. If $\langle u, t\rangle, \langle u, t'\rangle \in P$ then $\langle u, t \cup t'\rangle$ is a member of P which extends both $\langle u, t\rangle$ and $\langle u, t'\rangle$. Hence two members of P can be incompatible only if they contain different first components. Since $|[\omega]^{<\omega}| = \aleph_0$, P satisfies the c.a.c.

For $a \in A$ let $D_a = \{\langle u, t\rangle \in P \mid a \in t\}$. D_a is dense in P since for every member $\langle u, t\rangle$ of P $\langle u, t\rangle \leqslant \langle u, t \cup \{a\}\rangle \in D_a$. For $b \in B$ and $n < \omega$ let $D_{b,n} = \{\langle u, t\rangle \in P \mid |u \cap b| \geqslant n\}$. To see that $D_{b,n}$ is dense in P let $\langle u, t\rangle \in P$. By the hypothesis of the theorem t does not almost cover b, i.e., $|b \sim \bigcup t| = \aleph_0$. Let $v \subseteq b \sim \bigcup t$, $|v| = n$, then $\langle u, t\rangle \leqslant \langle u \cup v, t\rangle$ since $(u \cup v) \cap \bigcup t = (u \cap \bigcup t) \cup (v \cap \bigcup t) = u \cap \bigcup t \cup 0 = u \cap \bigcup t$, and $\langle u \cup v, t\rangle \in D_{b,n}$ since $(u \cup v) \cap b \supseteq v$ and $|v| = n$. Since $|\{D_a \mid a \in A\} \cup \{D_{b,n} \mid b \in B \wedge n < \omega\}| \leqslant \mathfrak{a}$ there is a subnet G of P such that $G \cap D_a \neq 0$ for $a \in A$ and $G \cap D_{b,n} \neq 0$ for $b \in B$ and $n < \omega$. Let w be the union of the first components of the members of G. For $a \in A$ let $\langle u, t\rangle \in G \cap D_a$, then $a \in t$. We shall see that $w \cap a \subseteq u \cap a$, and $u \cap a$ is finite since u is finite. Let $i \in w \cap a$, then, since $i \in w$, there is a $\langle u', t'\rangle \in G$ such that $i \in u'$. Since G is a net it contains a common extension $\langle u'', t''\rangle$ of $\langle u, t\rangle$ and $\langle u', t'\rangle$ and we have $i \in u''$. By the definition of $\leqslant$ $a \in t$ implies $u'' \cap a = u \cap a$. Since $i \in u'' \cap a$ also $i \in u \cap a$. Thus we have proved that $w \cap a \subseteq u \cap a$ which establishes the finiteness of $w \cap a$. To prove that for $b \in B$ $w \cap b$ is infinite let $b \in B$ and assume that $|w \cap b| < \aleph_0$. Let $n < \omega$ be such that $n > |w \cap b|$. Let $\langle u, t\rangle \in G \cap D_{b,n}$ then $|u \cap b| \geqslant n$ and since $w \supseteq u$ $|w \cap b| \geqslant n$, which is a contradiction. □

4.21 Exercise. Prove 4.20 for the case where $\mathfrak{a} = \aleph_0$ without using $A'_{\aleph_0}$. □

4.22Ac Corollary (Martin and Solovay 1970). *If $A_{\mathfrak{a}}$ holds then $2^{\mathfrak{a}} = 2^{\aleph_0}$. Thus Martin's axiom implies that $2^{\mathfrak{a}} = 2^{\aleph_0}$ for all $\aleph_0 \leqslant \mathfrak{a} < 2^{\aleph_0}$.*

Proof. If $A_{\mathfrak{a}}$ holds then $\mathfrak{a} < 2^{\aleph_0}$ (by 4.9) and hence, by V.5.25, there is a subset C of $\mathbf{P}(\omega)$ which consists of $\mathfrak{a}$ almost disjoint sets. We define a function F on $\mathbf{P}(\omega)$ into $\mathbf{P}(C)$ by $F(w) = \{a \in C \mid |w \cap a| < \aleph_0\}$. For every subset A of C there is, by 4.20 (where we take $B = C \sim A$), a $w \subseteq \omega$ such that $F(w) = A$. Thus F is a surjection, hence $2^{\aleph_0} = |\mathbf{P}(\omega)| \geqslant |\mathbf{P}(C)| = 2^{\mathfrak{a}}$. □

4.23Ac Corollary (Martin and Solovay 1970). *Martin's axiom implies that $2^{\aleph_0}$ is a regular cardinal.*

Proof. Assume that $2^{\aleph_0}$ is singular, then $\aleph_0 \leqslant \mathrm{cf}(2^{\aleph_0}) < 2^{\aleph_0}$; denote $\mathrm{cf}(2^{\aleph_0})$ with $\mathfrak{a}$. By 4.22 $2^{\mathfrak{a}} = 2^{\aleph_0}$, but by V:5.2(ii) $2^{\mathfrak{a}}$ is not cofinal with $\mathfrak{a}$, contradicting $\mathrm{cf}(2^{\aleph_0}) = \mathfrak{a}$. □

4.24 Exercises. (i) Let $T \subseteq \mathbf{P}(\omega)$ be a set such that $|T| \leqslant \mathfrak{a}$ and for every finite subset t of T $\bigcap t$ is infinite. If $A'_{\mathfrak{a}}$ holds then there is an infinite subset w of ω such that w is *almost included* in every member of T, i.e., $w \sim a$ is finite for all $a \in T$. (Hint. In 4.20 take $A = \{\omega \sim a \mid a \in T\}$, $B = \{\omega\}$).

(ii)Ac. For $a, b \subseteq \omega$ define $a \supset^* b$ if $|a \sim b| = \aleph_0$ while $|b \sim a| < \aleph_0$. Show that, by (i), if Martin's axiom holds then there is a sequence $\langle a_\alpha \mid \alpha < 2^{\aleph_0} \rangle$ of subsets of ω such that for all $\alpha < \beta < 2^{\aleph_0}$ $a_\alpha \supset^* a_\beta$. (Without Martin's axiom one can still prove the existence of such a sequence $\{a_\alpha \mid \alpha < \aleph_1\}$.) Let H be a splitting function on $\mathbf{P}(\omega)$, i.e., for every infinite subset a of ω $H(a)$ and $a \sim H(a)$ are infinite subsets of a. Let J be a function on $\mathbf{P}(\mathbf{P}(\omega))$ such that for all $W \subseteq \mathbf{P}(\omega)$ $J(W)$ is an infinite subset of ω which is almost included in every member of W, if there is such, and 0 otherwise. We define F on ${}^{\overset{2^{\aleph_0}}{\smile}}2$ into $\mathbf{P}(\omega)$ by $F(0) = \omega$, $F(u^\frown\langle 0 \rangle) = H(F(u))$ and $F(u^\frown\langle 1 \rangle) = F(u) \sim H(F(u))$, and if Length$(u)$ is a limit ordinal then $F(u) = J(\{F(u \upharpoonright \beta) \mid \beta < \text{Length}(u)\})$. Assuming Martin's axiom prove that if Length(u) = Length(v) and $u \neq v$ then $F(u)$ and $F(v)$ are almost disjoint subsets of ω. Use this also to give an alternative proof that if Martin's axiom holds then for $\mathfrak{a} < 2^{\aleph_0}$ $2^{\mathfrak{a}} = 2^{\aleph_0}$. □

4.25Ac Theorem. *Each of the following two statements $A''_{\mathfrak{a}}$ and $A'''_{\mathfrak{a}}$ (which are generalizations of the Baire category theorem* VI.3.8) *is equivalent to $A_{\mathfrak{a}}$.*

$A''_{\mathfrak{a}}$ *No compact Hausdorff space $\langle X, O \rangle$ which satisfies the* c.a.c. *is the union of $\leqslant \mathfrak{a}$ nowhere dense sets.*

$A'''_{\mathfrak{a}}$ *In every compact Hausdorff space $\langle X, O \rangle$ which satisfies the* c.a.c. *no open set U is included in the union of $\mathfrak{a}$ nowhere dense sets.*

Proof. $A'_{\mathfrak{a}} \to A'''_{\mathfrak{a}}$. Let P be the set of all closed subsets Y of U with non-void interior, and let $Y \leqslant Z$ be $Z \subseteq Y$. If $Y, Z \in P$ and Int$(Y) \cap$ Int$(Z)) \neq 0$ then Cl(Int$(Y) \cap$ Int(Z)) is in P and $Y, Z \leqslant$ Cl (Int$(Y) \cap$ Int(Z)), hence Y and Z are compatible. Therefore if T is a subset of P which consists of pairwise incompatible members then $\langle \text{Int}(Y) \mid Y \in T \rangle$ is a one–one function on T, and its range consists of pairwise disjoint non-void open sets. Since the space satisfies the c.a.c. the range of $\langle \text{Int}(Y) \mid Y \in T \rangle$ is countable and hence T is countable too. Thus P satisfies the c.a.c.

Let $\{W_j \mid j \in J\}, |J| \leqslant \mathfrak{a}$, be a family of nowhere dense subsets of X. For $j \in J$ let D_j be the set of all members of P which are disjoint from W_j. We shall see that D_j is dense in P. Let $Y \in P$, then Int(Y) is a non-void open set and since W_j is nowhere dense Int(Y) has a non-void open subset Y' disjoint from W_j. Since the space is regular by VI.2.14(ii) Y' has an open non-void subset Z such that Cl$(Z) \subseteq Y'$. Thus Cl$(Z) \geqslant Y$ and Cl$(Z) \in D_j$.

By $A'_{\mathfrak{a}}$ P has a subnet G such that $G \cap D_j \neq 0$ for $j \in J$. G is a set of closed sets such that for every finite subset H of G $\bigcap H \neq 0$ (since G is a net). Since the space is compact we have $\bigcap G \neq 0$. Let $C \in G \cap D_j$ then $C \cap W_j = 0$, and since $0 \neq \bigcap G \subseteq C$ we get $\bigcap G \cap W_j = 0$. Thus the non-void subset $\bigcap G$ of U is disjoint from $\bigcup_{j \in J} W_j$.

$A'''_{\mathfrak{a}} \to A''_{\mathfrak{a}}$ is trivial.

$A''_{\mathfrak{a}} \to A_{\mathfrak{a}}$. Let B be a Boolean algebra which satisfies the c.a.c. and let $\langle S_j \mid j \in J \rangle$ be a family of subsets of B such that $\Sigma S_j = 1$ for $j \in J$. Let W_j be the set of all ultrafilters in the Stone space X of B which are disjoint from S_j. We shall see that W_j is nowhere dense in X. A basic open set in X is of the form $\{F\in X \mid a\in F\}$ where $0\neq a\in B$. Since $\Sigma S_j=1$ we have $a\cdot b\neq 0$ for some $b\in S_j$. Thus $\{F\in X \mid a\cdot b\in F\}$ is an open subset of $\{F\in X \mid a\in F\}$ which is disjoint from W_j, and W_j is nowhere dense. By $A''_{\mathfrak{a}}$ there is a member F of X which belongs to no W_j, hence $F\cap S_j\neq 0$ for every $j\in J$, and thus $A_{\mathfrak{a}}$ holds.

4.26Ac Exercise. In a Boolean algebra B we say that an ultrafilter F *preserves* the sum $\Sigma S=c$ if either $s\in F$ for some $s\in S$ or else $c\notin F$, and we say that F preserves the product $\Pi S=c$ if either for some $s\in S$ $s\notin F$ or else $c\in F$.

(i) Prove that the set of all ultrafilters F which do not preserve a given sum or product in B is a nowhere dense set in the Stone space of B.

(ii) Use (i) to give an alternative proof of Theorem 4.6 (using VI.3.8).

(iii) Use (i) to prove Exercise 4.8 with the hypothesis $A_{\mathfrak{a}}(B)$ replaced by $A''_{\mathfrak{a}}$ and the assumption that B satisfies the c.a.c. □

4.27Ac Theorem (Martin and Solovay 1970). *$A_{\mathfrak{a}}$ implies that in a topological space $\langle X, O\rangle$ with a countable base the union of $\leqslant\mathfrak{a}$ meager sets is meager.*

Proof. It is clearly sufficient to prove that the union of a family $\{E_j \mid j\in J\}$, with $|J|\leqslant\mathfrak{a}$, of nowhere dense sets is meager. For this purpose we shall construct a sequence $\langle K_i \mid i\in\omega\rangle$ of basic open sets with the following properties (a) and (b). (a) For every $j\in J$ there is an $l\in\omega$ such that for all $i\geqslant l$ $K_i\cap E_j=0$. (b) For every basic open set G there are arbitrarily large i's such that $K_i\subseteq G$. We can set $H_l=\bigcup_{l\leqslant i<\omega}K_i$, H_l is obviously open and by (b) H_l is dense. $\bigcap_{l\in\omega}H_l$ is disjoint from $\bigcup_{j\in J}E_j$ since for each $j\in J$ if l is as in (a) above then $H_l\cap E_j=0$ hence $\bigcap_{l\in\omega}H_l\cap E_j=0$. Since $\bigcap_{l\in\omega}H_l$ is the intersection of $\aleph_0$ dense open sets its complement, which includes $\bigcup_{j\in J}E_j$, is meager.

Let $\langle G_i \mid i<\omega\rangle$ be a sequence consisting of all non-void basic open sets and in which every basic open set occurs infinitely often. We shall obtain the K's to satisfy (a) and (b) above by taking the subsequence of $\langle G_i \mid i<\omega\rangle$ obtained by taking only i's which belong to some infinite subset w of ω; using w, rather than ω, as the index set does obviously not invalidate any of the arguments we gave above. In order to satisfy (a) we require that for every given $j\in J$ for only finitely many $i\in w$ does $G_i\cap E_j\neq 0$, i.e., that $w\cap\{i\in\omega \mid G_i\cap E_j\neq 0\}$ be finite for every $j\in J$. In order to satisfy (b) we require that for every $n\in\omega$ for arbitrarily large $i\in w$ we have $G_i\subseteq G_n$, i.e., that $w\cap\{i\in\omega \mid G_i\subseteq G_n\}$ be infinite for every $n\in\omega$. If we set $A_j=\{i\in\omega \mid G_i\cap E_j\neq 0\}$ for $j\in J$ and $B_n=\{i\in\omega \mid G_i\subseteq G_n\}$ for $n\in\omega$ then our requirements are that $|A_j\cap w|<\aleph_0$ and $|B_n\cap w|=\aleph_0$ for $j\in J$ and $n\in\omega$. By 4.20 $A'_{\mathfrak{a}}$ implies the existence of such a w provided no finite union of A_j's almost covers any B_n; we shall now prove that this is indeed the case. Let J_0 be a finite subset of J. Clearly $\bigcup_{j\in J_0}A_j=\{i\in\omega \mid G_i\cap\bigcup_{j\in J_0}E_j\neq 0\}$. Since the union of finitely many nowhere dense sets is nowhere dense (as follows easily from VI.3.2) $\bigcup_{j\in J_0}E_j$ is nowhere dense and the open non-void set G_n has a non-void basic open subset G' disjoint from $\bigcup_{j\in J_0}E_j$. Since G' occurs infinitely often in the sequence $\langle G_i \mid i\in\omega\rangle$ the set

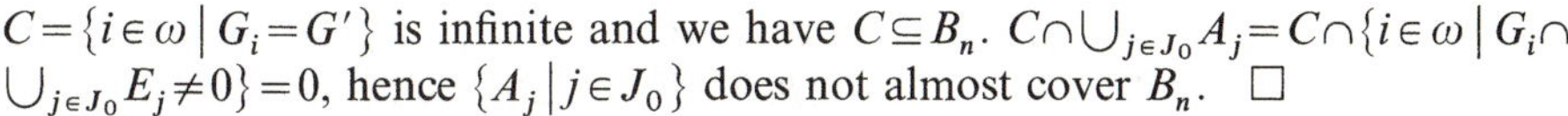
$C=\{i\in\omega \mid G_i=G'\}$ is infinite and we have $C\subseteq B_n$. $C\cap\bigcup_{j\in J_0}A_j=C\cap\{i\in\omega \mid G_i\cap \bigcup_{j\in J_0}E_j\neq 0\}=0$, hence $\{A_j \mid j\in J_0\}$ does not almost cover B_n. □

4.28Ac Corollary (Martin and Solovay 1970). *$A_{\mathfrak{a}}$ implies that in a topological space with a countable base the union and the intersection of $\leqslant\mathfrak{a}$ Baire sets are Baire sets.* □

4.29 Exercise (Shoenfield 1975). (i) Assume $A'_{\mathfrak{a}}$, let $\langle X, O\rangle$ be a topological space and let $x\in X$ be an accumulation point of the sequence $\langle x_n \mid n\in\omega\rangle$, i.e., for every neighborhood G of x the set $\{n\in\omega \mid x_n\in G\}$ is infinite. If the filter of all neighborhoods of x has a base of cardinality $\leqslant\mathfrak{a}$ (IV.4.4(vi)) then the sequence $\langle x_n \mid n\in\omega\rangle$ has a subsequence converging to x, i.e., there is an infinite subset w of ω such that for every neighborhood G of x there is an $n_0\in\omega$ such that $x_n\in G$ for all $n\in w$ such that $n\geqslant n_0$.

(ii)Ac Let us call a topological space *sequentially compact* if every sequence $\langle x_n \mid n\in\omega\rangle$ in it has a convergent subsequence. Prove that $A_{\mathfrak{a}}$ implies that the product of $\leqslant\mathfrak{a}$ compact metric spaces is sequentially compact.

Hints. (i) Let $\{G_j \mid j\in J\}$, $|J|\leqslant\mathfrak{a}$, be a base of open neighborhoods of x. Let $a_j=\{n \mid x_n\notin G_j\}$ for $j\in J$, $A=\{a_j \mid j\in J\}$ and $B=\{\omega\}$. Take w as in 4.20. (ii) Use (i). □

4.30 Theorem (Martin and Solovay 1970). *$A'_{\mathfrak{a}}$ implies that the union of $\leqslant\mathfrak{a}$ sets of real numbers of Lebesgue measure* 0 *is a set of measure* 0. *This is also true for the natural measures of the Cantor and Baire spaces* (*by* VII.3.18 *and* VII.3.19).

Proof. Let $\{E_j \mid j\in J\}$, where $|J|\leqslant\mathfrak{a}$, be a family of sets of measure 0 and let $E=\bigcup_{j\in J}E_j$. We shall show that for every real $\varepsilon>0$ there is an open set of measure $\leqslant\varepsilon$ which includes E; hence $\mu(E)=0$. Let P be the set of all open sets in the real line of measure $<\varepsilon$. For $p, q\in P$ we write $p\geqslant q$ if $p\supseteq q$. (Notice that this is unlike the natural definition of $p\geqslant q$ as $p\subseteq q$ or $p\leqslant q$.) Let us prove now that P satisfies the c.a.c. Let Q be a subset of P consisting of pairwise incompatible sets. For $0<\delta<\varepsilon$ let $Q_\delta=\{p\in Q \mid \mu(p)<\varepsilon-\delta\}$, where μ denotes the measure; it suffices to prove that $|Q_\delta|\leqslant\aleph_0$ since $Q=\bigcup_{n\in\omega}Q_{\varepsilon/(n+2)}$. For $p\in Q_\delta$ p is the union of countably many intervals hence, as easily seen, there is a subset q of p which is the union of finitely many intervals with rational endpoints such that $\mu(p\sim q)<\delta$. Let F be a mapping of Q_δ into $\mathbf{P}(\mathbb{R})$ where for each $p\in Q_\delta$ $F(p)\subseteq p$, $F(p)$ is a finite union of intervals with rational endpoints and $\mu(p\sim F(p))<\delta$. F is an injection since if for $p, r\in Q_\delta$ $F(p)=F(r)$ then $\mu(p\cup r)\leqslant\mu(p)+\mu(r\sim p)\leqslant\mu(p)+\mu(r\sim F(p))=\mu(p)+\mu(r\sim F(r))<(\varepsilon-\delta)+\delta=\varepsilon$, which is impossible since p and q are incompatible. Clearly $\mathrm{Rng}(F)$ is countable, hence $|Q_\delta|\leqslant\aleph_0$, which establishes that P satisfies the c.a.c.

For each $j\in J$ let $D_j=\{p\in P \mid p\supseteq E_j\}$. To see that D_j is dense let $p\in P$ and let q be an open set including E_j with $\mu(q)<\varepsilon-\mu(p)$, then $\mu(p\cup q)\leqslant\mu(p)+\mu(q)<\mu(p)+(\varepsilon-\mu(p))=\varepsilon$, hence $p\leqslant p\cup q\in D_j$. By $A'_{\mathfrak{a}}$ P has a subnet G such that $G\cap D_j\neq 0$ for every $j\in J$; let $t=\bigcup G$. t is an open set, and since $G\cap D_j\neq 0$ for every $j\in J$ $t\supseteq\bigcup_{j\in J}E_j$. We have to prove that $\mu(t)\leqslant\varepsilon$. Since $t=\bigcup G$, since G consists of open sets and since the space has a countable base $t=\bigcup_{i\in\omega}p_i$ for some $\{p_i \mid i\in\omega\}\subseteq G$.

We have $t=\bigcup_{i\in\omega}(p_i\sim\bigcup_{m<i}p_m)$ and the sets $p_i\sim\bigcup_{m<i}p_m$ are pairwise disjoint, hence $\mu(t)=\Sigma_{i=0}^{\infty}\mu(p_i\sim\bigcup_{m<i}p_m)$. If $\mu(t)>\varepsilon$ then for some $n<\omega$ $\varepsilon<\Sigma_{i=0}^{n}\mu(p_i\sim\bigcup_{m<i}p_m)=\mu(\bigcup_{i\leqslant n}(p_i\sim\bigcup_{m<i}p_m))=\mu(\bigcup_{i\leqslant n}p_i)$. Since G is a net G contains a member q which includes each one of $p_0,\ldots,p_n$, hence $\mu(q)\geqslant\mu(\bigcup_{i\leqslant n}p_i)>\varepsilon$, which contradicts $q\in p$. □

4.31Ac Corollary (Martin and Solovay 1970). *$A_{\mathfrak{a}}$ implies that the union of $\leqslant\mathfrak{a}$ Lebesgue measurable sets on the real line is Lebesgue measurable* (*and similarly for the natural measures in the Cantor and Baire spaces*).

Proof. Let $\{E_j \mid j\in J\}$, where $|J|\leqslant\mathfrak{a}$, be a family of Lebesgue measurable sets and let $E=\bigcup_{j\in J}E_j$. As is well known (see, for example, Halmos 1950) E has a measurable kernel C, i.e., $C\subseteq E$, C is measurable, and every measurable subset of $E\sim C$ is of measure 0. For each $j\in J$ we have $E_j\sim C\subseteq E\sim C$ hence $\mu(E_j\sim C)=0$. By 4.30 $\bigcup_{j\in J}(E_j\sim C)=E\sim C$ is a measurable set of measure 0, hence $E=C\cup(E\sim C)$ is measurable. □

Chapter IX
Infinite Combinatorics and Large Cardinals

The second part of this book has dealt until now with applications of set theory; now we return to set theory proper. The subjects which are treated in this chapter, namely trees, partitions and measurable cardinals, seem at first sight to be subjects which are interesting mostly because they can be applied to the study of various structures which occur in mathematics, and in particular to combinatorial structures and structures of mathematical analysis. However, it turns out that these subjects shed much light on the fundamental questions of "how far do the ordinals go on" and "what are the possible subsets of a given set". For this reason these subjects have become topics of central interest in set theory. Much of this chapter will deal with very large cardinal numbers such as the weakly compact, the ineffable and the measurable cardinals. These are cardinals whose existence cannot be proved in ZF (if ZF is consistent). We shall study the properties of these cardinals and prove also theorems of the kind: "if a large cardinal of type A exists then the least such cardinal is already greater than many large cardinals of type B".

Mathematical logic turned out to be a very useful tool in the study of large cardinals. As we do not go into logic in this book, and also for reasons of space, we can go here only through a small part of the theory of large cardinals while the deepest and the most beautiful theorems of it are not even mentioned here. For more information about this subject the reader is referred to the book of Drake 1974, to the articles of Boos 1975, Abramson, Harrington, Kleinberg and Zwicker 1977 and Kanamori and Magidor 1978 and to the many references listed in the biliographies of these works.

The present chapter starts with a short section on the concept of constructibility. While the topic of constructibility lies outside the scope of this book, this concept is so strongly tied with so many questions of set theory, and in particular with questions with which we deal in this chapter, that one just cannot avoid mentioning it.

1. The Axiom of Constructibility

As we have already mentioned in Chapter I the two main questions of set theory can be vaguely formulated as "how long do the ordinals go on?" and "which are the subsets of a given infinite set?". These questions are, to a large extent, unanswered by the axioms of ZF. We touched on the first question already in Chapter IV, where we discussed the existence of inaccessible cardinals, and it motivates also much of what we shall do in the present chapter. We touched on the second question when we discussed in Chapter V the hypotheses concerning the cardinals of power sets, namely the generalized continuum hypothesis and the gimel hypothesis.

To look deeper into the second question let us ask which are the sets which "must exist" once we take the class of ordinals as given. One may be tempted to answer that these are all the members of the class $\bigcup_{\alpha\in\mathrm{On}} R(\alpha)=\mathrm{Wf}$, but this is not the case. Already the set $R(\omega+1)$ contains, by II.6.8 and II.6.12, all the subsets of ω, and there is no reason for us to assume that all the subsets of ω which exist *must* exist once the ordinals exist; there may be many subsets of ω in a particular universe of set theory which are there but whose existence is not required by the axioms of ZFC. In fact, we have already mentioned that while we cannot prove in ZFC the existence of more than $\aleph_1$ subsets of ω, ω may have many more subsets. Having noticed that, we replace the function R, for our present purpose, by a function L which we shall define below. As in the case of R, all the members of $L(\alpha)$ will be subsets of $\bigcup_{\beta<\alpha} L(\beta)$, i.e., the members of $L(\alpha)$ consist of sets which are already known to have to exist, being members of "earlier" $L(\beta)$'s. Unlike the case of R, each member of $L(\alpha)$ will be obtained from $\bigcup_{\beta<\alpha} L(\beta)$ by a constructive process.

1.1 Definition. Df is a function on V defined as follows. First we define a function D on V by taking $D(a)$ to be the closure of the set $a\cup\{a, \epsilon_a, \epsilon_a^{-1}\}$ under the unary functions F_1, F_2, F_3 and the binary functions F_4, F_5, where $\epsilon_a=\{\langle x, y\rangle \mid x, y\in a \wedge x\in y\}$ and $\epsilon_a^{-1}=\{\langle y, x\rangle \mid x, y\in a\wedge x\in y\}$, and $F_1, \ldots, F_5$ are defined by

$$
\begin{aligned}
&F_1(u)=\mathrm{Dom}(u),\\
&F_2(u)=\{\langle\langle x, z\rangle, y\rangle \mid \langle\langle x, y\rangle, z\rangle\in u\},\\
&F_3(u)=\{\langle\langle z, x\rangle, y\rangle \mid \langle\langle x, y\rangle, z\rangle\in u\},\\
&F_4(u, v)=u\times v,\\
&F_5(u, v)=u\sim v.
\end{aligned}
$$

Now we set $\mathrm{Df}(a)=D(a)\cap\mathbf{P}(a)$.

For those among the readers who are familiar with mathematical logic let us remark that the reason for defining $\mathrm{Df}(a)$ the way we did is that $\mathrm{Df}(a)$ is exactly the set of all subsets of a which are definable in the structure $\langle a, \epsilon\rangle$ by a first-order formula from a finite number of parameters in a.

Notice that each member of $\mathrm{Df}(a)$ is a subset of a which is constructed in a finite number of steps which determine completely all the intermediate sets. The intermediate sets are not necessarily subsets of a but they consist of objects which are either members of a or else are obtained from members of a by one or more applications of pairing. Thus $\mathrm{Df}(a)$ is indeed very closely connected with a, and if we "know" a $\mathrm{Df}(a)$ will not contain any surprise.

1.2 Definition (Constructible sets—Gödel 1938). We define $L(\alpha)$ by recursion on α as follows. $L(0)=0$, $L(\alpha+1)=\mathrm{Df}(L(\alpha))$, and for a limit ordinal α $L(\alpha)=\bigcup_{\beta<\alpha}L(\beta)$. L is defined to be $\bigcup_{\alpha\in\mathrm{On}}L(\alpha)$ and the members of L are called *constructible* sets.

1.3 Proposition (Gödel 1938). *There is a definable relation which well-orders the class L.*

Hint of proof. One defines by recursion a relation $S(\alpha)$ which well-orders the set $L(\alpha)$, then one combines the $S(\alpha)$'s to obtain a well-ordering of L. To obtain the well-ordering $S(\alpha+1)$ of $L(\alpha+1)$ from the well-ordering $S(\alpha)$ of $L(\alpha)$ one proceeds as follows. From the well-ordering $S(\alpha)$ of $L(\alpha)$ one obtains a well-ordering of $D(L(\alpha))$ as in the last part of the proof of III.4.24; this yields a well-ordering of $L(\alpha+1)=\mathrm{Df}(L(\alpha))\subseteq D(L(\alpha))$. □

1.4 Exercise. Prove, without using the axiom of foundation, that $L\subseteq\mathrm{Wf}$. □

1.5 The Axiom of Constructibility. The *axiom of constructibility* is the statement that every set is constructible, i.e., that $V=L$. As is the case with Martin's axiom, this is not an axiom in the same sense that the axioms of ZF are axioms. While the axioms of ZF are assumptions about the sets which mathematicians usually accept either as being "true" or as constituting a reasonable basis for set theory, the axiom of constructibility is a pivotal assumption about the universe of sets which is usually neither accepted as being true nor even as being one of the basic assumptions of set theory. The axiom of constructibility should have better been called the constructibility hypothesis, but its present name is already in general use. When added as an axiom to ZF it yields a set theory which is usually understood to describe not the full universe of sets but a limited universe which consists of the constructible sets only. Gödel 1938 proved that this set theory is consistent (if ZF is consistent); in other words, one cannot prove in ZF (and hence also in ZFC) the existence of a non-constructible set.

The axiom of constructibility is an extremely strong hypothesis. By 1.4 and 1.3 it implies the axiom of foundation and the axiom of global choice. Gödel 1938 proved that it implies also the generalized continuum hypothesis. Jensen has developed methods of getting very detailed information about the constructible sets and used those methods to settle, under the assumption that the axiom of constructibility holds, many questions which are undecided in ZFC (see Devlin 1973 and Devlin and Johnsbraten 1974); we shall mention some of those results in § 2. The axiom of constructibility gives such an exact description of what all sets are that one of the most profound open problems in set theory is to find a natural statement of set theory which does not refer, directly or indirectly, to very large ordinals (as does a statement which asserts the existence of an inaccessible ordinal) and which is neither proved nor refuted by the axiom of constructibility. We required the statement to be "natural" in order to exclude statements such as the statements of number theory which are not decided by the axiom of constructibility as a result of Gödel's incompleteness theorem; the latter statements are especially constructed for this purpose and are not of a kind across which one may naturally come when one deals with set theory.

2. Trees

When we look at the general pattern of questions discussed in set theory we see that one main theme is the study of the structure of well ordered classes, or, what amounts to the same thing, the study of the structure of the class of the ordinals. What is special about the ordinals is that in going from smaller to larger ordinals no decisions have to be made as the only available direction is upward and it leads us from an ordinal to its successor and, at limit stages, to the appropriate limit ordinal. This does not at all make the theory of ordinals trivial; in Chapters II and IV we witnessed a rich and deep theory concerned with the ordinals. When we come to study other aspects of set theory and its applications we encounter processes and constructions where in going from one step to the next we have to choose among several possible directions. For example, when we construct a subset a of ω we often go through the following steps, where at each step we can make one out of two possible decisions: Shall we put 0 in a? Shall we put 1 in a?...The general framework for representing such a process is that of a tree, to be defined below. Most of the present section handles questions of whether a tree which satisfies certain conditions must have a branch which is as long as possible; these are the questions we address ourselves to in our study of the Aronszajn and Souslin trees. We are led also to studying Kurepa trees by questions concerning the relationship between the number of the choices we can make at each step of a process or a construction and the number of all possible outcomes of the process or the construction.

In our discussion of trees we shall encounter many concepts of set theory which are of independent interest. The most important among them are the concepts of the weakly compact ordinals and the ineffable ordinals, which are two species of large cardinals; we continue studying them also in the next section. We shall also discuss the combinatorial hypothesis $\diamondsuit$ and some of its consequences.

2.1 Definition. (i) A tree is a partially ordered set $\langle T, <\rangle$ such that for every $x \in T$ the set $pr_T(x) = \{y \in T \mid y < x\}$ of the *predecessors* of x is well ordered by $<$.

(ii) For a poset $\langle T, <\rangle$, a subset A of T which is totally ordered by $<$ is called a *chain*. A maximal chain (i.e., a chain A such that for no $x \in T \sim A$ is $A \cup \{x\}$ a chain) is called a *branch*. A *path* is a chain with no gaps, i.e., a chain A is a path if whenever $x, y \in A$, $z \in T$ and $x < z < y$ then also $z \in A$. In Fig. 9 the set

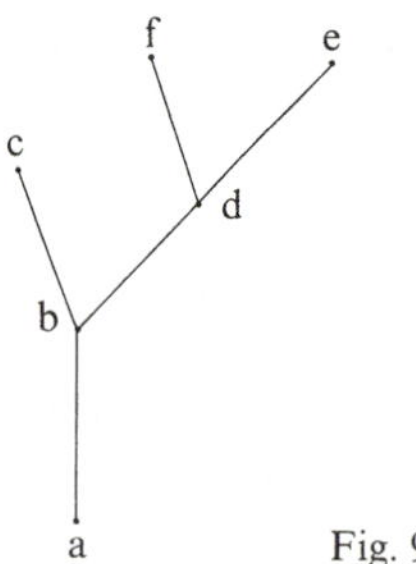

Fig. 9

$\{a, b, d, e, f\}$ is not a chain, the set $\{a, d, e\}$ is a chain but not a path, the set $\{b, d, e\}$ is a path but not a branch, and the set $\{a, b, d, e\}$ is a branch.

(iii) For a poset $\langle T, \prec \rangle$ we say that y is an *immediate successor* of x if $y \succ x$ and there is no z such that $y \succ z \succ x$. Members of T which have no predecessors are called *roots* of T.

2.2 Proposition. *If $\langle T, \prec \rangle$ is a tree then $\prec$ is well founded on T, and for every chain C in the tree $\langle C, \prec \rangle$ is a well ordered set.*

Proof. Since T is a set all we have to show is that every non-void subset S of T has a $\prec$-minimal member. Let z be a member of S. If z is not $\prec$-minimal in S then the set $\{x \in S \mid x \prec z\}$ is a non-void subset of the well ordered set $pr_T(z)$, hence it has a $\prec$-minimal member y. y is also a $\prec$-minimal member of S since if $x \in S$ and $x \prec y$ then, since $x \prec y \prec z$, also $x \prec z$ and $x \in \{x \in S \mid x \prec z\}$, contradicting the $\prec$-minimality of y.

If C is a chain in T then $\prec$ is well founded on C (by II.5.14(iv)), and since $\langle C, \prec \rangle$ is totally ordered and well founded it is well ordered. □

2.3 Definition. If $\langle T, \prec \rangle$ is a tree then, since $\prec$ is well founded on T, every member of T has a $\prec$-rank (II.5.12). We shall call this rank, somewhat unrigorously, the *T-rank* of x and denote it with $\rho_T(x)$. (Notice that the roots of T are exactly the members of T of T-rank 0). $\sup^+ \{\rho_T(x) \mid x \in T\}$ will be called the *length* of T. The set $\{x \in T \mid \rho_T(x) = \alpha\}$ will be called the *α-th level* of T and will be denoted with T_α. $T \mid \alpha$ will be used ambiguously for the set $\{x \in T \mid \rho_T(x) < \alpha\}$ and the tree $\langle \{x \in T \mid \rho_T(x) < \alpha\}, \prec \rangle$.

2.4 An Example of a Tree. For every set B and every ordinal α $\langle {}^{\underline{\alpha}}B, \subset \rangle$ is a tree of length α. For every $x \in {}^{\underline{\alpha}}B$ $\rho_T(x) = \text{Length}(x)$.

2.5 Proposition. *Let $\langle T, \prec \rangle$ be a tree.*

(i) *For every $x \in T$ $\rho_T(x)$ is the ordinal of $\langle \text{pr}_T(x), \prec \rangle$.*
(ii) *For every $\lambda < \rho_T(x)$ there is a unique $y \in T_\lambda$ such that $y \prec x$.*
(iii) *If κ is the length of $\langle T, \prec \rangle$ then for every $\lambda < \kappa$ $T_\lambda \neq 0$.*

Proof. (i) We shall see that $f = \rho_T 1 \text{pr}_T(x)$ is an isomorphism of $\langle \text{pr}_T(x), \prec \rangle$ onto $\langle \rho_T(x), < \rangle$. By II.5.13 f is an order preserving map of $\text{pr}_T(x)$ into $\rho_T(x)$, so all we have to prove is that f is a surjection. By II.5.14(vi) $\text{Rng}(f)$ is an ordinal β. We have, by the definition of $\rho_T(x)$,

$$\rho_T(x) = \sup^+ \{\rho_T(y) \mid y \in \text{pr}_T(x)\} = \sup^+ \text{Rng}(f) = \sup^+ \beta = \beta,$$

and thus

$$\text{Rng}(f) = \beta = \rho_T(x).$$

(ii) In the proof of (i) we saw that $\rho_T 1 \text{pr}_T(x)$ is a bijection of $\text{pr}_T(x)$ onto $\rho_T(x)$, hence for every $\lambda < \rho_T(x)$ there is a unique $y \in \text{pr}_T(x)$ such that $\rho_T(y) = \lambda$.

(iii) By II.5.13 $\rho_T[T]$ is an ordinal, hence $\rho_T[T]=\sup^+\rho_T[T]=\kappa$, and thus for every $\lambda<\kappa$ there is an $x\in T$ such that $\rho_T(x)=\lambda$. □

2.6 Proposition. *Let $\langle T,<\rangle$ be a tree and let C be a branch in T.*

(i) *If $y\in C$ and $z<y$ then also $z\in C$ (and hence every branch is a path).*

(ii) *If the length of C is λ then C has no members of T-rank $\geqslant\lambda$, and for every $\mu<\lambda$ C has a unique member of T-rank μ, which is therefore the μ-th member of C. Thus C is no longer than T.*

(iii)Ac *For every $x\in T$ T has a branch which contains x.*

(iv)Ac *If λ is less than the length κ of T then T has a branch of length $\geqslant\lambda$.*

Proof. (i) We shall prove that if $z\notin C$ then $C\cup\{z\}$ is a chain, contradicting the maximality of C. Since C is a chain it suffices to prove that z is comparable with every $x\in C$. For $x\in C$ we have $x<y$ or $x\geqslant y$, since C is totally ordered by $<$. If $x<y$ then $x, z\in \mathrm{pr}_T(y)$ and hence z and x are comparable; if $y\leqslant x$ then by $z<y\leqslant x$ z and x are comparable.

(ii) Let $g=\rho_T\upharpoonright C$. By II.5.12 g is an order preserving map, and by (i) and II.5.14(vi) $\mathrm{Rng}(g)$ is an ordinal. Thus g is an isomorphism of C on an ordinal which must be the length λ of C. For every $x\in C$ we have $\rho_T(x)=g(x)<\lambda$, and for every $\mu<\lambda$ there is a unique $x\in C$ such that $\rho_T(x)=g(x)=\mu$.

(iii) Let P be the set of all chains which contain x. The poset $\langle P,\subset\rangle$ is easily seen to satisfy the hypothesis of Zorn's lemma (P is non-void since $\{x\}\in P$), hence it has a maximal member S. S is obviously a branch which contains x.

(iv) By 2.5(iii) $T_\lambda\neq 0$, let $x\in T_\lambda$. By (iii) there is a branch S through x. By (ii) $\lambda=\rho_T(x)<$ the length of the branch S. □

2.7 Exercises. (i) A chain A in a tree $\langle T,<\rangle$ is a branch iff (a) for every $y\in A$ $\mathrm{pr}_T(y)\subseteq A$, and (b) there is no t such that $t>y$ for every $y\in A$.

(ii) Prove that $\langle T,<\rangle$ is a tree iff it is a well-founded poset in which every two incomparable members are incompatible.

(iii) Prove that a tree $\langle T,<\rangle$ with $|T|>1$ satisfies condition (h) of VIII.3.20 (with $<$ being equal to $>$) iff every $x\in T$ has at least two immediate successors. □

2.8 The Existence of Long Branches. Having seen in 2.6(iv) that a tree of length κ has branches of length $\geqslant\lambda$ for every $\lambda<\kappa$ it is natural to ask whether a given tree T of length κ has a branch of length κ. This is a worthwhile question since a tree represents a branching process and it is usually very important to know not just the least upper bound of the lengths of all the branches that the process can go along, but also the actual length of such a branch. In the uninteresting case where κ is a successor ordinal T has a branch of length κ, since every branch through an $x\in T_{\kappa-1}$ is such. In the case where κ is a limit ordinal such a branch does not have to exist, unless we make some special assumptions about the tree.

To obtain a tree of length κ which has no branches of length κ take $T=\{\langle\mu,\lambda\rangle\mid\mu<\lambda<\kappa\}$, and define $<$ by $\langle\mu,\lambda\rangle<\langle\mu',\lambda'\rangle$ iff $\lambda=\lambda'$ and $\mu<\mu'$; this tree is shown in Fig. 10. As easily seen, the branches of T are the sets $S_\lambda=\{\langle\mu,\lambda\rangle\mid\mu<\lambda\}$, where λ is fixed for each branch, and S_λ is clearly of length $\lambda<\kappa$.

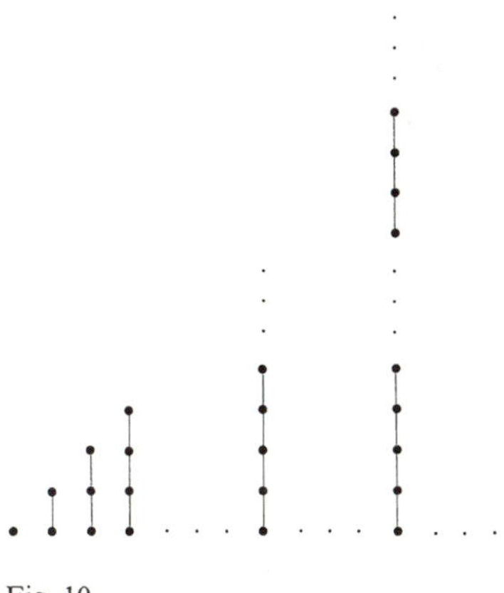

Fig. 10

This example of a tree of length κ with no branch of length κ can be viewed as a rather special and artificial example for the following reason. The T-rank of the member $\langle\mu,\lambda\rangle$ of T is μ since the function $\langle\langle\alpha,\lambda\rangle \mid \alpha<\lambda\rangle$ is obviously an isomorphism of $\langle\mu,<\rangle$ on $\langle \mathrm{pr}_T(\langle\mu,\lambda\rangle),<\rangle$. Since the only members of the tree which are $>\langle\mu,\lambda\rangle$ are of the form $\langle\beta,\lambda\rangle$ with $\beta<\lambda$ we can "go up from $\langle\mu,\lambda\rangle$" only to levels $\beta<\lambda$. This is one of the features which make T rather artificial. We would like instead to deal with trees in which one can go up from every member to every higher level below κ, and which have also other regularity properties, as will be introduced in the following definition.

2.9 Definition. (i) Let $\langle T,<\rangle$ be a tree of length κ. A member $x\in T$ is said to be a *blind alley* if the set $\{\rho_T(y) \mid y>x\}$ is bounded by an ordinal $\lambda<\kappa$, i.e., if all the members of T which are above x are of T-rank $<\lambda$, where $\lambda<\kappa$.

(ii) A tree $\langle T,<\rangle$ is said to be *normal* if

(a) T has a single root,

(b) every $x\in T$ has at least two immediate successors, and

(c) T has no blind alleys.

2.10 Proposition. *If κ is an ordinal of cofinality ω then every tree T of length κ which has no blind alleys has a branch of length κ.* □

As a consequence of 2.10, as far as normal trees are concerned, the question whether a branch of length κ must exist has a positive answer for κ's of cofinality ω. We shall now see that for all other infinite limit ordinals the answer is negative.

2.11 Proposition. *Let κ be an ordinal of cofinality $>\omega$, then there exists a normal tree T of length κ with no branch of length κ.*

Proof. The idea is that instead of constructing a tree of disconnected branches as in 2.8 we start here with a single root, connect to it κ paths such that for $\lambda<\kappa$ the λ-th path is of length λ. Each member of these paths is now clearly a blind alley, and to prevent that we connect κ new paths to each member of these paths, and then we again connect κ new paths to each new member, and so on for ω steps. The tree T which we get this way is best described as the set of all pairs $\langle\mu, h\rangle$ where $\mu<\kappa$ and h is a function on a finite subset of κ into κ which satisfies the following condition.

(1) If $h=\{\langle\delta_0,\lambda_0\rangle, \langle\delta_1,\lambda_1\rangle, \ldots, \langle\delta_{n-1},\lambda_{n-1}\rangle\}$ where $\delta_0<\delta_1<\cdots<\delta_{n-1}$, then if $n=0$ then $\mu=0$ and if $n>0$ then $\delta_0=1$, $\lambda_i>\delta_i$ for $i<n$, $\delta_{i+1}\leqslant\lambda_i$ for $i<n-1$, and $\delta_{n-1}\leqslant\mu<\lambda_{n-1}$.

The member $\langle\mu, h\rangle$ of the tree is to be construed as follows. $\langle\mu, h\rangle$ is in the level μ of T on a path which starts at the level δ_{n-1} and from which one has to get out at or below the level λ_{n-1}. To leave this path at the level δ, $\delta_{n-1}<\delta\leqslant\lambda_{n-1}$, one goes to the member $\langle\delta, h\cup\{\langle\delta,\lambda\rangle\}\rangle$ of T, where $\lambda>\delta$. Thus whenever one leaves a path and enters a new path one has to make at that time a commitment to leave the new path below or at some level $\lambda<\kappa$. Accordingly we define $\langle\mu', h'\rangle<\langle\mu, h\rangle$ if $\mu'<\mu$ and $h'=h\upharpoonright(\mu'+1)$. It is easily seen that $\langle T,<\rangle$ is a poset. For a given member $\langle\mu, h\rangle$ of T the mapping $\langle\langle\xi, h\upharpoonright(\xi+1)\rangle \mid \xi<\mu\rangle$ is obviously an isomorphism of $\langle\mu,<\rangle$ on $\langle \mathrm{pr}_T(\langle\mu, h\rangle),<\rangle$ and therefore T is a tree and the T-rank of $\langle\mu, h\rangle$ is μ. To see that T is normal we verify the following facts. $\langle 0,0\rangle$ is clearly the single root of T. Each member $\langle\mu, h\rangle$ of T has as successors all the members $\langle\mu+1, h\cup\{\langle\mu+1,\lambda\rangle\}\rangle$ with $\mu+1<\lambda<\kappa$, in addition to $\langle\mu+1, h\rangle$ if $\mu+1<\lambda_{n-1}$ (where h is as in (1)). Since $\mathrm{cf}(\kappa)>\omega$ this means that $\langle\mu, h\rangle$ has at least $\aleph_1$ immediate successors. No $\langle\mu, h\rangle$ in T is a blind alley since for every $\lambda>\mu$ we have $\langle\lambda, h\cup\{\langle\mu+1, \lambda+1\rangle\}\rangle>\langle\mu, h\rangle$ and $\langle\lambda, h\cup\{\langle\mu+1,\lambda+1\rangle\}\rangle$ is a member of T of T-rank λ. To see that T has no branch of length κ assume that $\{\langle\mu, h_\mu\rangle \mid \mu<\kappa\}$ is such a branch and $\langle\mu, h_\mu\rangle$ is its μ-th member. Since h_μ is finite, $h_\lambda\subseteq h_\mu$ for $\lambda<\mu$, and $\mathrm{cf}(\kappa)>\omega$ there is an ordinal $\mu_0<\kappa$ such that $h_\mu=h_{\mu_0}$ for all $\mu_0\leqslant\mu<\kappa$. (To see that define, for $k\in\omega$, $f(k)=$ the least μ such that $|h_\mu|=k$, if there is such a μ, and $f(k)=0$ otherwise; by IV.3.4 the range of f is bounded below κ, and we take for μ_0 any such bound.) Let $h_{\mu_0}=\{\langle\delta_0,\lambda_0\rangle, \ldots, \langle\delta_{n-1},\lambda_{n-1}\rangle\}$ be as in (1) then $\mu_0<\lambda_{n-1}$ and a branch through $\langle\mu_0, h_{\mu_0}\rangle$ must leave this path at a level not higher than λ_{n-1}, but the branch $\{\langle\mu, h_\mu\rangle \mid \mu<\lambda\}$ never leaves this path since $h_\mu=h_{\mu_0}$ for $\mu_0\leqslant\mu<\kappa$. Formally we obtain a contradiction from the fact that for $\mu\geqslant\lambda_{n-1}$ $\langle\mu, h_\mu\rangle=\langle\mu, h_{\mu_0}\rangle$ is a member of T, but this contradicts the requirement $\mu<\lambda_{n-1}$ of (1). □

2.12 The Need of Narrowness Conditions on Trees. 2.11 shows that for ordinals κ with $\mathrm{cf}(\kappa)>\omega$ even very nice trees T of length κ do not have to have branches of length κ. One may conjecture that the reason for why a tree of length κ does not have a branch of length κ is that the tree is too wide, and as in the case of the green trees, if we want T to have long branches we have to make sure it is well trimmed. The tree T we constructed in 2.11 is such that $T_\mu\subseteq\{\mu\}\times[\kappa\times\kappa]^{<\omega}$ and hence, by III.4.21(ii), $|T_\mu|\leqslant|\kappa|$. (It is, in fact, easily seen that for $0<\mu<\kappa$ $|T_\mu|=|\kappa|$.) Therefore, the least narrowness condition we might reasonably impose on a tree T in order to make it have branches of length κ is to require that $|T_\lambda|<|\kappa|$ for $\lambda<\kappa$. Accordingly, we define as follows.

2.13 Definition. We say that a tree $\langle T, <\rangle$ is a *κ-Aronszajn tree* if T is of length κ, for every $\lambda<\kappa$ $|T_\lambda|<|\kappa|$, and T has no branch of length κ. If there are no κ-Aronszajn trees, i.e., if every tree $\langle T, <\rangle$ of length κ in which $|T_\lambda|<|\kappa|$ for $\lambda<\kappa$ has a branch of length κ, we say that κ has the *tree property*.

2.14Ac Proposition. *Let T be a normal tree of length κ then for $\lambda<\kappa$ $|T_\lambda|\geq|\lambda|$. Therefore, if κ is not a cardinal then no κ-Aronazajn tree is normal.*

Hint of proof. Let $w\in T_\lambda$. For $\mu<\lambda$ let a_μ, b_μ, c_μ be such that $a_\mu\in T_\mu$, $a_\mu<w$, $b_\mu\in T_{\mu+1}$, $b_\mu>a_\mu$, $b_\mu\neq a_{\mu+1}$, $c_\mu\in T_\lambda$, $c_\mu>b_\mu$, then $\langle c_\mu \mid \mu<\lambda\rangle$ is an injection of λ into T_λ. □

2.15 Proposition. (i) *If κ is a singular cardinal with* $\mathrm{cf}(\kappa)>\omega$ *then there is a normal κ-Aronszajn tree.*

(ii) *If κ is a singular limit ordinal with* $\mathrm{cf}(\kappa)<|\kappa|$ *(and this includes all the singular alephs) then there is a κ-Aronszajn tree, and κ does not have the tree property.*

(iii) *If κ is a singular limit ordinal with* $\mathrm{cf}(\kappa)=|\kappa|$ *then κ has the tree property iff* $|\kappa|$ *has the tree property.*

Hint of proof. (i) Repeat the construction of the tree in 2.11, this time connecting only $\mathrm{cf}(\kappa)$ paths to each new member. (Accordingly, take h in (1) to be a finite subset of $\kappa\times\mathrm{cf}(\kappa)$).

(ii) Use the construction of the tree in 2.8 but admit only $\mathrm{cf}(\kappa)$ branches. One can easily modify this construction to obtain a tree which has a single root and two immediate successors of each member. Another way of constructing a tree as required is to use a construction similar to that used in (i), but to branch off not at every member but only at the members of $\mathrm{cf}(\kappa)$-many levels. This construction yields a tree with a single root and no blind alleys, but many of its members have only one immediate successor.

(iii) Let w be a cofinal subset of κ of order type $|\kappa|$, and let f be the function on $|\kappa|$ enumerating w. If T is a κ-Aronszajn tree then $\langle \bigcup_{\mu\in w} T_\mu, <\rangle$ is a $|\kappa|$-Aronszajn tree. In the other direction, if S is a $|\kappa|$-Aronszajn tree then one can easily construct a κ-Aronszajn tree T by stretching S to the length κ, i.e., by turning S_α to $T_{f(\alpha)}$ by filling in appropriately the levels in $\kappa\sim w$. □

Propositions 2.14 and 2.15 give us full answers to the questions of whether κ has the tree property and whether there are normal κ-Aronszajn trees for all singular limit ordinals κ, except that if $\mathrm{cf}(\kappa)=|\kappa|$ then the question of whether κ has the tree property is reduced to the question whether the regular ordinal $|\kappa|$ has the tree property. Therefore we shall turn now to regular ordinals κ, and this is where the deeper theory lies. First we shall see that for regular ordinals κ it makes no difference whether we look for normal κ-Aronszajn trees or for general κ-Aronszajn trees.

2.16 Proposition. *For every regular aleph κ if there is a κ-Aronszajn tree then there is a normal κ-Aronszajn tree.*

Hint of proof. Start with any κ-Aronszajn tree $\langle T, \lessdot\rangle$. First delete all blind alleys of T. Then, for every $\lambda<\kappa$ which is a limit ordinal or 0 and for every $x \in T_\lambda$, add to T a member x' such that $x' \lessdot x$ and $x' \gtrdot u$ for every $u \in \mathrm{pr}_T(x)$ so that $x'=y'$ iff $\mathrm{pr}_T(x)=\mathrm{pr}_T(y)$. Finally delete from T every member x which has no more than one immediate successor. □

2.17Ac König's Infinity Lemma (D. König 1926). *ω has the tree property. i.e., if $\langle T, \lessdot\rangle$ is a tree of length ω such that each level of T is finite then T has an infinite branch.*

Proof. Let us write $T[\geqslant a]$ for the set $\{b \in T \mid b \geqslant a\}$. We define by recursion on n a branch $\{a_n \mid n<\omega\}$ such that for every $n<\omega$ $a_n \in T_n$ and $T[\geqslant a_n]$ is infinite. Let $a_0 \in T_0$ be such that $T[\geqslant a_0]$ is infinite. There is such an a_0 since $T=\bigcup_{a \in T_0} T[\geqslant a]$ and if there were no such a_0 T would be a finite union of finite sets, and hence T and its length would be finite. Given $a_n \in T_n$ such that $T[\geqslant a_n]$ is infinite we notice that since $T[\geqslant a_n]=\{a_n\} \cup \bigcup_{a \in T_{n+1} \wedge a \geqslant a_n} T[\geqslant a]$ there must be an $a \in T_{n+1}$ such that $a \geqslant a_n$ and $T[\geqslant a]$ is infinite. We choose for a_{n+1} such an a. □

2.18 Exercise. Let A be a denumerable set of finite sets. Prove in ZF, i.e., in the absence of the axiom of choice, that if König's infinity lemma 2.17 holds then there is a choice function on A. □

2.19Ac Lemma. *For every infinite cardinal κ there is an ordered set $\langle Q, <\rangle$ such that* (a) *$|Q|=\kappa$ and* (b) *every $\alpha<\kappa^+$ can be embedded in every interval of Q, i.e., for all $u, v \in Q$ such that $u<v$ the interval $(u, v)=\{w \in Q \mid u<w<v\}$ has a subset B such that* $\mathrm{Ord}(\langle B, <\rangle)=\alpha$.

Proof. For $\kappa=\aleph_0$ we can take for $\langle Q, <\rangle$ the set of the rational numbers in their natural order. For a general κ take $Q \doteq {}^{\omega}\kappa$ and let $<$ be the left lexicographic order on ${}^{\omega}\kappa$, where we define an extension of a sequence to precede the sequence (i.e., for $u, v \in {}^{\omega}\kappa$ $u<v$ if $u^\frown\langle\kappa\rangle$ precedes $v^\frown\langle\kappa\rangle$ in the left lexicographic order of ${}^{\omega}\mathrm{On}$). $|Q|=\kappa$ by III.4.21(i). We prove (b) by induction on α. For $\alpha=0$ (b) is trivial. Let us be given $u, v \in {}^{\omega}\kappa$, $u<v$; we want to embed $\alpha>0$ in the open interval (u, v). If there is an $n<\mathrm{Length}(u), \mathrm{Length}(v)$ such that $u(n)<v(n)$ we set $w=v$ and by the definition of the order on ${}^{\omega}\kappa$ we have

(2) for every $\lambda<\kappa$ $u<w^\frown\langle\lambda\rangle<v$.

If there is no n as above then u is an extension of v. Let $\mathrm{Length}(v)=m$ and take w to be $v^\frown\langle u(m)+1\rangle$. then we have for all $\lambda<\kappa$ $u<w^\frown\langle\lambda\rangle<w<v$, and thus (2) holds in this case too. If α is a successor ordinal then, by the induction hypothesis, the interval $(w^\frown\langle 0\rangle, w^\frown\langle 1\rangle)$ has a subset b of order type $\alpha-1$. Let x be any member of $(w^\frown\langle 1\rangle, w^\frown\langle 2\rangle)$, such as $w^\frown\langle 2, 0\rangle$, then the order type of $b\cup\{x\}$ is clearly $(\alpha-1)+1=\alpha$, thus by (2) $b\cup\{x\}$ is a subset of (u, v) of order type α. If α is a limit ordinal let $\mathrm{cf}(\alpha)=\tau\leqslant\kappa$; then there exists an increasing sequence $\langle\beta_\nu \mid \nu<\tau\rangle$ of ordinals $<\alpha$ such that $\sup_{\nu<\tau}\beta_\nu=\alpha$. By the induction hypothesis the interval $(w^\frown\langle\nu\rangle, w^\frown\langle\nu+1\rangle)$

includes a set b_ν of order type β_ν. The set $b = \bigcup_{\nu<\tau} b_\nu$ is easily seen to be well ordered by $<$. Since the order type of b_ν is β_ν the order type of b is $\geqslant \alpha = \sup_{\nu<\tau}\beta_\nu$, hence b has a subset of order type α, which is, by (2), also a subset of (u, v). $\square$

2.20Ac Theorem (Aronszajn—see Kurepa 1938, Specker 1949). *If κ is a cardinal such that $\kappa^{\underline{\kappa}} = \kappa$ then there exists a κ^+-Aronszajn tree.*

Proof. Let Q be as in Lemma 2.19. T will consist of some, but not all, increasing sequences of members of Q of length $<\kappa^+$ and will be such that if $g \in T$ and $\beta <$ Length(g) then also $g \upharpoonright \beta \in T$. The relation $\prec$ on T is defined to be proper inclusion. If $g \in T$ is of length α then $\langle g\upharpoonright\beta \mid \beta<\alpha\rangle$ is an isomorphism of $\langle \alpha, <\rangle$ on the set of all predecessors of g in T. Hence T is a tree and the T-rank of g is equal to its length α. We shall now define T by recursion on α, thus we may assume that $T \mid \alpha$ is already defined. Concurrently we shall also prove by induction on α that the following (3)–(6) hold.

(3) $|T_\alpha| \leqslant \kappa$.

(4) If $\beta<\alpha$ and $h \in T_\alpha$ then $h\upharpoonright\beta \in T_\beta$.

(5) If $\beta<\alpha$ and $g \in T_\beta$ and $x \in Q$ is greater than some upper bound of Rng(g) then g has an extension h in T_α such that x is an upper bound of Rng(h).

(6) For every limit ordinal $\alpha<\kappa^+$, if cf$(\alpha)<\kappa$ and $h \in {}^\alpha Q$ is such that for all $\beta<\alpha$ $h\upharpoonright\beta \in T_\beta$ then $h \in T_\alpha$.

We define T_0 to consist only of the null sequence 0. (3)–(6) hold trivially for $\alpha=0$. If α is a successor ordinal we define T as the set of all ascending sequences $h \in {}^\alpha Q$ such that $h\upharpoonright(\alpha-1) \in T_{\alpha-1}$. Since by the induction hypothesis $|T_{\alpha-1}| \leqslant \kappa$, $|T_\alpha| \leqslant |T_{\alpha-1} \times Q| = |T_{\alpha-1}|\,|Q| = |T_{\alpha-1}|\kappa \leqslant \kappa^2 = \kappa$ and (3) holds for α. (4) is a trivial consequence of the induction hypothesis. To see that (5) holds we consider two cases. If $\beta<\alpha-1$ then let $y<x$ where y is an upper bound of Rng(g). Since 1 can be embedded in the interval (y, x) there is a z such that $y<z<x$. By the induction hypothesis g has an extension h' such that $h' \in T_{\alpha-1}$ and z is an upper bound of Rng(h'). Let $h = h' \cup \{\langle \alpha-1, z\rangle\}$ then h is as required in (5). If $\beta=\alpha-1$ put $h'=g$ and continue as above. (6) holds vacuously for α.

If α is a limit ordinal and cf$(\alpha)<\kappa$ we put in T_α every increasing sequence $h \in {}^\alpha Q$ such that for every $\beta<\alpha$ $h\upharpoonright\beta \in T_\beta$. (4) and (6) hold trivially for α. Let $\lambda = \text{cf}(\alpha)<\kappa$ and let $\langle \xi_\mu \mid \mu<\lambda\rangle$ be a continuous increasing sequence whose range is cofinal in α. The function F on T_α defined by $F(h) = \langle h\upharpoonright\xi_\mu \mid \mu<\lambda\rangle$ is obviously an injection of T_α into $\times_{\mu<\lambda} T_{\xi_\mu}$, hence $|T_\alpha| \leqslant \Pi_{\mu<\lambda}|T_{\xi_\mu}| \leqslant \Pi_{\mu<\lambda}\kappa = \kappa^\lambda = \kappa$, since $\kappa^{\underline{\kappa}} = \kappa$, thus (3) holds. To see that (5) holds let $\langle z_\mu \mid \mu<\lambda\rangle$ be an order preserving map of λ into the interval (y, x), where y is an upper bound of Rng(g) and $y<x$, Let $\langle \xi_\mu \mid \mu<\lambda\rangle$ be as above with $\xi_0=\beta$. We define $f_\mu \in T_{\xi_\mu}$ by recursion on μ so that z_μ is an upper bound of Rng(f_μ) and for $\mu<\nu$ $f_\mu \subseteq f_\nu$. Put $f_0=g$, since $y<z_0$ z_0 is an upper bound of f_0. For a successor ordinal μ we have $z_{\mu-1}<z_\mu$ and

$\xi_{\mu-1}<\xi_\mu$, and by the induction hypothesis $f_{\mu-1}\in T_{\xi_{\mu-1}}$ and $z_{\mu-1}$ is an upper bound of $\mathrm{Rng}(f_{\mu-1})$; therefore there is an $f_\mu\in T_{\xi_\mu}$ which extends $f_{\mu-1}$ and such that z_μ is an upper bound of f_μ. For a limit ordinal μ we have, since $\mu<\lambda<\kappa$, $\mathrm{cf}(\mu)<\kappa$. We define $f_\mu=\bigcup_{\nu<\mu}f_\nu$. Since $\langle\xi_\nu\mid\nu<\lambda\rangle$ is increasing and continuous we have Length $(f_\mu)=\bigcup_{\nu<\mu}\mathrm{Length}(f_\nu)=\sup_{\nu<\mu}\xi_\nu=\xi_\mu$. Since by IV.3.19 $\mathrm{cf}(\xi_\mu)=\mathrm{cf}(\mu)<\kappa$ we get, since (6) holds for ξ_μ, that $f_\mu\in T_{\xi_\mu}$. For $\nu<\mu$ z_ν is an upper bound of Rng (f_ν), and since $z_\mu>z_\nu$ also z_μ is an upper bound of $\mathrm{Rng}(f_\nu)$, hence z_μ is an upper bound of $\mathrm{Rng}(f_\mu)=\bigcup_{\nu<\mu}\mathrm{Rng}(f_\nu)$. We define now $h=\bigcup_{\mu<\lambda}f_\mu$. The length of h is $\bigcup_{\mu<\lambda}\xi_\mu=\alpha$, and since x is an upper bound of all the z_μ's x is an upper bound of $\mathrm{Rng}(h)$. To prove that $h\in T_\alpha$ we have to show, by the definition of T_α, that for every $\beta<\alpha$ $h\upharpoonright\beta\in T_\beta$. Let β be $<\alpha$, and let $\mu<\lambda$ be such that $\xi_\mu>\beta$, then $h\upharpoonright\xi_\mu=f_\mu$ and $h\upharpoonright\beta=(h\upharpoonright\xi_\mu)\upharpoonright\beta=f_\mu\upharpoonright\beta$. Since $f_\mu\in T_{\xi_\mu}$ $f_\mu\upharpoonright\beta\in T_\beta$, thus $h\upharpoonright\beta\in T_\beta$, which concludes the proof that h is as required in (5).

If $\mathrm{cf}(\alpha)=\kappa$ then for every $g\in T|\alpha$ and every $x\in Q$ which is greater than an upper bound of $\mathrm{Rng}(g)$ we take $h_{g,x}$ to be a member h of ${}^\alpha Q$ such that $h\supseteq g$, x is an upper bound of $\mathrm{Rng}(h)$, and for all $\beta<\alpha$ $h\upharpoonright\beta\in T_\beta$. The proof that such an h exists is exactly like the proof of the existence of an h as required by (5) in the case where $\mathrm{cf}(\alpha)<\kappa$, which we carried out above. We put $T_\alpha=\{h_{g,x}\mid g\in T|\alpha\wedge x$ is greater than an upper bound of $\mathrm{Rng}(g)\}$. It follows directly from the definition of T_α that (4) and (5) hold. (6) holds vacuously. $T_\alpha=\bigcup_{\beta<\alpha}\{h_{g,x}\mid g\in T_\beta\wedge x$ is an upper bound of $\mathrm{Rng}(g)\}$, hence, by the induction hypothesis, $|T_\alpha|\leqslant\Sigma_{\beta<\alpha}|T_\beta|\,|Q|\leqslant\Sigma_{\beta<\alpha}\kappa\leqslant\kappa|\alpha|=\kappa$, and thus (3) holds. □

2.21 The Tree Property for Accessible Regular Ordinals. Let us examine now whether the various uncountable accessible (i.e., not inaccessible) cardinals have or have not the tree property; for reasons which will become apparent later it is convenient to separate these cardinals from the inaccessible ones. First we shall state what is known to be provable in set theory without or with the generalized continuum hypothesis and the axiom of constructibility, and then we shall state what is known to be unprovable.

In ZFC we can apply Theorem 2.20 with $\kappa=\omega$. Since $\omega^{\underline{\omega}}=\omega$ there exists a normal ω_1-Aronszajn tree and ω_1 does not have the tree property. This is the only thing about the existence of κ-Aronszajn trees for regular uncountable κ's which we know to prove in set theory without any further assumptions. If κ is an inaccessible cardinal, or even just a weakly inaccessible cardinal such that $2^\lambda\leqslant\kappa$ for every cardinal $\lambda<\kappa$ (and we know that the existence of such cardinals is unprovable in set theory), then $\kappa^{\underline{\kappa}}=\kappa$ and by 2.20 κ^+ does not have the tree property. Once we assume the generalized continuum hypothesis we get $\kappa^{\underline{\kappa}}=\kappa$ for every regular cardinal κ (by V.5.16) and hence, by 2.20, there is a normal κ^+-Aronszajn tree for every regular κ, i.e., no cardinal successor of a regular cardinal has the tree property. Notice that as a consequence of the generalized continuum hypothesis the only regular accessible cardinals are the successor cardinals and $\aleph_0$ (since every weakly inaccessible cardinal is inaccessible). Out of these we have not yet dealt with the cardinal successors of singular cardinals. Theorem 2.20 is of no help in their case since for a singular cardinal κ we have $\kappa^{\underline{\kappa}}\geqslant\kappa^{\mathrm{cf}(\kappa)}>\kappa$ by V.5.10. If we assume the still stronger axiom of constructibility then, as proved by Jensen (see Devlin 1973, Ch. 9), for every successor cardinal κ there is a normal κ-Aronszajn tree. (Notice that if we assume the axiom of constructibility then we know exactly which ordinals have the tree property; it is typical of this axiom to decide questions one way or the other.)

To what extent are the results just presented the best possible? As to the cardinal successors of singular cardinals it is not known whether one can prove that they do not have the tree property even by means of the generalized continuum hypothesis; one cannot prove that they have the tree property since if the axiom of constructibility holds they fail to have this property. As to cardinal successors of regular cardinals it was shown by Silver (see Mitchell 1972) that in ZFC one cannot prove that these cardinals, other than ω_1, do not have the tree property (assuming that the existence of weakly compact cardinals—see 2.23 below—is consistent with ZFC); it is even consistent with ZFC that ω_2 has the tree property.

2.22 The Tree Property for Inaccessible Cardinals. The method used here to prove, by means of the generalized continuum hypothesis, that cardinal successors κ of regular cardinals have κ-Aronszajn trees, and other methods used to prove related results for all successor cardinals κ (without using the generalized continuum hypothesis) do not work for inaccessible cardinals κ. Therefore it was suspected for a while that those properties fail for inaccessible κ, i.e., for an inaccessible κ there is no κ-Aronszajn tree and κ has the tree property. The breakthrough was achieved by Hanf 1964 with respect to related properties taken from logic and was transferred by Tarski 1962 to trees (using earlier work of Erdös and Tarski 1961). The breakthrough consisted of proving that the properties under discussion, such as the existence of a κ-Aronszajn tree, which were already known to be true for many or all of the successor cardinals, were also true for the first inaccessible cardinal, the second inaccessible cardinal and many more inaccessible cardinals (if they exist at all). It was shown that if there are inaccessible cardinals κ for which these properties fail they must be very large cardinals even with respect to the inaccessible cardinals. This opened the gates for the discovery and study of the really large cardinals in set theory, the first of which we encounter in the next definition.

2.23 Definition. An ordinal is said to be *weakly compact* if it is inaccessible and has the tree property. (This term was chosen because of properties of such ordinals which we do not discuss in this book.) A weakly compact ordinal is obviously a cardinal.

2.24 Do Weakly Compact Cardinals Exist? We cannot prove in set theory the existence of a weakly compact cardinal since we cannot even prove the existence of an inaccessible cardinal (IV.3.28). Even if we assume the existence of an inaccessible cardinal we cannot prove from this assumption the existence of a weakly compact cardinal since we cannot prove from it even the existence of more than one inaccessible cardinal (this is shown by an argument very much like that in IV.3.28) and if there is only one inaccessible cardinal it does not have the tree property, being the first inaccessible cardinal. Using similar arguments one can show that we cannot derive the existence of a weakly compact cardinal even from the assumption that there are arbitrarily large inaccessible cardinals nor from stronger similar assumptions. On the other hand, if we assume the existence of a weakly compact cardinal κ we can derive from it the existence of many inaccessible cardinals; κ itself is inaccessible, it is not the first inaccessible cardinal since the latter does not have the tree property; for the same reason it is also not the second or third inaccessible cardinal, and so on. As we have mentioned in IV.3.28 with respect to the inaccessible cardinals, the fact that we cannot prove in ZFC the existence of a weakly compact cardinal, even if we assume the existence of inaccessible cardinals, should by no means be construed as evidence for the inexistence of weakly compact cardinals, or as a reason for their exclusion from set theory. As we shall see these cardinals are a source of some quite interesting mathematics.

The picture drawn here with respect to the weakly compact cardinals is very much the same for all other large cardinals, such as the Mahlo cardinals and the measurable cardinals discussed below.

To prove that weakly compact cardinals are very large we establish first one of their properties from which it is easy to obtain results concerning their size.

2.25Ac Lemma (Kunen 1977). *Let A be a set of alephs such that for all uncountable regular ordinals λ $A \cap \lambda$ is not a stationary subset of λ. Then there is a one–one regressive function g on A.*

Remark. For a singular cardinal κ and a subset $A \subseteq \kappa$ there is always a regressive function on A which obtains each value $<\kappa$ times; for a regular ordinal κ such a function exists iff A is not stationary (IV.4.41). The present lemma establishes, for a set A which consists of alephs only, the stronger condition for the existence of such a function which attains each value only once. The requirement that for every uncountable regular λ $A \cap \lambda$ be a non-stationary subset of λ is clearly necessary for the existence of a one–one regressive function in A, as is evident from IV.4.41.

Proof. Let $\gamma = \sup A$. We prove the lemma by induction on γ; therefore we assume that for every aleph $\delta < \gamma$ we are given a regressive one–one function g_δ on $A \cap (\delta+1)$. As the cases where γ is 0 or $\aleph_0$ are trivial we are left to deal with the following three cases.

Case a: γ is a successor cardinal, $\gamma = \delta^+$. Then we set $g = g_\delta \cup \{\langle \gamma, \delta \rangle\}$.

Case b: γ is singular. Let $\theta = \text{cf}(\gamma)$, let C be a closed unbounded subset of γ of order type θ which consists of cardinals $> \text{cf}(\gamma)$, and let $\langle \delta_\mu \mid \mu < \theta \rangle$ be the enumerating function of C. For $\alpha \in A$ we define

$$g(\alpha) = \begin{cases} g_{\delta_0}(\alpha) \dotplus 1 & \text{if} \quad \alpha < \delta_0, \\ \delta_\mu \dotplus g_{\delta_{\mu+1}}(\alpha) & \text{if} \quad \delta_\mu < \alpha < \delta_{\mu+1}, \\ \omega \bullet \mu & \text{if} \quad \alpha = \delta_\mu. \end{cases}$$

g is clearly a regressive one–one function.

Case c: γ is weakly inaccessible. By the hypothesis of the lemma $A\cap\gamma$ is a non-stationary subset of γ. Therefore there is a closed unbounded subset C of γ disjoint from A. Since the set of all alephs in γ is also closed and unbounded one can assume, without loss of generality, that C consists of alephs only (by IV.4.15). Let $\langle\delta_\mu \mid \mu<\gamma\rangle$ be the enumerating function of C. Define g on A by

$$g(\alpha)=\begin{cases} g_{\delta_0}(\alpha) & \text{if } \alpha<\delta_0, \\ \delta_\mu+g_{\delta_{\mu+1}}(\alpha) & \text{if } \delta_\mu<\alpha<\delta_{\mu+1}. \end{cases}$$

We do not have to define $g(\alpha)$ for $\alpha=\delta_\mu$ as we did above since by $\delta_\mu\in C$ and $C\cap A=0$ we have $\delta_\mu\notin A$. □

2.26Ac Theorem. *If κ is a weakly compact ordinal then for every stationary subset S of κ there is an uncountable regular ordinal $\lambda<\kappa$ such that $S\cap\lambda$ is a stationary subset of λ.*

Outline of proof (Kunen 1977). If S is a stationary subset of κ so is also $S\cap\{\aleph_\alpha \mid \alpha<\kappa\}$, hence we can assume that S consists of alephs. We assume that for every infinite regular $\lambda<\kappa$ $S\cap\lambda$ is not a stationary subset of λ and prove that S is not a stationary subset of κ. Let T be the set of all regressive one–one functions on sets of the form $S\cap\gamma$ for $\gamma<\kappa$. For the partial order on T we take proper inclusion. T is a tree. Let μ be the λ-th member of S, then T_λ consists of the one–one regressive functions on $S\cap\mu$, as easily seen. Since $\mu<\kappa$ the set $S\cap\mu$ clearly satisfies the hypothesis of Lemma 2.25 and hence there is a one–one regressive function on $S\cap\mu$ and thus $T_\lambda\neq 0$. $T_\lambda\subseteq{}^{\mu}\mu$, and since κ is inaccessible $|T_\lambda|\leqslant|\mu|^{|\mu|}<\kappa$. Since κ has the tree property T must have a branch W of length κ. $\bigcup W$ is obviously a one–one regressive function on S, and therefore S is not stationary, which is what we set out to prove. □

Is the property of weakly compact cardinals established in 2.26 equivalent to weak compactness? Or, more precisely, is it the case that

(7) κ is an inaccessible cardinal such that for every $S\subseteq\kappa$ which is stationary in κ there is an uncountable regular ordinal $\lambda<\kappa$ such that $S\cap\lambda$ is stationary in λ,

then κ is necessarily weakly compact? It follows from a theorem of Jensen (see Devlin 1977) that the axiom of constructibility implies that (7) is indeed equivalent to weak compactness; however, Kunen 1978a proved that in ZFC (7) does not imply weak compactness.

2.27 Definition (Mahlo 1911). An ordinal κ is called a *Mahlo ordinal* (or cardinal) if κ is inaccessible and the set of all regular ordinals $<\kappa$ is a stationary subset of κ.

2.28Ac Proposition. (i) *If κ is a Mahlo ordinal then also the set of all inaccessible ordinals $<\kappa$ is a stationary subset of κ.*

(ii) (Mahlo 1911). *If κ is a Mahlo ordinal then κ is the κ-th inaccessible ordinal and there are arbitrarily large $\lambda<\kappa$ such that λ is the λ-th inaccessible ordinal.*

Hint. Notice that the set $\{\lambda\in\kappa\cap Cn \mid (\forall\mu\in\lambda\cap Cn)2^{\mu}<\lambda\}$ of all strong limit cardinals $\lambda<\kappa$ is a closed unbounded subset of κ, and use IV.4.38(ii). □

2.29Ac Proposition (Hanf 1964, Keisler and Tarski 1964). *A weakly compact ordinal κ is a Mahlo ordinal, and the set of all Mahlo ordinals below κ is a stationary subset of κ. (Thus the least weakly compact ordinal, if it exists, is a lot bigger than the least Mahlo ordinal, which is, in turn, a lot bigger than the least inaccessible ordinal.)*

Proof. To prove that κ is Mahlo we shall show that the set $\{\lambda<\kappa \mid \lambda \text{ is regular}\}$ has a non-void intersection with every closed unbounded subset C of κ. For such a C there is, by 2.26, an uncountable regular ordinal $\lambda<\kappa$ such that $C\cap\lambda$ is a stationary subset of λ, thus $\sup(C\cap\lambda)=\lambda$. Since λ is a limit ordinal $<\kappa$ and C is a closed subset of κ we have $\lambda\in C$, thus $C\cap\{\lambda<\kappa \mid \lambda \text{ is regular}\}\neq 0$.

To prove the second half of the proposition assume, in order to obtain a contradiction, that there is a closed unbounded subset C of κ such that no member of C is a Mahlo ordinal. Let $S=\{\lambda<\kappa \mid \lambda \text{ is a strong limit cardinal}\}$. As easily seen, S is a closed unbounded subset of κ. Let $D=C\cap S\cap\{\lambda<\kappa \mid \lambda \text{ is regular}\}$; D is a stationary subset of κ by IV.4.38(ii). By 2.26 there is an uncountable regular $\lambda<\kappa$ such that $D\cap\lambda$ is a stationary subset of λ, hence $\lambda\in C\cap S$ (since $C\cap S$ is closed in κ). Since λ is regular and $\lambda\in S$ λ is inaccessible. $\{\mu<\lambda \mid \mu \text{ is regular}\}\supseteq D\cap\lambda$ hence $\{u<\lambda \mid \mu \text{ is regular}\}$ is stationary in λ, and λ is also Mahlo, which contradicts $\lambda\in C$. □

2.30Ac Exercise. Prove that if κ is a weakly compact ordinal then $\{\lambda<\kappa \mid \lambda$ is a Mahlo ordinal and the set of Mahlo ordinals $<\lambda$ is stationary in $\lambda\}$ is a stationary subset of κ. □

2.31 Stronger Narrowness Conditions. When we saw, at the beginning of the present section, that trees of length κ do not necessarily have branches of length κ we tried to make these trees have branches of length κ by imposing on them the narrowness requirement $|T_\lambda|<\kappa$ for $\lambda<\kappa$. We saw that if κ is ω or a weakly compact ordinal then this requirement does indeed imply the existence of a branch of length κ, but in all other cases we either found such trees which have no branch of length κ or else we were not able to decide whether such a tree T must have a branch of length κ. We shall now study the effect of a stronger narrowness requirement, namely that for some fixed $\mu<\kappa$ $|T_\lambda|<\mu$ for all $\lambda<\kappa$.

2.32Ac Proposition. *If κ is a limit ordinal, $\langle T,<\rangle$ is a tree of length κ and $\mathfrak{a}$ is a cardinal $<\mathrm{cf}(\kappa)$ such that for all $\lambda<\kappa$ $|T_\lambda|<\mathfrak{a}$ then $\langle T,<\rangle$ has a branch of length κ.*

Proof. Let

$$T'=\{x\in T \mid x \text{ is not a blind alley}\}.$$

Clearly T' is a subtree of T (i.e., if $x \in T'$, $y \in T$ and $y \prec x$ then also $y \in T'$), and hence for every $x \in T'$ $\rho_{T'}(x) = \rho_T(x)$ and for every $\lambda < \kappa$ $T'_\lambda = T' \cap T_\lambda$. We shall now prove that the length of T' is κ by showing that for every $\lambda < \kappa$ $T'_\lambda = T' \cap T_\lambda \neq 0$. Assume that $T' \cap T_\lambda = 0$ then all the members of T_λ are blind alleys. For each blind alley x let μ_x be the least ordinal $\mu < \kappa$ such that T_μ contains no member $y \succ x$. $|\{\mu_x | x \in T_\lambda\}| \leqslant |T_\lambda| < \mathfrak{a}$ and since $\mathfrak{a} < \mathrm{cf}(\kappa)$ $\{\mu_x \mid x \in T_\lambda\}$ is a bounded subset of κ; let $v < \kappa$ be an upper bound of this set. This is a contradiction since $T_v \neq 0$ and for every member $y \in T_v$ there is an $x \in T_\lambda$ such that $x \prec y$, contradicting $\mu_x \leqslant v$. An almost verbatim repetition of the same argument proves that in T' there are no blind alleys. If $\mathrm{cf}(\kappa) = \omega$ then T' has a branch of length κ by 2.10, so let us deal now with the case where $\mathrm{cf}(\kappa) > \omega$.

We say that $x \in T'$ is a *branching point* if there is a $z \in T'$ such that $z \neq x$ and $\mathrm{pr}_{T'}(z) = \mathrm{pr}_{T'}(x)$ (see Fig. 11). Let $\lambda < \kappa$ and $\mathrm{cf}(\lambda) \geqslant \mathfrak{a}$ and let $x \in T'_\lambda$. For $\alpha < \lambda$ let x_α

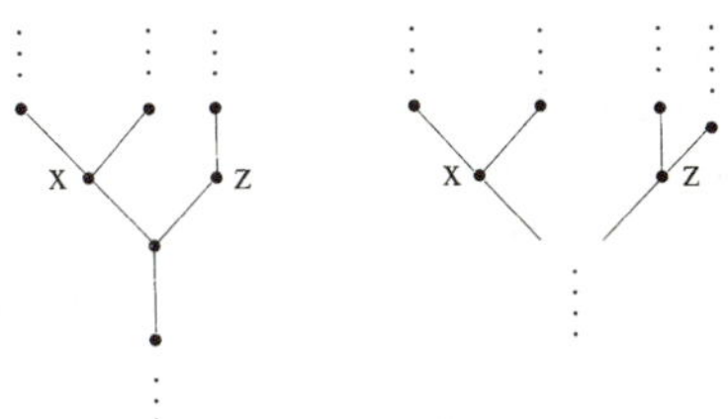

Fig. 11. Branching points x and z.

be the α-th member (in the order $\prec$) of $\mathrm{pr}_{T'}(x)$. If x_α is a branching point then, since T' has no blind alleys T'_λ must contain a member y such that $\mathrm{pr}_{T'}(y) \supseteq \mathrm{pr}_{T'}(x_\alpha)$ but $y \not\succ x_\alpha$; let us denote such a y with y_α. If $\alpha < \beta < \lambda$ and x_α and x_β are branching points then $y_\alpha \neq y_\beta$ since $y_\alpha \not\succ x_\alpha$ but $y_\beta \succ x_\alpha$. Thus the function

$$h_x = \langle y_\alpha \mid \alpha < \lambda \wedge x_\alpha \text{ is a branching point} \rangle$$

is an injection of a subset of λ into T'_λ, and therefore $|\mathrm{Dom}(h_x)| \leqslant |T'_\lambda| < \mathfrak{a}$. Since $\mathrm{cf}(\lambda) \geqslant \mathfrak{a}$ $\mathrm{Dom}(h_x)$ is a bounded subset of λ. Let $\mu_x < \lambda$ be the least upper bound of $\mathrm{Dom}(h_x)$, then $\mathrm{pr}_{T'}(x)$ contains no branching points of T'-rank $> \mu_x$. Let $\mu = \sup\{\mu_x \mid x \in T'_\lambda\}$ then since $|T'_\lambda| < \mathfrak{a}$ and $\mathrm{cf}(\lambda) \geqslant \mathfrak{a}$ we get $\mu < \lambda$, and since T' contains no blind alleys it has no branching points of T'-rank α for $\mu < \alpha < \lambda$. The set $\{\lambda < \kappa \mid \mathrm{cf}(\lambda) \geqslant \mathfrak{a}\}$ is a stationary subset of κ (by IV.4.38); let f be the function defined on it by $f(\lambda) =$ the least $\mu < \lambda$ such that there are no branching points in T' with T'-rank α for $\mu < \alpha < \lambda$. f is clearly a regressive function and hence, by IV.4.40, there is an unbounded subset w of $\{\lambda < \kappa \mid \mathrm{cf}(\lambda) \geqslant \mathfrak{a}\}$ on which the values of f are bounded by some $\eta < \kappa$. Thus T' has no branching points of T'-rank $> \eta$ (since if x were a branching point of T' of T'-rank $\alpha > \eta$ let $\lambda \in w$ be such that $\lambda > \alpha$ then $f(\lambda) \leqslant \eta < \alpha < \lambda$, contradicting the definition of f). Let $x \in T'_\eta$ and let $B = \{y \in T' \mid y \preccurlyeq x \text{ or } y \succ x\}$. Since x is not a blind alley in T' B contains members of every T'-rank below κ; since T' has no branching points of T'-rank $> \eta = \rho_{T'}(x)$ B is easily seen to be a branch of length κ. $\square$

2.33 Proposition. (i) *Every tree $\langle T, <\rangle$ of length κ such that $|T_\lambda| < \mathrm{cf}(\kappa)$ for every $\lambda < \kappa$ has a branch of length κ iff $\mathrm{cf}(\kappa)$ has the tree property.*

(ii) *For every limit ordinal κ with $\mathrm{cf}(\kappa) > \omega$ there is a tree $\langle T, <\rangle$ of length κ with no blind alleys such that $|T_\lambda| \leqslant \mathrm{cf}(\kappa)$ for every $\lambda < \kappa$ and T has no branch of length κ.*

Hint of proof. (i) Use the ideas of the proof of 2.15(iii). (ii) Use the ideas of the proof of 2.15(ii). □

2.34 An Intermediate Narrowness Requirement. In defining κ-Aronszajn trees we made the narrowness requirement that $|T_\lambda| < \kappa$ for $\lambda < \kappa$. Later we discussed the stronger narrowness requirement $|T_\lambda| < \mathfrak{a}$ for some $\mathfrak{a} < \kappa$. Let us look now at the intermediate requirement that $|T_\lambda| \leqslant |\lambda|$ for $0 < \lambda < \kappa$. When we try to construct a tree of length κ which meets given narrowness requirements but has no branch of length κ we have to make sure that the tree meets the narrowness requirements for T_λ with large $\lambda < \kappa$ but we do not need to pay attention to the narrowness requirements for T_λ with small λ, for a reason which is roughly as follows. Any tree which meets the narrowness requirements for large λ's and has no branch of length κ can be changed to such a tree which for small λ's has only one member in each level T_λ either by propping up the original tree on such a thin stalk or by trimming the original tree. If κ is not a cardinal then the intermediate narrowness requirement for $\lambda > |\kappa|$ is $|T_\lambda| \leqslant |\lambda| = |\kappa|$ and this does not imply the existence of a branch of length κ if $\mathrm{cf}(\kappa) > \omega$ (as we saw in 2.11). Similarly, if κ is a singular cardinal of cofinality $> \omega$ then for $\lambda \geqslant \mathrm{cf}(\kappa)$ the narrowness requirement $|T_\lambda| \leqslant |\lambda|$ again does not imply the existence of a branch of length κ (by 2.33(ii)). If κ is the cardinal successor of a cardinal ν then for $\lambda \geqslant \nu$ the narrowness requirement $|T_\lambda| \leqslant |\lambda|$ is equivalent to the narrowness requirement $|T_\lambda| < \kappa$ which we discussed in detail above. Therefore our present narrowness condition is a new interesting one only for weakly inaccessible ordinals κ. It is yet an open problem whether this narrowness condition (that $|T_\lambda| \leqslant |\lambda|$ for $0 < \lambda < \kappa$) implies the existence of a branch of length κ for κ which is a weakly inaccessible ordinal which is not weakly compact. (If κ is weakly compact then already the requirement that $|T_\lambda| < \kappa$ for $\lambda < \kappa$ suffices to establish the existence of a branch of length κ in T).

2.35Ac Exercise. (i) Prove that for every weakly inaccessible ordinal κ which is not weakly compact there is a tree $\langle T, <\rangle$ of length κ such that for all $\lambda < \kappa$ $|T_\lambda| \leqslant 2^{|\lambda|}$ and T has no branch of length κ.

(ii) Prove that if κ is a weakly inaccessible ordinal and $\langle T, <\rangle$ is a tree of length κ such that for every infinite cardinal $\lambda < \kappa$ $|T_\lambda| < \lambda$ then T has a branch of length κ.

Hint. (i) If κ is not inaccessible use a tree similar to that of 2.8. If κ is strongly inaccessible start with any normal κ-Aronszajn tree of length κ and stretch it upwards appropriately, filling in the new levels (see 3.39). □

2.36 Definition. (i) A subset A of a poset $\langle T, <\rangle$ is called an *antichain* if no two members of A are comparable, i.e., if for all $a, b \in A$ if $a \neq b$ then neither $a < b$ nor $b < a$. Notice that every level of a tree is an antichain.

(ii) A tree is said to be *κ-Souslin* if it is a tree of length κ which has no antichains of cardinality $|\kappa|$ and no branches of length κ. Since every level in a tree is an antichain every κ-Souslin tree is a κ-Aronszajn tree. The ω_1-Souslin trees are the κ-Souslin trees which were historically observed first and they are particularly interesting (see 2.40 below), therefore they are called just *Souslin trees*.

(iii) We say that an ordinal κ has the *Souslin property* if there is no κ-Souslin tree. If κ has the tree property it has, obviously, also the Souslin property. Thus ω and the weakly compact ordinals have the Souslin property.

We have obtained the concept of a κ-Souslin tree by strengthening the narrowness requirement of 2.13 not in the direction of a further restriction on $|T_\lambda|$ for $\lambda<\kappa$ but by extending the restriction $|T_\lambda|<\kappa$ from the levels to all antichains.

2.37Ac Proposition. (i) *For every singular limit ordinal κ such that* $\mathrm{cf}(\kappa)<|\kappa|$ *(and this includes all the singular alephs) there is a κ-Souslin tree and hence κ does not have the Souslin property. If* $\mathrm{cf}(\kappa)>\omega$ *there is a κ-Souslin tree with no blind alleys.*

(ii) *For every singular ordinal κ such that* $\mathrm{cf}(\kappa)=|\kappa|$ *there is a κ-Souslin tree iff there is a $|\kappa|$-Souslin tree and hence κ has the Souslin property iff $|\kappa|$ has this property.*

(iii) *For every singular limit ordinal κ if $\langle T,<\rangle$ is a tree of length κ in which every member has at least two immediate successors then T includes an antichain of cardinality $|\kappa|$.*

Hint of proof. (i) Use a tree similar to that of 2.8. If $\mathrm{cf}(\kappa)>\omega$ stretch a tree of length $\mathrm{cf}(\kappa)$ as in 2.11 to length κ.

(ii) Follow the proof of 2.15(iii).

(iii) Assume that $\langle T,<\rangle$ is a tree as in (iii). Prove first that if $U\subseteq T$ is a chain of length λ in T then T includes an antichain of cardinality $|\lambda|$ as in Fig. 12. Hence (iii) holds trivially if κ is not a cardinal.

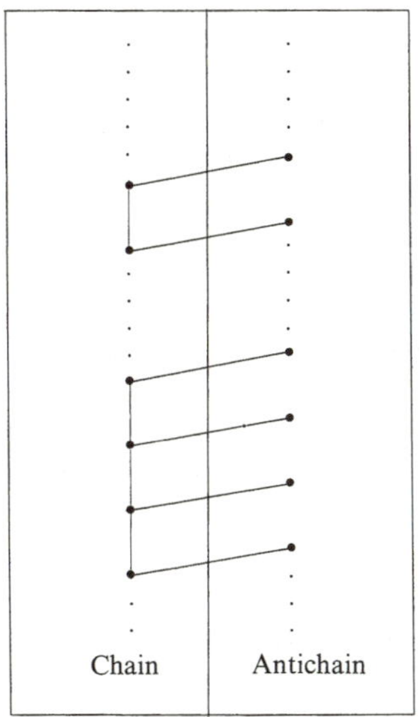

Fig. 12

Assume now that κ is a cardinal and prove that T has an antichain of cardinality κ in each one of the cases a–c below, which exhaust all possibilities. Define $h_T(x) = \sup\{\rho_T(y) \mid y \geqslant x\}$; clearly $h_T(x) = \kappa$ iff x is not a blind alley and $h_T(x) < \kappa$ if x is one. *Case a*: For all $\mu, \sigma < \kappa$ there is a blind alley $x \in T$ such that $\rho_T(x) \geqslant \mu$ and $h_T(x) \geqslant \rho_T(x) \dotplus \sigma$. *Case b*: For some $\mu, \sigma < \kappa$ every blind alley x of T-rank $\geqslant \mu$ satisfies $h_T(x) < \rho_T(x) \dotplus \sigma$, and for every $\lambda < \kappa$ there is a blind alley $x \in T$ such that $\lambda \leqslant \rho_T(x) < \lambda \dotplus \sigma$. *Case c*: For every $\sigma < \kappa$ there is a $\lambda < \kappa$ such that for every $x \in T$ if $\lambda \leqslant \rho_T(x) < \lambda \dotplus \sigma$ then x is not a blind alley. (Then by the proof of 2.14 there is a $\nu < \kappa$ such that $|T_\nu| \geqslant \mathrm{cf}(\kappa)$ and T_ν contains no blind alleys.) □

2.38 The Souslin Property for Regular Ordinals. 2.37 established that the singular alephs fail to have the Souslin property in a trivial way and that for the other singular ordinals the question whether they have the Souslin property has an easy answer once we know which regular ordinals have the Souslin property. Therefore the major question is which regular ordinals have that property (in addition to the regular ordinals which are known already to have the stronger tree property). Here the picture is similar to what we saw in 2.21 and 2.22 with respect to the tree property; for some regular ordinals it is known that ZFC does not decide the question whether the ordinal has the Souslin property and for other ordinals even this is not known, except that here we know even less than what we know in the case of the tree property, as the generalized continuum hypothesis is not known to yield any information about the Souslin property.

Jensen has proved (see Devlin 1973) that the axiom of constructibility implies that no regular ordinal, except ω and the weakly compact ones, has the Souslin property (and thus, by 2.15 and 2.37 an ordinal has the Souslin property iff it has the tree property). We shall see in 2.45 below a part of this proof when we shall use a certain hypothesis $\diamond$, which follows from the axiom of constructibility, to establish the existence of an ω_1-Souslin tree. (The consistency of the existence of such a tree was proved earlier by Jech 1967 and Tennenbaum 1968.) We shall also give in 2.41 below the proof of Solovay and Tennenbaum 1971 that the statement $A_{\aleph_1}$ (of VIII.4.11), which follows from Martin's axiom together with $2^{\aleph_0} > \aleph_1$, implies that ω_1 has the Souslin property. Since Martin and Solovay 1970 have proved that ZFC is consistent with Martin's axiom and $2^{\aleph_0} > \aleph_1$, we know that ZFC does not decide the question whether ω_1 has the Souslin property (since the axiom of constructibility decides it one way and $A_{\aleph_1}$ decides it the other way). Moreover, Jensen has proved (see Devlin and Johnsbraten 1974) that even by means of the generalized continuum hypothesis we cannot prove the existence of an ω_1-Souslin tree and thus even the generalized continuum hypothesis does not decide the question whether ω_1 has the Souslin property. As for regular ordinals above ω_1, other than the weakly compact ones, we know that they cannot be proved to have the Souslin property (since the axiom of constructibility implies that they do not have it) and we know also that (assuming the consistency with ZFC of the existence of weakly compact cardinals) we cannot prove in ZFC that cardinal successors of regular cardinals do not have the Souslin property (since by Silver's result mentioned in 2.18 one cannot even prove that those cardinals do not have the stronger tree property). It is, however, unknown whether one can prove in ZFC that cardinal successors of singular cardinals fail to have the Souslin property and whether one can prove by means of the generalized continuum hypothesis that some regular ordinal above ω_1 which is not weakly compact fails to have the Souslin property.

2.39 Exercises. (i) For every regular aleph κ, if there is a κ-Souslin tree then there is a normal κ-Souslin tree.

(ii) We can strengthen the narrowness condition of a κ-Souslin tree $\langle T, <\rangle$ by requiring that every antichain S of T should be of cardinality $<\mathfrak{a}$, where $\mathfrak{a}$ is some fixed cardinal $<\kappa$. Establish how the existence or inexistence of such a tree depends on the value of $\mathfrak{a}$.

Hints. (i) See 2.16. (ii) See 2.32 and 2.33. □

2.40Ac Exercise (Miller 1943). The question of whether κ-Souslin trees exist originated from the following hypothesis of Souslin 1920.

(8) If $\langle Q, <\rangle$ is an ordered set such that every family of pairwise disjoint open intervals of Q is countable then Q includes a countable set which is dense in Q (and, as a consequence which is not proved here, Q is isomorphic to some set of real numbers ordered by magnitude).

Prove that (8) holds iff there is no Souslin tree. (This is what makes the Souslin property especially interesting for ω_1. As we have mentioned, ZFC, even with the generalized continuum hypothesis added, does not decide whether ω_1 has the Souslin property and therefore it does not decide whether (8) holds).

Hint. Let $\langle T, <\rangle$ be a normal Souslin tree and assume that every member of T has $\aleph_0$ immediate successors (such a tree can be obtained from a normal Souslin tree by omitting all the members whose level is a successor ordinal). Let $<*$ be a total ordering of T such that for each $x \in T$ the set of all immediate successors of x is ordered by $<*$ in the order type of the rational numbers in their natural order. Define a total ordering $<$ of T by: $x<y$ iff either $x<y$ or else for the least λ such that $(\{x\}\cup \mathrm{pr}_T(x))\cap T_\lambda \neq (\{y\}\cup \mathrm{pr}_T(y))\cap T_\lambda$ if x_λ and y_λ are the unique members of T_λ such that $x_\lambda \leqslant x$ and $y_\lambda \leqslant y$ then $x_\lambda <* y_\lambda$. $\langle T, <\rangle$ satisfies the hypothesis, but not the conclusion, of (8).

Let $\langle Q, <\rangle$ be a counterexample to (8), i.e., $\langle Q, <\rangle$ is an ordered set such that every family of pairwise disjoint open intervals in it is countable but there is no countable subset of Q which is dense in it. Define a sequence $\langle I_\alpha \mid \alpha<\omega_1\rangle$ of non-void open intervals of Q as follows. For $\alpha<\omega_1$ let A_α be the set of all endpoints of the intervals I_β, $\beta<\alpha$. A_α is countable and hence not dense in Q; let $a, b \in Q$ be such that $a<b$, $(a, b)\neq 0$ and $[a, b]\cap A_\alpha=0$, and set $I_\alpha=(a, b)$. The set $T=\{I_\alpha \mid \alpha<\omega_1\}$, ordered by inverse proper inclusion, is a Souslin tree. □

We shall now see how two different set theoretical hypotheses, which we have already studied for other purposes, give opposite answers to the question of whether ω_1 has the Souslin property. First we shall deal with the hypothesis $A_{\aleph_1}$ of VIII.4.11, which is a consequence of Martin's axiom and $2^{\aleph_0}>\aleph_1$ (by VIII.4.9).

2.41Ac Theorem (Solovay and Tennenbaum 1971, Martin and Solovay 1970). *If $A_{\aleph_1}$ holds then there is no Souslin tree.*

Proof. If there is a Souslin tree then, by 2.39(i), there is such a tree $\langle T,<\rangle$ with no blind alleys. Each set $D_\beta=\bigcup_{\beta<\gamma<\omega_1}T_\gamma$ is therefore dense in $\langle T,<\rangle$. Since $\langle T,<\rangle$ is a Souslin tree it satisfies the countable antichain condition and hence, by $A_{\aleph_1}$ and VIII.4.15, there is a subnet G of T such that $G\cap D_\beta\neq 0$ for all $\beta<\omega_1$. Let $W=\{y\in T\mid \exists x\in G(y\leqslant x)\}\supseteq G$. Since G contains members with arbitrarily large T-rank W contains members of every T-rank, hence if W is a branch it is of length ω_1. We know already that if $y\in W$ and $z<y$ then also $z\in W$, thus in order to prove that W is a branch we have only to show that all its members are comparable. If $y_1, y_2\in W$ then there are $x_1, x_2\in G$ such that $y_1\leqslant x_1$ and $y_2\leqslant x_2$. Since G is a net there is a $z\in G$ such that $x_1, x_2\leqslant z$ and hence $y_1, y_2\leqslant z$. Since T is a tree y_1 and y_2 are comparable. Thus W is a branch of length ω_1 which is a contradiction. □

2.42 The Diamond Hypothesis (Jensen 1968). The following hypothesis, which is denoted by $\diamond$ and is called *diamond*, is a consequence of the axiom of constructibility.

$\diamond$ There exists a sequence $\langle s_\alpha\mid\alpha<\omega_1\rangle$ such that for all $\alpha<\omega_1$ $s_\alpha\subseteq\alpha$ and for every $A\subseteq\omega_1$ the set $\{\alpha<\omega_1\mid A\cap\alpha=s_\alpha\}$ is a stationary subset of ω_1.

This can be explained intuitively as follows. Without knowing the subset A of ω_1 we try, for each α, to guess $A\cap\alpha$ and the guess we make is s_α. $\diamond$ says that we can make such guesses so that whatever A will be the set of α's at which we made a correct guess is stationary.

2.43 Proposition. *If $\diamond$ holds then $2^{\aleph_0}=\aleph_1$.*

Proof. Let $A\subseteq\omega$, and let $\beta>\omega$ belong to the stationary subset $\{\alpha<\omega_1\mid A\cap\alpha=s_\alpha\}$ of ω_1, then, since $A\subseteq\beta$, $A=A\cap\beta=s_\beta$. Thus $\mathbf{P}(\omega)\subseteq\{s_\beta\mid\beta<\omega_1\}$, hence $|\mathbf{P}(\omega)|\leqslant\omega_1$. □

The inverse implication, i.e., that $2^{\aleph_0}=\aleph_1$ implies $\diamond$, does not hold since, as will be shown in 2.45, $\diamond$ implies the existence of a Souslin tree while, as we have mentioned in 2.38, the generalized continuum hypothesis does not imply the existence of a Souslin tree.

2.44 Proposition. *If $\diamond$ holds then there is a subset W of $\mathbf{P}(\omega_1)$ which consists of $2^{\aleph_1}$ almost disjoint stationary subsets of ω_1, i.e., if $B, C\in W$ then $B\cap C$ is countable.*

Remark. If the axiom of constructibility holds then it is the case that for every stationary subset D of ω_1, not just for ω_1 itself, there is a subset W of $\mathbf{P}(D)$ as above.

Proof. Let $\langle s_\alpha \mid \alpha<\omega_1\rangle$ be as in $\diamond$. We define a function F on $\mathbf{P}(\omega_1)$ by setting, for $A\subseteq\omega_1$, $F(A)=\{\alpha<\omega_1 \mid A\cap\alpha=s_\alpha\}$. By $\diamond$, $F(A)$ is a stationary subset of ω_1. If $A\neq B$ then for some $\gamma<\omega_1$ γ belongs to one of A and B but not to the other, thus $A\cap\delta\neq B\cap\delta$ for all $\delta>\gamma$ hence $F(A)\cap F(B)\subseteq\gamma+1$ and $F(A)$, $F(B)$ are almost disjoint. Thus $W=F[\mathbf{P}(\omega_1)]$ is as claimed. $\square$

2.45Ac Theorem (Jensen 1968). $\diamond$ *implies the existence of a Souslin tree.*

Proof. We intend to construct a normal tree $\langle T,<\rangle$ of length ω_1 whose members will be just the ordinals $<\omega_1$, i.e., $T=\omega_1$, and such that $|T_\lambda|\leqslant\aleph_0$ for $\lambda<\omega_1$. We shall prove that every antichain S in T is countable. This implies that also every branch in T is countable (since an uncountable chain yields an uncountable antichain—see the proof of 2.37(iii)) and thus T is a Souslin tree.

Suppose we have already constructed $\langle T,<\rangle$ as we have just mentioned. Let W be an antichain in T. By applying Zorn's lemma to the set of all antichains in T which include W, with proper inclusion as the partial order relation, we obtain a maximal antichain S which includes W. Once we prove that S is countable we know that W is countable too. We shall first see that

(9) for every $\alpha<\omega_1$ there is a $\beta<\omega_1$ such that $T|\beta\supseteq\alpha$, $\beta\supseteq T|\alpha$ and for every $\gamma\in T|\alpha$ there is a $\delta\in S\cap\beta$ such that $\delta\leqslant\gamma$ or $\gamma\leqslant\delta$.

Since for $\lambda<\alpha$ $|T_\lambda|\leqslant\aleph_0$ $T|\alpha$ is a countable subset of ω_1, hence $T|\alpha$ is bounded by some ordinal $\beta_0<\omega_1$. Similarly, since $T=\omega_1$ the regularity of ω_1 implies that there is a $\beta_1<\omega_1$ such that $T|\beta_1\supseteq\alpha$. Let h be the function on $T|\alpha$ given by

$$h(\gamma)=\text{the least }\delta\in S\text{ such that }\gamma\leqslant\delta\text{ or }\delta\leqslant\gamma$$

(there is such a δ by the maximality of S). $h[T|\alpha]$ is a countable subset of ω_1 and hence it is bounded by some $\beta_2<\omega_1$. Take $\beta=\max(\beta_0,\beta_1,\beta_2)$ then β satisfies (9). We define now a function F on ω_1 by

$$F(\alpha)=\text{the least }\beta\text{ which satisfies (9) with respect to }\alpha.$$

It is obvious from (9) that $\alpha<\alpha'\to F(\alpha)\leqslant F(\alpha')$ and that F is continuous (i.e., for a limit ordinal α $F(\alpha)=\bigcup_{\gamma<\alpha}F(\gamma)$). Since F is monotonic and continuous and $F(\alpha)\geqslant\alpha$ for every $\alpha<\omega_1$ the set A of α's for which $F(\alpha)=\alpha$ is a closed unbounded subset of ω_1 (as in IV.4.24). Let $\alpha\in A$; by (9) we have $T|\alpha=\alpha$ and we shall see that $S\cap\alpha$ is a maximal antichain in $T|\alpha$. That $S\cap\alpha$ is an antichain is obvious. To prove its maximality we have to show that for every $\gamma\in T|\alpha=\alpha$ there is a $\delta\in S\cap\alpha$ such that $\gamma\leqslant\delta$ or $\delta\leqslant\gamma$, but this is explicitly demanded in (9).

Suppose now that for some $\alpha\in A$ we guessed correctly, before constructing T, what $S\cap\alpha$ will be, and our guess is a certain subset s_α of $\alpha=T|\alpha$. Since $s_\alpha=S\cap\alpha$ s_α is, as we have seen, a maximal antichain in $T|\alpha$. We shall see that we could have constructed T_α so as to make s_α a maximal antichain in the whole of T, and since $S\supseteq S\cap\alpha=s_\alpha$ we would have $S=s_\alpha$ (since s_α is already maximal) and therefore S is countable. To achieve this purpose all we have to do is to see to it that T_α is constructed so that

(10) each member of T_α stands in the relation $\succ$ to some member of s_α.

Assume now that (10) holds and γ is any ordinal $<\omega_1$. If $\rho_T(\gamma)<\alpha$ then since s_α is a maximal antichain in $T|\alpha$ γ is comparable with some member of s_α. If $\rho_T(\gamma)\geqslant\alpha$ then $\gamma\succcurlyeq\mu$ for a unique $\mu\in T_\alpha$ and by (10) $\mu\succ\delta$ for some $\delta\in s_\alpha$, thus γ is again comparable with some member of s_α. Thus s_α is already a maximal antichain in the whole of T, and therefore $S=s_\alpha$.

Before we fully describe the formal details of the construction of T let us go through the main points of that construction. We choose $\langle s_\alpha \mid \alpha<\omega_1\rangle$ to be as in $\diamondsuit$. Now we construct T_α by recursion on α, therefore when we construct T_α at the α-th step $T|\alpha$ is already given. At this step we guess that $S\cap\alpha$ will be s_α, and if s_α is indeed a maximal antichain in $T|\alpha$ we construct T_α so that (10) will hold. Once we are through the construction of T we pick any maximal antichain S in T. By $\diamondsuit$ the set $\{\alpha<\omega_1 \mid S\cap\alpha=s_\alpha\}$ is a stationary subset of ω_1. Let A be the closed unbounded subset of ω_1 defined above for this set S then the set $B=A\cap\{\alpha<\omega_1 \mid S\cap\alpha=s_\alpha\}$ is non-void. For $\alpha\in B$ we have $S\cap\alpha=s_\alpha$, and since $\alpha\in A$ $s_\alpha=S\cap\alpha$ is a maximal antichain in $T|\alpha$ as we saw above. Since T_α satisfies (10) we have, as mentioned above, $S=s_\alpha$ and S is countable. Thus our construction is successful since for every maximal antichain S we have guessed $S\cap\alpha$ correctly at least at one appropriate α (i.e., at $\alpha\in A$).

As we construct T we simultaneously prove by induction on $\alpha<\omega_1$ that

(11) T_α is a countable subset of ω_1, and

(12) for all $\gamma<\alpha$ and $x\in T_\gamma$ there is a $y\in T_\alpha$ such that $y\succcurlyeq x$.

We set $T_0=\{0\}$, (11) and (12) hold trivially for $\alpha=0$. If α is a successor ordinal let $\beta=\alpha-1$. We construct T_α by putting two distinct ordinals out of $\omega_1\sim\bigcup_{\gamma<\alpha}T_\gamma$ on top of each member of T_β. Since by the induction hypothesis T_γ is countable for $\gamma<\alpha$ we have $|\omega_1\sim\bigcup_{\gamma<\alpha}T_\gamma|=\aleph_1$ so we can find in the set $\omega_1\sim\bigcup_{\gamma<\alpha}T_\gamma$ $2\cdot|T_\beta|\leqslant 2\cdot\aleph_0=\aleph_0$ ordinals for T_α ($|T_\beta|\leqslant\aleph_0$ by the induction hypothesis). T_α is obviously countable. If $\gamma<\alpha$ and $x\in T_\gamma$ than $\gamma\leqslant\beta$ and by the induction hypothesis there is a $z\in T_\beta$ such that $x\preccurlyeq z$. By the construction of T_α there is a $y\in T_\alpha$ such that $z\prec y$ hence $x\prec y$ and (12) is established.

If α is a limit ordinal then let us see first that for every $x\in T|\alpha$ there is in $T|\alpha$ a branch of length α which contains x. Let δ be the $T|\alpha$-rank of x. Since α is a limit ordinal $<\omega_1$ $\mathrm{cf}(\alpha)=\omega$ and hence there is an increasing sequence $\langle\delta_i \mid i<\omega\rangle$ of ordinals such that $\delta_0=\delta$ and $\sup_{i<\omega}\delta_i=\alpha$. We construct by recursion a sequence $\langle z_i \mid i\in\omega\rangle$ of members of $T|\alpha$ such that $z_0=x$, $\rho_{T|\alpha}(z_i)=\delta_i$ and $z_i\prec z_{i+1}$ for $i<\omega$. To obtain such a z_{i+1} notice that since $\delta_i<\delta_{i+1}<\alpha$ and $z_i\in T_{\delta_i}$ there is, by the induction hypothesis (12), a $z_{i+1}\in T_{\delta_{i+1}}$ such that $z_i\prec z_{i+1}$. The set $\{u\in T|\alpha| \text{ for some } i,\ u\preccurlyeq z_i\}$ is clearly a branch of length α which contains x. For each $x\in T|\alpha$ let b_x be some branch of length α through x. If s_α is not a maximal antichain in $T|\alpha$ we construct T_α by putting, for every $x\in T|\alpha$, a distinct ordinal $\lambda_x\in\omega_1\sim T|\alpha$

on top of b_x. Since $T|\alpha$ is countable, by the induction hypothesis, $\omega_1 \sim T|\alpha$ contains the $\leqslant\aleph_0$ ordinals needed for this purpose. Formally we extend the partial order $<$ on $T|\alpha$ to $T|(\alpha+1)$ by adding to it the set $\bigcup_{x\in T|\alpha} b_x \times \{\lambda_x\}$. (11) and (12) obviously hold. If s_α is a maximal antichain in $T|\alpha$ we construct T_α by putting an ordinal $\lambda_x \in \omega_1 \sim T|\alpha$ on top of b_x only for those $x \in T|\alpha$ which are $\geqslant$ some member of s_α. This is obviously possible and (11) clearly holds. To see that (12) holds let $x \in T|\alpha$ then since s_α is a maximal antichain in $T|\alpha$ there is a $y \in s_\alpha$ such that $y \leqslant x$ or $x \leqslant y$. If $y \leqslant x$ then $\lambda_x \in T_\alpha$ satisfies $\lambda_x \succ x$, and if $x \leqslant y$ then $\lambda_y \in T_\alpha$ satisfies $\lambda_y \succ y \geqslant x$, and thus (12) holds. In this case we have also to verify that (10) holds, but this follows directly from the way we constructed T.

Finally, we have to see that $T=\omega_1$ and that T is indeed normal. To make $T=\omega_1$ we take care that for every α when we construct T_α we take α to be one of the members of T_α, if it is not already in $T|\alpha$. That T is normal follows immediately from the construction of T_0 and $T_{\alpha+1}$ and from (12). □

2.46Ac Can Narrow Trees Have Many Branches? We saw that if a tree of length κ is sufficiently narrow it will have a branch of length κ. We shall see now that while the narrowness of a tree may assure the existence of a single branch of length κ it may prevent the existence of too many such branches. If for a tree T of length κ we have $|T_\lambda| \leqslant \mathfrak{a}$ for every $\lambda < \kappa$ then, by a rough upper estimate, T has at most $\mathfrak{a}^{|\kappa|}$ branches of length κ (since for every such branch b $\{\langle \rho_T(x), x\rangle \mid x \in b\} \in \bigtimes_{\lambda<\kappa} T_\lambda$), and if $\mathfrak{a} \leqslant |\kappa|$ then T has at most $|\kappa|^{|\kappa|} = 2^{|\kappa|}$ branches. If a tree satisfies the strong narrowness requirement that $|T_\lambda| < \mathfrak{a}$ for all $\lambda < \kappa$, where $\mathfrak{a} < \mathrm{cf}(\kappa)$, then, as follows easily from the proof of 2.32, T has less than $\mathfrak{a}$ branches of length κ. The question we ask now is how much do we have to relax the narrowness requirement in order to obtain trees of length κ which have $2^{|\kappa|}$ branches of length κ (as $2^{|\kappa|}$ is the maximal number possible as long as $|T_\lambda| \leqslant \kappa$ for $\lambda < \kappa$) or even at least κ^+ such branches.

In order not to widen too much the scope of this discussion we shall restrict it from now on to κ's which are cardinals. A very natural narrowness requirement on trees of length κ is the one we used in 2.13 to define κ-Aronszajn trees, namely that $|T_\lambda| < \kappa$ for $\lambda < \kappa$; let us call the trees of length κ which satisfy this requirement *κ-slim* trees. For a strong limit cardinal κ there is a κ-slim tree with 2^κ branches of length κ. To get it take for T the set ${}^{\underset{\sim}{\kappa}}2$ and for $<$ the proper inclusion relation $\subset$. For each $\lambda < \kappa$ $T_\lambda = {}^\lambda 2$ hence $|T_\lambda| = 2^{|\lambda|} < \kappa$ and thus T is κ-slim. The set of all branches of T of length κ is $\{\{f \upharpoonright \lambda \mid \lambda < \kappa\} \mid f \in {}^\kappa 2\}$ and its cardinality is therefore 2^κ. If the generalized continuum hypothesis holds then all limit cardinals κ are strong limit cardinals and by what we have just seen there are κ-slim trees with 2^κ branches of length κ. Without assuming the generalized continuum hypothesis there may be limit cardinals κ which are not strong limit cardinals and it is not known whether for such κ's there can be κ-slim trees with more than κ branches of length κ.

If κ is a successor cardinal let μ be its cardinal predecessor. If there is a κ-slim tree with $\mathfrak{b}$ branches of length κ for some $\mathfrak{b}$ such that $\kappa < \mathfrak{b} \leqslant 2^\kappa$ then we can easily construct such a tree T' which satisfies the stronger narrowness requirement

(13) $|T'_\lambda| \leqslant |\lambda|$ for every $0<\lambda<\kappa$

as follows. For T' we take the tree of Figure 13 obtained from T by adding at the bottom of T a totally ordered set of order type μ. This leads us to investigate trees which satisfy the narrowness requirement (13).

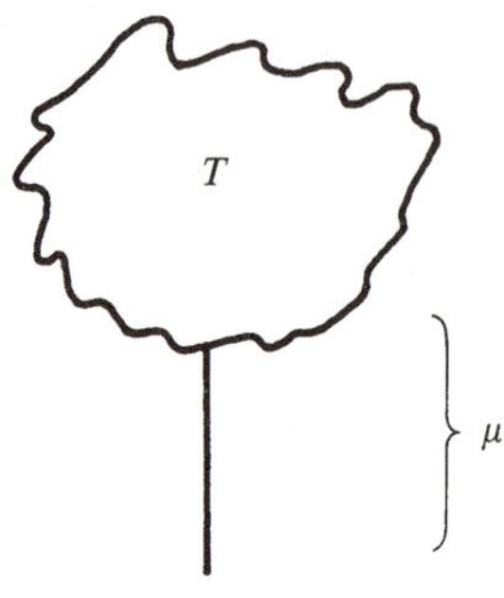

Fig. 13

2.47 Definition. For a cardinal κ, a tree T of length κ is called a *κ-Kurepa tree* if for all $\lambda<\kappa$ $|T_\lambda| \leqslant |\lambda| + \aleph_0$ and T has more than κ branches of length κ.

2.48Ac Proposition. *If κ is a cardinal and* $\text{cf}(\kappa)=\omega$ *then there is a κ-Kurepa tree with $\kappa^{\aleph_0}$ branches of length κ. (Notice that since* $\text{cf}(\kappa)=\omega$ $\kappa^{\aleph_0}>\kappa$ *by* V.5.10, *and if κ is a strong limit cardinal then* $\kappa^{\aleph_0}=2^\kappa$).

Hint of proof. Let T be the set of all sequences $z \in {}^{\underset{\smile}{\kappa}}2$ such that the set

$$\{\alpha \in \text{Dom}(z) \mid z(\alpha)=1\}$$

is finite, partially ordered by proper inclusion. To prove that T has at least $\kappa^{\aleph_0}$ branches of length κ use IV.2.25. □

2.49Ac Theorem (Baumgartner and Prikry 1976). *Let κ be a singular cardinal such that* $\text{cf}(\kappa)>\omega$ *and let T be a tree of length κ. If for some $\mu<$* $\text{cf}(\kappa)$ *the set*

$$\{\alpha<\kappa \mid |T_\alpha| \leqslant |\alpha|^{+\mu}\}$$

is a stationary subset of κ then T has at most $(\underset{\smile}{\kappa}^{\text{cf}(\kappa)})^{+\mu}$ branches of length κ. Therefore if $\underset{\smile}{\kappa}^{\text{cf}(\kappa)}=\kappa$ (and this holds, in particular, if κ is a strong limit cardinal) then there is no κ-Kurepa tree.

Proof. This is a more general version of Theorem V.5.6. The proof is exactly like that of V.5.6 with the following replacements. Replace $\mathfrak{a}$ by κ, T by the set of all branches of length κ of our present T, $x \cap \mathfrak{b}$ by the single member of $x \cap T_{\mathfrak{b}}$ (where x is a branch of T of length κ). Wherever an upper estimate for the cardinality of a subset of T, such as T_Q, is computed in the proof of V.5.6 from upper estimates for the cardinalities of the sets $\{y \cap \mathfrak{b} \mid y \in T_Q\}$ for $\mathfrak{b} \in Q$ we proceed now as follows. Here T_Q is a set of branches of length κ. We look at the subtree $T_Q^* = \langle \bigcup T_Q, < \rangle$. The same computation as in V.5.6 yields now an upper estimate on the cardinality of the set $\bar{T}_Q$ of all branches of T_Q^* of length κ. Since, clearly, $T_Q \subseteq \bar{T}_Q$ the same cardinal is also an upper estimate for $|T_Q|$. □

2.50Ac Exercise. The following exercises yield some information on what one might have considered as alternatives to the definition of a κ-Kurepa tree.

(i) If κ is an ordinal such that $\mathrm{cf}(\kappa) > \omega$ then there is a normal tree T such that $|T_\lambda| \leqslant |\lambda| + \aleph_0$ for every $\lambda < \kappa$ and T has exactly $|\kappa|$ branches of length κ.

(ii) If $\mathrm{cf}(\kappa) = \omega$ then every normal tree of length κ has at least $|\kappa|^{\aleph_0}$ branches of length κ.

(iii) If κ is a regular uncountable ordinal, and T is a tree such that $|T_\lambda| < |\lambda| + \aleph_0$ for every $\lambda < \kappa$ then T has less than κ branches.

Hints. (i), (ii). See the proof of 2.48. (iii) See the discussion in 2.46. □

2.51 κ-Kurepa Trees for Regular κ's. We shall see that while for some large cardinals κ κ-Kurepa trees are known not to exist, for many cardinals κ the question of the existence of a κ-Kurepa tree cannot be decided either way in ZFC and for other cardinals κ the only thing known is that the existence of a κ-Kurepa tree cannot be refuted in ZFC. We shall define below the concept of an ineffable cardinal. We shall prove in 3.37 and 3.40 that every ineffable cardinal is weakly compact and that the least ineffable cardinal, if there is such, is greater than many weakly compact cardinals, (remember that even the existence of an inaccessible cardinal, and a fortiori a weakly compact cardinal, is unprovable in ZFC). The only result concerning the existence of κ-Kurepa trees which we know how to prove in ZFC is 2.55 below which asserts that for every ineffable cardinal κ there is no κ-Kurepa tree. The axiom of constructibility, on the other hand, gives a full picture; as shown by Jensen and Solovay (see Devlin 1973) if that axiom holds then for every cardinal κ there is a κ-Kurepa tree iff κ is not ineffable. Silver 1971 has shown that for no particular regular ordinal μ (given by an appropriate set term such as ω_{13} or ω_{ω_1} or "the second inaccessible cardinal") can one prove in ZFC that there exists a μ^+-Kurepa tree. □

The next proposition will show that we can replace dealing with arbitrary κ-Kurepa trees by dealing with such trees which are subtrees of the full binary tree ${}^{\underset{\smile}{\kappa}}2$ of length κ.

2.52Ac Proposition. *For an aleph κ, there is a κ-Kurepa tree iff there is a family $S \subseteq P(\kappa)$ such that $|S| > \kappa$ and for every $\lambda < \kappa$ $|\{x \cap \lambda \mid x \in S\}| \leqslant |\lambda| + \aleph_0$.*

Proof. If there is a family S as above then let S' be the set of all characteristic functions on κ of the members of S. Put $T = \{f \upharpoonright \lambda \mid \lambda < \kappa \wedge f \in S'\}$, and let $<$ be proper inclusion. T is clearly a tree of length κ. $|T_\lambda| = |\{f \upharpoonright \lambda \mid f \in S'\}| = |\{x \cap \lambda \mid x \in S\}| \leqslant |\lambda| + \aleph_0$. T is a Kurepa tree since for every member of f of S $\{f \upharpoonright \lambda \mid \lambda < \kappa\}$ is a branch of T of length κ.

If $\langle T, <\rangle$ is a κ-Kurepa tree then let $\lhd$ be an arbitrary well-ordering of T and let $<$ be the ordering of T defined by $x<y$ if $\rho_T(x)<\rho_T(y)$ or if $\rho_T(x)=\rho_T(y)$ and $x \lhd y$. $<$ is clearly a well-ordering; we shall see that the order type of $\langle T, <\rangle$ is κ. Let $x\in T_\lambda$ for some $\lambda<\kappa$. For $\mu\leqslant\lambda$ let x_μ denote that $y\leqslant x$ such that $\rho_T(y)=\mu$. We clearly have for $\sigma<\mu\leqslant\lambda$ $x_\sigma<x_\mu$, hence $x_\lambda\geqslant$ the λ-th member of T; since this holds for every $\lambda<\kappa$ the order type of T is $\geqslant\kappa$. If $x\in T$ then for some $\lambda<\kappa$ $x\in T_\lambda$ and therefore $\{y\in T \mid y<x\}\subseteq\{y\in T\mid \rho_T(y)\leqslant\lambda\}=\bigcup_{\mu<\lambda}T_\mu$ hence

$$|\{y\in T\mid y<x\}|\leqslant\Sigma_{\mu\leqslant\lambda}|T_\mu|=\Sigma_{\mu\leqslant\lambda}(|\mu|+\aleph_0)\leqslant(|\lambda|+\aleph_0)^2<\kappa.$$

Thus there are less than κ members $<x$ in T and therefore the order type of $\langle T, <\rangle$ is exactly κ. If for every $\lambda<\kappa$ we replace in T the λ-th member in the ordering $<$ by the ordinal λ and denote the relation obtained from $<$ again by $<$, we get a Kurepa tree $T'=\langle\kappa, <\rangle$ isomorphic to $\langle T, <\rangle$ such that for every $\mu<\kappa$ $\rho_{T'}(\mu)\leqslant\mu$ (since by what we proved above, if $\mu\in T'_\lambda$ then μ is $\geqslant$ than the λ-th member of $\langle\kappa, <\rangle$ which is now λ, i.e., $\mu\geqslant\lambda=\rho_{T'}(\mu)$).

Let S be the set of all branches of T' of length κ, then $|S|$ is the same as the cardinality of the set of all branches of T of length κ, hence $|S|>\kappa$. Let $\lambda<\kappa$ and let, for $x\in S$, x_λ be that member y of x such that $\rho_{T'}(y)=\lambda$. If $\mu\in x\cap\lambda$ then since $\rho_{T'}(\mu)\leqslant\mu<\lambda$ we have $\mu<x_\lambda$. Obviously, also if $\mu<x_\lambda$ then $\mu\in x$. Thus $x\cap\lambda=\{\mu<\lambda\mid\mu<x_\lambda\}$ and therefore

$$|\{x\cap\lambda\mid x\in S\}|=|\{\{\mu<\lambda\mid\mu<x_\lambda\}\mid x\in S\}|\leqslant|\{x_\lambda\mid x\in S\}|\leqslant|T'_\lambda|=|T_\lambda|=|\lambda|+\aleph_0. \quad \square$$

Historical remark. The existence of an ω_1-Kurepa tree was first suggested by Kurepa, who suggested the existence of a subset S of $P(\omega_1)$ such that for every $\lambda<\omega_1$ $|\{x\cap\lambda\mid x\in S\}|\leqslant\aleph_0$ but $|S|>\aleph_1$ (see Ricabarra 1958).

2.53 Definition. (i) For a class X of ordinals a sequence $\langle A_\alpha\mid\alpha\in X\rangle$ such that $A_\alpha\subseteq\alpha$ for every $\alpha\in X$ is said to be *coherent* if for all $\alpha,\beta\in X$, if $\alpha<\beta$ then $A_\alpha=A_\beta\cap\alpha$. Obviously if $\langle A_\alpha\mid\alpha\in X\rangle$ is coherent and we set $A=\bigcup_{\alpha\in X}A_\alpha$ then we get $A_\alpha=A\cap\alpha$ for every $\alpha\in X$.

(ii) (Jensen—see Devlin 1973.) An ordinal κ is said to be *ineffable* if $\mathrm{cf}(\kappa)>\omega$ and for every sequence $\langle A_\alpha\mid\alpha<\kappa\rangle$ such that $A_\alpha\subseteq\alpha$ for $\alpha<\kappa$ there is a stationary subset X of κ such that the subsequence $\langle A_\alpha\mid\alpha\in X\rangle$ of $\langle A_\alpha\mid\alpha<\kappa\rangle$ is coherent. We shall see in 3.40 that the ineffable ordinals, if such exist, are weakly compact ordinals and the least of them is larger than many weakly compact ordinals.

2.54 Proposition. *Every ineffable ordinal is regular.*

Hint of proof. Use the fact that for every singular limit ordinal κ there is a regressive function f on κ into $\mathrm{cf}(\kappa)<\kappa$ such that for every $\lambda<\mathrm{cf}(\kappa)$ $f^{-1}[\{\lambda\}]$ is a bounded subset of κ. $\square$

2.55 Theorem (Jensen and Kunen—see Devlin 1973). *For an ineffable ordinal κ there is no κ-Kurepa tree.*

Proof. We shall prove the theorem by showing that there is no subset S of $P(\kappa)$ as in 2.52. Let S be a family of subsets of such that for every $\lambda<\kappa$ $|\{x\cap\lambda \mid x\in S\}|\leqslant |\lambda|+\aleph_0$. Let P be a pairing function on the ordinals such that all the alephs are critical points of P (the pairing functions of III.3.21 and IV.2.23(ii) are such by IV.2.23(iii) and IV.2.13). The class of all critical points of a binary function on On into On is always closed, as easily seen. Let C be the set of all infinite critical points of P below κ. Since κ is regular, by 2.54, C is unbounded in κ (since if $\omega<\alpha<\kappa$ and we define $\gamma_0=\alpha$ and $\gamma_{i+1}=\sup(\gamma_i\cup P[\gamma_i\times\gamma_i])$ then $\alpha<\sup_{i<\omega}\gamma_i\in C$). For $\lambda\in C$ $|\{x\cap\lambda \mid x\in S\}|\leqslant|\lambda|+\aleph_0=|\lambda|$, hence there is a surjection $\langle w_{\lambda,\beta} \mid \beta<\lambda\rangle$ of λ on $\{x\cap\lambda \mid x\in S\}$. We want to obtain a stationary subset X of κ such that

(14) for every $\beta<\kappa$ the sequence $\langle w_{\lambda,\beta} \mid \lambda\in X\sim(\beta+1)\rangle$, where β is fixed, is coherent.

To obtain such an X we define $Q_\lambda=\{P(\beta,\gamma) \mid \beta<\lambda\wedge\gamma\in w_{\lambda,\beta}\}$ for $\lambda\in C$ and $Q_\lambda=0$ for $\lambda\notin C$. Since every member of C is a critical point of P we get $Q_\lambda\subseteq\lambda$ for every $\lambda<\kappa$. Since κ is ineffable there is a stationary subset X of κ such that the sequence $\langle Q_\lambda \mid \lambda\in X\rangle$ is coherent; since C is a closed unbounded subset of κ we may assume, without loss of generality, that $X\subseteq C$ (otherwise we can replace X by $X\cap C$, by IV.4.38(ii)). To prove (14) we shall show that

(15) if $\lambda,\mu\in X$ and $\beta<\lambda<\mu$ then $w_{\lambda,\beta}=w_{\mu,\beta}\cap\lambda$.

Since $\lambda,\mu\in X$ and $\lambda<\mu$ we have $Q_\lambda=Q_\mu\cap\lambda$, and since $X\subseteq C$ and $\beta<\lambda$ we have $\gamma\in w_{\lambda,\beta}\leftrightarrow P(\gamma,\beta)\in Q_\lambda\leftrightarrow P(\gamma,\beta)\in Q_\mu\cap\lambda\leftrightarrow\gamma\in w_{\mu,\beta}\wedge\gamma<\lambda$, which establishes (15).

For $\beta<\kappa$ we define $w_\beta=\bigcup\{w_{\lambda,\beta} \mid \lambda>\beta\wedge\lambda\in X\}\subseteq\kappa$. We shall conclude our proof by showing that $S\subseteq\{w_\beta \mid \beta<\kappa\}$ which obviously implies $|S|\leqslant\kappa$ and hence S is not as in 2.52. Let x be any member of S and let g be the function on X defined by $g(\lambda)=$a $\beta<\lambda$ such that $x\cap\lambda=w_{\lambda',\beta}$. g is a regressive function on X hence, by IV.4.41, there is an $\alpha<\kappa$ such that $g^{-1}[\{\alpha\}]$ is an unbounded subset of X (and κ). Therefore

$$x=\bigcup\{x\cap\lambda \mid \lambda\in g^{-1}[\{\alpha\}]\}=\bigcup\{w_{\lambda,g(\lambda)} \mid \lambda\in g^{-1}[\{\alpha\}]\}=\bigcup\{w_{\lambda,\alpha} \mid \lambda\in g^{-1}[\{\alpha\}]\}$$

and since $\{w_{\lambda,\alpha} \mid \alpha<\lambda\in X\}$ is coherent and $g^{-1}[\{\alpha\}]$ is an unbounded subset of X we have

$$x=\bigcup\{w_{\lambda,\alpha} \mid \lambda\in g^{-1}[\{\alpha\}]\}=\bigcup\{w_{\lambda,\alpha} \mid \alpha<\lambda\in X\}=w_\alpha. \quad \square$$

2.56 An Undecided Statement About Coloring. We shall conclude this section with another example of a combinatorial statement such that $\diamond$ decides it one way, the version $A_{\aleph_1}$ of Martin's axiom (VIII.4.11) decides it the other way, and the generalized continuum hypothesis does not decide it. The statement we shall deal with is about the coloring of graphs. We dealt with this subject already in V.1.11(v); here we shall admit also an infinite number of colors. The idea is to try to color a graph with as few colors as possible. The statement we consider is "If G is a graph on ω_1 such that for some $\tau<\omega_1$ G has the property that for all $\alpha<\omega_1$ the order type of $\{\beta<\alpha \mid \{\beta, \alpha\} \in G\}$ is $\leqslant\tau$, then ω_1 has a coloring with $\aleph_0$ colors consistent with G". We shall prove that this follows from $A_{\aleph_1}$, which contradicts the continuum hypothesis, and that a counterexample to this statement, even for $\tau=\omega$, can be obtained by means of $\diamond$. This statement is consistent with the generalized continuum hypothesis (Avraham, Devlin and Shelah 1978, or see Shelah 1977); its negation is obviously consistent with the generalized continuum hypothesis since it follows from the axiom of constructibility via $\diamond$.

2.57Ac Proposition (Hajnal and Máté 1975). *$\diamond$ implies the existence of a graph G on ω_1 such that* (a) *for every $\alpha<\omega_1$ the set $G_\alpha=\{\beta<\alpha \mid \{\beta, \alpha\} \in G\}$ is either* 0 *or else it is of order type ω and* $\sup G_\alpha=\alpha$, *and* (b) *ω_1 has no countable coloring consistent with G.*

Proof. Let us call a coloring F of ω_1 such that $\mathrm{Rng}(F)\subseteq\omega$ an *ω-coloring*. What we are looking for is a graph G on ω_1 for which there is no ω-coloring consistent with it. We shall construct such a G in ω_1 steps by deciding at step α which of the β's below α to put in G_α, namely, which ordinals below α to connect to α by an edge of G. Our construction will essentially be a diagonalization; at the α-th step we shall take a certain function $f_\alpha\colon \alpha\rightarrow\omega$, guess that f_α is a restriction to α of an ω-coloring F of ω_1, and try to construct G_α so as to prevent F from being a coloring consistent with G. The idea is to connect α, for each color n, to an ordinal $\beta<\alpha$ which f_α colors with n (i.e., such that $f_\alpha(\beta)=n$), if there is such a β. If we would succeed in doing that, and if all the colors in $\mathrm{Rng}(F)$ would occur at ordinals $<\alpha$, then F could not be consistent with G since the color $F(\alpha)$ of α would occur at some $\beta<\alpha$ which was connected to α by an edge of G. Since we want requirement (a) of our proposition to hold too we cannot always connect α to β's of all different colors below α and instead we shall connect limit ordinals α only to as many β's below α as we can while obeying (a). If α is a successor ordinal or 0 we set $G_\alpha=0$. If α is a limit ordinal let $\langle\alpha_n \mid n\in\omega\rangle$ be an ascending sequence such that $\sup_{n<\omega}\alpha_n=\alpha$. Then we define the sequence $\langle B_n \mid n\in\omega\rangle$ by recursion as follows. $B_0=0$. For $n\in\omega$, if there is a $\beta<\alpha$ such that $\beta>\alpha_n$, β is greater than every member of B_n and $f_\alpha(\beta)=n$ then we denote this β with β_n and we set $B_{n+1}=B_n\cup\{\beta_n\}$; if there is no such β we set $B_{n+1}=B_n$. Let $B=\bigcup_{n<\omega}B_n$. If B is finite we set $G_\alpha=0$. If B is infinite then, by the way we defined the B_n's, the order type of B is obviously ω and $\sup B=\alpha$; in this case we set $G_\alpha=B$. Notice that if there are infinitely many colors which occur unboundedly high below α in the coloring f_α then G connects α to ordinals colored with each one of these colors (since if n is such a color then B_{n+1} contains β_n). This concludes the construction of a graph G which satisfies (a) provided we specify what the f_α's are. Since the s_α's of $\diamond$ (2.42) are rather good at guessing subsets of ω_1 we shall take f_α to be the function f on α into ω such that $s_\alpha=\{\omega\bullet\beta+f(\beta) \mid \beta<\alpha\}$ if there is such an f. If there is no such f it does not matter what we take for f_α.

To prove (b) suppose that F is an ω-coloring of ω_1 and let $W\subseteq\omega$ be the set of colors which occur for arbitrarily large ordinals $<\omega_1$ in the coloring F (i.e., $W=\{n<\omega \mid \sup\{\alpha<\omega_1 \mid F(\alpha)=n\}=\omega_1\}$). We shall now see that if ω_1 has a ω-coloring F consistent with G it has one for which W is infinite, hence we can assume that W is infinite. W is clearly non-void since $\mathrm{cf}(\omega_1)>\omega$; let $k\in W$. Let F' be the function on ω_1 defined by $F'(\alpha)=2\cdot F(\alpha)$ if $F(\alpha)\neq k$, and if $F(\alpha)=k$ then $F'(\alpha)=2l+1$ where l is such that for some $\beta<\omega_1$ α is the $\omega\cdot\beta+l$-th member of $F^{-1}[\{k\}]$. It is easy to verify that since F is consistent with G also F' is so and the set of all colors which occur unboundedly in F' includes all the odd numbers and is therefore infinite.

For each $n\in W$ let C_n be the set of all limit points of $F^{-1}[\{n\}]$ in ω_1; by IV.4.13(iii) C_n is a closed unbounded subset of ω_1. For $n\notin W$ the set $F^{-1}[\{n\}]$ is a bounded subset of ω_1; we set $C_n=\omega_1\sim\sup^+F^{-1}[\{n\}]$; also this C_n is clearly a closed unbounded subset of ω_1. Let $C=\bigcap_{n\in\omega}C_n\cap\{\omega^{\cdot\omega}\cdot\beta\mid\beta<\omega_1\}$; by IV.4.15 C is a closed unbounded subset of ω_1. By the definition of C, if $\alpha\in C$ then no color in $\omega\sim W$ occurs at α or above α in F, thus $F(\alpha)\in W$, and each color of W occurs unboundedly high below α in F. Let $S=\{\omega\cdot\beta\dot{+}F(\beta)\mid\beta<\omega_1\}$. By $\diamondsuit$ the set $Q=\{\alpha<\omega_1\mid s_\alpha=S\cap\alpha\}$ is a stationary subset of ω_1, hence $Q\cap C\neq 0$. Let $\alpha\in Q\cap C$. Since $\alpha\in C$ it is a multiple of $\omega^{\cdot\omega}$, hence for $\beta<\alpha$ $\omega\cdot\beta\dot{+}F(\beta)<\alpha$ and therefore $s_\alpha=S\cap\alpha=\{\omega\cdot\beta\dot{+}F(\beta)\mid\beta<\alpha\}$. Since f_α was defined to be that $f:\alpha\to\omega$ such that $s_\alpha=\{\omega\cdot\beta\dot{+}f(\beta)\mid\beta<\alpha\}$ we have $f_\alpha=F\upharpoonright\alpha$. Since $\alpha\in C$ we have, as mentioned above, $F(\alpha)\in W$ and each color of W occurs unboundedly high below α in F. By the definition of G_α, G connects α to ordinals colored by $f_\alpha=F\upharpoonright\alpha$ with each one of the infinitely many colors which occur unboundedly high below α, hence G connects α to ordinals colored by F with each one of the colors in W. Since also $F(\alpha)\in W$ the coloring F is not consistent with G. □

2.58Ac Proposition (Hajnal and Máté 1975). *The version $A_{\aleph_1}$ of Martin's axiom (VIII.4.11) implies that for every graph G on ω_1 for which there is an ordinal $\tau<\omega_1$ such that for all $\alpha<\omega_1$ $G_\alpha=\{\beta<\omega_1\mid\{\beta,\alpha\}\in G\}$ is of order type $\leqslant\tau$ there is a coloring F of ω_1 with $\aleph_0$ colors consistent with G.*

Outline of proof. Let G be a graph on ω_1 as in the proposition. Let P be the set of all colorings of finite subsets of ω_1 which are consistent with G, partially ordered by proper inclusion. To prove that P satisfies the countable antichain condition let W be an uncountable set of pairwise incompatible members of P. We can, by deleting members of W, assume that all the members of W have domains of the same cardinality n. Let q be a member of P with a domain of maximal cardinality which is included in $>\aleph_0$ members of W. We can assume that $q=\bigcap W$. By the maximality of q one can construct by recursion a sequence $\langle p_\alpha\mid\alpha<\omega_1\rangle$ of members of W such that if $\alpha<\beta<\omega_1$ then $\mathrm{Dom}(p_\alpha)\cap\mathrm{Dom}(p_\beta)=\mathrm{Dom}(q)$. One can also construct a subsequence $\langle r_\alpha\mid\alpha<\omega_1\rangle$ of $\langle p_\alpha\mid\alpha<\omega_1\rangle$ such that for all $\alpha<\beta<\omega_1$ $\max(\mathrm{Dom}(r_\alpha))<\min(\mathrm{Dom}(r_\beta\sim q))$. By IV.2.21 $\tau\#\tau\#\cdots\#\tau$ (n times) $<\omega_1$; take σ such that $\tau\#\tau\#\cdots\#\tau<\sigma<\omega_1$. Since r_σ is incompatible with r_α for $\alpha<\sigma$ there is an $\eta\in\mathrm{Dom}(r_\sigma)$ and a $\xi\in\mathrm{Dom}(r_\alpha)$ such that $\{\xi,\eta\}\in G$. For $\eta\in\mathrm{Dom}(r_\sigma)$ let $W_\eta=\{\alpha<\sigma\mid(\exists\xi\in\mathrm{Dom}(r_\alpha))(\{\xi,\eta\}\in G)\}$. We have $\sigma=\bigcup_{\eta\in\mathrm{Dom}(r_\sigma)}W_\eta$ hence, by IV.2.22(vii), there is an $\eta\in\mathrm{Dom}(r_\sigma)$ such that the order type of W_η is $>\tau$, hence the order type of G_η is $>\tau$, which contradicts our assumption on G.

Now apply $A_{\aleph_1}$ in the version $A'_{\aleph_1}$ of VIII.4.15(d) where for every $\alpha<\omega_1$

D_α of $A'_{\aleph_1}(P)$ (of VIII.4.14) is the set of all $p \in P$ such that $\alpha \in \mathrm{Dom}(p)$. If H is a subnet of P as G in $A'_{\aleph_1}(P)$ then $\bigcup H$ is a coloring of ω_1 as required. $\square$

3. *Partition Properties*

Our starting point is the familiar pigeon hole principle: if for a finite n $n+1$ balls are distributed among n holes at least one hole must contain more than one ball. The infinite version of this principle is even stronger: for a regular $\aleph_\alpha$ if $\aleph_\alpha$ balls are distributed among $<\aleph_\alpha$ holes then at least one of the holes must contain $\aleph_\alpha$ balls. These principles are trivial but we shall deal with related results which are deep and useful. We start getting non-trivial results when rather than distribute the members of some set A among different sets we distribute the pairs of the members of A. This can be illustrated by the following finite example, which is very easy but not outright trivial.

Suppose we are given a complete polygon A with $n \geqslant 6$ vertices (where by *complete* we mean that every two vertices of the polygon are joined by an edge) and that every edge of the polygon is colored red or black. (Thus we have distributed the unordered pairs of the vertices of A, which are the edges of A, between the set of the black edges and the set of the red edges.) We shall see that A includes an all-red or an all-black triangle. Let $A_0, \ldots, A_5$ be any six distinct vertices of the polygon. Among the edges A_0A_i, $1 \leqslant i \leqslant 5$, at least three must be of the same color. Suppose, without loss of generality, that the edges A_0A_1, A_0A_2 and A_0A_3 are red. If for some $1 \leqslant i < j \leqslant 3$ the edge A_iA_j is red then the triangle $A_0A_iA_j$ has red edges only. If all the edges A_iA_j with $1 \leqslant i < j \leqslant 3$ are black then the triangle $A_1A_2A_3$ has black edges only.

The question which this example answers can be extended to many problems in finite and infinite combinatorics. Since only infinite combinatorics is set theoretical, both in its flavor and in its applications, we shall limit ourselves to the infinite domain. In the present section we shall study situations like that of the example we gave generalized in several directions, such as coloring the faces of a complete n-dimensional polyhedron instead of the edges of a complete polygon, having $\mathfrak{a}$ vertices, where $\mathfrak{a}$ is any infinite cardinal, instead of just finitely many vertices, allowing also more than two colors, and looking not just for triangles of one color but for complete n-dimensional polyhedra with $\mathfrak{b}$ vertices, where $\mathfrak{b}$ is any cardinal. These questions are interesting not only for their own sake. The results which we shall prove have many applications; the most important one among those which will be proved here will supply us with additional information on the power function $\langle 2^{\mathfrak{a}} \mid \mathfrak{a} \in \mathrm{Cn} \rangle$. The study of these questions leads us to large cardinals and, as a consequence, we shall continue here our discussion of the weakly compact and the ineffable cardinals.

The first results in this area are due to Ramsey 1930 and we begin our treatment by presenting the simplest case of Ramsey's theorem. The general case will be formulated and proved in 3.7 below.

3.1 Theorem (Ramsey 1930). *Let A be a denumerable set and suppose we color each pair $\{x, y\}\subseteq A$, where $x\neq y$, either red or black then A has an infinite subset B such that all the pairs $\{x, y\}\subseteq B$, with $x\neq y$, are of the same color.*

Proof. We define for every finite sequence s of 0's and 1's a subset C_s of A and if $C_s\neq 0$ also a member $x_s\in C_s$ by recursion on the proper inclusion relation (restricted to ${}^{\underline{\omega}}2$) as follows. $C_0=A$. For every $s\in{}^{\underline{\omega}}2$ if $C_s\neq 0$ we take x_s to be an arbitrary member of C_s and we set

$$C_{s^\frown\langle 0\rangle}=\{y\in C_s \mid y\neq x_s\wedge\{x_s, y\}\text{ is black}\}\text{ and}$$
$$C_{s^\frown\langle 1\rangle}=\{y\in C_s \mid y\neq x_s\wedge\{x_s, y\}\text{ is red}\},$$

and if $C_s=0$ we set $C_{s^\frown\langle 0\rangle}=C_{s^\frown\langle 1\rangle}=0$. We have, clearly,

(1) if $C_s\neq 0$ then $C_s=\{x_s\}\cup C_{s^\frown\langle 0\rangle}\cup C_{s^\frown\langle 1\rangle}$, $C_{s^\frown\langle 0\rangle}\cap C_{s^\frown\langle 1\rangle}=0$ and $x_s\notin C_{s^\frown\langle 0\rangle}\cup C_{s^\frown\langle 1\rangle}$.

(1) obviously implies that

(2) for $s, t\in{}^{\underline{\omega}}2$ if $C_t\neq 0$ and $s\subset t$ then $C_t\subset C_s$, $x_s\notin C_t$ and $x_s\neq x_t$.

(3) If $C_t\neq 0$ and $s\subset t$ then either $t(\text{Length}(s))=0$ and then, since $x_t\in C_t\subseteq C_{s^\frown\langle 0\rangle}$ and by the definition of $C_{s^\frown\langle 0\rangle}$, $\{x_s, x_t\}$ is black, or else $t(\text{Length}(s))=1$ and $\{x_s, x_t\}$ is red.

Let $T=\{s\in{}^{\underline{\omega}}2 \mid C_s\neq 0\}$. By (1) and (2) $\langle T, \subset\rangle$ is a tree, and every initial of a member s of T is in T (i.e., for $n<\text{Length}(s)$ also $s{\restriction}n\in T$). Therefore for every $s\in T$ $\rho_T(s)=\text{Length}(s)<\omega$. It is easily seen, by induction on n, that

(4) $A=\{x_t \mid t\in{}^{\underline{n}}2\cap T\}\cup\bigcup_{s\in{}^n2}C_s$.

Since A is denumerable and both ${}^{\underline{n}}2$ and ${}^n2$ are finite there is an $s\in{}^n2$ such that $C_s\neq 0$, and hence the n-th level T_n of T is non-void. Thus $\text{Length}(T)=\omega$. Each level T_n of T is finite, being a subset of ${}^n2$, hence by König's infinity lemma (2.17) T has a branch W of Length ω. Let $w=\bigcup W$, then $w\in{}^\omega2$ and $W=\{w{\restriction}n \mid n\in\omega\}$. In w at least one of 0 and 1 must occur infinitely often; assume that it is 0 which occurs infinitely often in w and let $B=\{x_{w\restriction n} \mid n\in\omega\wedge w(n)=0\}\subseteq A$. By our assumption and (2) B is infinite. We claim that for all $y, z\in B$ if $y\neq z$ then $\{y, z\}$ is black. Any two different members of B must be of the form $x_{w\restriction m}$ and $x_{w\restriction n}$ where $m, n\in\omega$, $w(m)=w(n)=0$, and $m\neq n$. Assume, without loss of generality, that $m<n$ then since $(w{\restriction}n)(m)=w(m)=0$ we have, by (3), that $\{x_{w\restriction m}, x_{w\restriction n}\}$ is black. If it is 1 which occurs infinitely often in w then we obtain similarly an infinite subset B of A such that for all $y, z\in B$ if $y\neq z$ then $\{y, z\}$ is red. □

Ramsey's theorem turned out to be quite useful in several areas of mathematics.

We present here an elementary application to geometry of the simple case 3.1 of Ramsey's theorem.

3.2 Definition. Given a subset S of a Euclidean space ${}^n\mathbb{R}$ we say that x *can be seen from* y *in* S if the closed interval $[x, y]$ is included in S. A subset T of S is called *visually independent in* S if no point of T can be seen from any other point of T in S. S is called *locally star shaped at* x if x has a neighborhood W such that every member of $S \cap W$ can be seen in S from x.

3.3Ac Proposition. *Let S be a closed set in a Euclidean space ${}^n\mathbb{R}$. Then either S includes an infinite subset which is visually independent in S or else S is locally star shaped at every point of S.*

Proof. Suppose S is not locally star shaped at x, then each sphere $S(x, 2^{-n})$ with center x and radius 2^{-n} contains a member x_n of S which cannot be seen from x in S. The set $T = \{x_n \mid n < \omega\}$ is obviously infinite and $\lim_{n\to\infty} x_n = x$. Let us color a pair $\{y, z\} \subseteq T$ red if z can be seen from y in S and black otherwise. By Ramsey's theorem 3.1 T has an infinite subset P such that either all pairs $\{y, z\} \subseteq P$ are black or all of them are red. In the former case P is an infinite subset of S which is visually independent in S. In the latter case let us write P as $\{x_{i_0}, x_{i_1}, \ldots\}$ with $i_0 < i_1 < \ldots$ We have $[x_{i_0}, x_{i_n}] \subseteq S$ for all n. $\lim_{n\to\infty} x_{i_n} = \lim_{k\to\infty} x_k = x$ and since S is closed also $[x_{i_0}, x] \subseteq S$, contradicting our choice of x_{i_0}. □

3.4Ac Exercise. Let A and B be subsets of ${}^n\mathbb{R}$ neither of which includes an infinite subset which is visually independent in the respective set. Prove that also $A \cap B$ has this property. □

In order to formulate the general case of Ramsey's theorem we need the following definition.

3.5 Definition. (i) For a class A and $n < \omega$ $[A]^n$ denotes the set of all subsets of A of cardinality n, i.e., the set of all unordered n-tuples of members of A, and $[A]^{<\omega}$ denotes the class of all finite subsets of A.

(ii) (Erdös and Rado 1956). For cardinals $\mathfrak{a}, \mathfrak{b}, \mathfrak{c}$ and $n < \omega$ $\mathfrak{a} \to (\mathfrak{b})^n_{\mathfrak{c}}$ denotes the statement "for all sets A and C with $|A| = \mathfrak{a}$ and $|C| = \mathfrak{c}$ and every function $f: [A]^n \to C$ there is a subset B of A with $|B| = \mathfrak{b}$ such that $|f[[B]^n]| = 1$. Such a set B is said to be *homogeneous* with respect to f. We can regard f as a coloring of all unordered n-tuples of members of A by at most $\mathfrak{c}$ different colors, and then $\mathfrak{a} \to (\mathfrak{b})^n_{\mathfrak{c}}$ says that if we color all unordered n-tuples of members of a set A of cardinality $\mathfrak{a}$ by at most $\mathfrak{c}$ colors then A has a subset B of cardinality $\mathfrak{b}$ such that all unordered n-tuples of B are colored by the same color. Following tradition we shall refer to statements of the form $\mathfrak{a} \to (\mathfrak{b})^n_{\mathfrak{c}}$ as *partition relations* since a function $f: [A]^n \to C$ can be viewed as a partition of $[A]^n$ into $|C|$ sets.

Theorem 2.1 can now be written as $\aleph_0 \to (\aleph_0)^2_2$. Notice that if for some fixed sets A and C with $|A| = \mathfrak{a}$ and $|C| = \mathfrak{c}$ it is the case that for every function $f: [A]^n \to C$ there is a subset B of A such that B is homogeneous with respect to f, then $\mathfrak{a} \to (\mathfrak{b})^n_{\mathfrak{c}}$

holds (i.e., what we said above holds for all A and C such that $|A|=\mathfrak{a}$ and $|C|=\mathfrak{c}$); this can be easily shown since any counterexample to $\mathfrak{a}\to(\mathfrak{b})^n_{\mathfrak{c}}$ using sets A and C and a function $f:[A]^n\to C$ can be trivially transferred to any other sets A' and C' such that $|A'|=|A|$ and $|C'|=|C|$ by taking a corresponding function $f':[A']^n\to C'$.

3.6 Proposition. (i) *If* $\mathfrak{a}\to(\mathfrak{b})^n_{\mathfrak{c}}$ *holds and* $\mathfrak{a}'\geqslant\mathfrak{a}$, $\mathfrak{b}'\leqslant\mathfrak{b}$ *and* $\mathfrak{c}'\leqslant\mathfrak{c}$ *then also* $\mathfrak{a}'\to(\mathfrak{b}')^n_{\mathfrak{c}'}$.
(ii) *For an aleph* $\mathfrak{a}$ *and* $\mathfrak{b}\geqslant\aleph_0$, *if* $m<n$ *and* $\mathfrak{a}\to(\mathfrak{b})^n_{\mathfrak{c}}$ *then also* $\mathfrak{a}\to(\mathfrak{b})^m_{\mathfrak{c}}$.

Hint of proof. (ii) Let $|A|=\mathfrak{a}$ and let $f:[A]^m\to C$ with $|C|\leqslant\mathfrak{c}$. Well-order A and define $f':[A]^n\to C$ by $f'(u)=f$(the set of the first m members of u). □

3.7 Theorem (Ramsey 1930). *For* $n, k<\omega$ $\aleph_0\to(\aleph_0)^n_k$.

Proof. We prove the theorem by induction on n. For $n=1$ it is the trivial pigeon-hole principle. For $n>1$ we shall now assume that $\aleph_0\to(\aleph_0)^{n-1}_k$ and prove $\aleph_0\to(\aleph_0)^n_k$. In the proof of the induction step we shall make use of the idea of the proof of $\aleph_0\to(\aleph_0)^2_2$ in 3.1. There we constructed a sequence $\langle x_{w1m}\mid m\in\omega\rangle$ of members of A such that for every fixed $l<\omega$ all the pairs $\{x_{w1l},x_{w1m}\}$ with $m>l$ were of the same color. This enabled us to associate a color with each x_{w1l}, namely the common color of all pairs $\{x_{w1l},x_{w1m}\}$ with $m>l$. Using the simple pigeon-hole principle we obtained an infinite subset B of $\{x_{w1m}\mid m\in\omega\}$ such that all the members of B were associated with the same color. This set B turned out to be the required set. Here we shall aim at the construction of a sequence $\langle x_m\mid m<\omega\rangle$ of members of A such that for every n-tuple $\{x_{i_1},\ldots,x_{i_n}\}$ with $i_1<i_2<\cdots<i_n$ the color of the n-tuple depends only on its $n-1$ first members $x_{i_1},\ldots,x_{i_{n-1}}$. Then we shall be able to associate a color with each $n-1$-tuple $\{x_{i_1},\ldots,x_{i_{n-1}}\}$ and apply the induction hypothesis $\aleph_0\to(\aleph_0)^{n-1}_k$ to this coloring of $[\{x_m\mid m\in\omega\}]^{n-1}$.

By the remark at the end of Definition 3.5 we can choose A and C of Definition 3.5 to be ω and k, respectively. Given $f:[\omega]^n\to k$ we define a tree $\langle T,\supset\rangle$ whose members are non-void subsets of ω and whose partial order is the inverse of proper inclusion. We shall define by recursion on $m<\omega$ the m-th level T_m of T. We shall simultaneously prove by induction that

(4) T_m is finite,

(5) $\bigcup T_m=\omega\sim\{\operatorname{Min}D\mid D\in\bigcup_{l<m}T_l\}$,

(6) any two different members of T_m are disjoint, and

(7) for every $l<m$ and every member E of T_m there is a unique member of T_l which includes E.

We take ω itself to be the unique member of T_0; (4)–(7) obviously hold for $m=0$. Now we suppose that T_l is already defined for $l\leqslant m$ and that (4)–(7) hold for $l\leqslant m$, and we define T_{m+1} and prove (4)–(7) for $m+1$. Let $E\in T_m$. For $l<m$ there is, by (7), a unique $D\in T_l$ which includes E; we denote it with E_l, and we set $y_l=\operatorname{Min}E_l$ for

$l<m$ and $y_m=\operatorname{Min} E$. For every $u \in E \sim \{y_m\}$ let $g_u:[\{y_0,\ldots,y_m\}]^{n-1}$ be the function given by

$$g_u(\{t_1,\ldots,t_{n-1}\})=f(\{t_1,\ldots,t_{n-1},u\})$$

for all $\{t_1,\ldots,t_{n-1}\} \in [\{y_0,\ldots,y_m\}]^{n-1}$. (If $n-1>m+1$ then $[\{y_0,\ldots,y_m\}]^{n-1}=0$ and $g_u=0$). We define an equivalence relation $\approx_E$ on $E \sim \{y_m\}$ by setting $u \approx_E v$ iff $g_u=g_v$. Let Q_E be the set of all equivalence classes of $\approx_E$. Since the number of functions on $[\{y_0,\ldots,y_m\}]^{n-1}$ into k is finite Q_E is finite. $\bigcup Q_E=E \sim \{y_m\}=E \sim \{\operatorname{Min} E\}$. We set $T_{m+1}=\bigcup_{E \in T_m} Q_E$; thus T_{m+1} consists of non-void subsets of ω. By the induction hypothesis T_m is finite, hence T_{m+1} is finite.

$$\begin{aligned}\bigcup T_{m+1}=\bigcup_{E \in T_m}\bigcup Q_E=\bigcup_{E \in T_m}(E\sim\{\operatorname{Min} E\})&=\bigcup T_m\sim\{\operatorname{Min} E \mid E \in T_m\} \text{ by (6)}\\ &=\omega\sim\{\operatorname{Min} D \mid D \in \textstyle\bigcup_{l<m} T_l\}\sim\{\operatorname{Min} E \mid E \in T_m\} \text{ by (5)}\\ &=\omega\sim\{\operatorname{Min} D \mid D \in \textstyle\bigcup_{l<m+1} T_l\}, \text{ and (5) holds for } m+1.\end{aligned}$$

(6) for $m+1$ follows immediately from (6) for m and the definition of T_{m+1}. The uniqueness part of (7) for $m+1$ follows from (6) for $l \leqslant m$; the existence part of (7) for $m+1$ follows easily from the definition of T_{m+1} and (7) for m.

T is clearly a tree of length ω and by (4) T_m is finite for all $m<\omega$. Hence, by König's infinity lemma (2.17), T has an infinite branch $\{D_l \mid l<\omega\}$, where $D_l \in T_l$ for $l<\omega$. We denote $\operatorname{Min} D_l$ with x_l and $\{x_l \mid l<\omega\}$ with X. For an $n-1$-tuple $\{x_{i_1},\ldots,x_{i_{n-1}}\}$ with $i_1<i_2<\cdots<i_{n-1}$ put $m=i_{n-1}$. For all $l>m$ we have $x_l \in D_l \subseteq D_{m+1}$. Since D_{m+1} is an equivalence class of the relation $\approx_{D_m}$ we have that $f(\{x_{i_1},\ldots,x_{i_{n-1}},u\})=g_u(\{x_{i_1},\ldots,x_{i_{n-1}}\})$ does not depend on u as long as $u \in D_{m+1}$, hence $f(\{x_{i_1},\ldots,x_{i_{n-1}},x_{i_n}\})$ does not depend on i_n and we can denote this value with $g(\{x_{i_1},\ldots,x_{i_{n-1}}\})$. We have now $g:[X]^{n-1} \to k$. By the induction hypothesis there is an infinite subset B of X such that $g[[B]^{n-1}]=\{l\}$ for some $l \in k$. Therefore we have $f(\{x_{i_1},\ldots,x_{i_{n-1}},x_{i_n}\})=l$ for all $x_{i_1},\ldots,x_{i_{n-1}},x_{i_n} \in B$ (actually if $i_1<i_2<\cdots<i_{n-1}$ and $x_{i_1},\ldots,x_{i_{n-1}} \in B$ then $f(\{x_{i_1},\ldots,x_{i_{n-1}},u\})=l$ holds for every $u \in D_{i_{n-1}+1}$). $\square$

The question which comes up naturally now is whether we can extend the theorem $\aleph_0 \to (\aleph_0)_k^n$ to cardinals greater than $\aleph_0$, and the least such cardinal is $\aleph_1$. The answer we get is negative; not only does $\aleph_1 \not\to (\aleph_1)_2^2$ but even $2^{\aleph_0} \not\to (\aleph_1)_2^2$, where $\not\to$ means that $\to$ does not hold.

3.8Ac Proposition (Sierpinski 1933). $2^{\aleph_0} \not\to (\aleph_1)_2^2$.

Proof. Let $\langle \mathbb{R}, < \rangle$ be the set of the real numbers ordered by magnitude, and let $<^*$ be a relation which well-orders $\mathbb{R}$. Define $f:[\mathbb{R}]^2 \to 2$ as follows. For $\{x,y\} \subseteq \mathbb{R}$ let $f(\{x,y\})=0$ if both $<$ and $<^*$ order the pair $\{x,y\}$ in the same way, and $f(\{x,y\})=1$ if $<$ and $<^*$ order $\{x,y\}$ in opposite ways; i.e., if $x \neq y$ then $f(\{x,y\})= 0 \leftrightarrow (x<y \leftrightarrow x<^*y)$. We shall see that there is no set of cardinality $\aleph_1$ which is homogeneous with respect to f. Suppose that there is such a set B. If $f[[B]^2]=\{0\}$ then $x,y \in B \to (x<y \leftrightarrow x<^*y)$, and since $<^*$ is a well-ordering this means that the natural order of the reals well-orders the uncountable set B, which is impossible (since the rational numbers are dense in $\mathbb{R}$). If $f[[B]^2]=\{1\}$ we have $x, y \in B \rightarrow (x<y \leftrightarrow y<^*x)$ and thus $<$ is an inverse well-ordering of the uncountable set B, which is also impossible. □

3.9 A More General Partition Relation. As a consequence of 3.8 and 3.6 we have $\aleph_1 \not\rightarrow (\aleph_1)^2_2$, and all we can say right now about $\aleph_1$ is that $\aleph_1 \rightarrow (\aleph_0)^2_2$. This will be strengthened in 3.10 where it will be shown that for every aleph $\mathfrak{a}$ $\mathfrak{a} \rightarrow (\mathfrak{a}, \aleph_0)^2$. This means that if $f:[\mathfrak{a}]^2 \rightarrow 2$ then either there is a subset $B \subseteq \mathfrak{a}$ with $|B|=\mathfrak{a}$ such that $f[[B]^2]=\{0\}$ or else there is a subset $B \subseteq \mathfrak{a}$ with $|B|=\aleph_0$ such that $f[[B]^2]=\{1\}$. As the identity of the two colors does not matter we have in this case that for every coloring of $\mathfrak{a}$ with two colors there is either a homogeneous set of cardinality $\mathfrak{a}$ or else for each one of the two colors there is a denumerable set all of whose pairs are colored with that color.

3.10Ac Exercise (Dushnik and Miller 1941). For every aleph $\mathfrak{a}$ $\mathfrak{a} \rightarrow (\mathfrak{a}, \aleph_0)^2$.

Hint. Use induction on $\mathfrak{a}$. Let $f:[\mathfrak{a}]^2 \rightarrow 2$. First show that if for every $B \subseteq \mathfrak{a}$ with $|B|=\mathfrak{a}$ there is an $x \in B$ such that $|\{y \in B \mid f(\{x,y\})=1\}|=\mathfrak{a}$ then there is a denumerable $D \subseteq \mathfrak{a}$ such that $f[[D]^2]=\{1\}$. Now assume that there is no such D and let $B \subseteq \mathfrak{a}$ be such that $|B|=\mathfrak{a}$ and for every $x \in B$ $|\{y \in B \mid f(\{x,y\})=1\}|<\mathfrak{a}$. If $\mathfrak{a}$ is regular construct by recursion a sequence $\langle x_\alpha \mid \alpha<\mathfrak{a}\rangle$ such that $f(\{x_\alpha, x_\beta\})=0$ for $\alpha<\beta<\mathfrak{a}$. (Notice that this yields a proof of $\aleph_0 \rightarrow (\aleph_0)^2_2$ different from the one in 3.1). If $\mathfrak{a}$ is singular let $\lambda=\mathrm{cf}(\mathfrak{a})$. Let $e: B \rightarrow \mathfrak{a}$ be defined by $e(x)=|\{y \in B \mid f(\{x,y\})=1\}|$. Let A_ξ, $\xi<\lambda$ be subsets of B such that $\bigcup_{\xi<\lambda} A_\xi=B$, for all $\xi<\lambda$ e is bounded below $\mathfrak{a}$ on A_ξ, and for all $\xi<\eta<\lambda$ $\aleph_0 \leqslant |A_\xi|<|A_\eta|<\mathfrak{a}$. Since for $\xi<\lambda$ $|A_\xi|<\mathfrak{a}$ we may assume, by the induction hypothesis, that there is a subset B_ξ of A_ξ such that $|B_\xi|=|A_\xi|$ and $f[[B_\xi]^2]=\{0\}$. Define $g:\lambda \rightarrow \mathfrak{a}$ by recursion so that g is increasing and for $\alpha<\lambda$ $|B_{g(\alpha)}|>\Sigma_{\beta<\alpha}\Sigma_{x \in B_{g(\beta)}} e(x)$. Set

$$E=\bigcup_{\alpha<\lambda}(B_{g(\alpha)} \sim \bigcup_{\beta<\alpha}\bigcup_{x \in B_{g(\beta)}}\{y \in B \mid f(\{x,y\})=1\}).$$

then $|E|=\mathfrak{a}$ and $f[[E]^2]=\{0\}$. □

3.11 Exercise. If $\mathfrak{a} \rightarrow (\mathfrak{a})^m_2$ then $\mathfrak{a} \rightarrow (\mathfrak{a})^m_k$ for all $2 \leqslant k<\omega$.

Hint. Use induction on k. For a given k identify $k-1$ of the colors. □

3.12Ac Wider Application for the Proof of Ramsey's Theorem. In spite of the negative result of Proposition 3.8 we shall now go back to the proof of Ramsey's theorem 3.7 and see which positive results one can obtain by means of the same proof applied to other cardinals. For all alephs $\mathfrak{a}$ we have, by IV.3.9 $\mathfrak{a} \rightarrow (\mathfrak{a})^1_\mathfrak{c}$ for all $\mathfrak{c}<\mathrm{cf}(\mathfrak{a})$. We shall see that the proof of 3.7 provides an inductive procedure for increasing the superscript. Suppose our inductive procedure led us to $\mathfrak{b} \rightarrow (\mathfrak{a})^{n-1}_\mathfrak{c}$ and we want to obtain $\mathfrak{d} \rightarrow (\mathfrak{a})^n_\mathfrak{c}$ for as small a $\mathfrak{d}$ as possible. In the proof of 3.7 we obtained a sequence $\langle x_n \mid n<\omega\rangle$ such that

(8) for all $i_1<i_2<\cdots<i_n$ the color $f(\{x_{i_1},\ldots,x_{i_n}\})$ of $\{x_{i_1},\ldots,x_{i_n}\}$ depends only on $i_1,\ldots,i_{n-1}$.

We defined this color to be the color $g(\{x_{i_1},\ldots,x_{i_{n-1}}\})$ of $\{x_{i_1},\ldots,x_{i_{n-1}}\}$ and used $\aleph_0\to(\aleph_0)_k^{n-1}$ to obtain a subset B of $\{x_i \mid i\in\omega\}$ homogeneous with respect to the coloring g and hence also with respect to the original coloring f. Here we have at our disposal $\mathfrak{b}\to(\mathfrak{a})_{\mathfrak{c}}^{n-1}$ therefore we need a sequence x of length $\mathfrak{b}$ in order to obtain a homogeneous set of cardinality $\mathfrak{a}$. To put it formally, if $f\colon[\mathfrak{d}]^n\to\mathfrak{c}$ and $\langle x_i \mid i<\mathfrak{b}\rangle$ is such that for all $i_1<i_2<\cdots<i_n<\mathfrak{b}$ the color $f(x_{i_1},\ldots,x_{i_n}\})$ does not depend on x_{i_n} we define the coloring g of $[\{x_i \mid i<\mathfrak{b}\}]^{n-1}$ by $g(\{x_{i_1},\ldots,x_{i_{n-1}}\})=f(\{x_{i_1},\ldots,x_{i_{n-1}},x_{i_n}\})$ for $i_1<\cdots<i_{n-1}<i_n$. By $\mathfrak{b}\to(\mathfrak{a})_{\mathfrak{c}}^{n-1}$ $\{x_i \mid i<\mathfrak{b}\}$ has a subset B of cardinality $\mathfrak{a}$ homogeneous with respect to g, and hence also with respect to f, which establishes $\mathfrak{d}\to(\mathfrak{a})_{\mathfrak{c}}^n$.

Now we have to obtain a sequence $\langle x_i \mid i<\mathfrak{b}\rangle$ which satisfies (8). We shall construct it by means of a tree T as in the proof of 3.7, except that now it is a tree of length $\mathfrak{b}$. We remind the reader that the cardinal $\mathfrak{d}$ for which we prove $\mathfrak{d}\to(\mathfrak{a})_{\mathfrak{c}}^n$ is not yet determined. The tree T is going to be a tree of non-void subsets of $\mathfrak{d}$, partially ordered by inverse proper inclusion, such that the following (9)–(11) (which are analogous to (5)–(7) of 3.7) hold for all $\mu<\mathfrak{b}$.

(9) $\bigcup T_\mu=\mathfrak{d}\sim\{\mathrm{Min}\,D \mid D\in\bigcup_{\lambda<\mu}T_\lambda\}$.

(10) Any two different members of T_μ are disjoint.

(11) For every $\lambda<\mu$ and $E\in T_\mu$ there is a unique member $D\in T_\lambda$ such that $D\supset E$.

We define T_μ by recursion. $T_0=\{\mathfrak{d}\}$; this obviously satisfies (9)–(11). Given T_μ, let $E\in T_\mu$, and for every $\lambda\leqslant\mu$ let E_λ be the unique member of T_λ which includes E, and let $y_\lambda=\mathrm{Min}\,E_\lambda$. For every $u\in E\sim\{y_\mu\}$ let $g_u\colon[\{y_\lambda \mid \lambda\leqslant\mu\}]^{n-1}\to\mathfrak{c}$ be the function given by $g_u(\{t_1,\ldots,t_{n-1}\})=f(\{t_1,\ldots,t_{n-1},u\})$. We define an equivalence relation $\approx_E$ on $E\sim\{y_\mu\}$ by setting $u\approx_E v$ iff $g_u=g_v$. Let Q_E be the set of all equivalence classes of $\approx_E$; since the number of functions $g\colon[\{y_\lambda \mid \lambda\leqslant\mu\}]^{n-1}\to\mathfrak{c}$ is $\mathfrak{c}^{\max(\aleph_0,|\mu|)}$ we have $|Q_E|\leqslant\mathfrak{c}^{\max(\aleph_0,|\mu|)}$. We set now $T_{\mu+1}=\bigcup_{E\in T_\mu}Q_E$, hence

(12) $|T_{\mu+1}|\leqslant|T_\mu|\cdot\mathfrak{c}^{\max(\aleph_0,|\mu|)}$.

It is now easily seen, as in the proof of 3.7, that (9)–(11) hold also for $T_{\mu+1}$.

For a limit ordinal μ we set $T_\mu=\{\bigcap_{\lambda<\mu}h(\lambda) \mid h\in \times_{\lambda<\mu}T_\lambda\}\sim\{0\}$, which is the set of all non-void intersections along all the branches of $T|\mu$. To see that (9) holds, let $u\in\mathfrak{d}\sim\{\mathrm{Min}\,D \mid D\in\bigcup_{\lambda<\mu}T_\lambda\}$ then, since for $\lambda<\mu$ (9) holds by the induction hypothesis, we have $u\in\bigcup T_\lambda$, and hence for some $E_\lambda\in T_\lambda$ $u\in E_\lambda$, and therefore $u\in\bigcap_{\lambda<\mu}E_\lambda\in T_\mu$. In the other direction, if $u\in\bigcup T_\mu$ we have, by the definition of T_μ, $u\in\bigcap_{\lambda<\mu}h(\lambda)$ for some $h\colon\mu\to V$ such that $h(\lambda)\in T_\lambda$ for $\lambda<\mu$. Since (9) holds for $\lambda<\mu$ we have $u\notin\{\mathrm{Min}\,D \mid D\in\bigcup_{\xi<\lambda}T_\xi\}$ for every $\lambda<\mu$ hence

$u \in \mathfrak{d} \sim \{\mathrm{Min} D \mid D \in \bigcup_{\lambda<\mu} T_\lambda\}$. It follows easily from the definition of T_μ that also (10) and (11) hold and that

$$(13)\qquad |T_\mu| \leqslant \Pi_{\lambda<\mu} |T_\lambda|.$$

What we need now is a branch of T of length $\mathfrak{b}$. Suppose that $\langle D_\lambda \mid \lambda<\mathfrak{b}\rangle$, where $D_\lambda \in T_\lambda$ for $\lambda<\mathfrak{b}$, is such a branch, then we set $x_\lambda = \mathrm{Min} D_\lambda$ for $\lambda<\mathfrak{b}$. Let $i_1 < i_2 < \cdots < i_n < \mathfrak{b}$ and let $i_{n-1} < j < \mathfrak{b}$ then we have $i_n, j \geqslant i_{n-1}+1$, and since $\{D_\lambda \mid \lambda<\mathfrak{b}\}$ is a branch we have $D_{i_n}, D_j \subseteq D_{i_{n-1}+1}$ hence $x_{i_n}, x_j \in D_{i_{n-1}+1}$. By the definition of $T_{i_{n-1}+1}$ both x_{i_n} and x_j belong to the same equivalence class with respect to $\approx_{D_{i_{n-1}}}$, hence $f(\{x_{i_1}, \ldots, x_{i_{n-1}}, x_{i_n}\}) = f(\{x_{i_1}, \ldots, x_{i_{n-1}}, x_j\})$, which is what is needed to conclude the proof.

Our final task is to see under which conditions we can guarantee the existence of a branch of T of length $\mathfrak{b}$. One way of obtaining such a branch is to continue the construction of the tree to the level $\mathfrak{b}$ itself, and if $T_\mathfrak{b}$ turns out to be non-void we can pick any member D of $T_\mathfrak{b}$ and then the set $\{D_\lambda \mid \lambda<\mathfrak{b}\}$, where D_λ is the unique member of T_λ which includes D, is a branch of length $\mathfrak{b}$. For $T_\mathfrak{b}$ to be non-void it suffices, by (9), to have

$$(14)\qquad \mathfrak{d} > |\{\mathrm{Min} E \mid E \in \bigcup_{\lambda<\mathfrak{b}} T_\lambda\}|.$$

$|\{\mathrm{Min} E \mid E \in \bigcup_{\lambda<\mathfrak{b}} T_\lambda\}| \leqslant |\bigcup_{\lambda<\mathfrak{b}} T_\lambda| \leqslant \Sigma_{\lambda<\mathfrak{b}} |T_\lambda|$, so (14) holds if

$$(15)\qquad \mathfrak{d} > \Sigma_{\lambda<\mathfrak{b}} |T_\lambda|.$$

We shall now prove by induction on λ that for all $\lambda<\mathfrak{b}$ $|T_\lambda| \leqslant 2^{|\lambda|+\aleph_0+\mathfrak{c}}$. For $\lambda=0$ $|T_\lambda|=1$. For $\lambda+1$ we have, by (12) and the induction hypothesis, $|T_{\lambda+1}| \leqslant |T_\lambda| \cdot \mathfrak{c}^{|\lambda|+\aleph_0} \leqslant 2^{|\lambda|+\aleph_0+\mathfrak{c}} \cdot (2^\mathfrak{c})^{|\lambda|+\aleph_0} = (2^{|\lambda|+\aleph_0+\mathfrak{c}})^2 = 2^{|\lambda|+\aleph_0+\mathfrak{c}}$ by III.3.29(ii), III.3.13, and III.3.6. For a limit ordinal λ we have, by (13).

$$|T_\lambda| \leqslant \Pi_{\mu<\lambda} |T_\mu| \leqslant \Pi_{\mu<\lambda} 2^{|\mu|+\aleph_0+\mathfrak{c}} = 2^{\Sigma_{\mu<\lambda}|\mu| + |\lambda|\cdot\aleph_0 + |\lambda|\cdot\mathfrak{c}} = 2^{|\lambda|+\aleph_0+\mathfrak{c}}.$$

If $\aleph_0, \mathfrak{c} < \mathfrak{b}$ we have $|T_\lambda| \leqslant 2^{\underline{\mathfrak{b}}}$ for all $\lambda<\mathfrak{b}$, and since $\mathfrak{b} \leqslant 2^{\underline{\mathfrak{b}}}$ we have $\Sigma_{\lambda<\mathfrak{b}} |T_\lambda| \leqslant 2^{\underline{\mathfrak{b}}}$. Thus (15) holds if $\mathfrak{d} \geqslant (2^{\underline{\mathfrak{b}}})^+$. This establishes the following theorem (for the case $\mathfrak{b} = \aleph_0$ use 3.7 and 3.6).

3.13Ac Theorem (Erdös and Rado 1956). *If* $\mathfrak{a}$, $\mathfrak{b}$ *are alephs,* $n \geqslant 2$, $\mathfrak{c}<\mathfrak{b}$, *and* $\mathfrak{b} \to (\mathfrak{a})^{n-1}_\mathfrak{c}$ *then* $(2^{\underline{\mathfrak{b}}})^+ \to (\mathfrak{a})^n_\mathfrak{c}$. □

3.14Ac Definition (The beth function). For a cardinal $\mathfrak{a}$ and an ordinal α let $\beth_\alpha(\mathfrak{a})$ be defined by: $\beth_0(\mathfrak{a}) = \mathfrak{a}$, $\beth_{\alpha+1}(\mathfrak{a}) = 2^{\beth_\alpha(\mathfrak{a})}$, and for a limit ordinal α $\beth_\alpha(\mathfrak{a}) = \sup_{\beta<\alpha} \beth_\beta(\mathfrak{a})$.

3.15Ac Corollary. (i) *If* $\mathfrak{a}$ *is an aleph,* $n<\omega$ *and* $\mathfrak{c} < \mathrm{cf}(\mathfrak{a})$ *then* $\beth_n(2^{\underline{\mathfrak{a}}})^+ \to (\mathfrak{a})^{n+2}_\mathfrak{c}$.

(ii) *If* $\mathfrak{a}$ *is an aleph,* $n<\omega$ *and* $\mathfrak{c} \leqslant \mathfrak{a}$ *then* $\beth_n(\mathfrak{a})^+ \to (\mathfrak{a}^+)^{n+1}_\mathfrak{c}$.

Proof. (i) Since $\mathfrak{c}<\mathrm{cf}(\mathfrak{a})$ we have $\mathfrak{a}\to(\mathfrak{a})^1_{\mathfrak{c}}$, hence, by 3.13, $(2^{\mathfrak{a}})^+\to(\mathfrak{a})^2_{\mathfrak{c}}$, which is what the corollary claims for $n=0$. Now we proceed by induction on n as follows. Applying 3.13 to $\beth_n(2^{\mathfrak{a}})^+\to(\mathfrak{a})^{n+2}_{\mathfrak{c}}$ we obtain $\beth_{n+1}(2^{\mathfrak{a}})^+=(2^{\beth_n(2^{\mathfrak{a}})})^+\to(\mathfrak{a})^{n+3}_{\mathfrak{c}}$.

(ii) This is obtained by substituting $\mathfrak{a}^+$ for $\mathfrak{a}$ in (i). □

3.16Ac Theorem (Erdös and Tarski 1961). *If $\mathfrak{a}$ is a weakly compact cardinal, $0<n<\omega$ and $\mathfrak{c}<\mathfrak{a}$ then $\mathfrak{a}\to(\mathfrak{a})^n_{\mathfrak{c}}$.*

Proof. We continue the discussion of 3.12, assuming now that $\mathfrak{b}$ is a weakly compact cardinal. Instead of choosing $\mathfrak{d}$ so big as to secure a branch of length $\mathfrak{b}$ we use now the tree property of $\mathfrak{b}$. We have shown that $|T_\lambda|\leqslant 2^{|\lambda|+\aleph_0+\mathfrak{c}}$ for $\lambda<\mathfrak{b}$, and since $\mathfrak{b}$ is a strong limit cardinal we have $|T_\lambda|<\mathfrak{b}$ for $\lambda<\mathfrak{b}$. Therefore, by the tree property of $\mathfrak{b}$, the tree T has a branch of length $\mathfrak{b}$, provided the length of T is $\geqslant\mathfrak{b}$. To get this we choose $\mathfrak{d}=\mathfrak{b}$ and we shall show that $T_\mu\neq 0$ for $\mu<\mathfrak{b}$. Since $\mathfrak{b}$ is regular and $|T_\lambda|<\mathfrak{b}$ for $\lambda<\mathfrak{b}$ we have, for $\lambda<\mathfrak{b}$, $\mathfrak{d}=\mathfrak{b}>\Sigma_{\lambda<\mu}|T_\lambda|=|\bigcup_{\lambda<\mu}T_\lambda|=|\{\mathrm{Min}\,D\mid D\in\bigcup_{\lambda<\mu}T_\lambda\}|$. Therefore the right-hand side of (9) is non-void and hence $T_\mu\neq 0$. Thus we have shown that

(16) if $\mathfrak{b}\to(\mathfrak{a})^{n-1}_{\mathfrak{c}}$ then $\mathfrak{b}\to(\mathfrak{a})^n_{\mathfrak{c}}$.

Since a weakly compact cardinal $\mathfrak{a}$ is regular we have $\mathfrak{a}\to(\mathfrak{a})^1_{\mathfrak{c}}$. Applying (16) $n-1$ times we get $\mathfrak{a}\to(\mathfrak{a})^n_{\mathfrak{c}}$. □

Having generalized Ramsey's theorem in 3.15 and 3.16 let us turn to generalize Sierpinski's negative result 3.8. This generalization will establish that, on the whole, the results of 3.15 cannot be improved. We begin with 3.18, which is a straightforward generalization of 3.8.

3.17Ac Lemma. *For an aleph κ let $<$ be the left lexicographic ordering of ${}^\kappa 2$. No subset of ${}^\kappa 2$ is ordered by $<$, or by its inverse relation $>$, in the order type κ^+.*

Proof. If a subset A of ${}^\kappa 2$ is of order type κ^+ let f be an isomorphism of κ^+ onto A. We shall prove by induction on μ that

(17) for every $\mu\leqslant\kappa$ there is an $\alpha_\mu<\kappa^+$ such that for every $\beta\geqslant\alpha_\mu$ $f(\beta)\upharpoonright\mu=f(\alpha_\mu)\upharpoonright\mu$.

Taking $\mu=\kappa$ in (17) we get for every $\beta\geqslant\alpha_\kappa$ $f(\beta)=f(\beta)\upharpoonright\kappa=f(\alpha_\kappa)\upharpoonright\kappa=f(\alpha_\kappa)$, which is a contradiction since f is an injection.

For $\mu=0$ (17) obviously holds for $\alpha_\mu=0$. Suppose now that (17) holds for μ. If $f(\beta)(\mu)=0$ for every $\beta\geqslant\alpha_\mu$ then set $\alpha_{\mu+1}=\alpha_\mu$, and for every $\beta\geqslant\alpha_{\mu+1}$ we have $f(\beta)\upharpoonright\mu=f(\alpha_\mu)\upharpoonright\mu$, $f(\beta)(\mu)=0=f(\alpha_\mu)(\mu)$, hence $f(\beta)\upharpoonright(\mu+1)=f(\alpha_{\mu+1})\upharpoonright(\mu+1)$. If $f(\beta)(\mu)=1$ for some $\beta\geqslant\alpha_\mu$ then let $\alpha_{\mu+1}$ be the least such β; we have now $f(\alpha_{\mu+1})(\mu)=1$. For $\beta\geqslant\alpha_{\mu+1}$ we have, since $\alpha_{\mu+1}\geqslant\alpha_\mu$, $f(\beta)\upharpoonright\mu=f(\alpha_\mu)\upharpoonright\mu=f(\alpha_{\mu+1})\upharpoonright\mu$. Since f is an isomorphism $f(\beta)\geqslant f(\alpha_{\mu+1})$ and since $f(\beta)$ and $f(\alpha_{\mu+1})$ coincide up to μ we must have $f(\beta)(\mu)\geqslant f(\alpha_{\mu+1})(\mu)=1$, hence $f(\beta)(\mu)=1$. Since $f(\beta)\upharpoonright\mu=f(\alpha_{\mu+1})\upharpoonright\mu$ and $f(\beta)(\mu)=1=f(\alpha_{\mu+1})(\mu)$ we have $f(\beta)\upharpoonright(\mu+1)=f(\alpha_{\mu+1})\upharpoonright(\mu+1)$. If μ is a limit ordinal take $\alpha_\mu=\sup_{\lambda<\mu}\alpha_\lambda$. Since κ^+ is regular and $\mu\leqslant\kappa$ we have

$\alpha_\mu < \kappa^+$. For $\beta \geqslant \alpha_\mu$ and $\lambda < \mu$ we have, since $\alpha_{\lambda+1} \leqslant \alpha_\mu$ and by (17) for λ, $f(\beta)(\lambda) = f(\alpha_{\lambda+1})(\lambda) = f(\alpha_\mu)(\lambda)$, thus $f(\beta) \upharpoonright \mu = f(\alpha_\mu) \upharpoonright \mu$.

If a subset A of ${}^\kappa 2$ is of order type κ^+ in the inverse ordering $>$ we obtain a contradiction similarly, interchanging 0 and 1 in the proof. □

3.18Ac Proposition (Sierpinski 1933). *For every aleph* $\mathfrak{a}$ $2^{\mathfrak{a}} \nrightarrow (\mathfrak{a}^+)^2_2$.

Hint of proof. Follow the proof of 3.8, using 3.17. □

3.19Ac Proposition (Erdös, Hajnal and Rado 1965). *If* $\mathfrak{a}$ *is a regular aleph*, $n > 2$, *and* $\mathfrak{b} \nrightarrow (\mathfrak{a})^{n-1}_2$ *then* $2^{\mathfrak{b}} \nrightarrow (\mathfrak{a})^n_2$.

Outline of proof. Let $g: [\mathfrak{b}]^{n-1} \to 2$ be a counterexample to $\mathfrak{b} \to (\mathfrak{a})^{n-1}_2$. Let $<$ be the left lexicographic ordering of ${}^{\mathfrak{b}}2$ and let $<^*$ be a well-ordering of ${}^{\mathfrak{b}}2$. For $u \in [{}^{\mathfrak{b}}2]^n$ let $p(u) \in {}^n n$ be the permutation p of n such that the i-th member of u in the ordering $<^*$ is the $p(i)$-th member of u in the ordering $<$. For a subset D of ${}^{\mathfrak{b}}2$ and $v \in {}^{\underset{\smile}{\mathfrak{b}}}2$ let $\text{Left}(D, v) = \{s \in D \mid s \supseteq v^\frown\langle 0\rangle\}$, $\text{Right}(D, v) = \{s \in D \mid s \supseteq v^\frown\langle 1\rangle\}$, and $\text{Thru}(D, v) = \{s \in D \mid s \supseteq v\} = \text{Left}(D, v) \cup \text{Right}(D, v)$. If both $\text{Left}(D, v)$ and $\text{Right}(D, v)$ are non-void we say that v is a *junction* of D, and if both are of cardinality $\mathfrak{a}$ we say that v is an $\mathfrak{a}$-*junction* of D. It is easily seen by induction on n that if $u \in [{}^{\mathfrak{b}}2]^n$ then u has exactly $n-1$ junctions. Let p_0 be the identity permutation on n, let p_1 be given by $p_1(i) = n-1-i$ for $i < n$ and let p_2 be any fixed permutation of n other than p_0 and p_1 (this is where we use the assumption that $n > 2$). Let $f: [{}^{\mathfrak{b}}2]^n \to 2$ be given by: $f(u) = 1$ iff $p(u) = p_2$ or else $p(u)$ is one of p_0, p_1, $|J(u)| = n-1$, and $g(J(u)) = 1$, where $J(u) = \{\text{Length}(v) \mid v \text{ is a junction of } u\}$. We shall prove that this coloring f of $[{}^{\mathfrak{b}}2]^n$ is a counterexample to $2^{\mathfrak{b}} \to (\mathfrak{a})^n_2$; in order to do it we assume, to obtain a contradiction, that D is a subset of ${}^{\mathfrak{b}}2$ of cardinality $\mathfrak{a}$ homogeneous with respect to f. Without loss of generality we can assume that the order type of D in the well-ordering $<^*$ is $\mathfrak{a}$. We consider two cases.

Case a. For every $v \in {}^{\underset{\smile}{\mathfrak{b}}}2$ if $|\text{Thru}(D, v)| = \mathfrak{a}$ then some $v' \in {}^{\underset{\smile}{\mathfrak{b}}}2$ which includes v is an $\mathfrak{a}$-junction of D. In this case we can easily obtain $v_0, \ldots, v_{n-1} \in {}^{\underset{\smile}{\mathfrak{b}}}2$ such that for $i < j < n$ v_i is incompatible with v_j, $v_i < v_j$ in the left lexicographic ordering of ${}^{\underset{\smile}{\mathfrak{b}}}2$, and $|\text{Thru}(D, v_i)| = \mathfrak{a}$. For any permutation p of n it is now easy to construct a set $u = \{u_0, \ldots, u_{n-1}\} \subseteq D$ such that $u_i \supseteq v_i$ for $i < n$ and $p(\{u_0, \ldots, u_{n-1}\}) = p$, contradicting the homogeneity of D.

Case b. There is a $v \in {}^{\underset{\smile}{\mathfrak{b}}}2$ such that $|\text{Thru}(D, v)| = \mathfrak{a}$ and for every $v' \supseteq v$ in ${}^{\underset{\smile}{\mathfrak{b}}}2$ at most one of $\text{Left}(D, v')$ and $\text{Right}(D, v')$ is of cardinality $\mathfrak{a}$. In this case one can easily construct a sequence $\langle v_\alpha \mid \alpha < \mathfrak{a}\rangle$ of junctions of D such that $\alpha < \beta \to v_\alpha \subset v_\beta$ and $|\text{Thru}(D, v_\alpha)| = \mathfrak{a}$. By deleting a subset of $\{v_\alpha \mid \alpha < \mathfrak{a}\}$ one can obtain an increasing sequence $\langle v'_\alpha \mid \alpha < \mathfrak{a}\rangle$ of junctions of D such that $|\text{Left}(D, v'_\alpha)| = \mathfrak{a}$ for all α or $|\text{Right}(D, v'_\alpha)| = \mathfrak{a}$ for all α. Without loss of generality assume that $|\text{Right}(D, v'_\alpha)| = \mathfrak{a}$ for all α. Since v'_α is a junction of D there is an $s_\alpha \in \text{Left}(D, v'_\alpha)$, and we have $\beta < \alpha \to s_\beta < s_\alpha$. By a further pruning of the s_α's we can obtain a subset e of $\mathfrak{a}$ such that $|e| = \mathfrak{a}$ and for $\alpha, \beta \in e$ $\beta < \alpha \to s_\beta <^* s_\alpha$. Let $S' = \{s_\alpha \mid \alpha \in e\}$. By the definition of S' $p(u)=p_0$ for every $u\in[S']^n$. If $u=\{s_{i_1},\ldots,s_{i_n}\}\in[S']^n$ and $i_1<i_2<\cdots<i_n$ then $J(u)=\{\text{Length}(v'_{i_1}),\ldots,\text{Length}(v'_{i_{n-1}})\}$; thus $J(u)$ has $n-1$ members and we get, by $p(u)=p_0$ and the definition of f, $f(u)=g(J(u))=g(\{\text{Length}(v'_{i_1}),\ldots,\text{Length}(v'_{i_{n-1}})\})$. Therefore if $Q=\{\text{Length}(v'_\alpha)\mid\alpha\in e\}$ we get that $|Q|=\mathfrak{a}$ and Q is homogeneous with respect to g, which contradicts our choice of g. □

3.20Ac Corollary. *For all alephs* $\mathfrak{a}$ $\beth_n(\mathfrak{a})\nrightarrow(\mathfrak{a}^+)_2^{n+1}$

Proof. For $n=0$ this is trivial. For $n=1$ this is 3.18. For $n\geqslant 2$ the corollary is proved by induction on n, using 3.19. □

Corollary 3.20 shows that in 3.15(ii) the $\beth_n(\mathfrak{a})^+$ on the left-hand side of the arrow cannot be decreased. As to 3.15(i), the case not treated also in 3.15(ii) is that of a limit cardinal $\mathfrak{a}$, and for this case the picture is not uniform. If $\mathfrak{a}$ is weakly compact then 3.15(i) is indeed improved to $\mathfrak{a}\to(\mathfrak{a})_{\mathfrak{c}}^{n+2}$ by 3.16; for other weakly inaccessible cardinals $\mathfrak{a}$ 3.15(i) cannot be improved (see next exercise). For a singular $\mathfrak{a}$ the situation is more complicated; while 3.15(i) can be improved in some cases for the subscript 2 it cannot be improved if the subscript is a large finite number (see Erdös, Hajnal, Mate and Rado 1979).

3.21Ac Exercise. (i) If $\mathfrak{a}>\aleph_0$ is regular but not weakly compact then for $n<\omega$ $\beth_n(2^{\underline{\mathfrak{a}}})\nrightarrow(\mathfrak{a})_2^{n+2}$.

(ii) If $\mathfrak{a}$ is singular, $\text{cf}(\mathfrak{a})>\omega$, and $\text{cf}(\mathfrak{a})$ is not weakly compact then $2^{\underline{\mathfrak{a}}}\nrightarrow(\mathfrak{a})_2^2$.

(iii) If $\mathfrak{a}$ is singular, $\text{cf}(\mathfrak{a})=\omega$ or $\text{cf}(\mathfrak{a})$ is weakly compact, and $2^{\underline{\mathfrak{a}}}=2^{\mathfrak{b}}$ for some $\mathfrak{b}<\mathfrak{a}$ then $2^{\underline{\mathfrak{a}}}\nrightarrow(\mathfrak{a})_2^2$.

For the remaining case of a singular $\mathfrak{a}$ such that $\text{cf}(\mathfrak{a})=\omega$ or $\text{cf}(\mathfrak{a})$ is weakly compact and $2^{\underline{\mathfrak{a}}}>2^{\mathfrak{b}}$ for all $\mathfrak{b}<\mathfrak{a}$ it has been proved that 3.15(i) can be improved to $2^{\underline{\mathfrak{a}}}\to(\mathfrak{a})_2^2$ (Shelah 1975).

(iv) If $\mathfrak{a}$ is a singular aleph then $2^{\underline{\mathfrak{a}}}\nrightarrow(\mathfrak{a})_3^2$.

Hints. (i) By Theorem 3.34 since $\mathfrak{a}$ is uncountable and not weakly compact $\mathfrak{a}\nrightarrow(\mathfrak{a})_2^2$; let f be a coloring of $[\mathfrak{a}]^2$ which establishes this fact. Let A_λ, $\lambda<\mathfrak{a}$ be pairwise disjoint sets such that $|A_\lambda|=2^{|\lambda|}$, and let f_λ be a coloring of $[A_\lambda]^2$ which establishes $2^{|\lambda|}\nrightarrow(\lambda^+)_2^2$. Define a coloring g on $[\bigcup_{\lambda<\mathfrak{a}}A_\lambda]^2$ by $g(\{p,q\})=f_\lambda(\{p,q\})$ if $p,q\in A_\lambda$, and $g(\{p,q\})=f(\{\lambda,\mu\})$ if $p\in A_\lambda$, $q\in A_\mu$, $\lambda\neq\mu$. This proves $2^{\underline{\mathfrak{a}}}\nrightarrow(\mathfrak{a})_2^2$. Now use 3.19 to obtain (i).

(ii) Proceed as in (i) with the following difference. Let $\mathfrak{a}=\Sigma_{\lambda<\text{cf}(\mathfrak{a})}\mathfrak{b}_\lambda$, where $\mathfrak{b}_\lambda<\mathfrak{a}$ for $\lambda<\text{cf}(\mathfrak{a})$. Choose for these λ's A_λ's such that $|A_\lambda|=2^{\mathfrak{b}_\lambda}$.

(iii) Use 3.18 for $\mathfrak{b}$.

(iv) Proceed as in (ii) but define $g(\{p,q\})=f_\lambda(\{p,q\})$ if $p,q\in A_\lambda$ and $g(\{p,q\})=2$ otherwise. □

3.22 Definition (Erdös, Hajnal and Rado 1965). A statement much weaker than $\mathfrak{a}\to(\mathfrak{b})_{\mathfrak{c}}^n$, which is useful where $\mathfrak{a}\to(\mathfrak{b})_{\mathfrak{c}}^n$ fails, is the following one, denoted by $\mathfrak{a}\to[\mathfrak{b}]_{\mathfrak{c}}^n$: If A and C are sets of the respective cardinalities $\mathfrak{a}$ and $\mathfrak{c}$ and $f:[A]^n\to C$

then A has a subset B of cardinality $\mathfrak{b}$ such that $f[[B]^n]$ is not all of C, i.e., while B is not necessarily homogeneous in the sense that all of the members of $[B]^n$ have the same color, it at least avoids the other extreme where the members of $[B]^n$ have all possible colors. Clearly $\mathfrak{a} \to (\mathfrak{b})^n_2$ iff $\mathfrak{a} \to [\mathfrak{b}]^n_2$.

3.23 Exercise. Prove the following, where n may also be infinite.

(i) If $\mathfrak{c} \leqslant \mathfrak{d}$ and $\mathfrak{a} \to [\mathfrak{b}]^n_{\mathfrak{c}}$ then also $\mathfrak{a} \to [\mathfrak{b}]^n_{\mathfrak{d}}$.

(ii) If $\mathfrak{a} \geqslant \mathfrak{b}$ and $\mathfrak{d} > \mathfrak{b}^n$ then $\mathfrak{a} \to [\mathfrak{b}]^n_{\mathfrak{d}}$. □

3.24Ac Partition Relations With an Infinite Superscript. Until now we have considered only partition relations $\mathfrak{a} \to (\mathfrak{b})^n_{\mathfrak{c}}$ where n is finite; what can we say about the case of an infinite n? In that case we get very strong negative answers. Not only that, by using the axiom of choice, we can prove that for every $\mathfrak{a}$ $\mathfrak{a} \not\to (\aleph_0)^{\aleph_0}_2$ but, as will be shown in 3.25, we have even $\mathfrak{a} \not\to [\aleph_0]^{\aleph_0}_{2^{\aleph_0}}$ for every $\mathfrak{a}$. This is a strongest negative result for the superscript $\aleph_0$ since by 3.23(ii) we have $\mathfrak{a} \to [\aleph_0]^{\aleph_0}_{(2^{\aleph_0})^+}$ for all infinite $\mathfrak{a}$.

3.25Ac Proposition. *For all cardinals* $\mathfrak{a}$ $\mathfrak{a} \not\to [\aleph_0]^{\aleph_0}_{2^{\aleph_0}}$, *i.e., there is a coloring with* $2^{\aleph_0}$ *colors of the set of all denumerable subsets of* $\mathfrak{a}$ *such that for every denumerable subset* B *of* $\mathfrak{a}$ *the denumerable subsets of* B *are colored with all possible colors.*

Outline of proof. First prove $\aleph_0 \not\to [\aleph_0]^{\aleph_0}_{2^{\aleph_0}}$. Let $\{A_\alpha \mid \alpha < 2^{\aleph_0}\} = [\omega]^{\aleph_0}$ and let $\Phi_{\alpha,\beta}$ denote the requirement "A_α has a subset of color β". Let $P: 2^{\aleph_0} \times 2^{\aleph_0} \to 2^{\aleph_0}$ be a bijection and let Q_1, Q_2 be the functions on $2^{\aleph_0}$ which are the inverses of P, i.e., $P(Q_1(\gamma), Q_2(\gamma)) = \gamma$ for all $\gamma < 2^{\aleph_0}$. We color $[\omega]^{\aleph_0}$ in $2^{\aleph_0}$ steps by coloring at the α-th step a single member of $[\omega]^{\aleph_0}$ in order to satisfy $\Phi_{Q_1(\alpha), Q_2(\alpha)}$.

To prove $\mathfrak{a} \not\to [\aleph_0]^{\aleph_0}_{2^{\aleph_0}}$ color the members of $[\mathfrak{a}]^{\aleph_0}$ by recursion. When handling $A \in [\mathfrak{a}]^{\aleph_0}$, if some $B \subseteq A$ is already colored then color A arbitrarily, and if no $B \subseteq A$ is colored then color A and all its denumerable subsets by a coloring as in the first part of the proof. □

3.26Ac Proposition (Erdös and Hajnal 1966). *For all infinite cardinals* $\mathfrak{a}$ $\mathfrak{a} \not\to [\mathfrak{a}]^{\aleph_0}_{\mathfrak{a}}$.

Hint of partial proof. We shall prove the proposition only for the case of a regular $\mathfrak{a}$ since this case, unlike the general case, follows easily from earlier theorems in this book, and this case is a good theorem in itself. For $\mathfrak{a} = \aleph_0$ the proposition follows from 3.25. For a regular $\mathfrak{a} > \aleph_0$ let $\bigcup_{\lambda < \mathfrak{a}} A_\lambda$ be a decomposition of $\{\alpha \in \mathfrak{a} \mid \mathrm{cf}(\alpha) = \omega\}$ to $\mathfrak{a}$ pairwise disjoint stationary subsets of $\mathfrak{a}$ (IV.4.48). Let $f: [\mathfrak{a}]^{\aleph_0} \to \mathfrak{a}$ be given by: $f(B)$ = that λ for which $\sup B \in A_\lambda$, if $\mathrm{cf}(\sup B) = \omega$, and $f(B) = 0$ otherwise. This f establishes $\mathfrak{a} \not\to [\mathfrak{a}]^{\aleph_0}_{\mathfrak{a}}$. □

3.27 More on Partition Relations With an Infinite Superscript. What we have shown does not close the lid on the subject of partition relations with an infinite superscript. First, the proofs of 3.25 and 3.26 used the axiom of choice. There are axioms which contradict the axiom of choice and which imply partition relations with an infinite superscript (see Kleinberg 1973). Second, we know that for an arbitrary coloring $f:[\mathfrak{a}]^{\aleph_0}\to 2$ we cannot, in general, find an infinite homogeneous subset of $\mathfrak{a}$, but this changes if we insist that f be a particularly nice coloring. $[\mathfrak{a}]^{\aleph_0}$ has the following natural topology. When we regard 2 as a discrete topological space we obtain a power topology on ${}^{\mathfrak{a}}2$. Using a natural bijection of ${}^{\mathfrak{a}}2$ on $\mathbf{P}(\mathfrak{a})$ (III.3.32) we transfer this topology to the set $\mathbf{P}(\mathfrak{a})$, which includes $[\mathfrak{a}]^{\aleph_0}$. We say that $f:[\mathfrak{a}]^{\aleph_0}\to 2$ is a Borel or analytic coloring of $[\mathfrak{a}]^{\aleph_0}$ if $f^{-1}[\{0\}]$ is a Borel or analytic set, respectively, in the subspace $[\mathfrak{a}]^{\aleph_0}$ of $\mathbf{P}(\mathfrak{a})$. It is known that a homogeneous set exists for every Borel coloring f of $[\omega]^{\aleph_0}$ (Galvin and Prikry 1973) and even for every analytic coloring f of $[\omega]^{\aleph_0}$ (Silver 1970). □

We shall now present, as an application of the Erdös–Rado Theorem (3.15), the proof of V.5.2(v), which gives us some new information on the behavior of the power function $\langle 2^{\mathfrak{a}} \mid \mathfrak{a}\in \mathrm{Cn}\rangle$ at singular cardinal $\mathfrak{a}$. We shall first deal with a few preliminary concepts and propositions.

3.28 Definition. F is said to be a *set function* on A if $F:A\to \mathbf{P}(A)$. A subclass B of A is said to be *independent with* respect to F if for all $x, y\in B$ if $x\neq y$ then $x\notin F(y)$.

3.29Ac Proposition (Piccard 1937). *Let $\mathfrak{a}$ be a regular cardinal and $\mathfrak{b}<\mathfrak{a}$. If f is a set function on a set A of cardinality $\mathfrak{a}$ such that $|f(x)|<\mathfrak{b}$ for all $x\in A$ then A has a subset B of cardinality $\mathfrak{a}$ independent with respect to f. (This holds also for a singular $\mathfrak{a}$—Hajnal 1961.)*

Proof. Let $D\subseteq A$ be called a *monopoly* if $|D|<\mathfrak{a}$ and for almost all $x\in A$ (i.e., for all $x\in A$ except $<\mathfrak{a}$ many) $f(x)\cap D\neq 0$. To see that every set of pairwise disjoint monopolies is of cardinality $<\mathfrak{b}$ let E be such a set with $|E|=\mathfrak{b}$.

(18) $$\{x\in A \mid f(x)\cap D=0 \text{ for some } D\in E\}=\textstyle\bigcup_{D\in E}\{x\in A\mid f(x)\cap D=0\}.$$

Since the members of E are monopolies the right-hand side of (18) is a union of $|E|=\mathfrak{b}$ sets of cardinality $<\mathfrak{a}$, and by the regularity of $\mathfrak{a}$ the cardinality of the set in (18) is $<\mathfrak{a}$. Since $|A|=\mathfrak{a}$ there is a $y\in A$ which is not in (18), i.e., for all $D\in E$ $f(y)\cap D\neq 0$. $f(y)\supseteq\bigcup_{D\in E}(f(y)\cap D)$, and since distinct D's in E are assumed to be disjoint and $f(y)\cap D\neq 0$ for $D\in E$, we get $|f(y)|\geqslant\Sigma_{D\in E}|f(y)\cap D|\geqslant\Sigma_{D\in E}1=|E|\cdot 1=\mathfrak{b}$, contradicting our hypothesis that $|f(x)|<\mathfrak{b}$. By Zorn's lemma there is a maximal set E of pairwise disjoint monopolies; as we saw $|E|<\mathfrak{b}$, and, since $\mathfrak{a}$ is regular and $|D|<\mathfrak{a}$ for $D\in E$, also $|\bigcup E|<\mathfrak{a}$.

By Zorn's lemma there is a maximal subset B of A which is disjoint from $\bigcup E$ and is independent with respect to f. We shall conclude the proof by showing that $|B|=\mathfrak{a}$. Suppose that $|B|<\mathfrak{a}$. B is not a monopoly by the maximality of E (since $B\cap\bigcup E=0$), hence

(19) $$|\{x\in A\mid f(x)\cap B=0\}|=\mathfrak{a}.$$

By the regularity of $\mathfrak{a}$ $|\bigcup_{x\in B}f(x)|<\mathfrak{a}$. Using this, together with (19) and $|B|<\mathfrak{a}$, we get

(20) $\{x\in A \mid f(x)\cap B=0\}\sim\bigcup_{x\in B}f(x)\sim B\sim\bigcup E\neq 0.$

If y belongs to the left-hand side of (20) then $B\cup\{y\}$ is an independent set, contradicting the maximality of B. □

3.30 Definition. Let κ be a regular ordinal and let $A=\langle A_\alpha \mid \alpha<\kappa\rangle$ be a sequence of sets. A subset W of $\times_{\alpha<\kappa}A_\alpha$ is said to be an *almost disjoint transversal system* for A if for all $f, g\in W$ if $f\neq g$ then there is a $\mu<\kappa$ such that $f(\lambda)\neq g(\lambda)$ for every $\mu\leqslant\lambda<\kappa$. $T(A)$ is defined to be $\sup\{|W| \mid W$ is an almost disjoint transversal system for $A\}$.

3.31Ac Lemma. *Let κ be an ordinal of cofinality $>\omega$ and let $<$ be the relation on ${}^\kappa\mathrm{On}$ defined by: $f_1<f_2$ iff $\{\alpha<\kappa \mid f_1(\alpha)<f_2(\alpha)\}$ includes a closed unbounded subset of κ. The relation $<$ satisfies the first of the two requirements for a well founded relation (that every non-void set have an $<$-minimal member*—II.5.1(a)) *but it is not left-narrow. Nevertheless the relation $<$ admits a rank function* (II.5.14).

Proof. Let F be the filter of all subsets of κ which include a closed unbounded subset of κ. For $f, g\in{}^\kappa\mathrm{On}$ we write $f\approx_F g$ if $\{\alpha<\kappa \mid f(\alpha)=g(\alpha)\}\in F$. $\approx_F$ is an equivalence relation, hence by II.7.13 (and its proof) there is a function H on ${}^\kappa\mathrm{On}$ such that for $f, g\in{}^\kappa\mathrm{On}$ $H(f)=H(g)$ iff $f\approx_F g$, and $H(f)\subseteq\{h \mid h\approx_F f\}$. $H(f)$ is essentially the $\approx_F$-equivalence class of f, except that it does not contain all the functions equivalent to f, and maybe not f itself, in order that it be not a proper class. We define a relation $\ll$ by $u\ll v$ iff $u, v\in\mathrm{Rng}(H)$ and there are $f\in u$ and $g\in v$ such that $f<g$. As easily seen, for all $f', g'\in{}^\kappa\mathrm{On}$ $H(f')\ll H(g')$ iff $f'<g'$.

To prove that $\ll$ is a well founded relation it suffices, by II.5.3(ii), to show that there is no infinite descending $\ll$-sequence and that $\ll$ is left-narrow. Assume that $\langle u_i \mid i<\omega\rangle$ is such that $u_{i+1}\ll u_i$ for $i<\omega$. Let f_i be a member of u_i, then since $H(f_{i+1})=u_{i+1}\ll u_i=H(f_i)$ we have $f_{i+1}<f_i$, hence $W_i=\{\alpha<\kappa \mid f_{i+1}(\alpha)<f_i(\alpha)\}\in F$. By IV.4.17 F is $\aleph_1$-complete hence $\bigcap_{i<\omega}W_i\in F$. Let $\beta\in\bigcap_{i<\omega}W_i$ then we have $f_{i+1}(\beta)<f_i(\beta)$ for all $i<\omega$, which is a contradiction.

To see that $\ll$ is left-narrow let $u\in\mathrm{Rng}(H)$ and $g\in u$. If $H(f)\ll u$ then $\{\alpha<\kappa \mid f(\alpha)<g(\alpha)\}\in F$. Let $f':\kappa\to\mathrm{On}$ be defined by $f'(\alpha)=f(\alpha)$ if $f(\alpha)<g(\alpha)$ and $f'(\alpha)=0$ otherwise. Clearly $f'\approx_F f$, hence $H(f)=H(f')$, and for all $\alpha<\kappa$ $f'(\alpha)<\max(g(\alpha), 1)$. Therefore $\{v \mid v\ll u\}\subseteq H[\times_{\alpha<\kappa}\max(g(\alpha), 1)]$, hence $\ll$ is left-narrow.

Since $\ll$ is well founded there is, by II.5.12, a function $\rho_\ll: V\to\mathrm{On}$ such that $\rho_\ll(u)=\sup^+\{\rho_\ll(v) \mid v\ll u\}$. We define $\rho_<$ on ${}^\kappa\mathrm{On}$ by $\rho_<(f)=\rho_\ll(H(f))$. Since $H[\{g\in{}^\kappa\mathrm{On} \mid g<f\}]=\{v \mid v\ll H(f)\}$ we have also $\rho_<(f)=\sup^+\{\rho_<(g) \mid g<f\}$. This implies, as easily seen (compare II.5.14(ii)) that every non-void subset of ${}^\kappa\mathrm{On}$ has a $<$-minimal member. □

3.32Ac Proposition (Galvin and Hajnal 1975). *Let κ be a regular aleph. For every aleph $\mathfrak{c}$ and every function $g:\kappa\to\mathrm{On}$ if $h:\kappa\to\mathrm{On}$ is the function defined by $h(\alpha)=\mathfrak{c}^{+g(\alpha)}$ then we have $T(h)\leqslant(\mathfrak{c}^\kappa)^{+\rho_<(g)}$, where T and $\rho_<$ are as in* 3.30 *and* 3.31.

Proof. We shall prove by induction on $\rho_<(g)$ that for every almost disjoint transversal system W for h we have

(21) $\quad |W| \leqslant (\mathfrak{c}^\kappa)^{+\rho_<(g)}.$

If $\rho_<(g)=0$ then the set $B=\{\alpha \mid g(\alpha)=0\}$ is a stationary subset of κ. For $\alpha \in B$ $h(\alpha)=\mathfrak{c}$. Since B is unbounded in κ the function E on W defined by $E(f)=f\upharpoonright B$ is an injection of W into ${}^B\mathfrak{c}$ (as the members of W are almost disjoint). Thus $|W|=|\mathrm{Rng}(E)|\leqslant|{}^B\mathfrak{c}|=\mathfrak{c}^\kappa$, which establishes (21).

If $\rho_<(g)=\gamma>0$ we define g_f and h_f for every $f\in W$ to be the unique functions in ${}^\kappa\mathrm{On}$ which satisfy, for all $\alpha<\kappa$,

$$h_f(\alpha)=\mathfrak{c}^{+g_f(\alpha)}=\max(\mathfrak{c},|f(\alpha)|).$$

Since $\rho_<(g)>0$ we have for almost all $\alpha<\kappa$ (i.e., for all but an insignificant set of α's) $g(\alpha)>0$, hence $h(\alpha)>\mathfrak{c}$, $h_f(\alpha)<h(\alpha)$, $g_f(\alpha)<g(\alpha)$ and therefore $\rho_<(g_f)<\rho_<(g)=\gamma$. Thus $W=\bigcup_{\mu<\gamma}W_\mu$, where

$$W_\mu=\{f\in W \mid \rho_<(g_f)=\mu\}.$$

Since $\gamma\leqslant(\mathfrak{c}^\kappa)^{+\gamma}=(\mathfrak{c}^\kappa)^{+\rho_<(g)}$ (21) is established once we prove that for $\mu<\gamma$ $|W_\mu|\leqslant(\mathfrak{c}^\kappa)^{+\gamma}$; we shall actually prove that for $\mu<\gamma$ $|W_\mu|\leqslant(\mathfrak{c}^\kappa)^{+\mu+1}$.

We define a set function J on W_μ by

$$J(f)=\{e\in W_\mu \mid \forall\alpha(e(\alpha)\leqslant f(\alpha))\}.$$

$J(f)$ is obviously an almost disjoint transversal system for $\langle f(\alpha)+1 \mid \alpha<\kappa\rangle$, and since $|f(\alpha)+1|\leqslant\mathfrak{c}^{+g_f(\alpha)}$ and $\rho_<(g_f)=\mu<\gamma$ we get, by the induction hypothesis, $|J(f)|\leqslant(\mathfrak{c}^\kappa)^{+\mu}$. Now we assume $|W_\mu|\geqslant(\mathfrak{c}^\kappa)^{+\mu+2}$ and obtain a contradiction, which establishes $|W_\mu|\leqslant(\mathfrak{c}^\kappa)^{+\mu+1}$. We apply 3.29, where we take $\mathfrak{a}=(\mathfrak{c}^\kappa)^{+\mu+2}$, $\mathfrak{b}=(\mathfrak{c}^\kappa)^{+\mu+1}$, and A is a subset of W_μ of cardinality $(\mathfrak{c}^\kappa)^{+\mu+2}$, we obtain a subset of W_μ of cardinality $(\mathfrak{c}^\kappa)^{+\mu+2}$ which is independent with respect to the set function J. Since $(\mathfrak{c}^\kappa)^{+\mu+2}>2^\kappa$ there is a sequence $\langle f_\xi \mid \xi<(2^\kappa)^+\rangle$ of members of W_μ such that for $\xi<\eta<(2^\kappa)^+$ $f_\xi\notin J(f_\eta)$, i.e., there is an $\alpha<\kappa$ such that $f_\xi(\alpha)>f_\eta(\alpha)$. Let us color every pair $\{\xi,\eta\}$, where $\xi<\eta<(2^\kappa)^+$, with the least color α such that $f_\xi(\alpha)>f_\eta(\alpha)$. By 3.15(ii) $(2^\kappa)^+\to(\kappa^+)^2_\kappa$, hence there is a subset B of $(2^\kappa)^+$ of cardinality κ^+ all of whose pairs are colored with the same color. If $\langle\xi_0,\xi_1,\ldots\rangle$ is the enumeration of B we get $f_{\xi_0}(\alpha)>f_{\xi_1}(\alpha)>f_{\xi_2}(\alpha)>\ldots$, which is a contradiction since B is infinite. $\square$

3.33Ac Theorem (Galvin and Hajnal 1975—*this is* V.5.2(v)). *If $\kappa>\omega$ is a regular ordinal and for all $\alpha<\kappa$ $2^{\mathfrak{c}^{+\alpha}}<\mathfrak{c}^{+(2^\kappa)^+}$ then also $2^{\mathfrak{c}^{+\kappa}}<\mathfrak{c}^{+(2^\kappa)^+}$.*

Proof. We shall apply Proposition 3.32 for the case where g is the function on κ defined by $\mathfrak{c}^{+g(\alpha)}=2^{(\mathfrak{c}^{+\alpha})}$. Therefore the function h of 3.32 is given by $h(\alpha)=2^{(\mathfrak{c}^{+\alpha})}$. For $\alpha<\kappa$ let $A_\alpha={}^{(\mathfrak{c}^{+\alpha})}2$, and let $F\colon{}^{(\mathfrak{c}^{+\kappa})}2\to\bigtimes_{\alpha<\kappa}A_\alpha$ be defined by $F(v)=\langle v\upharpoonright\mathfrak{c}^{+\alpha} \mid \alpha<\kappa\rangle$. Since κ is a limit ordinal F is clearly an injection, $\mathrm{Rng}(F)$ is an almost disjoint transversal system for $\langle A_\alpha \mid \alpha<\kappa\rangle$, and $|\mathrm{Rng}(F)|=|^{(\mathfrak{c}^{+\kappa})}2|=2^{\mathfrak{c}^{+\kappa}}$. For $\alpha<\kappa$ $|A_\alpha|=2^{(\mathfrak{c}^{+\alpha})}=h(\alpha)$, hence also h has an almost disjoint transversal system of cardinality $2^{\mathfrak{c}^{+\kappa}}$, and thus $T(h)\geqslant 2^{(\mathfrak{c}^{+\kappa})}$.

Since we assume $\mathfrak{c}^{+g(\alpha)}=2^{(\mathfrak{c}^{+\alpha})}<\mathfrak{c}^{+(2^\kappa)^+}$ we have $g(\alpha)<(2^\kappa)^+$ for $\alpha<\kappa$, and by the regularity of $(2^\kappa)^+$ there is a bound $\beta<(2^\kappa)^+$ of $\{g(\alpha)\mid\alpha<\kappa\}$. By the definition of $\rho_<(g)$ in 3.31 $\rho_<(g)=\rho_\ll(H(g))$. The relation $\ll$ of 3.31 is transitive, as easily seen, hence $\{v\mid v\ll H(g)\}$ is a $\ll^{-1}$-closed class, and by II.5.14(vi) $\{\rho_\ll(v)\mid v\ll H(g)\}$ is On or an ordinal. In the proof of 3.31 we saw that $\{v\mid v\ll H(g)\}\subseteq H[\times_{\alpha<\kappa}\max(g(\alpha),1)]$, hence

$$(22)\qquad \begin{aligned}&|\{\rho_\ll(v)\mid v\ll H(g)\}|\leqslant|\{v\mid v\ll H(g)\}|\leqslant|H[\times_{\alpha<\kappa}\max(g(\alpha),1)]|\\&\leqslant\Pi_{\alpha<\kappa}\max(|g(\alpha)|,1)\leqslant(2^\kappa)^\kappa=2^\kappa.\end{aligned}$$

Since $\{\rho_\ll(v)\mid v\ll H(g)\}$ is On or an ordinal (22) implies that $\{\rho_\ll(v)\mid v\ll H(g)\}$ is an ordinal $<(2^\kappa)^+$, hence

$$\rho_<(g)=\rho_\ll(H(g))=\sup{}^+\{\rho_\ll(v)\mid v\ll H(g)\}\leqslant\{\rho_\ll(v)\mid v\ll H(g)\}<(2^\kappa)^+.$$

Applying 3.32 we get $2^{(\mathfrak{c}^{+\kappa})}\leqslant T(h)\leqslant(\mathfrak{c}^\kappa)^{+\rho_<(g)}<(\mathfrak{c}^\kappa)^{+(2^\kappa)^+}$. □

A simple generalization of Proposition 3.32 yields many additional results, including Silver's 5.2(iv) (Galvin and Hajnal 1975). □

Let us move up from the ordinary cardinals with which we dealt mostly until now to large cardinals. We proved in 3.7 and 3.16 that $\aleph_0$ and all weakly compact cardinals $\mathfrak{a}$ have the property that $\mathfrak{a}\to(\mathfrak{a})^n_{\mathfrak{c}}$ for $n<\omega$ and $\mathfrak{c}<\mathfrak{a}$. We shall now see that already the weakest non-trivial case of this partition relation characterizes these cardinals.

3.34Ac Theorem (Erdös and Tarski 1961, Hanf 1964a, Monk and Scott 1964). *For all uncountable cardinals $\mathfrak{a}$ the following conditions are equivalent.*

(a) *$\mathfrak{a}$ is weakly compact.*

(b) *Every ordered set $\langle A,<\rangle$ of cardinality $\mathfrak{a}$ has a subset B such that $|B|=\mathfrak{a}$ and B is either well-ordered by $<$ or inversely well ordered by $<$ (i.e., B is well-ordered by $>$).*

(c) $\mathfrak{a}\to(\mathfrak{a})^2_2$.

Proof. (a)→(c). This is a particular case of 3.16.

(c)→(b). Let $<^*$ be a well-ordering of A. Define a coloring $f:[A]^2\to 2$ as follows. $f(\{a,b\})=0$ if both $a<b$ and $a<^*b$ or both $b<a$ and $b<^*a$, and $f(\{a,b\})=$ $=1$ otherwise. Let B be a subset of A of cardinality $\mathfrak{a}$ which is homogeneous with respect to f, i.e., $f[[B]^2]=\{i\}$ for some $i<2$. Since $B\subseteq A$, B is well ordered by $<^*$. If $i=0$ then $<$ coincides with $<^*$ on B and hence also $<$ well-orders B. If $i=1$ then for all $x,y\in B$ $x<^*y$ iff $y<x$ and thus $<$ inversely well-orders B.

(b) → (a). We assume (b), and first we prove that $\mathfrak{a}$ is regular. Suppose $\mathfrak{a}$ is singular, then $\mathfrak{a}$ is the union $\bigcup_{\alpha<\lambda} A_\alpha$ of $\lambda<\mathfrak{a}$ pairwise disjoint sets A_α of cardinality $<\mathfrak{a}$. Let us define an ordering $<^*$ on $\mathfrak{a}$ as follows. For $\xi, \eta \in \mathfrak{a}$ $\xi <^* \eta$ iff ξ and η belong to the same A_α and $\xi<\eta$ or if $\xi \in A_\alpha$, $\eta \in A_\beta$, and $\beta<\alpha$ (notice that here ξ comes before η if it is in the later set A_α). Let $B \subseteq \mathfrak{a}$ and $|B|=\mathfrak{a}$; we shall see that B is neither well ordered nor inversely well ordered by $<^*$. Since $|B|=\mathfrak{a}$ while the union of finitely many A_α's is of cardinality $<\mathfrak{a}$ $B \cap A_\alpha \neq 0$ for infinitely many α's; let $\langle \alpha_n \mid n<\omega \rangle$, be the first ω such α's. For $n \in \omega$ let β_n be the least ordinal in $B \cap A_{\alpha_n}$. By the definition of $<^*$ we have $\cdots <^* \beta_2 <^* \beta_1 <^* \beta_0$, and thus B is not well ordered by $<^*$. If B were inversely well ordered by $<^*$ then, since by the definition of $<^*$ $<^*$ and $<$ coincide on each A_α, each $B \cap A_\alpha$ were both well-ordered and inversely well-ordered by $<^*$ and hence each $B \cap A_\alpha$ were finite. Thus $|B|=|\bigcup_{\alpha<\lambda}(B \cap A_\alpha)| \leqslant \Sigma_{\alpha<\lambda} |B \cap A_\alpha| \leqslant \Sigma_{\alpha<\lambda} \aleph_0 = |\lambda| \cdot \aleph_0 < \mathfrak{a}$, contradicting $|B|=\mathfrak{a}$.

Now let us prove that $\mathfrak{a}$ is inaccessible. If this is not the case then for some $\mathfrak{b}<\mathfrak{a}$ $2^{\mathfrak{b}} \geqslant \mathfrak{a}$. Let $<^*$ be the left lexicographic ordering of ${}^{\mathfrak{b}}2$, and since $\mathfrak{a} \leqslant 2^{\mathfrak{b}}$ let $A \subseteq {}^{\mathfrak{b}}2$ be of cardinality $\mathfrak{a}$. By 3.17 A has no subset of cardinality $>\mathfrak{b}$ which is well ordered or inversely well ordered by $<^*$.

Finally let us prove that $\mathfrak{a}$ has the tree property. Let $\langle T, \prec \rangle$ be a tree of length $\mathfrak{a}$ such that for every $\lambda<\mathfrak{a}$ the set T_λ of the members of T of T-rank λ is of cardinality $<\mathfrak{a}$. Since $T=\bigcup_{\lambda<\mathfrak{a}} T_\lambda$, the T_λ's are pairwise disjoint, and $1 \leqslant |T_\lambda| < \mathfrak{a}$ for $\lambda<\mathfrak{a}$ we have $|T|=\Sigma_{\lambda<\mathfrak{a}}|T_\lambda|$ and hence $|T|=\mathfrak{a}$. Let $<$ be any order on T. We define another order $<^*$ on T as follows. For $x \in T_\lambda$ and $\mu \leqslant \lambda$ we denote with x_μ the unique member z of T_μ such that $z \preccurlyeq x$. For $x, y \in T$ we define $x <^* y$ if $x \prec y$ or if x and y are incomparable with respect to $\prec$ and for the least ordinal μ such that $x_\mu \neq y_\mu$ $x_\mu < y_\mu$. By (b) T has a subset B of cardinality $\mathfrak{a}$ which is well ordered or inversely well ordered by $<^*$. If B is well ordered by $<^*$ then, since $|B|=\mathfrak{a}$, we can assume that the order type of $\langle B, <^* \rangle$ is $\mathfrak{a}$ (otherwise we delete some of the members of B to obtain the order type $\mathfrak{a}$). Let

$$S=\{x \in T \mid |\{y \in B \mid y \succcurlyeq x\}|=\mathfrak{a}\}.$$

We shall see that S is a branch of T of length $\mathfrak{a}$. Since, obviously, if $x \in S$ and $z \prec x$ then also $z \in S$, it suffices to prove that for every $\lambda<\mathfrak{a}$ $|S \cap T_\lambda|=1$. We shall do it by deriving contradictions from each one of the assumptions $S \cap T_\lambda=0$ and $|S \cap T_\lambda| \geqslant 2$. We have, for $\lambda<\mathfrak{a}$,

$$(23) \qquad B=\bigcup_{\mu<\lambda}(T_\mu \cap B) \cup \bigcup_{x \in T_\lambda}\{y \in B \mid y \succcurlyeq x\}.$$

For $\mu<\lambda$ $|T_\mu \cap B| \leqslant |T_\mu| < \mathfrak{a}$. If $S \cap T_\lambda=0$ then for every $x \in T_\lambda$ $\{y \in B \mid y \succcurlyeq x\}<\mathfrak{a}$. Thus the right-hand side of (23) is a union of $<\mathfrak{a}$ sets of cardinality $<\mathfrak{a}$ and since $\mathfrak{a}$ is regular we get $|B|<\mathfrak{a}$, which is a contradiction. If $|S \cap T_\lambda| \geqslant 2$ let $x, z \in S \cap T_\lambda$, $x <^* z$. By the definition of $<^*$ we have, as easily seen, also $y <^* z$ for every $y \succcurlyeq x$. Since $x \in S$ $|\{y \in B \mid y \succcurlyeq x\}|=\mathfrak{a}$, and thus B has at least $\mathfrak{a}$ members which are $<^* z$ while $z \in B$, contradicting our choice of B as a set of order type $\mathfrak{a}$. If $<^*$ inversely well orders B we prove in essentially the same way the existence of a branch S of T of length $\mathfrak{a}$, which shows that $\mathfrak{a}$ has the tree property. □

3.35Ac Exercise. (i) A field of sets is said to be $\mathfrak{a}$-*complete* if it is closed under union and intersection of $<\mathfrak{a}$ sets. An $\mathfrak{a}$-complete field G of subsets of a set X is said to be $\mathfrak{a}$-*generated* by a subset H of G if every member of G is obtained from members of H by the operations of complementation with respect to X and of union and intersection of $<\mathfrak{a}$ sets, i.e., no proper subset of G which includes H is an $\mathfrak{a}$-complete field of subsets of X. Prove that for such G and H if $\mathfrak{a}$ is inaccessible and $|H|=\mathfrak{a}$ then also $|G|=\mathfrak{a}$.

(ii) (Erdös and Tarski 1961, Monk and Scott 1964). Prove that the following conditions on $\mathfrak{a}$ are equivalent.

(a) $\mathfrak{a}$ is weakly compact.

(b) In every $\mathfrak{a}$-complete field G of sets which is $\mathfrak{a}$-generated by a subset of G of cardinality $\mathfrak{a}$ every $\mathfrak{a}$-complete filter F can be extended to an $\mathfrak{a}$-complete ultrafilter U.

(c) $\mathfrak{a}$ is regular and (b) holds for every field of subsets of $\mathfrak{a}$ as in (b).

Hint. (ii) (a)$\rightarrow$(b). Since $\mathfrak{a}$ is weakly compact $\mathfrak{a}$ is inaccessible by definition (2.23), hence if G and F are as in (b) we can write $G=\{b_\lambda \mid \lambda<\mathfrak{a}\}$. For $b\subseteq\mathfrak{a}$ let $b^{(0)}=b$, $b^{(1)}=\mathfrak{a}\sim b$. Let the tree $\langle T, \subset\rangle$ be defined by $T\subseteq {}^{\underline{\mathfrak{a}}}2$ and for all $t\in {}^{\underline{\mathfrak{a}}}2$ $t\in T$ iff for every $b\in F$ $\bigcap_{\mu<\text{Length}(t)} b_\mu^{(t(\mu))}\cap b\neq 0$. A branch of length $\mathfrak{a}$ in T determines an $s\in {}^{\mathfrak{a}}2$ such that $s\upharpoonright\lambda\in T$ for every $\lambda<\mathfrak{a}$. U is the ultrafilter generated by the set $\{\bigcap_{\mu<\lambda} b_\mu^{(s(\mu))} \mid \lambda<\mathfrak{a}\}$.

(c)$\rightarrow$(a). If $f:[\mathfrak{a}]^2\rightarrow 2$ let G be the field of subsets of $\mathfrak{a}$ $\mathfrak{a}$-generated by the set $\mathfrak{a}\cup\{\{\gamma\in\mathfrak{a} \mid \alpha<\gamma\wedge f(\{\alpha,\gamma\})=0\} \mid \alpha\in\mathfrak{a}\}\subseteq\mathbf{P}(\mathfrak{a})$. Let U be an $\mathfrak{a}$-complete ultrafilter in G containing all complements of singletons. Define by recursion a sequence $\langle\lambda_\alpha \mid \alpha<\mathfrak{a}\rangle$ such that for $\alpha>\beta$ $\lambda_\beta\neq\lambda_\alpha$ and $f(\{\lambda_\beta,\lambda_\alpha\})=0$ iff $\{\gamma\in\mathfrak{a} \mid f(\{\lambda_\beta,\gamma\})=0\}\in U$. For $i<2$ let $W_i=\{\lambda_\beta \mid \{\gamma \mid f(\{\lambda_\beta,\gamma\})=i\}\in U\}$. $W_0\cup W_1=\{\lambda_\alpha \mid \alpha\in\mathfrak{a}\}$, hence one of W_0, W_1 is a homogeneous set of cardinality $\mathfrak{a}$.

(b)$\rightarrow$(c). It clearly suffices to prove from (b) that $\mathfrak{a}$ is regular. For $\beta<\mathfrak{a}$ and $i<2$ let $B_{\beta,i}=\{f\in {}^{\mathfrak{a}}2 \mid f(\beta)=i\}$. Let G be the field of subsets of ${}^{\mathfrak{a}}2$ which is $\mathfrak{a}$-generated by the $B_{\beta,i}$'s, and let U be any $\mathfrak{a}$-complete ultrafilter in G. If $\mathfrak{a}$ were singular then G and U were $\mathfrak{a}^+$-complete; let $g\in {}^{\mathfrak{a}}2$ be such that $B_{\beta,g(\beta)}\in U$ for all $\beta<\mathfrak{a}$, then $\bigcap_{\beta<\mathfrak{a}} B_{\beta,g(\beta)}=\{g\}\in U$ and U were principal. Let $F=\{D\subseteq {}^{\mathfrak{a}}2 \mid |{}^{\mathfrak{a}}2\sim D|\leqslant\mathfrak{a}\}\subseteq G$, then F is an $\mathfrak{a}$-complete filter which cannot be extended to any principal ultrafilter. □

3.36 Ineffable Ordinals. In 2.53 we defined the concept of an ineffable ordinal and in 2.54 we proved that the ineffable ordinals are regular. We rephrase here Definition 2.53 in terms of characteristic functions instead of subsets. Accordingly, if X is a set of ordinals and $\langle f_\alpha \mid \alpha\in X\rangle$ is a sequence of functions such that $\text{Dom}(f_\alpha)=\alpha$ for $\alpha\in X$ then we say that $\langle f_\alpha \mid \alpha\in X\rangle$ is a *coherent* sequence if for $\beta,\alpha\in X$ if $\beta<\alpha$ then $f_\beta=f_\alpha\upharpoonright\beta$. If $\langle f_\alpha \mid \alpha\in X\rangle$ is coherent then $F=\bigcup_{\alpha\in X} f_\alpha$ is a function on sup$X$ and we have $f_\alpha=F\upharpoonright\alpha$ for every $\alpha\in X$. An ordinal κ is ineffable if $\text{cf}(\kappa)>\omega$ and for every sequence $\langle f_\alpha \mid \alpha<\kappa\rangle$ such that $f_\alpha:\alpha\rightarrow 2$ for $\alpha<\kappa$ there is a stationary subset X of κ such that the subsequence $\langle f_\alpha \mid \alpha\in X\rangle$ is coherent. In 3.37 and 3.40 we shall show that every ineffable ordinal is weakly compact and that the least ineffable ordinal, if there is such, is greater than many weakly compact ordinals.

Thus the concept of an ineffable ordinal is a new concept of a large cardinal, still larger than that of a weakly compact cardinal.

3.37Ac Theorem (Kunen—see Devlin 1973). *For every infinite ordinal $\kappa>\omega$ κ is ineffable iff κ is a cardinal such that $\kappa\to(\text{stationary})^2_2$, i.e., iff for every coloring $f:[\kappa]^2\to 2$ there is a stationary subset S of κ which is homogeneous with respect to f. As a consequence, every ineffable ordinal is weakly compact* (by 3.34).

Proof. Assume that κ is ineffable. Let $f:[\kappa]^2\to 2$ then we define, for $\alpha<\kappa$, $f_\alpha=\langle f(\{\beta,\alpha\})\mid\beta<\alpha\rangle$. Since κ is ineffable there is a stationary subset X of κ such that $\langle f_\alpha\mid\alpha\in X\rangle$ is coherent. Let $F=\bigcup_{\alpha\in X}f_\alpha$, then for $\beta<\kappa$ $F(\beta)$ is the color $f(\{\beta,\alpha\})$ of every $\{\beta,\alpha\}$ for $\alpha\in X$, $\alpha>\beta$. Since $X=X\cap F^{-1}[\{0\}]\cup X\cap F^{-1}[\{1\}]$ and X is stationary, at least one of the sets $X\cap F^{-1}[\{0\}]$ and $X\cap F^{-1}[\{1\}]$ must be stationary (IV.4.35), let $i<2$ be such that $X\cap F^{-1}[\{i\}]$ is a stationary subset of κ, then we have $f[[X\cap F^{-1}[\{i\}]]^2]=\{i\}$. Thus we have $\kappa\to(\text{stationary})^2_2$.

Now we assume $\kappa\to(\text{stationary})^2_2$ and we prove that κ is ineffable. Since κ clearly satisfies $\kappa\to(\kappa)^2_2$, κ is weakly compact (by 3.34) and hence regular. Let $\langle f_\alpha\mid\alpha<\kappa\rangle$ be such that $f_\alpha:\alpha\to 2$ for $\alpha<\kappa$. Let $<_{\text{lex}}$ denote the left lexicographic order of ${}^{\underline{\kappa}}2$ where $f<_{lex}h$ also when $f\subset h$. We define $g:[\kappa]^2\to 2$ as follows. For $\beta<\alpha<\kappa$ let $g(\{\beta,\alpha\})=0$ if $f_\beta<_{\text{lex}}f_\alpha$, and $g(\{\beta,\alpha\})=1$ otherwise. Since $\kappa\to(\text{stationary})^2_2$ there is a stationary subset S of κ homogeneous with respect to g. Assume that $g[[S]^2]=\{0\}$ then for $\alpha,\beta\in S$ if $\alpha<\beta$ then $f_\alpha<_{\text{lex}}f_\beta$. We shall now define by recursion a normal function $j\in{}^\kappa\kappa$ such that for every $\lambda<\kappa$ and $\mu<\lambda$ the values of $f_\alpha(\mu)$ are the same for all $\alpha\geqslant j(\lambda)$ such that $\alpha\in S$, i.e.,

(24) if $\alpha,\beta\geqslant j(\lambda)$ and $\alpha,\beta\in S$ then $f_\alpha\upharpoonright\lambda=f_\beta\upharpoonright\lambda$.

Set $j(0)=0$, then (24) holds trivially for $\lambda=0$. For a given λ, if there is a $\gamma\geqslant j(\lambda)$, $\gamma\in S$, such that $f_\gamma(\lambda)=1$ we take $j(\lambda+1)$ to be the least such γ. For $\alpha\geqslant j(\lambda+1)$ we have, by (24), $f_{j(\lambda+1)}\upharpoonright\lambda=f_\alpha\upharpoonright\lambda$, and by $j(\lambda+1)\leqslant\alpha$ $f_{j(\lambda+1)}\leqslant_{lex}f_\alpha$; therefore $1=f_{j(\lambda+1)}(\lambda)\leqslant f_\alpha(\lambda)$ and hence $f_\alpha(\lambda)=1=f_{j(\lambda+1)}(\lambda)$ and $f_\alpha\upharpoonright(\lambda+1)=f_{j(\lambda+1)}\upharpoonright(\lambda+1)$. If there is no $\gamma\geqslant j(\lambda)$ such that $\gamma\in S$ and $f_\gamma(\lambda)=1$ then we have $f_\alpha(\lambda)=0$ for every $\alpha\in S\sim j(\lambda)$. In this case we take $j(\lambda+1)$ to be $j(\lambda)+1$ and we get again, for every $\alpha\geqslant j(\lambda+1)$, $f_\alpha(\lambda)=f_{j(\lambda+1)}(\lambda)$, which establishes (24) for $\lambda+1$. For a limit ordinal λ define $j(\lambda)=\sup_{\mu<\lambda}j(\mu)$; since κ is regular $j(\lambda)>\kappa$. (24) for λ follows now easily from the induction hypothesis that (24) holds for $\mu<\lambda$.

Let C be the set of the fixed points of j; by IV.4.12(ii) and IV.4.24 C is a closed unbounded subset of κ. Since S is stationary also $C\cap S$ is stationary (IV.4.38(ii)). For every $\lambda\in C\cap S$ we have $j(\lambda)=\lambda$ and hence by (24) if also $\alpha\in S$ and $\alpha>\lambda$ then $f_\alpha\upharpoonright\lambda=f_\lambda\upharpoonright\lambda=f_\lambda$ and thus the sequence $\langle f_\alpha\mid\alpha\in C\cap S\rangle$ is coherent.

If $g[[S]^2]=\{1\}$ we interchange 0 and 1 in the proof and arrive at the same conclusion. □

3.38Ac Exercise. Prove that for an uncountable cardinal κ the following conditions are equivalent.

(a) κ is ineffable.

(b) If $\langle f_\alpha \mid \alpha<\kappa\rangle$ is such that $f_\alpha: \alpha \to \alpha$ for $\alpha<\kappa$ then there is a stationary subset X of κ such that the subsequence $\langle f_\alpha \mid \alpha \in X\rangle$ is coherent.

(c) For every $\lambda<\kappa$ $\kappa \to (\text{stationary})^2_\lambda$, i.e., if $f:[\kappa]^2 \to \lambda$ then there is a stationary subset A of κ which is homogeneous with respect to f.

(d) For every ordering $<^*$ of κ κ has a stationary subset A such that $<$ and $<^*$ coincide on A or $<$ and $<^*$ are inverses of each other on A.

Hint of proof. (a) $\to$ (b). Consider the sets $A_\alpha = \{P(\beta,\sigma) \mid f_\alpha(\beta)=\sigma\}$, where P is a pairing function on κ. For sufficiently many of the A_α's $A_\alpha \subseteq \alpha$. (b) $\to$ (c) $\to$ (d) $\to$ (a). See the proof of 3.37. □

While for every ineffable cardinal κ and every $\lambda<\kappa$ $\kappa \to (\text{stationary})^2_\lambda$ it is not the case that $\kappa \to (\text{stationary})^3_2$ for every ineffable κ, if there are ineffable cardinals. In particular, the least ineffable cardinal, if such exists, does not satisfy $\kappa \to (\text{stationary})^3_2$ (Baumgartner 1975). □

Now we proceed to prove that the least ineffable cardinal, if such exists, is bigger than many weakly compact cardinals.

3.39Ac Lemma. *If κ is a regular ordinal and $\langle T, <\rangle$ is a tree of length κ such that $|T_\lambda|<\kappa$ for $\lambda<\kappa$ then there is an increasing function $G:\kappa \to \kappa$ and a bijection F of T onto $S \subseteq {}^{\underline{\kappa}}2$ such that*

(25) $\quad F[T_\lambda] \subseteq {}^{G(\lambda)}2$ *for $\lambda<\kappa$, and*

(26) $\quad$ *F is an isomorphism of $\langle T, <\rangle$ onto $\langle S, \subset\rangle$.*

Hint of proof. Use recursion on λ to define $G(\lambda)$ and $F1T_\lambda$. To go from λ to $\lambda+1$ let α be the least non-zero cardinal such that $2^\alpha \geqslant \sup_{x \in T_\lambda} |\{y \in T_{\lambda+1} \mid y \succcurlyeq x\}|$ and for $x \in T_\lambda$ let H_x be an injection of $\{y \in T_{\lambda+1} \mid y \succcurlyeq x\}$ into ${}^\alpha 2$. Set $G(\lambda+1)=G(\lambda)\dotplus\alpha$ and for $y \in T_{\lambda+1}$ $F(y)=F(x)^\frown H_x(y)$, where x is the member of T_λ such that $x<y$. For a limit ordinal proceed similarly, with T_λ taking the role of $T_{\lambda+1}$ and the set of all branches of $T|\lambda$ of length λ taking the role of T_λ. □

3.40Ac Theorem. *If κ is an ineffable ordinal then the set W of all weakly compact ordinals below κ is a stationary subset of κ. Moreover, W is a "fat" stationary subset of κ in the sense that if $\langle A_\alpha \mid \alpha<\kappa\rangle$ is a sequence such that $A_\alpha \subseteq \alpha$ for $\alpha<\kappa$ then the stationary subset X of κ such that $\langle A_\alpha \mid \alpha \in X\rangle$ is coherent can always be taken to be a subset of W.*

Hint of proof (Shelah). Let $A_\alpha \subseteq \alpha$ for $\alpha<\kappa$. Let $J:\kappa \to {}^{\underline{\kappa}}2$ be such that for every strong limit cardinal $\lambda<\kappa$ $J[\lambda] \doteq {}^{\underline{\lambda}}2$. For $\alpha<\kappa$ let $B_\alpha \subseteq \alpha$ be as follows. If α is singular B_α is a cofinal subset of α of order type $\text{cf}(\alpha)$. If α is regular but not inaccessible then $B_\alpha=0$. If α is inaccessible but not weakly compact let T^α be an α-Aronszajn tree as in 3.39 and let $B_\alpha=J^{-1}[T^\alpha]$. If α is weakly compact let $B_\alpha=A_\alpha$. Since κ is ineffable there is a stationary subset Y of κ such that $\langle B_\alpha \mid \alpha \in Y\rangle$

is coherent. $Y_1 = Y \cap \{\alpha \mid \alpha \text{ is singular}\}$ is an insignificant set, since the function $\langle \mathrm{cf}(\alpha) \mid \alpha \in Y_1 \rangle$ is regressive on Y_1 and otherwise Y_1 would have a stationary subset Y_1' on which $\mathrm{cf}(\alpha)$ is fixed and, as a consequence, on which no two B_α's cohere. $Y_2 = \{\alpha \mid \alpha \text{ is not a strong limit cardinal}\}$ is an insignificant set. $Y_3 = Y \cap \{\alpha \mid \alpha \text{ is inaccessible but not weakly compact}\}$ contains at most one member since if $\alpha < \beta$, $\alpha, \beta \in Y_3$, and B_α and B_β cohere then $T^\alpha = T^\beta \mid \alpha$ and thus T^α has branches of length α. The set $X = Y \sim (Y_1 \cup Y_2 \cup Y_3)$ is as required by the theorem. □

3.41 Definition. (i) Let K be a set of ordinals. We say that A is a *limit* of the sequence $\langle A_\alpha \mid \alpha \in K \rangle$ if for every x there is a $\beta \in K$ such that for all $\beta < \alpha \in K$ $x \in A_\alpha$ iff $x \in A$.

(ii) Let $W \subseteq P(\kappa)$. We say that $B \subseteq \kappa$ is *consistent with respect to* W if for every $X \in W$ either "$B \subseteq X$ above some ordinal $< \kappa$". i.e., $B \sim X$ is a bounded subset of κ, or else "B is disjoint from X above some ordinal $< \kappa$", i.e., $B \cap X$ is a bounded subset of κ.

(iii)Ac For a cardinal $\mathfrak{a}$ we denote with $\log \mathfrak{a}$ the least cardinal $\mathfrak{c}$ such that $2^{\mathfrak{c}} \geqslant \mathfrak{a}$.

3.42 Exercise. Prove that the following two statements about $\mathfrak{a}$ and an aleph κ are equivalent.

(a) For every set W of cardinality $\mathfrak{a}$ every sequence $\langle A_\alpha \mid \alpha < \kappa \rangle$ of subsets of W has a subsequence of length κ which has a limit (i.e., there is a subset B of κ such that $|B| = \kappa$ and $\langle A_\alpha \mid \alpha \in B \rangle$ has a limit).

(b) For every subset W of $\mathbf{P}(\kappa)$ of cardinality $\mathfrak{a}$ there is a subset B of κ of cardinality κ which is consistent with respect to W.

Hint. To prove (a)→(b) set $A_\alpha = \{X \in W \mid \alpha \in X\}$, then the B of (a) is the B whose existence is asserted in (b). □

3.43Ac Exercise. Let κ be an aleph and let $\Phi(\kappa, \mathfrak{a})$ be the statement of Exercise 3.42. Notice that in 3.42 B is required to be of cardinality κ, not just a cofinal subset of κ; this makes a difference if κ is singular. Prove the following.

(i) If $\mathfrak{b} < \mathfrak{a}$ then $\Phi(\kappa, \mathfrak{a})$ implies $\Phi(\kappa, \mathfrak{b})$.

(ii) If κ is regular then $\Phi(\kappa, \kappa)$ holds iff $\kappa = \aleph_0$ or κ is weakly compact.

(iii) If κ is regular, uncountable and not weakly compact then $\Phi(\kappa, \mathfrak{a})$ holds iff $\mathfrak{a} < \log \kappa$.

(iv) $\Phi(\kappa, \mathfrak{a})$ fails for all $\mathfrak{a} \geqslant 2^\kappa$.

(iii) settles the question of $\Phi(\kappa, \mathfrak{a})$ for regular uncountable κ's which are not weakly compact. (i), (ii) and (iv) settle $\Phi(\kappa, \mathfrak{a})$ for $\kappa = \aleph_0$ and weakly compact κ's, provided $2^\kappa = \kappa^+$. If $2^\kappa > \kappa^+$ all we can say is that it is consistent with ZFC that $2^{\aleph_0} > \aleph_1$ and $\Phi(\omega, \mathfrak{a})$ for all $\mathfrak{a} < 2^{\aleph_0}$ (it follows from (iv)) and it is also consistent with ZFC that $\Phi(\omega, \aleph_1)$ does not hold ($\Phi(\omega, \aleph_1)$ fails in the model of Cohen 1966 where $2^{\aleph_0} > \aleph_1$).

(v) Martin's axiom (VIII.4.11) implies $\Phi(\omega, \mathfrak{a})$ for every $\mathfrak{a} < 2^{\aleph_0}$.

(vi) If κ is singular then $\Phi(\kappa, \mathfrak{a})$ implies $\Phi(\mathrm{cf}(\kappa), \mathfrak{a})$.

(vii) If κ is singular and $\mathrm{cf}(\kappa)$ is neither $\aleph_0$ nor weakly compact then $\Phi(\kappa, \mathfrak{a})$ holds iff $\mathfrak{a} < \log \mathrm{cf}(\kappa)$.

(viii) If κ is singular, and $\mathrm{cf}(\kappa)$ is $\aleph_0$ or weakly compact then $\Phi(\kappa, \mathfrak{a})$ holds for $\mathfrak{a} < \mathrm{cf}(\kappa)$, and also for $\mathfrak{a} = \mathrm{cf}(\kappa)$ if $2^{\mathrm{cf}(\kappa)} < \kappa$, and $\Phi(\kappa, \mathfrak{a})$ fails for $\mathfrak{a} \geqslant 2^{\mathrm{cf}(\kappa)}$.

(ix) If $\Psi(\kappa, \mathfrak{a})$ is like the statement of 3.42 except that κ may be any limit ordinal and B is only required to be an unbounded subset of κ then $\Psi(\kappa, \mathfrak{a})$ holds iff $\Phi(\mathrm{cf}(\kappa), \mathfrak{a})$ does.

Hints. (ii) Use the ideas of the proof of 3.37.

(iii) If $\mathfrak{a} < \log\kappa$ then $2^{\mathfrak{a}} < \kappa$ and since κ is regular $\Phi(\kappa, \mathfrak{a})$ holds by simple cardinality considerations. If $\log\kappa = \kappa$ use (ii). If $\log\kappa < \kappa$ let $\mathfrak{a} = \log\kappa$ then $\mathfrak{a} < \kappa \leqslant 2^{\mathfrak{a}}$. Since $\mathfrak{a} < \kappa$ and κ is regular, if $\langle A_\alpha \mid \alpha \in B \rangle$ has a limit A then $A_\alpha = A$ for sufficiently large α's and since $2^{\mathfrak{a}} \geqslant \kappa$ we can choose A_α's so that this does not happen.

(iv) Since there are at most 2^κ different candidates for B as in 3.42(b) we can put in W sufficiently many subsets of κ as counterexamples.

(v) For W as in 3.42(b) first construct a subset T of $\mathbf{P}(\omega)$ such that for every $X \in W$ $X \in T$ or $\omega \sim X \in T$ and such that the intersection of any finite number of members of T is infinite. Then use VIII.4.24.

(vi) Assume that $\Phi(\mathrm{cf}(\kappa), \mathfrak{a})$ fails to hold and prove that also $\Phi(\kappa, \mathfrak{a})$ does not hold.

(vii) Use (iii) and (vi) in one direction, and the idea of the proof of (iii) in the other direction.

(viii) We mention only the case where $2^{\mathrm{cf}(\kappa)} < \kappa$ and $\mathfrak{a} = \mathrm{cf}(\kappa)$. Assume that $\langle A_\mu \mid \mu < \kappa \rangle$ is a counterexample to $\Phi(\kappa, \mathfrak{a})$. Let $\langle \kappa_\xi \mid \xi < \mathfrak{a} \rangle$ be a function enumerating a cofinal subset of κ which consists of successor cardinals and such that $\kappa_0 > 2^{\mathrm{cf}(\kappa)}$. For $\xi < \mathfrak{a}$ $2^{\mathfrak{a}} < \kappa_0 \leqslant \kappa_\xi$ and κ_ξ is regular, hence $\Phi(\kappa_\xi, \mathfrak{a})$ holds in the strong sense that there is a subset B_ξ of κ_ξ such that all the A_α's for $\alpha \in B_\xi$ are the same set; let us denote this set with A'_ξ. $\langle A'_\xi \mid \xi < \mathfrak{a} \rangle$ is a counterexample to $\Phi(\mathfrak{a}, \mathfrak{a})$, contradicting (ii). □

3.44 Definition. Let $\langle A_\alpha \mid \alpha < \mu \rangle$ be a sequence of subsets of a set W. A *flip* of $\langle A_\alpha \mid \alpha < \mu \rangle$ (with respect to W) is a sequence $\langle B_\alpha \mid \alpha < \mu \rangle$ such that for each $\alpha < \mu$ either $B_\alpha = A_\alpha$ or $B_\alpha = W \sim A_\alpha$.

3.45Ac Exercise (Abramson, Harrington, Kleinberg and Zwicker 1977). For an aleph κ prove the following.

(i) κ is inaccessible iff for every $\mu < \kappa$ every sequence in ${}^\mu\mathbf{P}(\kappa)$ has a flip $\langle B_\alpha \mid \alpha < \mu \rangle$ such that $|\bigcap_{\alpha<\mu} B_\alpha| = \kappa$.

(ii) κ is weakly compact iff every sequence $\langle A_\alpha \mid \alpha < \kappa \rangle$ in ${}^\kappa\mathbf{P}(\kappa)$ has a flip $\langle B_\alpha \mid \alpha < \kappa \rangle$ such that for all $\mu < \kappa$ $|\bigcap_{\alpha<\mu} B_\alpha| = \kappa$.

(iii) κ is ineffable iff every sequence in ${}^\kappa\mathbf{P}(\kappa)$ has a flip $\langle B_\alpha \mid \alpha < \kappa \rangle$ such that its diagonal intersection $D_{\alpha<\mu} B_\alpha$ is a stationary subset of κ.

Remark. The flipping properties of the cardinals can be regarded as partition relations which deal with partitions of the set κ itself (rather than of $[\kappa]^n$) but, as a counterweight to this relaxation, one looks for subsets of κ which are homogeneous with respect to many given partitions. A sequence $\langle A_\alpha \mid \alpha<\mu\rangle$ in ${}^\mu\mathbf{P}(\kappa)$ can be regarded as a sequence of μ 2-partitions of κ (i.e., of partitions of κ into 2 sets), the α-th partition being $\{A_\alpha, \kappa\sim A_\alpha\}$. Thus (i) asserts that κ is inaccessible iff for every sequence of $<\kappa$ 2-partitions of κ there is a set of cardinality κ which is homogeneous for each one of the partitions. (iii) asserts that κ is ineffable if for every sequence of κ 2-partitions of κ there is a stationary subset B of κ such that for each $\mu<\kappa$ $B\sim\mu$ is homogeneous for each one of the first μ partitions.

Hints. (ii) If κ is weakly compact consider the tree $T\subseteq{}^{\kappa}2$ such that for $\mu<\kappa$ and $f\in{}^\mu 2$ $f\in T$ iff $|\bigcap_{\alpha<\mu}A_\alpha^{(f(\alpha))}|=\kappa$, where $A_\alpha^{(0)}=A_\alpha$, $A_\alpha^{(1)}=\kappa\sim A_\alpha$. In the other direction use (i), and given a tree $\langle T,<\rangle$ of length κ such that $|T_\lambda|<\kappa$ for $\lambda<\kappa$, assume $T=\kappa$ and let $A_\alpha=\{\gamma\in T \mid \gamma \not\geq \alpha\}$.

(iii) If $A_\alpha\subseteq\alpha$ for $\alpha<\kappa$ let $A'_\beta=\{\alpha<\kappa \mid \beta\in A_\alpha\}$ for $\beta<\kappa$. Take a flip $\langle B_\beta \mid \beta<\kappa\rangle$ of $\langle A'_\beta \mid \beta<\kappa\rangle$ as in (iii), then $D=D_{\beta<\kappa}B_\beta$ is such that the sequence $\langle A_\alpha \mid \alpha\in D\rangle$ is coherent. The other direction is proved by going through this proof backwards. □

3.46Ac Proposition (Baumgartner 1975). *Let κ be an inaccessible cardinal and let $F:[\kappa]^2\to\{0,1\}$ then there is a stationary subset A of κ such that $F[[A]^2]=\{0\}$ or else there is a closed unbounded subset C of κ and a sequence $\langle S_\alpha \mid \alpha\in C\rangle$ such that for $\alpha\in C$ S_α is an unbounded subset of α and $F[[S_\alpha\cup\{\alpha\}]^2]=\{1\}$.*

3.47Ac Corollary (Baumgartner 1975). *If κ is inaccessible then $\kappa\to$ (stationary, $<\kappa$), i.e., if $F:[\kappa]^2\to\{0,1\}$ then there is a stationary subset A of κ homogeneous for F, or else for both $i=0,1$ and for each $\mathfrak{a}<\kappa$ there is a subset A of κ such that $|A|=\mathfrak{a}$ and $F[[A]^2]=\{i\}$.* □

Hint of proof of 3.46. For $\alpha<\kappa$ construct $S_\alpha\subseteq\alpha$ by scanning all the ordinals $<\alpha$ and putting $\beta<\alpha$ in S_α if $F(\{\beta,\alpha\})=1$ and $F(\{\gamma,\beta\})=1$ for all $\gamma\in S_\alpha\cap\beta$. If for a stationary subset B of κ $\alpha\in B$ $\sup S_\alpha<\alpha$ we may assume, by Fodor's theorem IV.4.40, that there is a $\lambda<\kappa$ such that $\sup S_\alpha=\lambda$ for all $\alpha\in B$, and since λ has $2^{|\lambda|}<\kappa$ subsets we may assume that for some $W\subseteq\lambda$ $S_\alpha=W$ for all $\alpha\in B$. We have $F[[B\sim\lambda]^2]=\{0\}$, since if $\gamma,\delta\in B\sim\lambda$ and $\gamma<\delta$ then if $F(\{\gamma,\delta\})=1$ we would have $\gamma\in S_\delta$. □

4. *Measurable Cardinals*

Vitali 1905 showed that there is no non-trivial translation-invariant countably additive measure on all sets of real numbers. (A measure μ on $\mathbf{P}(\mathbb{R})$ is said to be *translation-invariant* if for all $A\subseteq\mathbb{R}$ and $u\in\mathbb{R}$ $\mu(\{x+u \mid x\in A\})=\mu(A)$). Then the following questions came up. If we give up the requirement that the measure be translation-invariant, is there a non-trivial countably additive measure on all sets of reals? Is there such a measure for all subsets of some general set X? These questions are the origin of the subject of the measurable cardinals. The cardinals of those infinite sets X for which some measures as above, of special natural kinds, exist are called real-valued-measurable cardinals and measurable cardinals. It turns out that the most important concept here is that of the measurable cardinals. Measurable cardinals are very large cardinals; the least measurable cardinal (if there are any measurable cardinals) is greater than many weakly compact and even ineffable cardinals. The concept of a measurable cardinal occupies a much more central role in the theory of large cardinals than those played by the weakly compact and the ineffable cardinals; the full use of the power of this concept is way beyond the scope of this book.

4.1 Definition (Infinite sums of reals). Let $\langle a_i \mid i \in I \rangle$ be an indexed family of non-negative real numbers. We say that $\Sigma_{i \in I} a_i$ *converges* if the set $\{\Sigma_{i \in J} a_i \mid J$ is a finite subset of $I\}$ is a bounded set of reals; if $\Sigma_{i \in I} a_i$ does not converge we say that it *diverges*. If $\Sigma_{i \in I} a_i$ converges then $\Sigma_{i \in I} a_i$ is defined to be the least upper bound of the set $\{\Sigma_{i \in J} a_i \mid J$ is a finite subset of $I\}$. We assume the properties of this operation Σ to be known to the reader.

4.2Ac Proposition. *Let $\langle a_i \mid i \in I \rangle$ be an indexed family of non-negative real numbers such that $\Sigma_{i \in I} a_i$ converges, then the set $K = \{i \in I \mid a_i > 0\}$ is countable and $\Sigma_{i \in K} a_i = \Sigma_{i \in I} a_i$.*

Proof. Let $\Sigma_{i \in I} a_i = a$, then for every $n < \omega$ the set $K_n = \{i \in I \mid a_i \geqslant \frac{1}{n}\}$ has at most $n \cdot a$ members. Clearly $K = \bigcup_{n \in \omega} K_n$ and thus K is countable, being the union of $\aleph_0$ finite sets. The equality $\Sigma_{i \in K} a_i = \Sigma_{i \in I} a_i$ follows directly from the definition of Σ. □

4.3 Definition. (i) By a *measure* on the power set $\mathbf{P}(X)$ of a set X we shall mean, throughout this section, a function μ on $\mathbf{P}(X)$ into the closed unit interval $[0, 1]$ such that

(a) $\mu(X) = 1, \quad \mu(0) = 0$,
(b) if $A, B \subseteq X$ and $A \cap B = 0$ then $\mu(A \cup B) = \mu(A) + \mu(B)$, and
(c) For every $x \in X \quad \mu(\{x\}) = 0$.

(ii) A measure on $\mathbf{P}(X)$ is said to be $\mathfrak{m}$-*additive* if for every family $\langle A_i \mid i \in I \rangle$ of $< \mathfrak{m}$ pairwise disjoint subsets of $X \quad \mu(\bigcup_{i \in I} A_i) = \Sigma_{i \in I} \mu(A_i)$. We shall also use the term *countably additive* for $\aleph_1$-additive (not for $\aleph_0$-additive!).

(iii) A measure μ on $\mathbf{P}(X)$ is called *two-valued* if for every $A \subseteq X \quad \mu(A) = 0$ or $\mu(A) = 1$.

(iv) When we deal with a fixed measure μ on $\mathbf{P}(X)$ we shall mean by "$\Phi(x)$ holds for almost all $x \in X$" that the set of all x's in X for which $\Phi(x)$ holds is of measure 1.

Remark. The requirement that $\mu(X) = 1$ is not essential for what we shall do; we could have let $\mu(X)$ be any positive real. Indeed, demanding $\mu(X) = 1$ imposes no restriction at all since from every measure μ on $\mathbf{P}(X)$ which gives X a positive finite value we can pass to an essentially equivalent measure μ' for which $\mu'(X) = 1$ by setting $\mu'(A) = \mu(A)/\mu(X)$ for every $A \subseteq X$. The requirement that $\mu(\{x\}) = 0$ for every $x \in X$ is adopted in order to streamline our treatment and to avoid cases which are trivial from our point of view. We assume that the simple properties of a measure are known to the reader (see § VII.3).

4.4 Proposition. *If for a set X there is an $\mathfrak{m}$-additive measure on $\mathbf{P}(X)$ then there is an $\mathfrak{m}$-additive measure on $\mathbf{P}(Y)$ for every set Y which is equinumerous with X.* □

4.5Ac Proposition. *If a measure μ on $\mathbf{P}(X)$ is $\mathfrak{m}$-additive and $Y \subseteq X$, $|Y| < \mathfrak{m}$, then $\mu(Y) = 0$; hence $|X| \geqslant \mathfrak{m}$.* □

4.6 Proposition. *If for an aleph $\mathfrak{m}$ μ is an $\mathfrak{m}$-additive measure on $\mathbf{P}(X)$ and $\langle A_i \mid i \in I \rangle$ is a family of subsets of X such that $|I| < \mathfrak{m}$ then either $\Sigma_{i \in I} \mu(A_i)$ diverges or else $\Sigma_{i \in I} \mu(A_i) \geqslant \mu(\bigcup_{i \in I} A_i)$.*

Proof. Let $<$ be a relation which well-orders I. Since $\langle A_i \sim \bigcup_{j<i} A_j \mid i \in I \rangle$ is a family of pairwise disjoint sets we have, by the $\mathfrak{m}$-additivity of μ,

$$\mu(\textstyle\bigcup_{i \in I} A_i) = \mu(\bigcup_{i \in I}(A_i \sim \bigcup_{j<i} A_j)) = \Sigma_{i \in I} \mu(A_i \sim \bigcup_{j<i} A_j) \leqslant \Sigma_{i \in I} \mu(A_i). \quad \square$$

4.7 Proposition. *If μ is a measure on $\mathbf{P}(X)$ and $\langle A_i \mid i \in I \rangle$ is a family of pairwise disjoint subsets of X then $\Sigma_{i \in I} \mu(A_i)$ converges and $\Sigma_{i \in I} \mu(A_i) \leqslant \mu(\bigcup_{i \in I} A_i)$. (Notice that we deal here with the general case where we do not assume that μ is $\mathfrak{m}$-additive for some $\mathfrak{m} > |I|$).* □

4.8 Proposition. (i) *Let μ be an $\mathfrak{m}$-additive measure on $\mathbf{P}(X)$, then $\{A \subseteq X \mid \mu(A) = 0\}$ is a non-principal $\mathfrak{m}$-complete ideal on X and $\{A \subseteq X \mid \mu(A) = 1\}$ is a non-principal $\mathfrak{m}$-complete filter on X. (Thus the intersection of $< \mathfrak{m}$ sets of measure 1 is a set of measure 1). The measure μ is two-valued iff the ideal $\{A \subseteq X \mid \mu(A) = 0\}$ is a prime ideal, iff the filter $\{A \subseteq X \mid \mu(A) = 1\}$ is an ultrafilter.*

(ii) *If I is a non-principal $\mathfrak{m}$-complete prime ideal on X (i.e., I is a prime ideal containing all singletons) then the function μ on $\mathbf{P}(X)$ defined by $\mu(A) = 0$ if $A \in I$ and $\mu(A) = 1$ otherwise is a two-valued $\mathfrak{m}$-additive measure on $\mathbf{P}(X)$.*

(iii) *If μ is an $\mathfrak{m}$-additive two-valued measure on $\mathbf{P}(X)$ and $\langle A_i \mid i \in I \rangle$ is a family of subsets of X with $|I| < \mathfrak{m}$ then $\mu(\bigcap_{i \in I} A_i) = 0$ iff for some $i \in I$ $\mu(A_i) = 0$.* □

Remark. 4.8(i) and 4.8(ii) establish the interchangability of the concepts of an $\mathfrak{m}$-additive two-valued measure on $\mathbf{P}(X)$ and of a non-principal $\mathfrak{m}$-complete ultrafilter on X. During the rest of this section we shall move freely back and forth between these two concepts.

4.9 On the Existence of Countably Complete Measures on Power Sets. We asked at the beginning of this section if there is a countably complete measure on the set of all sets of reals. Once we do not require translation-invariance it does not matter whether the measure is on the set of all sets of reals or on the power set of any other set of cardinality $2^{\aleph_0}$ (by 4.4). A natural extension of our question is to ask for which cardinals $\mathfrak{a}$ is there a countably additive measure on power sets of sets of cardinality $\mathfrak{a}$, or, for that matter, on $\mathbf{P}(\mathfrak{a})$ itself, if $\mathfrak{a}$ is an aleph. Let us first notice that by 4.5 there is no such measure on $\aleph_0$. Once there is such a measure μ on an aleph $\mathfrak{a}$ there is such a measure μ' on every aleph $\mathfrak{b}>\mathfrak{a}$, where we set for $A \subseteq \mathfrak{b}$ $\mu'(A)=\mu(A\cap\mathfrak{a})$. Therefore what we should be looking for is the least aleph $\mathfrak{a}$ such that there is a countably additive measure on $\mathfrak{a}$.

4.10 Theorem. (i) *If* $\mathfrak{a}$ *is the least aleph such that there is a countably additive measure on* $\mathbf{P}(\mathfrak{a})$ *then every countably additive measure on* $\mathbf{P}(\mathfrak{a})$ *is* $\mathfrak{a}$*-additive.*

(ii) *If* $\mathfrak{a}$ *is the least aleph such that there is a countably additive two-valued measure on* $\mathbf{P}(\mathfrak{a})$ *then every countably additive two-valued measure on* $\mathbf{P}(\mathfrak{a})$ *is* $\mathfrak{a}$*-additive.*

Proof. (i) Let μ be a countably additive measure on $\mathbf{P}(\mathfrak{a})$. Suppose μ is not $\mathfrak{a}$-additive and let $\mathfrak{b}<\mathfrak{a}$ be the least aleph such that there is a sequence $\langle A_\sigma \mid \sigma<\mathfrak{b}\rangle$ of pairwise disjoint subsets of $\mathfrak{a}$ with

(1) $\qquad \Sigma_{\sigma\in\mathfrak{b}}\mu(A_\sigma)\neq\mu(\bigcup_{\sigma\in\mathfrak{b}}A_\sigma).$

Since μ is countably additive $\mathfrak{b}\geqslant\aleph_1$. By 4.7 we have $\Sigma_{\sigma\in\mathfrak{b}}\mu(A_\sigma)<\mu(\bigcup_{\sigma\in\mathfrak{b}}A_\sigma)$. Let $u=\mu(\bigcup_{\sigma\in\mathfrak{b}}A_\sigma)-\Sigma_{\sigma\in\mathfrak{b}}\mu(A_\sigma)>0$. We shall now define a $\mathfrak{b}$-additive measure ν on $\mathbf{P}(\mathfrak{b})$, and since $\mathfrak{b}\geqslant\aleph_1$ ν is also countably additive, contradicting the minimality of $\mathfrak{a}$.

For $Y\subseteq\mathfrak{b}$ define

(2) $\qquad \nu(Y)=\frac{1}{u}(\mu(\bigcup_{\sigma\in Y}A_\sigma)-\Sigma_{\sigma\in Y}\mu(A_\sigma)).$

$\nu(Y)\geqslant 0$ by 4.7.

$$\nu(\mathfrak{b})=\tfrac{1}{u}(\mu(\bigcup_{\sigma<\mathfrak{b}}A_\sigma)-\Sigma_{\sigma<\mathfrak{b}}\mu(A_\sigma))=\tfrac{1}{u}\cdot u=1.$$

Now let us prove that ν is $\mathfrak{b}$-additive. For some $\mathfrak{c}<\mathfrak{b}$ let $\langle Y_\lambda \mid \lambda\in\mathfrak{c}\rangle$ be a family of pairwise disjoint subsets of $\mathfrak{b}$, and let $Y=\bigcup_{\lambda\in\mathfrak{c}}Y_\lambda$. $\mu(\bigcup_{\sigma\in Y}A_\sigma)= \mu(\bigcup_{\lambda\in\mathfrak{c}}\bigcup_{\sigma\in Y_\lambda}A_\sigma)=\Sigma_{\lambda\in\mathfrak{c}}\mu(\bigcup_{\sigma\in Y_\lambda}A_\sigma)$, since $\{\bigcup_{\sigma\in Y_\lambda}A_\sigma \mid \lambda\in\mathfrak{c}\}$ is a family of pairwise disjoint sets (because the A_σ's and Y_λ's are pairwise disjoint) and since $\mathfrak{b}$ is the minimal cardinal for which (1) holds. By the associativity of infinite addition of non-negative real numbers we have $\Sigma_{\sigma\in Y}\mu(A_\sigma)=\Sigma_{\lambda\in\mathfrak{c}}\Sigma_{\sigma\in Y_\lambda}\mu(A_\sigma)$, and thus we have, by (2),

$$\begin{aligned}\nu(Y)&=\tfrac{1}{u}(\Sigma_{\lambda\in\mathfrak{c}}\mu(\bigcup_{\sigma\in Y_\lambda}A_\sigma)-\Sigma_{\lambda\in\mathfrak{c}}\Sigma_{\sigma\in Y_\lambda}\mu(A_\sigma))\\&=\tfrac{1}{u}(\Sigma_{\lambda\in\mathfrak{c}}(\mu(\bigcup_{\sigma\in Y_\lambda}A_\sigma)-\Sigma_{\sigma\in Y_\lambda}\mu(A_\sigma)))\\&=\Sigma_{\lambda\in\mathfrak{c}}\tfrac{1}{u}(\mu(\bigcup_{\sigma\in Y_\lambda}A_\sigma)-\Sigma_{\sigma\in Y_\lambda}\mu(A_\sigma))=\Sigma_{\lambda\in\mathfrak{c}}\nu(Y_\lambda).\end{aligned}$$

Finally, if $\sigma\in\mathfrak{b}$ then $\nu(\{\sigma\})=\frac{1}{u}(\mu(A_\sigma)-\mu(A_\sigma))=0$. Thus γ is a $\mathfrak{b}$-complete measure on $\mathbf{P}(\mathfrak{b})$.

(ii) Follow the proof of (i), using two-valued measures instead of general ones. □

As we have mentioned, if $\mathfrak{a}$ is the least cardinal such that there is a countably additive, general or two-valued, measure on $\mathbf{P}(\mathfrak{a})$ then we have such a measure also on each $\mathbf{P}(\mathfrak{b})$ for $\mathfrak{b} > \mathfrak{a}$. However, not all these cardinals will have the interesting properties of $\mathfrak{a}$. Therefore we shall rather discuss the narrower classes of those cardinals which share with $\mathfrak{a}$ the properties of the conclusions of the two parts of Theorem 4.10.

4.11 Definition. A cardinal $\mathfrak{a}$ is said to be *real-valued-measurable* if $\mathfrak{a} > \aleph_0$ and there is an $\mathfrak{a}$-additive measure on $\mathbf{P}(A)$ for some set A of cardinality $\mathfrak{a}$; and $\mathfrak{a}$ is said to be *measurable* if $\mathfrak{a} > \aleph_0$ and there is an $\mathfrak{a}$-additive two-valued measure on $\mathbf{P}(A)$ for some set A of cardinality $\mathfrak{a}$. Obviously, every measurable cardinal is real-valued-measurable; as will follow from 4.18 and the discussion in 4.15 there may be real-valued-measurable cardinals which are not measurable.

By Theorem 4.10 the least aleph $\mathfrak{a}$ for which there is a countably additive measure on $\mathbf{P}(\mathfrak{a})$ is real-valued-measurable, and the least aleph $\mathfrak{a}$ for which there is a countably additive two-valued measure on $\mathbf{P}(\mathfrak{a})$ is measurable.

4.12 Exercise. Prove the generalization of 4.10(i) that for every aleph $\mathfrak{d} > \aleph_0$ if $\mathfrak{a}$ is the least aleph for which there is a $\mathfrak{d}$-additive measure on $\mathbf{P}(\mathfrak{a})$ then every $\mathfrak{d}$-additive measure on $\mathbf{P}(\mathfrak{a})$ is $\mathfrak{a}$-additive, and hence $\mathfrak{a}$ is real-valued-measurable. Prove also the analogous generalization of 4.10(ii). □

4.13 Proposition (Banach 1930). *Every real-valued-measurable cardinal is regular.*

Proof. Let μ be an $\mathfrak{a}$-additive measure on $\mathbf{P}(\mathfrak{a})$. If $\mathfrak{a}$ is singular then $\mathfrak{a} = \bigcup_{\xi < \mathfrak{b}} A_\xi$ for some $\mathfrak{b} < \mathfrak{a}$ and $|A_\xi| < \mathfrak{a}$ for $\xi \in \mathfrak{b}$. Since $|A_\xi| < \mathfrak{a}$ $\mu(A_\xi) = 0$ for $\xi \in \mathfrak{b}$, by 4.5. Hence, by 4.6, $\mu(\mathfrak{a}) = \mu(\bigcup_{\xi \in \mathfrak{b}} A_\xi) \leqslant \Sigma_{\xi \in \mathfrak{b}} \mu(A_\xi) = 0$, which contradicts $\mu(\mathfrak{a}) = 1$. □

4.14Ac Theorem (Banach 1930, Ulam 1930). *Every real-valued-measurable cardinal is a limit cardinal.*

Proof. Suppose μ is an $\mathfrak{a}$-additive measure on a successor cardinal $\mathfrak{a} = \lambda^+$. Consider a matrix A with λ rows and λ^+ columns as follows. In the 0-th column we inscribe, in any order, the ordinals $< \lambda$, in the next column we inscribe the ordinals $< \lambda + 1$ and so forth; for every $\sigma < \lambda^+$ we inscribe in the σ-th column the ordinals $< \lambda + \sigma$. This is possible since for every $\sigma < \lambda^+$ $|\lambda + \sigma| = \lambda$. Let $A(\xi, \eta)$ denote the ordinal inscribed in the ξ-th row and the η-th column. For every ordinal $\sigma < \lambda^+$ σ occurs at least in all columns from the σ-th column rightwards, hence the set $\{\eta < \lambda^+ \mid (\exists \xi < \lambda)(A(\xi, \eta) = \sigma)\}$ of the columns in which σ occurs is of measure 1 (since its complement is included in $\sigma \cup \{\sigma\}$ which is of measure 0 by 4.5). Let $B_{\xi, \sigma}$ be the set of all the places in the ξ-th row in which σ occurs, i.e., $B_{\xi, \sigma} = \{\eta < \lambda^+ \mid A(\xi, \eta) = \sigma\}$. By the λ^+-additivity of μ and since $B_{\xi, \sigma} \cap B_{\eta, \sigma} = 0$ for $\xi \neq \eta$ we have

$$\Sigma_{\xi < \lambda} \mu(B_{\xi, \sigma}) = \mu(\bigcup_{\xi < \lambda} B_{\xi, \sigma}) = \mu(\{\eta < \lambda^+ \mid (\exists \xi < \lambda)(A(\xi, \eta) = \sigma\}) = 1.$$

therefore

(3) for some $\xi<\lambda \quad \mu(B_{\xi,\sigma})>0$,

i.e., σ occurs in the ξ-th row in a set of positive measure. In the ξ-th row we have $\lambda^+=\bigcup_{\sigma<\lambda^+}B_{\sigma,\xi}$ and since $B_{\xi,\sigma}\cap B_{\xi,\tau}=0$ for $\sigma\neq\tau$ we have, by 4.7 and 4.2, that for at most $\aleph_0$ σ's is $\mu(B_{\xi,\sigma})>0$; let S_ξ be the set of those σ's. We have $|S_\xi|\leqslant\aleph_0$ and $\mu(B_{\xi,\sigma})>0$ iff $\sigma\in S_\xi$. S_ξ is the set of σ's which occur with positive measure in the ξ-th row. Since there are only λ rows there are only $\lambda\cdot\aleph_0=\lambda$ σ's which occur with positive measure in any row, which contradicts the assertion in (3) that every $\sigma<\lambda^+$ occurs with positive measure in some row. □

4.15Ac On the Size of the Real-valued-Measurable Cardinals. By 4.13 and 4.14 every real-valued-measurable cardinal is weakly inaccessible. Since we cannot prove in ZFC the existence of weakly inaccessible cardinals we cannot prove in ZFC the existence of real-valued-measurable cardinals. The real-valued-measurable cardinals, if they exist at all, are not merely weakly inaccessible cardinals. Solovay 1971 proved that every real-valued-measurable cardinal $\mathfrak{a}$ has the property that

(4) for every stationary subset S of $\mathfrak{a}$ there is a regular (and even a weakly inaccessible) ordinal $\lambda<\mathfrak{a}$ such that $S\cap\lambda$ is a stationary subset of λ.

This is the property we proved for the weakly compact cardinals in 2.26. As a consequence, a real-valued-measurable cardinal $\mathfrak{a}$ is not the first weakly inaccessible cardinal, it has $\mathfrak{a}$ weakly inaccessible cardinals below it and is not even the least weakly inaccessible cardinal $\mathfrak{b}$ which has $\mathfrak{b}$ weakly inaccessible cardinals below it. Moreover, we can follow 2.27–2.30 to supply some additional information on how big a real-valued-measurable cardinal is. First we follow 2.27 and define a cardinal $\mathfrak{b}$ to be *weakly Mahlo* if $\mathfrak{b}$ is infinite and regular and the set of all regular ordinals $<\mathfrak{b}$ is a stationary subset of $\mathfrak{b}$. Such a $\mathfrak{b}$ is obviously weakly inaccessible. As in 2.28 one can prove that also the set of all weakly inaccessible ordinals below a weakly Mahlo cardinal $\mathfrak{b}$ is a stationary subset of $\mathfrak{b}$, and thus $\mathfrak{b}$ is the $\mathfrak{b}$-th weakly inaccessible ordinal and is even the $\mathfrak{b}$-th weakly inaccessible ordinal $\mathfrak{c}$ such that $\mathfrak{c}$ is the $\mathfrak{c}$-th weakly inaccessible ordinal. As in 2.29 one can prove that a real-valued-measurable cardinal $\mathfrak{a}$ is a weakly Mahlo cardinal and the set of all weakly Mahlo cardinals below it is a stationary subset of $\mathfrak{a}$; as in 2.30 one can also prove that the set $\{\mathfrak{b}<\mathfrak{a}\mid\mathfrak{b}$ is a weakly Mahlo cardinal and the set of all weakly Mahlo cardinals $<\mathfrak{b}$ is a stationary subset of $\mathfrak{b}\}$ is a stationary subset of $\mathfrak{a}$. All this shows how big a real-valued-measurable cardinal is with respect to the weakly inaccessible cardinals. A real-valued-measurable cardinal is not necessarily big with respect to the inaccessible cardinals or even with respect to $2^{\aleph_0}$; Solovay 1971 has shown that (assuming the existence of a measurable cardinal is consistent) it is consistent with ZFC to assume the existence of real-valued-measurable cardinals $\leqslant 2^{\aleph_0}$. If there is a real-valued-measurable cardinal $\leqslant 2^{\aleph_0}$ then also $2^{\aleph_0}$ is very big compared to the weakly inaccessible and the weakly Mahlo cardinals.

4.16Ac Theorem (Ulam 1930). *Every real-valued-measurable cardinal is either* $\leqslant 2^{\aleph_0}$ *or else measurable* (*but not both, since by* 4.18 *every measurable cardinal is inaccessible*).

Proof. Let $\mathfrak{a}$ be a real-valued-measurable cardinal and let μ be an $\mathfrak{a}$-additive measure on $\mathbf{P}(\mathfrak{a})$. We shall first assume

(5) For every $A \subseteq \mathfrak{a}$ such that $\mu(A)>0$ there is a subset B of A such that $\frac{1}{3}\mu(A) \leqslant \mu(B) \leqslant \frac{2}{3}\mu(A)$.

Let F_0 and F_1 be functions, defined on all subsets of $\mathfrak{a}$ of positive measure, which satisfy $F_0(A) \subseteq A$, $\frac{1}{3}\mu(A) \leqslant \mu(F_0(A)) \leqslant \frac{2}{3}\mu(A)$, and $F_1(A)=A \sim F_0(A)$. For every $s \in {}^{\omega}2$ we define $B_s \subseteq \mathfrak{a}$ by $B_0=\mathfrak{a}$, $B_{s^\frown\langle 0\rangle}=F_0(B_s)$, and $B_{s^\frown\langle 1\rangle}=F_1(B_s)$. Clearly $\bigcup_{s \in {}^n2} B_s=\mathfrak{a}$ and for all $s, s' \in {}^n2$ if $s \neq s'$ then $B_s \cap B_{s'}=0$. Since for all $A \subseteq \mathfrak{a}$ $\mu(F_0(A))$, $\mu(F_1(A)) \leqslant \frac{2}{3}\mu(A)$ we get, by induction on n, that for $s \in {}^n2$ $\mu(B_s) \leqslant (\frac{2}{3})^n$. For $t \in {}^{\omega}2$ we define $C_t=\bigcap_{n<\omega} B_{t\restriction n}$; as easily seen $\bigcup_{t \in {}^{\omega}2} C_t=\mathfrak{a}$. Since $C_t \subseteq B_{t\restriction n}$ for each $n<\omega$ we get $\mu(C_t) \leqslant \mu(B_{t\restriction n}) \leqslant (\frac{2}{3})^n$ for every $n<\omega$, hence $\mu(C_t)=0$. If $\mathfrak{a}>2^{\aleph_0}$ we get, by 4.6, that

$$1=\mu(\mathfrak{a})=\mu(\textstyle\bigcup_{t \in {}^{\omega}2} C_t) \leqslant \Sigma_{t \in {}^{\omega}2}\, \mu(C_t)=\Sigma_{t \in {}^{\omega}2}\, 0=0,$$

which is a contradiction. Thus (5) implies that $\mathfrak{a} \leqslant 2^{\aleph_0}$. If (5) fails we have

(6) There is a subset A of $\mathfrak{a}$ of positive measure such that for all $B \subseteq A$ $\mu(B)<\frac{1}{3}\mu(A)$ or $\mu(B)>\frac{2}{3}\mu(A)$.

For a set A as in (6) we have $|A|=\mathfrak{a}$, since if $|A|<\mathfrak{a}$ then by 4.5 we would have $\mu(A)=0$. We shall now define a two-valued measure v on $\mathbf{P}(A)$ and thereby establish that $\mathfrak{a}$ is measurable. For $B \subseteq A$ we set $v(B)=0$ if $\mu(B)<\frac{1}{3}\mu(A)$ and $v(B)=1$ if $\mu(B)>\frac{2}{3}\mu(A)$. By 4.8(ii) we shall know that v is an $\mathfrak{a}$-additive two-valued measure on $\mathbf{P}(A)$ once we prove that the set $I=\{B \subseteq A \mid v(B)=0\}=\{B \subseteq A \mid \mu(B)<\frac{1}{3}\mu(A)\}$ is an $\mathfrak{a}$-complete prime ideal which contains all singletons. It is immediate from the definition of I that every subset of a member of I, and every singleton, is in I. It is also obvious that $A \notin I$. By (6), if $B \subseteq A$ is not in I then $\mu(B)>\frac{2}{3}\mu(A)$ and hence $\mu(A \sim B)<\frac{1}{3}\mu(A)$ and $A \sim B \in I$. Now let us prove that if $B, C \in I$ then also $B \cup C \in I$. Since $B, C \in I$ $\mu(B \cup C) \leqslant \mu(B)+\mu(C)<\frac{1}{3}\mu(A)+\frac{1}{3}\mu(A)=\frac{2}{3}\mu(A)$. By (6) $\mu(B \cup C)<\frac{2}{3}\mu(A)$ implies $\mu(B \cup C)<\frac{1}{3}\mu(A)$, thus $B \cup C \in I$. Now that we know that I is an ideal we know that the union of finitely many members of I is again in I.

To prove the $\mathfrak{a}$-completeness of I it suffices, by IV.4.6(iv), to show that if $\mathfrak{b}<\mathfrak{a}$ and $\{B_\alpha \mid \alpha \in \mathfrak{b}\} \subseteq I$ consists of pairwise disjoint sets then $\bigcup_{\alpha \in \mathfrak{b}} B_\alpha \in I$. Since μ is $\mathfrak{a}$-additive and the B_α's are pairwise disjoint we have $\mu(\bigcup_{\alpha \in \mathfrak{b}} B_\alpha)=\Sigma_{\alpha \in \mathfrak{b}}\mu(B_\alpha)=$ the least upper bound of $\{\Sigma_{\alpha \in J}\, \mu(B_\alpha) \mid J$ is a finite subset of $\mathfrak{b}\}=$ the least upper bound of $\{\mu(\bigcup_{\alpha \in J} B_\alpha) \mid J$ is a finite subset of $\mathfrak{b}\}$. Since for every finite subset J of $\mathfrak{b}$ $\bigcup_{\alpha \in J} B_\alpha \in I$, we have $\mu(\bigcup_{\alpha \in J} B_\alpha)<\frac{1}{3}\mu(A)$; thus $\frac{1}{3}\mu(A)$ is an upper bound of the set $\{\mu(\bigcup_{\alpha \in J} B_\alpha) \mid J$ is a finite subset of $\mathfrak{b}\}$ and hence the least upper bound $\mu(\bigcup_{\alpha \in \mathfrak{b}} B_\alpha)$ of this set is $\leqslant \frac{1}{3}\mu(A)$. By (6) $\mu(\bigcup_{\alpha \in \mathfrak{b}} B_\alpha)<\frac{1}{3}\mu(A)$, hence $\bigcup_{\alpha \in \mathfrak{b}} B_\alpha \in I$. □

7Ac Exercise. Let μ be an $\mathfrak{a}$-additive measure on some set $\mathbf{P}(A)$ where $\mathfrak{a}>2^{\aleph_0}$. ere are a sequence $\langle A_i \mid i<\lambda\rangle$, where $0<\lambda\leqslant\omega$, of pairwise disjoint subsets A of positive measure and a sequence $\langle \mu_i \mid i<\lambda\rangle$, where for $i<\lambda$ μ_i is a two-ued $\mathfrak{a}$-additive measure on $\mathbf{P}(A_i)$, such that for every subset B of A $\mu(B)=$ $_{<\lambda}\mu(A_i)\cdot\mu_i(B\cap A_i)$. □

By 4.16 and 4.17 if we want to investigate real-valued-measurable cardinals ove $2^{\aleph_0}$, or even if we want to investigate, for $\mathfrak{a}>2^{\aleph_0}$, real-valued $\mathfrak{a}$-additive asures on sets $\mathbf{P}(A)$, it suffices to deal with measurable cardinals and with two-ued measures. This is where our attention turns now.

8Ac Theorem (Ulam 1930). *Every measurable cardinal* $\mathfrak{a}$ *is inaccessible.*

oof. By 4.13 $\mathfrak{a}$ is regular. We still have to prove that $\mathfrak{a}$ is a strong limit cardinal. ppose that for some $\mathfrak{b}<\mathfrak{a}$ $2^{\mathfrak{b}}\geqslant\mathfrak{a}$, then by 4.4 there is an $\mathfrak{a}$-additive two-valued asure μ on $\mathbf{P}(A)$ for a subset A of ${}^{\mathfrak{b}}2$. For $\alpha\in\mathfrak{b}$ and $i<2$ let $A_{\alpha,i}=\{h\in A \mid h(\alpha)=i\}$ n $A_{\alpha,0}\cap A_{\alpha,1}=0$, $A_{\alpha,0}\cup A_{\alpha,1}=A$. Define $f(\alpha)$ to be that $i<2$ for which $\mu(A_{\alpha,i})=1$. have therefore, for each $\alpha\in\mathfrak{b}$ $\mu(A_{\alpha,f(\alpha)})=1$, and since $\mathfrak{b}<\mathfrak{a}$ we have $\bigcap_{\alpha\in\mathfrak{b}}A_{\alpha,f(\alpha)})=1$. But $\bigcap_{\alpha\in\mathfrak{b}}A_{\alpha,f(\alpha)}=\bigcap_{\alpha\in\mathfrak{b}}\{h\in A \mid h(\alpha)=f(\alpha)\}=\{f\}\cap A\subseteq\{f\}$, l thus $\bigcap_{\alpha\in\mathfrak{b}}A_{\alpha,f(\alpha)}$ has at most one member and its measure is 0, which is a ntradiction. □

9Ac Exercise. Prove that every measurable cardinal $\mathfrak{a}$ has the tree property l is therefore weakly compact.

nt. If $\langle T,<\rangle$ is a tree and $T=\mathfrak{a}$ obtain a branch of length $\mathfrak{a}$ by picking in each el T_λ that x for which $\mu(\{y\in T \mid y\geqslant x\})=1$. □

0Ac Exercise. Let $\mathfrak{a}$ be a measurable cardinal. Let $|A|=\mathfrak{a}$ and let $F\subseteq\mathbf{P}(A)$ be h that $|F|=\mathfrak{a}$ and for every subset G of F if $|G|<\mathfrak{a}$ then $\bigcap G\neq 0$. Then there is an dditive measure v on $\mathbf{P}(A)$ such that $v(B)=1$ for every $B\in F$. Here we allow also asures v such that $v(\{x\})=1$ for some $x\in A$.

mark. While the statement that $\mathfrak{a}$ is measurable tells us that there is some dditive measure on the power set $\mathbf{P}(A)$ of a set A of cardinality $\mathfrak{a}$ the present rcise gives us some freedom in the choice of such a measure. As mentioned ove, the measure may concentrate on a single point x (i.e., $v(\{x\})=1$). Obviously s does not happen if the set F contains the complements of all singletons.

nt. Let μ be an $\mathfrak{a}$-additive measure on $\mathbf{P}(\mathfrak{a})$. Let $F=\{B_\alpha \mid \alpha\in\mathfrak{a}\}$; define $h:\mathfrak{a}\to A$ $h(\alpha)\in\bigcap_{\beta<\alpha}B_\beta$. Obtain an $\mathfrak{a}$-complete ultrafilter U on A by setting $B\in U$ iff $\alpha<\mathfrak{a} \mid h(\alpha)\in B\})=1$. □

1Ac Exercise (Abramson, Harrington, Kleinberg and Zwicker 1977). An inite cardinal κ is measurable iff for every ordinal λ every sequence $\langle A_\alpha \mid \alpha<\lambda\rangle$ of osets of κ has a flip $\langle B_\alpha \mid \alpha<\lambda\rangle$ such that the intersection of any number less than

κ of B_α's has cardinality κ. (See 3.44 and 3.45. 4.21 together with 3.45 give another proof that every measurable cardinal is weakly compact.) $\square$

4.22 Proposition. *Let μ be a κ-additive two-valued measure on the power set $\mathbf{P}(\kappa)$ of the aleph κ. The following three conditions are equivalent.*

(a) *Whenever $\langle B_\lambda \mid \lambda<\kappa\rangle$ is a sequence of subsets of κ of measure* 1 *also their diagonal intersection $D_{\lambda<\kappa} B_\lambda$ is of measure* 1.

(b) *Every regressive function on κ is constant on a set of measure* 1.

(c) *If $\langle A_\alpha \mid \alpha<\kappa\rangle$ is a sequence such that $A_\alpha \subseteq \alpha$ for $\alpha<\kappa$ then there is a subset B of κ of measure* 1 *such that the subsequence $\langle A_\alpha \mid \alpha \in B\rangle$ of $\langle A_\alpha \mid \alpha<\kappa\rangle$ is coherent.*

Proof. (a)$\to$(c). Let $\langle A_\gamma \mid \gamma<\kappa\rangle$ be a sequence such that $A_\gamma \subseteq \gamma$ for $\gamma<\kappa$. We use now a method which may be called "interchanging the index and the variable". From the statement $\alpha \in A_\gamma$ in which α is the variable and γ is an index we go over, for $\alpha<\kappa$, to the sets $C_\alpha=\{\gamma<\kappa \mid \alpha \in A_\gamma\}$, and then $\alpha \in A_\gamma$ is equivalent to $\gamma \in C_\alpha$, where γ is now the variable and α is the index. Let B_α be that set among C_α and $\kappa \sim C_\alpha$ which has measure 1. By (a) $D=D_{\delta<\kappa} B_\delta$ has measure 1. We shall prove that $\langle A_\gamma \mid \gamma \in D\rangle$ is coherent; for this purpose all we have to show is that if $\beta, \gamma \in D$ and $\beta, \gamma>\alpha$ then if $\alpha \in A_\beta$ then also $\alpha \in A_\gamma$. If $\alpha \in A_\beta$ then, by the definition of C_α, $\beta \in C_\alpha$, and since $\beta \in D_{\delta<\kappa} B_\delta$ we have $B_\alpha=C_\alpha$ (since otherwise $B_\alpha=\kappa \sim C_\alpha$ and then $\beta \in D_{\delta<\kappa} B_\delta$ and $\alpha<\beta$ would imply $\beta \in B_\alpha=\kappa \sim C_\alpha$, contradicting $\beta \in C_\alpha$). Since also $\gamma \in D_{\delta<\kappa} B_\delta$ and $\alpha<\gamma$ we have $\gamma \in B_\alpha=C_\alpha$; thus, by the definition of C_α, $\alpha \in A_\gamma$.

(c)$\to$(b). Let f be a regressive function on κ. Set $A_0=0$ and $A_\alpha=\{f(\alpha)\}$ for $0<\alpha<\kappa$. By (c) there is a subset B of measure 1 such that the sequence $\langle A_\alpha \mid \alpha \in B\rangle$ is coherent. Let α and β be non-zero members of B and $\alpha<\beta$, then since $f(\alpha) \in A_\alpha$ and $f(\alpha)<\alpha<\beta$ also $f(\alpha) \in A_\beta=\{f(\beta)\}$ and hence $f(\alpha)=f(\beta)$. Thus f is constant on the set $B \sim \{0\}$ which is of measure 1.

(b)$\to$(a). Let $\langle B_\lambda \mid \lambda<\kappa\rangle$ be a sequence of subsets of κ of measure 1, and assume that $\mu(D_{\lambda<\kappa} B_\lambda)=0$. Define $f: \kappa \to \kappa$ as follows. If $\alpha \notin D_{\lambda<\kappa} B_\lambda$ then for some $\lambda<\alpha$ $\alpha \notin B_\lambda$, let $f(\alpha)$ be the least such λ, and for $\alpha \in D_{\lambda<\kappa} B_\lambda$ let $f(\alpha)=0$. f is a regressive function, therefore there is a subset E of κ of measure 1 and an ordinal $\gamma<\kappa$ such that $f[E]=\{\gamma\}$. Therefore, for every $\alpha \in E \sim D_{\lambda<\kappa} B_\lambda$ we have, by the definition of f and since $f(\alpha)=\gamma$, $\alpha \notin B_\gamma$; thus $B_\gamma \cap (E \sim D_{\lambda<\kappa} B_\lambda)=0$, which is impossible since both B_γ and $E \sim D_{\lambda<\kappa} B_\lambda$ are of measure 1. $\square$

4.23 Exercise. Give an alternative proof of Proposition 4.22 by giving direct proofs of (a)$\to$(b), (b)$\to$(c) and (c)$\to$(a).

Hint. (a)$\to$(b). Let f be a regressive function on κ. If (b) does not hold for f then $\mu(D_{\lambda<\kappa}(\kappa \sim f^{-1}[\{\lambda\}]))=1$, but if $\alpha \in D_{\lambda<\kappa}(\kappa \sim f^{-1}[\{\lambda\}]) \sim \{0\}$ then $f(\alpha) \geqslant \alpha$, which is a contradiction.

(b)$\to$(c). Define $A \subseteq \kappa$ by $\beta \in A \leftrightarrow \mu(\{\alpha<\kappa \mid \beta \in A_\alpha\})=1$. Let $f(\alpha)=0$ if $A_\alpha=A \cap \alpha$, and $f(\alpha)=$ the least λ which belongs to one of the sets A, A_α but not to both, otherwise. Use (b) with respect to f to get a contradiction, unless

$$\mu(\{\alpha<\kappa \mid A_\alpha=A \cap \alpha\})=1.$$

(c)$\to$(a). Given $\langle B_\alpha \mid \alpha<\kappa\rangle$ such that $\mu(B_\alpha)=1$ for $\alpha<\kappa$, use the method of interchanging the index and the variable and define $A_\beta=\{\alpha<\beta \mid \beta\in B_\alpha\}$. If $D\subseteq\kappa$, $\mu(D)=1$, and $\{A_\beta \mid \beta\in D\}$ is coherent then $D\subseteq D_{\alpha<\kappa}B_\alpha$. $\square$

4.24 Definition. (i) For a cardinal κ, a two-valued measure on $\mathbf{P}(\kappa)$ is said to be a *normal measure* if it is a κ-additive measure which satisfies the conditions of 4.22.

(ii) Let A be an uncountable set and $<$ a well-ordering of A of order type $|A|$. A two-valued measure μ on $\mathbf{P}(A)$ is called *normal with respect to* $<$ if it is $|A|$-additive and it satisfies conditions analogous to those of 4.22. Thus an $|A|$-additive two-valued measure μ on $\mathbf{P}(A)$ is normal with respect to $<$ iff every function $f:A\to A$ which is regressive with respect to $<$ (i.e., f is such that $f(x)<x$ for every $x\in A$ other than the least member of A) is constant on a subset of A of measure 1.

4.25 Proposition. *If κ is an uncountable cardinal and μ is a normal measure on $\mathbf{P}(\kappa)$ then every closed unbounded subset C of κ is of measure* 1, *and hence every subset of κ of measure* 1 *is stationary.*

Proof. Let $f:\kappa\to\kappa$ be defined by $f(\alpha)=0$ if $\alpha\in C$ and $f(\alpha)=\sup(C\cap\alpha)$ if $\alpha\notin C$. Since C is closed $\sup(C\cap\alpha)\in C$ or $\sup(C\cap\alpha)=0$, therefore if $\alpha>0$ and $\alpha\notin C$ we have $f(\alpha)=\sup(C\cap\alpha)<\alpha$. Thus f is a regressive function and hence f has a constant value γ on a set E of measure 1. We shall now obtain a contradiction from the assumption that $\mu(C)=0$. If $\mu(C)=0$ then since $\mu(E)=1$ also $\mu(E\sim C)=1$. Since C is unbounded it contains a member $\delta>\gamma$, and since $\mu(E\sim C)=1$ $E\sim C$ contains a member $\alpha>\delta$. Since $\alpha\notin C$ $f(\alpha)=\sup(\alpha\cap C)\geqslant\sup\{\delta\}=\delta>\gamma$, which is a contradiction since $f(\alpha)=\gamma$ for $\alpha\in E$. $\square$

4.26Ac Exercise. For a measurable cardinal κ let μ be a two-valued normal measure on $\mathbf{P}(\kappa)$, then for every stationary subset S of κ $S\cap\alpha$ is a stationary subset of α for almost every $\alpha<\kappa$.

Hint. Assume otherwise, then let $A_\alpha=$a closed unbounded subset of α disjoint from S, if there is such, and $A_\alpha=0$ otherwise. Let $B\subseteq\kappa$ be such that $\mu(B)=1$ and $\langle A_\alpha \mid \alpha\in B\rangle$ is coherent. $\bigcup_{\alpha\in B}A_\alpha$ is a closed unbounded subset of κ disjoint from S. $\square$

4.27Ac Theorem. *For every measurable cardinal κ there is a normal two-valued measure on* $\mathbf{P}(\kappa)$.

Proof. Let κ be a measurable cardinal and let μ be a κ-additive measure on a set K of cardinality κ. We say that a function $f:K\to\kappa$ is *incompressible* if f is not constant on a set of measure 1 but every function $g:K\to\kappa$ such that $g(\alpha)<f(\alpha)$ for almost all α is constant on a set of measure 1. (By 4.24, using 4.22(b), a two-valued measure μ on $\mathbf{P}(\kappa)$ is normal iff the identity function $\langle\lambda \mid \lambda<\kappa\rangle$ on κ is incompressible.) First let us prove the existence of an incompressible function f. We define on ${}^K\kappa$ a relation $<^*$ by setting $f<^*g$ if $\mu(\{x\in K \mid f(x)<g(x)\})=1$. Let

us show now that every non-void subset of ${}^{\kappa}\kappa$ has a $<^*$-minimal member. By II.5.3(ii) it suffices to prove that there is no sequence $\langle g_i \mid i<\omega\rangle$ of members of ${}^{K}\kappa$ such that $g_{i+1}<^* g_i$ for all $i<\omega$. Suppose $\langle g_i \mid i<\omega\rangle$ were such a sequence and $B_i=\{x\in K \mid g_{i+1}(x)<g_i(x)\}$ then $\mu(B_i)=1$ for $i<\omega$. Since $\kappa>\aleph_0$ and μ is κ-additive we have $\mu(\bigcap_{i\in\omega}B_i)=1$. If $\alpha\in\bigcap_{i\in\omega}B_i$ we have $g_{i+1}(\alpha)<g_i(\alpha)$ for all $i<\omega$, contradicting the fact that On is well ordered. Let $A\subseteq{}^{K}\kappa$ be the set of all functions which are constant on no set of measure 1. $A\neq 0$ since every injection of K into κ is in A. By what we have just shown A has a minimal member with respect to $<^*$. Each such minimal member of A is an incompressible function.

Given an incompressible function $f: K\to\kappa$ we shall now define a two-valued measure ν on $\mathbf{P}(\kappa)$ by identifying the ordinal $\lambda<\kappa$ with the subset $f^{-1}[\{\lambda\}]$ of K. Thus, for a subset A of κ we define $\nu(A)=\mu(f^{-1}[A])$. We have $\nu(\kappa)=\mu(f^{-1}[\kappa])=\mu(K)=1$; if $|I|<\kappa$ and $\langle A_i \mid i\in I\rangle$ is a family of pairwise disjoint subsets of κ then $\nu(\bigcup_{i\in I}A_i)=\mu(f^{-1}[\bigcup_{i\in I}A_i])=\mu(\bigcup_{i\in I}f^{-1}[A_i])=\Sigma_{i\in I}\mu(f^{-1}[A_i])=\Sigma_{i\in I}\nu(A_i)$, since if $A_i\cap A_j=0$ also $f^{-1}[A_i]\cap f^{-1}[A_j]=0$; and for $\lambda<\kappa$ $\nu(\{\lambda\})=\mu(f^{-1}[\{\lambda\}])=0$ since f is not constant on a set of measure 1. Therefore ν is a two-valued measure on κ; we shall now prove that ν is normal.

Let g be a regressive function on κ. We shall see that since f is incompressible and gf is a further compression of f g must be constant on a subset of κ of measure 1. Since g is regressive we have for every $x\in K$ $f(x)=0$ or $(gf)(x)=g(f(x))<f(x)$. Since every value, including 0, is obtained by f only on a set of measure 0 we have $(gf)(x)<f(x)$ for almost all x, and since f is incompressible gf has a constant value λ on a set $D\subseteq K$ of measure 1. Let $E=g^{-1}[\{\lambda\}]$, then $\nu(E)=\mu(f^{-1}[E])=\mu(f^{-1}[g^{-1}[\{\lambda\}]])=\mu((gf)^{-1}[\{\lambda\}])\geqslant\mu(D)=1$, since $D\supseteq(gf)^{-1}[\{\lambda\}]$. Thus g has the constant value λ on the set E for which $\nu(E)=1$, which establishes the normality of ν. □

4.28Ac Corollary. *Every measurable cardinal is ineffable.*

Proof. To prove that a measurable cardinal κ is ineffable we have to show that if a sequence $\langle A_\alpha \mid \alpha<\kappa\rangle$ is such that $A_\alpha\subseteq\alpha$ for $\alpha<\kappa$ then there is a stationary subset B of κ such that the subsequence $\langle A_\alpha \mid \alpha\in B\rangle$ is coherent. By 4.27 there is a two-valued normal measure μ on $\mathbf{P}(\kappa)$. By the definition 4.24 of a normal measure there is a subset B of κ such that $\mu(B)=1$ and the subsequence $\langle A_\alpha \mid \alpha\in B\rangle$ is coherent. B is a stationary subset of κ by 4.25. □

4.29Ac Theorem. *If κ is a measurable cardinal then the set of all ineffable cardinals below κ is of measure* 1 *with respect to each normal measure μ on* $\mathbf{P}(\kappa)$, *and therefore this set is a stationary subset of κ.*

Proof. Let μ be a two-valued normal measure on $\mathbf{P}(\kappa)$ and let I be the set of all ineffable cardinals below κ. Assume that $\mu(I)=0$ to prove the theorem by contradiction. If $\beta\in\kappa\sim I$ then there is a sequence $A^\beta=\langle A^\beta_\alpha \mid \alpha<\beta\rangle$ such that $A^\beta_\alpha\subseteq\alpha$ for $\alpha<\beta$ which shows that β is not ineffable. $\bigcup_{Q\in\mathbf{P}(\alpha)}\{\beta<\kappa \mid A^\beta_\alpha=Q\}=\kappa\sim(\alpha+1)\sim I$, and since for $\alpha<\kappa$ $\mathbf{P}(\alpha)$ has $<\kappa$ members $\mu(\{\beta<\kappa \mid A^\beta_\alpha=Q\})=1$ for exactly one $Q\in\mathbf{P}(\alpha)$; we denote this Q with A_α and get $A^\beta_\alpha=A_\alpha$ for almost all β. Since μ is a normal measure there is a subset E of κ such that $\mu(E)=1$ and the sequence $\langle A_\alpha \mid \alpha \in E\rangle$ is coherent. Let $A=\bigcup_{\alpha\in E} A_\alpha$, then $A_\alpha = A\cap\alpha$ for $\alpha\in E$. Since for $\beta\in\kappa\sim I$ $\langle A_\alpha^\beta \mid \alpha<\beta\rangle$ is a witness that β is not ineffable the set $W_\beta=\{\alpha<\beta \mid A_\alpha^\beta = A\cap\alpha\}$ is a non-stationary subset of β (since $\langle A_\alpha^\beta \mid \alpha\in W_\beta\rangle$ is a coherent sequence). Therefore there is a closed and unbounded subset C_β of β such that for $\alpha\in C_\beta$ $A\cap\alpha\neq A_\alpha^\beta$. Since μ is a normal measure there is a subset D of κ such that $\mu(D)=1$ and $\langle C_\beta \mid \beta\in D\rangle$ is coherent; let $C=\bigcup_{\beta\in D} C_\beta$, then $C_\beta = C\cap\beta$ for all $\beta\in D$. We shall now prove that C is a closed unbounded subset of κ. To see that C is closed let $Y\subseteq C$ be such that $\sup Y<\kappa$. Since $\mu(D)=1$ D is unbounded in κ, so let $\beta\in D$ be such that $\beta>\sup Y$. Since $\beta\in D$ $C\cap\beta=C_\beta$, and since $\sup Y<\beta$ we get $Y\subseteq C\cap\beta\subseteq C_\beta$. C_β is a closed subset of β and $\sup Y<\beta$ therefore $\sup Y\in C_\beta = C\cap\beta\subseteq C$, which shows that C is closed. To see that C is unbounded let $\lambda<\kappa$ and let $\beta\in D\sim(\lambda+1)$, then $\beta>\lambda$ and $C_\beta=C\cap\beta$. C_β is unbounded in β and $\lambda<\beta$, hence there is an $\alpha\geqslant\lambda$ such that $\alpha\in C_\beta = C\cap\beta\subseteq C$ and thus C is unbounded in κ. By 4.25 $\mu(C)=1$. Since also $\mu(E)=1$ we get $\mu(C\cap E)=1$, hence $C\cap E\neq 0$. Let $\alpha\in C\cap E$. Since $\alpha\in C$ we have for almost all β's $\alpha\in C_\beta$ and, by the definition of the C_β's, for almost all β's $A\cap\alpha\neq A_\alpha^\beta$. Since $\alpha\in E$ we have $A\cap\alpha=A_\alpha$, thus for almost all β's $A_\alpha\neq A_\alpha^\beta$, which contradicts the definition of A_α. □

4.30Ac Theorem. *Let κ be a measurable cardinal and let μ be a normal two-valued measure on* $\mathbf{P}(\kappa)$.

(i) *If $\lambda<\kappa$ and $F:[\kappa]^n\to\lambda$ then there is an $A\subseteq\kappa$ of measure* 1 *which is homogeneous with respect to F, i.e.,* $|F[[A]^n]|=1$.

(ii) *For $\lambda<\kappa$ and $F:[\kappa]^{<\omega}\to\lambda$ there is an $A\subseteq\kappa$ of measure* 1 *such that* $|F[[A]^n]|=1$ *for all* $n<\omega$.

Remark. We cannot demand $|F[[A]^{<\omega}]|=1$, since if we define $F:[\kappa]^{<\omega}\to\omega$ by $F(a)=|a|$ then $F[[A]^{<\omega}]=\omega$ for all infinite A.

Proof. (i) By induction on n. For $n=1$ this is just the κ-additivity of μ. We assume now that (i) holds for n and prove it for $n+1$. We are given $F:[\kappa]^{n+1}\to\lambda$. For every $\gamma<\kappa$ let $F_\gamma:[\kappa]^n\to\lambda$ be defined by $F_\gamma(a)=F(a\cup\{\gamma\})$ if $\gamma\notin a$, and $F_\gamma(a)=0$ if $\gamma\in a$. By the induction hypothesis there is a subset A_γ of κ such that $\mu(A_\gamma)=1$ and $|F_\gamma[[A_\gamma]^n]|=1$; let us denote $F_\gamma[[A_\gamma]^n]$ by $\{h(\gamma)\}$, then $h(\gamma)<\lambda$ for $\gamma<\kappa$. Since $\kappa=\bigcup_{\sigma<\lambda} h^{-1}[\{\sigma\}]$ and $\lambda<\kappa$ we have $\mu(h^{-1}[\{\sigma\}])=1$ for some $\sigma<\lambda$; let $E=h^{-1}[\{\sigma\}]$, then $\mu(E)=1$ and $h[E]=\{\sigma\}$. Let $A=E\cap D_{\gamma<\kappa} A_\gamma$; since μ is normal $\mu(A)=1$; we shall prove that $F[[A]^{n+1}]=\{\sigma\}$, which terminates the induction step.

Let $\{\xi_1,\ldots,\xi_{n+1}\}\in[A]^{n+1}$, $\xi_1<\xi_2<\cdots<\xi_{n+1}$, then by the definition of the diagonal intersection, $\{\xi_2,\ldots,\xi_{n+1}\}\subseteq A_{\xi_1}$, and hence $h(\xi_1)=F_{\xi_1}(\{\xi_2,\ldots,\xi_{n+1}\})=F(\{\xi_1,\ldots,\xi_{n+1}\})$. But since $\xi_1\in A\subseteq E$ $h(\xi_1)=\sigma$, thus $F(\{\xi_1,\ldots,\xi_{n+1}\})=h(\xi_1)=\sigma$.

(ii) Given F as in (ii) let $F_n=F\restriction[\kappa]^n$, then by (i) there is a set $A_n\subseteq\kappa$ such that $\mu(A_n)=1$ and $|F_n[[A_n]^n]|=1$. Let $A=\bigcap_{n\in\omega} A_n$. By the κ-additivity of μ $\mu(A)=1$, and since $F[[A]^n]=F_n[[A]^n]\subseteq F_n[[A_n]^n]$ we get $|F_n[[A]^n]|=1$. □

4.31Ac Ramsey Cardinals and the Axiom of Constructibility. Since on every measurable cardinal there is a normal measure it is an immediate consequence of 4.20 that every measurable cardinal satisfies the partition relation $\kappa \to (\kappa)_\lambda^{<\omega}$ for every $\lambda < \kappa$ (i.e., for every $F:[\kappa]^{<\omega} \to \lambda$ there is a subset A of κ of cardinality κ which is homogeneous with respect to each $F \upharpoonright [\kappa]^n$). From the existence of κ's satisfying this partition relation, called *Ramsey cardinals*, and even from the existence of cardinals κ satisfying $\kappa \to (\aleph_1)_2^{<\omega}$, one can prove that there are non-constructible sets (Scott 1961, Rowbottom 1971) and many other strong consequences along this line. Thus, if one assumes the axiom of constructibility then there are no Ramsey cardinals and, a fortiori, no measurable cardinals. The axiom of constructibility also implies that there are no real-valued-measurable cardinals since the continuum hypothesis $2^{\aleph_0} = \aleph_1$ implies that every real-valued-measurable cardinal is measurable (by 4.16 and 4.14). The assumption that measurable cardinals exist offers a significantly richer and more interesting universe than that offered by the axiom of constructibility which contradicts it. Therefore, many mathematicians adopt the former assumption, rather than the latter, as part of their mental image of set theory.

4.32 The Dependence of the Normality of a Measure on the Order. In 4.24 we defined, for an arbitrary set A, the concept of the normality of a two-valued measure on $\mathbf{P}(A)$ with respect to a given well-ordering of A of order type $|A|$. A two-valued measure which is normal with respect to one well-ordering of a set is not necessarily normal with respect to another well-ordering of the same set. For example, if μ is any two-valued normal measure on $\mathbf{P}(\kappa)$ and D is the set of all limit ordinals below κ then $\mu(D) = 1$, by 4.25; and if we take any other well-ordering of κ in which all the limit ordinals of κ become successors (such as $\{\langle \omega \cdot \alpha + n,\ \omega \cdot \beta + m \rangle \mid \alpha < \beta < \kappa \vee \alpha = \beta < \kappa \wedge (n < m \wedge m > 1 \vee n = 1 \wedge m = 0)\}$) then μ is not normal with respect to that well-ordering since $\mu(D) = 1$ while D is not stationary in that well ordering.

4.33 Definition. Let K be a set and μ a two-valued measure on $\mathbf{P}(K)$. μ is called *prenormal* if there is an ordering $<$ of K of order type $|K|$ such that μ is normal with respect to $<$.

4.34Ac Theorem. *Let μ be a two-valued $|K|$-additive measure on the power-set $\mathbf{P}(K)$ of a set K. μ is prenormal iff*

(7) *for every $G:[K]^2 \to 2$ there is a homogeneous subset A of K of measure* 1.

Proof. If μ is prenormal order K by a relation $<$ of order type $|K|$ such that μ is normal with respect to $<$. By 4.30 (7) holds.

Now assume (7) and denote $|K|$ with κ. In the proof of 4.27 we essentially made an arbitrary two-valued κ-additive measure μ normal on $\mathbf{P}(\kappa)$ by taking an incompressible function f, identifying all the members of $f^{-1}[\{\alpha\}]$ with each other to get a single object a_α (for α's such that $f^{-1}[\{\alpha\}] \neq 0$), and ordering the a_α's by the order of the α's. Here we want to order K without identifying different members, therefore the proof of the prenormality of μ is just a repetition of the relevant part of the proof of 4.27, once we have an incompressible function $f: K \to \kappa$ which is a bijection. Let us start with any incompressible function $g: K \to \kappa$. Let h be the coloring of $[K]^2$ defined by $h(\{x, y\})=0$ if $g(x)=g(y)$ and $h(\{x, y\})=1$ otherwise. By (7) there is a subset A of K of measure 1 which is homogeneous with respect to h. We cannot have $h[[A]^2]=\{0\}$ since this would mean that g has a constant value on the set A, contradicting our choice of g as an incompressible function; therefore $h[[A]^2]=\{1\}$, which means that on A g is a one–one function. Let B be a subset of A of measure 0 such that $|B|=\kappa$ (every splitting of A into two complementary sets of cardinality κ yields such a set B). By $B \subseteq \kappa \sim (A \sim B)$ and $|B|=\kappa$ we have $|K \sim (A \sim B)|=\kappa$, and since g is one–one on A and $A \supseteq B$ we have $\kappa \supseteq \kappa \sim g[A \sim B] \supseteq g[B]$ and $|\kappa \sim g[A \sim B]|=\kappa$. Thus $K \sim (A \sim B) \approx \kappa \sim g[A \sim B]$. Let j be any bijection of $K \sim (A \sim B)$ on $\kappa \sim g[A \sim B]$, then define $f = g \upharpoonright (A \sim B) \cup j$. f is obviously a bijection of K on κ (by I.6.30). Since f agrees with g on the set $A \sim B$, which is of measure 1, f is also an incompressible function, which is what we need to complete the proof. □

4.35Ac Exercise. Let K be a set of cardinality κ and let μ be a two-valued κ-additive measure on $\mathbf{P}(K)$. The following conditions on μ are equivalent.

(a) μ is prenormal.

(b) For every ordering $<$ of K of order type κ and every function $f: K \to \kappa$ which is not constant on a set of measure 1 there is a subset A of K of measure 1 such that f is strictly increasing or strictly decreasing on A.

(c) Every function $f: K \to \kappa$ is either constant on a set of measure 1 or else one–one on a set of measure 1.

(d) For every family $\langle A_\alpha \mid \alpha \in \kappa \rangle$ of pairwise disjoint non-void subsets of K of measure 0 there is a choice set $B \subseteq K$ (i.e., $|B \cap A_\alpha|=1$ for every $\alpha \in \kappa$) of measure 1.

(e) There is a one–one incompressible function f on some subset S of K of measure 1.

Hint. Prove (a) → (b) → (c) → (e) → (a) by following closely the proof of 4.34. (d) is immediately equivalent to (c). □

4.36Ac Exercises. Let μ be a two-valued prenormal measure on $\mathbf{P}(\kappa)$, where κ is a measurable cardinal.

(i) μ is normal iff every closed unbounded subset C of κ is of measure 1.

(ii) μ is normal iff for every stationary subset S of κ $S \cap \alpha$ is a stationary subset of α for almost all $\alpha < \kappa$.

Hints. (i) If μ is not normal then there is a regressive function g on κ which is not constant on any set of measure 1. By 4.35 there is an $A \subseteq \kappa$ of measure 1 such that g is one–one on A. Since every closed unbounded set has measure 1 A must be stationary, and by Fodor's theorem IV.4.40 g cannot be one–one on A.

(ii) If C is a closed unbounded subset of κ then $\{\alpha \mid C \cap \alpha$ is a stationary subset of $\alpha\} \subseteq C$, hence $\mu(C)=1$. Therefore by (i) μ is normal. □

4.37 Proposition. *For every measurable cardinal κ there is a two-valued κ-additive measure ν on $\mathbf{P}(\kappa)$ which is not prenormal.*

Outline of proof. Let μ be a κ-additive two-valued measure on $\mathbf{P}(\kappa)$. We construct a measure ν as claimed by the proposition on the set $K=[\kappa]^2$, which is equinumerous to κ. For $A \subseteq K$ let

$$\nu(A)=1 \quad \text{iff} \quad \mu(\{\alpha \mid \mu(\{\beta \mid \{\alpha, \beta\} \in A\})=1\})=1,$$

i.e., $\nu(A)=1$ iff for almost every α it is the case that for almost every β $\{\alpha, \beta\} \in A$. Define $f:[K]^2 \to 2$ as follows. For $\alpha<\beta<\kappa$ and $\gamma<\delta<\kappa$ let $f(\{\{\alpha, \beta\}, \{\gamma, \delta\}\})=0$ if $\beta<\gamma$ or $\delta<\alpha$, and $f(\{\{\alpha, \beta\}, \{\gamma, \delta\}\})=1$ if $\beta \geqslant \gamma$ and $\delta \geqslant \alpha$. As easily seen, there is no homogeneous subset A of K with $\nu(A)=1$. $\square$

Appendix X
The Eliminability and Conservation Theorems

1.1 On the Syntax of the Language. To handle theorems about provability we have to give a rigorous description of the underlying logic. First we shall discuss a few syntactical concepts.

We recall that the only primitive quantifier in our language is $\forall$, while $\exists x$ is defined as $\neg \forall x \neg$. By the *scope* of a particular occurrence of a quantifier $\forall x$ in a formula Φ we mean that part of the formula to which this quantification applies, i.e., the part of Φ which is inside the pair of parentheses whose left parenthesis is immediately to the right of this $\forall x$. For example, the scope of $\forall y$ in $\exists x(\forall y\,(y \notin x) \vee \exists t(t \in x))$ is $y \notin x$.

In class terms $\{x \mid \Phi(x)\}$ $\{x|$ also acts as a quantifier in that it causes all the free occurrences of x in $\Phi(x)$ to become bound. Therefore whenever we mention quantifiers in the appendix we also mean $\{x \mid$. The scope of this "quantifier" $\{x \mid$ in $\{x \mid \Phi(x)\}$ is $\Phi(x)$.

Two formulas Φ and Ψ are said to be *alphabetic variants* of each other if each one of them is obtained from the other by replacing bound variables by other variables in a way which does not violate the usual rules for such replacements. We say that a variable y is *substitutable* for a set or class variable x in a formula Φ if no free occurrence of x occurs in Φ within the scope of a quantifier $\forall y$ or $\{y \mid$ on y. (Every occurrence of a class variable is considered to be free.) We say that a class term τ is substitutable for x in Φ if no free occurrence of x occurs in Φ within the scope of a quantifier on a variable free in τ.

1.2 The System P*. We shall now present the logical and basic extralogical axioms and rules of inference of the extended language of set theory. The deductive system which consists of these axioms and rules of inference is denoted with P*.

The sentential calculus. (a) Every tautology is an axiom.

(b) The rule of *Modus Ponens*: from Φ and $\Phi \rightarrow \Psi$ infer Ψ.

First-order predicate calculus with equality for sets. (c) Particularization: If y is substitutable for x in $\Phi(x)$ then $\forall x\,\Phi(x) \rightarrow \Phi(y)$ is an axiom.

(d) The rule of generalization: If x is not free in Ψ then from $\Psi \rightarrow \Phi(x)$ infer $\Psi \rightarrow \forall x \Phi(x)$.

(e) The axioms of equality: $x=x$, $x=y\to y=x$, $x=y\wedge y=z\to x=z$, $x=y\wedge x\in z\to y\in z$, $x=y\wedge z\in x\to z\in y$.

Free variable calculus for classes. (f) The rule of substitution: If τ is a class variable, or a class term which is substitutable for A in $\Phi(A)$, then from $\Phi(A)$ infer $\Phi(\tau)$.

The axioms of equality for classes, which should have been listed here, follow from the other axioms and rules of inference.

The basic extralogical axioms and rule of inference (I.4.3 and I.4.2). (g) The axiom of class comprehension: $y\in\{x\mid\Phi(x)\}\leftrightarrow\Phi'(y)$, where $\Phi'(x)$ is any alphabetic variant of $\Phi(x)$ in which y is substitutable for x.

(h) The axiom of extensionality: $A=B\leftrightarrow\forall x(x\in A\leftrightarrow x\in B)$.

(i) The axiom of membership: $A\in B\leftrightarrow\exists x(x=A\wedge x\in B)$.

(j) The rule of substitution of sets for classes: If x is substitutable for A in $\Phi(A)$ then from $\Phi(A)$ infer $\Phi(x)$.

1.3 (I.4.5) The Eliminability Theorem. *Let $\Phi(A_1,\ldots,A_n)$ be a formula of the extended language of set theory with no class variables other than indicated. Let $\{x_1\mid\Psi_1(x_1)\},\cdots,\{x_n\mid\Psi_n(x_n)\}$ be class terms which do not contain class variables. Then there is a formula ϕ of the basic language such that*

$$P^*\vdash\Phi(\{x_1\mid\Psi_1(x_1)\},\cdots,\{x_n\mid\Psi_n(x_n)\})\leftrightarrow\phi.$$

Proof. Let Λ be a formula of the extended language without class variables. By the *weight* of Λ we mean the pair $\langle b(\Lambda), l(\Lambda)\rangle$ where $l(\Lambda)$ is the *length* of Λ, i.e., the number of occurrences of symbols in Λ, and $b(\Lambda)$ is the *term-count* of Λ which is the number of occurrences of class terms in Λ, where each occurrence of a term $\{x\mid\Psi(x)\}$ is counted twice if it is of the form $\{x\mid\Psi(x)\}=\tau$ or $\tau=\{x\mid\Psi(x)\}$, where τ is any variable or class term, and is counted three times if it is of the form $\{x\mid\Psi(x)\}\in\tau$, where τ is as above. For example, the term-count of $y\in\{x|\{y\mid y\neq y\}\in x\}$ is 4 since $\{y\mid y\neq y\}$ contributes 3 and $\{x|\{y\mid y\neq y\}\in x\}$ contributes 1. We order the weights in the left lexicographic ordering.

To prove the theorem let us denote $\Phi(\{x_1\mid\Psi_1(x_1)\},\cdots,\{x_n\mid\Psi_n(x_n)\})$ with Γ. We shall prove the existence of ϕ such that $P^*\vdash\Gamma\leftrightarrow\phi$ by induction on the weight of Γ. If the weight of Γ is $\langle 0, l\rangle$ then Γ contains no class terms and hence Γ is in the basic language. In this case we take ϕ to be Γ itself and we obviously have $P^*\vdash\Gamma\leftrightarrow\phi$. If Γ is an atomic formula then Γ is of one of the forms in the left-hand side of Table 1. By axioms (g), (h) and (i) and rule (j) of 1.2 each formula in the left-hand side of Table 1 is equivalent in P^* to the corresponding formula in the right-hand side of Table 1, which is of lower weight; by the induction hypothesis the formula on the right-hand side is equivalent in P^* to some formula ϕ of the basic language, hence also the formula in the left-hand side is equivalent in P^* to the same formula ϕ.

Every non-atomic formula Γ is built up from shorter atomic formulas $\Gamma_1,\ldots,\Gamma_k$ by an operation $\Delta(\Gamma_1,\ldots,\Gamma_k)$ which applies to $\Gamma_1,\ldots,\Gamma_k$ sentential connectives and quantifiers of the form $\forall x$. By the induction hypothesis there are formulas $\phi_1, \ldots, \phi_k$ of the basic language such that $P^* \vdash \Gamma_i \leftrightarrow \phi_i$ for $i=1, \ldots, k$, hence $P^* \vdash \Delta(\Gamma_1, \ldots, \Gamma_k) \leftrightarrow \Delta(\phi_1, \ldots, \phi_k)$. Since $\Delta(\Gamma_1, \ldots, \Gamma_k)$ is Γ and $\Delta(\phi_1, \ldots, \phi_k)$ is a formula of the basic language our proof is done. □

Table 1

Atomic formula	*Equivalent formula of lower weight*
$y \in \{x \mid \Lambda(x)\}$	$\Lambda'(y)$ where $\Lambda'(x)$ is an alphabetic variant of $\Lambda(x)$
$\{y \mid \Lambda(y)\} \in \{z \mid \Theta(z)\}$	$\exists x(x=\{y \mid \Lambda(y)\} \wedge x \in \{z \mid \Theta(z)\})$
$\{y \mid \Lambda(y)\} \in z$	$\exists x(x=\{y \mid \Lambda(y)\} \wedge x \in z)$
$\{y \mid \Lambda(y)\} = \{z \mid \Theta(z)\}$	$\forall x(x \in \{y \mid \Lambda(y)\} \leftrightarrow x \in \{z \mid \Theta(z)\})$
$z=\{y \mid \Lambda(y)\}$ or $\{y \mid \Lambda(y)\}=z$	$\forall x(x \in z \leftrightarrow x \in \{y \mid \Lambda(y)\})$

1.4 (I.4.6(i)). The Conservation Theorem—*Part* (i). *For every formula ϕ of the basic language $P^* \vdash \phi$ iff $P \vdash \phi$, where P is the theory in the basic language whose only axiom is the axiom of extensionality* (I.2.1).

Proof. All the logical axioms and rules of inference of P are also axioms and rules of P^*. The only extralogical axiom of P is the axiom of extensionality, which follows from axiom (h), $A=B \leftrightarrow \forall x(x \in A \leftrightarrow x \in B)$, of P^* by substitution of y and z for A and B (by rule (j) of P^*). Therefore, if $P \vdash \phi$ then every proof of ϕ in P is also a proof of ϕ from the axiom of extensionality in P^*, and since the axiom of extensionality is a theorem of P^* we have $P^* \vdash \phi$.

To prove the other direction of the theorem let us represent every proof of a formula Γ in P^* as a finite tree T of formulas as in Fig. 14. T has a single bottom formula, which is the formula Γ which T proves. Each formula Ψ in the tree has at most two formulas, called the *predecessors* of Ψ, immediately above it. The same formula may occur in several different places in the tree, but we shall refer to the

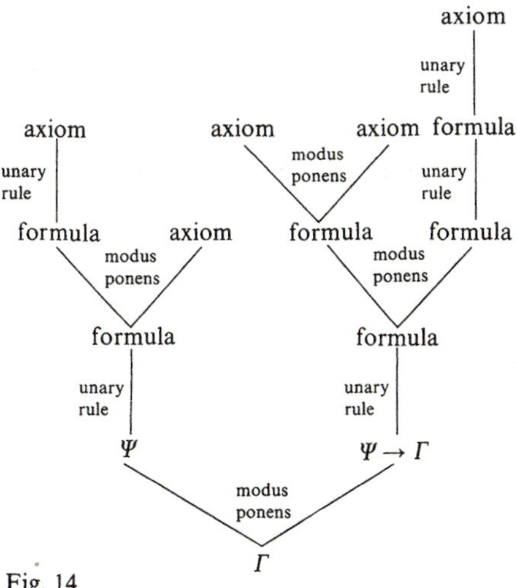

Fig. 14

different occurrences of the formula as different formulas. A formula with no predecessors must be an axiom. A formula Ψ which has predecessors must follow from them by one of the rules of inference, i.e., by one of the unary rules (d), (f), (j) if Ψ has a single predecessor, and by the binary rule (b) of modus ponens if Ψ has two predecessors. We shall say that a formula Φ is *above* a formula Ψ in the tree if Φ can be reached from Ψ by going up a branch. The part of the tree above a formula Ψ, including Ψ itself, is obviously a proof of Ψ; it will be called *the proof of Ψ in the tree.*

A proof in P^* of a formula Γ will be called *normal* if it is the case that whenever we go down along a branch from an axiom downwards we first encounter only applications of the rules (f) and (j) of substitution and then only applications of the rules (b) of modus ponens and (d) of generalization. We shall show that if $P^* \vdash \Gamma$ then Γ has a normal proof.

An application of one of the rules of substitution such that in the part of the tree above it there is an application of one of the rules (b) and (d) is called a *special application.* We shall prove by induction on l that if Γ has a proof with l special applications then Γ has also a normal proof. If $l=0$ the proof is already normal. Now we shall deal with the case where $l=1$.

Let the number of applications of the rules (b) and (d) above the special application be k. Since the application is special $k>0$. We shall prove by induction on k that ϕ has a normal proof. Let us look at the rule which is applied immediately above the special application. If this rule is modus ponens then the part of the proof consisting of these two applications is as in Fig. 15, where τ is a class term or a set variable which is substitutable for A in $\Phi(A)$. If τ is also substitutable for A in $\Psi(A)$ then we replace this part by the part in Fig. 16. In the proof of $\Psi(\tau) \to \Phi(\tau)$ in the new tree there are k' applications of the rules (b) and (d) above the application of substitution with $k'<k$ (as the application of modus ponens is now below the application of substitution). If $k'=0$ then the proof of $\Psi(\tau) \to \Phi(\tau)$ in the new tree is normal; otherwise $\Psi(\tau) \to \Phi(\tau)$ has a normal proof by the induction hypothesis. Similarly, also $\Psi(\tau)$ has a normal proof. Replacing, if necessary, the proofs of $\Psi(\tau) \to \Phi(\tau)$ and $\Psi(\tau)$ in the tree by normal proofs we obtain a normal proof of ϕ since the only special application in the original proof is eliminated. If τ is not substitutable in $\Phi(A)$ it is because there are free variables $x_1, \ldots, x_m$ of τ (or τ itself if τ is a set variable) such that A occurs in the scopes of quantifiers on x_i, for $i=1, \ldots, m$, in Ψ. Let $y_1, \ldots, y_m$ be different variables which do not occur in

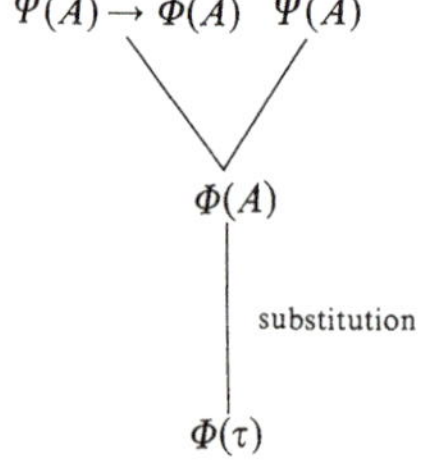

Fig. 15

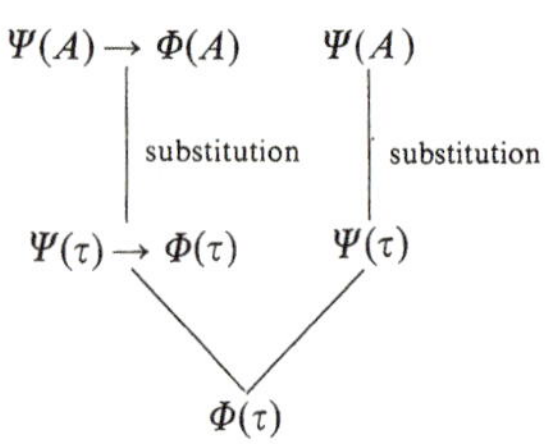

Fig. 16

$\Phi(A)$, $\Psi(A)$ or τ. Writing τ as $\tau(x_1, \ldots, x_m)$ it is obvious that $\tau(y_1, \ldots, y_m)$ is substitutable in both $\Psi(A)$ and $\Phi(A)$, and for every $1 \leqslant i \leqslant m$ x_i is substitutable for y_i in $\Phi(\tau(x_1, \ldots, x_{i-1}, y_i, \ldots, y_m))$. The part of the proof in Fig. 15 can now be replaced by the part in Fig. 17.

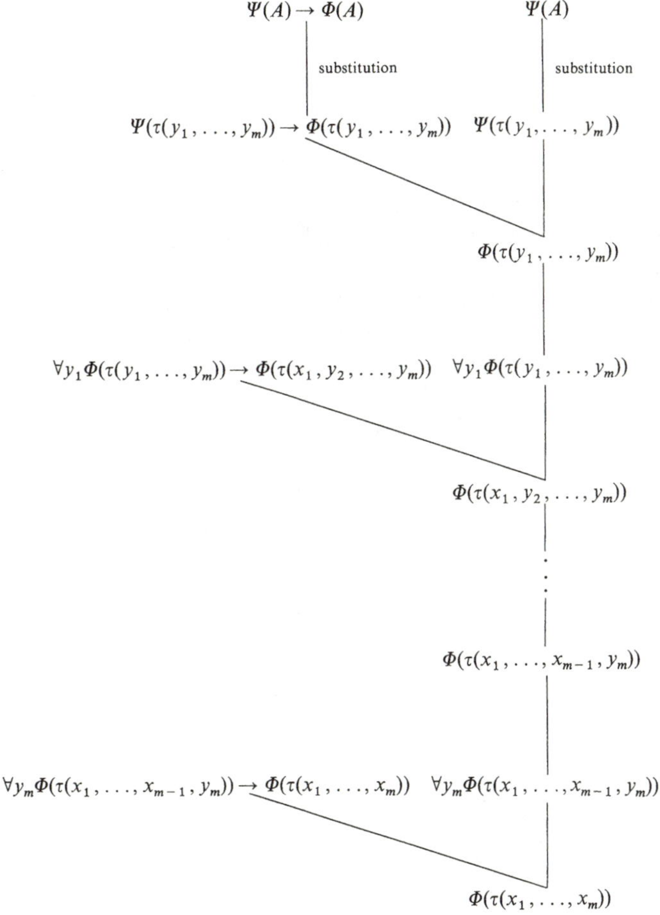

Fig. 17

Applying the induction hypothesis on k to the proofs of $\Psi(\tau(y_1, \ldots, y_m)) \rightarrow \Phi(\tau(y_1, \ldots, y_m))$ and $\Psi(\tau(y_1, \ldots, y_m))$ in the new tree we get, again, a normal proof of Γ.

If the rule applied immediately above the special application is the rule of generalization the relevant part of the proof is as in Fig. 18, where τ is substitutable for A in $\Psi(A) \rightarrow \forall x \Phi(A, x)$ and x is not free in $\Psi(A)$. If x is not free in τ we can replace the part of the proof in Fig. 18 by the part in Fig. 19 in which the application of substitution has $<k$ applications of rules (b) and (d) above it in the new tree, and thus the proof of $\Psi(\tau) \rightarrow \Phi(\tau, x)$ in the new tree is either already normal or else can be replaced by a normal proof, by the induction hypothesis on k. If x is free in τ

Fig. 18

Fig. 19

we shall write $\tau(x)$ for τ, and since $\tau(x)$ is supposed to be substitutable for A in $\Psi(A) \to \forall x \Phi(A, x)$ A does not occur at all in $\Phi(A, x)$ and we can write $\Phi(x)$ for $\Phi(A, x)$. Let y be a variable which does not occur in $\Psi(A)$, $\Phi(x)$ or $\tau(x)$. The part of the proof in Fig. 18 can be replaced by the part described in Fig. 20. As above, in the new tree the part of the proof above the application of substitution contains $<k$ applications of rules (b) and (d) therefore the proof of $\Psi(\tau(y)) \to \Phi(x)$ in the new tree is either already normal or else can be replaced by a normal proof, by the induction hypothesis. Thus Γ has a normal proof when $l=1$.

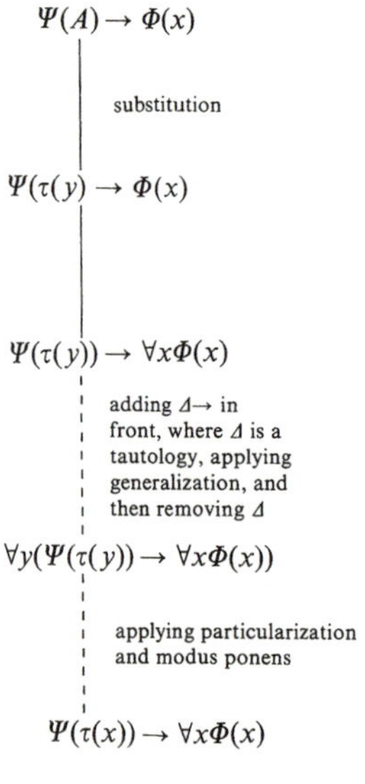

$\Psi(A) \to \Phi(x)$
substitution
$\Psi(\tau(y)) \to \Phi(x)$
$\Psi(\tau(y)) \to \forall x\Phi(x)$
adding $\Delta \to$ in front, where Δ is a tautology, applying generalization, and then removing Δ
$\forall y(\Psi(\tau(y)) \to \forall x\Phi(x))$
applying particularization and modus ponens
$\Psi(\tau(x)) \to \forall x\Phi(x)$

Fig. 20

If $l>1$ then out of the l special applications one must be such that there is no special application above it; let Φ be the formula proved by this special application. The proof of Φ in the tree contains only one special application, namely the one used to obtain Φ. By what we have shown for $l=1$ Φ has a normal proof. Replacing the proof of Φ in the tree by a normal proof we get a proof of Φ with $l-1$ special applications. By the induction hypothesis Φ has a normal proof.

Once we have a normal proof of the formula ϕ of the theorem we go on changing it as follows. Going down a branch we first encounter an axiom, then zero or more substitutions, and then zero or more applications of rules (b) and (d). Let Ψ be the formula resulting from all the substitutions in the axiom. If Ψ still contains class variables $A_1, \ldots, A_k$ then by k applications of the substitution rule (f) we substitute for these class variables a class term with no parameters, say $\{x \mid x=x\}$. We carry out the same substitutions in each branch of the proof, and in the part of the proof below these substitutions we replace all the class variables by the same class term. ϕ remains unchanged since it contains no class variables. The new tree obtained is still a proof since in the lower part of the proof, where there are no substitutions, a class term with no parameters does not function differently from a class variable, and having no free variables it does not interfere with generalization.

Now the proof of ϕ has the form that at the top of every branch there is an axiom, then rules of substitution are applied to it till a formula with no class variables is obtained, which we shall call a *source formula*, and then only rules (b) and (d) are applied.

For a formula Δ with no class variables we define the *basification* $b(\Delta)$ of Δ to be the formula of the basic language, with the same free variables as Δ, which is defined by induction on the length of Δ as follows. If Δ is of the form $x=y$ or $x \in y$ then $b(\Delta)=\Delta$. If Δ is of the form $y \in \{x \mid \Phi(x)\}$ then $b(\Delta)$ is $\phi(y)$, where $\phi(x)$ is an alphabetic variant of $b(\Phi(x))$ such that y is substitutable for x in $\phi(x)$. If Δ is of the form $\{x \mid \Phi(x)\}=\{y \mid \Psi(y)\}$ or $\{x \mid \Phi(x)\}=y$ or $y=\{x \mid \Phi(x)\}$ then $b(\Delta)$ is $b(\forall z(z \in \{x \mid \Phi(x)\} \leftrightarrow z \in \{y \mid \Psi(y)\}))$, or $b(\forall z(z \in y \leftrightarrow z \in \{x \mid \Phi(x)\}))$, respectively, where z is a variable which is not free in $\Phi(x)$ or $\Psi(y)$ and is not y. If Δ is of the form $\{x \mid \Phi(x)\} \in y$ or $\{x \mid \Phi(x)\} \in \{y \mid \Psi(y)\}$ then $b(\Delta)$ is $b(\exists z(z=\{x \mid \Phi(x)\} \wedge z \in y))$ or $b(\exists z(z=\{x \mid \Phi(x)\} \wedge z \in \{y \mid \Psi(y)\}))$, respectively, for an appropriate variable z. If Δ is $\neg\, \Phi$, $\Phi \wedge \Psi$ or $\forall x \Phi$ then $b(\Delta)$ is $\neg\, b(\Phi)$, $b(\Phi) \wedge b(\Psi)$ or $\forall x b(\Phi)$, respectively.

We shall now see that if we go through the part of the proof of ϕ from the source formulas downwards and basify every formula in it we get a skeleton of a proof of ϕ in P. This is a skeleton of a proof in the sense that it can be easily completed to become a proof. It is easily seen that since ϕ is in the basic language $b(\phi)=\phi$. Now let us see that the basification of a source formula is always a theorem of P. A source formula Ψ is always the result of zero or more substitutions in an axiom Γ. If Γ is a tautology then Ψ is, obviously, also a tautology, and so is also $b(\Psi)$. If Γ is an instance of particularization then also Ψ is such, and thus Ψ is of the form $\forall x \Phi(x) \rightarrow \Phi(y)$. $b(\Psi)$ is therefore $\forall x b(\Phi(x)) \rightarrow b(\Phi(y))$. We cannot say that $b(\Phi(x))(y)=b(\Phi(y))$ or even that y is substitutable for x in $b(\Phi(x))$ since when we compare how $b(\Phi(x))$ and $b(\Phi(y))$ are obtained changes of bound quantifiers, required in the basification of expressions of the form $u \in \{v \mid \Theta(v)\}$, may have been done differently in the two cases. However, looking at the construction of $b(\Phi(x))$ we can see that by replacing the bound variable y in $b(\Phi(x))$, if it happens to occur there, by another variable we obtain a formula $\chi(x)$ equivalent in P to $b(\Phi(x))$ such that y is substitutable for x in $\chi(x)$ and $\chi(y)$ is an alphabetic variant of $b(\Phi(y))$ (this can be formally established by induction on the length of $\Phi(x)$). In such a case $\chi(x) \leftrightarrow b(\Phi(x))$ and $\chi(y) \leftrightarrow b(\Phi(y))$ are theorems of logic and hence also of P.

Thus $b(\Psi)$, which is $\forall x b(\Phi(x)) \rightarrow b(\Phi(y))$ follows in P from $\forall x\chi(x) \rightarrow \chi(y)$, which is an instance of particularization. If Γ is an axiom (e) of equality then Γ is in the basic language and b does not change it. If Γ is an axiom of class comprehension (g) then Ψ is also such and it is of the form $y \in \{x \mid \Phi(x)\} \leftrightarrow \Phi'(y)$, where $\Phi'(x)$ is an alphabetic variant of $\Phi(x)$ such that y is substitutable for x in it, and $b(\Psi)$ is $\phi(y) \leftrightarrow b(\Phi'(y))$, where $\phi(x)$ is an alphabetic variant of $b(\Phi(x))$ such that y is substitutable for x in $\phi(x)$. Since $\Phi(x)$ and $\Phi'(x)$ are alphabetic variants it is easily seen that also $b(\Phi(x))$ and $b(\Phi'(x))$ are alphabetic variants, hence $\phi(x)$ and $b(\Phi'(x))$ are alphabetic variants. As we mentioned above, this implies that $\phi(y)$ and $b(\Phi'(y))$ are alphabetic variants hence $b(\Psi)$, which is $\phi(y) \rightarrow b(\Phi'(y))$, is provable in P. If Γ is the axiom (h) of extensionality or the axiom (i) of membership one can show in a similar way that $b(\Psi)$ is provable in P.

We still have to know that when we basify all the formulas in the proof of ϕ from the source formulas downwards, if in the original proof a formula Ψ is obtained from a formula Γ or from formulas Γ and Δ by a rule of inference then also $b(\Psi)$ follows from $b(\Gamma)$, or from $b(\Gamma)$ and $b(\Delta)$. Straightforward checking shows that this is indeed the case. □

1.5 (I.4.6(iii)) The Conservation Theorem—*Part* (iii). *If T is a theory in the basic language which includes P and T^* is a theory in the extended language which includes P^* such that for every formula ϕ of the basic language $T^* \vdash \phi$ iff $T \vdash \phi$, $\Psi(A)$ is a formula of the extended language which has no class variables other than A, and R is a set of formulas of the basic language such that each member of R is equivalent to some formula $\Psi(\{x \mid \Gamma(x)\})$ for some formula Γ with no class variables, and for every formula $\chi(x)$ of the basic language R contains a formula equivalent to $\Psi(\{x \mid \chi(x)\})$, then for every formula ϕ of the basic language we have $T^* \cup \{\Psi(A)\} \vdash \phi$ iff $T \cup R \vdash \phi$.*

Proof. If $T \cup R \vdash \phi$ then a proof of ϕ from $T \cup R$ uses only finitely many members $\psi_1, \ldots, \psi_k$ of R. We have $T \cup \{\psi_1, \ldots, \psi_k\} \vdash \phi$, hence $T \vdash \psi_1 \wedge \ldots \wedge \psi_k \rightarrow \phi$. By the assumption of the theorem also $T^* \vdash \psi_1 \wedge \cdots \wedge \psi_k \rightarrow \phi$. By our assumption about R each formula ψ_i is provable from $T^* \cup \{\Psi(A)\}$, hence $T^* \cup \{\Psi(A)\} \vdash \phi$.

Now suppose that $T^* \cup \{\Psi(A)\} \vdash \phi$. As we saw in the proof of 1.4 ϕ follows from source formulas, which are substitution instances of the logical and extra-logical axioms of $T^* \cup \{\Psi(A)\}$, by a proof which uses only the rules of modus ponens and generalization. Let the substitution instances of $\Psi(A)$ used in the proof of ϕ be $\Psi(\{x_1 \mid \Gamma_1(x_1)\}), \ldots, \Psi(\{x_n \mid \Gamma_n(x_n)\}), \Psi(y_1), \ldots, \Psi(y_m)$. Since each formula $\Psi(y_i)$ follows from a formula $\Psi(x)$ by generalization and particularization we have

$$T^* \cup \{\Psi(\{x_1 \mid \Gamma_1(x_1)\}), \ldots, \Psi(\{x_n \mid \Gamma_n(x_n)\}), \Psi(x)\} \vdash \phi.$$

Since $P^* \vdash x = \{y \mid y \in x\}$ and since, by 1.3, each formula $\Gamma_i(x_i)$ is equivalent in P^* to a formula $\phi_i(x_i)$ of the basic language, we have

$$T^* \cup \{\Psi(\{x_1 \mid \phi_1(x_1)\}), \ldots, \Psi(\{x_n \mid \phi_n(x_n)\}), \Psi(\{y \mid y \in x\})\} \vdash \phi.$$

By our assumption about R, R contains formulas $\chi_1, \ldots, \chi_n, \chi$ such that χ_i is equivalent in P^* to $\Psi(\{x_i \mid \phi_i(x_i)\})$ and χ is equivalent in P^* to $\Psi(\{y \mid y \in x\})$, hence $T^* \cup \{\chi_1, \ldots, \chi_n, \chi\} \vdash \phi$. Thus we have $T^* \vdash \chi_1 \wedge \cdots \wedge \chi_n \wedge \chi \to \phi$, hence $T \vdash \chi_1 \wedge \cdots \wedge \chi_n \wedge \chi \to \phi$, hence $T \cup \{\chi_1, \ldots, \chi_n, \chi\} \vdash \phi$, hence $T \cup R \vdash \phi$. □

Bibliography

The abbreviations of the names of the journals are usually those used by Mathematical Reviews.

Information about later editions and printings and English translations of the works listed in this bibliography can be found in the bibliography of Fraenkel, Bar-Hillel and Levy 1973.

Abramson, F. G., Harrington, L. A., Kleinberg, E. M., Zwicker, W. S.

1977 Flipping properties: A unifying thread in the theory of large cardinals. Ann. Math. Logic *12*, 25–58

Ackermann, W.

1937 Die Widerspruchsfreiheit der allgemeinen Mengenlehre. Math. Ann. *114*, 305–315

Alexandroff, P. S.

1916 Sur la puissance des ensembles mesurables B. C. R. Acad. Sci. Paris *162*, 323–325

Alexandroff, P. S., Urysohn, P.

1929 Mémoire sur les éspaces topologiques compacts. Verh. Nederl. Akad. Wetensch. Sec. I, *14*, 1–96

Avraham, U., Devlin, K., Shelah, S.

1978 The consistency with CH of some consequences of Martin's axiom plus $2^{\aleph_0} > \aleph_1$. Israel J. Math. *31*, 19–33

Bachmann, H.

1955 Transfinite Zahlen. 204 pp.

Baire, R.

1899 Sur les fonctions de variables reells. Ann. Mat. Pura Appl., Ser. IIIa, *3*, 1–122

Banach, S.

1930 Über additive Massfunktionen in abstrakten Mengen. Fund. Math. *15*, 97–101

Baumgartner, J. E.

1975 Ineffability properties of cardinals I. Infinite and finite sets, vol. I (Hajnal, Rado, Sos, editors), 109–130

Baumgartner, J. E., Prikry, K.

1976 On a theorem of Silver. Discrete Math. *14*, 17–21

Bendixson, I.

1883 Quelques théorèmes de la théorie des ensembles de points. Acta Math. *2*, 415–429

Bernays, P.

1937 A system of axiomatic set theory I. J. Symbolic Logic *2*, 65–77

1941 A system of axiomatic set theory II. J. Symbolic Logic *6*, 1–17

1942 A system of axiomatic set theory III. J. Symbolic Logic *7*, 65–89

1958 Axiomatic set theory. 226 pp.

Bernstein, F.
1905 Untersuchungen aus der Mengenlehre. Math. Ann. *61*, 117–155

Bloch, G.
1953 Sur les ensembles stationaires de nombres ordinaux et les suites distinguées de fonctions regressives. C. R. Acad. Sci. Paris *236*, 265

Bolzano, B.
1851 Paradoxien des Unendlichen

Boole, G.
1847 The mathematical analysis of logic. 82 pp
1854 An investigation of the laws of thought. 424 pp

Boos, W.
1975 Lectures on large cardinal axioms. Logic conference, Kiel 1974. Lecture Notes Math. *499*, 25–88

Borel, É,
1898 Leçons sur la theorie des fonctions

Bukovsky, L.
1965 The continuum problem and the powers of alephs. Comment. Math. Univ. Carolinae *6*, 181–197

Burali-Forti, C.
1897 Una questione sui numeri transfiniti. Rend. Circ. Mat. Palermo *11*, 154–164, 260

de Bruin, *see* under D.

Cantor, G.
1878 Ein Beitrag zur Mannigfaltigkeitslehre. J. f. Mathematik *84*, 242–258; also in Cantor 1932, 119–138
1883 Über unendliche lineare Punktmannigfaltigkeiten, Nr. 5. Math. Ann. *21*, 545–586; also in Cantor 1932, 165–204
1887 Mitteilungen zur Lehre vom Transfiniten I. Z. Philos. philos. Kritik, N.S. *91*, 81–125, 252–270; also in Cantor 1932, 378–439
1892 Über eine elementare Frage der Mannigfaltigkeitslehre. Jahresber. Deutsch. Math. Verein *1*, 75–78; also in Cantor 1932, 278–281
1895 Beiträge zur Begründung der transfiniten Mengenlehre I. Math. Ann. *46*, 481–512; also in Cantor 1932, 282–311
1897 Beiträge zur Begründung der transfiniten Mengenlehre II. Math. Ann. *49*, 207–246; also in Cantor 1932, 312–356
1899 Letters to Dedekind. Also in Cantor 1932, 443–451
1932 Gesammelte Abhandlungen mathematischen und philosophischen Inhalts. 486 pp.

Carruth, P. W.
1942 Arithmetic of ordinals with applications to the theory of ordered abelian groups. Bull. Amer. Math. Soc. *48*, 262–271

Cartan, H.
1937 Théorie des filtres. C. R. Acad. Sci. Paris *205*, 595–598

Cohen, P. J.
1963 The independence of the continuum hypothesis. Proc. Nat. Acad. Sci. U.S.A. *50*, 1143–1148
1966 Set theory and the continuum hypothesis. 160 pp.

De Bruin, N. G., Erdös, P.
1951 A colour problem for infinite graphs and a problem in the theory of relations. Nederl. Akad. Wetensch. Proc. Sec. A 54 = Indag. Math. *13*, 369–373

Dedekind, R.
1888 Was sind und was sollen die Zahlen? 58 pp.
1932 Gesammelte mathematische Werke, vol. III

Devlin, K. J.
1973 Aspects of constructibility. Lecture Notes Math. *354*, 240 pp.
1977 Constructibility. In: Handbook of Mathematical Logic. Barwise, J. (ed.), pp. 453–489. Amsterdam: North-Holland

Devlin, K. J., Jensen, R. B.
1976 Marginalia to a theorem of Silver. Logic Conference Kihl 1974. Lecture Notes Math. *499*, 115–142

Devlin, K. J., Johnsbraten, H.
1974 The Souslin problem. Lecture Notes Math. *405*, 132 pp.

Drake, F. R.
1974 Set theory, an introduction to large cardinals. 351 pp.

Dushnik, B., Miller, E. W.
1941 Partially ordered sets. Amer J. Math. *63*, 600–610

Easton, W. B.
1964 Powers of regular cardinals. Ph.D. thesis, Princeton Univ. 66 pp. *See* Easton 1970
1970 Powers of regular cardinals. Ann. Math. Logic *1*, 139–178

Erdös, P., Hajnal, A.
1966 On a problem of B. Jónsson. Bull. Acad. Polon. Sci. Sér. Sci. Math. Astronom. Phys. *14*, 19–23

Erdös, P., Hajnal, A., Máté, A., Rado, R.
1979 Combinatorial set theory—partition relations for cardinals. To appear

Erdös, P., Hajnal, A., Rado, R.
1965 Partition relations for cardinal numbers. Acta Math. Acad. Sci. Hungar. *16*, 93–196

Erdös, P., Rado, R.
1956 A partition calculus in set theory. Bull. Amer. Math. Soc. *62*, 427–489

Erdös, P., Tarski, A.
1961 On some problems involving inaccessible cardinals. Essays on the foundations of mathematics. Bar-Hillel Y. et al. (ed.), pp. 50–82

Feferman, S.
1965 Some applications of the notions of forcing and generic sets. Fund. Math. *56*, 325–345.

Feferman, S., Levy, A.
1963 Independence results in set theory by Cohen's method II (abstract). Notices Amer. Math. Soc. *10*, 592

Felgner, U.
1971 Comparison of the axioms of local and universal choice. Fund. Math. *71*, 43–62

Fodor, G.
1956 Eine Bemerkung zur Theorie der regressiven Funktionen. Acta Sci. Math. (Szeged) *17*, 139–142
1966 On stationary sets and regressive functions. Ibid. *27*, 105–110

Fraenkel, A. A.
1922 Zu den Grundlagen der Cantor-Zermeloschen Mengenlehre. Math. Ann. *86*, 230–237

Fraenkel, A. A., Bar-Hillel, Y., Levy, A.
1973 (with the collaboration of D. van Dalen) Foundations of set theory. 404 pp.

Frege, G.
1884 Die Grundlagen der Arithmetik. Eine logischmathematische Untersuchung über den Begriff der Zahl. 119 pp. § 68 (*see* also Frege 1893, § 42)
1893 Grundgesetze der Arithmetik, begriffsschriftlich abgeleitet. Vol. I. 254 pp.

Gaifman, H.
1975 Global and local choice functions. Israel J. Math. *22*, 257–265

Galvin, F., Hajnal, A.
1975 Inequalities for cardinal powers. Ann. of Math. (2) *101*, 491–498

Galvin, F., Prikry, K.
1973 Borel sets and Ramsey's theorem. J. Symbolic Logic *38*, 193–198

Gitik, M.
1979 All uncountable cardinals can be singular, to appear

Gödel, K.
1938 The consistency of the axiom of choice and of the generalized continuum hypothesis. Proc. Nat. Acad. Sci. U.S.A. *24*, 556–557
1939 Consistency proof for the generalized continuum hypothesis. Ibid *25*, 220–224
1965 Remarks before the Princeton bicentennial conference on problems in mathematics. The undecidable. Davis, M. (ed), pp. 84–88

Hajnal, A.
1961 Proof of a conjecture of S. Ruziewicz. Fund. Math. *50*, 123–128

Hajnal, A., Máté, A.
1975 Set mappings, partitions and chromatic numbers. Logic colloquium '73. Rose, H. E., Shepherdson, J. C. (eds.), pp. 347–379

Halmos, P. R.
1950 Measure theory. 304 pp.

Halpern, J. D.
1964 The independence of the axiom of choice from the Boolean prime ideal theorem. Fund. Math. *55*, 57–66

Halpern, J. D., Levy, A.
1971 The Boolean prime ideal theorem does not imply the axiom of choice. Axiomatic set theory. Amer. Math. Soc. Proc. Symposia Pure Math. *13*, vol. I, 83–134

Hanf, W.
1964 Incompactness in languages with infinitely long expressions. Fund. Math. *53*, 309–324
1964a On a problem of Erdös and Tarski. Ibid, 325–334

Hartogs, F.
1915 Über das Problem der Wohlordnung. Math. Ann. *76*, 438–443

Hausdorff, F.
1904 Der Potenzbegriff in der Mengenlehre. Jahresber. Deutsch. Math.-Verein *13*, 569–571
1908 Grundzüge einer Theorie der geordnetn Mengen. Math. Ann. *65*, 435–505
1914 Grundzüge der Mengenlehre. 476 pp.
1916 Die Mächtigkeit der Borelschen Mengen. Math. Ann. *77*, 430–437
1919 Über halbstetige Funktionen und deren Verallgemeinerung. Math. Z. *5*, 292–309

Hechler, S. H.
1973 Powers of singular cardinals and a strong form of the negation of the generalized continuum hypothesis. Z. Math. Logik Grundlagen Math. *19*, 83–84

Hessenberg, G.
1906 Grundbegriffe der Mengenlehre. Abh. der Friesschen Schule. N.S.[1], Heft 4. 220 pp.

Hilbert, D.
1900 Mathematische Probleme. Nachr. Göttingen, 253–297

Hilbert, D., Bernays, P.
1939 Grundlagen der Mathematik. Vol. II. 498 pp.

Jacobsthal, E.
1909 Über den Aufbau der transfiniten Arithmetik. Math. Ann. *66*, 145–194

Jech, T.
1966 Interdependence of weakened forms of the axiom of choice. Comment. Math. Univ. Carolinae *7*, 359–371
1967 Non provability of Souslin's hypothesis. Ibid. *8*, 291–305
1974 Properties of the gimel function and a classification of singular cardinals. Fund. Math. *81*, 57–64

Jensen, R. B.
1966 Independence of the axiom of dependent choices from the countable axiom of choice (abstract). J. Symbolic Logic *31*, 294
1967 Consistency results of ZF (abstract). Notices Amer. Math. Soc. *14*, 137
1968 Souslin's hypothesis is incompatible with $V=L$ (abstract). Notices Amer. Math. Soc. *15*, 935

Jourdain, P. E. B.
1908 On the multiplication of alephs. Math. Ann. *65*, 506–512
1908a On infinite sums and products of cardinal numbers. Quart. J. pure appl. Math. *39*, 375–384

Kanamori, A., Magidor, M.
1978 The evolution of large cardinal axioms in set theory. In: Lecture Notes in Mathematics, Vol. 699, Higher Set Theory, edited by Müller, G. H. and Scott, D. S., pp. 99–275

Keisler, H. J., Tarski, A.
1964 From accessible to inaccessible cardinals. Fund. Math. *53*, 225–308

Kelley, J. L.
1950 The Tychonoff product theorem implies the axiom of choice. Fund. Math. *37*, 75–76

Kleinberg, E. M.
1973 Infinitary combinatorics. Cambridge Summer School in Mathematical Logic. Lecture Notes Math. *337*, 361–418

König, D.
1926 Sur les correspondances multivoques des ensembles. Fund. Math. *8*, 114–134

König, J.
1905 Zum Kontinuumproblem. Math. Ann. *60*, 177–180

Kunen, K.
1977 Combinatorics. In: Handbook of Mathematical Logic. Barwise, J. (ed.), pp. 371–401. Amsterdam: North-Holland
1978 Saturated ideals. J. Symbolic Logic *43*, 65–76

Kuratowski, K.
1921 Sur la notion d'ordre dans la théorie des ensembles. Fund. Math. *2*, 161–171
1922 Sur l'opération A de l'Analysis Situs. Fund. Math. *3*, 192–195
1925 Sur l'état actuel de l'axiomatique de la théorie des ensembles. Ann. Soc. Polon. Math. *3*, 146–147
1966 Topology. Vol. I. 560 pp.

Kurepa, G.
1938 Ensembles linéaires et une classe de tableaux ramifiées (tableaux de M. Aronszajn). Publ. Math. Univ. Belgrade *6/7*, 129–160

Läuchli, H.
1971 Coloring infinite graphs and the Boolean prime ideal theorem. Israel J. Math. *9*, 422–429

Lebesgue, H.
1905 Sur les fonctions représentables analytiquement. J. de Math. (6) *1*, 139–216

Levi, B.
1902 Intorno alla teoria degli aggregati. R. Istituto Lombardo di Sci. e Lettere, Rendic. (2) *35*, 863–868

Levy, A.
1964 The interdependence of certain consequences of the axiom of choice. Fund. Math. *54*, 135–157

1965 Definability in axiomatic set theory I. Logic, methodology and philosophy of science. Proc. of the 1964 Intern. Congr. Bar-Hillel, Y. (ed.), pp. 127–151

1976 The role of classes in set theory. Sets and Classes; on the work by Paul Bernays, pp. 173–215; also in Fraenkel, Bar-Hillel and Levy 1973, pp. 119–153

Lindenbaum, A., Tarski, A.

1926 Communication sur les recherches de la theorie des ensembles. C. R. Soc. Sci. lettres Varsovie, Cl. III *19*, 299–330

Łoś, J., Ryll-Nardzewski, C.

1951 On the application of Tychonoff's theorem in mathematical proofs. Fund. Math. *38*, 233–237

1954 Effectiveness of the representation theory for Boolean algebras. Ibid. *41*, 49–56

Lusin, N.

1917 Sur la classification de M. Baire, C.R. Acad. Sci. Paris *164*, 91–94

1925 Sur les ensembles non mesurables B et l'emploi de la diagonale de Cantor. Ibid. *181*, 95–96

1927 Sur les ensembles analytiques. Fund. Math. *10*, 1–95

Lusin, N., Sierpinski, W.

1923 Sur un ensemble non mesurable B. J. de Math. (9) *2*, 53–72

Mac Neille, H. M.

1937 Partially ordered sets. Trans. Amer. Math. Soc. *42*, 416–460

Magidor, M.

1977 On the singular cardinals problem I. Israel J. Math. *28*, 1–31

Mahlo, P.

1911 Über lineare transfinite Mengen. Berich. Verh. Sächsischen Akad. Wissensch. Leipzig, Math. Phys. Kl. *63*, 187–225

Martin, D. A., Solovay, R.

1970 Internal Cohen extensions. Ann. Math. Logic *2*, 143–178

Miller, E. W.

1943 A note on Souslin's problem. Amer. J. Math. *65*, 673–678

Milner, E. C., Rado, R.

1965 The pigeon-hole principle for ordinal numbers. Proc. London Math. Soc. (3) *15*, 750–768

Mirimanoff, D.

1917 Les antinomies de Russell et de Burali-Forti et le problème fondamental de la théorie des ensembles. L'Enseignement Math. *19*, 37–52

Mitchell, W. J.

1972 Aronszajn trees and the independence of the transfer property. Ann. Math. Logic *5*, 21–46

Monk, D., Scott, D.

1964 Additions to some results of Erdös and Tarski. Fund. Math. *53*, 335–343

Montague, R.

1955 Well-founded relations; generalizations of principles of induction and recursion (abstract). Bull. Amer. Math. Soc. *61*, 442

1961 Fraenkel's addition to the axioms of Zermelo. Essays on the foundations of mathematics. Bar-Hillel, Y. et al. (eds.), pp. 91–114

Mostowski, A.

1948 On the principle of dependent choices. Fund. Math. *35*, 127–130

1949 An undecidable arithmetical statement. Ibid. *36*, 143–164

Mycielski, J.

1961 Some remarks and problems on the colouring of infinite graphs and the theorem of Kuratowski. Acta Math. Acad. Sci. Hungar. *12*, 125–129

von Neumann, *see* under V.

Peirce, C. S.
1880 On the algebra of logic. Amer. J. Math. *3*, 15–57
1885 On the algebra of logic, a contribution to the philosophy of notation. Ibid. *7*, 180–202

Piccard, Sophie
1937 Sur un probleme de M. Ruziewicz de la theorie des relations pour les nombres cardinaux $m < \aleph_\Omega$. C.R. Soc. Sci. lettres Varsovie, Cl. III *30*, 12–18

Pincus, D.
1972 Zermelo–Fraenkel consistency results by Fraenkel–Mostowski methods. J. Symbolic Logic *37*, 721–743
1974 Cardinal representatives. Israel J. Math. *18*, 321–344

Quine, W. V.
1963 Set theory and its logic. 359 pp.

Ramsey, F. P.
1930 On a problem of formal logic. Proc. London Math. Soc. (2) *30*, 264–286

Rasiowa, H., Sikorski, R.
1950 A proof of the completeness theorem of Gödel. Fund. Math. *37*, 193–200

Ricabarra, R. A.
1958 Conjuntos ordinados y ramificados

Rowbottom, F.
1971 Some strong axioms of infinity incompatible with the axiom of constructibility. Ann. Math. Logic *3*, 1–44

Rubin, H.
1960 Two propositions equivalent to the axiom of choice only under both the axioms of extensionality and regularity (abstract). Notices Amer. Math. Soc. *7*, 380

Rubin, H., Rubin, J. E.
1963 Equivalents of the axiom of choice. 134 pp.

Rubin, H., Scott, D.
1954 Some topological theorems equivalent to the Boolean prime ideal theorem (abstract). Bull. Amer. Math. Soc. *60*, 389

Russell, B.
1903 The principles of mathematics I
1906 On some difficulties in the theory of transfinite numbers and order types. Proc. London Math. Soc. (2) *4*, 29–53

Schönflies, A.
1913 Entwickelung der Mengenlehre und ihrer Anwendungen. 388 pp.

Schröder, E.
1891 Algebra der Logik. Vol. II. 606 pp. § 47

Scott, D.
1955 Definitions by abstraction in axiomatic set theory (abstract). Bull. Amer. Math. Soc. *61*, 442
1961 Measurable cardinals and constructible sets. Bull. Acad. Polon. Sci. Ser. Sci. Math. Astronom. Phys. *9*, 521–524
1974 Axiomatizing set theory. Axiomatic set theory. Amer. Math. Soc. Proc. Symposia Pure Math. *13*, vol. II, 207–214

Shelah, S.
1975 Notes on partition calculus. Infinite and finite sets. Hajnal, Rado and Sos (eds.), vol. III, pp. 1257–1276
1977 Whitehead groups may not be free, even assuming CH I. Israel J. Math. *28*, 193–204

Shoenfield, J. R.
1975 Martin's axiom. Amer. Math. Monthly *82*, 610–617

Sierpiński, W.
1928 Sur une decomposition d'ensembles. Monatshefte Math. Phys. *35*, 239–242
1933 Sur un probleme de la theorie des relations. Ann. Scuola Norm. Sup. Pisa, Sec. 2, *2*, 285–287
1958 Cardinal and ordinal numbers

Sierpiński, W., Tarski, A.
1930 Sur une propriété caractéristique des nombres inaccessibles. Fund. Math. *15*, 292–300

Sikorski, R.
1948 A theorem on extension of homomorphisms. Ann. Soc. Polon. Math. *21*, 332–335

Silver, J.
1970 Every analytic set is Ramsey. J. Symbolic Logic *35*, 60–64
1971 The independence of Kurepa's conjecture and two cardinal conjectures in model theory. Axiomatic set theory. Amer. Math. Soc. Proc. Symposia Pure Math. *13*, part I, 383–390
1975 On the singular cardinals problem. Proc. International Congress of Mathematicians, Vancouver 1974, vol. I, pp. 265–268

Skolem, T.
1923 Einige Bemerkungen zur Axiomatischen Begründung der Mengenlehre. Wiss. Vorträge 5. Kongress skandinav. Mathematiker in Helsingfors 1922, pp. 217–232

Smith, E. C., Tarski, A.
1957 Distributivity and completeness in Boolean algebras. Trans. Amer. Math. Soc. *84*, 230–257

Solovay, R. M.
1967 A nonconstructible Δ_3^1 set of integers. Trans. Amer. Math. Soc. *127*, 50–75
1971 Real-valued measurable cardinals. Axiomatic set theory. Amer. Math. Soc. Proc. Symposia Pure Math. *13*, part I, 397–428
1974 Strongly compact cardinals and the GCH. Proceedings of the Tarski Symposium. Amer. Math. Soc. Proc. Symposia Pure Math. *25*, 365–372

Solovay, R. M., Tennenbaum, S.
1971 Iterated Cohen extensions and Souslin's problem. Ann. of Math. (2) *94*, 201–245

Souslin, M.
1917 Sur une définition des ensembles mesurables B sans nombres transfinis. C.R. Acad. Sci. Paris *164*, 88–91
1920 Problème 3. Fund. Math. *1*, 223

Specker, E.
1949 Sur une problème de Sikorski. Colloq. Math. *2*, 9–12
1954 Verallgemeinerte Kontinuumshypothese und Auswahlaxiom. Archiv der Math. *5*, 332–337

Stone, M. H.
1934 Boolean algebras and their application to topology. Proc. Nat. Acad. Sci. U.S.A. *20*, 197–202
1936 The theory of representations for Boolean algebras. Trans. Amer. Math. Soc. *40*, 37–111

Szpilrajn, E.
1930 Sur l'expansion de l'ordre partiel. Fund. Math. *16*, 386–389

Tarski, A.
1924 Sur quelques théorèmes qui équivalent à l'axiome du choix. Fund. Math. *5*, 147–154
1924a Sur les ensembles finis. Ibid. *6*, 45–95
1925 Quelques théorèmes sur les alephs. Ibid. *7*, 1–14
1928 Sur la décomposition des ensembles en sous ensembles presque disjoints. Ibid. *12*, 188–205
1929 Sur la décomposition des ensembles en sous ensembles presque disjoints. Ibid. *14*, 205–215
1930 Une contribution à la theorie de la mesure. Ibid. *15*, 42–50

1954 Prime ideal theorems for Boolean algebras and the axiom of choice; Prime ideal theorems for set algebras and ordering principles; Prime ideal theorems for set algebras and the axiom of choice (3 abstracts). Bull. Amer. Math. Soc. *60*, 390–391

1954a Theorems on the existence of successors of cardinals and the axiom of choice. Nederl. Akad. Wetensch. Proc. Ser. A 57 = Indag. Math. *16*, 26–32

1955 General principles of induction and recursion; The notion of rank in axiomatic set theory and some of its applications (2 abstracts). Bull. Amer. Math. Soc. *61*, 442–443

1962 Some problems and results relevant to the foundations of set theory. Logic, methodology, and philosophy of science; proceedings of the 1960 international congress. Nagel et al. (eds.), pp. 125–135

Tennenbaum, S.

1968 Souslin's problem. Proc. Nat. Acad. Sci. U.S.A. *56*, 60–63

Tychonoff, A.

1935 Über einen Funktionenraum. Math. Ann. *111*, 762–766

Ulam, S.

1929 Concerning functions of sets. Fund. Math. *14*, 231–233

1930 Zur Masstheorie in der allgemeinen Mengenlehre. Ibid. *16*, 140–150

Veblen, O.

1908 Continuous increasing functions of finite and transfinite ordinals. Trans. Amer. Math. Soc. *9*, 280–292

Vitali, G.

1905 Sul problema della misura dei gruppi di punti di una retta. Bologna

von Neumann, J.

1923 Zur Einführung der transfiniten Zahlen. Acta Sci. Math. (Szeged) *1*, 199–208

1925 Eine Axiomatisierung der Mengenlehre. J. fur. Math. *154*, 219–240

1928 Die Axiomatisierung der Mengenlehre. Math. Z. *27*, 669–752

1928a Über die Definition durch transfinite Induktion und verwandte Fragen der allgemeinen Mengenlehre. Math. Ann. *99*, 373–391

Whitehead, A. N.

1902 On cardinal numbers. Amer. J. Math. *24*, 367–394

Whitehead, A. N., Russell, B.

1912 Principia Mathematica. Vol. II, 772 pp.

Wiener, N.

1914 A simplification of the logic of relations. Proc. Cambridge Philos. Soc. *17*, 387–390

Zermelo, E.

1904 Beweis, dass jede Menge wohlgeordnet werden kann. Math. Ann. *59*, 514–516

1908 Untersuchungen über die Grundlagen der Mengenlehre I. Ibid. *65*, 261–281

1909 Sur les ensembles finis et le principe de l'induction complète. Acta Math. *32*, 185–193

1930 Über Grenzzahlen und Mengenbereiche. Fund. Math. *16*, 29–47

1935 Grundlagen einer allgemeinen Theorie der mathematischen Satzsysteme. Ibid. *25*, 136–146

Zorn, M.

1935 A remark on method in transfinite algebra. Bull. Amer. Math. Soc. *41*, 667–670

1944 Idempotency of infinite cardinals. Univ. California Publ. Math., N.S. *2*, 9–12

Additional Bibliography

Baumgartner, J. E.
1976 Almost disjoint sets, the dense set problem and the partition calculus. Ann. Math. Logic 10, 401–439.

Birkhof, G.
1933 On the combination of subalgebras. Proc. Cambridge Philosophical Soc. 29, 441–464.

Cantor, G.
1883a Fondements d'une théorie générale des ensembles. Acta Mathematica 2, 381–408.

Felgner, U. (ed.)
1979 Mengenlehre. 331 pp.

Hausdorff, F.
1909 Die Graduierung nach dem Endverlauf. Abhandlungen d. Königlich Sächlischen Gesellschaft d. Wiss., Math.-Phys. Klasse, 295–334.

Jech, T.
1978 Set Theory. 621 pp.

Jourdain, P. E. B.
1904 On the Transfinite Cardinal Numbers of Well-ordered Aggregates. Philosophical Magazine (6) 7, 61–75.

Kanamori, A.
1994 The Higher Infinite. 536 pp.

Kunen, K.
1980 Set Theory. 313 pp.

Moore, G. H.
1982 Zermelo's Axiom of Choice, Its Origins, Development and Influence. 410 pp.

Moschovakis, Y. N.
1980 Descriptive Set Theory. 637 pp.

Noether, E. and Cavailles, J. (eds.)
1937 Briefwechsel Cantor-Dedekind. 60 pp.

Rang, B. and Thomas, W.
1981 Zermelo's Discovery of the "Russell Paradox." Historia Mathematica 8, 15–22.

Russell, B.
1908 Mathematical Logic as based on the Theory of Types. Amer. J. Math. 30, 222–262.

Shelah, S.
1994 Cardinal Arithmetic. 481 pp.

Tarski, A.
1938 Eine äquivalente Formulierung des Auswahlaxioms. Fund. Math. 30, 197-201

Tukey, J.
1940 Convergence and Uniformity in Topology. Annals of Math. Studies, 2. 90pp.

Truss, J.
1973 Convex sets of cardinals. Proc. London Math. Soc. (3) 27, 577–599.

Index of Notation

Numbers in italics denote the *definition* (or introduction) of the notion.

Use of Variables

The list below gives the most common use of the different letters, but in some places the same letters are used for different objects.

$x, y, z, \ldots$	sets 4, 12
φ, ϕ, χ	formulas of the basic language 10
$\Phi, \Psi, \Gamma, \Theta, \Lambda$	formulas of the extended language 10
$A, B, C, \ldots$	classes (including sets) 12
τ	terms
$\alpha, \beta, \gamma, \delta, \ldots$	ordinals 52
i, j, k, l, m, n	natural numbers 56
$\mathfrak{a}, \mathfrak{b}, \mathfrak{c}, \ldots, \mathfrak{m}, \mathfrak{n}, \ldots$	cardinals 83

Latin Alphabetical Notations

(A)	a certain operation in a topological space *224*
$A_\mathfrak{a}$	Martin's axiom for $\mathfrak{a}$ in Boolean algebras *279*, 308, 310, 317–320
$A'_\mathfrak{a}$	Martin's axiom for $\mathfrak{a}$ in posets *282*
$A''_\mathfrak{a}, A'''_\mathfrak{a}$	Martin's axiom for $\mathfrak{a}$ in compact Hausdorff spaces *285*
$A_\mathfrak{a}(B)$	Martin's axiom for $\mathfrak{a}$ in the Boolean algebra B *278*, 282
$A'_\mathfrak{a}(P)$	Martin's axiom for $\mathfrak{a}$ in the poset P *281*
Ac	axiom of choice *24*
AC_ω	axiom of choice for denumerable sets *167*, 168
AC_{wo}	axiom of choice for well-orderable sets *169*–171
Bd( )	the boundary of *212*
Card(σ)	the cardinality of the order type σ *113*, 114, 119
cf	the cofinality of *132*
Cl( )	the closure of *202*
Cn	the class of all cardinals *83*, 86–88
Concat(w)	the concatenation of all the members of the sequence w *236*
$d(,)$	the distance function of a metric space *201*
$D_{\alpha<\Omega}$	diagonal intersection *146*
DC_ω	axiom of ω dependent choices *167*, 168–170
DC_{ω_1}	axiom of ω_1 dependent choices *171*
Df(a)	the set of all subsets of a first-order definable from parameters in a *290*
Dom	the domain of *26*
F_σ	the set of countable unions of closed sets *215*
G_δ	the set of countable intersections of open sets *215*
$H_\mathfrak{a}$	the class of all x such that $\lvert \mathrm{Tc}(x) \rvert < \mathfrak{a}$ *136*
$\mathrm{HC}_\mathfrak{a}$	the class of all sets hereditarily of cardinality $< \mathfrak{a}$ *136*

Other Notations

Chapter I

Chapter II

Chapter III

Chapter IV

Chapter V

Chapter VI

Chapter VII

Chapter VIII

Chapter IX

Index

Numbers in italics denote the *definition* (or introduction) of the notion.

Appendix: Corrections and Additions

The number of asterisks in front of each item indicates the importance of the item. An item without asterisk corrects a misprint which the reader would have interpreted correctly anyway. An item with a single asterisk corrects a mistake which the reader could correct himself, with some effort, or adds something to the text. An item with two asterisks is an item which corrects a serious misprint or omission.

A minus sign in front of the line number means that this is the number of the line counted from the bottom of the page.

P. 5, l. 3	Erase one l in auxilliary
* P. 6, l. 21	Between Russell 1903 and) insert ,Zermelo — see Rang and Thomas 1981
P. 7, l. 15	In by Russell 1906 replace 6 by 7
* P. 10, l. -7	Replace added now by now added
P. 12, l. 2	Replace §1 by §3
P. 13, l. 15	Replace §1 by §3
P. 19, l. -2	In $\tau(u, x_1, \ldots, x_n, \ldots, x_n)$ erase one $\ldots, x_n$
P. 20, l. 18	Replace by by be
P. 22, l. 5	Replace 1874 by 1847
**P. 23, l. 8	In $(\exists u \in z)$ replace $\exists$ by $\forall$
P. 23, l. -15	Replace **of Schema** by **Schema of**
P. 23, l. -5	Insert) between 2.23 and .
* P. 25, l. 4	In $\{\{x, y\}\}$ erase the first $\{$
P. 25, l. 10	In (6.3) replace 3 by 2
P. 26, l. -14	In $\langle xy \rangle \in F$ insert , after x
* P. 26, l. -5	In of A on B. replace . by , or that F *maps* A *onto* B.
* P. 26, l. -1	In also *onto* B insert maps A after also
P. 27, l. -13	In the superscript $^{n=1}$ of $\bigwedge_{i=1}^{n=1}$ replace $=$ by $-$
* P. 27, l -8	After is in A. add This notation is not the standard one for this operation.
* P. 35, l. -7	In $xRy \leftarrow$ replace $\leftarrow$ by $\rightarrow$
* P. 50, l. 7	After **Theorem** insert (Cantor 1899)
* P. 53, l. -6	After **Proposition** insert (Cantor 1899)
P. 54, l. 5	In $\alpha \cup \{a\}$ replace a by α
P. 56, l. 13	In *a limit number* replace *number* by *ordinal*
P. 56, l. -14	In a limit number replace number
P. 61, l. -9	In with includes A replace with by which
* P. 63, l. 5	In $\bigcup_{i=1}^{k} R_1$- replace R_1 by R_i
* P. 67, l. -8	In satisfies (5) replace (5) by 5.12
* P. 68, l. -1	Replace $\rho^{\in \mid \mathrm{Wf}(x)}$ by $\rho^{\in \mid \mathrm{Wf}}(x)$
P. 70, l. 6	Replace **Comulative** by **Cumulative**
P. 78, l. -15	Insert ; between $z \notin a$ and we

P. 80, l. 16 In of the finite replace of by on

P. 81, l. 1 In natural number m replace number by numbers

**P. 83, l. 14 In $f(k) \in f(n)$ replace f by g in both places

* P. 85, l. 2 In Cantor 1895 replace 1895 by 1883a — see also Noether and Cavailles 1937, p. 55

P. 87, l. 19 In *a cardinal c* replace c by $\mathfrak{c}$

* P. 87, l. -6 After **Corollary** insert (Cantor 1899)

* P. 91, l. -14 After (iii) insert (Beppo Levi 1902)

* P. 94, l. -1 Replace Hessenberg 1906 by Jourdain 1904

P. 99, l. 7 In $a < 2^{\mathfrak{a}}$ replace a by $\mathfrak{a}$

P. 99, l. 10 In Hartog's theorem replace Hartog's by Hartogs'

* P. 100, l. 20 Below this line add the line

(vi) (Truss 1973). If $\mathfrak{a}+\mathfrak{c} \leqslant \mathfrak{b}+\mathfrak{c}$ then there is a $\mathfrak{d} \leqslant \mathfrak{b}$ such that $\mathfrak{a}+\mathfrak{c} = \mathfrak{d}+\mathfrak{c}$.

P. 105, l. -17 In in the sum replace in by is

P. 109, l. 9 Delete - in n-variables

P. 110, l. 20 In $= \aleph_\alpha \cdot \aleph_0, = \aleph_\alpha$ delete ,

**P. 118, l. -8 At the end of the line add

We read "σ *times* τ because this is the order of the multiplicands, but the meaning of this multiplication is better described by "τ *times* σ" since $\sigma \bullet \tau$ is the order type of an ordered set which consists of τ copies of σ, i. e., copies of σ ordered, among themselves, in the order type τ.

P. 121, l. 1 Replace monotonous by monotonic

P. 121, l. 3 Replace monotonous by monotonic

* P. 121, l. 11 In $\omega \cdot \omega(\sup\{n|n < \omega\})$ insert $=$ after $\omega \cdot \omega$

P. 126, l. 9 In all the function replace function by functions

* P. 127, l. -13 In $\beta' = \alpha^{\cdot\gamma_0} \bullet \delta_0 +$ replace $^{\cdot\gamma_0}$ by $^{\cdot\gamma_0'}$

P. 129, l. 11 In expansion in powers replace expansion by expansions

P. 129, l. 12 Replace idealogical by ideological

**P. 131, l. -8 In for $\beta, \alpha < \kappa$ replace α by γ

P. 132, l. 2 In increasing). And replace . And by , and

* P. 136, l. 10 Between **Exercise** and . insert (partly in Tarski 1938)

P. 136, l. 13 In well founded set x replace set by sets

* P. 136, l. 20 In for every cardinal insert infinite after every

* P. 136, l. -9 In $\sup_{\lambda<\mathrm{cf}(\omega_\gamma)}$ replace $_{\omega_\gamma}$ by $_{\omega\cdot\gamma}$

* P. 136, l. -3 In **3.17 Exercise** replace **3.17** by **3.17Ac**

**P. 137, l. 8 In if $p \subseteq q$ or replace $\subseteq$ by $\subset$

P. 137, l. 12 Replace determinated by determined

**P. 140, l. 5 In $|R(\omega) \dotplus \alpha| = \beth_\alpha$ replace $R(\omega) \dotplus \alpha$ by $R(\omega \dotplus \alpha)$

P. 141, l. 22 In be effected by replace effected by affected

P. 143, l. 11 In substracting from A replace substracting by subtracting

P. 149, l. 19 In we have only discussed delete only

P. 152, l. 20 In the subset of replace subset by subsets

**P. 157, l. -7 Replace

$\langle \delta_\xi \mid \xi < \Omega \rangle$ is a normal function on Ω, and hence, by 4.24, by
As a consequence of (ii) we have: (a) $\delta_\xi > 0$ for $\xi > 0$, and since for $g_\xi(\gamma) > 0$ we have $g_\xi(\gamma) = f_\gamma(\xi) \geqslant \xi$, we have $\delta_\xi \geqslant \xi$, (b) $\delta_\xi < \delta_\eta$ for $\xi < \eta$, and (c) the function $\langle \delta_\xi \mid \xi < \Omega \rangle$ is continuous. As in the proof of 4.24 it follows that

P. 157, l. -7 In $\delta_\xi = \xi\rangle$ replace $\rangle$ by $\}$

P. 158, l. 14 In an extention ZFGC replace extention by extension

* P. 158, l. -9 Erase Beppo Levy 1902;

* P. 158, l. -8 Between 1904 and) insert and Moore 1982

* P. 160, l. 7 Replace (Russell 1906). by
(Zermelo 1908 and Russell 1908).

* P. 160, l. 11 After (ii) insert (Russell 1907)

* P. 160, l. 13 In member b. insert of after member

* P. 161, l. -5. In (Hausdorff 1914) replace 1914 by 1909

* P. 162, l. -6 Replace Zorn 1944 by Tukey 1940

P. 169, l. -5 Replace $AC_{\omega o}$ by AC_{wo}

**P. 171, l. 1 Insert AC_{wo} between Also and does

**P. 179, l. -5 In $\leftrightarrow \Phi(x)$ replace x by y

* P. 181, l. 4 Between the present section. and We start insert
For an advanced treatment of cardinal exponentiation see Shelah 1994.

* P. 181, l. -14 In we estimate $2^{\mathfrak{c}+\kappa}$ is insert , between $2^{\mathfrak{c}+\kappa}$ and is

* P. 183, l. 19 Between of ZFC and . insert
(assuming the existence of an inaccessible cardinal is consistent with ZFC)

* P. 188, l. 13 In $\prod_{\alpha<\lambda} \mathfrak{d}_\alpha^{\mathfrak{b}}$ $\prod_{\alpha<\lambda}$ insert $\leqslant$ in the space between $\prod_{\alpha<\lambda} \mathfrak{d}_\alpha^{\mathfrak{b}}$ a

* P. 188, l. 13 Insert below this line the folowing rows.

5.12.1Ac Exercise. (i) For $0 \leqslant m < \omega$ and for every ordinal α

$$\aleph_m^{\aleph_\alpha} = \begin{cases} 2^{\aleph_\alpha} & \text{if } \aleph_m \leqslant 2^{\aleph_\alpha} \\ \aleph_m & \text{otherwise} \end{cases}$$

(ii) Conclude that for the example of Magidor 1977 in 5.8 one could not use cardinals smaller than those in 5.8.

**P. 188, l. -14 In then if $\mathrm{cf}(\mathfrak{a}^{\mathfrak{b}})$ move the arc from under $\mathfrak{b}$ to under $\mathfrak{a}$

* P. 188, l. -6 Replace for all $n < \omega$, $2^{\aleph_n} = \aleph_{\omega+n+1}$. by
$2^{\aleph_0} > \aleph_\omega$ and $2^{\aleph_1} > 2^{\aleph_0}$.

* P. 190, l. 5 Below this line insert the following lines

5.18.1Ac Exercise. (i) Prove the following equalities for all infinite cardinals $\mathfrak{b}$.
For a successor cardinal α

$$\beth_\alpha{}^{\mathfrak{b}} = \begin{cases} \beth_\alpha & \text{if } \mathfrak{b} \leqslant \beth_{\alpha-1} \\ 2^{\mathfrak{b}} & \text{if } \mathfrak{b} \geqslant \beth_{\alpha-1} \end{cases}$$

and for a limit ordinal α and for $\alpha = 0$

$$\beth_\alpha{}^{\mathfrak{b}} = \begin{cases} \beth_\alpha & \text{if } \mathfrak{b} < \mathrm{cf}(\beth_\alpha) \\ \mathfrak{c}, \text{ where } \beth_\alpha < \mathfrak{c} \leqslant \beth_{\alpha+1} & \text{if } \mathrm{cf}(\beth_\alpha) \leqslant \mathfrak{b} \leqslant \beth_\alpha \\ 2^{\mathfrak{b}} & \text{if } \mathfrak{b} \geqslant \beth_\alpha \end{cases}$$

(ii) Show that 5.17 is a direct consequence of (i).
(iii) Assume the generalized continuum hypothesis, and use 5.17 to give an example where $\mathfrak{a} < \mathfrak{b}$ and $\mathfrak{c} < \mathfrak{d}$ and still $\mathfrak{a}^{\mathfrak{c}} = \mathfrak{b}^{\mathfrak{d}}$.

P. 193, l. 14 In a set A replace a by an infinite
**P. 194, l. -12 In $2^{\mathfrak{a}}$ add an arc under the $\mathfrak{a}$ like the one two lines below
* P. 195 Below the last line add the line
For additional information on almost disjoint sets see Baumgartner [1976].
P. 199. l. 7 In to Cantors discovery replace Cantors by Cantor's
P. 206, l. 13 In open set A, B replace set by sets
P. 208, l. 19 Delete the s at the end of observations
P. 210, l. -7 In an k_0 such replace an by a
* P. 213, l. 5 In where $\omega = \min(\frac{q}{2}, 2^{-i}$ replace ω by w
* P. 213, l. 12 In open set $S(t,p) \sim C_i$ replace C_i by $\mathrm{Cl}(C_i)$
P. 218, l. 9 Between also and the insert $0 \neq 1$ and
* P. 227, l. 7 Replace perfect separable metric by Polish
**P. 230, l. 5 Between Polish space and is homeomorphic insert with no isolated points
**P. 230, l. 11-12. Replace these lines by
the W_k's be pairwise disjoint and that there be exactly 2^m such, by splitting sets if necessary.
* P. 232, l. 9 Replace (a)→(c) and the idea of the proof of (a)→(b). by (a)→(b) and the idea of the proof of (a)→(c).
* P. 232, l. -9 In hence $F(t) = \bigcap_{n \in \omega}$ insert { before $F(t)$ and } after
P. 242, l. 8 In (ii For add) after (ii
P. 245, l. 23 In while we have replace while by when
P. 248, l. -7 In a field of set replace set by sets
* P. 254, l. 1 In Boolean algebra insert B between algebra and .
P. 254, l. -2 Insert (in front of in the sense
P. 257, l. -6 In (Stone 1934) insert Birkhof 1933, between (and Stone
* P. 260, l. 13 Replace "1935)." by "1935, see also Tychonoff 1930)."
P. 264, l. 10 In $W \subseteq B \sum W$ insert some space between $W \subseteq B$ and $\sum$
P. 269, l. 12 In P is a Boolean algebra. insert) between algebra and .
* P. 271, l. -5 Below this line insert the following line
This is commonly called the *countable chain condition* or *ccc*.
P. 280, l. 6 In the c.a.c. were insert , between c.a.c. and were
P. 281, l. 17 In If $|J| = \leqslant \mathfrak{a}$ and delete the =
**P. 281, l. 18 In P has a subset G replace subset by subnet
**P. 282, l.18 Before Let $p_o \in P$ insert
Let C be a function on $P \times P$ such that if p, q are compatible members of P then $C(p,q) \succcurlyeq p, q$.
**P. 282, l. 19 After $j \in J$ insert , and C
At the end of the line insert
To see that P' satisfies the c.a.c. it suffices to show that if p, q are incompatible

in P' then they are also incompatible in P. Let $p, q \in P'$ be compatible in P, then $C(p,q) \succcurlyeq p, q$. Since P' is closed under C also $C(p,q) \in P'$, hence p, q are compatible in P'.

* P. 287, l. -10 between for and $p, r \in Q_\delta$ insert distinct
**P. 287, l. -8 In p and q are replace q by r
* P. 295, l. -10 Replace **2.10** by **2.10Ac**
* P. 296, l. -13 In $\{\langle \mu, h_\mu \rangle | \mu < \lambda$ replace λ by κ
P. 297, l. 6 Replace the first a in κ-Aronazajn by s
P. 303, l. -6 In Kunen 1978a proved Erase the a in 1978a
* P. 304, l. 22 In $\{u < \lambda \mid \mu \text{ is regular}\}$ replace u by μ
P. 312, l. -13 In than $\gamma \leqslant \beta$ replace than by then
* P. 317, l. -14 In such that $x \cap \lambda = w_{\lambda'\beta}$. replace $w_{\lambda'\beta}$ by $w_{\lambda,\beta}$
P. 321, l. -17 Insert segment between initial and of a
P. 325, l. -17 In 1}). Let Replace) by |
P. 325, l. -5 In by IV.3.9 insert , after IV.3.9
* P. 326, l. -10 In is $\mathfrak{c}^{\max(\aleph_0,|\mu|)}$ we insert $\leqslant$ in front of $\mathfrak{c}^{\max(\aleph_0,|\mu|)}$
P. 330, l. 17 In Mate change a and e to á and é
**P. 338, l. 4 in *iff κ is a cardinal* insert regular between a and cardinal
**P. 338, l. 7 After ineffable. insert κ is regular by 2.54.
**P. 338, l. 14 After we assume insert that κ is regular and
**P. 338, l. 14–15 Erase the sentence starting with Since κ and ending with hence regular.
* P. 342, l. -14 Replace $\alpha \in B \sup S_\alpha < \alpha$ by $\sup S_\alpha < \alpha$ for all $\alpha \in B$
* P. 346, l. -7 Replace from the σ-th column rightwards, by to the right of the σ-th column,
P. 347, l. 4 In $\lambda^+ = \bigcup_{\sigma<\lambda^+} B_{\sigma,\xi}$ replace $_{\sigma,\xi}$ by $_{\xi,\sigma}$
* P. 349, l. 13 In $A_{\alpha,\iota} = \{h \in A | h(\alpha) = i\}$ replace $_\iota$ by $_i$
* P. 352, l. 1 In of ${}^\kappa\kappa$ has replace ${}^\kappa$ by ^{K}
**P. 352, l. -6 In let I be replace I by I'
**P. 352, l. -5 In $\mu(I) = 0$ replace I by I'
**P. 352, l. -4 Insert before If $\beta \in \kappa \sim I$

By IV4.48(ii) the set $J = \{\beta < \kappa | \text{cf}(\beta) \leqslant \omega\}$ is not stationary. Hence, by 4.25, $\mu(J) = 0$. Therefore $\mu(I' \cup J) = 0$. Let $I = I' \cup J$ then $\mu(I) = 0$.

P. 362, Fig. 20 In the second formula from the top $\Psi(\tau(y) \to F(x)$ insert) between $\Psi(\tau(y)$ and $\to$

P. 368 Fraenkel 1922. Add at the end Also in Felgner 1979, 49–56.
P. 370 Gödel 1939. Add at the end Also in Felgner 1979, 257–261.
P. 372 Mahlo 1911. Add at the end Also in Felgner 1979, 209–222.
P. 373 Russell 1906. Replace 1906 by 1907.
P. 373 Scott, 1961. Add at the end Also in Felgner 1979, 234–237.
P. 374 Skolem 1923. Add at the end Also in Felgner 1979, 57–72.
P. 374 Specker 1954. Add at the end Also in Felgner 1979, 159–165..
P. 375 Ulam 1930. Add at the end Also in Felgner 1979, 223-233.

P. 375 von Neumann 1923. Add at the end Also in Felgner 1979, 92–101.
P. 375 Zermelo 1908. Add at the end Also in Felgner 1979, 28–48.

P. 383, right column. In the list of Baumgartner, J. E. add —[1976] 195
P. 384, left column. Insert Birkhof, G. [1933] 257
In the list of Cantor, G. add — [1883a] 85
In Cantor, G. [1895] erase 85
P. 384, right column. Cavailles, J. see Noether and Cavailles J. 1937
P. 385, left column. In the list of countable add — chain condition 271.
P. 386, right column. In the list of Hausdorff, F. add —[1909] 161
In Hausdorff [1914] 26, 37, 135, 161, 199 delete 161,
In Hessenberg G. [1906] 37, 94, 97, 130 delete 94
P. 387, left column. In the list of Jourdain, P.E.B. add — [1904] 94
P. 388, left column. In the list of Magidor, M. add 188 to — [1977] 182, 186
P. 388, left column. In the list of Mate, A. in both places where Mate is written change a and e to á and é
P. 388, left column. Add Moore, G.H. [1982] 158
P. 388, right column. Insert Noether, E. and Cavailles, J. [1937] 85
P. 389, right column. In the list of Russell, B. replace [1906] by [1907] and add —[1908] 160
P. 390, right column. Insert Rang, B. and Thomas, W. [1981] 6
In the list of Tarski, A. add — [1938] 136
P. 391, left column.
Insert Thomas, W. — see Rang, B. and Thomas, W. [1981] 6
Insert Truss, J. [1973] 100
Insert Tukey, J. [1940] 162
P. 391, right column.
After (Zermelo, E.) [1908] 18, 19, 21, 23, 24, 107 add , 160
In the list of Zorn, M. delete 162 from — [1944] 162, 163